CO_2 Adsorbents

CO_2 Adsorbents comprehensively explores the materials and technologies used for adsorbing carbon dioxide. It covers materials synthesis, characterization, adsorption mechanisms, applications, modeling, and potential environmental and industrial impacts to help readers find the best adsorbent for their needs. Through its coverage of advanced research and technological developments in the field, such as novel adsorbent materials, innovative synthesis methods, and advancements in scalable production techniques, the book aims to contribute to the understanding of CO_2 adsorption processes, materials, and their practical applications in addressing the global challenge of climate change and CO_2 emissions reduction.

1. Presents an overview of different adsorbent materials used for CO_2 capture and the methods of synthesis, modification, and functionalization to enhance their CO_2 adsorption performance
2. Describes characterization techniques used to assess the structural, morphological, and surface properties of CO_2 adsorbents
3. Explores the fundamental mechanisms behind CO_2 adsorption of various materials, the interactions between CO_2 molecules and the adsorbent surface, and the kinetics and thermodynamics of CO_2 adsorption processes
4. Discusses through real-world case studies the applications of CO_2 adsorbents in various industries, such as natural gas purification, carbon capture and storage, biogas upgrading, and air purification and their potential in mitigating greenhouse gas emissions and addressing environmental concerns
5. Analyzes the environmental impact and feasibility of using CO_2 adsorbents in real-world scenarios and the challenges and benefits of integrating adsorption technologies into existing industrial processes
6. Provides insights into potential improvements, challenges, and directions for further research and development

Through its practical relevance, advanced insights, and focus on sustainability and innovation, this text serves as a valuable resource for researchers, engineers, and professionals working in the field of carbon capture and utilization, including those in chemical, environmental, and related engineering disciplines.

Ahad Ghaemi earned a BSc at Razi University, Kermanshah, Iran, in 2003 and a PhD at Iran University of Science and Technology (IUST), Tehran, Iran, in 2009. He is a chemical engineering professor at IUST. His research interests include CO_2 capture technologies, gas adsorption, and modeling and simulation of chemical engineering processes. He is an expert in CO_2 capture.

Zohreh Khoshraftar earned a PhD in applied chemistry in 2018, with a focus on various aspects of nanotechnology, green chemistry, material synthesis, and pesticides. In 2022, she earned a postdoctorate in chemical engineering at Iran University of Science and Technology (IUST), Tehran, Iran. During her postdoctorate, her research primarily focused on CO_2 adsorption, CO_2 absorption, dye removal, and heavy metals removal.

Fereshteh Pouresmaeil earned a BSc at Iran University of Science and Technology (IUST), Tehran, Iran, in 2013 and a PhD at Gubkin Russian State University of Oil and Gas, Moscow, Russia, in 2021. She is a postdoctoral researcher at IUST, and her research focuses on gas and heavy metal adsorption.

CO_2 Adsorbents

Advances in Materials, Technologies, and Applications for a Sustainable Future

Edited by
Ahad Ghaemi, Zohreh Khoshraftar, and
Fereshteh Pouresmaeil

CRC Press
Taylor & Francis Group
Boca Raton London New York

CRC Press is an imprint of the
Taylor & Francis Group, an **informa** business

Designed cover image: © 2026 Ahad Ghaemi, Zohreh Khoshraftar, and Fereshteh Pouresmaeil

First edition published 2026
by CRC Press
2385 NW Executive Center Drive, Suite 320, Boca Raton FL 33431

and by CRC Press
4 Park Square, Milton Park, Abingdon, Oxon, OX14 4RN

CRC Press is an imprint of Taylor & Francis Group, LLC

ISBN: 978-1-032-75919-7 (hbk)
ISBN: 978-1-032-75920-3 (pbk)
ISBN: 978-1-003-47619-1 (ebk)

DOI: 10.1201/9781003476191

Typeset in Times
by Apex CoVantage, LLC

In the Name of God

Contents

Preface

Carbon dioxide (CO_2) is considered the most important greenhouse gas with the largest impact on climate change. The gradual increase in the atmospheric concentration of CO_2 due to the burning of fossil fuels is becoming a serious environmental problem. CO_2 capture and sequestration from point source emissions has been recognized as a potential way to stabilize CO_2 in the atmosphere.

Various processes, such as liquid solvent absorption, cryogenic techniques, membrane separation, solid sorbents, and pressure and temperature swing adsorption, have been proposed for the separation and recovery of CO_2 emitted by power plants. To date, most commercial CO_2 capture plants use mine-based processes and wet scrubbing systems, but these processes are energy intensive due to the large amount of water needed in these systems. Another problem associated with these processes is amine degradation by oxidation, leading to the corrosion of process equipment.

The goal of reducing CO_2 emissions on an industrial scale is the development of cost-effective techniques for the separation and capture of CO_2. Adsorption is considered to be one potential option because of the low energy requirement, cost-effectiveness, and the ease of applicability over a relatively wide range of temperatures and pressures.

The main aim of this book is to recognize the potential of the types of various adsorbents in the removal of CO_2 by considering the economic cases, structural morphology, mass transfer, mechanism, and synthesis methods for helping the readers find the best adsorbent. In addition, we aim to comprehensively explore the field of materials and technologies used for adsorbing CO_2. The book's scope covers various aspects related to CO_2 adsorption, including materials synthesis, characterization, adsorption mechanisms, applications, modeling, and potential environmental and industrial impacts.

In addition, it provides an overview of different types of adsorbent materials used for CO_2 capture, such as zeolites, clays, minerals, natural, metal-organic frameworks (MOFs), activated carbons, and porous polymers. It discusses the methods of synthesis, modification, and functionalization of these materials to enhance their CO_2 adsorption performance.

Further, it describes the characterization techniques used to assess the structural, morphological, and surface properties of CO_2 adsorbents. Techniques such as X-ray diffraction (XRD), thermal gravity (TGA), scanning electron microscopy (SEM), and surface area analysis (BET) are relevant in this context. Furthermore, it explores the fundamental mechanisms behind CO_2 adsorption on various materials. In addition, it discusses the interactions between CO_2 molecules and the adsorbent surface, including physical adsorption, chemisorption, and cooperative effects.

This book covers the kinetics and thermodynamics of CO_2 adsorption processes. It explains the factors influencing adsorption rates, isotherms, and the role of temperature and pressure. Identification of absorbent behavior includes endothermic and exothermic, single- or multiple-layer absorption, temperature effects, and other important factors. Moreover, it discusses the applications of CO_2 adsorbents in various industries, such as natural gas purification, carbon capture and storage, biogas upgrading, and air purification. It highlights the potential of these materials in mitigating greenhouse gas emissions and addressing environmental concerns.

The book includes recent research and technological developments in the field, such as novel adsorbent materials, innovative synthesis methods, and advancements in scalable production techniques.

It analyzes the environmental impact and feasibility of using CO_2 adsorbents in real-world scenarios. It discusses the challenges and benefits of integrating adsorption technologies into existing industrial processes. It provides insights into the future prospects of CO_2 adsorption technologies, including potential improvements, challenges, and directions for further research and development.

Overall, the book aims to serve as a comprehensive reference for researchers, engineers, students, and professionals working in the field of carbon capture and storage. It intends to contribute to the understanding of CO_2 adsorption processes, materials, and their practical applications in addressing the global challenge of climate change and CO_2 emissions reduction.

Ahad Ghaemi

Contributors

Hossein Alizadeh
School of Chemical Engineering
Noshirvani University of Technology
Babol, Mazandaran, Iran

Azam Aslani
Department of Chemical Engineering
University of Guilan
Rasht, Iran

Fatemeh Bahmanzadegan
School of Chemical, Petroleum and Gas Engineering
Iran University of Science and Technology
Narmak, Tehran, Iran

Ahad Ghaemi
School of Chemical, Petroleum and Gas Engineering
Iran University of Science and Technology
Narmak, Tehran, Iran

Amir Kazemi
Research Laboratory of Inorganic Chemistry and Environment
Department of Chemistry
Iran University of Science and Technology
Tehran, Iran

Zohreh Khoshraftar
School of Chemical, Petroleum and Gas Engineering
Iran University of Science and Technology
Narmak, Tehran, Iran

Faranak Manteghi
Research Laboratory of Inorganic Chemistry and Environment
Department of Chemistry
Iran University of Science and Technology
Tehran, Iran

Hadiseh Masoumi
School of Chemical, Petroleum and Gas Engineering
Iran University of Science and Technology
Narmak, Tehran, Iran

Farzad Moazami
School of Chemical, Petroleum and Gas Engineering
Iran University of Science and Technology
Narmak, Tehran, Iran

Shanli Nezami
School of Chemical, Petroleum and Gas Engineering
Iran University of Science and Technology
Narmak, Tehran, Iran

Reza Norouzbeigi
School of Chemical, Petroleum and Gas Engineering
Iran University of Science and Technology
Narmak, Tehran, Iran

Hassan Pashaei
School of Chemical, Petroleum and Gas Engineering
Iran University of Science and Technology
Narmak, Tehran, Iran

Fereshteh Pouresmaeil
School of Chemical, Petroleum and Gas Engineering
Iran University of Science and Technology
Narmak, Tehran, Iran

Mahyar Ashourzadeh Pordsari
School of Chemical, Petroleum and Gas Engineering
Iran University of Science and Technology
Narmak, Tehran, Iran

Mohsen Shayanmehr
School of Chemical, Petroleum and Gas Engineering
Iran University of Science and Technology
Narmak, Tehran, Iran

1 CO_2 Capture and Adsorption Technologies

Hassan Pashaei, Hossein Alizadeh, and Ahad Ghaemi

1.1 INTRODUCTION

In recent years, environmental warming issues, because of the diffusion of greenhouse gases (GHGs), have become a global problem [1, 2]. According to the different climate models, global warming will increase the temperature of the earth closely by 1.4–5.8°C until the first of the 22nd century [3]. To fulfill important global objectives, developing a technically practical CO_2 uptake process that allows for the decrease of GHGs from the flue gases is necessary [4]. The CO_2 capture with amine solution has gained widespread acceptance [5]. Amine absorption is more energy efficient than other technologies for CO_2 removal from natural gas, synthetic, and refinery gas industries [6]. Primary and secondary alkanolamines have a high reaction rate with CO_2 but have a lower CO_2 absorption capacity and need a high energy value for regeneration compared to tertiary amines [7]. On the contrary, tertiary amines have high CO_2 absorption capacity and excellent stability and are easily regenerated [8]. In this regard, Pz as a new cyclic diamine could be an impressive amine applicant for CO_2 removal [9]. One of the most remarkable subjects in the adsorption processes is mass transfer flux. There are two physical and chemical adsorption methods related to the adsorbents [10]. In physical adsorption, molecules of CO_2 are going to be dissolved in the solvent, but no reaction is taking place between CO_2 and the solvent. Therefore, the value of adsorbed gas enhanced linearly with the increase of partial pressure based on Henry's law. In chemical adsorption, the reaction rate is a significant specification of the mass transfer. Chemical adsorption is characterized by properties such as high CO_2 recovery capacity, high heat of adsorption, and excellent selectivity. It is particularly suitable for processes operated at atmospheric pressure. But there are some problems like adsorbent demolition, high energy need for the regeneration, and corrosion [11]. The post-combustion method of carbon dioxide adsorption is a process in which CO_2 is captured from flue gas streams after the combustion of fossil fuels, such as coal or natural gas. The absorption process typically involves the use of a liquid solvent, such as monoethanolamine (MEA), which reacts with the CO_2 in the flue gas to form a compound that can be separated and captured. Some of the advantages of the post-combustion method include the following: (1) It can be retrofitted to existing power plants and industrial facilities. (2) It can capture up to 90% of the CO_2 in the flue gas stream. Some disadvantages of this method

DOI: 10.1201/9781003476191-1

include the following: (1) It can be energy-intensive and expensive to implement and operate. The liquid solvents used in the process can be corrosive and require careful handling and disposal. (2) It can produce a concentrated stream of CO_2 that must be transported and stored, which can be a challenge.

CO_2 adsorption involves capturing carbon dioxide molecules on solid materials, primarily for carbon capture and storage (CCS) applications. Key factors influencing this process include adsorbent properties (e.g., surface area, pore structure), temperature, and pressure. Recent studies highlight the effectiveness of porous carbon derived from biomass, zeolites, and metal-organic frameworks (MOFs) in enhancing CO_2 uptake. Modifications like introducing functional groups or optimizing pore sizes can significantly improve adsorption capacity and selectivity toward CO_2 over other gases like N_2 or CH_4 [12].

CO_2 adsorption is a critical area of research for carbon capture technologies. Various materials, including porous carbons, zeolites, and MOFs, are studied for their ability to effectively adsorb CO_2. Factors influencing adsorption capacity include pore structure, surface chemistry, and operating conditions like temperature and pressure. For instance, the presence of functional groups can enhance CO_2 uptake, while larger pore sizes improve adsorption efficiency under high pressures [13]. Recent advancements focus on optimizing these materials to increase selectivity and reduce energy costs in CO_2 separation processes.

The most effective materials for CO_2 adsorption include:

1. MOFs: Known for their high surface areas and tunable porosity, MOFs like MOF-210 exhibit exceptional CO_2 uptake capacities, surpassing 2400 mg/g [14].
2. Zeolites: These crystalline aluminosilicates have well-defined micropores, enhancing selectivity and capacity for CO_2 adsorption [15].
3. Porous Carbons: Biomass-derived porous carbons demonstrate high adsorption capacities due to their large surface area and tunable surface chemistry, with some achieving up to 10.51 mmol/g [16].
4. Composite Materials: Combining MOFs with other materials, such as carbon nanotubes (CNTs) or functionalized polymers, can significantly improve CO_2 capture efficiency [14].

MOFs generally outperform zeolites in CO_2 adsorption efficiency due to their higher surface areas and tunable pore structures. MOFs can achieve surface areas up to 7000 m^2/g, significantly exceeding zeolites' maximum of around 1500 m^2/g, which enhances their adsorption capacity. For instance, MOF-5 exhibits CO_2 adsorption capacities nearly double that of zeolites [17]. Additionally, MOFs allow for chemical modification to optimize selectivity and binding strength, making them more adaptable for specific applications compared to the more rigid structure of zeolites.

Using composite materials for CO_2 adsorption offers several benefits:

1. Enhanced Adsorption Capacity: Composites often exhibit increased CO_2 uptake compared to individual components. For example, combining HKUST-1 with activated natural zeolite improved CO_2 adsorption capacity by approximately 12% [18].

2. Improved Stability: Composites can enhance the stability of adsorbents, particularly in humid environments. For instance, incorporating CNTs into MOFs increases hydrophobicity and structural integrity.
3. Synergistic Effects: The combination of different materials can create additional active sites and optimize pore structures, leading to better adsorption performance. This is evident in composites like Cu-BDC with MCM-41, which achieved higher capacities than either material alone [18].
4. Tailored Properties: Composites allow for customization of surface chemistry and pore size, improving selectivity for CO_2 over other gases [19].

The structure of mesoporous silica significantly influences its CO_2 adsorption capacity through several factors:

1. Pore Size and Volume: Larger and well-defined mesopores facilitate enhanced gas diffusion and increased surface area for adsorption. For instance, SBA-15 exhibits superior CO_2 uptake due to its open structure compared to other types like KIT-6, which has a more complex arrangement.
2. Surface Chemistry: The presence of reactive silanol (Si-OH) groups on the surface provides active sites for CO_2 interaction. Functionalization with amines, such as 1-methylpiperazine, further enhances adsorption by creating stronger interactions with CO_2 molecules, leading to higher capacities.
3. Modification Effects: Tailoring the silica with organic molecules can improve performance. Modified mesoporous silicas have shown significant increases in CO_2 adsorption compared to unmodified counterparts, highlighting the importance of structural characteristics and functional group content on adsorption efficiency.

Temperature significantly affects CO_2 adsorption capacity, generally showing an inverse relationship. As temperature increases, the adsorption capacity tends to decrease due to enhanced thermal energy that promotes desorption. For instance, studies indicate that CO_2 excess adsorption on materials like kaolinite decreases linearly with rising temperatures [20]. Similarly, lower temperatures enhance the adsorption capacity of molecular sieves, with specific capacities dropping dramatically as temperatures rise from −80°C to 80°C [21]. This trend is attributed to the exothermic nature of CO_2 absorption processes, where higher temperatures reduce the physical solubility of CO_2 in solvents [22].

The pre-combustion method of carbon dioxide adsorption involves capturing CO_2 before the combustion process takes place. In this method, the fossil fuel is converted into a mixture of CO_2 and hydrogen through a process called gasification or pyrolysis. The CO_2 is then separated from the hydrogen and captured. Advantages of this method include the following:

1. It can capture CO_2 at a higher concentration than the post-combustion method, reducing the energy required for separation and capture.
2. It can produce hydrogen as a valuable by-product, which can be used as a clean fuel source.
3. It can be integrated into new power plants and industrial facilities.

Disadvantages of this method include the following:

1. It requires significant changes to the design and operation of power plants and industrial facilities, which can be costly and time-consuming.
2. It can produce additional pollutants, such as nitrogen oxides and sulfur oxides, which must be managed and controlled.
3. It can be less flexible than the post-combustion method in terms of the types of fuels and feedstocks that can be used.

The oxy-fuel method of carbon dioxide absorption involves burning fossil fuels in a pure oxygen environment rather than in air. This produces a flue gas stream that is primarily composed of CO_2 and water vapor, which can be easily separated and captured.

Advantages of this method include the following:

1. It can capture CO_2 at a very high concentration, reducing the energy required for separation and capture.
2. It can be applied to a wide range of fossil fuels, including coal, natural gas, and biomass.
3. It produces a relatively pure stream of CO_2 that is easier to transport and store.

Disadvantages of this method include the following:

1. It requires a significant amount of energy to produce the pure oxygen required for combustion.
2. It can be more expensive to implement and operate than other carbon capture methods.
3. It can produce higher levels of pollutants, such as nitrogen oxides, which must be managed and controlled.
4. It can be challenging to integrate into existing power plants and industrial facilities, which may require significant modifications.

Several CO_2 capture techniques, such as chemical and physical solvent absorption, membrane purification, cryogenic distillation, catalytic conversion, and adsorption processes, have been proposed for the separation and regeneration of CO_2 emitted into the atmosphere [23]. The most widely applied technique for CO_2 removal is the absorption of various solvent types [24, 25]. Choosing an economically solvent system includes two essential criteria: high CO_2 selectivity over other gases such as CH_4 and N_2 and high CO_2 solubility. Furthermore, there is minimal solvent loss, low volatility, and a low energy requirement for solvent regeneration [26]; high thermal stability, high reaction rate, long-term stability, and high thermal stability are significant in the CO_2 capture process [27].

According to the Intergovernmental Panel on Climate Change (IPCC), global GHG emissions must be reduced by 50% to 80% by 2050 to avoid the dramatic consequences of global warming [28, 29]. Scenarios from the International Energy Agency indicate that the potential for reduced CO_2 emissions through enhanced

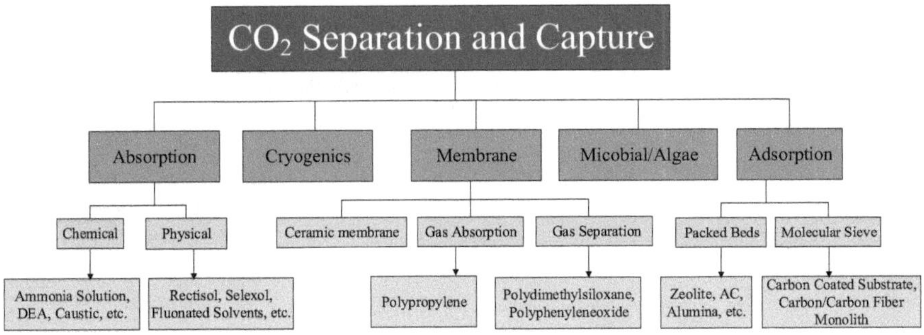

FIGURE 1.1 Schematic of carbon dioxide adsorption and separation methods.

energy efficiency and increased renewable energy production is limited. According to the IPCC, a delay in CO_2 emission reductions can lead to dramatic consequences, and a new strategy for reducing CO_2 emissions as soon as possible is required. CO_2 capture is a technology with the potential for large reductions in CO_2 emissions within 10 to 20 years. Therefore, the strategy for reducing global CO_2 emissions must be a combination of (1) increased energy efficiency, (2) more renewable energy production, and (3) a wide implementation. By establishing stronger incentives favoring energy efficiency and renewable energy, global CO_2 emissions can be reduced by approximately 70% by 2050 compared to emissions today. Therefore, in the present study, we have tried to study all CO_2 emission methods and use the best method with optimal performance for different scenarios. Figure 1.1 shows a schematic representation of the adsorption methods, and Table 1.1 presents the advantages and disadvantages of carbon dioxide absorption technology.

1.2 CO_2 CAPTURE TECHNOLOGIES

Because fossil fuel-burning power plants, especially those that use coal, contribute significantly to global CO_2 emissions, reducing CO_2 emissions has become a top research issue. Carbon dioxide adsorption technology has a great role in reducing the adverse effects of GHG emissions on the environment. Therefore, CO_2 uptake technologies such as post-combustion technology, pre-combustion technology, oxy-fuel combustion, chemical looping combustion (CLC), cryogenic separation, membrane-based absorption, absorption-based CO_2 absorption, direct air absorption, and absorption are used biologically. Because of its "end-of-pipe" feature, post-combustion technology is currently seen as the most practical way to capture CO_2 from existing coal-fired power plants. Despite being energy-intensive, chemical absorption, often known as scrubbing, is the method that is most likely to be used in the near future. In recent times, membrane-based CO_2 separation methods have become a competitive substitute for traditional chemical absorption techniques. The purpose of the aforementioned technologies is to reduce carbon dioxide emissions from various industrial environments, especially in the energy and production sectors. In this research, a summary of new carbon dioxide absorption technologies has been used [30].

TABLE 1.1
Advantages and Disadvantages of Carbon Dioxide Capture Technology

CO_2 Capture Technologies	Advantages	Disadvantages
Post-combustion capture	1. Easy installation and retrofitting 2. Higher efficiency 3. Flexible operation	1. High energy consumption 2. High corrosion 3. Limited capacity
Pre-combustion capture	1. High purity of CO_2 2. Reduced energy consumption 3. Reduced corrosive gases	1. High complexity 2. Hydrogen production 3. Limited fuel options
Oxy-fuel combustion	1. High purity of CO_2 2. Low cost	1. Energy consumption 2. Safety 3. Limited fuel options
Chemical looping combustion	1. High purity of CO_2 2. Flexible fuel options 3. Reduced NO_x emissions	1. Complexity 2. Temperature requirements 3. Limited commercial experience
Cryogenic separation	1. High purity of CO_2 2. Low energy consumption	1. High capital cost 2. High operating cost
Membrane-based CO_2 capture	1. Low energy consumption 2. High selectivity 3. Low capital cost	1. Low permeability 2. Limited scalability 3. Maintenance requirements
Adsorption-based CO_2 capture	1. High selectivity 2. Flexible operating conditions 3. Low energy consumption	1. Limited capacity 2. High cost 3. Limited scalability
Direct air capture	1. Flexibility 2. Low emissions	1. High energy consumption 2. High cost 3. Limited scalability
Biological capture	1. Low cost 2. Scalability	1. Low efficiency 2. Limited lifespan 3. Limited scalability

1.2.1 Post-Combustion Capture

Large-point sources of CO_2 emissions can benefit from CCS, according to a 2005 [31], IPCC report. Captured CO_2 can be utilized as feedstock for industrial operations or compressed, transported, and stored in mineral carbonates, seas, or geological formations. Three main approaches—post-combustion, pre-combustion, and oxy-fuel combustion [32]—can be used to achieve CCS. After fuels have burned entirely, CO_2 is extracted from flue gas via post-combustion capture. Post-combustion capture appears to be a more practical option for present coal-fired power plants because of its "end-of-pipe" technology, whereas traditional coal-fired power plants now burn fuel directly in a boiler to generate power [33]. As a result post-combustion carbon capture (PCC) technologies play an important role in the reduction of CO_2 emissions to address climate challenges. This process is usually simulated in process simulation software based on first-principle models, which calculate physical properties directly

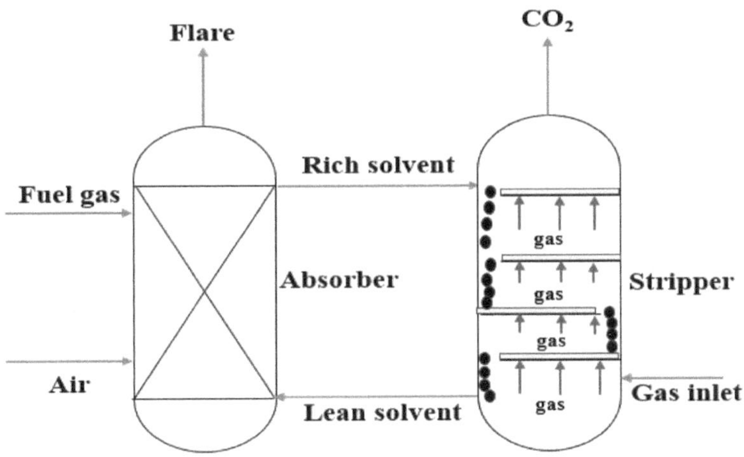

FIGURE 1.2 Schematic of the chemisorption process for post-combustion CO_2 adsorption.

from basic physical quantities such as mass and temperature. Figure 1.2 shows a schematic of the chemical adsorption process for the absorption of CO_2 after combustion.

1.2.2 Pre-Combustion Capture

Carbon dioxide can be captured in power stations, with an integrated gasification combined cycle process (pre-combustion capture). Pre-combustion carbon capture involves the removal of all or part of the carbon content of fuel before burning it. The basic idea with pre-combustion de-carbonization is to convert a fossil fuel to a synthesis gas that contains mostly hydrogen and carbon dioxide. In the pre-combustion process, the fuel gas produced contains a high concentration of CO_2 and comes out at a total pressure of $2.5e^{5.0}$ MPa. And the separated hydrogen can be used in a gas turbine or a fuel cell [34]. Figure 1.3 shows a view of carbon dioxide absorption by pre-combustion adsorption method.

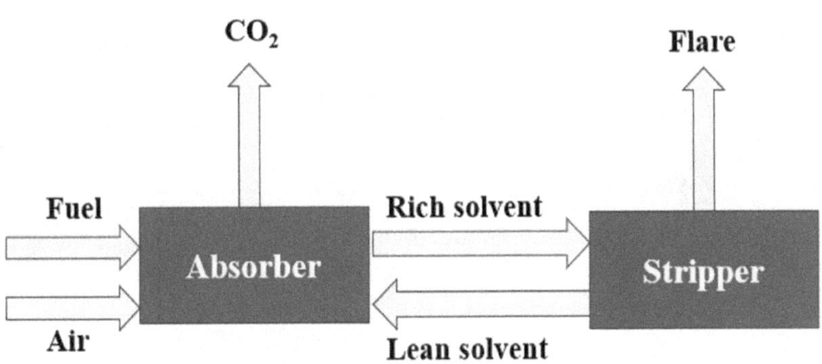

FIGURE 1.3 Pre-combustion adsorption method view of CO_2 adsorption.

1.2.3 Oxy-Fuel Combustion

Oxy-fuel combustion technology is an effective way to reduce CO_2, and this has been established by various researchers working in this area [35]. The oxy-fuel combustion technology for power generation with CCS is illustrated as a flow chart shown in Figure 1.4. The basis of oxy-fuel combustion technology relies on using pure oxygen to replace air as the intake charge to react with fuel for combustion. This ensures that exhaust emissions mainly contain CO_2 and water vapor (H_2O). In the meantime, other harmful emissions related to nitrogen elements, such as NO_x, are significantly reduced. Moreover, the other significant advantage of oxy-fuel combustion is that CO_2 emissions can be almost eliminated from the exhaust. This can be achieved by recycling or separating from water vapor after the condensation. As a result, nearly pure CO_2 can be compressed, but the purity of CO_2 should depend on a trade-off between efficiency losses and operational costs during purification and the safety demands of transportation and storage. The main difference between oxy-fuel combustion and conventional combustion can be explained mathematically using Eqs. (1.1) and (1.2). Unlike conventional air combustion, oxy-fuel combustion conception replaces air with oxygen, which is utilized to react with fuel directly, leading to the final products, which merely contain CO_2 and H_2O. Figure 1.4 shows the general flow chart of oxy-fuel combustion technology.

For the conventional air combustion:

$$\mathrm{C}_x\mathrm{H}_y\mathrm{O}_z + \left(X + \frac{Y}{4} - \frac{Z}{2}\right)(\mathrm{O}_2 + 3.773\,\mathrm{N}_2) \rightarrow \mathrm{XCO}_2 + \frac{Y}{2}\mathrm{H}_2\mathrm{O} + 3.773\left(X + \frac{Y}{4} - \frac{Z}{2}\right)\mathrm{N}_2 \tag{1.1}$$

For the oxy-fuel combustion

$$\mathrm{C}_x\mathrm{H}_y\mathrm{O}_z + \left(X + \frac{Y}{4} - \frac{Z}{2}\right)\mathrm{O}_2 \rightarrow \mathrm{XCO}_2 + \frac{Y}{2}\mathrm{H}_2\mathrm{O} \tag{1.2}$$

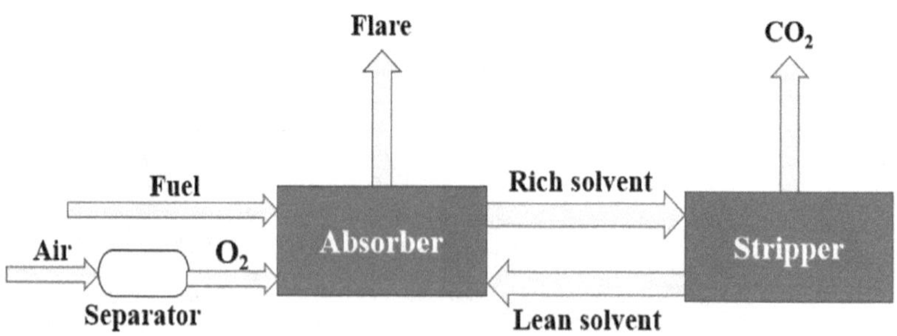

FIGURE 1.4 A general flow chart of oxy-fuel combustion technology.

Oxy-fuel technology enhances CO_2 adsorption by combusting fossil fuels in pure oxygen and recycled CO_2, resulting in flue gas with high CO_2 concentration. This increases the driving force for mass transfer during carbon capture, leading to improved efficiency and reduced energy requirements compared to traditional air combustion methods. Studies indicate that oxy-fuel combustion can achieve over 90% CO_2 removal with energy savings of up to 84% [36]. Additionally, innovative solvents like ionic liquids are being explored for their effectiveness in capturing CO_2 from oxy-fuel processes, further optimizing energy consumption and purity of captured CO_2.

1.2.4 Chemical Looping Combustion

The CCS process generates energy from the burning of existing hydrocarbon fuel in such a way that it separates the existing CO_2 from the flow of flue gases and stores it under pressure and outside the atmosphere. Three methods, oxy-fuel combustion, post-combustion, and pre-combustion, are used to absorb CO_2 in large industries such as power plants and petrochemicals. These methods have received much attention and development in the past three decades [37]. Since the goal of all methods is to reduce the release of CO_2 into the environment, the use of energy has become a bigger problem [38]. Therefore, to avoid the energy problem, power plants and large industries do not use CO_2 absorption technology. Therefore, to avoid this problem, CLC has emerged as a new achievement in the CO_2 absorption process with less energy consumption. It seems that the use of CLC as the most economical absorption technology is associated with 100% absorption of carbon dioxide [38].

CLC is a new and suitable combustion technology for the CO_2 absorption process. To carry out the aforementioned process, small metal particles called oxygen should flow between the air and fuel reactor and be added to the reactor to perform fuel combustion. Therefore, the oxygen of the metal (Me) is reduced to a less oxidized state (MeO_{a-1}), although this process is called a reduction reaction. To reach the original point(MeO_a) through the airflow, the reduced oxygen must be introduced into the air and transferred to the reactor chamber. This is called the oxidation reaction, and this process continues in another fuel reactor to form the second ring. So, with this process, CLC is created between fuel and pure oxygen. As a result, the flue gases will be free of N_2 and consist only of carbon dioxide (CO_2) and water vapor, and water vapor can be reduced and carbon dioxide (CO_2) can be stored away from the atmosphere. The advantage of using CLC is the spontaneous absorption of carbon dioxide (CO_2) without reducing energy in the separation processes [39].

1.2.5 Cryogenic Separation Method

The cryogenic method of carbon capture technology uses natural gas to provide cold energy to capture CO_2. Cryogenic CO_2 capture is used in oxy-fuel combustion technology as well as in PCC technology to separate CO_2 from flue gas. With cryogenic CO_2 capture, it is possible to produce high-purity CO_2 of up to 99.17%. This method includes a few processes, such as compression, expansion, separation, and

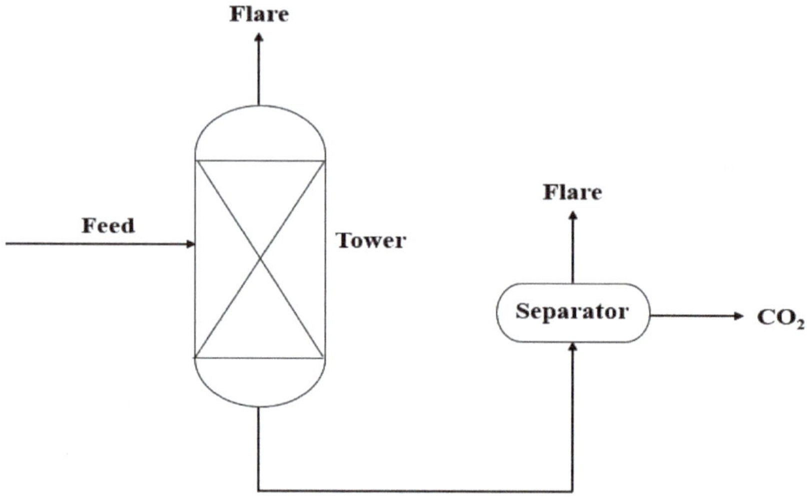

FIGURE 1.5 Schematic of carbon dioxide absorption by cryogenic separation method.

cooling. The cryogenic method is less preferred because of its high operational cost. The cryogenic method used in PCC is carried out using various methods [25]. Figure 1.5 shows a schematic of carbon dioxide absorption by the cryogenic separation method.

1.2.6 Membrane-Based CO_2 Capture

A wide range of processes have been extended to CO_2 removal from gas streams including adsorption, cryogenics absorption, and especially membrane processes [5]. With membrane separation, high-purity streams are strict to get, exclusively on the scale of CO_2 removal from power plants. Membrane separation technology offers many advantages due to its inherent attributes such as energy-saving and continuous operation compared to traditional methods like absorption and adsorption processes [40].

Facilitated transport membranes, in addition to exhibiting "solution diffusion" characteristics, contain an active transport mechanism that increases the permeability and selectivity of the membrane material. The target species reversibly reacts with either a fixed or mobile carrier present in the membrane that then diffuses across the membrane driven by a concentration gradient in the complex rather than a gradient in the permeate. Unlike non-facilitated transport membranes, the main characteristic of these membranes is represented by the decrease of both CO_2 permeance and selectivity with increasing CO_2 feed partial pressure. A schematic representation of the mechanisms for a fixed-carrier amine-based facilitated transport membrane is provided in Figure 1.6. As shown, the CO_2 is absorbed and reacts to form bicarbonate, the form in which it transverses the membrane. Upon reaching the other side, it back-reacts to form CO_2 and is then released on the permeate side [41].

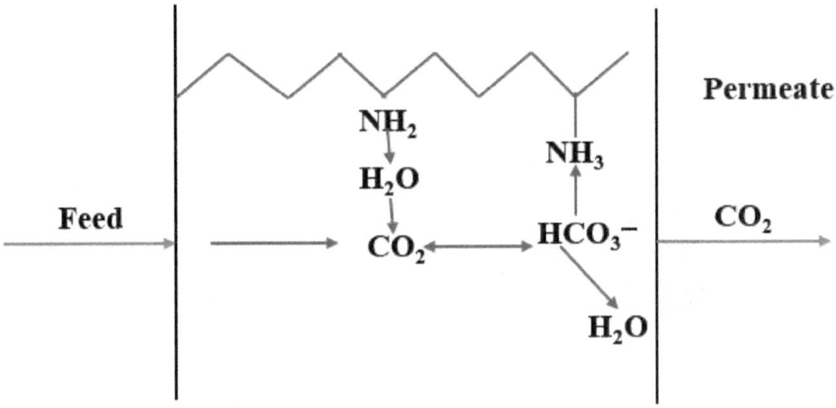

FIGURE 1.6 Schematic of the mechanisms for a fixed-carrier amine-based facilitated transport membrane.

1.2.7 Adsorption-Based CO_2 Capture

Adsorption has been examined, but low CO_2 selectivity and poor capacity restrict the potential for usage for CO_2 removal. Up until now, adsorption and absorption methods have been the most economical for separating CO_2 from bulk flows. That's why this section focuses on the expansion of more effective adsorption and absorption technology [42]. CO_2 absorption into physical or chemical solvents [43] is a well-developed technology that has been used for many commercial processes [44], including ammonia production and gas treatment [45]. There has been a lot of research on this technology over the past 50 years, mainly on developing an understanding of the specific properties of the solvent. While a significant portion of work has been published on specific amines, little research has been done on comprehension or indicating complex mixtures, which are often the most effective technologies. The main purpose of this study was to explore the review of CO_2 capture using adsorption technologies [46]. For this purpose, we tried, in addition to reviewing the literature, to investigate the effect of operational parameters such as temperature, flow rate, pressure, and adsorbent loading on the CO_2 removal efficiency and CO_2 capture capacity. Figure 1.7 shows a schematic of CO_2 absorption. Depending on the conditions, multiple adsorbents have been used in Table 1.2.

1.2.7.1 Temperature Effect on CO_2 Adsorption

The effect of different temperatures on the CO_2 adsorption capacity for activated alumina (AA) and AA/NaOH, AC/NaOH, NaOH [47], and AA/Pz [49] adsorbents is shown in Figure 1.2. As can be seen from the trend for AA, AA/NaOH [59], AA/Pz [60], and AC, the increase in temperature from 293 to 353 K reduced the CO_2 adsorption capacity for the adsorbents, and the maximum value of CO_2 adsorption capacity was observed at 293 K. Also, the result indicates that the

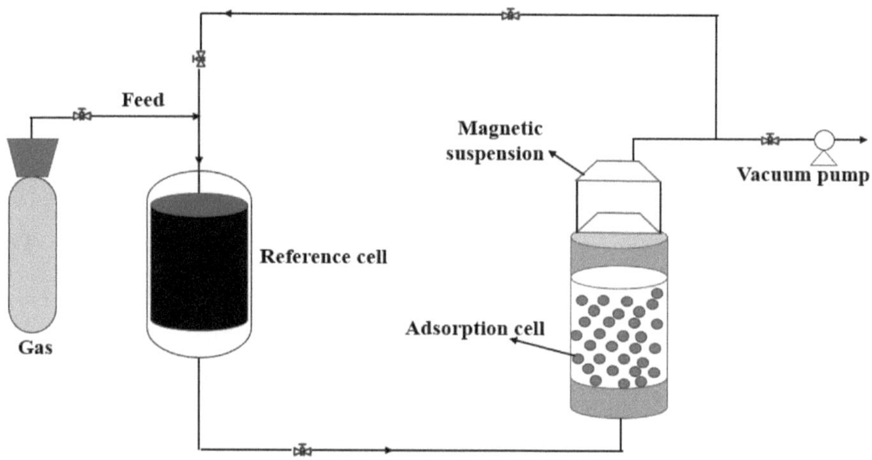

FIGURE 1.7 Schematic of CO_2 adsorption.

TABLE 1.2
CO_2 Adsorption with Different Adsorbents

Sorbent	Process	Temperature (K)	Capacity (mmol/g)	Ref.
NaOH	Fixed-bed reactor	298–338	2.82–29.75	[47]
Graphene oxide/TiO_2 nanocomposite	Fixed-bed reactor	298	0.77–0.95	[48]
AC/NaOH	Fixed-bed reactor	293–353	0.48–8.69	[49]
Molecular sieve 13X	Pressure swing adsorption	298	2.80–3.60	[50]
Active carbon–KOH	Pressure swing adsorption	298	—	[51]
NaOH-granular coconut shell AC	Fixed-bed reactor	332	27.10	[52]
NaOH/CaO	Temperature swing	350	3.00	[53]
Graphene	—	195	0.80	[54]
Na_2CO_3	Thermogravimetric adsorption	323–343–70	2.30	[55]
Na_2CO_3/Al_2O_3	Fixed-bed reactor	323–353	—	[56]
NaY zeolites/Si/Al	Batch reactor	323	1.55–6.89	[57]
K_2CO_3/Al_2O_3	Fixed-bed reactor	328–537	0.44–0.63	[58]

CO_2 adsorption process is highly exothermic and that CO_2 adsorption capacity decreases with increasing temperature. However, a trend in NaOH indicates that the optimum temperature for CO_2 removal is 308 K. To describe the perceived trend in this figure, it can be noted that, below 308 K, an increase in temperature leads to an increase in the carbonation reactions (i.e., forward reactions

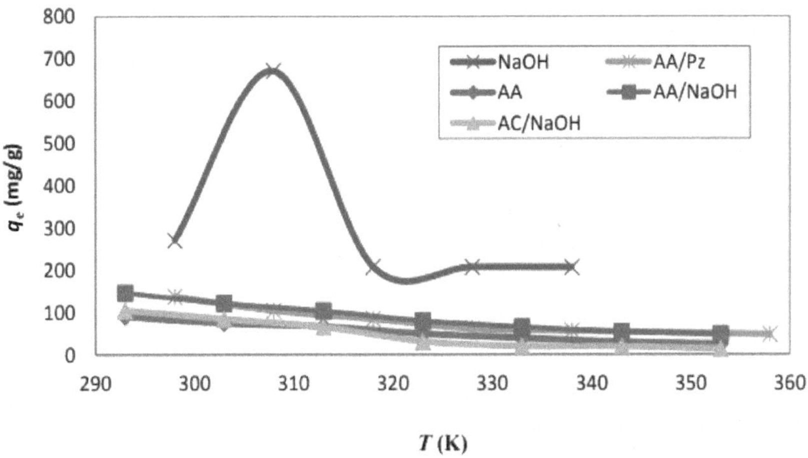

FIGURE 1.8 Effect of temperature on CO_2 adsorption capacity.

in Eqs. (1.3) and (1.4)) [28]. Nevertheless, above 308 K, the reverse $NaHCO_3$ decomposition reaction becomes thermodynamically favorable and slows down the overall carbonation reaction. Consequently, the optimum temperature for the present CO_2 removal system is 308 K. Also, Figure 1.8 indicates that NaOH has better CO_2 adsorption capacity than other adsorbents [29, 61]. There are two main steps for this reason. First, NaOH loading leads to an increase in CO_2 removal rate due to direct growth in the reactant amount in the reaction system, along with an increase in the gas–solid contact surface. The second is the pH effect. In the beginning, CO_2 gas instantly reacts with hydroxide ions (OH^-) and is converted to carbonate (CO_3^{-2}). With increasing solution conversion, the concentrations of [Amine]H^+ and HCO^{-3} rise; this can be described via a base catalyst mechanism. At the lower conversion, the carbamate is formed rapidly and then decreases slowly at the higher conversion. When the pH value becomes low enough, the amine conversion to HCO^{-3} becomes higher, postponing the formation of [amine]COO^-. Therefore, due to the conversion of amine to amine carbamate and protonated amine, the concentration of free amine becomes low, but the value of amine is still important for the absorption of CO_2.

$$2NaOH_{(g)} + CO_{2(g)} \leftrightarrow Na_2CO_{3(S)} + H_2O_{(g)} \quad \Delta H_{298K} = -127.5\,\text{Kj} \tag{1.3}$$

$$NaOH_{(g)} + CO_{2(g)} \leftrightarrow NaHCO_{3(S)} \quad \Delta H_{298K} = -131.5\,\text{Kj} \tag{1.4}$$

1.2.7.2 Pressure Effect on CO_2 Adsorption

The effects of different adsorption pressures, including 2, 4, 6, and 8 bars, on CO_2 adsorption by AA and AA/NaOH [28], AC/NaOH [49], NaOH [62], and AA/Pz [23, 63, 64] adsorbents are presented in Figure 1.9. The adsorption capacity at 2 bar is

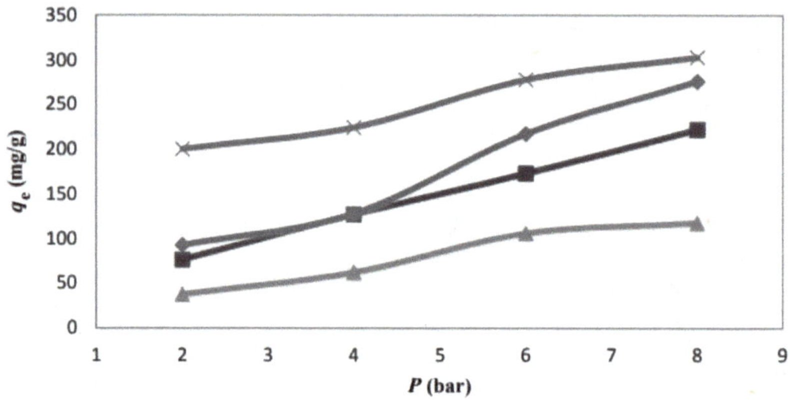

FIGURE 1.9 Effect of pressure on CO_2 adsorption capacity.

low, and it gradually increases with increasing pressure from 2 to 8 bar. Some graphs such as AC/NaOH show that maximum adsorption capacity was achieved at 6 bars, and, after this pressure increase, there is no impressive effect on CO_2 adsorption capacity. Nevertheless, the maximum CO_2 adsorption capacity for all adsorbents is near the pressure of 8 bar.

1.2.8 Direct Air Capture

The use of direct air capture (DAC) technology in the separation of carbon dioxide (CO_2) can help reduce the emission of CO_2 in environmental communities and develop technologies that have a strong need for CO_2 to obtain raw materials. As a result, it eliminates the need for difficult methods such as storage and transportation and creates technologies such as converting CO_2 into alcohol, fuel into air, and the effective growth of plants in greenhouses. Also, by using the development and solving of the problems of DAC technology, materials can be provided to simplify other carbon absorption methods such as pre-combustion absorption, biogas, and other fuel gases [61].

The technology of DAC is based on the existence of an optimal and cost-effective adsorbent, which is based on the relationship and ratio of gases and vapors with synthetics, optimal absorption and energy, and selectivity of CO_2. These things can be determined by the concentration of CO_2 and atmospheric oxide significantly combined. Also, the most important thing about absorbents is their ability to create physical and chemical stability. The new DAC mostly uses solid chemical adsorbents based on organic amines, in which functional groups in chemical adsorbents are placed on surfaces such as cellulose [65], polymer networks [66], and porous silica. Therefore, these adsorbents need a temperature above 100°C to be made, which may limit some activities due to the high temperature. The DAC based on physisorption is a creative work, but it has limitations due to the lack of suitable physical adsorbents that have good selectivity toward CO_2 and also maintain excellent stability against competing vapors and gases. Among the other advantages of physisorption, it is a good absorber of CO_2 and uses less energy for recovery, which makes it compatible with the environment. As a result of continuing research and development based on the discovery of various types of adsorbents, these limitations can be reduced and be useful for increasing the efficiency of DAC for CO_2 absorption and separation [67].

1.2.9 Biological Capture

Nowadays, the use of microalgae technology (biological process) in the process of CO_2 absorption has been prioritized over chemical and physical methods due to its low energy consumption and stability in different environments. But biological processes have a lower rate of CO_2 absorption than chemical processes, and the reason is that biological methods are sensitive to the types of pollution that exist in flue gases [23, 25, 68]. But the amount of production output in the biological method is superior to the chemical and physical methods that have more CO_2 flow. Therefore, it is beneficial for the economic sustainability of the whole process. There are differences in the production of biofuels, one of which is the efficiency of microalgae (biological process) to absorb CO_2. As a result, the reliability of biological processes and the commercial development of CO_2 absorption processes depend on the use of microalgae masses [69, 70]. In recent years, the production of microalgae on an industrial scale has become widespread due to the use and expansion of microalgae in food products and the improvement of technology. Therefore, some improvements such as better energy efficiency, mass transfer capacity in reactors, and the production of new downstream products have been revealed [71]. Therefore, these developments have shown the better use of microalgae in absorbing CO_2 more seriously than in previous years. As a result, for the rapid growth of microalgae and the development of the biological process, factors such as light intensity, temperature, and pH have a direct effect.

1.3 CO_2 ADSORPTION PROCESS

CO_2 adsorption is an important technology for separating CO_2 from different industrial processes and atmospheric air, which helps the world reduce GHG emissions and resist climate change. For development and progress in this field, research and development for better efficiency, economic efficiency, and scalability of the CO_2 adsorption process are necessary. In addition, the CO_2 adsorption process is investigated by methods such as physical adsorption, chemical adsorption, pre-combustion adsorption, and post-combustion adsorption. The process of adsorption with amine solutions is one of the most common methods for adsorbing CO_2 on high scales using alkyl amine solutions such as MEA and diethanolamine or other solutions with a base nature [72–74]. Although this method has disadvantages such as high corrosion rate of equipment due to the evaporation process, oxidative degradation of adsorbent, high energy consumption for amine reprocessing, and reduction of its effect over time due to low thermal stability and formation of foam [75]. To avoid the disadvantages mentioned about the use of amine adsorbents, one of the alternative techniques for removing CO_2 is adsorption using solid porous adsorbents. In addition, CO_2

adsorption using solid adsorbents is a practical method because of its versatility and efficiency, and it reduces the potential energy costs for reproduction in comparison with the absorption process by amine solutions [76, 77]. Generally, CO_2 adsorption using solid adsorbents is being either physically or chemically developed, but, before designing an adsorption process, the choice of a suitable adsorbent with high CO_2 adsorption capacity is a very important issue that should be taken into consideration. In the adsorption processes, various adsorbents, such as porous minerals, activated carbon, AA, bases, and amine-based oxides, are used to adsorb CO_2. However, some of these adsorbents, such as amine-based adsorbents, need high energy and have the potential for destruction through oxidation, which results in corrosion [78, 79].

One of the highest, and most investigated, acid gas removal technologies is an absorption process that uses a chemical solvent [80]. Processes like this are usually used in natural gas and ammonia production [81, 82]. There are multiple types of this flow sheet [74, 75], including an isothermal and a temperature process. In many standard absorption processes, the variation of temperature swing, a gas flow including CO_2 enters a column's bottom. The treated gas and CO_2 are released, which exit the absorber's top. A solvent enters the top of the column and counter-currently contacts the gas phase in different absorbers such as the packed column, tray column, and bubble column. The CO_2 removal efficiency of absorption was calculated in Eq. (1.5) [23, 68]:

$$CO_2 \text{ removal efficiency}(\%) = \frac{y_{co_2,in} - y_{co_2,out}}{y_{co_2,in}} \times 100 \tag{1.5}$$

CO_2 adsorption technology involves capturing carbon dioxide from gas streams using solid adsorbents. Key methods include pressure swing adsorption (PSA) and temperature swing adsorption (TSA), which are favored for their energy efficiency and lower operational costs compared to traditional absorption techniques. Advanced materials, such as carbonaceous and MOFs, are being developed to enhance adsorption capacity and reduce regeneration energy requirements. Recent advancements focus on optimizing pore structures and integrating various technologies to improve performance in industrial applications, particularly for post-combustion CO_2 capture. Recent advancements in CO_2 adsorption technologies focus on enhancing efficiency and reducing costs. Key developments include:

1. Innovative Adsorbents: MOFs and advanced carbonaceous materials are gaining attention for their high capacity and selectivity in CO_2 capture [83].
2. Integrated Technologies: Combining different adsorption methods, such as temperature and PSA, has shown improved performance and energy efficiency [84].
3. Sustainable Regeneration: TSA is favored for its lower energy requirements, enabling integration with renewable energy sources [84].
4. Hybrid Systems: New hybrid processes that combine membrane and adsorption technologies are being explored to optimize CO_2 capture efficiency while minimizing operational costs [85].

TABLE 1.3
Some of the Best Absorbent Materials for Absorbing Carbon Dioxide (CO_2)

Carbonaceous Adsorbents	Non-carbonaceous Adsorbents
1. Carbon nanotube adsorbents	1. Metal-organic framework (MOF) adsorbents
2. Carbon molecular sieve adsorbents	2. Zeolits-based adsorbents
3. Graphene adsorbents	3. Silica material-based adsorbents

1.3.1 Physical Adsorption

1.3.1.1 Physical Adsorbents

The absorption of CO_2, according to the physical adsorbents, is divided into two categories: carbonaceous and non-carbonaceous materials, which, compared to chemical adsorbents, consume little energy, and the reason for the low energy consumption is not forming a bond between the CO_2 and the surface of the adsorbent. Also, less energy is used to recover CO_2. Therefore, physical adsorbents such as activated carbon have given power to amine absorption technology and low energy consumption due to their low selectivity to CO_2/nitrogen and good overcoming of this issue [86]. On the other hand, some materials, such as zeolites, which have good selectivity for CO_2, do not perform well if they absorb a small amount of CO_2 in wet conditions. Finally, the design and discovery of materials with high selectivity for CO_2, good strength, acceptable scalability, and low energy consumption are prioritized for carbon dioxide separation. Some of the best adsorbent materials used for CO_2 absorption are given in Table 1.3.

1.3.1.1.1 Carbonaceous Materials Adsorbents

These materials are mainly composed of carbon and several existing materials that can evaluate and estimate factors such as thermal and chemical strength, heat, and electrical conductivity [87]. There are different types of carbon materials, such as CNTs, graphene, and molecular carbon sieves, that are cheap, are available from natural sources, have a large pore volume, and have lightweight.

1.3.1.1.1.1 CNT Adsorbents CNT materials are a new member of the carbon group which has aroused high attention as a novel type of adsorbent. Many theoretical and experimental studies are dedicated to examining the adsorption performance of CO_2 on the CNT adsorbents. CNTs can act as an efficient adsorbent for CO_2 separation. Cinke and co-workers (2003) [88] investigated the CO_2 uptake on the surface of purified single-walled CNT adsorbent at different temperatures ranging from 0°C to 200°C. They reported that the surface area of the adsorbent is 1587 m^2/g with 1.55 cm^3/g pore volume. The study extended to compare the performance of purified single-walled CNT adsorbent to activated carbon adsorbent at different temperatures. Figure 1.10 highlights that the nanotube adsorbent exhibited CO_2 adsorption capacity twice more than that of activated carbon.

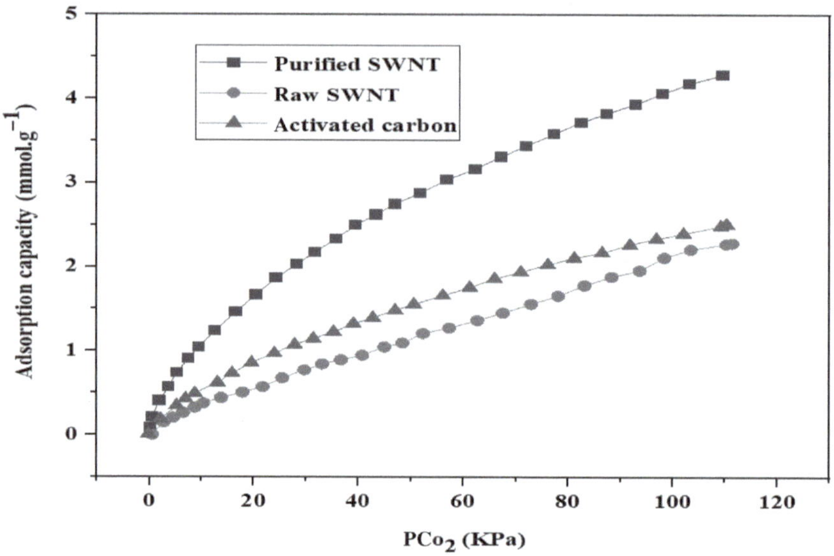

FIGURE 1.10 Comparison of carbon dioxide uptake of activated carbon and carbon nanotube adsorbents at 308 K.

Lu's research group [89] studied the modifications of CNT materials and stated that the CNT adsorbents are compatible adsorbents at low temperatures. They modified the surface of CNTs using 3-aminopropyltriethoxysilane (APTS) solution where the physicochemical properties enhanced to entrap more CO_2 molecules. The adsorption performance of modified CNTs was 43.3 mg/g at 293 K with 97.42 m^2/g. In addition, they stated that the carbon dioxide uptake for both pure and modified CNTs decreases as the temperature is elevated. The same group extended the study to compare the performance of CNTs, zeolite, and granular activated carbon under the same modification conditions to report that the CNT adsorbents exhibit the greatest CO_2 uptake. Moreover, they studied the impact of functionalized multi-walled CNTs with APTS at various adsorption temperatures. They stated that increasing the temperature of the process has a negative impact on the adsorption performance while elevating moisture content would improve the CO_2 uptake, which reflected the exothermic process of the adsorption. The estimated CO_2 adsorption capacity for the proposed adsorbent (APTES-CNT) was 2.59 mmol/g at 25°C.

1.3.1.1.1.2 Carbon Molecular Sieve Adsorbents Carbon molecular sieves (CMSs) are a special member of microporous carbon adsorbents with molecule-sized pores. The preparation process of CMS adsorbent includes four steps: (1) carbonization of raw material, (2) surface activation, (3) deposition usually via chemical vapor, and (4) subsequent carbonization of aromatic molecules. Foley [90] mentioned that the efficient CMS adsorbent for CO_2 adsorption should offer narrow pore size distribution, which can result in high adsorption performance and selectivity. Burchell et al. [91] modified a porous monolithic carbon fiber composite molecular sieve with a carbon dioxide adsorption capacity of more than 2.27 mmol/g at 303 K and atmospheric pressure; this carbon dioxide (CO_2) uptake decreased as the temperature increased. In addition, Alcañiz-Monge et al. [92] studied the CO_2 uptake on the surface of CMS adsorbents in the form of particles and monoliths and tested a full study about the impacts of pore blockage to conclude that the decomposition of the surface during the carbonization process results in the formation of narrow microporosity, which is compatible for CO_2 molecules. Bikshapathi and co-workers [93] studied the preparation of CMS adsorbents using two different raw materials (a hierarchal web of carbon micro- and nanofibers and activated carbon fibers). They reported that the adsorption of the CMS that was prepared based on the hierarchal web of carbon micro- and nanofibers was superior to those synthesized based on activated carbon fibers. The carbon dioxide uptake was 10 mg/g of the inlet concentration of 5000 ppm. Silvestre-Albero and co-workers [94] studied the CMS adsorbents and reported that these adsorbents have at least the same adsorption capacity or more than other carbon-based adsorbents.

According to the results of CMS adsorbents, these adsorbents have at least the same or greater adsorption capacity than other carbon-based adsorbents. The ion exchange process is another technology to modify the molecular sieve carbon adsorbent surface.

1.3.1.1.1.3 Graphene Adsorbents Graphene is a new class of carbonaceous materials with acceptable adsorption capacity and has recently taken massive attention. It is basically a flat single layer of sp^2 hybridized carbon atoms, densely packed into an ordered two-dimensional honeycomb network. Before 2011, as far as we can tell, there was no research about the deployment of graphene/graphite as carbon dioxide (CO_2) adsorbents. However, since 2012, many papers researches have been conducted to investigate the employing of graphene/graphite as CO_2 adsorbent owing to the large active surface area and low preparation cost. The investigations are generally in three research directions: (1) proposing new structures, (2) surface modifications, and (3) synthesizing hybrid materials. Several studies employed various treatments to modify the graphene surface and propose a high surface area and tunable pore volume [95].

1.3.1.2 Non-Carbonaceous Dry Adsorbents

1.3.1.2.1 MOF Adsorbents

MOF materials are a new generation of solid adsorbent materials that are produced by the combination of metal ions linked by coordination. The flexibility of manipulation of both metal ions and the organic linkers offers copious options to manage the pore size, shape, and the potential of the adsorbing surface, which improves the selectivity, CO_2 loading, and adsorption/desorption kinetics. Millward et al. [96] and Liu et al. [97] stated that the MOF adsorbent has many features such as strong 3D structures, large active surface area, controlled pore shapes, and tunable pore surface properties. All the mentioned features nominate MOF adsorbents ideal for CO_2 capture. Almost, all the metals and a large range of organic species can be deployed to produce MOFs. Two important criteria for selecting a proper MOF adsorbent for CO_2 removal are that the adsorbent pores have to be harmonious with the CO_2 molecule's diameter. Besides, the adsorbent should come with polar (OH^-, $N=N^-$, NH^{-2}, and $-N = C(R)^-$), where pores have a larger CO_2 loading owing to quadrupole moments of carbon dioxide (CO_2) molecules. Therefore, considering these criteria in the design of the MOF adsorbents can result in a huge improvement in CO_2 adsorption.

1.3.1.2.2 Zeolite-Based Adsorbents

Zeolites are microporous crystalline framework materials that exist naturally and can be synthesized in the research laboratory. Zeolites possess regular pore sizes ranging from 0.5 to 1.2 nm, consisting of a chain of channels to capture the gas molecules [98]. It has been therefore widely deployed in gas separation technologies. Zeolites have been broadly studied for carbon dioxide removal in the interest of their molecular sieving impact and the robust dipole–quadrupole (electrostatic) interactions between CO_2 and alkali metal cations in the zeolite frameworks [99]. Li, Al, and/or Na cations of zeolites affect the CO_2 heat of adsorption, where the heat of adsorption increases in the act of rising monovalent charge density [100]. Zeolites attract strongly to gases with high quadrupole moments like carbon dioxide ($-14.29 \times 10{-}40$ C m^2) via the electric field produced by the cations of the zeolites [101]. The efficiency of zeolite adsorbents relates to several parameters like structure and composition of the framework, cationic shape, purity, size and shape of the molecules, and molecular polarity. There are several common types of zeolites for CO_2 adsorption such as zeolite A, X, and Y or natural zeolites like chabazites, clintopiles, ferrierites, and mordenites [102]. Out of the mentioned, zeolites, 13X, and chabazite were reported as efficient adsorbents for carbon dioxide from nitrogen [103]. Recently, the development of zeolite-based adsorbents following two research aspects covering (1) manipulating the content composition and the structure of the zeolite framework and (2) cationic exchange.

Siriwardane and others [102] stated that the natural zeolites with a high content of sodium showed high carbon dioxide adsorption. Harlick and Tezel [104] conducted a comprehensive evaluation study for some synthetic zeolites covering 5A, 13X, NaY, ZSM-5, and HiSiV-3000 for CO_2 uptake at a pressure ranging from 0 to 1.5 bar. Their results showed that the adsorption capacities enhanced in the act of increasing the pressure as shown in Figure 1.11, where 13X zeolite reached the highest carbon dioxide adsorption followed by NaY, HZSM-5-30, HiSiV-3000, and HY-5, respectively. This is by virtue of the low ratio of Si/Al of 13X with the cations that cause a strong reaction with carbon dioxide molecules.

1.3.1.2.3 Silica Material-Based Adsorbents

Silica-based materials are another generation of non-carbonaceous dry adsorbents for CO_2 capture, which characterize a high surface area, large pore size, and great mechanical stability [105]. Silica is generally employed as support on which other materials are added for carbon dioxide (CO_2) removal. Therefore, most of the research efforts on silica-based adsorbents are mainly interested in modifying various types of silicas and deploying appropriate amines types. Heretofore, many studies reported

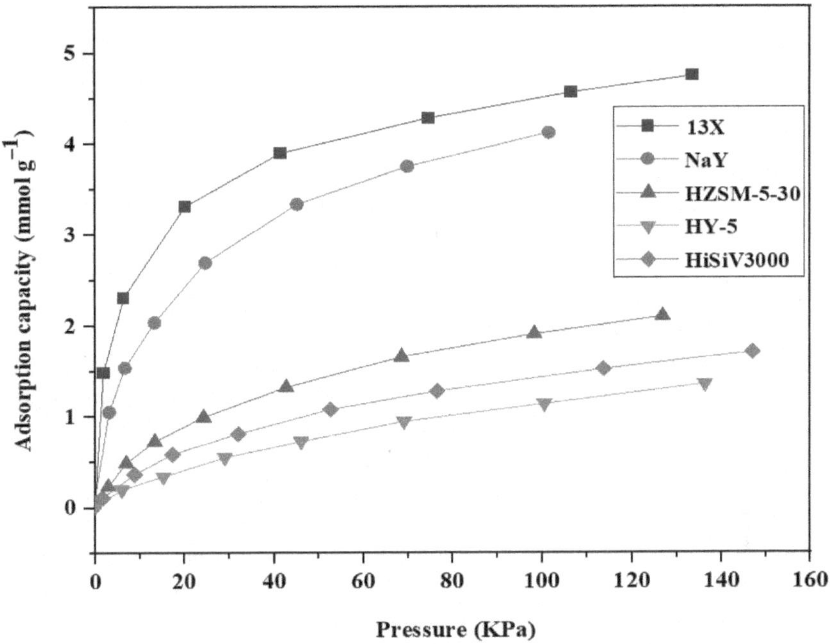

FIGURE 1.11 Comparison of carbon dioxide adsorption isotherms for fresh zeolite at 22°C based on ZSM-5 structure with (Si/Al > 1000).

the use of silica material-based adsorbents for CO_2. Three driving aspects should be taken in the process of production of silica-based material adsorbent: (1) using new preparation technologies to design amine-modified silica materials, (2) advancing new surface structures, and (3) evaluating carbon dioxide adsorption performances of various silica composites. There are a wide range of silica-based adsorbents, and recent studies show interest in the development of mesoporous silica materials for effective carbon dioxide (CO_2) removal [106].

1.3.2 Chemical Adsorption

The adsorption approach utilizing amine-based solvents (typically 30 hydroamines) [107], including ethanolamine, diethanolamine, diglycolamine, and MEA, has attracted widespread industrial attention [108] due to its high efficiency for capturing CO_2 through chemical reactions [109]. CO_2 absorption using amine-based solvents produces carbamate as the end product. The amine-based absorption process has been extensively used since the 1970s, and it is considered a first-generation technology for carbon capture [110]. Compared to conventional CO_2 separation using amine-based solvents, separation of CO_2 via adsorption by solid materials is a more sustainable technology. It is now widely adaptable due to its simplicity and effectiveness [111]. To overcome the poor gas selectivity for CO_2 in physisorbents, chemical grafting or coating is widely performed on the surface of the porous materials by incorporating basic groups that effectively interact with the acidic CO_2 gas molecules [112]. Among different basic groups, the amine is a commonly used functionality for the surface modification of CO_2 adsorbents [66]. The CO_2 gas molecule forms a chemical bond with the adsorption sites in chemisorption. Usually, CO_2 adsorption at elevated temperatures (above 140°C) is primarily governed by chemisorption [113]. Both chemisorption and physisorption can occur in the temperature range of 25–140°C. The heat of adsorption for chemisorbents can vary between −60 and −100 kJ/mol depending on the chemical functionality and bonding nature [45].

1.3.2.1 CO_2 Chemisorption Using Amine-Functionalized Porous Carbon Materials

Even though porous carbon-based materials which adsorb CO_2 molecules via physisorption exhibit remarkable CO_2 capture capacities and easy regenerability, the selectivity of CO_2 over other gas molecules is poor, and the presence of moisture hinders the CO_2 capture performance [114]. To overcome these significant drawbacks, research groups have suggested increasing the basic nature of the adsorbent surface to make the capture of acidic CO_2 molecules relatively easier. Therefore, the amine functionalization of porous carbon materials has become a common practice to enhance the basicity of adsorbents [115]. Such a process could enhance the separability of CO_2 in the presence of other gases owing to the highly favorable CO_2–amine interactions [116]. Generally, amine-functionalized porous carbon materials can be categorized into two major types based on their preparation methods [117]: (1) amine-impregnated porous sorbents where the amine groups and the carbon support interact via Van der Waals forces [118] and (2) amine-grafted porous supports where the amine species and the support are covalently bonded.

Interestingly, amine functionalization on porous carbon supports offers some advantages over absorption of CO_2 using alkanolamines, such as reduced corrosion problems due to decreased contact between amines and the equipment [117], solids being easy to handle, and, most importantly, reduced energy consumption for regeneration due to the lower heat capacity of solid sorbents compared to that of water [119].

Previous studies have reported that the CO_2 adsorption capacity of amine-grafted sugar cane bagasse depends on the amine content and hydrophobicity of the matrix fiber since the presence of hydroxyl groups in the carbon support enhances CO_2 chemisorption via changing the mechanism into a more favorable route, requiring less activation energy. Additionally, it was noted that the amine-grafted carbon supports with mild alkalinity could exhibit higher CO_2 adsorption capacities [120]. Deng and Park [121] reported that the PEI-grafted CNT demonstrates enhanced CO_2 capture performance using spacers containing an optimal or moderate chain length. Furthermore, Gunathilake et al. [122] reported that the CO_2 uptake increased from 2.84 to 3.85 mmol/g when the amidoxime content was increased from 10% to 40% at 120°C. Similar observations were reported by Zhodi et al. [123] for amine-grafted CNT/hollow silica particle composites. It was also reported that the CO_2 adsorption capacity decreases with increased grafted amine loading due to pore blockage by larger amine groups that prevent effective CO_2 diffusion at low temperatures [124].

1.3.3 Pre-Combustion Adsorption

Pre-combustion adsorption relates to the treatment of a synthetic gas composed principally of CO_2 and hydrogen. Gasification of coal or reforming of natural gas with oxygen leads to a mixture composed principally of$CO_2 + H_2$ (on dry) and then, after conversion of CO_2 with steam, to a mixture of $CO_2 + H_2$. CO_2 can then be separated using a physical absorbent (which is less expensive as an investment and less penalizing for the efficiency than if the exhaust gases are treated, because the gas is under pressure and there is no nitrogen); CO_2 is sent to the compression unit while hydrogen is used as input to a combined cycle to produce electricity. The integrated gasification combined cycle (IGCC) concept has not so far been presented by the American DOE as being the "clean coal" power station of the future in a context in which something else other than electricity has to be produced from coal (e.g., synthetic fuels). However, this system is still handicapped by low availability and high investment and production costs. The optimum separation process will be chosen as a function of the partial pressure of CO_2 in the gas to be treated. In the case of pre-combustion adsorption, the partial pressure of CO_2 depends on the IGCC process chosen for coal or the pressure of the reforming process chosen for natural gas. However, the increase in pressure is not a free operation, and it must be assured that the gain in capture due to the high pressure is not cancelled out by the additional energy that has to be supplied.

Pre-combustion technology for CO_2 adsorption involves capturing carbon dioxide from syngas before combustion in power generation. Key methods include the Selexol process, which uses a glycol-based solvent to absorb CO_2 and H_2S, achieving up to 95% capture efficiency. Other techniques like the Purisol process utilize *N*-methyl-2-pyrrolidone for high-pressure applications, while the calcium looping process offers a cost-effective solution by cycling CaO and CO_2. This technology is particularly beneficial in IGCC systems, where CO_2 concentrations are higher, making capture more efficient and economically viable [125].

1.3.4 Post-Combustion Adsorption

Post-combustion adsorption consists of treating exhaust gases on the output side of the PC or natural gas combined cycle (NGCC). As mentioned earlier, technologies based on chemical absorption appear to be best adapted to this separation. Other technologies, adsorption, membranes, and cryogeny, are even less suitable for post-combustion adsorption than for pre-combustion capture, mainly for the following two reasons: a much lower partial pressure of CO_2 in post-combustion exhaust gases than in synthetic gas originating from a gasifier or a reformer: about 50 mbars in NGCC exhaust gases and 100 mbars in PC exhaust gases compared with 8 bars (for the Puertollano-type IGCC) to more than 16 bars (for the Texaco IGCC-slurry and NGCC before reforming) in synthetic gas—the presence of larger quantities of dust, impurities such as SO_x and NO_x and incondensable gases, particularly oxygen (which do not exist or exist only in very small quantities in synthetic gas). The choice of the absorbent is guided by the partial pressure of CO_2 in the gas to be treated. As mentioned earlier, the very low levels of CO_2 partial pressures in exhaust gases impose chemical absorption. Chemical absorption (processes with amines, Benfield process with potassium carbonate) is an industrial technology and has been used for decades, but in the past it has only been used to treat very clean gas mixtures containing no or few impurities such as dust, SO_x, NO_x, oxygen (ammonia industry and fertilizer industry, separation of CO_2 from natural gas), or containing only very small quantities of them. Therefore, chemical absorption will have to satisfy two challenges before it can be applied to the capture of CO_2 in exhaust gases from coal power stations, first related to its strong energy demand for the regeneration of the absorbent (increasing with decreasing of CO_2 partial pressure) and second degradation of the absorbent by impurities contained in the exhaust gases to be treated. Degradation of chemical solvents will be treated subsequently in the chapter concerning technical limitations. CO_2 adsorption technology for post-combustion capture is a promising method for reducing GHG emissions from power plants and industrial sources. This technology utilizes solid sorbents to capture CO_2 from flue gas, offering energy savings compared to traditional amine-based absorption methods. Recent advancements focus on developing novel adsorbent materials with enhanced capacity and stability while minimizing regeneration energy requirements. Effective gas–solid contact systems and innovative regeneration strategies are also critical for optimizing the overall efficiency and cost-effectiveness of CO_2 capture processes. Post-combustion CO_2 adsorption technology focuses on capturing carbon dioxide from flue gases after fossil fuel combustion. This method is gaining traction due to its potential for energy savings compared to traditional amine-based absorption methods. Key advancements include the development of novel adsorbent materials that enhance adsorption capacity and reduce regeneration energy requirements [126]. Effective gas–solid contact systems and innovative regeneration strategies are crucial for optimizing performance and cost-effectiveness in CO_2 capture processes [127]. Overall, adsorption offers a promising alternative for significant emission reductions in industrial applications [128].

1.3.5 Temperature Swing Adsorption

TSA is carried out either in direct mode, in which a hot stream (steam, CO_2) is used as a purge gas in direct contact with the adsorbent particles, or in indirect mode, in which a heat exchanger is used to provide the required heat for regeneration. The regeneration energy requirement in TSA comprises sensible heat required to heat the adsorbent to the target regeneration temperature, reaction heat to drive the endothermic desorption reaction, and latent heat of vaporization if the sorbent has adsorbed water as well [129]. This regeneration mode has been applied in the different reactor configurations with pros and cons as will be discussed in what follows. The various research works have focused on reducing the overall heat requirement in this TSA mode through sorbent development with improved adsorption capacity and lower specific heat capacity but, more importantly, with lower heat of reaction. Others focused on reducing the energy requirement through heat integration in the process [130].

1.3.5.1 Indirect TSA

Indirect TSA is done by heating or cooling the adsorbent by means of a heat exchanger in the reactor, as illustrated in Figure 1.12. This heating mode was widely

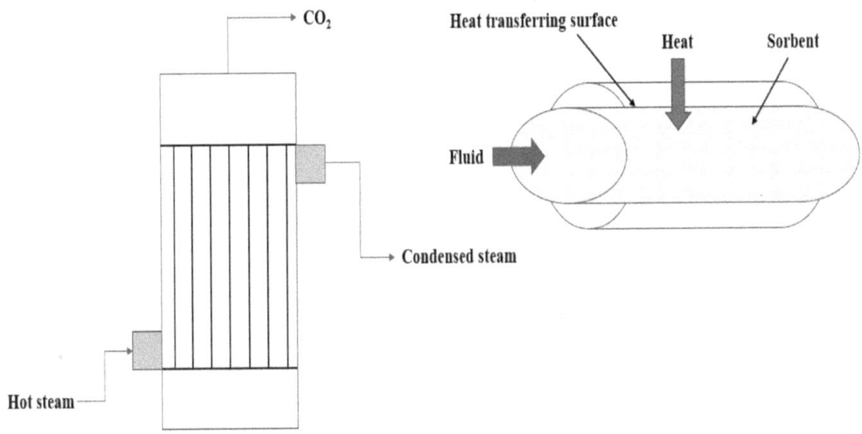

FIGURE 1.12 Indirect temperature swing adsorption.

used in early works in a fixed-bed configuration due to its simplicity and the achieved high CO_2 purity if no purging is applied through the bed [131]. Its major drawback remains the very low tube-to-bed heat transfer coefficient in a fixed-bed configuration making such a heating mode inherently unfeasible in a conventional fixed bed, and it shows a very low heat transfer coefficient in the fixed bed, although some gas purge could improve the convective heat transfer [132]. But remains limited though against the values achieved in a fluidized bed that could be up to an order of magnitude larger. This advantage makes the fluidized bed favored for indirect TSA, not only for supplying heat in the regeneration but also for improving the control of temperature in the adsorption to maximize the working adsorption capacity and improve the capture efficiency [133]. Nevertheless, a good example of a successful experience of indirect heating in a fixed bed was demonstrated in a hollow polymeric fiber that enabled cycling fast TSA in the reactor by pumping hot/cold fluid through the hollow structure discussed earlier.

1.3.5.2 Direct TSA

Direct TSA is the mode of regeneration in which the adsorbent is heated directly by means of a hot gas stream, preferably steam (Figure 1.13). This TSA mode has a much better heating rate than the indirect one, thus decreasing the heating time. The main advantage remains the capability of achieving sufficiently high purity of CO_2 with this mode explaining the widespread application of such an option to the different reactor configurations discussed in the previous section. The addition of steam reduces the CO_2 concentration, which further increases the driving force for desorption, reducing the size of the temperature swing required. However, this also has some drawbacks such as (1) the need for an additional unit operation for water removal from the captured CO_2 before being compressed for transportation and storage; (2) the used sorbent should tolerate the presence of water and in many cases a drying step is needed before starting a new cycle (this depends on the nature

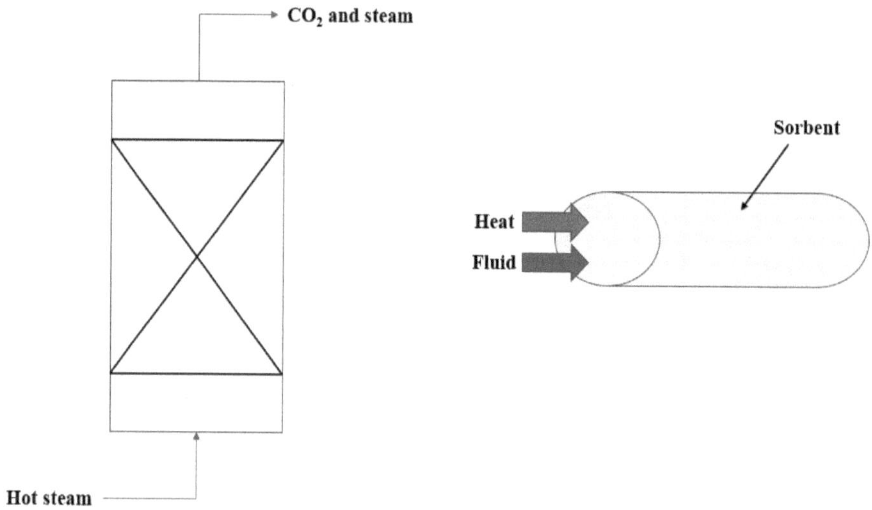

FIGURE 1.13 Direct temperature swing adsorption.

of sorbent and operating conditions in the regeneration step). Such additional steps increase the process complexity, the energy penalty, and costs [134].

1.3.6 Electrothermal Swing Adsorption

Electrothermal swing adsorption (ESA) is considered another potential mode to reduce the energy penalty of adsorption-based CO_2 capture. The ESA mode is conducted by heating the adsorbent beds by means of Joule heating inside the particles by passing an electric current [135]. This in-situ heating of the particles enables a fast heat transfer rate in comparison to a conventional TSA mode and also provides better desorption kinetics [136]. The essential feature for the adsorbent to work under an ESA mode is its electrical conductivity. Activated carbon fiber was considered a potential sorbent that works in the ESA mode of regeneration [137]. The ESA mode is, however, applicable only in a fixed bed configuration where the long cooling time counteracts the advantage of the in-situ fast heating, thus hindering the prospects of such a mode in implementation at the industrial scale. From an economic point of view, the main drawback is that electrical energy is many times more costly than the low-grade heat typically used for regeneration in the TSA processes.

1.3.7 Vacuum Swing Adsorption

Another widely used regeneration approach is by varying the pressure, commonly known as vacuum swing adsorption (VSA) or PSA. In post-combustion CO_2, only VSA makes sense as the PSA requires pressurizing the incoming large quantities of the flue gas from the source, making the process uneconomical. VSA is predominantly applied to a fixed-bed reactor configuration with physisorption-based sorbents such as zeolites or activated carbon [138]. However, as emphasized in the previous section, the VSA regeneration mode can achieve a very short cycle, but a low-pressure drop across the bed in adsorption is a prerequisite, thus favoring the structured advanced fixed bed. Additionally, a high CO_2 capture rate requires deep vacuum levels, making this regeneration mode suitable only for industrial applications producing flue gases with high CO_2 partial pressure [139]. The first proposal of VSA regeneration mode comprises four steps: pressurization with feed gas, adsorption, forward blowdown, and reverse evacuation.

1.3.8 Pressure Swing Adsorption

CCS is a key tool in the global commitment to tackle climate change. Absorption, whether chemical or physical, is commonly regarded as the most mature technology for CO_2 capture. Nevertheless, absorption suffers from some drawbacks, such as high energy requirements and corrosion of process equipment. Adsorption is considered a promising alternative, especially when the regeneration process is carried out through a pressure reduction (i.e., PSA), with the potential for reducing energy penalty, environmental impact, and cost of CO_2 capture [140]. A significant research effort has been already undertaken to develop PSA technology for CO_2 capture applications. A wide range of adsorbent materials has been synthesized and studied [141]. However, material science is still very active, and newly discovered adsorbents are currently under investigation. With regard to the engineering of the process, PSA cycles have been studied in depth [142]. Some key criteria to define the most efficient way to use and regenerate adsorbents can be already pinpointed. Further work is ongoing, focusing on decreasing the relative energy consumption and on process optimization. The literature is less comprehensive with regard to the integration of PSA processes into the power plants and the relative system performance. Some analyses at a system level have been published lately [143], touching upon these topics and giving some indications of the strengths and weaknesses of PSA in the contexts investigated.

The performance of a PSA-based plant was demonstrated to be competitive with that associated with other decarbonization technologies, chemical absorption in the first instance [144]. Table 1.4 reports the relative simulation outputs of a large advanced supercritical (ASC) coal-fired power plant implementing CO_2 capture by PSA. To the authors' knowledge, this is the only post-combustion full-plant analysis available in the literature. The performance is evaluated in terms of net electric efficiency (η_{el})

TABLE 1.4
Performances of Power Plants Implementing Post-Combustion CO_2 Capture

Case	CO_2 Capture Technology	R_{co2} (%)	η_{el} (HLV) (%)	Ref
ASC-absorption	Abs. (MEA)	90.0	34.2	[144]
ASC-PSA	PSA	90.2	34.8	[144]

The nomenclature used refers to CO_2 recovery (R_{CO2}) and net electric efficiency (η_{el}).

and CO_2 recovery (the CO_2 purity is not reported, but it is implied that the systems achieve the 95% specification requested for safe transport and storage of CO_2). For comparison the simulation outputs of the same plant using a common amine absorption process as CO_2 capture technology are also shown. Steam is used for solvent regeneration in the absorption process, while only electric power needs to be supplied to the PSA process. In spite of the different energy consumption patterns, the final net electric efficiency is similar for the two cases tested, with PSA displaying a slightly lower energy penalty. Regarding the CO_2 separation performance, both technologies allow for meeting the CO_2 separation targets (i.e., 90% CO_2 recovery and 95% CO_2 purity). A serious issue envisaged regards the size of the PSA unit. Common PSA processes would result in an unfeasible footprint. Taking into account some conservative design criteria, concerning the minimum fluidization velocity and the maximum pressure drop in the adsorption columns, the PSA unit would need an unacceptable number of columns to process the whole flue gas flow rate. Further studies must focus on an intensification of the process. Structured adsorbents, allowing higher superficial velocities with reduced pressure drops, are interesting from this perspective but necessitate further investigation [145]. Otherwise, different process frameworks are probably to be considered (e.g., moving bed adsorption reactors). Another issue, which has often been disregarded in the literature, concerns the detrimental effect of water on typical physisorbents, zeolites in the first instance. Since the presence of water is unavoidable in flue gases, methods for effective removal must prove to perform satisfactorily integrated in the complex arrangement of a power plant, while keeping as low as possible the additional power consumption. The issue would no longer apply if water-resistant adsorbents were demonstrated to be applicable.

1.4 CARBON CAPTURE, UTILIZATION, AND STORAGE

To achieve the prevention of carbon dioxide emissions in the environment, a portfolio of technologies has been proposed, one of which is carbon capture, utilization, and storage (CCUS) [146]. CCUS comprises CCS and CCU, both of which involve the capture of CO_2 from various sources such as large stationary power plants [37]. However, they differ in the destination of the captured CO_2: (1) in CCS, CO_2 is transported and injected into underground storage sites; (2) CO_2 is converted to commercial products in CCU. The energy trilemma including energy security, energy affordability, and environmental sustainability makes CCUS the most probable technology for reducing emissions from power plants and carbon-intensive industries [147]. In addition to special CCUS rules, other regulations governing environmental and climate protection involve CCUS. Generally, these notices gradually increase the significance of CCUS as a priority technology for a sustainable future. The development of the CCUS regulatory framework can be divided into three stages, and each stage has similar objectives.

1.5 CONCLUSION

This chapter is an introduction to the various technologies and methods of surface adsorption for carbon dioxide absorption. Surface adsorption is based on mass transfer to absorb carbon dioxide on the surface. Surface adsorption is actually due to the presence of molecules and particles of materials on the surface with high surface energy. Today, methods such as pre-combustion, post-combustion, oxy-fuel combustion, chemical ring combustion, cryogenic separation, and membrane adsorption are used to prevent gas emissions. In addition, a series of advantages and disadvantages of the previous methods have been mentioned in this chapter, the most important advantages of surface adsorption technology can be high efficiency, low cost, reduced energy consumption, and high properties of CO_2, and its most important disadvantages can be high corrosion from acid production, low capacity, and high operating cost. Also, the method of using and storing carbon absorption is mostly used when it is to prevent GHG emissions on a larger scale. In addition, various methods such as physical adsorption, chemical adsorption, pre-combustion adsorption, post-combustion adsorption, TSA, PSA, and VSA can be mentioned, with the post-combustion adsorption method being the most used. Also, in this chapter, adsorbents NaOH, graphene oxide/TiO_2 nanocomposite, AC/NaOH, active carbon–KOH, Na_2CO_3 / Al_2O_3, and K_2CO_3 / Al_2O_3 have been used.

REFERENCES

1. López, A., et al., Characterization of MIPA and DIPA aqueous solutions in relation to absorption, speciation and degradation. *Journal of Industrial and Engineering Chemistry*, 2015. **21**: p. 428–435.
2. Ramazani, R., et al., Characterization of monoethanolamine+ potassium lysinate blend solution as a new chemical absorbent for CO_2 capture. *International Journal of Greenhouse Gas Control*, 2016. **51**: p. 29–35.
3. Bajpai, A. and M.K. Mondal, Equilibrium solubility of CO_2 in aqueous mixtures of DEA and AEEA. *Journal of Chemical & Engineering Data*, 2013. **58**(6): p. 1490–1495.
4. Stec, M., et al., Density of unloaded and CO_2-loaded aqueous solutions of piperazine and 2-amino-2-methyl-1-propanol and their mixtures from 293.15 to 333.15 K. *Physics and Chemistry of Liquids*, 2016. **54**(4): p. 475–486.
5. Pashaei, H., H. Mashhadimoslem, and A. Ghaemi, Modeling and optimization of CO_2 mass transfer flux into Pz-KOH-CO_2 system using RSM and ANN. *Scientific Reports*, 2023. **13**(1): p. 4011.
6. Rochelle, G.T., Amine scrubbing for CO_2 capture. *Science*, 2009. **325**(5948): p. 1652–1654.
7. Khan, A.A., G. Halder, and A. Saha, Experimental investigation of sorption characteristics of capturing carbon dioxide into piperazine activated aqueous 2-amino-2-methyl-1-propanol solution in a packed column. *International Journal of Greenhouse Gas Control*, 2016. **44**: p. 217–226.
8. Benamor, A. and A. Aroussi. Modeling analysis of CO_2 solubility in solutions of DEA, MDEA and their mixtures using Kent-Eisenberg and Deshmukh-Mather models. In *3rd International Conference on Chemical, Biological and Environment Sciences (ICCEBS'2013)*. Kuala Lumpur, Malaysia. 2013. p. 80e85.
9. Rochelle, G., et al., Aqueous piperazine as the new standard for CO_2 capture technology. *Chemical Engineering Journal*, 2011. **171**(3): p. 725–733.
10. Etemad, E., A. Ghaemi, and M. Shirvani, Rigorous correlation for CO_2 mass transfer flux in reactive absorption processes. *International Journal of Greenhouse Gas Control*, 2015. **42**: p. 288–295.
11. Mortazavi-Manesh, S., M. Satyro, and R.A. Marriott, Modelling carbon dioxide solubility in ionic liquids. *The Canadian Journal of Chemical Engineering*, 2013. **91**(4): p. 783–789.

12. Boer, D.G., J. Langerak, and P.P. Pescarmona, Zeolites as selective adsorbents for CO_2 separation. *ACS Applied Energy Materials*, 2023. **6**(5): p. 2634–2656.
13. Khosrowshahi, M.S., et al., The role of surface chemistry on CO_2 adsorption in biomass-derived porous carbons by experimental results and molecular dynamics simulations. *Scientific Reports*, 2022. **12**(1): p. 8917.
14. Ren, F. and W. Liu, Review of CO_2 adsorption materials and utilization technology. *Catalysts*, 2023. **13**(8): p. 1176.
15. Reddy, M.S.B., et al., Carbon dioxide adsorption based on porous materials. *RSC Advances*, 2021. **11**(21): p. 12658–12681.
16. Hoang, T.-D., et al., Carbon-based synthesized materials for CO_2 adsorption and conversion: Its potential for carbon recycling. *Recycling*, 2023. **8**(4): p. 53.
17. Krishna, R. and J.M. van Baten, A comparison of the CO_2 capture characteristics of zeolites and metal–organic frameworks. *Separation and Purification Technology*, 2012. **87**: p. 120–126.
18. Lestari, W.W., et al., Composite material consisting of HKUST-1 and Indonesian activated natural zeolite and its application in CO_2 capture. *Open Chemistry*, 2019. **17**(1): p. 1279–1287.
19. Fathalian, F., et al., Efficient CO_2 adsorption using chitosan, graphene oxide, and zinc oxide composite. *Scientific Reports*, 2024. **14**(1): p. 3186.
20. Kang, G., et al., Effect of pressure and temperature on CO_2/CH_4 competitive adsorption on kaolinite by Monte Carlo simulations. *Materials*, 2020. **13**(12): p. 2851.
21. Wang, S., et al., Adsorption behaviour of molecular sieve and activated carbon for CO_2 adsorption at cold temperatures. *Carbon Neutrality*, 2022. **1**(1): p. 16.
22. Gul, A. and U.T. Un, Effect of temperature and gas flow rate on CO_2 capture. *European Journal of Sustainable Development Research*, 2022. **6**(2): p. em0181.
23. Pashaei, H., M.N. Zarandi, and A. Ghaemi, Experimental study and modeling of CO_2 absorption into diethanolamine solutions using stirrer bubble column. *Chemical Engineering Research and Design*, 2017. **121**: p. 32–43.
24. Rezaei, B. and S. Riahi, Prediction of CO_2 loading of amines in carbon capture process using membrane contactors: A molecular modeling. *Journal of Natural Gas Science and Engineering*, 2016. **33**: p. 388–396.
25. Heydarifard, M., et al., Reactive absorption of CO_2 into Piperazine aqueous solution in a stirrer bubble column: Modeling and experimental. *International Journal of Greenhouse Gas Control*, 2018. **79**: p. 91–116.
26. Amiri, M. and S. Shahhosseini, Optimization of CO_2 capture from simulated flue gas using K_2CO_3/Al_2O_3 in a micro fluidized bed reactor. *Energy & Fuels*, 2018. **32**(7): p. 7978–7990.
27. Goli, A., et al., An overview of biological processes and their potential for CO_2 capture. *Journal of Environmental Management*, 2016. **183**: p. 41–58.
28. Mohammad, N.K., A. Ghaemi, and K. Tahvildari, Hydroxide modified activated alumina as an adsorbent for CO_2 adsorption: Experimental and modeling. *International Journal of Greenhouse Gas Control*, 2019. **88**: p. 24–37.
29. Behroozi, A.H., N. Akbarzad, and A. Ghaemi, CO 2 reactive absorption into an aqueous blended MDEA and TMS solution: Experimental and modeling. *International Journal of Environmental Research*, 2020. **14**: p. 347–363.
30. Wang, Y., et al., A review of post-combustion CO_2 capture technologies from coal-fired power plants. *Energy Procedia*, 2017. **114**: p. 650–665.
31. Metz, B., et al., *IPCC special report on carbon dioxide capture and storage*. 2005: Cambridge: Cambridge University Press.
32. Figueroa, J.D., et al., Advances in CO_2 capture technology—the US Department of Energy's Carbon Sequestration Program. *International Journal of Greenhouse Gas Control*, 2008. **2**(1): p. 9–20.

33. Merkel, T.C., et al., Power plant post-combustion carbon dioxide capture: An opportunity for membranes. *Journal of Membrane Science*, 2010. **359**(1–2): p. 126–139.
34. Lee, H.J., et al., Gas hydrate formation process for pre-combustion capture of carbon dioxide. *Energy*, 2010. **35**(6): p. 2729–2733.
35. Huang, X., et al., Opportunities and challenges of oxy-fuel combustion. *Oxy-fuel Combustion*, 2018: p. 1–12.
36. Talei, S., et al., Oxyfuel combustion makes carbon capture more efficient. *ACS Omega*, 2024. **9**(3): p. 3250–3261.
37. Cuéllar-Franca, R.M. and A. Azapagic, Carbon capture, storage and utilisation technologies: A critical analysis and comparison of their life cycle environmental impacts. *Journal of CO_2 Utilization*, 2015. **9**: p. 82–102.
38. Adánez-Rubio, I., et al., Chemical looping combustion of biomass: CLOU experiments with a Cu-Mn mixed oxide. *Fuel Processing Technology*, 2018. **172**: p. 179–186.
39. Abuelgasim, S., W. Wang, and A. Abdalazeez, A brief review for chemical looping combustion as a promising CO_2 capture technology: Fundamentals and progress. *Science of the Total Environment*, 2021. **764**: p. 142892.
40. Momeni, M., et al., Hybrid membranes for carbon capture. In Sustainable Agriculture Reviews 38: *Carbon Sequestration Vol. 2 Materials and Chemical Methods*. 2020: Cham: Springer International Publishing. p. 85–120.
41. Khalilpour, R., et al., Membrane-based carbon capture from flue gas: A review. *Journal of Cleaner Production*, 2015. **103**: p. 286–300.
42. Ashraf Talesh, S.S., et al., Adsorption properties of carbon dioxide and methane by synthesized fine particles of SAPO-34 molecular sieve. *Iranian Journal of Chemistry and Chemical Engineering (IJCCE)*, 2010. **29**: p. 41–49.
43. Ghaemi, A., A. Hemmati, and H. Mashhadimoslem, Non-equilibrium modeling of CO 2 reactive-absorption process using sodium hydroxide–ammonia–water solution in a packed bed column. *Journal of the Iranian Chemical Society*, 2021. **18**: p. 2303–2314.
44. Rahmandoost, E., B. Roozbehani, and M.H. Maddahi, Experimental studies of CO_2 capturing from the flue gases. *Iranian Journal of Oil and Gas Science and Technology*, 2014. **3**(4): p. 1–15.
45. Afandiyeva, L., et al., Investigation of organic complexes of imidazolines based on synthetic oxy-and petroleum acids as corrosion inhibitors. *Iranian Journal of Chemistry and Chemical Engineering (IJCCE)*, 2018. **37**(3): p. 73–79.
46. Saeidi, M., A. Ghaemi, and K. Tahvildari, CO_2 capture exploration on potassium hydroxide employing response surface methodology, isotherm and kinetic models. *Iranian Journal of Chemistry and Chemical Engineering (IJCCE)*, 2020. **39**(5): p. 255–267.
47. Saeidi, M., et al., Exploiting response surface methodology (RSM) as a novel approach for the optimization of carbon dioxide adsorption by dry sodium hydroxide. *Journal of the Chinese Chemical Society*, 2018. **65**(12): p. 1465–1475.
48. Nazari Kudahi, S., A. Noorpoor, and N.M. Mahmoodi, Adsorption performance indicator for power plant CO_2 capture on graphene oxide/TiO_2 nanocomposite. *Iranian Journal of Chemistry and Chemical Engineering (IJCCE)*, 2019. **38**(3): p. 293–307.
49. Karbalaei Mohammad, N., et al., Experimental investigation and modeling of CO_2 adsorption using modified activated carbon. *Iranian Journal of Chemistry and Chemical Engineering (IJCCE)*, 2020. **39**(1): p. 177–192.
50. Siriwardane, R.V., et al., Adsorption of CO_2 on molecular sieves and activated carbon. *Energy & Fuels*, 2001. **15**(2): p. 279–284.
51. Derylo-Marczewska, A., et al., Adsorption of chlorophenoxy pesticides on activated carbon with gradually removed external particle layers. *Chemical Engineering Journal*, 2017. **308**: p. 408–418.
52. Tan, Y., et al., Adsorption of carbon dioxide by sodium hydroxide-modified granular coconut shell activated carbon in a fixed bed. *Energy*, 2014. **77**: p. 926–931.

53. Siriwardane, R.V., et al., Novel regenerable sodium-based sorbents for CO_2 capture at warm gas temperatures. *Energy & Fuels*, 2007. **21**(4): p. 2088–2097.
54. Ghosh, A., et al., Uptake of H_2 and CO_2 by graphene. *The Journal of Physical Chemistry C*, 2008. **112**(40): p. 15704–15707.
55. Lee, J.B., et al., Sodium-based dry regenerable sorbent for carbon dioxide capture from power plant flue gas. *Industrial & Engineering Chemistry Research*, 2008. **47**(13): p. 4465–4472.
56. Kazemi, H., S. Shahhosseini, and M. Amiri, Optimization of CO_2 capture process using dry sodium-based sorbents. *Iranian Journal of Chemistry and Chemical Engineering (IJCCE)*, 2021. **40**(4): p. 1179–1194.
57. Alinezhad, H., et al., Application of MEA, TEPA and morpholine grafted NaY zeolite as CO_2 capture. *Iranian Journal of Chemistry and Chemical Engineering (IJCCE)*, 2021. **40**(2): p. 581–592.
58. Amiri, M., S. Shahhosseini, and A. Ghaemi, Optimization of CO_2 capture process from simulated flue gas by dry regenerable alkali metal carbonate based adsorbent using response surface methodology. *Energy & Fuels*, 2017. **31**(5): p. 5286–5296.
59. Ghaemi, A., Z. Jafari, and E. Etemad, Prediction of CO_2 mass transfer flux in aqueous amine solutions using artificial neural networks. *Iranian Journal of Chemistry and Chemical Engineering (IJCCE)*, 2020. **39**(4): p. 269–280.
60. Pashaei, H., et al., Experimental modeling and optimization of CO_2 absorption into piperazine solutions using RSM-CCD methodology. *ACS Omega*, 2020. **5**(15): p. 8432–8448.
61. Ghaemi, A. and A.H. Behroozi, Comparison of hydroxide-based adsorbents of Mg (OH) 2 and Ca (OH) 2 for CO_2 capture: Utilization of response surface methodology, kinetic, and isotherm modeling. *Greenhouse Gases: Science and Technology*, 2020. **10**(5): p. 948–964.
62. Abdollahi, S. and Z. Zarei, Reduction of CO_2 emission and production costs by using pozzolans in Lamerd cement factory. *Iranian Journal of Chemistry and Chemical Engineering (IJCCE)*, 2018. **37**(1): p. 223–230.
63. Fashi, F., A. Ghaemi, and P. Moradi, Piperazine-modified activated alumina as a novel promising candidate for CO_2 capture: Experimental and modeling. *Greenhouse Gases: Science and Technology*, 2019. **9**(1): p. 37–51.
64. Pashaei, H., et al., Experimental investigation of the effect of nano heavy metal oxide particles in Piperazine solution on CO_2 absorption using a stirrer bubble column. *Energy & Fuels*, 2018. **32**(2): p. 2037–2052.
65. Zhao, M., S. Ou, and C.-D. Wu, Improvement of the CO_2 capture capability of a metal–organic framework by encapsulating dye molecules inside the mesopore space. *Crystal Growth & Design*, 2017. **17**(5): p. 2688–2693.
66. Wang, J., et al., Application of polyethylenimine-impregnated solid adsorbents for direct capture of low-concentration CO_2. *AIChE Journal*, 2015. **61**(3): p. 972–980.
67. Kumar, A., et al., Direct air capture of CO_2 by physisorbent materials. *Angewandte Chemie International Edition*, 2015. **54**(48): p. 14372–14377.
68. Pashaei, H., A. Ghaemi, and M. Nasiri, Modeling and experimental study on the solubility and mass transfer of CO 2 into aqueous DEA solution using a stirrer bubble column. *RSC Advances*, 2016. **6**(109): p. 108075–108092.
69. Pashaei, H. and A. Ghaemi, CO_2 absorption into aqueous diethanolamine solution with nano heavy metal oxide particles using stirrer bubble column: Hydrodynamics and mass transfer. *Journal of Environmental Chemical Engineering*, 2020. **8**(5): p. 104110.
70. Besagni, G. and F. Inzoli, The effect of liquid phase properties on bubble column fluid dynamics: Gas holdup, flow regime transition, bubble size distributions and shapes, interfacial areas and foaming phenomena. *Chemical Engineering Science*, 2017. **170**: p. 270–296.

71. Chu, F., et al., Mass transfer and energy consumption for CO_2 absorption by ammonia solution in bubble column. *Applied Energy*, 2017. **190**: p. 1068–1080.
72. Ghaemi, A., et al., Kinetics and absorption rate of CO_2 into partially carbonated ammonia solutions. *Chemical Engineering Communications*, 2011. **198**(10): p. 1169–1181.
73. Niknafs, H., A. Ghaemi, and S. Shahhosseini, Dynamic heat and mass transfer modeling and control in carbon dioxide reactive absorption process. *Heat and Mass Transfer*, 2015. **51**: p. 1131–1140.
74. Norouzbahari, S., S. Shahhosseini, and A. Ghaemi, Modeling of CO_2 loading in aqueous solutions of piperazine: Application of an enhanced artificial neural network algorithm. *Journal of Natural Gas Science and Engineering*, 2015. **24**: p. 18–25.
75. Ghaemi, A., S. Shahhosseini, and M.G. Maragheh, Nonequilibrium dynamic modeling of carbon dioxide absorption by partially carbonated ammonia solutions. *Chemical Engineering Journal*, 2009. **149**(1–3): p. 110–117.
76. Naeem, S., A. Ghaemi, and S. Shahhosseini, Experimental investigation of CO 2 capture using sodium hydroxide particles in a fluidized bed. *Korean Journal of Chemical Engineering*, 2016. **33**: p. 1278–1285.
77. Naeem, S., S. Shahhosseini, and A. Ghaemi, Simulation of CO_2 capture using sodium hydroxide solid sorbent in a fluidized bed reactor by a multi-layer perceptron neural network. *Journal of Natural Gas Science and Engineering*, 2016. **31**: p. 305–312.
78. Younas, M., et al., Feasibility of CO 2 adsorption by solid adsorbents: A review on low-temperature systems. *International Journal of Environmental Science and Technology*, 2016. **13**: p. 1839–1860.
79. Pal, A., et al., Experimental investigation of CO_2 adsorption onto a carbon based consolidated composite adsorbent for adsorption cooling application. *Applied Thermal Engineering*, 2016. **109**: p. 304–311.
80. Jahanmiri, A.H. and A. Eghbalpoor, Gas absorption with chemical reaction in turbulent flow. *Iranian Journal of Chemistry and Chemical Engineering (IJCCE)*, 1993. **12**(2): p. 85–90.
81. Kazemi, S., A. Ghaemi, and K. Tahvildari, Chemical absorption of carbon dioxide into aqueous piperazine solutions using a stirred reactor. *Iranian Journal of Chemistry and Chemical Engineering (IJCCE)*, 2020. **39**(4): p. 253–267.
82. Najafi, P., H. Ramezanipour Penchah, and Ů. Ghaemi, Improving CO_2/N_2 and CO_2/H_2 selectivity of hypercrosslinked carbazole-based polymeric adsorbent for environmental protection. *Journal of Chemical and Petroleum Engineering*, 2020. **54**(2): p. 311–321.
83. Cui, J., Recent advances in post-combustion CO_2 capture via adsorption methods. *Highlights in Science, Engineering and Technology*, 2022. **6**: p. 172–181.
84. Das, A., et al., Advancements in adsorption based carbon dioxide capture technologies-A comprehensive review. *Heliyon*, 2023. **9**: p. e22341.
85. Nwabueze, Q.A. and S. Leggett, Advancements in the application of CO_2 capture and utilization technologies—A comprehensive review. *Fuels*, 2024. **5**(3): p. 508–532.
86. Li, J.-R., et al., Carbon dioxide capture-related gas adsorption and separation in metal-organic frameworks. *Coordination Chemistry Reviews*, 2011. **255**(15–16): p. 1791–1823.
87. Bilalis, P., et al., Non-covalent functionalization of carbon nanotubes with polymers. *RSC Advances*, 2014. **4**(6): p. 2911–2934.
88. Cinke, M., et al., CO_2 adsorption in single-walled carbon nanotubes. *Chemical Physics Letters*, 2003. **376**(5–6): p. 761–766.
89. Su, F., C. Lu, and H.-S. Chen, Adsorption, desorption, and thermodynamic studies of CO_2 with high-amine-loaded multiwalled carbon nanotubes. *Langmuir*, 2011. **27**(13): p. 8090–8098.
90. Foley, H.C., Carbogenic molecular sieves: Synthesis, properties and applications. *Microporous Materials*, 1995. **4**(6): p. 407–433.

91. Burchell, T., et al., A novel process and material for the separation of carbon dioxide and hydrogen sulfide gas mixtures. *Carbon*, 1997. **35**(9): p. 1279–1294.
92. Alcañiz-Monge, J., J.P. Marco-Lozar, and M.Á. Lillo-Ródenas, CO_2 separation by carbon molecular sieve monoliths prepared from nitrated coal tar pitch. *Fuel Processing Technology*, 2011. **92**(5): p. 915–919.
93. Bikshapathi, M., et al., Preparation of carbon molecular sieves from carbon micro and nanofibers for sequestration of CO_2. *Chemical Engineering Research and Design*, 2011. **89**(9): p. 1737–1746.
94. Silvestre-Albero, J., et al., Ultrahigh CO 2 adsorption capacity on carbon molecular sieves at room temperature. *Chemical Communications*, 2011. **47**(24): p. 6840–6842.
95. Kumar, H.P. and M.A. Xavior, Graphene reinforced metal matrix composite (GRMMC): A review. *Procedia Engineering*, 2014. **97**: p. 1033–1040.
96. Millward, A.R. and O.M. Yaghi, Metal– organic frameworks with exceptionally high capacity for storage of carbon dioxide at room temperature. *Journal of the American Chemical Society*, 2005. **127**(51): p. 17998–17999.
97. Liu, Y., Z.U. Wang, and H.C. Zhou, Recent advances in carbon dioxide capture with metal-organic frameworks. *Greenhouse Gases: Science and Technology*, 2012. **2**(4): p. 239–259.
98. Chester, A.W. and E.G. Derouane, *Zeolite Characterization and Catalysis*. Vol. 360. 2009: Springer.
99. Zhang, J., R. Singh, and P.A. Webley, Alkali and alkaline-earth cation exchanged chabazite zeolites for adsorption based CO_2 capture. *Microporous and Mesoporous Materials*, 2008. **111**(1–3): p. 478–487.
100. Dunne, J., et al., Calorimetric heats of adsorption and adsorption isotherms. 2. O_2, N_2, Ar, CO_2, CH_4, $C2H_6$, and SF_6 on NaX, H-ZSM-5, and Na-ZSM-5 zeolites. *Langmuir*, 1996. **12**(24): p. 5896–5904.
101. Coriani, S., et al., On the molecular electric quadrupole moment and the electric-field-gradient-induced birefringence of CO_2 and CS_2. *Chemical Physics Letters*, 2000. **326**(3–4): p. 269–276.
102. Siriwardane, R.V., M.-S. Shen, and E.P. Fisher, Adsorption of CO_2, N_2, and O_2 on natural zeolites. *Energy & Fuels*, 2003. **17**(3): p. 571–576.
103. Inui, T., Y. Okugawa, and M. Yasuda, Relationship between properties of various zeolites and their carbon dioxide adsorption behaviors in pressure swing adsorption operation. *Industrial & Engineering Chemistry Research*, 1988. **27**(7): p. 1103–1109.
104. Harlick, P.J. and F.H. Tezel, An experimental adsorbent screening study for CO_2 removal from N_2. *Microporous and Mesoporous Materials*, 2004. **76**(1–3): p. 71–79.
105. Qin, C., et al., Effect of support material on the performance of K_2CO_3-based pellets for cyclic CO_2 capture. *Applied Energy*, 2014. **136**: p. 280–288.
106. Hakiki, A., et al., Synthesis and characterization of mesoporous silica SBA-15 functionalized by mono-, di-, and tri-amine and its catalytic behavior towards Michael addition. *Materials Chemistry and Physics*, 2018. **212**: p. 415–425.
107. Tu, R., et al., Bio-tar-derived porous carbon with high gas uptake capacities. *Renewable Energy*, 2021. **167**: p. 82–90.
108. Benedetti, V., et al., CO_2 Adsorption study on pure and chemically activated chars derived from commercial biomass gasifiers. *Journal of CO_2 Utilization*, 2019. **33**: p. 46–54.
109. Prasetyo, I., et al., Nanoporous carbon prepared from palm kernel shell for CO_2/CH_4 separation. *Waste and Biomass Valorization,* 2020. **11**: p. 5599–5606.
110. Hu, H., et al., Functionalization of multi-walled carbon nanotubes with phenylenediamine for enhanced CO 2 adsorption. *Adsorption*, 2017. **23**: p. 73–85.
111. Auta, M., et al., Diethanolamine functionalized waste tea activated carbon for CO_2 adsorption. In *International Conference on Chemical, Environmental and Biological Science*. Dubai. 2015.

112. Kulkarni, V., D. Panda, and S.K. Singh, Direct air capture of CO_2 over amine-modified hierarchical silica. *Industrial & Engineering Chemistry Research*, 2023. **62**(8): p. 3800–3811.
113. Matito-Martos, I., et al., Potential of CO_2 capture from flue gases by physicochemical and biological methods: A comparative study. *Chemical Engineering Journal*, 2021. **417**: p. 128020.
114. Das, D. and B. Meikap, Role of amine-impregnated activated carbon in carbon dioxide capture. *Indian Chemical Engineer*, 2021. **63**(4): p. 435–447.
115. Gómez-Pozuelo, G., et al., CO_2 adsorption on amine-functionalized clays. *Microporous and Mesoporous Materials*, 2019. **282**: p. 38–47.
116. Cabriga, C.K.C., et al., Evaluation of biochar derived from the slow pyrolysis of rice straw as a potential adsorbent for carbon dioxide. *Biomass Conversion and Biorefinery*, 2021: p. 1–8.
117. Ohs, B., M. Krödel, and M. Wessling, Adsorption of carbon dioxide on solid amine-functionalized sorbents: A dual kinetic model. *Separation and Purification Technology*, 2018. **204**: p. 13–20.
118. Gunawardene, O.H., et al., Carbon Dioxide capture through physical and chemical adsorption using porous carbon materials: A review. *Atmosphere*, 2022. **13**(3): p. 397.
119. Peng, H.-L., et al., Chitosan-derived mesoporous carbon with ultrahigh pore volume for amine impregnation and highly efficient CO_2 capture. *Chemical Engineering Journal*, 2019. **359**: p. 1159–1165.
120. Luo, S., et al., Preparation and characterization of amine-functionalized sugarcane bagasse for CO_2 capture. *Journal of Environmental Management*, 2016. **168**: p. 142–148.
121. Deng, M. and H.G. Park, Spacer-assisted amine-coiled carbon nanotubes for CO_2 capture. *Langmuir*, 2019. **35**(13): p. 4453–4459.
122. Gunathilake, C., et al., Amidoxime-functionalized microcrystalline cellulose–mesoporous silica composites for carbon dioxide sorption at elevated temperatures. *Journal of Materials Chemistry A*, 2016. **4**(13): p. 4808–4819.
123. Zohdi, S., M. Anbia, and S. Salehi, Improved CO_2 adsorption capacity and CO_2/CH_4 and CO_2/N_2 selectivity in novel hollow silica particles by modification with multi-walled carbon nanotubes containing amine groups. *Polyhedron*, 2019. **166**: p. 175–185.
124. Tian, Y., et al., Surface-microporous graphene for CO_2 adsorption. *Catalysis Today*, 2020. **356**: p. 514–518.
125. Chen, Z. A review of pre-combustion carbon capture technology. In *2022 7th International Conference on Social Sciences and Economic Development (ICSSED 2022)*. 2022: Atlantis Press.
126. Raganati, F., F. Miccio, and P. Ammendola, Adsorption of carbon dioxide for post-combustion capture: A review. *Energy & Fuels*, 2021. **35**(16): p. 12845–12868.
127. Allangawi, A., et al., Carbon capture materials in post-combustion: Adsorption and absorption-based processes. *C*, 2023. **9**(1): p. 17.
128. Zanco, S.E., et al., Postcombustion CO_2 capture: A comparative techno-economic assessment of three technologies using a solvent, an adsorbent, and a membrane. *ACS Engineering Au*, 2021. **1**(1): p. 50–72.
129. Zhang, W., et al., Parametric study on the regeneration heat requirement of an amine-based solid adsorbent process for post-combustion carbon capture. *Applied Energy*, 2016. **168**: p. 394–405.
130. Park, J., et al., One-dimensional modeling of a turbulent fluidized bed for a sorbent-based CO_2 capture process with solid–solid sensible heat exchange. *Energy*, 2019. **168**: p. 1168–1180.
131. Bonjour, J., J.-B. Chalfen, and F. Meunier, Temperature swing adsorption process with indirect cooling and heating. *Industrial & Engineering Chemistry Research*, 2002. **41**(23): p. 5802–5811.

132. Ntiamoah, A., et al., CO_2 capture by temperature swing adsorption: Use of hot CO_2-rich gas for regeneration. *Industrial & Engineering Chemistry Research*, 2016. **55**(3): p. 703–713.
133. Mondino, G., et al., Moving bed temperature swing adsorption for CO_2 capture from a natural gas combined cycle power plant. *International Journal of Greenhouse Gas Control*, 2019. **85**: p. 58–70.
134. Dhoke, C., et al., Review on reactor configurations for adsorption-based CO_2 capture. *Industrial & Engineering Chemistry Research*, 2021. **60**(10): p. 3779–3798.
135. Petkovska, M., et al., Temperature-swing gas separation with electrothermal desorption step. *Separation Science and Technology*, 1991. **26**(3): p. 425–444.
136. Lillia, S., et al., A comprehensive modeling of the hybrid temperature electric swing adsorption process for CO_2 capture. *International Journal of Greenhouse Gas Control*, 2018. **74**: p. 155–173.
137. Zhao, Q., et al., CO_2 capture using a novel hybrid monolith (H-ZSM5/activated carbon) as adsorbent by combined vacuum and electric swing adsorption (VESA). *Chemical Engineering Journal*, 2019. **358**: p. 707–717.
138. Shen, C., et al., Two-stage VPSA process for CO_2 capture from flue gas using activated carbon beads. *Industrial & Engineering Chemistry Research*, 2012. **51**(13): p. 5011–5021.
139. Krishnamurthy, S., V. R. Rao, S. Guntuka, P. Sharratt, R. Haghpanah, A. Rajendran, M. Amanullah, I. A. Karimi, and S. Farooq. CO_2 capture from dry flue gas by vacuum swing adsorption: A pilot plant study. *AIChE Journal*, 2014. **60**(5): p. 1830–1842.
140. Glier, J.C. and E.S. Rubin, Assessment of solid sorbents as a competitive post-combustion CO_2 capture technology. *Energy Procedia*, 2013. **37**: p. 65–72.
141. Hedin, N., et al., Adsorbents for the post-combustion capture of CO_2 using rapid temperature swing or vacuum swing adsorption. *Applied Energy*, 2013. **104**: p. 418–433.
142. Abanades, J., et al., Emerging CO_2 capture systems. *International Journal of Greenhouse Gas Control*, 2015. **40**: p. 126–166.
143. Liu, Z. and W.H. Green, Analysis of adsorbent-based warm CO_2 capture technology for integrated gasification combined cycle (IGCC) power plants. *Industrial & Engineering Chemistry Research*, 2014. **53**(27): p. 11145–11158.
144. Riboldi, L. and O. Bolland, Evaluating Pressure Swing Adsorption as a CO_2 separation technique in coal-fired power plants. *International Journal of Greenhouse Gas Control*, 2015. **39**: p. 1–16.
145. Thiruvenkatachari, R., et al., A site trial demonstration of CO_2 capture from real flue gas by novel carbon fibre composite monolith adsorbents. *International Journal of Greenhouse Gas Control*, 2015. **42**: p. 415–423.
146. Wang, M. and E. Oko, *Special Issue on Carbon Capture in the Context of Carbon Capture, Utilisation and Storage (CCUS)*. 2017: Springer. p. 1–4.

2 Mass Transfer, Mechanism, and Diffusion on CO_2 Adsorption

Mahyar Ashourzadeh Pordsari, Fatemeh Bahmanzadegan, and Ahad Ghaemi

2.1 INTRODUCTION

Understanding mass transfer and diffusion processes is essential for optimizing adsorbents across various applications, such as environmental remediation and industrial processes. The efficiency of adsorption processes is largely governed by mass transfer mechanisms, which include external mass transfer, intraparticle diffusion, and surface diffusion. These mechanisms determine adsorption's kinetics and adsorbent materials' overall effectiveness. The external mass transfer involves the movement of adsorbate molecules from the bulk solution to the adsorbent's surface and is influenced by factors like concentration gradients and system hydrodynamics. For instance, a higher initial adsorbate concentration can accelerate the external mass transfer rate by increasing the driving force for diffusion. Additionally, the design aspects of the adsorption bed, such as its height and configuration, can influence external mass transfer, though this impact may be minimal under certain conditions [1]. After the adsorbates reach the surface, intraparticle diffusion becomes the rate-limiting step, especially in systems with high adsorbate concentrations and larger particle sizes. Intraparticle diffusion involves the movement of adsorbate molecules into the pores of the adsorbent, with pore size and structure playing a significant role. In microporous materials, surface diffusion often limits the adsorption rate, while molecular and Knudsen diffusion dominates in macroporous materials [2, 3]. Various kinetic models, such as the intraparticle diffusion model and the pseudo-first-order model, are utilized to analyze adsorption kinetics and identify rate-limiting steps [4, 5]. These models aid in the development of more effective adsorbents by elucidating how intraparticle diffusion influences overall adsorption kinetics [6]. The structural properties of adsorbents, such as pore size distribution, are also critical in influencing diffusion characteristics, with larger mesopores facilitating faster diffusion, while smaller micropores may impede diffusion due to steric effects [7]. Thus, optimizing pore size distribution and surface characteristics can significantly enhance

DOI: 10.1201/9781003476191-2

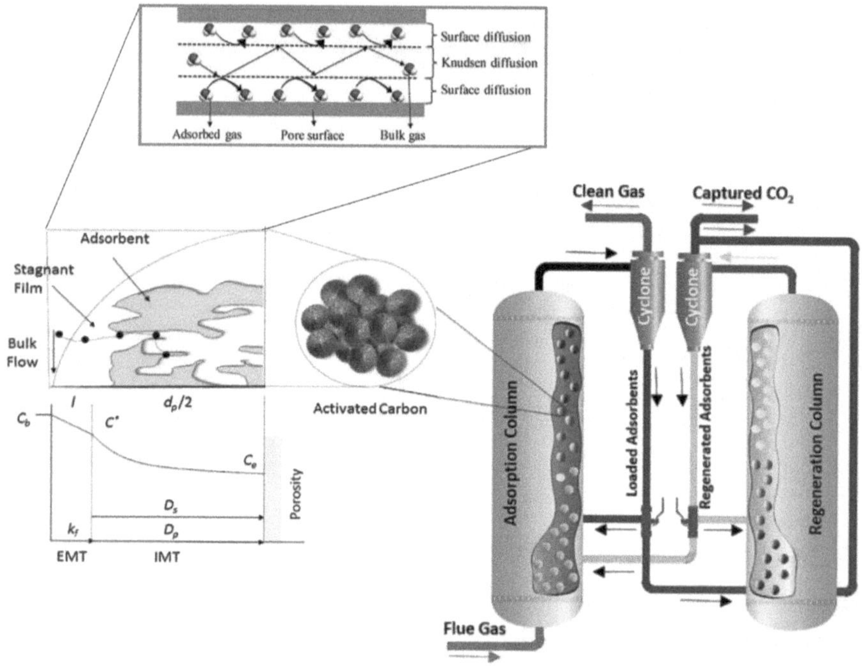

FIGURE 2.1 Diagram of CO_2 adsorption emphasizes mass transfer.

adsorption performance [8]. Figure 2.1 illustrates the various diffusion mechanisms, including surface diffusion, Knudsen diffusion, and bulk gas diffusion, which govern the transport of CO_2 molecules to the adsorbent's active sites. It focuses on the detailed mass transfer pathway, showing how CO_2 molecules diffuse from the bulk gas phase through a stagnant film into the porous structure of the adsorbent (e.g., activated carbon). It also emphasizes key parameters like film mass transfer coefficients (MTCs) (k_f) and diffusion coefficients (D_s). The industrial setup demonstrates the cyclic process of adsorption and regeneration. Flue gas enters the adsorption column, where CO_2 is captured on the adsorbent's surface, while clean gas exits.

2.2 MECHANISMS OF CO_2 ADSORPTION

The nature of weak interactions between CO_2 molecules and adsorbent surfaces is primarily characterized by van der Waals forces and other noncovalent interactions, which lead to relatively low adsorption energies. These weak interactions are crucial in determining the efficiency of CO_2 capture in various materials. One significant aspect of these interactions is the low adsorption energy associated with CO_2 on surfaces like boron nitride (BN). Research indicates that the adsorption energy of CO_2 on neutral BN sheets is approximately 4.46 kcal/mol, indicative of weak interactions primarily governed by van der Waals forces [9]. This finding is consistent with other studies that highlight the weak binding nature of CO_2 to various adsorbents, where the interaction energies are often in the range of 0.217 to 0.39 eV at high coverage, suggesting the stabilization of weakly bound states [10]. In adsorbents where weak forces are the primary mechanism for adsorption, surface area is a critical factor in determining a material's adsorption capacity [11]. Materials with higher specific surface areas offer more active sites for CO_2 molecules to adhere to, directly influencing their adsorption efficacy [12]. For example, porous materials like metal-organic frameworks (MOFs) exhibit remarkably high surface areas, often surpassing 2000 m^2/g, facilitating substantial CO_2 uptake [13]. Similarly, research on sulfur-doped nanoporous carbons has shown that increasing surface area significantly enhances CO_2 adsorption, with BET surface areas ranging from 837 to 2865 m^2/g [14]. Porosity, especially the distribution of pore sizes, is also a crucial factor in CO_2 adsorption [15]. Microporous materials are particularly effective in capturing CO_2 due to their ability to confine CO_2 molecules within their small pores. For instance, studies on functionalized 3D covalent organic frameworks have demonstrated that a microporous structure promotes CO_2 adsorption and separation, with porosity values ranging from 36.18% to 56.20% [16]. Additionally, the presence of ultramicropores (<0.7 nm) has been shown to enhance CO_2 adsorption, as these tiny pores effectively trap CO_2 molecules [17]. The interaction between the adsorbate and adsorbent is further influenced by porosity, as a more significant number of pores can increase the number of active sorption sites, thereby boosting overall adsorption capacity. Moreover, the synergy between surface area and porosity is vital for optimizing CO_2 adsorption. High porosity enables the accommodation of more CO_2 molecules, while a high surface area ensures sufficient active sites for effective interaction. For instance, synthesizing porous benzimidazole-linked polymers has shown that combining high surface area with porosity leads to significant CO_2 uptake, with one polymer achieving 235 mg/g at 273 K and 1 bar [18]. However, it is essential to recognize that excessive porosity can result in pore blockage, which may reduce adsorption efficiency [19]. Therefore, a balanced approach in designing materials with optimal surface area and porosity is crucial for enhancing CO_2 adsorption performance. Activated carbon is renowned for its high surface area and porosity, making it an efficient adsorbent for various gases and liquids. For example, it has been shown that the physisorption of oxygen on activated carbon is enhanced at lower temperatures, leading to an increased concentration of surface oxygen species and, consequently, greater reactivity in oxidation reactions [20, 21]. The adsorption properties of activated carbon can also be customized by selecting specific activation agents and carbonization conditions, which affect its surface area and pore structure [22]. This versatility allows activated carbon to be used in various applications, from air purification to energy storage [23, 24]. Zeolites, another class of adsorbents, exhibit unique physisorption characteristics due to their well-defined microporous structures [25, 26]. Research has shown that hierarchical ZSM-5 zeolite exhibits distinct nitrogen physisorption isotherms, highlighting its microporous nature and the absence of mesopores in conventional samples [27]. The interaction of gases with zeolites is also influenced by temperature; for instance, it has been found that the adsorption of hydrogen sulfide on faujasite zeolites is primarily governed by physisorption at elevated temperatures, indicating that the strength of these interactions varies significantly with thermal conditions [28]. MOFs are distinguished by their exceptional porosity and tunable structures, making them promising gas storage and separation candidates [29]. Studies on the physisorption of hydrogen in MOFs reveal that the presence of coordinatively unsaturated metal sites enhances hydrogen adsorption through stronger polarization effects [30]. Additionally, the performance of MOFs in gas adsorption is influenced by temperature and pressure. For example, the microporosity of carbon-based materials plays a significant role in hydrogen physisorption at low temperatures, where interactions are primarily governed by van der Waals forces [31].

The impact of temperature and pressure on physisorption is significant. Generally, increasing temperature reduces the extent of physisorption due to the increased kinetic energy of adsorbate molecules, which can overcome the weak van der Waals forces binding them to the adsorbent [32]. Conversely, increasing pressure enhances physisorption by increasing the concentration of adsorbate molecules near the adsorbent, thereby promoting more interactions [33]. Understanding this relationship is essential for designing effective adsorption systems in various industrial applications, such as gas storage and environmental remediation.

The interaction between CO_2 and adsorbents can be broadly classified into two main mechanisms: physisorption and chemisorption. Weaker van der Waals forces drive physisorption, while chemisorption involves the formation of stronger chemical bonds, typically associated with electron transfer or the creation of new chemical species [34]. The surface characteristics of the adsorbent significantly influence the efficiency of CO_2 capture. For example, surface hydroxyl groups and basic metal oxides can enhance adsorption capacity by facilitating the formation of chemical bonds with CO_2 [35]. This effect is particularly pronounced in materials such as amine-functionalized adsorbents, where $-NH_2$ groups can form carbamate species with CO_2, thereby substantially increasing the adsorption capacity [36–38]. The interaction between CO_2 and these functional groups is marked by the formation of strong covalent bonds, which are more favorable than the weaker interactions typical of physisorption [39]. Additionally, the chemical state of the adsorbent plays a crucial role in the adsorption process. For instance, oxygen vacancies in metal oxides like cerium oxide (CeO_2) have been shown to enhance CO_2 adsorption by providing additional active sites for interaction. The carbonate species formed during CO_2 adsorption on CeO_2 surfaces suggest a preference for specific configurations that are energetically favorable [40, 41]. Similar behavior is observed in other materials, such as layered double hydroxides, where the chemical nature of the support material significantly impacts CO_2 adsorption performance [42]. Furthermore, the surface morphology and the presence of defects or broken bonds on the adsorbent surface can affect the adsorption mechanisms, which can reduce the activation energy required for CO_2 to bond with the surface, thereby facilitating the adsorption process [43]. Metal oxides, such as magnesium ferrite, have demonstrated effective adsorption of heavy metals and gases via chemisorption mechanisms. Amine-functionalized magnesium ferrite, for example, exhibits considerable adsorption capacity due to the stable bonds formed between the amine groups and the target species [44]. The surface chemistry of these metal oxides can be engineered to enhance their adsorption properties. By introducing functional groups like amines, the number of active sites available for chemisorption can be increased, thereby improving the overall adsorption capacity [45]. Additionally, the interaction between CO_2 and the metal oxide surface may result in the formation of carbamate species, underscoring the significance of surface chemistry in determining the efficiency of chemisorption processes [46].

Amine-functionalized adsorbents have become increasingly important due to their selective adsorption of CO_2 over other gases such as N_2. The chemisorption process in these materials often involves the formation of carbamate ions through the reaction between CO_2 and amine groups [47]. Research suggests that the adsorption isotherms of these materials can be effectively modeled using the dual-site Langmuir model, which accounts for the presence of both physisorption and chemisorption sites [48]. The binding strength of CO_2 to these functionalized surfaces is significantly influenced by the type and density of amine groups, as well as the overall surface area of the adsorbent [49]. For instance, studies have shown that increasing the number of amine groups on silica support enhances CO_2 uptake, highlighting the critical role of surface chemistry in optimizing adsorption performance [50].

Ionic liquids (ILs) are increasingly recognized for their potential in gas capture applications due to their unique properties, including negligible vapor pressure and high thermal stability. Chemisorption in ILs typically involves the interaction of CO_2 with basic anions or cations, leading to the formation of stable reaction products [51, 52]. Task-specific ionic liquids (TSILs) functionalized with amine groups have shown significant CO_2 chemisorption capabilities, with reported capacities of up to 0.5 moles of CO_2 per mole of TSIL under ambient conditions [53]. The efficiency of these ILs in capturing CO_2 is heavily influenced by their surface chemistry, particularly the basicity of the anions, which enhances their interaction with CO_2 [54]. Moreover, the design of ILs can be optimized to improve their reactivity and stability, thereby enhancing their overall performance in chemisorption applications [55, 56].

2.3 THERMODYNAMICS OF CO_2 ADSORPTION

Enthalpy (ΔH) and entropy (ΔS) are key thermodynamic parameters that elucidate the energy transformations and disorders associated with the adsorption process [57]. In CO_2 capture, the enthalpy of adsorption typically reflects the strength of the interaction between CO_2 molecules and the adsorbent surface [58]. Research indicates that the isosteric heat of adsorption decreases with increased surface loading, suggesting the presence of energetically diverse adsorption sites and possible lateral interactions among CO_2 molecules [59]. This observation is particularly significant for materials like amine-impregnated silica, where chemical adsorption mechanisms prevail, resulting in the formation of carbamate species during the process [60]. Enthalpic contributions to adsorption can be considerable, with values around 90 kJ/mol reported for various amines, indicating strong CO_2–adsorbent interactions [61]. Entropy also plays a crucial role in adsorption. The entropy change during CO_2 adsorption is affected by the coverage degree on the adsorbent surface. For instance, primary amines show higher adsorption efficiencies than secondary amines, attributed to entropic factors at low coverages [62]. Furthermore, thermodynamic analysis of framework deformation in zeolites during CO_2 adsorption highlights that structural changes can induce significant entropic effects, which are vital for understanding adsorption behavior under varying conditions [63]. The balance between enthalpic and entropic contributions is essential, as high adsorption enthalpy may complicate adsorbent regeneration, whereas favorable entropy changes can improve CO_2 capture efficiency [64]. Additionally, the temperature dependence of CO_2 adsorption significantly affects both enthalpy and entropy. Higher temperatures can accelerate the adsorption kinetics, leading to a quicker attainment of thermodynamic equilibrium [65]. However, while low temperatures are thermodynamically advantageous for adsorption, they may also slow the adsorption rate due to kinetic constraints, as seen in mesoporous materials studies [66]. This interplay between temperature, enthalpy, and entropy underscores the complexity of the CO_2 adsorption process and the necessity for meticulous optimization of operational conditions to enhance capture efficiency. Examining the impact of temperature involves the calculation of the thermodynamic equilibrium constant at various temperatures utilizing Eq. (2.1).

$$\Delta G^0 = -RTLnK_d \tag{2.1}$$

The Vant Hoff equation determined the thermodynamic and enthalpy parameters associated with adsorption, denoted as ΔH^0. The slope of the natural logarithm of the distribution coefficient (ln K_d) curve is expressed as a function of the reciprocal of temperature ($1/T$) [67].

$$K_d = \left(\frac{P_i - P_e}{P_i}\right) \times \left(\frac{V}{w}\right) \tag{2.2}$$

2.3.1 Gibbs Free Energy and Spontaneity

Gibbs free energy (ΔG) is a fundamental thermodynamic parameter that determines the spontaneity of processes, including adsorption phenomena. In the context of CO_2 adsorption, ΔG assesses the feasibility of the adsorption process. A negative ΔG value indicates that the process is spontaneous, occurring without the need for external energy input [68, 69]. The spontaneity of CO_2 adsorption can be quantitatively evaluated through the Gibbs free energy change, which is influenced by the changes in enthalpy (ΔH) and entropy (ΔS) during adsorption. This relationship is described by the Gibbs free energy equation (2.3):

$$\Delta G^0 = \Delta H^0 - T\Delta S^0 \tag{2.3}$$

where T represents the absolute temperature [70]. A negative ΔG value signifies that the process is thermodynamically favorable. For instance, research has demonstrated that various adsorbents exhibit spontaneous CO_2 adsorption with negative ΔG values, reflecting the exothermic nature of the process [71]. Additionally, the efficiency of CO_2 adsorption is often linked to the magnitude of ΔG; a more negative ΔG generally indicates higher adsorption efficiency, suggesting stronger interactions between the adsorbent and CO_2 molecules. For example, studies show that materials such as biochar and MOFs exhibit enhanced adsorption capacities when ΔG values are significantly negative, indicating their effectiveness in capturing CO_2 from gas

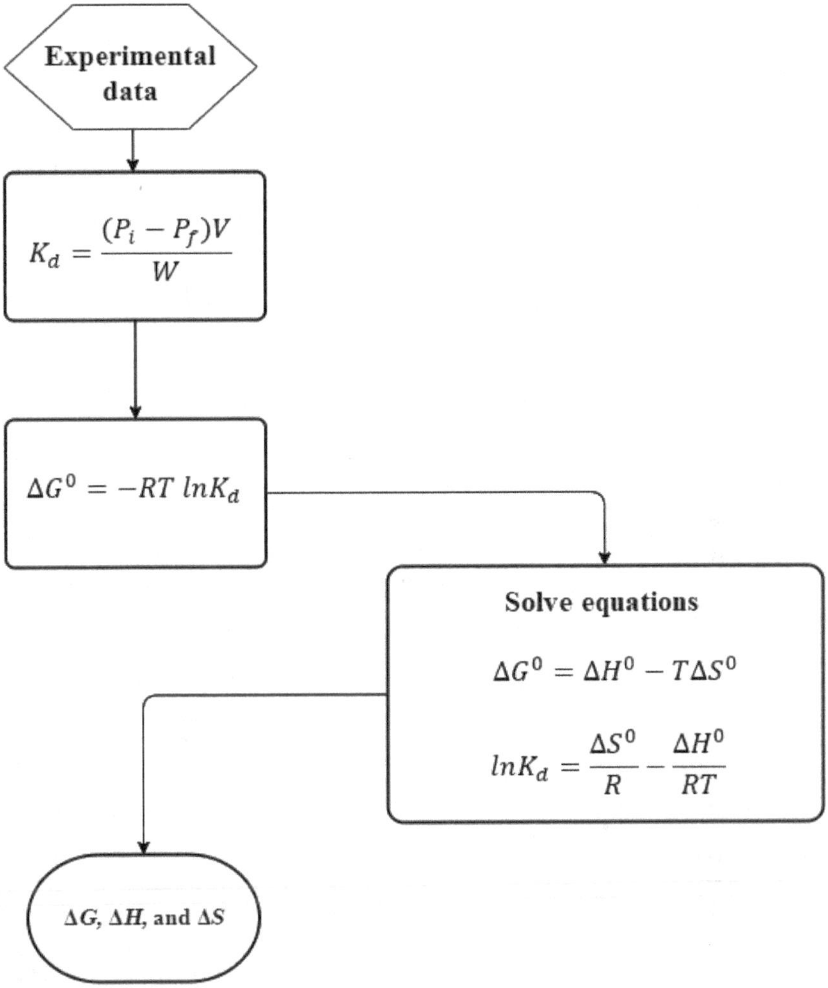

FIGURE 2.2 Algorithm for obtaining crucial thermodynamic parameters.

mixtures [72–74]. Adsorption isotherm models, such as Langmuir and Freundlich, are frequently used to analyze the relationship between ΔG and adsorption capacity, providing further insight into the efficiency of the adsorption process [75]. Figure 2.2 exhibits an algorithm for obtaining crucial thermodynamic parameters.

Bakhtiari et al. [57] used several types of alumina for CO_2 adsorption. The enthalpy for physical adsorption processes is below 20 kJ/mol, whereas chemical adsorption, involving chemical bonding, exhibits enthalpy values greater than 40 kJ/mol. The enthalpy values for the adsorption processes of both adsorbents are negative, indicating the exothermic nature of surface adsorption accompanied by energy release. Specifically, the enthalpy values for the Al_2O_3-A and 60% PEG/Al_2O_3-A adsorbents are 17.838 and 9.760 kJ/mol, respectively. These results confirm that

CO_2 adsorption by both adsorbents is characterized as a physical adsorption process$\left(\Delta H^\circ < 20 \text{ kJ/mol}\right)$ and exothermic, as evidenced by the negative enthalpy values [57].

2.3.2 Isotherms and Thermodynamic Models

Understanding adsorption mechanisms and optimizing materials for carbon capture critically rely on isotherm models. Each isotherm offers distinct insights into the adsorption properties of different adsorbents, which is crucial for advancing CO_2 capture technologies. The Langmuir isotherm model operates under the assumption of monolayer adsorption on a surface with a finite number of identical sites. This model is particularly effective for describing adsorbents with uniform adsorption energies and a limited number of active sites. Research indicates that many adsorbents, such as activated carbon and molecularly imprinted polymers, align well with the Langmuir model, suggesting a strong CO_2 affinity at specific sites. The role of isotherm models in understanding adsorption mechanisms cannot be overstated, as they provide crucial insights into the adsorption properties of different adsorbents, thereby playing a key role in advancing CO_2 capture technologies [76, 77]. For example, studies have shown that CO_2 capture on certain superactivated hydrochars conforms closely to the Langmuir isotherm, implying effective monolayer adsorption [78].

In contrast, the Freundlich isotherm model is suitable for heterogeneous surfaces where adsorption occurs on a non-uniform distribution of sites, facilitating multilayer adsorption. This model is especially relevant when the adsorption process is reversible and not limited to monolayer formation. The Freundlich model has been utilized to describe CO_2 adsorption on various materials, revealing diverse binding sites with varying affinities for CO_2 [79, 80]. For instance, molecularly imprinted polymers often exhibit adsorption behavior that aligns better with the Freundlich isotherm, indicating a more intricate interaction with CO_2 compared to the Langmuir model alone [81]. The BET isotherm extends the Langmuir model to accommodate multilayer adsorption, making it particularly useful for analyzing porous materials. It assumes that the initial layer of adsorbate molecules forms via a chemical reaction with the adsorbent, while subsequent layers are formed through physical adsorption. The BET model has been extensively applied to study CO_2 adsorption on porous materials, including activated carbons and MOFs [82, 83]. Research demonstrates that the BET model effectively characterizes CO_2 adsorption, providing valuable insights into the surface area and porosity of adsorbents, which are essential for optimizing their performance in carbon capture applications [84, 85]. Table 2.1 provides mentioned isotherms with their mathematical equation and description of their parameters.

Fathalian et al. investigated CO_2 adsorption on a novel composite adsorbent comprising chitosan (CTS), graphene oxide (GO), and zinc oxide (ZnO). Equilibrium adsorption data were analyzed using various isotherm models. The fitted isotherm curves at 298 K are also presented in Figure 2.3. The results indicate a decrease in the Langmuir model constant (q_L) with increasing temperature, confirming the exothermic nature of the CO_2 adsorption process. Similarly, the Freundlich model constant (K_F), which reflects the adsorbate's affinity for the solid sorbent, also decreases

TABLE 2.1
Isotherm Models Used Frequently in CO_2 Adsorption Modeling

Isotherm Model	Mathematical Equation	Parameters
Langmuir	$q_e = \frac{q_m K_L P_e}{1 + K_L P_e}$	K_L: Langmuir constant q_m: Maximum capacity
Freundlich	$q_e = K_F P_e^{1/n_F}$	K_F: Freundlich constant n_F: Freundlich exponent
Brunauer–Emmett–Teller (BET)	$\frac{P}{V(P^0 - P)} = \frac{1}{CV_m} + \frac{C-1}{CV_m} * \frac{P}{P^0}$	V_m: Monolayer adsorption amount V : adsorption amount P: equilibrium pressure

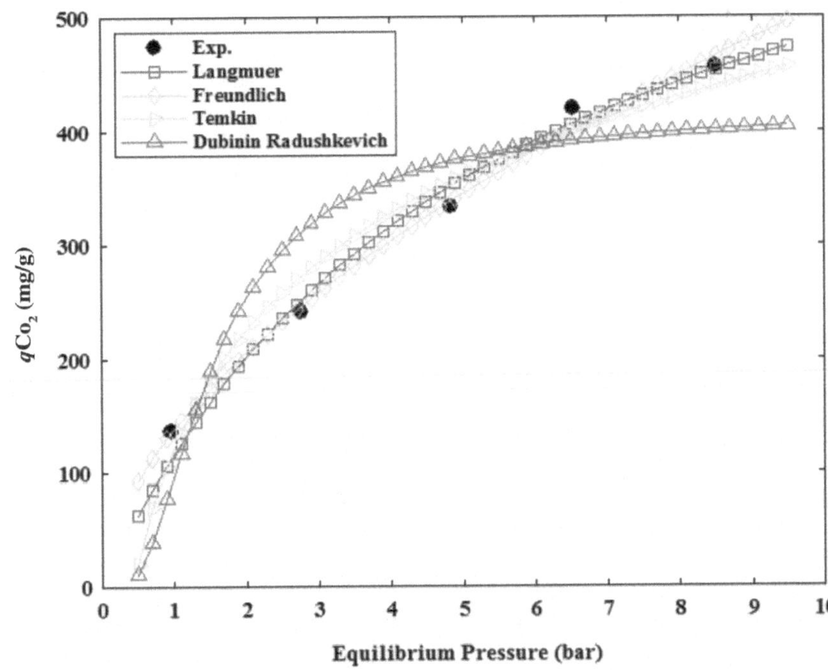

FIGURE 2.3 The fitted curves of the CO_2 adsorption isotherm at 25°C [86]. Copyright © 2024, Farnoush Fathalian et al.

with rising temperature. This behavior suggests that a physisorption mechanism predominantly governs CO_2 adsorption by the CTS/GO/ZnO composite. The parameter n_F of the Freundlich model, which lies between 1 and 2, further supports the favorable nature of the CO_2 capture process. Moreover, the Dubinin–Radushkevich (D–R) model constant (ω), representing the free energy of adsorption, was found to be below 8 kJ/mol. This indicates a physical adsorption mechanism, as ω values between 8 and 16 kJ/mol would signify chemisorption. These findings align with the physisorption nature of CO_2 adsorption previously established. The Freundlich model was identified as the most suitable among the different isotherm models evaluated, demonstrating the highest R^2 value. This suggests that the CTS/GO/ZnO composite has a heterogeneous surface where multilayer CO_2 adsorption occurs. A high Freundlich constant indicates a strong adsorption capacity, while a lower exponent value corresponds to a more linear adsorption isotherm, further validating the composite's efficiency as a CO_2 adsorbent [86].

2.4 MASS TRANSFER IN CO_2 ADSORPTION

MTCs quantify the rate at which CO_2 molecules move from the bulk phase to the adsorbent surface and subsequently into the adsorbent's pores. The values of these coefficients can vary significantly depending on the type of adsorbent, the operational conditions, and the specific adsorption mechanisms involved. MTCs can be determined using various models, with the linear driving force (LDF) model being one of the most commonly employed. This model simplifies the mass transfer process by assuming that the driving force for mass transfer is proportional to the concentration difference between the bulk gas phase and the adsorbent surface. The overall MTC, denoted (k), can be calculated using the equation:

$$\frac{1}{k} = \frac{1}{k_\phi} + \frac{1}{k_\pi} \tag{2.4}$$

where k_ϕ is the film MTC, and k_π is the pore MTC. This equation highlights the contributions of both external and internal mass transfer resistances to the overall mass transfer process [87].

The MTC in CO_2 adsorption is influenced by several factors, including the characteristics of the adsorbent material, temperature, and the physical properties of the gas being adsorbed. For example, studies have indicated that the MTC for CO_2 adsorption on activated carbon ranges between 0.00186 and 0.0052 s^{-1}, with an increase over time correlating with temperature changes and desorption rates [88]. This trend suggests that higher temperatures increase the mass transfer coefficient, enhancing the desorption rate but potentially reducing the overall adsorption capacity. Different adsorbents display varying mass transfer coefficients due to their distinct structural and chemical properties. For instance, comparisons between activated carbon and zeolite 13X reveal that, while the mass transfer coefficients for activated carbon align with values reported in the literature, zeolite 13X exhibits different kinetics, indicating that the rate-limiting step for CO_2 adsorption may differ between these materials [89].

The literature extensively documents the impact of adsorbent particle size and pore structure on mass transfer coefficients. Smaller particle sizes generally improve mass transfer rates by increasing the available surface area for adsorption and reducing the diffusion path length for CO_2 molecules. This relationship underscores the need to optimize adsorbent characteristics, particularly in industrial applications where space and efficiency are critical [90]. Additionally, adsorbents with larger pore sizes promote faster mass transfer by reducing diffusion resistance. However, as adsorption increases, intraparticle diffusion resistance can become significant, necessitating a detailed analysis of both surface and pore diffusion mechanisms [91]. The interaction between these mechanisms can be effectively modeled using various kinetic models, such as the LDF model, which has been shown to describe adsorption kinetics across different adsorbents accurately [92].

Beyond pore structure and particle size, the chemical properties of the adsorbent also affect MTCs. Incorporating functional groups or modifying adsorbents can improve CO_2 adsorption capacity and mass transfer efficiency. Research on amine-modified zeolites and MOFs has shown that these materials can achieve higher adsorption capacities and faster mass transfer rates compared to traditional adsorbents [19, 93]. The interaction between CO_2 and the adsorbent surface also influences the heat of adsorption, a critical factor in determining the overall efficiency of the adsorption process [94].

A demonstration of this mechanism is depicted in Figure 2.4. Understanding these diffusion processes is crucial for designing adsorbents with optimized pore structures to enhance mass transfer rates [95].

Operational parameters such as temperature, pressure, and gas composition add further complexity to calculating MTCs. Higher temperatures generally increase the kinetic energy of gas molecules, potentially enhancing mass transfer rates. However, the impact of temperature on MTCs can be complex, as it may also affect the adsorption capacity and selectivity of the adsorbent [97, 98]. Similarly, variations in pressure can alter the driving forces for mass transfer, influencing the overall performance of the adsorption system [99]. Furthermore, the interaction between CO_2 and other components in the gas mixture can complicate mass transfer dynamics.

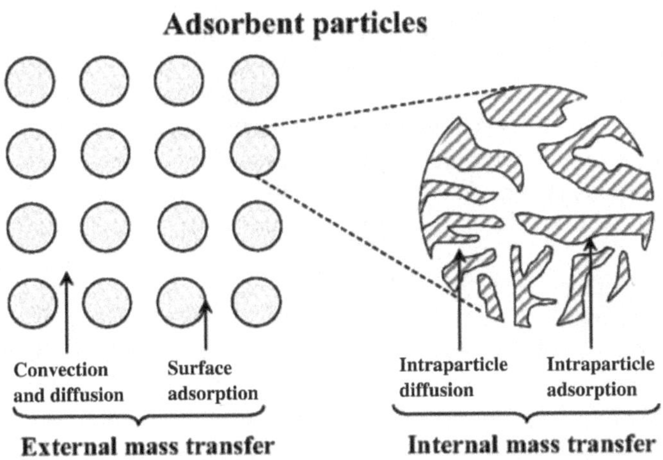

FIGURE 2.4 Different mass transfer and diffusion mechanisms involved in the adsorption process [96],

For example, the presence of water vapor has been found to influence the MTCs for CO_2 adsorption on zeolite 13X due to the competitive adsorption of water, which can reduce the effective surface area available for CO_2 [100].

Different adsorbents exhibit a range of MTCs due to their unique structural and chemical properties. For instance, activated carbon and zeolites have been extensively studied for CO_2 adsorption. Research shows that the MTCs for CO_2 adsorption on activated carbon can vary widely, with values reported in the literature ranging from 0.0145 s^{-1} for carbon molecular sieves to higher values for activated carbon, depending on the specific conditions and configurations used in experiments [89, 101]. The performance of these adsorbents is often assessed through breakthrough curves, which provide insights into the dynamics of mass transfer during adsorption [102, 103]. Variations in MTCs can be attributed to differences in pore structure, surface area, and the degree of heterogeneity in the adsorbent materials. In particular, the presence of micropores in activated carbon enhances adsorption capacity but may also introduce significant intraparticle diffusion resistance, potentially limiting the overall mass transfer rate [99].

Table 2.2 presents an overview of the correlations employed to estimate MTCs in the CO_2 adsorption process.

External mass transfer resistance refers to the barriers encountered by CO_2 molecules as they migrate from the bulk gas phase to the surface of the adsorbent. This resistance is influenced by factors such as the flow dynamics around the adsorbent particles and the thickness of the surrounding boundary layer. For instance, in fixed-bed adsorption systems, external mass transfer can significantly limit performance, particularly at low interstitial velocities, where increased axial dispersion diminishes the effectiveness of gas–solid interactions [113]. Similarly, in microporous adsorbents like NaY zeolite, external mass transfer resistance becomes pronounced, especially when the boundary layer thickness is considerable [114]. This type of resistance is pivotal as it impacts overall adsorption kinetics, determining the rate at which CO_2 is extracted from the gas phase. Furthermore, external mass transfer resistance can critically affect adsorption kinetics in large adsorbent particles or poorly distributed gas flow scenarios. For example, in zeolite-based adsorbents, larger particle sizes and associated film diffusion constraints often limit the adsorption kinetics [115]. Conversely, in fiber sorbents treated with polyethyleneimine (PEI), internal mass transfer limitations are more significant due to the fibers' unique structural properties, which introduce considerable intraparticle diffusion resistance [116]. Internal mass transfer resistance pertains to the movement of CO_2 molecules within the adsorbent's porous structure after reaching the surface. This resistance is governed by the adsorbent's pore structure, size, and connectivity. For instance, research on silica-supported poly(ethylenimine) hollow fiber sorbents has highlighted significant internal mass transfer limitations, particularly under varying temperature conditions, which impact adsorption capacity [115]. This underscores the need to optimize the internal structure of adsorbents to facilitate more efficient CO_2 diffusion through their pores.

The interplay between external and internal mass transfer resistances is evident in various studies. For example, the LDF model has been employed to analyze CO_2 adsorption dynamics, demonstrating that external film resistance is often less

TABLE 2.2
Summary of Correlations Used for Predicting Mass Transfer Coefficients for CO_2 Adsorption Process

Correlation	Equation	Typical Applications	Ref
Wilson–Geankoplis Correlation	• Based on Reynolds (Re) and Schmidt (Sc) numbers: $Sh = 1.09\times(Re^{0.33}Sc^{0.33})$ • Valid for laminar flow through packed beds	• Used for gas–solid adsorption in packed beds • Estimating external mass transfer coefficients	[104]
Higbie's Penetration Theory	• Assumes mass transfer occurs over a characteristic contact time: $k_f = \frac{2D}{\pi t_c^{0.5}}$ • t_c: contact time • D: diffusion coefficient	• Used for unsteady-state mass transfer • CO_2 adsorption onto surfaces with short contact times (e.g., fluidized beds)	[105]
Froessling Equation	• Empirical correlation for mass transfer around spherical particles: $Sh = 2 + 0.6Re^{0.5}Sc^{0.33}$ • Relates Sherwood, Reynolds, and Schmidt numbers	• External mass transfer in spherical adsorbent particles • Packed bed CO_2 adsorption systems	[106]
Wakao–Funazkri Correlation	• Describes combined film and pore diffusion: $Sh = 2 + 1.1Re^{0.6}Sc^{0.33}$	• Predicting mass transfer in packed beds with combined film and pore diffusion resistance	[107]
Geldart's Correlation	• Applies to fluidized beds: $Sh = 2.7 + 0.27Re^{0.52}Sc^{0.33}$	• Mass transfer in fluidized beds • Describes transfer around particles in fluidized adsorption	[108]
Gilliland Correlation	• For packed bed and column systems: $k_f d_p = \frac{D}{0.65 + 1.1Re^{0.7}}$	• Predicting mass transfer in adsorption columns • External diffusion in packed beds	[109]
Leva's Correlation	• Used for large particles in gas streams: $Sh = 0.81Re^{0.5}Sc^{0.33}$	• External mass transfer for large adsorbent particles • Fixed-bed and fluidized bed adsorption	[110]
Clift–Gauvin Correlation	• For small particles in fluidized beds: $Sh = 2 + 0.4Re^{0.5}Sc^{0.33}$	• Small particle mass transfer in fluidized beds • Often applied in CO_2 capture fluidized systems	[111]
Onda Correlation	• For mass transfer in packed beds: $Sh = 0.5\left(\frac{\epsilon}{1-\epsilon}\right)^{0.5}(ReSc)^{\frac{1}{3}}$ • ϵ: bed porosity	• Adsorption and desorption in packed bed systems • Used in large-scale industrial adsorption processes	[112]

significant than internal resistance in microporous materials like NaY zeolite [114]. Also, breakthrough curves for structured adsorbents have shown sharper profiles than traditional adsorbents, indicating reduced external mass transfer resistance due to better flow distribution and thinner films [87]. Moreover, the design and modification of adsorbents can affect mass transfer characteristics. For instance, hydrophobic modifications of ceramic membranes have improved CO_2 capture efficiency by minimizing external mass transfer resistance and enhancing the interaction between the gas and the adsorbent surface [117]. Similarly, composite materials, such as amine-grafted polymers, have been reported to enhance both external and internal mass transfer properties, resulting in improved CO_2 adsorption capacities [118].

2.5 DIFFUSION IN CO_2 ADSORPTION

Knudsen diffusion is a significant transport mechanism in small pores, particularly those in the nanometer range, where the mean free path of gas molecules exceeds the pore diameter. This phenomenon is characterized by the dominance of molecular-wall collisions over molecular–molecular collisions, leading to unique diffusion behaviors that differ from those observed in larger pores or bulk gas environments (Figure 2.5).

As the pore size decreases, the Knudsen number (Kn), the ratio of the mean free path of the gas molecules to the pore diameter, becomes critical in determining the flow regime. In small pores, particularly when the Kn exceeds 1, Knudsen diffusion becomes the primary mechanism for gas transport. This is because the mean free path of gas molecules is comparable to or larger than the pore dimensions, resulting in a significant increase in the frequency of collisions between gas molecules and the pore walls [120–122]. For instance, it has been shown that as the pore diameter approaches the nanometer scale, the contribution of Knudsen diffusion to gas

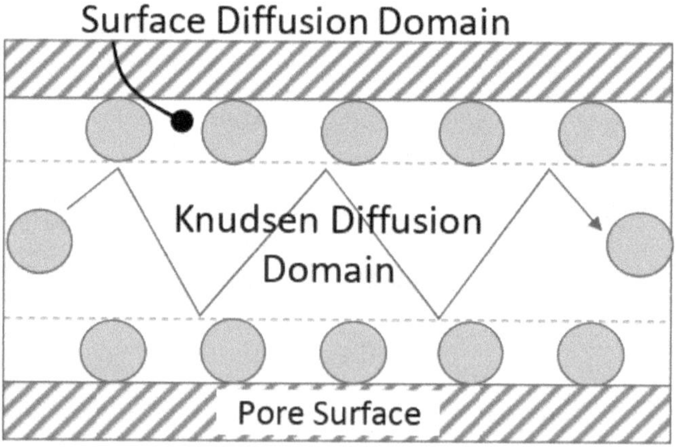

FIGURE 2.5 Mechanism of Knudsen diffusion in pore structure [119],

transport becomes non-negligible, necessitating its consideration in simulations and models of gas flow in nanoporous materials [123, 124].

Various factors, including the size of the gas molecules and the nature of the pore walls, influence the dynamics of Knudsen diffusion. Studies have indicated that lighter gas molecules tend to diffuse more rapidly than heavier ones through nanoporous membranes, which is a characteristic feature of Knudsen diffusion [125, 126]. This behavior is attributed to the kinetic and thermodynamic favorability of lighter gases in navigating through the constricted pathways of nanopores, as observed in experiments with graphene oxide membranes [126]. Additionally, the effective diffusivity of gases in nanopores can be significantly affected by the pore structure and surface interactions, which can either enhance or hinder the diffusion process [127, 128].

Molecular diffusion plays a critical role in the transport of fluids through porous media, particularly in larger pores and bulk phases. Various factors influence the diffusion dynamics, including pore size, molecular weight, and the interactions between the diffusing species and the pore walls. In larger pores, the diffusion coefficients tend to be higher due to reduced hindrance from the pore walls, allowing for more efficient transport of molecules (Figure 2.6).

Research indicates that, as the pore size increases, the self-diffusion coefficients of molecules also increase. For instance, it demonstrated that larger pore sizes facilitate diffusion by decreasing the average interaction energy between the diffusing molecules and the pore walls, thus enhancing the overall diffusivity [129]. This observation aligns with findings from those who noted that the diffusivity of residue fractions through polycarbonate membranes decreases as the molecular weight increases and the pore size decreases, highlighting the significance of both molecular size and pore size in diffusion processes [130].

Studies on MOFs and zeolites elucidate the relationship between molecular diffusion and pore size. Haldoupis et al. found that materials with pore diameters larger than the kinetic diameter of the diffusing species allow for rapid diffusion, while smaller pores can restrict movement and lead to lower diffusivity [95]. This is echoed in the work of those who noted that the diffusion of small molecules like CO_2 within zeolite frameworks is highly dependent on the pore size and shape, which can dictate the performance of these materials in applications such as chemical separations [131].

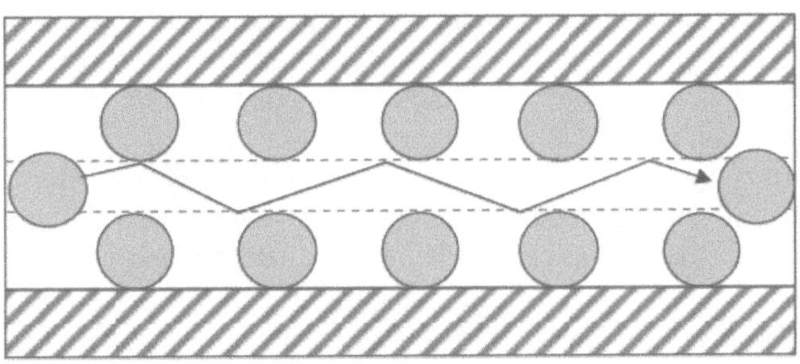

FIGURE 2.6 Mechanism of molecular diffusion in pore structure [119],

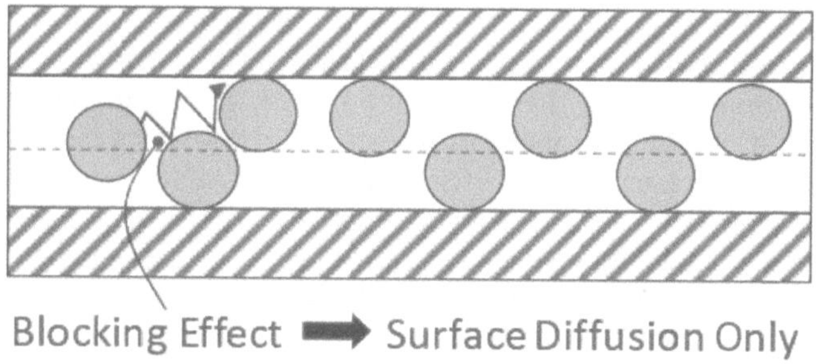

FIGURE 2.7 Mechanism of surface diffusion in pore structure [119].

Surface diffusion is fundamentally characterized by the hopping mechanism, where adsorbate molecules transition between adjacent adsorption sites on the surface. This process is energetically activated, requiring a certain amount of energy to overcome the potential barriers associated with moving from one site to another (Figure 2.7). The hopping model is widely accepted in the literature, as it effectively captures the essence of surface diffusion in various contexts, including gas adsorption in nanopores and on solid surfaces [132, 133]. The diffusion coefficient, which quantifies the rate of this process, is influenced by factors such as the adsorbate concentration, the nature of the adsorbent surface, and the temperature [3].

The dynamics of surface diffusion can be significantly affected by the surface coverage of the adsorbate. For instance, at low surface coverage, the diffusion behavior tends to exhibit a higher degree of heterogeneity, as the available adsorption sites are less populated, allowing for more pronounced hopping events between vacant sites [134]. Conversely, as the surface coverage increases, the interactions between adsorbed molecules can lead to a decrease in the effective diffusion coefficient, as the presence of neighboring adsorbates can hinder movement due to steric effects or repulsive interactions [135]. This interplay between surface coverage and diffusion dynamics underscores the complexity of adsorption processes and highlights the need for models that can accurately capture these relationships.

Moreover, the geometric and energetic characteristics of the adsorbent surface play a crucial role in determining diffusion behavior. For example, surfaces with varying curvatures can exhibit different diffusion rates, as observed in studies involving carbon nanotubes where the curvature of the surface was found to influence the lateral mobility of adsorbates [136]. Similarly, the presence of surface defects or variations in surface energy can create localized regions of enhanced or reduced diffusion rates, further complicating the overall diffusion landscape [137].

2.5.1 Pore Structure and Distribution

Pore size plays a pivotal role in determining the adsorption capacity and kinetics of CO_2. Smaller pores can restrict the diffusion of CO_2 molecules, leading to slower adsorption rates. For instance, it highlights that adsorbents with smaller pore volumes and sizes can hinder mass transfer, thus reducing CO_2 adsorption capacity [138]. This is corroborated by the findings of those who observed that increased loading of PEI on silica xerogels resulted in longer times to reach equilibrium CO_2 adsorption due to diffusion limitations within the filled pores [139]. Conversely, larger pores facilitate better mass transfer by allowing CO_2 molecules to access a greater surface area, thus enhancing adsorption kinetics [140]. Figure 2.8 illustrates the various pore size ranges according to the IUPAC classification.

The shape of the pores also significantly affects mass transfer dynamics. Different pore geometries can lead to variations in the adsorption behavior due to the way they influence the flow paths of CO_2 molecules (Figure 2.9). For example, discuss how pore connectivity and tortuosity impact the sorption of fluids, emphasizing that a well-connected pore network can enhance mass transfer rates [141]. Additionally, it

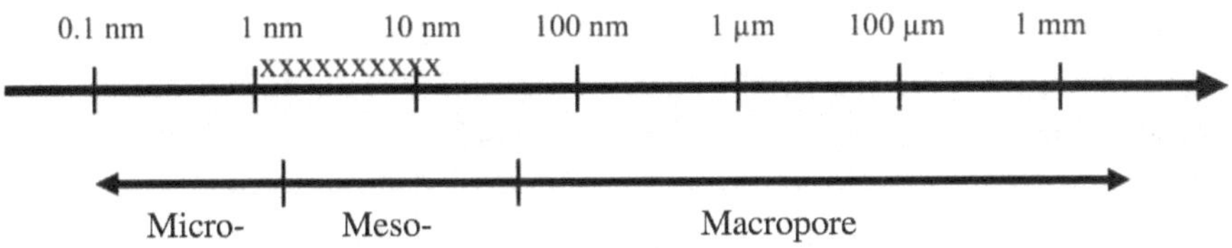

FIGURE 2.8 Pore size classification of IUPAC.

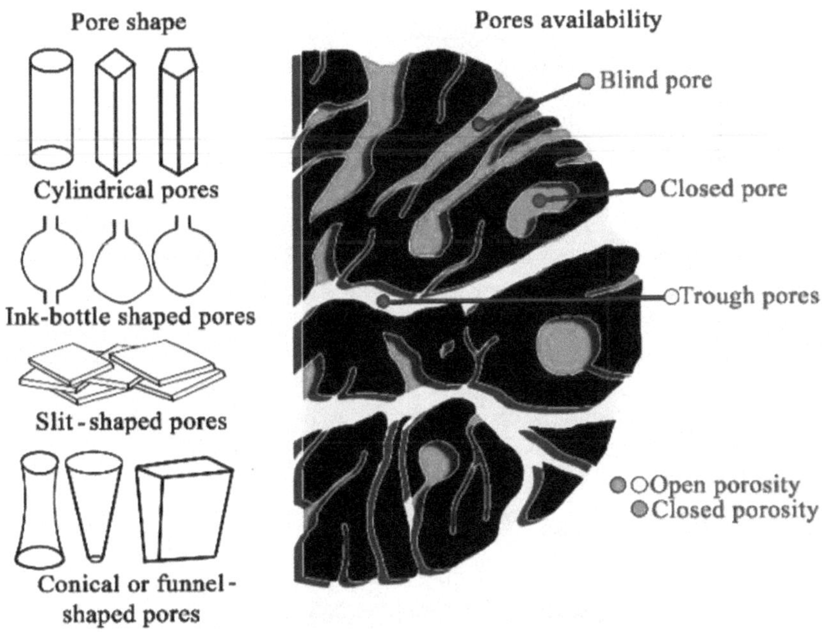

FIGURE 2.9 Various pore shapes present in adsorbent structures [143], © 2021 by the authors. Licensee MDPI, Basel, Switzerland.

illustrates that the interaction strength between CO_2 and the adsorbent framework can vary with pore shape, affecting the adsorption isotherms and overall efficiency [142].

Moreover, the distribution of pore sizes within an adsorbent material can create a hierarchical structure that optimizes mass transfer. Zhang et al. demonstrated that hierarchical porous silica with varied pore sizes can reduce mass transfer resistance, allowing for more efficient CO_2 capture compared to traditional materials with uniform pore sizes [140]. This is particularly important as it allows for a balance between high surface area and effective pore volume, which are essential for maximizing adsorption performance [144].

2.5.2 Temperature and Pressure Effects

The influence of temperature and pressure on diffusion rates in CO_2 adsorbents is critical to optimizing carbon capture technologies. Both parameters significantly affect the adsorption capacity and kinetics of CO_2 on various adsorbent materials, including activated carbons, zeolites, and amine-functionalized supports.

Temperature plays a dual role in the adsorption process. Generally, an increase in temperature tends to decrease the adsorption capacity of CO_2 due to the thermodynamic nature of the adsorption process. For instance, it demonstrated that CO_2 capture capacity decreased with increasing temperature for adsorbents with low PEI loading, indicating that higher temperatures weaken the interaction between CO_2 and the adsorbent due to increased molecular movement [145]. Similarly, it confirmed that CO_2 adsorption progressively decreased with rising temperatures, aligning with thermodynamic expectations [146]. However, certain materials, such as PEI-loaded mesoporous silica, exhibited enhanced CO_2 capture at elevated temperatures due to kinetic effects, suggesting that the diffusion of CO_2 molecules can be facilitated under specific conditions [147]. This highlights the complexity of temperature's influence, where the nature of the adsorbent and the specific conditions can lead to varying outcomes.

Pressure also significantly impacts the diffusion rates and adsorption capacity of CO_2. Higher pressures generally enhance the adsorption process by increasing the concentration of CO_2 in the vicinity of the adsorbent, thereby promoting diffusion into the porous structure. For instance, activated carbons exhibit considerable CO_2 loading at high pressures, which is essential for effective carbon capture [148]. Additionally, studies have shown that the effective diffusivity of CO_2 is positively correlated with pressure, as increased pressure enhances the driving force for diffusion into the adsorbent's pores [149]. Conversely, at lower pressures, the adsorption capacity may be limited, and the diffusion rates can be affected by the strong adsorption of CO_2 on high-energy sites within the adsorbent [150].

2.5.3 Adsorbent Material Properties

Hydrophilicity, or the affinity of a material to interact with water, significantly affects the adsorption capacity of CO_2 adsorbents. Materials with high hydrophilicity can attract water molecules, which may compete with CO_2 for adsorption sites, thereby reducing the overall CO_2 uptake. For instance, amine-functionalized adsorbents, known for their high CO_2 affinity, can exhibit varying degrees of hydrophilicity depending on the specific amines used and their loading on the support material.

TABLE 2.3
The Hydrophilicity of Different Amines Utilized for CO_2 Adsorbents

Amine Type	Hydrophilicity	Description
Monoethanolamine (MEA)	High	MEA is highly hydrophilic due to the presence of both amine and hydroxyl groups, which interact strongly with water.
Diethanolamine (DEA)	High	Like MEA, DEA has strong hydrophilic characteristics due to its two hydroxyl groups, leading to a higher affinity for water.
Methyldiethanolamine (MDEA)	Moderate	MDEA has a reduced hydrophilicity compared to MEA and DEA due to the presence of a methyl group, which provides some hydrophobic character.
Piperazine (PZ)	Low to moderate	Piperazine shows moderate hydrophilicity because of its cyclic structure, though it can still form hydrogen bonds with water.
Triethanolamine (TEA)	High	TEA has three hydroxyl groups, making it very hydrophilic, leading to significant water uptake.
Tetraethylenepentamine (TEPA)	High	TEPA contains multiple amine groups, contributing to its strong hydrophilicity and high water affinity.
2-Amino-2-methyl-1-propanol (AMP)	Moderate	AMP has a secondary amine and an alcohol group, leading to moderate hydrophilicity, though less than MEA or DEA.
Diisopropanolamine (DIPA)	Moderate	The isopropanol groups make DIPA less hydrophilic than MEA and DEA, providing a balance between water solubility and hydrophobicity.
N-Methylethanolamine (MMEA)	Moderate	MMEA is moderately hydrophilic due to the single hydroxyl group and the methyl group providing a degree of hydrophobicity.
Ethylenediamine (EDA)	High	EDA is highly hydrophilic because of two amine groups, which strongly attract water.

This balance between hydrophilicity and CO_2 affinity is critical. At the same time, some moisture can enhance CO_2 capture by facilitating the formation of carbamate species; excessive water can hinder adsorption kinetics by blocking access to active sites [151, 152]. Table 2.3 presents the hydrophilicity of various amine types utilized for adsorbent functionalization and describes their hydrophilic properties.

Surface functionality, which includes specific chemical groups on the adsorbent surface, also plays a pivotal role in mass transfer dynamics. Functional groups such as amines can enhance the adsorption capacity through chemical interactions with CO_2, forming stable carbamate bonds. The type and density of these functional groups can dictate the rate of CO_2 adsorption, as they influence both the intrinsic reaction kinetics and the diffusion of CO_2 molecules to the active sites [153, 154]. For example, using amine-functionalized materials has been shown to improve CO_2 uptake significantly, as these materials can provide high adsorption rates and favorable desorption characteristics, which are essential for efficient regeneration processes [155].

The adsorbent's hydrophilicity and surface functionality influence the kinetics of CO_2 adsorption and desorption. For instance, the adsorption process can be limited by the rate at which CO_2 molecules diffuse through the boundary layer surrounding the adsorbent particles and the rate at which they enter the pores. The presence of hydrophilic groups can alter the diffusion pathways and the energy barriers associated with CO_2 transport, thereby affecting the overall kinetics of the process [156, 157]. Additionally, these characteristics impact the regeneration of adsorbents, which is a critical aspect of their lifecycle. Adsorbents with high thermal stability and appropriate surface functionalities can be regenerated more efficiently, leading to lower operational costs and enhanced sustainability of CO_2 capture technologies [155, 158].

2.6 MODELING OF MASS TRANSFER AND DIFFUSION

2.6.1 Theoretical Models

Fick's laws of diffusion are fundamental principles that describe the movement of particles from regions of higher concentration to areas of lower concentration. These laws are pivotal in various scientific fields, including chemistry, physics, and engineering, particularly in gas adsorption processes such as CO_2 capture and storage.

Fick's first law states that a diffusing substance's flux is proportional to its concentration's negative gradient. Mathematically, it can be expressed as:

$$J = -D\frac{\partial C}{\partial x} \tag{2.5}$$

where J is the diffusion flux, D is the diffusion coefficient (m^2/s), and $\frac{\partial C}{\partial x}$ is the concentration gradient. This law is applicable under steady-state conditions where the concentration does not change with time [159, 160]. In contrast, Fick's second law deals with unsteady-state diffusion, which occurs when the concentration of a substance changes over time and is given by:

$$\frac{\partial C}{\partial t} = D\frac{\partial^2 C}{\partial x^2} \tag{2.6}$$

Applying Fick's laws to CO_2 adsorbents is significant, especially in materials designed for carbon capture. For instance, in the study by Yue et al., the classical Fick's diffusion model was employed to simulate the adsorption kinetics of nitrogen-doped porous carbons, demonstrating a rapid CO_2 adsorption rate that reached 90% of its maximum capacity within approximately 3.5 minutes [161]. Similarly, it reported that Fick's diffusion model effectively correlated with their adsorption kinetics data, achieving a high squared correlation coefficient ($R^2 \approx 0.91$) for nitrogen-doped porous carbons derived from graphene [162]. Moreover, Fick's second law, which accounts for unsteady-state diffusion, is often utilized in scenarios where diffusion processes limit the reaction kinetics. Highlighted that in their study of CO_2 adsorption in a tetraethylenepentamine thin film, the diffusivity of CO_2 was the limiting factor in the kinetics, allowing the application of Fick's second law to describe the system adequately [163]. This is further supported by, who applied Fick's first law to model CO_2 efflux in soil, demonstrating its versatility across different contexts of CO_2 diffusion [164]. Furthermore, the diffusion coefficients derived from Fick's laws provide critical insights into the efficiency of CO_2 capture materials. For example, the diffusion time constant for CO_2 in nitrogen-doped porous carbons was calculated to be significantly higher than that of previously reported biomass-derived carbons, indicating a superior CO_2 diffusion rate during the adsorption process [162]. This highlights the potential for optimizing adsorbent materials based on their diffusion characteristics, which could enhance performance in CO_2 capture applications. Table 2.4 summarizes the various theoretical models applied to the CO_2 adsorption process.

2.6.2 Computational Approaches

Numerical methods, particularly computational fluid dynamics (CFD) simulations, have become increasingly important in modeling the adsorption processes of CO_2 on various adsorbents. For instance, Saini et al.'s work on simulating breakthrough curves during CO_2 adsorption from biogas in a fixed-bed column illustrates the importance of numerical simulations in predicting the performance of adsorbents under operational conditions [165]. The breakthrough curve is a crucial parameter that indicates an adsorbent's saturation point and helps determine the optimal operating conditions for maximum CO_2 capture efficiency. Similarly, Zulkifli et al. employed CFD simulations to analyze CO_2 and methane adsorption at various temperatures for MOFs, providing insights into the dynamic adsorption processes and enhancing the understanding of mass transfer phenomena in these materials [166].

Modeling mass transfer in adsorption processes often involves applying various kinetic models to describe the adsorption behavior of CO_2 on different adsorbents. For example, the LDF model has been successfully applied to describe CO_2 adsorption kinetics on zeolite 13X, as demonstrated by Karka et al. [167]. This model simplifies the complex interactions during adsorption by assuming a linear relationship between the driving force and the mass transfer rate, thus facilitating the analysis of experimental data. Additionally, the use of statistical physics models, as highlighted by Aouaini et al., provides a more comprehensive understanding of the multilayer adsorption phenomena, which is particularly relevant for zeolite-based adsorbents [168]. Moreover, the development of software tools for simulating adsorption processes has facilitated the integration of experimental data with theoretical models. For instance, the work of Sinha et al. on temperature vacuum swing adsorption processes for direct air capture of CO_2 utilized modeling studies to assess the economic viability of different adsorbents [169]. By combining experimental findings with numerical simulations, researchers can optimize the design of adsorption systems, leading to more efficient and cost-effective solutions for CO_2 capture.

The impact of various factors on CO_2 adsorption performance, such as temperature, pressure, and the physical properties of adsorbents, can also be effectively modelled

TABLE 2.4
Summary of Different Theoretical Models Used for CO_2 Adsorption Process

Theoretical Model	Key Features	Typical Applications
Fick's Law of Diffusion	• Describes diffusion due to concentration gradients • Based on the proportionality between flux and gradient	• Predicting diffusion rates in porous materials • Simple diffusion in homogeneous media
Maxwell–Stefan Model	• Accounts for multi-component diffusion • Describes diffusion in gas mixtures and porous systems • Based on driving forces and friction	• Multi-component adsorption systems • CO_2 and gas mixture diffusion in zeolites, MOFs
Linear Driving Force (LDF) Model	• Simplifies mass transfer using a linear relationship • Assumes proportionality between the driving force and mass transfer rate	• Kinetic analysis of adsorption systems • Estimating overall mass transfer rates in packed beds
Knudsen Diffusion Model	• Relevant when the pore size is comparable to the mean free path of gas molecules • Diffusion is controlled by collisions with pore walls	• Diffusion in microporous adsorbents (e.g., activated carbon, zeolites) • Low-pressure adsorption
Dusty Gas Model (DGM)	• Combines Knudsen and Fickian diffusion • Considers bulk and surface diffusion, convection, and interaction with pore walls	• Diffusion in complex porous networks • Adsorption in microporous and mesoporous materials
Surface Diffusion Model	• Describes the movement of adsorbed species along the surface of the adsorbent • Driven by surface concentration gradients	• Surface diffusion in materials like MOFs, zeolites • CO_2 diffusion on high surface area adsorbents
Pore Diffusion Model	• Describes gas diffusion inside the pores of adsorbents • Incorporates pore geometry and structure effects	• Diffusion within porous adsorbents like MOFs, zeolites, and activated carbon • Predicting CO_2 capture performance in porous materials

using numerical methods. For example, the study investigated the adsorption kinetics of CO_2 on amine-impregnated mesoporous silica, exploring how temperature variations affect the adsorption capacity and kinetics [66]. This type of analysis is essential for understanding adsorbents' operational limits and developing strategies to enhance their performance under varying conditions. Furthermore, integrating machine learning techniques into modelling adsorption processes is an emerging trend that holds promise for advancing the field. By leveraging large datasets from experimental studies, machine learning algorithms can identify patterns and correlations that may not be evident through traditional modeling approaches. This can lead to the discovery of new adsorbent materials and optimization strategies for existing systems, as highlighted in recent reviews on the development of functionalized nanoporous materials for CO_2 capture [170].

2.7 CONCLUSION

This chapter has thoroughly examined the mechanisms, mass transfer dynamics, diffusion processes, and simulation methods that underpin effective CO_2 adsorption. The complex nature of CO_2 adsorption onto solid adsorbents is driven by various factors, including surface chemistry, pore architecture, and specific adsorbate–adsorbent interactions. By differentiating between physisorption and chemisorption, we have highlighted the essential aspects of binding affinity, selectivity, and energy profiles that support efficient CO_2 capture. Moreover, understanding the roles of mass transfer and diffusion within porous adsorbents underscores the significance of structural characteristics, particularly pore size and geometry, in enhancing diffusion pathways for effective adsorption.

Through a holistic investigation of these interconnected aspects, this chapter offers a foundational framework for advancing the development of innovative adsorbent materials. Researchers can design materials with superior adsorption performance by optimizing mass transfer properties, refining diffusion pathways, and leveraging thermodynamic insights. These advancements are crucial for scaling up carbon capture technologies and effectively addressing the global challenge of reducing CO_2 emissions.

REFERENCES

[1] Y. Zhu, J. Wu, H. Wang, J. Wang, H. Shen, and Z. Ying, "Interference Effect of Experimental Parameters on the Mercury Removal Mechanism of Biomass Char under an Oxy-Fuel Atmosphere," *ACS Omega,* vol. 6, no. 50, pp. 35124–35133, 2021, https://doi.org/10.1021/acsomega.1c06038.

[2] V. J. Inglezakis, M. M. Fyrillas, and J. Park, "Variable Diffusivity Homogeneous Surface Diffusion Model and Analysis of Merits and Fallacies of Simplified Adsorption Kinetics Equations," *Journal of Hazardous Materials,* vol. 367, pp. 224–245, 2019, https://doi.org/10.1016/j.jhazmat.2018.12.023.

[3] V. J. Inglezakis, M. Balsamo, and F. Montagnaro, "A Fractal-Based Correlation for Time-Dependent Surface Diffusivity in Porous Adsorbents," *Processes,* vol. 8, no. 6, p. 689, 2020, https://doi.org/10.3390/pr8060689.

[4] P. S. Kumar *et al.*, "Adsorption of Basic Dye Onto Raw and Surface-modified Agricultural Waste," *Environmental Progress & Sustainable Energy,* vol. 33, no. 1, pp. 87–98, 2013, https://doi.org/10.1002/ep.11756.

[5] V. Russo, R. Tesser, D. Masiello, M. Trifuoggi, and M. D. Serio, "Further Verification of Adsorption Dynamic Intraparticle Model (ADIM) for Fluid–solid Adsorption Kinetics in Batch Reactors," *Chemical Engineering Journal,* vol. 283, pp. 1197–1202, 2016, https://doi.org/10.1016/j.cej.2015.08.066.

[6] H. M. Mohamed, N. K. Soliman, D. A. Abdelrheem, A. A. Ramadan, A. H. Elghandour, and S. A. Ahmed, "Adsorption of Cd2+ and Cr3+ Ions from Aqueous Solutions by Using Residue of Padina Gymnospora Waste as Promising Low-Cost Adsorbent," *Heliyon,* vol. 5, no. 3, p. e01287, 2019, https://doi.org/10.1016/j.heliyon.2019.e01287.

[7] Q. Zhu, G. D. Moggridge, and C. D'Agostino, "Adsorption of Pyridine from Aqueous Solutions by Polymeric Adsorbents MN 200 and MN 500. Part 2: Kinetics and Diffusion Analysis," *Chemical Engineering Journal,* vol. 306, pp. 1223–1233, 2016, https://doi.org/10.1016/j.cej.2016.07.087.

[8] Z. Li, K. Wang, Y. Lou, P. Xia, L. Shao, H. Hu, and W. Gao, "Effect of Water on Effective Pore Structures for Medium-Rank Coal: Based on the Prefreezing Nitrogen Adsorption–Desorption Experiment," *ACS Omega,* vol. 8, no. 45, pp. 42379–42389, 2023.

[9] Q. Sun, Z. Li, D. J. Searles, Y. Chen, G. Q. Lu, and A. Du, "Charge-Controlled Switchable CO_2 Capture on Boron Nitride Nanomaterials," *Journal of the American Chemical Society,* vol. 135, no. 22, pp. 8246–8253, 2013, https://doi.org/10.1021/ja400243r.

[10] S. Ralser, A. Kaiser, M. Probst, J. Postler, M. Renzler, D. K. Böhme, and P. Scheier, "Experimental Evidence for the Influence of Charge on the Adsorption Capacity of Carbon Dioxide on Charged Fullerenes," *Physical Chemistry Chemical Physics,* vol. 18, no. 4, pp. 3048–3055, 2016, https://doi.org/10.1039/c5cp06587a.

[11] P. Mehrmohammadi and A. Ghaemi, "Investigating the Effect of textural Properties on CO_2 Adsorption in Porous Carbons via Deep Neural Networks Using Various Training Algorithms," *Scientific Reports,* vol. 13, no. 1, p. 21264, 2023.

[12] F. Bahmanzadegan and A. Ghaemi, "Exploring the Effect of Zeolite's Structural Parameters on the CO_2 Capture Efficiency Using RSM and ANN Methodologies," *Case Studies in Chemical and Environmental Engineering,* vol. 9, p. 100595, 2024.

[13] J. Liu, P. K. Thallapally, B. P. McGrail, D. R. Brown, and J. Li, "Progress in Adsorption-Based CO_2 Capture by Metal–organic Frameworks," *Chemical Society Reviews,* vol. 41, no. 6, pp. 2308–2322, 2012, https://doi.org/10.1039/c1cs15221a.

[14] D. Saha, G. Orkoulas, J. Chen, and D. K. Hensley, "Adsorptive Separation of CO_2 in Sulfur-Doped Nanoporous Carbons: Selectivity and Breakthrough Simulation," *Microporous and Mesoporous Materials,* vol. 241, pp. 226–237, 2017, https://doi.org/10.1016/j.micromeso.2016.12.015.

[15] M. Helmi, Z. Khoshdouni Farahani, A. Hemmati, and A. Ghaemi, "Facile Synthesis of Persian Gum–graphene Oxide Composite as a Novel Adsorbent for CO_2 Capture: Characterization and Optimization," *Scientific Reports,* vol. 14, no. 1, p. 5511, 2024.

[16] S. Wei *et al.*, "Functionalized 3D Covalent Organic Frameworks for High-Performance CO_2 Capture and Separation Over N_2," *Advanced Theory and Simulations,* vol. 5, no. 12, 2022, https://doi.org/10.1002/adts.202200588.

[17] C. M. Ghimbeu, J. Górka, V. Simone, L. Simonin, S. Martinet, and C. Vix-Guterl, "Insights on the Na+ Ion Storage Mechanism in Hard Carbon: Discrimination between the Porosity, Surface Functional Groups and Defects," *Nano Energy,* vol. 44, pp. 327–335, 2018, https://doi.org/10.1016/j.nanoen.2017.12.013.

[18] M. G. Rabbani and H. M. El-Kaderi, "Synthesis and Characterization of Porous Benzimidazole-Linked Polymers and Their Performance in Small Gas Storage and Selective Uptake," *Chemistry of Materials*, vol. 24, no. 8, pp. 1511–1517, 2012, https://doi.org/10.1021/cm300407h.

[19] S. M. Yusof and L. P. Teh, "Bifunctional Materials for CO_2 Adsorption: Short Review," *Journal of Chemical Engineering and Industrial Biotechnology,* vol. 7, no. 2, pp. 15–19, 2021, https://doi.org/10.15282/jceib.v7i2.7021.

[20] S. Q. Wang *et al.*, "Simultaneous Removal of SO_2 and NOx from Flue Gas by Low-Temperature Adsorption Over Activated Carbon," *Scientific Reports,* vol. 11, no. 1, 2021, https://doi.org/10.1038/s41598-021-90532-9.

[21] H. Mashhadimoslem, M. Vafaeinia, M. Safarzadeh, A. Ghaemi, F. Fathalian, and A. Maleki, "Development of Predictive Models for Activated Carbon Synthesis from Different Biomass for CO_2 Adsorption Using Artificial Neural Networks," *Industrial & Engineering Chemistry Research,* vol. 60, no. 38, pp. 13950–13966, 2021.

[22] P. M. Ejikeme, A. I. Okoye, M. C. Menkiti, and O. D. Onukwuli, "Activated Carbons from Gambeya Albida and Telfairia Occidentalis Seed Shells: Effect of Activating Agent on the Removal of Pb(II) Ions from Aqueous Solutions," *Chemical Science Transactions,* 2015, https://doi.org/10.7598/cst2015.1055.

[23] A. Ghaemi, H. Mashhadimoslem, and P. Zohourian Izadpanah, "NiO and MgO/activated Carbon as an Efficient CO_2 Adsorbent: Characterization, Modeling, and Optimization," *International Journal of Environmental Science and Technology,* vol. 19, no. 2, pp. 727–746, 2022.

[24] H. Mashhadimoslem, M. Safarzadeh Khosrowshahi, M. Jafari, A. Ghaemi, and A. Maleki, "Adsorption Equilibrium, Thermodynamic, and Kinetic Study of $O_2/N_2/CO_2$ on Functionalized Granular Activated Carbon," *ACS Omega,* vol. 7, no. 22, pp. 18409–18426, 2022.

[25] F. Fashi, A. Ghaemi, and A. H. Behroozi, "Piperazine Impregnation on Zeolite 13X as a Novel Adsorbent for CO_2 Capture: Experimental and Modeling," *Chemical Engineering Communications,* vol. 208, no. 8, pp. 1104–1120, 2021.

[26] F. Bahmanzadegan, M. A. Pordsari, and A. Ghaemi, "Improving the Efficiency of 4A-Zeolite Synthesized from Kaolin by Amine Functionalization for CO_2 Capture," *Scientific Reports,* vol. 13, no. 1, p. 12533, 2023.

[27] Z. Wan, W. Wu, W. Chen, H. Yang, and D. Zhang, "Direct Synthesis of Hierarchical ZSM-5 Zeolite and Its Performance in Catalyzing Methanol to Gasoline Conversion," *Industrial & Engineering Chemistry Research,* vol. 53, no. 50, pp. 19471–19478, 2014, https://doi.org/10.1021/ie5036308.

[28] A. Starke, C. Pasel, C. Bläker, T. Eckardt, J. Zimmermann, and D. Bathen, "Investigation of the Adsorption of Hydrogen Sulfide on Faujasite Zeolites Focusing on the Influence of Cations," *ACS Omega,* vol. 7, no. 48, pp. 43665–43677, 2022.

[29] H. J. Esfahani, A. Ghaemi, and S. Shahhosseini, "Improving CO_2 Adsorption Efficiency of an Amine-modified MOF-808 through the Synthesis of Its Graphene Oxide Composites," *Scientific Reports,* vol. 14, no. 1, p. 18871, 2024.

[30] M. T. Kapelewski *et al.*, "Record High Hydrogen Storage Capacity in the Metal–Organic Framework Ni_2(*m*-Dobdc) at Near-Ambient Temperatures," *Chemistry of Materials,* vol. 30, no. 22, pp. 8179–8189, 2018, https://doi.org/10.1021/acs.chemmater.8b03276.

[31] S. Giraudet, Z. Zhu, X. Yao, and G. Q. Lu, "Ordered Mesoporous Carbons Enriched with Nitrogen: Application to Hydrogen Storage," *The Journal of Physical Chemistry C,* vol. 114, no. 18, pp. 8639–8645, 2010, https://doi.org/10.1021/jp101119r.

[32] R. Jalab, "Biodegradable Polysaccharide Grafted Polyacrylamide Inhibitor for Corrosion in CO_2-Saturated Saline Solution," *Heliyon,* vol. 9, no. 10, p. e20304, 2023, https://doi.org/10.1016/j.heliyon.2023.e20304.

[33] S. Pinjari, T. Bera, and E. Kjeang, "Room Temperature Hydrogen Storage Enhancement in Copper-Doped Zeolitic Imidazolate Frameworks with Trioctylamine," *Sustainable Energy & Fuels,* vol. 7, no. 13, pp. 3142–3151, 2023, https://doi.org/10.1039/d3se00277b.

[34] R. A. Tayade and A. C. Kanojiya, "The Case Study of Isothermal Adsorption of Phenol, O-Cresol on Natural Charcoal's and Applications," *International Journal for Research in Applied Science and Engineering Technology,* vol. 10, no. 12, pp. 1718–1731, 2022, https://doi.org/10.22214/ijraset.2022.48281.

[35] M. A. Pordsari, M. Moradi, and A. Ghaemi, "Fabrication of a Cost-effective Metal Oxide-based Adsorbent from Industrial Waste Slag for Efficient CO_2 Separation under Flue Gas Conditions," *Journal of CO_2 Utilization,* vol. 88, p. 102930, 2024.

[36] Y. Liu, J. Liu, W. Yao, W. Cen, H. Wang, X. Weng, and Z. Wu, "The Effects of Surface Acidity on CO_2 Adsorption Over Amine Functionalized Protonated Titanate Nanotubes," *RSC Advances,* vol. 3, no. 41, p. 18803, 2013, https://doi.org/10.1039/c3ra42597e.

[37] Y.T. Sim and A.H. Ruhaimi, "Recent Progress on (3-Aminopropyl)triethoxysilane (APTES) Functionalized-Adsorbent for CO_2 Capture," *Journal of Physics Conference Series,* vol. 2259, no. 1, p. 012008, 2022, https://doi.org/10.1088/1742-6596/2259/1/012008.
[38] A. Torkashvand, M. R. Moradi, and A. Ghaemi, "Amine Grafting of Carbazole-based Hypercrosslinked Polymer as an Adsorbent to Enhance CO_2 Capture," *Case Studies in Chemical and Environmental Engineering,* vol. 8, p. 100472, 2023.
[39] D. P. Bezerra, R. S. Oliveira, R. S. Vieira, C. L. Cavalcante, and D. C. S. Azevedo, "Adsorption of CO_2 on Nitrogen-Enriched Activated Carbon and Zeolite 13X," *Adsorption,* vol. 17, no. 1, pp. 235–246, 2011, https://doi.org/10.1007/s10450-011-9320-z.
[40] A. Symington, R. M. Harker, M. T. Storr, M. Molinari, and S. C. Parker, "Thermodynamic Evolution of Cerium Oxide Nanoparticle Morphology Using Carbon Dioxide," *The Journal of Physical Chemistry C,* vol. 124, no. 42, pp. 23210–23220, 2020, https://doi.org/10.1021/acs.jpcc.0c07437.
[41] P. Albrecht, D. e. Jiang, and D. R. Mullins, "CO_2 Adsorption as a Flat-Lying, Tridentate Carbonate on $CeO_2(100)$," *The Journal of Physical Chemistry C,* vol. 118, no. 17, pp. 9042–9050, 2014, https://doi.org/10.1021/jp501201b.
[42] N. Tang, T. He, J. Liu, L. Li, H. Shi, W. Cen, and Z. Ye, "New Insights Into CO_2 Adsorption on Layered Double Hydroxide (LDH)-Based Nanomaterials," *Nanoscale Research Letters,* vol. 13, no. 1, 2018, https://doi.org/10.1186/s11671-018-2471-z.
[43] M. Stillings, Z. K. Shipton, and R. Lunn, "Mechanochemical Processing of Silicate Rocks to Trap CO_2," *Nature Sustainability,* vol. 6, no. 7, pp. 780–788, 2023, https://doi.org/10.1038/s41893-023-01083-y.
[44] M. Irfan *et al.*, "Kinetics and Adsorption Isotherms of Amine-Functionalized Magnesium Ferrite Produced Using Sol-Gel Method for Treatment of Heavy Metals in Wastewater," *Materials,* vol. 15, no. 11, p. 4009, 2022, https://doi.org/10.3390/ma15114009.
[45] S. Radchatawin, "Acid Dye Removal from Wastewaters Using Rice Husk Ash Functionalized with Organic Amine Groups as Adsorbent," *Asean Journal of Scientific and Technological Reports,* vol. 27, no. 1, pp. 102–110, 2023, https://doi.org/10.55164/ajstr.v27i1.250741.
[46] P. López-Aranguren, S. Builes, J. Fraile, L. F. Vega, and C. Domingo, "Understanding the Performance of New Amine-Functionalized Mesoporous Silica Materials for CO_2 Adsorption," *Industrial & Engineering Chemistry Research,* vol. 53, no. 40, pp. 15611–15619, 2014, https://doi.org/10.1021/ie502945r.
[47] D. Lee, C. Zhang, and H. Gao, "Amine-Functionalized Porous Polymer Network for Highly Selective Absorption of CO_2 Over N_2," *Macromolecular Chemistry and Physics,* vol. 216, no. 5, pp. 489–494, 2014, https://doi.org/10.1002/macp.201400504.
[48] C. Chen, D. Chen, S. Xie, H. Quan, X. Luo, and L. Guo, "Adsorption Behaviors of Organic Micropollutants on Zirconium Metal–organic Framework UiO-66: Analysis of Surface Interactions," *ACS Applied Materials & Interfaces,* vol. 9, no. 46, pp. 41043–41054, 2017.
[49] S. Builes and L. F. Vega, "Effect of Immobilized Amines on the Sorption Properties of Solid Materials: Impregnation Versus Grafting," *Langmuir,* vol. 29, no. 1, pp. 199–206, 2012, https://doi.org/10.1021/la3038507.
[50] H. A. Araj, D. Bahamón, K. S. K. Reddy, L. F. Vega, and G. N. Karanikolos, "A Systematic Approach to Understanding and Optimizing the CO_2 Capture Performance of Triamine-Functionalized Mesoporous Silica with Amine Blends Using Molecular Simulations," *The Journal of Physical Chemistry C,* vol. 127, no. 15, pp. 7410–7424, 2023, https://doi.org/10.1021/acs.jpcc.2c08453.
[51] S. Baj, A. Siewniak, A. Chrobok, T. Krawczyk, and A. Sobolewski, "Monoethanolamine and Ionic Liquid Aqueous Solutions as Effective Systems for CO_2 Capture," *Journal of Chemical Technology & Biotechnology,* vol. 88, no. 7, pp. 1220–1227, 2012, https://doi.org/10.1002/jctb.3958.

[52] G. Cui, J. Wang, and S. Zhang, "Active Chemisorption Sites in Functionalized Ionic Liquids for Carbon Capture," *Chemical Society Reviews,* vol. 45, no. 15, pp. 4307–4339, 2016, https://doi.org/10.1039/c5cs00462d.

[53] S. V. Yuvaraj *et al.*, "Solvation Mechanism of Task-specific Ionic Liquids in Water: A Combined Investigation Using Classical Molecular Dynamics and Density Functional Theory," *The Journal of Physical Chemistry B,* vol. 119, no. 40, pp. 12894–12904, 2015.

[54] X. Suo *et al.*, "CO_2 Chemisorption Behavior in Conjugated Carbanion-Derived Ionic Liquids via Carboxylic Acid Formation," *Journal of the American Chemical Society,* vol. 144, no. 47, pp. 21658–21663, 2022, https://doi.org/10.1021/jacs.2c09189.

[55] F. Li, S. Zeng, Y. Bai, H. Dong, H. Wang, X. Ji, and X. Zhang, "Efficient and Reversible Chemisorption of Carbon Dioxide with Dianionic-Functionalized Ionic Liquid-Based Solvents," *Energy & Fuels,* vol. 34, no. 7, pp. 8526–8533, 2020, https://doi.org/10.1021/acs.energyfuels.0c00728.

[56] X. Suo *et al.*, "CO_2 Chemisorption Behavior of Coordination-Derived Phenolate Sorbents," *Chemsuschem,* vol. 14, no. 14, pp. 2854–2859, 2021, https://doi.org/10.1002/cssc.202100666.

[57] J. H. Bakhtiari, K. Naderi, A. A. Shafiei, A. Bazyari, A. Ghaemi, and L. T. Thompson, "Tailoring Porosity: Unveiling the Ideal Alumina Support (Aerogel, Xerogel, Cryogel) for High-performance CO_2 Capture with Polyethylene Glycol," *Journal of Environmental Chemical Engineering,* vol. 12, no. 6, p. 114551, 2024.

[58] M. Ahmadi, F. Bahmanzadegan, M. Qasemnazhand, A. Ghaemi, and H. Ramezanipour Penchah, "Experimental, RSM Modelling, and DFT Simulation of CO_2 Adsorption on Modified Activated Carbon with LiOH," *Scientific Reports,* vol. 14, no. 1, p. 13595, 2024.

[59] F. Raganati, R. Chirone, and P. Ammendola, "CO_2 Capture by Temperature Swing Adsorption: Working Capacity as Affected by Temperature and CO_2 Partial Pressure," *Industrial & Engineering Chemistry Research,* vol. 59, no. 8, pp. 3593–3605, 2020, https://doi.org/10.1021/acs.iecr.9b04901.

[60] C. Quan, X. Jia, K. Zhang, Y. Zhang, S. R. Naqvi, N. A. S. Amin, and N. Gao, "Amine-impregnated Silica Zeolite from Microalgae Ash at Different Calcination Temperatures For CO_2 Capture," *International Journal of Energy Research,* vol. 46, no. 2, pp. 1220–1233, 2021, https://doi.org/10.1002/er.7241.

[61] M. A. Alkhabbaz, P. Bollini, G. S. Foo, C. Sievers, and C. W. Jones, "Important Roles of Enthalpic and Entropic Contributions to CO_2 Capture from Simulated Flue Gas and Ambient Air Using Mesoporous Silica Grafted Amines," *Journal of the American Chemical Society,* vol. 136, no. 38, pp. 13170–13173, 2014, https://doi.org/10.1021/ja507655x.

[62] C. Hayichelaeh, L. Reuvekamp, W. Dierkes, A. Blume, J. Noordermeer, and K. Sahakaro, "Enhancing the Silanization Reaction of the Silica-silane System by Different Amines in Model and Practical Silica-filled Natural Rubber Compounds," *Polymers,* vol. 10, no. 6, p. 584, 2018.

[63] M. Pera-Titus, M. Palomino, S. Valencia, and F. Rey, "Thermodynamic Analysis of Framework Deformation in Na, Cs-Rho Zeolite Upon CO_2 Adsorption," *Physical Chemistry Chemical Physics,* vol. 16, no. 44, pp. 24391–24400, 2014, https://doi.org/10.1039/c4cp03409k.

[64] D. M. Venturi *et al.*, "Increased CO_2 Affinity and Adsorption Selectivity in MOF-801 Fluorinated Analogues," *ACS Applied Materials & Interfaces,* vol. 14, no. 36, pp. 40801–40811, 2022, https://pubs.acs.org/doi/10.1021/acsami.2c07640.

[65] J. Wang, B. Wang, Z. Wen, N. Zhao, T. Li, and W. Zhao, "Adsorption Process Optimization and Adsorbent Evaluation Based on Langmuir Isotherm Model," *Langmuir,* vol. 39, no. 46, pp. 16404–16414, 2023.

[66] A. Zhao, A. Samanta, P. Sarkar, and R. Gupta, "Carbon Dioxide Adsorption on Amine-Impregnated Mesoporous SBA-15 Sorbents: Experimental and Kinetics Study," *Industrial & Engineering Chemistry Research,* vol. 52, no. 19, pp. 6480–6491, 2013, https://doi.org/10.1021/ie3030533.

[67] K. Naderi, F. Bahmanzadegan, B. Parizad, A. Ghaemi, and S. Shahhosseini, "Comprehensive Analysis of O_2 and N_2 Adsorption on Kaolin-Derived Zeolites: Experimental and Computational Insights," *Industrial & Engineering Chemistry Research,* vol. 63, no. 37, pp. 16061–16076, 2024.

[68] P. Sampranpiboon, P. Charnkeitkong, and X. S. Feng, "Determination of Thermodynamic Parameters of Zinc (II) Adsorpton on Pulp Waste as Biosorbent," *Advanced Materials Research,* vol. 931–932, pp. 215–219, 2014, https://doi.org/10.4028/www.scientific.net/amr.931–932.215.

[69] R. F. Abbas, H. K. Hami, and N. I. Mahdi, "Removal of Doxycycline Hyclate by Adsorption Onto Cobalt Oxide at Three Different Temperatures: Isotherm, Thermodynamic and Error Analysis," *International Journal of Environmental Science and Technology,* vol. 16, no. 10, pp. 5439–5446, 2018, https://doi.org/10.1007/s13762-018-2079-y.

[70] F. S. Taheri, A. Ghaemi, A. Maleki, and S. Shahhosseini, "High CO_2 Adsorption on Amine-Functionalized Improved Mesoporous Silica Nanotube as an Eco-Friendly Nanocomposite," *Energy & Fuels,* vol. 33, no. 6, pp. 5384–5397, 2019, https://doi.org/10.1021/acs.energyfuels.9b00703.

[71] S. S. Thakur and G. S. Chauhan, "Gelatin–Silica-Based Hybrid Materials as Efficient Candidates for Removal of Chromium(Vi) from Aqueous Solutions," *Industrial & Engineering Chemistry Research,* vol. 53, no. 12, pp. 4838–4849, 2014, https://doi.org/10.1021/ie401997g.

[72] T. Guo *et al.*, "Thermodynamics of CO_2 Adsorption on Cellulose-derived Biochar Prepared in Ionic Liquid," *The Canadian Journal of Chemical Engineering,* vol. 99, no. 9, pp. 1940–1961, 2021, https://doi.org/10.1002/cjce.23940.

[73] Z. Zhang, Y. Zhao, Q. Gong, Z. Li, and J. Li, "MOFs for CO_2 Capture and Separation from Flue Gas Mixtures: The Effect of Multifunctional Sites on Their Adsorption Capacity and Selectivity," *Chemical Communications,* vol. 49, no. 7, pp. 653–661, 2013, https://doi.org/10.1039/c2cc35561b.

[74] M. Senthilkumaran, C. Saravanan, L. Eswaran, P. Puthiaraj, and P. M. Mareeswaran, "Selective Carbon Dioxide Capture Using Silica-Supported Polyaminals," *Chemistryselect,* vol. 4, no. 29, pp. 8534–8541, 2019, https://doi.org/10.1002/slct.201901581.

[75] A. J. Gunorubon and N. Chukwunonso, "Kinetics, Equilibrium and Thermodynamics Studies of Fe^{3+} Ion Removal from Aqueous Solutions Using Periwinkle Shell Activated Carbon," *Advances in Chemical Engineering and Science,* vol. 08, no. 02, pp. 49–66, 2018, https://doi.org/10.4236/aces.2018.82004.

[76] A. I. Sultana, C. Chambers, M. M. N. Ahmed, P. Pathirathna, and M. T. Reza, "Multifunctional Loblolly Pine-Derived Superactivated Hydrochar: Effect of Hydrothermal Carbonization on Hydrogen and Electron Storage with Carbon Dioxide and Dye Removal," *Nanomaterials,* vol. 12, no. 20, p. 3575, 2022, https://doi.org/10.3390/nano12203575.

[77] D. Cantador-Fernandez, D. Esquivel, J. R. Jiménez, and J. M. Fernández-Rodríguez, "Use of Periodic Mesoporous Organosilica–Benzene Adsorbent for CO_2 Capture to Reduce the Greenhouse Effect," *Materials,* vol. 17, no. 11, p. 2669, 2024, https://doi.org/10.3390/ma17112669.

[78] J. Liu, Y. Li, X. An, C. Shen, Q. Xie, and D. Liang, "Activated Carbon Fiber Derived from Wasted Coal Liquefaction Residual for CO_2 Capture," *Environmental Research,* vol. 215, p. 114197, 2022.

[79] N. M. F. M. Yasin, M. S. Hossain, H. P. S. A. Khalil, M. Zulkifli, A. Al-Gheethi, A. J. Asis, and A. N. A. Yahaya, "Treatment of Palm Oil Refinery Effluent Using Tannin as a Polymeric Coagulant: Isotherm, Kinetics, and Thermodynamics Analyses," *Polymers,* vol. 12, no. 10, p. 2353, 2020, https://doi.org/10.3390/polym12102353.

[80] K. Zhi *et al.*, "Preparation and Evaluation of Molecularly Imprinted Polymer for Selective Recognition and Adsorption of Gossypol," *Journal of Molecular Recognition,* vol. 31, no. 3, 2017, https://doi.org/10.1002/jmr.2627.

[81] A. O. Gryshchenko and C. S. Bottaro, "Development of Molecularly Imprinted Polymer in Porous Film Format for Binding of Phenol and Alkylphenols from Water," *International Journal of Molecular Sciences,* vol. 15, no. 1, pp. 1338–1357, 2014, https://doi.org/10.3390/ijms15011338.
[82] K. Ignatowicz, "Chloroorganic Pesticide Adsorption from Water Solution Using Sewage Sludge Vermicompost as a Natural Sorbent," *Journal of Ecological Engineering,* vol. 19, no. 4, pp. 123–128, 2018, https://doi.org/10.12911/22998993/89667.
[83] C. Altintas, I. Erucar, and S. Keskin, "High-throughput Computational Screening of the Metal Organic Framework Database for CH_4/H_2 Separations," *ACS Applied Materials & Interfaces,* vol. 10, no. 4, pp. 3668–3679, 2018.
[84] P. Vyawahare, H. Tun, M. W. Vaughn, and C. C. Chen, "From Langmuir Isotherm to Brunauer–Emmett–Teller Isotherm," *Aiche Journal,* vol. 68, no. 3, 2021, https://doi.org/10.1002/aic.17523.
[85] N. N. Zulkefli, A. M. I. Noor Azam, M. S. Masdar, and W. N. R. W. Isahak, "Adsorption–desorption Behavior of Hydrogen Sulfide Capture on a Modified Activated Carbon Surface," *Materials,* vol. 16, no. 1, p. 462, 2023.
[86] F. Fathalian, H. Moghadamzadeh, A. Hemmati, and A. Ghaemi, "Efficient CO_2 Adsorption Using Chitosan, Graphene Oxide, and Zinc Oxide Composite," *Scientific Reports,* vol. 14, no. 1, p. 3186, 2024.
[87] F. Rezaei, A. Mosca, P. A. Webley, J. Hedlund, and P. Xiao, "Comparison of Traditional and Structured Adsorbents for CO_2 Separation by Vacuum-Swing Adsorption," *Industrial & Engineering Chemistry Research,* vol. 49, no. 10, pp. 4832–4841, 2010, https://doi.org/10.1021/ie9016545.
[88] A. Çağlar and C. Yamali, "Analysis of Heat and Mass Transfer in the Adsorbent Bed of a Thermal Wave Adsorption Cooling Cycle," *Computational Thermal Sciences an International Journal,* vol. 5, no. 2, pp. 97–106, 2013, https://doi.org/10.1615/computthermalscien.2013006349.
[89] S. Punpee, P. Tongpadungrod, T. Suttikul, and C. Phalakornkule, "Characteristics of CO_2 Adsorption and Desorption on Activated Carbon in Comparison with Zeolite 13X and Carbon Molecular Sieve and Applications in Biogas Upgrading Using Vacuum Pressure Swing Adsorption," *Journal of Chemical Technology & Biotechnology,* vol. 98, no. 11, pp. 2677–2690, 2023, https://doi.org/10.1002/jctb.7320.
[90] F. V. Yusubov, R. K. Babaev, E. A. Akhundov, and E. A. Mamedov, "Optimization of Processes in Oil and Gas Industry," *Chemistry and Technology of Fuels and Oils,* vol. 48, no. 2, pp. 156–161, 2012, https://doi.org/10.1007/s10553-012-0352-1.
[91] Y. Seida, N. Sonetaka, K. E. Noll, and E. Furuya, "Determination of Pore and Surface Diffusivities from Single Decay Curve in CSBR Based on Parallel Diffusion Model," *Water,* vol. 14, no. 22, p. 3629, 2022, https://doi.org/10.3390/w14223629.
[92] I. Essouradjane, J. R. J, and V. S. P, "CO_2 Emission Control Using Selective Oxidation of Catalytic Zeolite13x," *International Journal for Research in Applied Science and Engineering Technology,* vol. 10, no. 12, pp. 660–663, 2022, https://doi.org/10.22214/ijraset.2022.47980.
[93] X. Zhao, "Application of Metal-Organic Frameworks for CO_2 Capture," *Highlights in Science Engineering and Technology,* vol. 90, pp. 123–128, 2024, https://doi.org/10.54097/1ab8v510.
[94] F. Benevides *et al.*, "Estimation of the Heat of Absorption of CO_2 by IR Imaging of the Surface of Solid Absorbents," 2014, https://doi.org/10.21611/qirt.2014.210.
[95] E. Haldoupis, S. Nair, and D. S. Sholl, "Efficient Calculation of Diffusion Limitations in Metal Organic Framework Materials: A Tool for Identifying Materials for Kinetic Separations," *Journal of the American Chemical Society,* vol. 132, no. 21, pp. 7528–7539, 2010, https://doi.org/10.1021/ja1023699.

[96] L. Zhou, Z. Qu, T. Ding, and J. Miao, "Lattice Boltzmann Simulation of the Gas-solid Adsorption Process in Reconstructed Random Porous Media," *Physical Review E,* vol. 93, no. 4, p. 043101, 2016.

[97] N. Guilhaume, D. Bianchi, R. A. Wandawa, Y. Wang, and Y. Schuurman, "Study of CO_2 and H_2O Adsorption Competition in the Combined Dry/Steam Reforming of Biogas," *Catalysis Today,* vol. 375, pp. 282–289, 2021, https://doi.org/10.1016/j.cattod.2020.04.058.

[98] P. Goel, "An Insight into the CO_2 Adsorption and Temperature Evolution of a Zeolite Bed in a Fluidized Bed Reactor Using a Two-Fluid Modeling Approach," *Industrial & Engineering Chemistry Research,* vol. 63, no. 23, pp. 10343–10352, 2024, https://doi.org/10.1021/acs.iecr.4c00435.

[99] A. B. H. Jrad, A. Zagnani, A. Mhimid, and M. E. Ganaoui, "Feasibility Study of Cold Production Using Activated Carbon/CO_2 Pair," *International Journal of Heat and Technology,* vol. 37, no. 2, pp. 625–632, 2019, https://doi.org/10.18280/ijht.370234.

[100] M. Hefti and M. Mazzotti, "Postcombustion CO_2 Capture from Wet Flue Gas by Temperature Swing Adsorption," *Industrial & Engineering Chemistry Research,* 2018, https://doi.org/10.1021/acs.iecr.8b03580.

[101] B. Shirani and M. Eić, "Kinetic Behaviours of Carbon Dioxide and Carbon Monoxide on Carbon Molecular Sieve," *The Canadian Journal of Chemical Engineering,* vol. 94, no. 10, pp. 2023–2034, 2016, https://doi.org/10.1002/cjce.22585.

[102] M. K. A. Mesfer, M. Danish, M. A. Khan, I. H. Ali, M. Hasan, and A. E. Jery, "Continuous Fixed Bed CO_2 Adsorption: Breakthrough, Column Efficiency, Mass Transfer Zone," *Processes,* vol. 8, no. 10, p. 1233, 2020, https://doi.org/10.3390/pr8101233.

[103] H. A. Kindi, "Simulation on the Breakthrough Curve during CO_2 Adsorption from Biogas in a Fixed Bed Column," *Asean Journal of Chemical Engineering,* vol. 23, no. 3, p. 318, 2023, https://doi.org/10.22146/ajche.82470.

[104] S. Deridder and G. Desmet, "New Insights in the Velocity Dependency of the External Mass Transfer Coefficient in 2D and 3D Porous Media for Liquid Chromatography," *Journal of Chromatography A,* vol. 1227, pp. 194–202, 2012.

[105] S. Nedeltchev, U. Jordan, and A. Schumpe, "Correction of the Penetration Theory Applied to the Prediction of kLa in a Bubble Column with Organic Liquids," *Chemical Engineering & Technology: Industrial Chemistry-Plant Equipment-Process Engineering-Biotechnology,* vol. 29, no. 9, pp. 1113–1117, 2006.

[106] M. Ochowiak *et al.*, "The D 2-Law of Droplet Evaporation When Calculating the Droplet Evaporation Process of Liquid Containing Solid State Catalyst Particles," *Energies,* vol. 15, no. 20, p. 7642, 2022.

[107] V. Safari, F. Haghighat, C.-s. Lee, P. Blondeau, R. S. Popescu, and N. Lakdawala, "A Systematic Approach for Evaluation of Gas-phase Filter Model," *HVAC&R Research,* vol. 19, no. 6, pp. 705–714, 2013.

[108] J. Wang, "A Review of Eulerian Simulation of Geldart A Particles in Gas-fluidized Beds," *Industrial & Engineering Chemistry Research,* vol. 48, no. 12, pp. 5567–5577, 2009.

[109] L. P. McMaster and E. R. Gilliland, "An Investigation of the Influence of Porosity Variations on the Performance of Liquid Phase Packed Bed Chemical Reactors at Low Reynolds Numbers," *Chemical Engineering Science,* vol. 27, no. 12, pp. 2265–2280, 1972.

[110] M. Keyser, M. Conradie, M. Coertzen, and J. Van Dyk, "Effect of Coal Particle Size Distribution on Packed Bed Pressure Drop and Gas Flow Distribution," *Fuel,* vol. 85, no. 10–11, pp. 1439–1445, 2006.

[111] W. R. Goossens, "Review of the Empirical Correlations for the Drag Coefficient of Rigid Spheres," *Powder Technology,* vol. 352, pp. 350–359, 2019.

[112] Y. Djebbar and R. Narbaitz, "Improved Onda Correlations for Mass Transfer in Packed Towers," *Water Science and Technology,* vol. 38, no. 6, pp. 295–302, 1998.

[113] L. F. A. S. Zafanelli, A. Henrique, M. Karimi, A. r. E. Rodrigues, and J. A. C. Silva, "Single- And Multicomponent Fixed Bed Adsorption of CO_2, CH_4, and N_2 in Binder-Free Beads of 4A Zeolite," *Industrial & Engineering Chemistry Research,* vol. 59, no. 30, pp. 13724–13734, 2020, https://doi.org/10.1021/acs.iecr.0c01911.

[114] L. H. d. Oliveira, J. G. Meneguin, E. A. d. Silva, B. Maria Angélica Simões Dornellas de, P. A. Arroyo, W. M. Grava, and J. F. d. Nascimento, "Linear Driving Force Model in Carbon Dioxide Capture by Adsorption," *Applied Mechanics and Materials,* vol. 830, pp. 38–45, 2016, https://doi.org/10.4028/www.scientific.net/amm.830.38.

[115] Y. Fan, Y. Labreche, R. P. Lively, C. W. Jones, and W. J. Koros, "Dynamic CO_2 Adsorption Performance of Internally Cooled Silica-supported Poly(ethylenimine) Hollow Fiber Sorbents," *Aiche Journal,* vol. 60, no. 11, pp. 3878–3887, 2014, https://doi.org/10.1002/aic.14615.

[116] S. Sangsuradet, "Mass Transport, Kinetic Model, and Application of CO_2 Adsorption on Zeolite 5A Granules," *Clean—Soil Air Water,* vol. 52, no. 1, 2023, https://doi.org/10.1002/clen.202200406.

[117] H. Hongming, H. Fu, K. Xue, H. Zhang, and H. Chen, "Hydrophobic Modification of Ceramic Membrane and Its Application in CO_2 Capture," *Energy,* vol. 27, 2022, p. 2965, https://doi.org/10.46855/energy-proceedings-10192.

[118] F. Rezaei, R. P. Lively, Y. Labreche, G. Chen, Y. Fan, W. J. Koros, and C. W. Jones, "Aminosilane-Grafted Polymer/Silica Hollow Fiber Adsorbents for CO_2 Capture from Flue Gas," *ACS Applied Materials & Interfaces,* vol. 5, no. 9, pp. 3921–3931, 2013, https://doi.org/10.1021/am400636c.

[119] A. Caravella, P. F. Zito, A. Brunetti, E. Drioli, and G. Barbieri, "A Novel Modelling Approach to Surface and Knudsen Multicomponent Diffusion through NaY Zeolite Membranes," *Microporous and Mesoporous Materials,* vol. 235, pp. 87–99, 2016.

[120] F. Javadpour, "Nanopores and Apparent Permeability of Gas Flow in Mudrocks (Shales and Siltstone)," *Journal of Canadian Petroleum Technology,* vol. 48, no. 08, pp. 16–21, 2009, https://doi.org/10.2118/09-08-16-da.

[121] T. Wang, S. Tian, G. Li, and P. Zhang, "Analytical Model for Real Gas Transport in Shale Reservoirs with Surface Diffusion of Adsorbed Gas," *Industrial & Engineering Chemistry Research,* vol. 58, no. 51, pp. 23481–23489, 2019, https://doi.org/10.1021/acs.iecr.9b05630.

[122] C. Strangfeld, "Quantification of the Knudsen Effect on the Effective Gas Diffusion Coefficient in Partially Saturated Pore Distributions," *Advanced Engineering Materials,* vol. 23, no. 10, 2021, https://doi.org/10.1002/adem.202100106.

[123] A. M. Allan and G. Mavko, "The Effect of Adsorption and Diffusion on the Gas Permeability of Kerogen," In *SEG International Exposition and Annual Meeting*, pp. SEG-2012. SEG, 2012, https://doi.org/10.1190/segam2012–0076.1.

[124] Q. Zhang, W. Wang, Y. Kade, B. Wang, and L. M. Xiong, "Analysis of Gas Transport Behavior in Organic and Inorganic Nanopores Based on a Unified Apparent Gas Permeability Model," *Petroleum Science,* vol. 17, no. 1, pp. 168–181, 2019, https://doi.org/10.1007/s12182-019-00358-4.

[125] K. Wu, X. Li, C. Guo, C. Wang, and Z. Chen, "A Unified Model for Gas Transfer in Nanopores of Shale-gas Reservoirs: Coupling Pore Diffusion and Surface Diffusion," *Spe Journal,* vol. 21, no. 05, pp. 1583–1611, 2016.

[126] S. Y. Yoo, J. S. Roh, J. Y. Kim, W. Kim, H. B. Park, and H. W. Kim, "Gas Diffusion Through Nanoporous Channels of Graphene Oxide and Reduced Graphene Oxide Membranes," *ACS Applied Nano Materials,* vol. 5, no. 5, pp. 7029–7035, 2022, https://doi.org/10.1021/acsanm.2c00974.

[127] Y. Yin, Z. Qu, C. Zhu, and J. Zhang, "Visualizing Gas Diffusion Behaviors in Three-dimensional Nanoporous Media," *Energy & Fuels,* vol. 35, no. 3, pp. 2075–2086, 2021.

[128] W. Zhou, J. Zhu, H. Wang, and D. Kong, "Transport Diffusion Behaviors and Mechanisms of CO_2/CH_4 in Shale Nanopores: Insights from Molecular Dynamics Simulations," *Energy & Fuels,* vol. 36, no. 19, pp. 11903–11912, 2022, https://doi.org/10.1021/acs.energyfuels.2c02197.

[129] Z. Liu *et al.*, "Understanding the Diffusion Properties of Sulfur-Containing Compounds in Mesoporous Alumina: A Molecular Dynamics Study," *Industrial & Engineering Chemistry Research,* vol. 61, no. 8, pp. 3023–3030, 2022, https://doi.org/10.1021/acs.iecr.1c04934.

[130] Z. Chen, J. Gao, S. Zhao, Z. Xu, and X. Chen, "Effect of Composition and Configuration on Hindered Diffusion of Residue Fractions Through Polycarbonate Membranes," *AIChE Journal,* vol. 59, no. 4, pp. 1369–1377, 2012, https://doi.org/10.1002/aic.13884.

[131] D. Selassie, D. K. Davis, J. L. Dahlin, E. Feise, G. Haman, D. S. Shall, and D. Kohen, "Atomistic Simulations of CO_2 and N_2 Diffusion in Silica Zeolites: The Impact of Pore Size and Shape," *The Journal of Physical Chemistry C,* vol. 112, no. 42, pp. 16521–16531, 2008, https://doi.org/10.1021/jp803586m.

[132] K. Wu, X. Li, C. Wang, W. Yu, and Z. Chen, "A Model for Surface Diffusion of Adsorbed Gas in Nanopores of Shale Gas Reservoirs," In *Offshore Technology Conference*, pp. OTC-25662. OTC, 2015, https://doi.org/10.4043/25662-ms.

[133] W. Ren, G. Li, S. Tian, M. Sheng, and L. Geng, "Adsorption and Surface Diffusion of Supercritical Methane in Shale," *Industrial & Engineering Chemistry Research,* vol. 56, no. 12, pp. 3446–3455, 2017, https://doi.org/10.1021/acs.iecr.6b04432.

[134] M. Zheng, C. Xu, H. Huang, Z. Ye, and X. Chen, "Adsorption Diffusion Behaviour of 4,6-DMDBT from Diesel Fuel," *The Canadian Journal of Chemical Engineering,* vol. 95, no. 3, pp. 605–612, 2016, https://doi.org/10.1002/cjce.22690.

[135] R. Ocampo-Pérez, R. Leyva-Ramos, and É. Padilla-Ortega, "Equilibrium and Kinetic Adsorption of Organic Compounds Onto Organobentonite: Application of a Surface Diffusion Model," *Adsorption Science & Technology,* vol. 29, no. 10, pp. 1007–1024, 2011, https://doi.org/10.1260/0263–6174.29.10.1007.

[136] Q. Jia, B. J. Venton, and K. H. DuBay, "Structure and Dynamics of Adsorbed Dopamine on Solvated Carbon Nanotubes and in a CNT Groove," *Molecules,* vol. 27, no. 12, p. 3768, 2022, https://doi.org/10.3390/molecules27123768.

[137] M. Michailov, "Computational Study of Stripe Alloy Formation on Stepped Surfaces," *Physical Review B,* vol. 80, no. 3, 2009, https://doi.org/10.1103/physrevb.80.035425.

[138] S. Zhang *et al.*, "Design and Optimization of Solid Amine CO_2 Adsorbents Assisted by Machine Learning," *ACS Sustainable Chemistry & Engineering,* vol. 10, no. 39, pp. 13185–13193, 2022, https://doi.org/10.1021/acssuschemeng.2c04492.

[139] G. G. Kaya and H. Deveci, "Polietilenimin Ile Modifiye Edilmiş Silika Kserojellerin Kullanımı Ile CO_2 Tutumu," *Konya Journal of Engineering Sciences,* vol. 9, no. 4, pp. 1109–1118, 2021, https://doi.org/10.36306/konjes.1016120.

[140] G. Zhang, P. Zhao, and Y. Xu, "Development of Amine-Functionalized Hierarchically Porous Silica for CO_2 Capture," *Journal of Industrial and Engineering Chemistry,* vol. 54, pp. 59–68, 2017, https://doi.org/10.1016/j.jiec.2017.05.018.

[141] S. Gautam and D. R. Cole, "Effects of Pore Connectivity on the Sorption of Fluids in Nanoporous Material: Ethane and CO_2 Sorption in Silicalite," *Chemengineering,* vol. 5, no. 3, p. 55, 2021, https://doi.org/10.3390/chemengineering5030055.

[142] H. S. Cho *et al.*, "Physicochemical Understanding of the Impact of Pore Environment and Species of Adsorbates on Adsorption Behaviour," *Angewandte Chemie,* vol. 60, no. 37, pp. 20504–20510, 2021, https://doi.org/10.1002/anie.202107897.

[143] K. Sarna-Boś, K. Skic, J. Sobieszczański, P. Boguta, and R. Chałas, "Contemporary Approach to the Porosity of Dental Materials and Methods of Its Measurement," *International Journal of Molecular Sciences,* vol. 22, no. 16, p. 8903, 2021.

[144] F. Granados-Correa, J. Bonifacio-Martínez, H. Hernández-Mendoza, and S. Bulbulian, "Capture of CO_2 on Γ-Al_2O_3 Materials Prepared by Solution-Combustion and Ball-Milling Processes," *Journal of the Air & Waste Management Association,* vol. 66, no. 7, pp. 643–654, 2016, https://doi.org/10.1080/10962247.2016.1161673.
[145] X. Wan, X. Lu, J. Liu, Y. Pan, and H. Xiao, "Impregnation of PEI in Novel Porous $MgCO_3$ for Carbon Dioxide Capture from Flue Gas," *Industrial & Engineering Chemistry Research,* vol. 58, no. 12, pp. 4979–4987, 2019, https://doi.org/10.1021/acs.iecr.8b06153.
[146] L. A. Darunte, A. D. Oetomo, K. S. Walton, D. S. Sholl, and C. W. Jones, "Direct Air Capture of CO_2 Using Amine Functionalized MIL-101(Cr)," *ACS Sustainable Chemistry & Engineering,* vol. 4, no. 10, pp. 5761–5768, 2016, https://doi.org/10.1021/acssuschemeng.6b01692.
[147] J. R. Hoffman, A. E. Baumann, and C. M. Stafford, "Thickness Dependent CO_2 Adsorption of Poly (Ethyleneimine) Thin Films for Direct Air Capture," *Chemical Engineering Journal,* vol. 481, p. 148381, 2024.
[148] L. Castro-Gonzales, K. Kimmerle, E. Schippert, and D. Fickinger, "Finding and Evaluating of Adequate Adsorbents for the Adsorption of CO_2 from Humid Gas Streams," *Chemistryselect,* vol. 1, no. 11, pp. 2834–2841, 2016, https://doi.org/10.1002/slct.201600094.
[149] W. Guang, X. Liu, Z. Zhang, and P. Luo, "Effects of Sub- And Supercritical CO_2 on Coal Diffusivity and Surface Thermodynamics," *Energy & Fuels,* vol. 36, no. 7, pp. 3737–3748, 2022, https://doi.org/10.1021/acs.energyfuels.1c04408.
[150] T. Le, A. Striolo, and D. R. Cole, "CO_2–C_4H_{10} Mixtures Simulated in Silica Slit Pores: Relation between Structure and Dynamics," *The Journal of Physical Chemistry C,* vol. 119, no. 27, pp. 15274–15284, 2015, https://doi.org/10.1021/acs.jpcc.5b03160.
[151] S. Choi, J. H. Drese, P. Eisenberger, and C. W. Jones, "Application of Amine-Tethered Solid Sorbents for Direct CO_2 Capture from the Ambient Air," *Environmental Science & Technology,* vol. 45, no. 6, pp. 2420–2427, 2011, https://doi.org/10.1021/es102797w.
[152] L. A. Darunte, K. S. Walton, D. S. Sholl, and C. W. Jones, "CO_2 Capture via Adsorption in Amine-Functionalized Sorbents," *Current Opinion in Chemical Engineering,* vol. 12, pp. 82–90, 2016, https://doi.org/10.1016/j.coche.2016.03.002.
[153] J. A. K. Jovellana and B. B. Pajarito, "Carbon Dioxide Adsorbent Preparation by Coating Amine-Functionalized Pectin Onto Zeolites," *Key Engineering Materials,* vol. 801, pp. 179–184, 2019, https://doi.org/10.4028/www.scientific.net/kem.801.179.
[154] M. N. A. Tahari, A. H. Lahuri, Z. Ghazali, S. Samidin, S. S. Sulhadi, N. Dzakaria, and M. A. Yarmo, "Application of Octadecylamine-Based Adsorbent on Carbon Dioxide Capture," *Materials Science Forum,* vol. 1010, pp. 367–372, 2020, https://doi.org/10.4028/www.scientific.net/msf.1010.367.
[155] K. W. Tan, A. H. Ruhaimi, C. N. C. Hitam, and M. R. Taib, "A Brief on Tetraethylenepentamine (TEPA) Functionalized-Adsorbents in CO_2 Capture Application," *Journal of Physics Conference Series,* vol. 2259, no. 1, p. 012002, 2022, https://doi.org/10.1088/1742-6596/2259/1/012002.
[156] Q. Ma, L. Zhao, H. Su, Z. Chen, and Q. Xu, "Numerical Study of Mass Transfer and Desorption Behaviors in Deformable Porous Media Using a Coupling Lattice Boltzmann Model," *Physical Review E,* vol. 102, no. 2, p. 023309, 2020.
[157] Z. Peng, S. Liu, Y. Long, M. Xiao, and H. Feng, "Lattice Boltzmann Simulation of the Kinetics Process of Methane Diffusion with the Adsorption–Desorption Hysteresis Effect in Coal," *ACS Omega,* vol. 8, no. 34, pp. 31135–31144, 2023.
[158] Đ. V. Quang, T. A. Hatton, and M. R. Abu-Zahra, "Thermally Stable Amine-Grafted Adsorbent Prepared by Impregnating 3-Aminopropyltriethoxysilane on Mesoporous Silica for CO_2 Capture," *Industrial & Engineering Chemistry Research,* vol. 55, no. 29, pp. 7842–7852, 2016, https://doi.org/10.1021/acs.iecr.5b04096.

[159] P. Staszczuk, M. Płanda-Czyż, M. Blachnio, and V. V. Kutarov, "Studies of Liquids Diffusion in the Chosen Material Samples," *Journal of Thermal Analysis and Calorimetry,* vol. 86, no. 1, pp. 261–265, 2006, https://doi.org/10.1007/s10973-005-9979-z.

[160] J. Kärger and D. M. Ruthven, "Diffusion in Nanoporous Materials: Fundamental Principles, Insights and Challenges," *New Journal of Chemistry,* vol. 40, no. 5, pp. 4027–4048, 2016, https://doi.org/10.1039/c5nj02836a.

[161] L. Yue *et al.*, "Efficient CO_2 Adsorption on Nitrogen-Doped Porous Carbons Derived from d-Glucose," *Energy & Fuels,* vol. 32, no. 6, pp. 6955–6963, 2018, https://doi.org/10.1021/acs.energyfuels.8b01028.

[162] L. An *et al.*, "Novel Nitrogen-Doped Porous Carbons Derived from Graphene for Effective CO_2 Capture," *Industrial & Engineering Chemistry Research,* vol. 58, no. 8, pp. 3349–3358, 2019, https://doi.org/10.1021/acs.iecr.8b06122.

[163] D. D. Miller and S. S. C. Chuang, "Control of CO_2 Adsorption and Desorption Using Polyethylene Glycol in a Tetraethylenepentamine Thin Film: An in Situ ATR and Theoretical Study," *The Journal of Physical Chemistry C,* vol. 120, no. 44, pp. 25489–25504, 2016, https://doi.org/10.1021/acs.jpcc.6b09506.

[164] Z. Zhao, F. Shi, and F. Guan, "Effects of Plastic Mulching on Soil CO_2 Efflux in a Cotton Field in Northwestern China," *Scientific Reports,* vol. 12, no. 1, 2022, https://doi.org/10.1038/s41598-022-08793-x.

[165] A. K. Saini, T. Radu, K. Paritosh, V. Kumar, N. Pareek, D. Tripathi, and V. Vivekanand, "Bioengineered Bioreactors: A Review on Enhancing Biomethane and Biohydrogen Production by CFD Modeling," *Bioengineered,* vol. 12, no. 1, pp. 6418–6433, 2021.

[166] M. Z. A. Zulkifli, A. Mukhtar, M. F. M. Fadli, A. M. S. Bahari, A. Matsumoto, and H. Misran, "CFD Simulation of CO_2 and Methane Adsorption at Various Temperature for MOF-5 Using Dual-Site and Single-Site Langmuir Model," *CFD Letters,* vol. 13, no. 10, pp. 1–10, 2021, https://doi.org/10.37934/cfdl.13.10.110.

[167] S. Karka, S. Kodukula, S. V. Nandury, and U. Pal, "Polyethylenimine-modified Zeolite 13X for CO_2 Capture: Adsorption and Kinetic Studies," *ACS Omega,* vol. 4, no. 15, pp. 16441–16449, 2019.

[168] F. Aouaini, N. Bouaziz, W. F. Alfwzan, N. Khemiri, Z. M. Elqahtani, and A. B. Lamine, "Adsorption of CO_2 on ZSM-5 Zeolite: Analytical Investigation via a Multilayer Statistical Physics Model," *Applied Sciences,* vol. 12, no. 3, p. 1558, 2022, https://doi.org/10.3390/app12031558.

[169] A. Sinha, L. A. Darunte, C. W. Jones, J. Park, and Y. Kawajiri, "Systems Design and Economic Analysis of Direct Air Capture of CO_2 Through Temperature Vacuum Swing Adsorption Using MIL-101(Cr)-PEI-800 and Mmen-Mg_2 (dobpdc) MOF Adsorbents," *Industrial & Engineering Chemistry Research,* vol. 56, no. 3, pp. 750–764, 2017, https://doi.org/10.1021/acs.iecr.6b03887.

[170] N. Gargiulo, F. Pepe, and D. Caputo, "CO_2 Adsorption by Functionalized Nanoporous Materials: A Review," *Journal of Nanoscience and Nanotechnology,* vol. 14, no. 2, pp. 1811–1822, 2014, https://doi.org/10.1166/jnn.2014.8893.

3 CO_2 Adsorbent Characterizations

Fatemeh Bahmanzadegan, Mahyar Ashourzadeh Pordsari, and Ahad Ghaemi

3.1 INTRODUCTION

Precisely characterizing CO_2 adsorbents is essential for developing efficient materials in carbon capture and storage (CCS) technologies. Key factors influencing adsorbent performance include chemical composition, structural properties, and the mechanisms governing CO_2 interaction at the adsorbent surface. A deep understanding of these parameters is critical for optimizing material design and improving CO_2 capture efficiency. To assess the surface area, porosity, and pore volume, the Brunauer–Emmett–Teller (BET) method is widely used, providing insight into the adsorbent's textural properties. The crystalline structure of adsorbents is typically evaluated through X-ray diffraction (XRD), revealing the phase composition and crystallinity. Chemical composition is often determined using Fourier-transform infrared (FTIR) spectroscopy, Raman spectroscopy, and energy-dispersive X-ray (EDX) spectroscopy. Finally, the morphological characteristics are examined using scanning electron microscopy (SEM) and transmission electron microscopy (TEM), which offer high-resolution imaging of surface and particle structures. This chapter delves into these characterization methods, providing a comprehensive overview of the tools used to analyze CO_2 adsorbents. Figure 3.1 illustrates a range of advanced characterization techniques essential for comprehensively analyzing the physical, thermal, structural, surface, and chemical properties of CO_2 adsorbent materials.

3.2 MORPHOLOGICAL CHARACTERIZATION

3.2.1 FE-SEM Analyzes

Field-emission scanning electron microscopy (FE-SEM) is a powerful and advanced imaging technique used to examine the surface morphology, microstructure, and topographical features of materials with sub-nanometer resolution [1]. In the context of CO_2 adsorbents, FE-SEM is essential for providing detailed insights into the surface characteristics and structural properties of materials such as zeolites [2, 3], activated carbons (ACs) [4], and metal-organic frameworks (MOFs) [5]. These characteristics—particle size, pore distribution, surface roughness, and crystal architecture—are crucial in the adsorption mechanisms determining the material's performance in CO_2 capture [6]. FE-SEM operates by directing a focused beam of

DOI: 10.1201/9781003476191-3

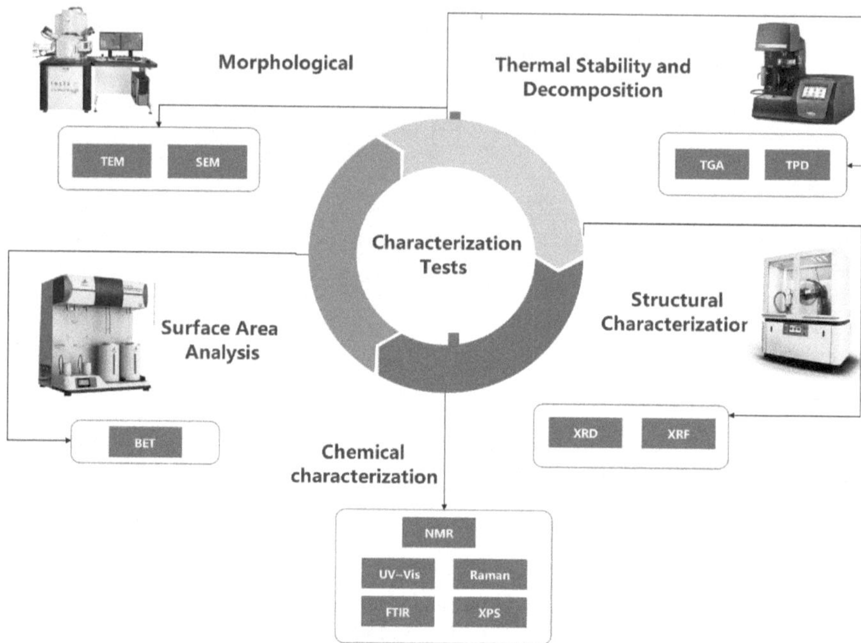

FIGURE 3.1 Characterization techniques used for analyzing CO_2 adsorbents.

high-energy electrons onto the surface of the analyzed material. Unlike conventional SEM, which uses a thermionic emitter to produce electrons, FE-SEM employs a field-emission electron gun, which generates a much more focused and intense electron beam. The field-emission process relies on the quantum mechanical phenomenon where electrons are emitted from the surface of a sharp metal tip under the influence of a strong electric field. This results in the emission of electrons with minimal spread, providing significantly higher spatial resolution compared to conventional SEM. FE-SEM is widely used to characterize CO_2 adsorbents to provide detailed information about their surface structure, which plays a key role in determining their adsorption efficiency. Understanding these surface characteristics allows researchers to optimize adsorbent materials for better CO_2 capture performance by correlating structural features with adsorption behavior. The surface morphology of CO_2 adsorbents, such as the size and shape of particles, roughness, and the presence of pores, significantly affects the adsorption process. For example, in zeolites and MOFs, the surface texture determines how gas molecules interact with the adsorbent material, influencing the diffusion and adsorption of CO_2 [7]. High surface roughness can enhance adsorption by increasing the surface area for gas interaction. Pore size and distribution are fundamental to CO_2 adsorption [8]. Materials like AC and zeolites rely on their porous structures to trap CO_2 molecules. FE-SEM provides high-resolution imaging of these pores, allowing for a detailed examination of the pore network. This information is critical in designing materials with the appropriate pore sizes for selective adsorption of CO_2 over other gases. Additionally, the

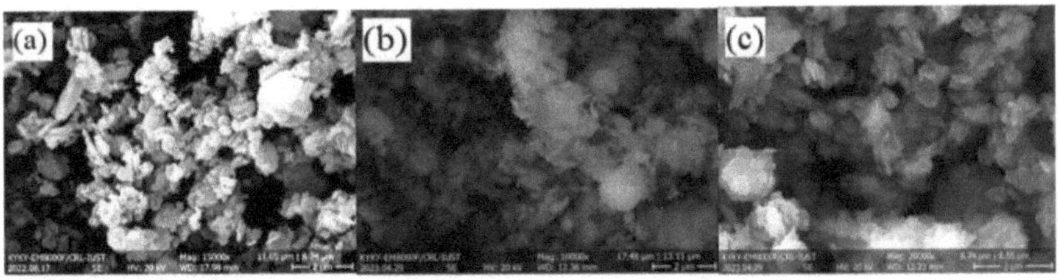

FIGURE 3.2 SEM images of (a) zeolite 4A, (b) zeolite 4A modified with 10% DEA, and (c) zeolite 4A modified with 15% TEPA [14], copyright © 2023, Nature.

pore size can affect the kinetics of the adsorption process, as small pores are ideal for trapping CO_2 molecules, while larger pores may be less effective. FE-SEM is often employed to study surface modifications of CO_2 adsorbents, such as the grafting of functional groups or the incorporation of metal particles. Surface functionalization is a common strategy to improve the selectivity and adsorption capacity of CO_2 adsorbents. FE-SEM can visually confirm these modifications, showing how functional groups or nanoparticles are distributed on the material's surface. Several types of zeolites have been extensively evaluated for their potential to capture CO_2 emissions from industrial processes. Zeolites such as ZSM-5 [9], 4A [10], 13X [11], Na-X [12], and β-zeolite [13] have demonstrated strong potential for CO_2 adsorption due to their distinct pore structures, large surface areas, and excellent thermal stability. These characteristics make them highly suitable for CO_2 capture applications. The adsorption capacity of zeolites can be further enhanced through amine modification. In a study by Bahmanzadegan and Ghaemi [8] zeolite was synthesized from kaolin and impregnated with two different amines. Among the various loadings, zeolite 4A with 10% diethanolamine (DEA) and 15% tetraethylenepentamine (TEPA) achieved the best performance. In Figure 3.2a, the bright and dark regions highlight variations in the surface topography and composition of zeolite 4A, with bright areas indicating raised surfaces and dark areas representing lower regions or pores.

The size of the particles affects the surface area-to-volume ratio, which is directly related to the adsorption capacity. Smaller particles typically provide a larger surface area for CO_2 adsorption, which enhances the material's overall performance. The particle shape also influences the packing density of adsorbents in columns or beds, impacting the flow of CO_2 gas through the material. The crystallinity and presence of defects in adsorbent materials influence their performance in CO_2 capture. This information helps correlate the surface features with the observed changes in adsorption behavior. AC is a highly microporous material known for its extensive surface area, making it practical for removing pollutants. Studies have demonstrated that CO_2 adsorption by modified AC, treated with active substances, significantly surpasses unmodified AC at ambient temperatures. Modifying AC with metal hydroxides enhances its surface area and adsorption capacity. Among the most effectual modifiers, potassium hydroxide (KOH) has been widely used to improve AC's ability to capture more significant amounts of CO_2. This modification enhances the material's porosity and increases the number of active sites, allowing for more efficient

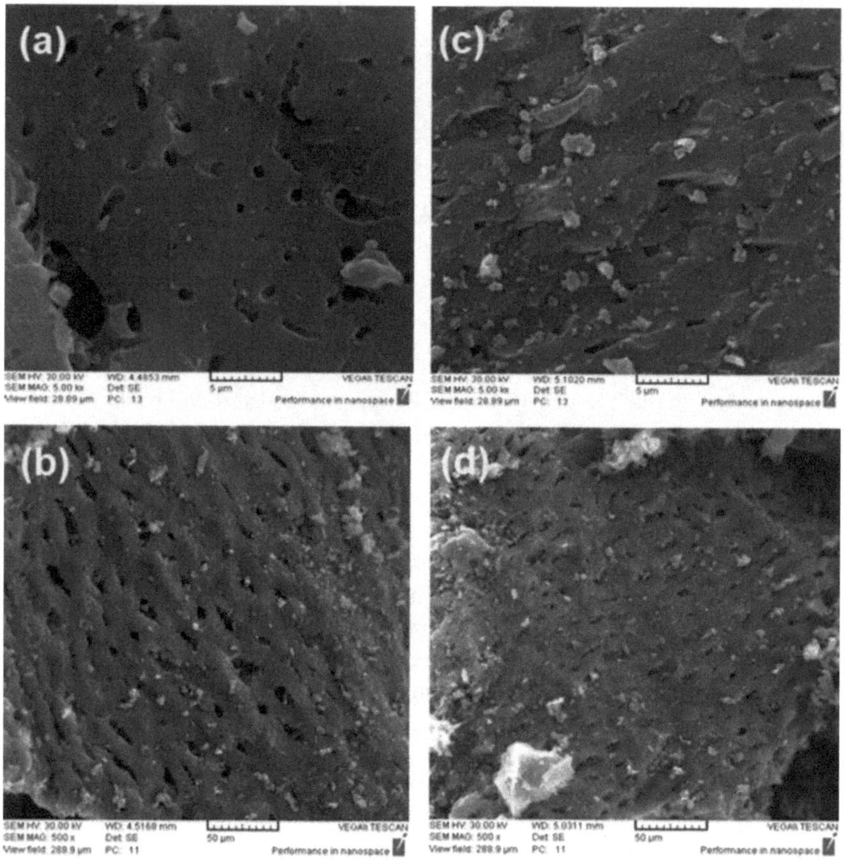

FIGURE 3.3 SEM analysis: (a) AC at 5 μm magnification, (b) AC at 50 μm magnification, (c) LiOH-modified AC at 5 μm magnification, and (d) LiOH-modified AC at 50 μm magnification [15], copyright © 2024, Nature.

interactions between the adsorbent and CO_2 molecules. Consequently, KOH-modified AC has shown substantial promise in improving CO_2 capture performance in various applications. Ahmadi et al. synthesized modified AC and examined its properties for CO_2 adsorption. Figure 3.3 displays the morphological structure of unmodified AC alongside AC modified with 24% lithium hydroxide (24Li-AC) at 50 and 500 μm magnifications. These images reveal the porous nature of the AC surface, illustrating a consistent distribution of pores with varying sizes. The modification with LiOH enhances the porosity, contributing to a more uniform and well-defined pore structure, which is crucial for improving the material's adsorption capacity. The comparison highlights the structural differences introduced by LiOH modification, which may enhance the effectiveness of the AC in various adsorption applications, particularly in capturing CO_2. The results indicate that the unmodified adsorbent has lower porosity than the modified sample, suggesting that LiOH penetrates the AC structure, enhancing surface area and porosity. The 24Li-AC sample exhibits highly

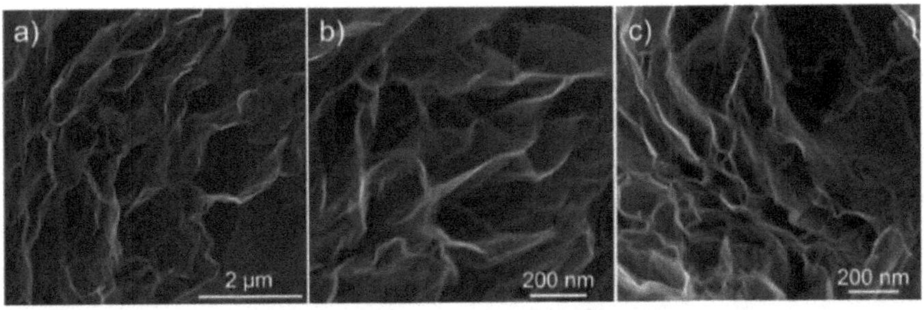

FIGURE 3.4 SEM of HPGC-850 (a, b) and HPGC-950 (c) [18], copyright (2014), Elsevier.

fractured surfaces with well-defined pore sizes, making it particularly suitable for CO_2 adsorption [15].

In some cases, CO_2 adsorbents are designed with hierarchical structures, where different porosity levels (micro-, meso-, and macropores) are combined to optimize the material for CO_2 capture [16]. FE-SEM can image these complex structures at different scales, providing a detailed understanding of how the hierarchical arrangement of pores contributes to enhanced adsorption performance. This hierarchical structuring often improves the capacity and kinetics of CO_2 capture by providing pathways for faster gas diffusion and multiple adsorption sites. It can be used to evaluate the stability of CO_2 adsorbents over time, particularly after exposure to gases or during adsorption–desorption cycles [17].

Xia et al. synthesized hierarchical porous graphene-based carbons (HPGCs) via graphite oxide's straightforward carbon dioxide activation. As depicted in Figure 3.4a and b, the material exhibits thin, layered sheets with a curled and wrinkled structure. This morphology is likely the result of graphene sheet restacking, which occurs after the oxygen-containing groups are eliminated from the graphite oxide surface. When subjected to a higher oxidation temperature of 950°C, HPGC-950 (Figure 3.4c) demonstrates a comparable morphology, though with noticeable perforations or "holey" defects within the curved layers. These structural changes at elevated temperatures suggest modifications in surface properties and porosity that may influence the material's performance in applications like adsorption [18]. Researchers can identify structural changes or degradation that may impact the adsorbent's long-term performance by comparing images of the material before and after use. Surface damage, pore collapse, or the loss of structural integrity can all be observed through FE-SEM imaging, providing insights into the material's durability under operational conditions.

3.2.2 TEM Analysis

Transmission electron microscopy (TEM) is an indispensable tool for investigating the microstructure of materials at the nanoscale. The adsorbent is typically dispersed in a suitable liquid medium and deposited onto a TEM grid. Negative staining can be used to enhance contrast. TEM utilizes various imaging techniques, including

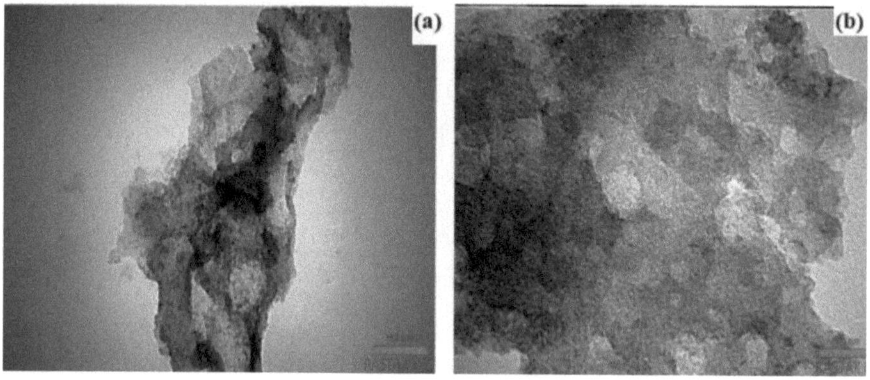

FIGURE 3.5 TEM of (a) GAC-pure and (b) modified GAC-10–500 [4], copyright © 2022 American Chemical Society.

bright-field imaging, dark-field imaging, and high-resolution imaging. Each method offers specific advantages for visualizing different features of the adsorbent. By providing high-resolution images, TEM offers invaluable insights into the morphology, size distribution, and structural features of CO_2 adsorbents. TEM enables precise measurement of adsorbent particle size and distribution [19]. This information is essential for understanding the adsorbent's surface area and porosity, directly influencing its CO_2 adsorption capacity. It allows for detailed visualization of the adsorbent's morphology, including its shape, surface roughness, and defects or agglomerates. These morphological characteristics play a crucial role in the adsorbent's performance. TEM provides insights into the intricate pore structure of adsorbents [20]. Researchers can determine the pore volume and surface area, by examining the size, shape, and distribution of pores, which are key parameters for CO_2 adsorption. In addition, TEM can detect and analyze structural defects such as dislocations, grain boundaries, and stacking faults within the adsorbent material. These defects can significantly impact the adsorbent's mechanical properties and adsorption behavior.

Mashhadimoslem et al. [4] conducted an in-depth study on the adsorption behavior of CO_2 using both granular activated carbon (GAC) and chemically functionalized GAC. Their investigation was performed across various temperatures—298, 308, and 318 K—while subjecting the materials to pressures reaching up to 10 bar. By evaluating the adsorption performance under these controlled conditions, they were able to assess the impact of temperature and functionalization on CO_2 uptake, providing insights into the efficiency and potential applications of these materials in high-pressure CO_2 capture environments. TEM images (Figure 3.5) reveal crystallinity formation within the porous framework, validating the structures of both GAC-pure and GAC-10–500 on the 50–100 nm scale. For GAC-10–500, the interconnected, wormlike porosity is particularly evident. By examining the distribution of CO_2 molecules within the adsorbent's pores, valuable insights can be obtained regarding the adsorption mechanism and the adsorbent's selectivity for CO_2 [4]. So, TEM analysis provides a comprehensive understanding of the structural and morphological properties of CO_2 adsorbents.

3.3 SURFACE AREA AND POROSITY ANALYSIS

3.3.1 BET Analysis

The BET analysis is a cornerstone technique for determining the specific surface area of porous and solid materials, widely employed in the study of CO_2 adsorbents [21]. This method is based on the physical adsorption of gas molecules, typically nitrogen, onto the surface of a solid under controlled conditions. By analyzing the adsorption data at various pressures, critical surface properties such as specific surface area and pore structure can be elucidated [22]. The BET model extends the Langmuir theory, which assumes monolayer adsorption, by accounting for multilayer adsorption. This theory assumes that (1) adsorption occurs in successive layers, beginning with a monolayer on the solid surface. Subsequent layers are adsorbed onto previously adsorbed molecules. (2) The model assumes that gas molecules in different layers do not interact with each other. The adsorption in one layer is independent of the adsorption in others. (3) A dynamic equilibrium exists between gas molecules adsorbed on the surface and those desorbed back into the gas phase. The BET equation, derived from these assumptions, is applied to experimental adsorption data to extract key material parameters, including specific surface area. The BET equation is shown in Equation (3.1)

$$\frac{1}{V\left(\frac{P_0}{P}-1\right)}=\frac{C-1}{V_mC}\cdot\frac{P}{P_0}+\frac{1}{V_mC} \tag{3.1}$$

where V is the volume of adsorbed gas, V_m is the monolayer adsorption capacity (volume of gas required to form a monolayer), P_0 is the saturation pressure of the adsorbing gas, P is the partial pressure of the gas, and C is a constant related to the enthalpy of adsorption in the first layer relative to subsequent layers. From this equation, the monolayer adsorption capacity V_m can be determined, which, combined with the molecular cross-sectional area of the adsorbed gas, allows the calculation of the material's specific surface area. The BET analysis typically requires cooling the sample to cryogenic temperatures, often to liquid nitrogen temperature (77 K), to enhance nitrogen adsorption. The solid sample is placed in a sample cell, where nitrogen gas is introduced gradually. As the gas pressure increases, the amount of gas adsorbed on the surface rises, and the volume of adsorbed gas at each pressure point is recorded. Once the system nears saturation pressure (P_0), adsorption levels off, indicating that multilayer adsorption has occurred. This information is plotted to create an adsorption isotherm, with the volume of adsorbed gas on the y-axis and relative pressure ($\frac{P}{P_0}$) on the x-axis.

The adsorption–desorption isotherms derived from BET analysis provide essential insights into the surface and pore characteristics of the material. According to the International Union of Pure and Applied Chemistry (IUPAC), these isotherms are categorized into six distinct types, as illustrated in Figure 3.6. Type I isotherms are characterized by a rapid rise in adsorption at low pressures, reflecting the formation of a monolayer on the surface. This behavior indicates strong binding interactions between

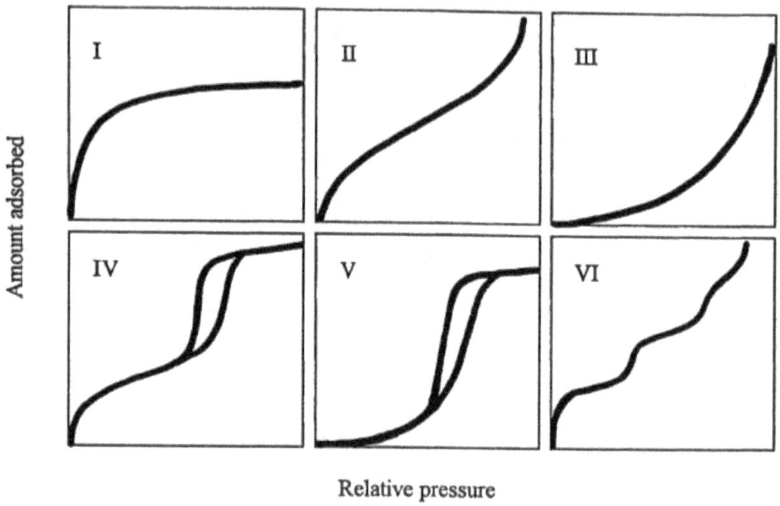

FIGURE 3.6 The types of IUPAC classification of adsorption isotherms [23], copyright (1988), Elsevier.

the adsorbate molecules and the microporous structure of the material. Type II isotherms represent nonporous or macroporous materials where adsorption occurs in multiple layers. The initial phase corresponds to the formation of a monolayer, followed by multilayer adsorption as pressure increases, typical of materials with substantial surface area and pore volume. Type III isotherms reflect weak interactions between the adsorbent and the adsorbate, with the curve indicating that adsorbate–adsorbate interactions are more significant than adsorbent–adsorbate interactions. This behavior is commonly observed in nonporous materials with low adsorption forces. Type IV isotherms, generally associated with mesoporous materials (pore sizes of 2–50 nm), display an initial monolayer phase to multilayer adsorption, followed by a sharp rise at higher pressures due to capillary condensation within mesopores. This type is crucial for studying materials utilized in gas storage and catalysis. Type V isotherms, similar to Type III, also indicate weak adsorbent–adsorbate interactions but display capillary condensation in later stages, suggesting the presence of mesopores. Finally, Type VI isotherms are characteristic of uniform, nonporous materials, indicating stepwise multilayer adsorption and reflecting a highly regular surface with minimal pore structure. Microporous materials, such as those exhibiting Type I isotherms, generally have a strong affinity for CO_2 due to their small pore sizes, which enhance gas–solid interactions. In contrast, mesoporous materials (Type IV isotherms) provide larger pore volumes that facilitate gas diffusion and improve adsorption kinetics. For mesoporous materials, capillary condensation enhances CO_2 adsorption, particularly at elevated pressures, making them suitable for applications in gas separation and storage.

MOFs present considerable benefits for CO_2 capture, primarily owing to their exceptionally high surface areas and customizable pore architectures. These materials allow for precise control over pore size, shape, and functionality, making them highly adaptable for optimizing gas adsorption processes [24, 25]. Kazemi et al. synthesized

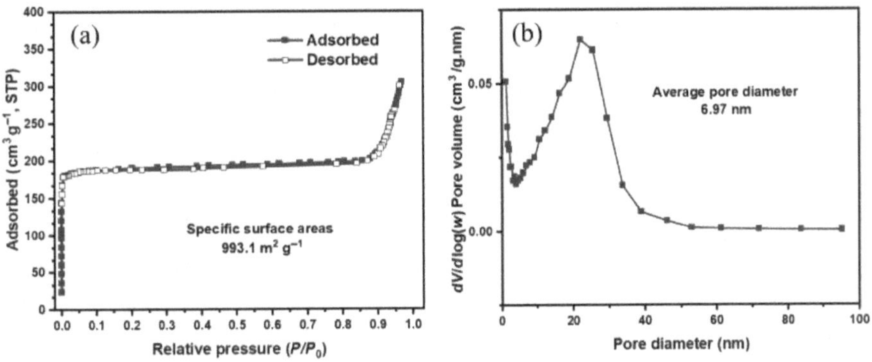

FIGURE 3.7 (a) The BET surface area and (b) the BJH pore diameter distribution of UiO-66-NH_2 [6], copyright © 2023, Nature.

a MOF specifically for CO_2 adsorption. Figure 3.7 presents the surface area (as determined by the BET method) and pore diameter (assessed using the Barrett, Joyner, and Halenda (BJH) method) of UiO-66-NH_2 produced through a sonochemical approach. The analysis demonstrates that the samples display a Type I isotherm with hysteresis, which indicates the material's microporous structure. This isotherm suggests strong adsorption at low pressures, reinforcing the presence of micropores that contribute to enhanced gas capture capabilities. The UiO-66-NH_2 sample, synthesized through the solvothermal method, exhibited a BET surface area of 876 m^2/g and a pore volume of 0.38 cm^3/g, both of which are critical parameters that underscore its potential for efficient CO_2 adsorption [6].

3.4 CHEMICAL CHARACTERIZATION

3.4.1 Ultraviolet–Visible (UV–Vis) Spectroscopy

Ultraviolet–visible (UV–Vis) spectroscopy is a powerful analytical technique widely utilized for the qualitative and quantitative analyzing of various substances based on their light absorption characteristics. This method operates on the principle that molecules absorb specific wavelengths of light, leading to electronic transitions within the molecules. The Beer–Lambert Law underpins this technique, establishing a linear relationship between absorbance and concentration, which is crucial for accurate measurements in trace analysis [26]. This method offers valuable insights into the interaction dynamics between CO_2 molecules and different adsorbent materials. The importance of UV–Vis spectroscopy stems from its capacity to track alterations in electronic states and molecular interactions, which are essential for elucidating the underlying adsorption mechanisms. A key application of UV–Vis spectroscopy in this field is determining the isosteric heat of adsorption (Q_{st}) for CO_2. Research has shown that in situ UV–Vis optical transmission spectroscopy is an effective method for measuring Q_{st} in microporous polymers such as PIM-1, which are recognized for their selective CO_2 adsorption capabilities. This measurement is particularly

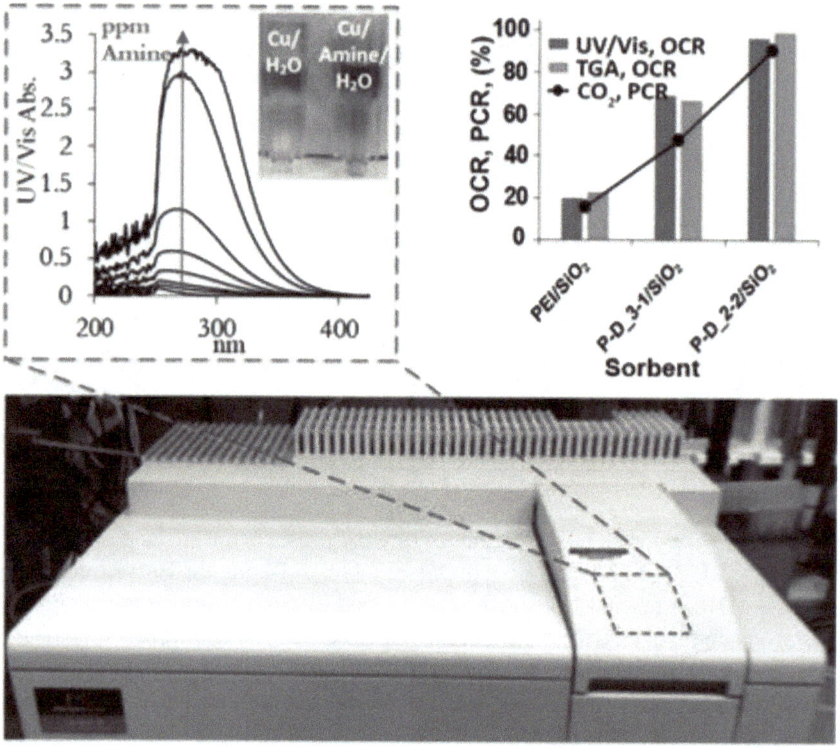

FIGURE 3.8 UV–Vis spectroscopic measurement of samples with different amine concentrations [28], copyright (2018) Wiley Online Library

significant as Q_{st} is directly related to the strength of the interaction between CO_2 and the adsorbent [27]. Moreover, the utility of UV–Vis spectroscopy extends to the characterization of various sorbents and catalysts. For example, a rapid UV–Vis technique has been introduced to assess the stability of immobilized amine sorbents used in CO_2 capture, demonstrating the method's efficiency in quantifying amine complexes formed during the adsorption process. As shown in Figure 3.8, increasing the amine content of adsorbents leads to higher UV–Vis absorbance. This approach is instrumental in evaluating sorbent performance under diverse conditions, which is crucial for optimizing CO_2 capture technologies [28].

In addition to stability assessments, UV–Vis spectroscopy has been utilized to investigate the structural properties of adsorbents. For instance, studies have used UV–Vis spectroscopy to examine MOFs with photoresponsive motifs, showcasing how light-induced isomerization can modulate CO_2 adsorption properties. This highlights the versatility of UV–Vis spectroscopy not only in quantifying adsorption but also in unraveling the mechanisms underlying the interaction between CO_2 and adsorbents [29]. Beyond traditional adsorbents, exploring MOFs and their photothermal properties has unveiled new opportunities for CO_2 capture. Recent work has reported on developing Co / Al_2O_3 Co/Al_2O_3 composites that exhibit high solar-to-fuel efficiency for CO_2 methanation, with UV–Vis light irradiation enhancing CO_2 adsorption. This synergy between light activation and adsorption processes underscores the potential of integrating UV–Vis spectroscopy with advanced materials to achieve enhanced CO_2 capture performance [30]. Furthermore, the interaction of CO_2 with nanomaterials has become a significant area of research, particularly concerning boron nitride (BN) nanomaterials. It has been demonstrated that introducing extra electrons to BN nanomaterials significantly enhances CO_2 adsorption, a phenomenon effectively monitored using UV–Vis spectroscopy. This underscores the critical role of electronic properties in modulating adsorption behavior, which is essential for the design of next-generation adsorbents [31].

3.4.2 Raman Spectroscopy

Raman spectroscopy is a highly effective analytical method that leverages the inelastic scattering of monochromatic light, commonly emitted from a laser, to offer comprehensive information on molecular vibrations, facilitating the identification and characterization of materials. This technique has become increasingly important for studying the adsorption of CO_2 on diverse substrates, especially within the realms of environmental and energy research. Raman spectroscopy delivers in-depth insights into molecular interactions, structural alterations, and the dynamics of CO_2 adsorption processes, which are essential for advancing efficient CO_2 capture technologies. A key advantage of Raman spectroscopy in the study of CO_2 adsorption lies in its capacity to deliver real-time, in situ measurements. For example, Mallick et al. utilized time-resolved Raman spectroscopy to examine CO_2 adsorption on titanium dioxide (TiO_2) surfaces during photocatalytic CO_2 reduction. Their results identified specific Raman bands corresponding to TiO_2, indicating the formation of reactive intermediates upon interaction with CO_2 as depicted in Figure 3.9. This underscores the technique's effectiveness in monitoring dynamic processes in real time [32].

Raman spectroscopy offers the capability to distinguish between various phases of adsorbed CO_2, which is essential for understanding the molecular-level adsorption process. For instance, coherent anti-Stokes Raman scattering has been used to differentiate between different phases of CO_2 adsorbed in mesoporous glass, successfully identifying the initial monolayer of adsorbed CO_2 as well as subsequent polymolecular layers. This ability to resolve distinct adsorption phases is critical for optimizing materials designed for CO_2 capture [33]. Beyond structural insights, Raman spectroscopy has played a significant role in examining the competitive adsorption of CO_2 alongside other gases, such as water vapor. For example, studies have explored the co-adsorption of CO_2 and H_2O, demonstrating that H_2O can displace adsorbed CO_2—a finding with substantial implications for the design of CO_2 capture systems in humid conditions. Understanding competitive adsorption is vital for developing materials that can maintain efficiency under varied operational conditions [34]. Raman spectroscopy is also extensively applied in characterizing the interactions between CO_2 and different adsorbents. In one particular study, Raman bands were observed at 1062 and 1085 cm^{-1}, which were attributed to the stretching vibrations of interlayer carbonate anions in hydrotalcites. These findings suggest that the material

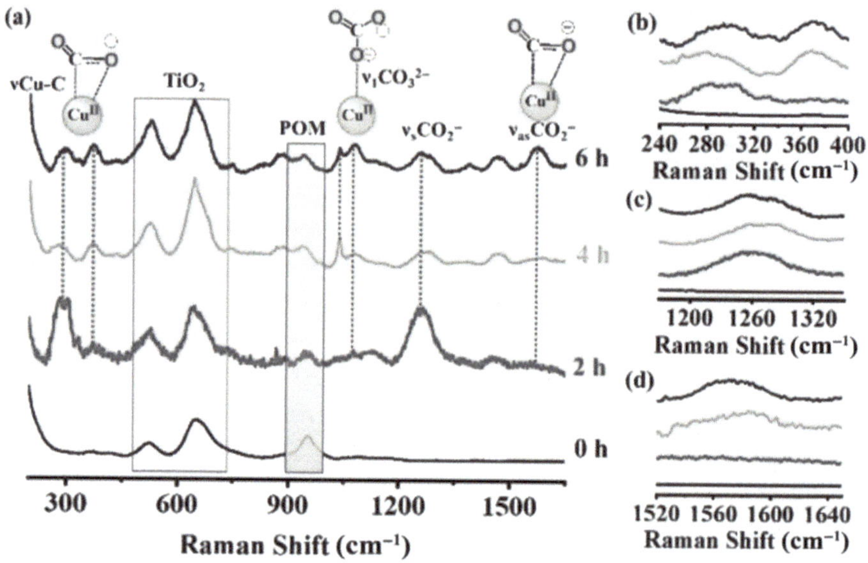

FIGURE 3.9 (a) Time-resolved quasi-in situ Raman analysis conducted during the photochemical reduction of Cux [POM]@TiO_2 to investigate CO_2 adsorption on the surface. (b–d) Highlighted regions of the Raman spectra emphasize various binding modes of $[CO_2]^-$ and $[CO_3]^{2-}$ with copper ions present on the surface [32], copyright (2024), Wiley Online Library.

undergoes structural regeneration upon CO_2 adsorption, confirming the reformation of its characteristic carbonate framework.

Additionally, the presence of K_2CO_3 was shown to facilitate the formation of $K_2Mg(CO_3)_2$, as evidenced by shifts in the IR spectra [35]. Similarly, Raman spectroscopy has been used to monitor CO_2 absorption in monoethanolamine solutions, revealing the formation of various chemical species during the absorption process. This capacity to detect specific molecular interactions is critical for refining CO_2 capture technologies [36, 37]. Furthermore, Raman spectroscopy has been employed to study the solubility and structural changes of CO_2 in different media. For example, it has been shown that Raman spectroscopy can quantify the solubility of CO_2 in the H_2O–CO_2 system under varying pressures, establishing a linear correlation between solubility and Raman peak intensity. Understanding this relationship is crucial for designing effective CO_2 capture systems, as it sheds light on how CO_2 behaves under different environmental conditions [38]. In addition, the technique has proven instrumental in investigating the formation of clathrate hydrates and semi-clathrates involving CO_2. Studies have noted that Raman shifts of CO_2 hydrates occur at specific frequencies, allowing for the differentiation between hydrate structures formed by CO_2 and other gases like N_2. Moreover, Raman spectroscopy has been employed to elucidate structural changes in semi-clathrates upon CO_2 encapsulation, providing valuable insights into the stability and efficiency of these materials for CO_2 capture [39, 40].

3.4.3 Fourier-Transform Infrared (FTIR) Spectroscopy

FTIR spectroscopy serves as a powerful analytical technique frequently utilized for detailed molecular characterization of materials. By analyzing the vibrational modes of chemical bonds, FTIR allows researchers to identify functional groups, assess material composition, and monitor chemical changes during processes like adsorption. This technique has become integral to the investigation of CO_2 adsorption processes, offering critical insights into the molecular interactions and underlying mechanisms. FTIR is particularly adept at identifying different surface species that arise during CO_2 adsorption and is invaluable in detecting subtle variations in molecular vibrations linked to the formation of carbonate species, thereby shedding light on the mechanisms governing adsorption. Recent studies have revealed the significant influence of surface functionalization on CO_2 adsorption capacity. For instance, a study revealed that carbon materials engineered from jute sticks exhibited substantial CO_2 adsorption capacity, even though their surface area was relatively low. This enhanced performance was attributed to the presence of hydroxyl and aromatic functional groups, which played a crucial role in facilitating CO_2 interactions, as confirmed by FTIR spectroscopy analysis. These functional groups enhance CO_2 adsorption by facilitating interactions such as acid–base and π–π interactions [41]. Similarly, FTIR analysis in another study emphasized the importance of characterizing zeolites and MOFs to better understand their CO_2 adsorption properties. However, this research primarily focused on developing suitable materials for CO_2 capture and storage, rather than on specific functional groups [42]. FTIR spectroscopy has also been employed to evaluate the effectiveness of various adsorbents in capturing CO_2. For instance, FTIR was used to examine the chemical functionalities of chromium terephthalate (MIL-101) MOFs, linking these properties to their CO_2 adsorption performance [43]. Similarly, in a study by Bahmanzadegan et al., FTIR was utilized to characterize the differences between 4A-zeolite, 4A-15%TEPA, and 4A-10%DEA adsorbents before and after amine functionalization (Figure 3.10), demonstrating how these modifications influence CO_2 adsorption efficiency [14]. The information derived from FTIR spectra is vital for understanding the chemical environment of the adsorbents, which is crucial for optimizing their design for CO_2 capture.

Furthermore, FTIR spectroscopy is crucial in understanding the interactions between CO_2 and adsorbents, particularly in assessing modified materials. For instance, FTIR analysis has shown the presence of functional groups in amine-modified biochar that enhance CO_2 adsorption, demonstrating the efficacy of these modifications in flue gas applications [44]. Similarly, another study employed in situ FTIR spectroscopy to investigate CO_2 adsorption dynamics on photocatalytic materials, providing insights into CO_2 capture mechanisms. However, the primary focus was on photocatalytic reduction rather than direct CO_2 adsorption [45]. Additionally, FTIR has been extensively used to study carbonate transformations during CO_2 adsorption. For example, in situ FTIR spectroscopy was utilized to examine the formation and desorption of carbonate species in K-promoted hydrotalcite materials, revealing that CO_2 adsorption led to the formation of polydentate carbonates, which were linked to a decrease in capacity during cycling [46]. This highlights the importance of understanding the chemical transformations that occur during CO_2 adsorption, as they can significantly affect the long-term viability of adsorbents.

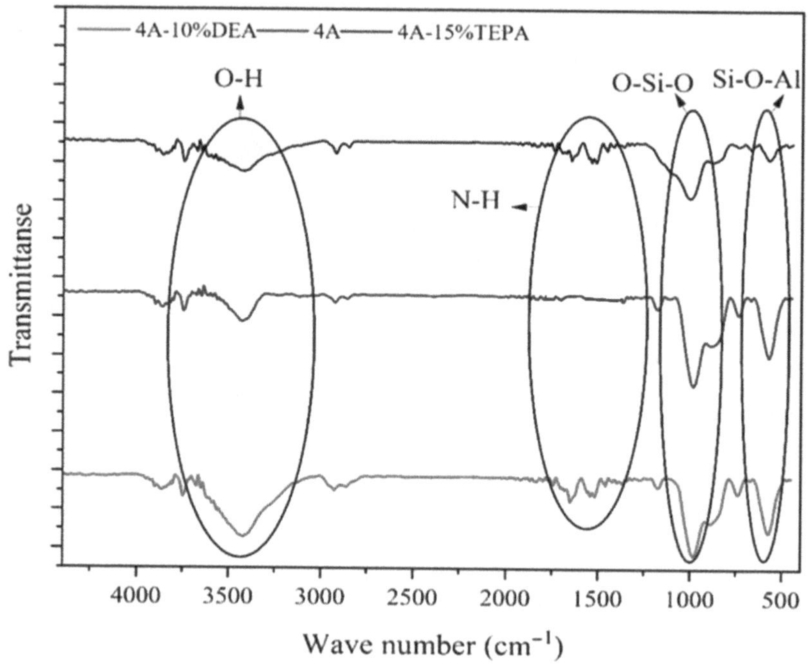

FIGURE 3.10 FT-IR spectra of zeolite 4A (middle line graph), zeolite 4A-TEPA (top line graph), and (bottom line graph) zeolite 4A-DEA [14], copyright © 2023, Nature.

3.4.4 X-Ray Photoelectron Spectroscopy

X-ray photoelectron spectroscopy (XPS) is an essential analytical method extensively utilized in the investigation of materials designed for CO_2 adsorption. This technique offers comprehensive insights into the elemental composition, chemical states, and electronic states of the materials under examination, making it invaluable for characterizing adsorbents in carbon capture applications. XPS operates by irradiating a sample with X-rays, leading to the emission of photoelectrons. By analyzing the kinetic energy of these emitted electrons, the binding energies can be determined, allowing for the inference of the chemical states of the elements within the sample [47–49]. XPS has been employed in the study of CO_2 adsorption on various materials, including nitrogen-doped graphitic carbons and MOFs. Research indicates that nitrogen doping improves the CO_2 adsorption capacity of carbon materials by increasing surface alkalinity, thereby enhancing interactions with CO_2 molecules [50]. XPS analysis of nitrogen-doped graphitic structures has shown an increase in the O 1s peak, signifying the presence of adsorbed CO_2 species, which is crucial for understanding the underlying adsorption mechanisms [47]. Furthermore, XPS has played a pivotal role in elucidating chemisorption processes in MOFs. For instance, XPS characterization of CO_2 interactions with diamine-appended MOFs confirmed the formation of carbamic acid species during adsorption, as illustrated

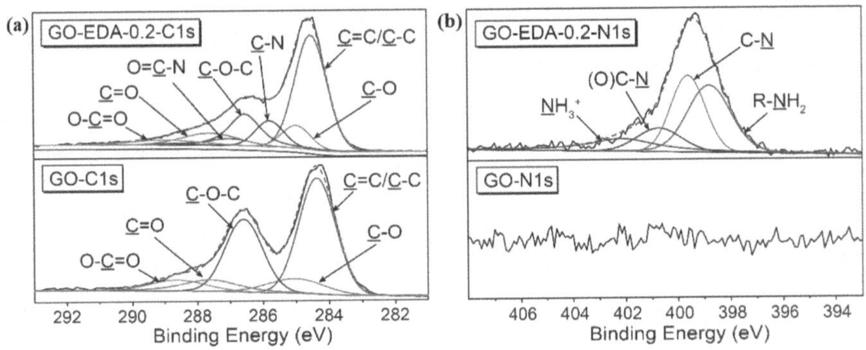

FIGURE 3.11 XPS of GO and GO–EDA–0.2 in the (a) C 1s region and (b) N 1s region [51], copyright © 2019, American Chemical Society.

in Figure 3.11 [51]. This finding is significant as it reveals the chemical transformations occurring during adsorption, providing valuable insights into the efficiency and selectivity of various adsorbent materials. The capability of XPS to distinguish between different chemical states enables researchers to refine the adsorbent design for improved CO_2 capture [49].

The influence of functional groups in improving CO_2 adsorption is further substantiated by research on amine-functionalized materials. Studies have demonstrated that the incorporation of branched amines markedly enhances CO_2 adsorption capacities in mesoporous silica [52]. In these investigations, XPS plays a crucial role in verifying the presence of functional groups and assessing their effectiveness in promoting CO_2 capture [53]. Additionally, XPS enables the analysis of chemical states during CO_2 adsorption, allowing researchers to observe changes in the electronic structure of the adsorbent. For example, it has been reported that the amine functionalization of silica significantly modifies its electronic properties, as evidenced by XPS measurements [54]. Moreover, the integration of amines into zeolites and other materials has enhanced CO_2 capture efficiency, with XPS providing vital information on the chemical states of nitrogen and its interaction with CO_2 molecules [55]. Beyond carbon-based materials, the adsorption of CO_2 on metal oxides has also been extensively examined using XPS. For instance, research by Kwon et al. indicated that the basicity of BaO influences its interaction with CO_2, although the relationship between surface activity and basicity remains intricate [56]. This complexity is further highlighted by the work of Wakuta et al., who demonstrated that incorporating alkaline earth metal oxides can enhance CO_2 adsorption capacities, as shown through XPS analysis of surface states [57]. XPS serves not only to characterize adsorbents but also to shed light on the underlying adsorption mechanisms. For example, a study on gold nanoparticles supported by titania revealed that CO_2 adsorption results in the formation of carbonate species, as confirmed by XPS analysis [58]. These findings are crucial for developing catalysts capable of efficiently converting CO_2 into valuable products.

3.4.5 Nuclear Magnetic Resonance Spectroscopy

Nuclear magnetic resonance (NMR) spectroscopy is an analytical technique used to determine the structure and dynamics of molecules by exploiting the magnetic properties of certain atomic nuclei. When nuclei such as hydrogen-1 or carbon-13 are placed in a strong magnetic field, they align with or against it. A pulse of radiofrequency energy causes these nuclei to absorb energy and transition between energy levels. When the energy is released, it provides a signal that reveals the chemical environment of the nuclei, known as the chemical shift [59]. Solid-state NMR techniques, such as ^{13}C and ^{17}O NMR spectroscopy, have been particularly instrumental in investigating the adsorption of CO_2 in porous materials like MOFs [60]. Researchers have employed solid-state NMR spectroscopy to identify specific surface sites on materials like MgO nanosheets that are active for CO_2 adsorption [61]. As shown in Figure 3.12, the ^{13}C NMR spectrum peaks around 167 ppm for NS-773 and 166 ppm for NS-1073, indicating the formation of unidentate carbonate species on the surface,

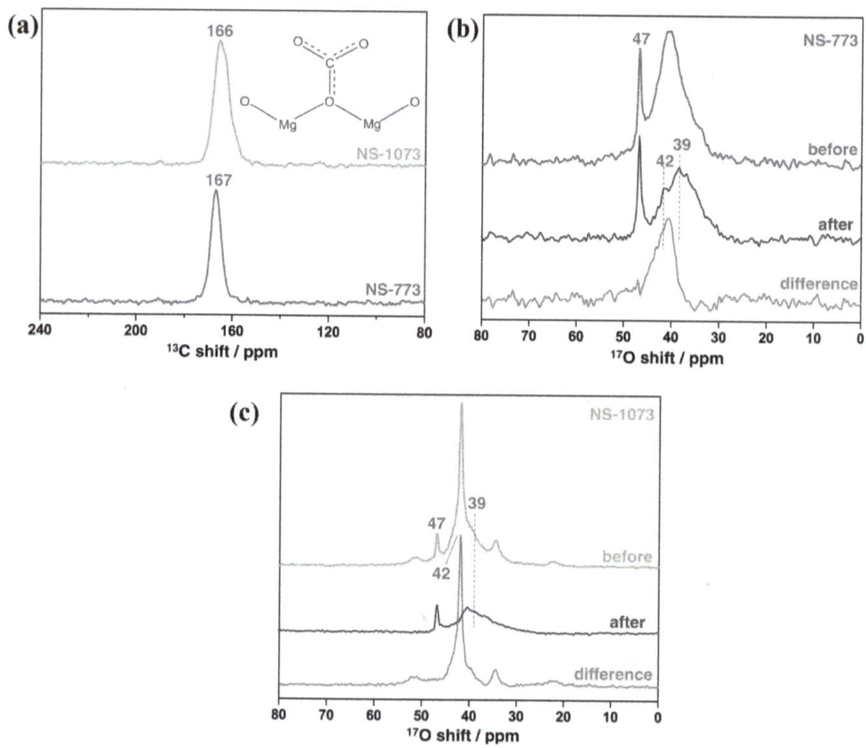

FIGURE 3.12 ^{13}C and ^{17}O single-pulse MAS NMR data of MgO nanosheets with adsorbed CO_2. (a) 13C MAS NMR spectra of NS-773 and NS-1073 after $13CO_2$ chemisorption at 9.4 T, with a MAS rate of 16 kHz and a 40 s recycle delay. 17O NMR spectra of NS-773 (b) and NS-1073 (c) before and after $13CO_2$ chemisorption, along with the difference spectrum at 9.4 T. MAS rate: 14 kHz, recycle delay: 5 s [61], copyright © 2022, Nature.

suggesting chemisorption of CO_2. Despite NS-1073 having a smaller surface area, its CO_2 adsorption capacity is 55% higher than NS-773, likely due to differences in surface species concentrations. The ^{17}O NMR analysis reveals that the surface oxygen species contributing to a peak at 42 ppm are involved in CO_2 binding, converting into carbonate species upon adsorption. The difference spectra for both samples confirm that the peaks at 39 and 42 ppm are essential for CO_2 adsorption, with most adsorption occurring at these sites.

By utilizing techniques like ^{17}O solid-state NMR spectroscopy, researchers have been able to pinpoint the sites where CO_2 molecules are adsorbed, providing valuable insights into the mechanisms of CO_2 capture on different surfaces [62]. Furthermore, NMR spectroscopy has been used to monitor unique switching transitions in flexible framework materials like DUT-49, providing valuable information on the local structure, dynamics, and flexibility of both the host framework and adsorbed guests like CO_2 [63]. This demonstrates the versatility of NMR spectroscopy in studying not only the adsorption of CO_2 but also the structural changes that occur within the materials during the adsorption process. In the realm of CO_2 capture research, NMR spectroscopy has been crucial in elucidating the chemisorption mechanisms of CO_2 in various materials. Studies have shown that solid-state NMR techniques can distinguish between different chemisorption products, such as carbamic acid pairs and ammonium carbamate chains, providing insights into the dominant mechanisms of CO_2 chemisorption in specific materials [62]. Forse et al. [64] used multinuclear NMR spectroscopy experiments to differentiate between various CO_2 chemisorption products in diamine-appended MOFs, highlighting the prevalence of ammonium carbamate chain formation in these materials. Additionally, Sin et al. [60] employed ^{13}C solid-state NMR spectroscopy to examine the adsorption of CO_2 in MOFs and explore the impact of flexibility on selectivity. This study underscores the established use of solid-state NMR spectroscopy in analyzing adsorbed CO_2 in porous materials like MOFs. Moreover, Ilkaeva et al. [65] demonstrated the innovative use of solid-state NMR to create isotherms of CO_2 adsorbed on an amine-modified silica sorbent, showcasing the versatility of NMR techniques in studying CO_2 adsorption processes. Xiao et al. [66] highlighted the utility of solid-state NMR spectroscopy in exploring host-guest interactions and dynamic properties of adsorbed small molecules, including CO_2, on various MOFs. Furthermore, Inukai et al. [67] utilized in situ magic angle spinning NMR spectroscopy to investigate the dynamics of CO_2 adsorption in porous coordination polymers controlled by molecular rotor dynamics, showcasing the effectiveness of NMR spectroscopy in studying the local mobility and diffusion of adsorbed CO_2. Additionally, NMR spectroscopy has been utilized to investigate the adsorption selectivity of binary mixtures like CO_2/CH_4 on materials such as SNU-9, highlighting its role in determining the adsorption behavior of different gases [68]. Moreover, NMR spectroscopy has been instrumental in studying the adsorption properties of different materials modified for CO_2 capture. By combining NMR spectroscopy with other characterization techniques like gas sorption analysis and XRD, researchers have been able to gain a comprehensive understanding of the adsorption mechanisms and selectivity of materials for CO_2 capture [69]. This integrated approach has allowed for the identification of cooperative adsorption mechanisms involving interactions between the material structure and the adsorbed CO_2 molecules, shedding light on the complex processes involved in CO_2 capture.

3.5 THERMAL STABILITY AND DECOMPOSITION

Thermal stability and thermal decomposition are crucial for the performance of CO_2 adsorbents. Thermal stability refers to the ability of CO_2 adsorbents to retain their structural integrity and adsorption capacity at high temperatures.

Heydari-Gorji and Sayari [70] found that PEI-modified adsorbents show high thermal stability at moderate temperatures but experience significant degradation when exposed to dry CO_2 at elevated temperatures. Similarly, Quang et al. [71] demonstrated that amino-functionalized mesoporous silica is a thermally stable adsorbent with a high capacity for CO_2 loading and optimal temperatures for adsorption and regeneration. In contrast, thermal decomposition in CO_2 adsorbents refers to the breakdown of the material when exposed to high temperatures, which results in a loss of adsorption capacity. Wang et al. [72] studied the adsorption and regeneration of polyethylenimine-impregnated mesoporous carbon spheres, highlighting the significance of thermal stability in sustaining high equilibrium adsorption capacity under varying conditions.

3.5.1 Thermogravimetric Analysis

Thermogravimetric analysis (TGA) is a widely used analytical technique that involves measuring a material's mass change as a function of temperature or time. TGA is valuable in characterizing materials, especially in CO_2 adsorbents, due to its ability to provide insights into the thermal stability, decomposition behavior, and composition of the materials [73]. In the context of CO_2 adsorbents, TGA plays a crucial role in characterizing these materials by providing data that can be correlated with CO_2 adsorption curves, aiding in understanding the adsorption mechanisms and efficiency of the adsorbents [74]. One of the key advantages of TGA in CO_2 adsorbent characterization is its ability to determine the thermal stability of the materials. By subjecting the CO_2 adsorbents to controlled temperature increases, TGA can identify the temperatures at which decomposition or weight loss occurs, providing information on the stability of the adsorbents under different conditions [75]. This is essential in assessing the suitability of the adsorbents for practical applications, such as CCS. Moreover, TGA enables the evaluation of the decomposition process of solid-state materials, including CO_2 adsorbents, by analyzing the weight loss as a function of temperature or time. This information is valuable in understanding the composition of the adsorbents, such as the content of ashes, volatiles, and fixed carbon, which are essential parameters in determining the adsorption capacity and efficiency of the materials [76]. Several studies have delved into the characterization and assessment of different materials for CO_2 capture through TGA analysis. Sultana et al. [77] explored using CaO nanoparticles coated with ZrO_2 layers, demonstrating enhanced CO_2 capture stability compared to uncoated counterparts. This study highlighted the importance of nanostructured materials in improving the efficiency and longevity of CO_2 capture processes. Similarly, Gorbounov et al. investigated the application of nanoporous carbon derived from biomass combustion ash for CO_2 adsorption, aligning the adsorbent characterization data with TGA results to validate CO_2 adsorption curves [74]. Furthermore, Samuel et al. focused on petroleum

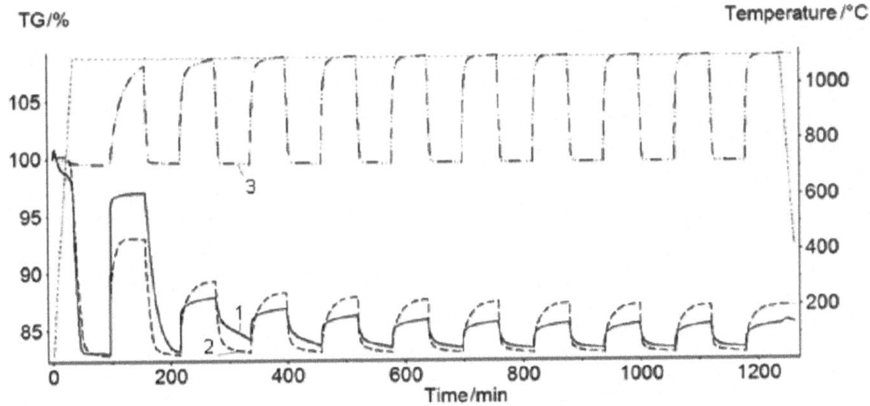

FIGURE 3.13 TGA results for ten cycles of calcination and carbonation conducted at a CO_2 pressure of 0.5 and a temperature of 1100°C, SrO/HA fine granules (curve 1, solid line), SrO/Al_2O_3 fine granules (curve 2, dashed line), and SrO/Al_2O_3 coarse granules (curve 3, dot-dashed line) [79], copyright © 2016 American Chemical Society.

asphaltene-based activated nanoporous carbon for CO_2 capture, emphasizing the significance of thorough characterization through spectroscopy and TGA to confirm material synthesis [78]. Miccio et al. studied high-temperature CO_2 capture using strontium oxide sorbents, employing various analytical techniques alongside TGA to assess microstructural changes influencing CO_2 capture efficiency. These studies underscore the multidimensional approach required to evaluate CO_2 adsorbent materials comprehensively, integrating TGA with other analytical methods to gain a holistic understanding of material performance (Figure 3.13) [79].

Moreover, Jo et al. [80] delved into the sorption kinetics of CaO-based sorbents quantitatively using experimental data from TGA, emphasizing the importance of TGA in elucidating the CO_2 sorption behavior of materials. Papalas et al. [81] evaluated calcium-based sorbents derived from natural ores and industrial wastes for high-temperature CO_2 capture, showcasing the versatility of TGA in assessing the CO_2 sorption activity of different materials. These studies collectively highlight TGA as a pivotal tool in the characterization and performance evaluation of diverse CO_2 adsorbent materials, enabling researchers to make informed decisions regarding material selection for carbon capture applications. Furthermore, Valverde et al. [82] explored calcium looping for post-combustion CO_2 capture, utilizing multicyclic TGA tests to demonstrate the feasibility of sorbent regeneration and reactivation under specific conditions. Petit et al. [83] investigated liquidlike POSS-based hybrid nanomaterials for CO_2 capture, employing TGA along with other analytical techniques to probe the interactions between materials and CO_2. This research underscores the importance of TGA in elucidating the thermodynamic behavior of materials during CO_2 adsorption processes, providing valuable insights into material performance under varying conditions. In the quest for efficient CO_2 adsorbents, researchers have explored a wide array of materials ranging from nanoporous carbon to activated porous carbons. Wu and Zhu [84] studied the behavior of $CaTio_3$ /nano-CaO as a CO_2 reactive adsorbent, showcasing the stable CO_2 sorption capacity achieved through material optimization and TGA analysis. Similarly, Jalilov et al. [85] investigated asphalt-derived activated porous carbons for CO_2 capture, emphasizing the high surface areas and CO_2 uptake performance of the materials through TGA assessments. These studies underscore the diverse material platforms available for CO_2 capture and the pivotal role of TGA in evaluating their adsorption capabilities and stability over multiple cycles. Moreover, synthesizing and modifying materials for CO_2 capture have been a focal point of research endeavors. Chen et al. [86] developed a new solid adsorbent based on an amine polycarboxylate ionic liquid, assessing its CO_2 adsorption capacity using TGA to elucidate the material's performance under specific conditions. Diyuk et al. [87] investigated the adsorption of CO_2 onto AC sorbents with different surface modifications, highlighting the utility of TGA in studying the adsorption behavior of functionalized materials.

3.5.2 Temperature-Programmed Desorption

Temperature-programmed desorption (TPD) is a technique utilized in catalysis and materials science to investigate the desorption kinetics of adsorbed species from material surfaces. TPD involves heating a sample at a controlled rate while monitoring the desorbed species, providing valuable information on the desorption process [88]. In the context of CO_2 adsorbent characterization, TPD offers several advantages. For instance, in CO_2-TPD experiments, different desorption peaks correspond to CO_2 adsorbed on weak, medium-strength, and strong basic sites, allowing for identifying and quantifying these sites on the adsorbent material [89, 90]. This information is crucial for understanding the interaction between CO_2 molecules and the adsorbent surface, which is essential for optimizing adsorption. Moreover, TPD has been used to characterize the basic sites and metal dispersion on catalyst surfaces, which are crucial for the adsorption and activation of feed gases in catalytic processes. By enhancing basic sites and improving metal dispersion, as observed in H_2-TPD and CO_2-TPD experiments, the efficiency of CO_2 adsorbents can be improved, leading to better performance in capturing and storing CO_2 [91]. Additionally, TPD has been utilized to quantify the proportion of specific functional groups, such as amines, on adsorbent surfaces, providing insights into the surface chemistry and reactivity of the material [92].

TPD has been instrumental in studying the adsorption characteristics of CO_2 and other gases on different catalysts and materials. For instance, TPD investigations have been conducted on cobalt–manganese oxide catalysts to analyze the adsorption of CH_3OH, H_2O, H_2, CO_2, and CO, providing valuable information on the adsorption behavior of these compounds (Figure 3.14) [93]. Additionally, TPD has been utilized to measure the adsorption of CO_2 and C_3H_8 in both single-component and mixture forms within zeolite crystals, demonstrating its versatility in studying adsorption processes in complex systems [94]. Moreover, TPD has been applied to study the adsorption of CO_2 on nitrogen-doped graphitic carbon materials, revealing insights into the chemisorption of CO_2 on these surfaces [95]. The technique has also been utilized to investigate the adsorption and dissociation of carbon dioxide on

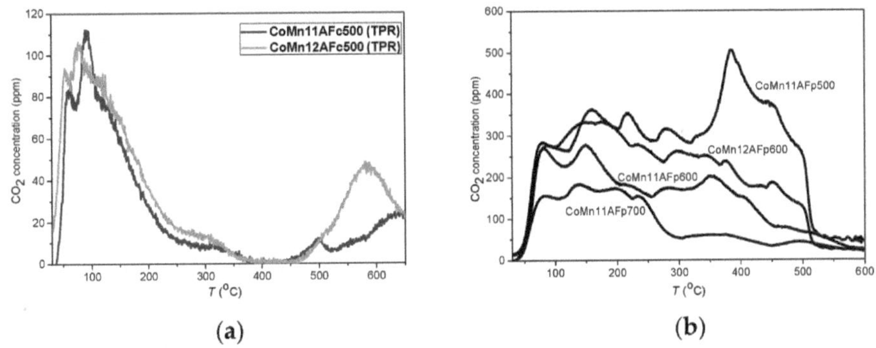

FIGURE 3.14 The TPD of CO_2 adsorption at 25°C on (a) CoMn1xAFc500 and (b) CoMn11AFpi, with *i* values of 500, 600, and 700°C for CoMn12AFp600 [93], copyright © 2016 MDPI (For Graph (a), CoMn11 . . . peaks slightly above 110 ppm, while CoMn12 . . . peaks at slightly under 110 ppm.)

metal surfaces, shedding light on the interaction mechanisms between CO_2 and the surface atoms [96]. These studies highlight the significance of TPD in elucidating the adsorption behavior of CO_2 on various substrates.

Furthermore, TPD has been employed in characterizing catalysts used in CO_2 reforming processes. For example, TPD has been used to analyze the effects of promoters on catalyst structures and performance for carbon dioxide reforming of methane, showcasing its importance in catalyst development and optimization [91]. Similarly, TPD has been utilized to assess the adsorption and activation of feed gases in nickel-based catalysts, emphasizing its role in understanding the surface properties of catalyst materials [97]. In the realm of astrochemistry, TPD has been a valuable tool for studying gas–surface interactions, including adsorption–desorption processes and surface diffusion, providing crucial insights into the behavior of molecules in interstellar environments [98]. Additionally, TPD has been instrumental in investigating the adsorption of gases on various surfaces, such as transition–metal surfaces, elucidating the energetics and structures of adsorbed molecules.

3.6 STRUCTURAL CHARACTERIZATION

3.6.1 X-Ray Diffraction

XRD is a widely used non-destructive analytical technique employed to determine the crystallographic structure of materials. In the realm of CO_2 adsorbents, XRD is essential for revealing structural characteristics that significantly influence adsorption efficiency. This technique provides critical information regarding crystal lattice parameters, phase composition, crystallinity, and grain size—factors vital for optimizing materials such as zeolites, MOFs, and other solid adsorbents for effective CO_2 capture. XRD operates on the principle that X-rays are scattered by the atoms within the crystal lattice when directed at a crystalline material. This scattering generates a diffraction pattern that reflects the material's internal structure. Bragg's Law describes the interaction of X-rays with the periodic arrangement of atoms, resulting in constructive interference and the formation of diffraction peaks at specific angles. The equation is Bragg's Law:

$$2dsin\theta = n\lambda \tag{3.2}$$

where d is the interplanar spacing between atomic layers, θ is the incident angle of the X-ray beam, n is an integer (the order of diffraction), and λ is the wavelength of the X-rays. Constructive interference occurs when the path difference between X-rays reflected from adjacent planes is an integer multiple of the wavelength. The resulting diffraction pattern, typically depicted as intensity versus scattering angle (2θ), provides insights into the sample's crystal structure and phase composition.

In a standard XRD experiment, the sample is exposed to a monochromatic X-ray beam, commonly generated by copper or molybdenum radiation. As the X-rays interact with the crystal structure, they scatter at various angles, producing diffraction peaks recorded by a detector. The position and intensity of these peaks yield valuable information: the peak position corresponds to the interplanar spacing (d) and is used to calculate lattice parameters; peak intensity indicates the distribution of atoms within the crystal lattice, providing insights into the atomic arrangement and preferred orientation of grains, while peak broadening can signal the presence of structural defects, strain, or small crystallite sizes. The Scherrer equation (Eq. 3.3) can be utilized to estimate crystallite size from the full width at half maximum (FWHM) of the diffraction peaks:

$$D = \frac{K\lambda}{\beta\cos\theta} \tag{3.3}$$

In the formula, D represents the crystallite size, K is the shape factor (typically 0.9), λ denotes the wavelength of the X-rays, β indicates the FWHM of the peak, and θ is the Bragg angle. XRD is essential for the detailed characterization of CO_2 adsorbent materials. Its structural information optimizes adsorbent properties, such as porosity, crystallinity, and phase purity. These properties directly affect the efficiency of CO_2 capture.

Activated alumina (AA) offers significant advantages in CO_2 adsorption due to its high surface area, porosity, and stability. To further enhance its performance, AA can be modified using chemical treatments. For instance, the introduction of sodium hydroxide (NaOH) and potassium hydroxide (KOH) solutions increases its carbon dioxide adsorption capacity [100]. Karbalaei et al. investigated the enhancement of carbon dioxide adsorption capacity in AA through modification withNaOH and KOH solutions. The XRD analysis of the modified adsorbents, 30NaOH-AA and 30KOH-AA (Figure 3.15), revealed distinct structural changes. For 30NaOH-AA, peaks at 14.45°, 28.54°, 38.51°, 49.46°, 67.44°, and 72.38° indicated the formation of aluminum oxide hydroxide. In the case of 30KOH-AA, peaks at 26.78° and 43.52° were attributed to the formation of aluminum hydroxide. These structural modifications were linked to improved CO_2 adsorption performance [99].

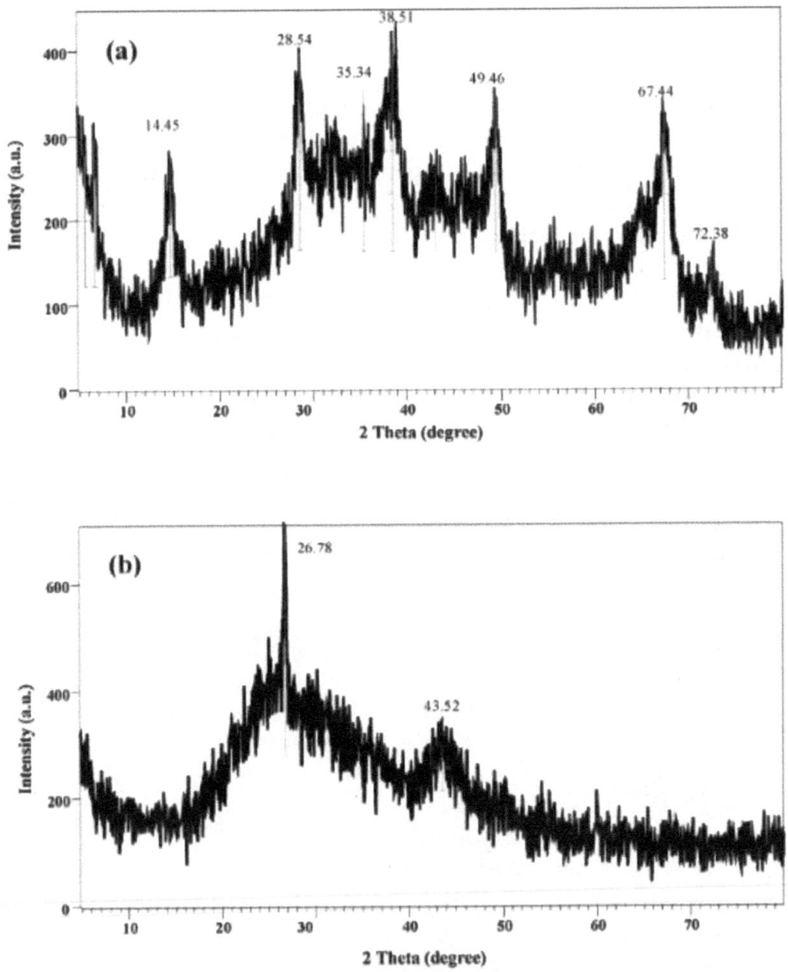

FIGURE 3.15 The XRD profiles for the adsorbents: (a) 30NaOH-AA and (b) 30KOH-AA [99],

3.6.2 X-Ray Fluorescence

X-ray fluorescence (XRF) is a highly effective, non-destructive analytical technique that provides quantitative and qualitative information about the elemental composition of materials. XRF is based on the fundamental principle that when a material is irradiated with high-energy X-rays, the atoms within the material become excited and emit secondary X-rays (fluorescence). This emission occurs because the X-ray photons from the incident beam displace electrons from the inner shells of the atoms, particularly the K or L shells. This displacement leaves vacancies quickly filled by electrons from higher energy levels. As these electrons drop to lower energy states, they release energy in the form of secondary (or fluorescent) X-ray photons. The energy of the emitted X-rays is characteristic of the specific elements present in the material, allowing for precise elemental analysis. In the context of CO_2 adsorbents, XRF is critical in identifying and quantifying the chemical elements that make up the adsorbent, such as silicon, aluminum, iron, and other metal or non-metal components. The elemental composition directly influences the material's properties, such as adsorption capacity, stability, and efficiency in capturing CO_2 molecules. XRF is essential for determining the elemental composition of CO_2 adsorbents, especially for materials like zeolites and MOFs, which rely on specific frameworks of metals (e.g., aluminum, iron, zinc) and non-metals (e.g., silicon, oxygen) for their function [101]. In zeolites, for example, the silicon-to-aluminum (Si/Al) ratio is a critical factor that influences the material's acid sites and overall adsorption capacity [102]. XRF can accurately measure this ratio, effectively tuning the material's properties. Additionally, XRF can detect the presence of impurities or undesirable elements that might reduce the adsorbent's efficiency or stability. Wang et al. synthesized zeolite 4A from kaolin for CO_2 adsorption. The main components of raw kaolin are SiO_2, Al_2O_3, and trace amounts of Fe_2O_3. Specifically, the weight percentages of SiO_2, Al_2O_3, and Fe_2O_3 in raw kaolin are 55.466%, 39.240%, and 1.415%, respectively.

3.7 CONCLUSION

This chapter thoroughly explores various analytical techniques used to characterize the morphological properties, surface chemistry, surface area, and zeta potential of CO_2 adsorbents. It highlights the effectiveness of FE-SEM and TEM for examining these adsorbents' morphological, textural, and topographical characteristics. For assessing chemical composition, techniques like Raman spectroscopy, FTIR spectroscopy, and EDX spectroscopy were emphasized as crucial tools, often utilized in combination to ensure accurate identification of chemical structures. Furthermore, the chapter emphasizes the significance of structural properties—such as internal surface area, total pore volume, and pore size distribution—as these parameters play a vital role in determining the adsorbent's efficiency in capturing CO_2 molecules. Understanding these aspects is essential for optimizing the performance of adsorbents in carbon capture applications.

REFERENCES

[1] H. Masoumi, A. Ghaemi, and H. G. Gilani, "Synthesis of Polystyrene-based Hyper-cross-linked Polymers for Cd (II) Ions Removal from Aqueous Solutions: Experimental and RSM Modeling," *Journal of Hazardous Materials,* vol. 416, p. 125923, 2021.

[2] F. Fashi, A. Ghaemi, and A. H. Behroozi, "Piperazine Impregnation on Zeolite 13X as a Novel Adsorbent for CO_2 Capture: Experimental and Modeling," *Chemical Engineering Communications,* vol. 208, no. 8, pp. 1104–1120, 2021.

[3] M. Helmi, A. Hemmati, and A. Ghaemi, "Clinoptilolite and MCM-41 Impregnated with Chitosan as a Green Novel Adsorbent for CO_2 Capture," *Case Studies in Chemical and Environmental Engineering,* vol. 9, p. 100764, 2024.

[4] H. Mashhadimoslem, M. Safarzadeh Khosrowshahi, M. Jafari, A. Ghaemi, and A. Maleki, "Adsorption Equilibrium, Thermodynamic, and Kinetic Study of $O_2/N_2/CO_2$ on Functionalized Granular Activated Carbon," *ACS Omega,* vol. 7, no. 22, pp. 18409–18426, 2022.

[5] F. S. Taheri, A. Ghaemi, and A. Maleki, "High Efficiency and Eco-friendly TEPA-functionalized Adsorbent with Enhanced Porosity for CO_2 Capture," *Energy & Fuels,* vol. 33, no. 11, pp. 11465–11476, 2019.

[6] A. Kazemi, F. Moghadaskhou, M. A. Pordsari, F. Manteghi, A. Tadjarodi, and A. Ghaemi, "Enhanced CO_2 Capture Potential of UiO-66-NH_2 Synthesized by Sonochemical Method: Experimental Findings and Performance Evaluation," *Scientific Reports,* vol. 13, no. 1, p. 19891, 2023.

[7] M. Khajeh and A. Ghaemi, "Nanoclay Montmorillonite as an Adsorbent for CO_2 Capture: Experimental and Modeling," *Journal of the Chinese Chemical Society,* vol. 67, no. 2, pp. 253–266, 2020.

[8] F. Bahmanzadegan and A. Ghaemi, "Exploring the Effect of Zeolite's Structural Parameters on the CO_2 Capture Efficiency Using RSM and ANN Methodologies," *Case Studies in Chemical and Environmental Engineering,* vol. 9, p. 100595, 2024.

[9] T. Yamazaki, M. Katoh, S. Ozawa, and Y. Ogino, "Adsorption of CO_2 over Univalent Cation-exchanged ZSM-5 Zeolites," *Molecular Physics,* vol. 80, no. 2, pp. 313–324, 1993, https://doi.org/10.1080/00268979300102281.

[10] D. Panda, E. A. Kumar, and S. K. Singh, "Introducing Mesoporosity in Zeolite 4A Bodies for Rapid CO_2 Capture," *Journal of CO_2 Utilization,* vol. 40, p. 101223, 2020, https://doi.org/10.1016/j.jcou.2020.101223.

[11] J.-S. Lee, J.-H. Kim, J.-T. Kim, J.-K. Suh, J.-M. Lee, and C.-H. Lee, "Adsorption Equilibria of CO_2 on Zeolite 13X and Zeolite X/activated Carbon Composite," *Journal of Chemical & Engineering Data,* vol. 47, no. 5, pp. 1237–1242, 2002, https://doi.org/10.1021/je020050e.

[12] A. Villarreal, G. Garbarino, P. Riani, E. Finocchio, B. Bosio, J. Ramirez, and G. Busca, "Adsorption and Separation of CO_2 from N_2-rich Gas on Zeolites: Na-X Faujasite vs Na-mordenite," *Journal of CO_2 Utilization,* vol. 19, pp. 266–275, 2017, https://doi.org/10.1016/j.jcou.2017.03.021.

[13] P. Li and F. H. Tezel, "Adsorption Separation of N_2, O_2, CO_2 and CH_4 Gases by β-zeolite," *Microporous and Mesoporous Materials,* vol. 98, no. 1–3, pp. 94–101, 2007, https://doi.org/10.1016/j.micromeso.2006.08.016.

[14] F. Bahmanzadegan, M. A. Pordsari, and A. Ghaemi, "Improving the Efficiency of 4A-zeolite Synthesized from Kaolin by Amine Functionalization for CO_2 Capture," *Scientific Reports,* vol. 13, no. 1, p. 12533, 2023.

[15] M. Ahmadi, F. Bahmanzadegan, M. Qasemnazhand, A. Ghaemi, and H. Ramezanipour Penchah, "Experimental, RSM Modelling, and DFT Simulation of CO_2 Adsorption on Modified Activated Carbon with LiOH," *Scientific Reports,* vol. 14, no. 1, p. 13595, 2024.

[16] G. Nazir, A. Rehman, and S.-J. Park, "Sustainable N-doped Hierarchical Porous Carbons as Efficient CO_2 Adsorbents and High-performance Supercapacitor Electrodes," *Journal of CO_2 Utilization,* vol. 42, p. 101326, 2020.

[17] J. A. Chagas, G. O. Crispim, B. P. Pinto, R. A. San Gil, and C. J. Mota, "Synthesis, Characterization, and CO_2 Uptake of Adsorbents Prepared by Hydrothermal Carbonization of Chitosan," *ACS Omega,* vol. 5, no. 45, pp. 29520–29529, 2020.

[18] K. Xia, X. Tian, S. Fei, and K. You, "Hierarchical Porous Graphene-based Carbons Prepared by Carbon Dioxide Activation and Their Gas Adsorption Properties," *International Journal of Hydrogen Energy,* vol. 39, no. 21, pp. 11047–11054, 2014.

[19] H. Masoumi, A. Ghaemi, and H. G. Gilani, "Evaluation of Hyper-cross-linked Polymers Performances in the Removal of Hazardous Heavy Metal Ions: A Review," *Separation and Purification Technology,* vol. 260, p. 118221, 2021.

[20] M. Azizi, S. Sedaghat, K. Tahvildari, P. Derakhshi, and A. Ghaemi, "Synthesis of Silver Nanoparticles Using Peganum Harmala Extract as a Green Route," *Green Chemistry Letters and Reviews,* vol. 10, no. 4, pp. 420–427, 2017.

[21] F. S. Taheri, A. Ghaemi, A. Maleki, and S. Shahhosseini, "High CO_2 Adsorption on Amine-functionalized Improved Mesoporous Silica Nanotube as an Eco-friendly Nanocomposite," *Energy & Fuels,* vol. 33, no. 6, pp. 5384–5397, 2019.
[22] Z. Khoshraftar, A. Ghaemi, and H. Mashhadimoslem, "Evaluation of the Silica Gel Adsorbent Potential for Carbon Dioxide Capture: Experimental and Modeling," *Iranian Journal of Chemical Engineering (IJChE),* vol. 18, no. 4, pp. 64–80, 2021.
[23] M. D. Donohue and G. L. Aranovich, "Classification of Gibbs Adsorption Isotherms," *Advances in Colloid and Interface Science,* vol. 76, pp. 137–152, 1998.
[24] H. J. Esfahani, A. Ghaemi, and S. Shahhosseini, "Improving CO_2 Adsorption Efficiency of an Amine-modified MOF-808 through the Synthesis of Its Graphene Oxide Composites," *Scientific Reports,* vol. 14, no. 1, p. 18871, 2024.
[25] A. Kazemi *et al.*, "Unveiling the Power of Defect Engineering in MOF-808 to Enhance Efficient Carbon Dioxide Adsorption and Separation by Harnessing the Potential of DFT Analysis," *Chemical Engineering Journal,* vol. 494, p. 153049, 2024.
[26] Z. Chen, "Application of UV-vis Spectroscopy in the Detection and Analysis of Substances," In *3rd International Conference on Biotechnology, Life Science and Medical Engineering,* vol. 3, pp. 131–136, 2024, https://doi.org/10.62051/mw4daz69.
[27] F. A. A. Nugroho, C. Xu, N. Hedin, and C. Langhammer, "UV–Visible and Plasmonic Nanospectroscopy of the CO_2 Adsorption Energetics in a Microporous Polymer," *Analytical Chemistry,* vol. 87, no. 20, pp. 10161–10165, 2015, https://doi.org/10.1021/acs.analchem.5b03108.
[28] W. C. Wilfong, B. W. Kail, Q. Wang, and M. L. Gray, "Novel Rapid Screening of Basic Immobilized Amine Sorbent/Catalyst Water Stability by a UV/Vis/Cu_{2+} Technique," *Chemsuschem,* vol. 11, no. 23, pp. 4114–4122, 2018, https://doi.org/10.1002/cssc.201801851.
[29] Y. Jiang, P. Tan, S. C. Qi, X. Liu, J. Yan, F. Fan, and L. B. Sun, "Metal–Organic Frameworks with Target-Specific Active Sites Switched by Photoresponsive Motifs: Efficient Adsorbents for Tailorable CO_2 Capture," *Angewandte Chemie,* vol. 58, no. 20, pp. 6600–6604, 2019, https://doi.org/10.1002/anie.201900141.
[30] X. Chen, Q. Li, M. Zhang, J. Li, S. Cai, J. Chen, and H. Jia, "MOF-Templated Preparation of Highly Dispersed Co/Al_2O_2 Composite as the Photothermal Catalyst with High Solar-to-Fuel Efficiency for CO_2 Methanation," *ACS Applied Materials & Interfaces,* vol. 12, no. 35, pp. 39304–39317, 2020, https://doi.org/10.1021/acsami.0c11576.
[31] Q. Sun, Z. Li, D. J. Searles, Y. Chen, G. Q. Lu, and A. Du, "Charge-Controlled Switchable CO_2 Capture on Boron Nitride Nanomaterials," *Journal of the American Chemical Society,* vol. 135, no. 22, pp. 8246–8253, 2013, https://doi.org/10.1021/ja400243r.
[32] L. Mallick, K. Samanta, and B. Chakraborty, "Post-synthetic Metalation on the Ionic TiO_2 Surface to Enhance Metal-CO_2 Interaction During Photochemical CO_2 Reduction," *Chemistry—A European Journal,* vol. 30, no. 38, 2024, https://doi.org/10.1002/chem.202400428.
[33] V. G. Arakcheev, A. N. Bekin, and V. B. Morozov, "Spectroscopic Characterization of Adsorbate Confined in Small Mesopores: Distinction of First Surface-adsorbed Layer, Polymolecular Layers, and Liquid Clusters," *Journal of Raman Spectroscopy,* vol. 49, no. 12, pp. 1945–1952, 2018, https://doi.org/10.1002/jrs.5491.
[34] N. Guilhaume, D. Bianchi, R. A. Wandawa, Y. Wang, and Y. Schuurman, "Study of CO_2 and H_2O Adsorption Competition in the Combined Dry/Steam Reforming of Biogas," *Catalysis Today,* vol. 375, pp. 282–289, 2021, https://doi.org/10.1016/j.cattod.2020.04.058.
[35] L. Sun, Y. Yang, H. Ni, D.-F. Liu, Z. Sun, P. Li, and J. Yu, "Enhancement of CO_2 Adsorption Performance on Hydrotalcites Impregnated with Alkali Metal Nitrate Salts and Carbonate Salts," *Industrial & Engineering Chemistry Research,* vol. 59, no. 13, pp. 6043–6052, 2020, https://doi.org/10.1021/acs.iecr.9b05700.

[36] M. H. W. N. Jinadasa, K. Chandra, and M. Halstensen, "System Development for on-Line Monitoring Using Raman Spectroscopy for CO_2 Absorption by MEA," In *Proceedings of the 59th Conference on Simulation and Modelling (SIMS 59)*, 2018, https://doi.org/10.3384/ecp18153328.
[37] M. Akram *et al.*, "Application of Raman Spectroscopy to Real-Time Monitoring of CO_2 Capture at PACT Pilot Plant; Part 1: Plant Operational Data," *International Journal of Greenhouse Gas Control,* vol. 95, p. 102969, 2020, https://doi.org/10.1016/j.ijggc.2020.102969.
[38] M.-C. Caumon, J. Sterpenich, A. Randi, and J. Pironon, "Measuring Mutual Solubility in the H 2 O–CO 2 System Up to 200 Bar and 100°C by in Situ Raman Spectroscopy," *International Journal of Greenhouse Gas Control,* vol. 47, pp. 63–70, 2016, https://doi.org/10.1016/j.ijggc.2016.01.034.
[39] X. Zang, N. Zhang, X. Zhou, L. Wan, and D. Liang, "Experimental Investigation of the Hydrate-Based Gas Separation of Synthetic Flue Gas with 5A Zeolite," *Energies,* vol. 13, no. 17, p. 4556, 2020, https://doi.org/10.3390/en13174556.
[40] B. Chazallon, M. Ziskind, Y. Carpentier, and C. Focsa, "CO_2 Capture Using Semi-Clathrates of Quaternary Ammonium Salt: Structure Change Induced by CO_2 and N_2 Enclathration," *The Journal of Physical Chemistry B,* vol. 118, no. 47, pp. 13440–13452, 2014, https://doi.org/10.1021/jp507789z.
[41] A. Hanif *et al.*, "CO_2 Adsorption on Pore-Engineered Carbons Derived from Jute Sticks," *Chemistry—An Asian Journal,* vol. 18, no. 17, 2023, https://doi.org/10.1002/asia.202300481.
[42] N. K. Daud and N. Nurul Huda Insyirah Mohammad, "CO_2 Adsorption on Zeolites and Fe-Mof for Carbon Capture and Storage (CCS)," *Journal of Chemical Engineering and Industrial Biotechnology,* vol. 8, no. 1, pp. 27–32, 2022, https://doi.org/10.15282/jceib.v8i1.7795.
[43] F. Yulia, V. J. Utami, N. Nasruddin, and A. Zulys, "Synthesis, Characterizations, and Adsorption Isotherms of CO_2 on Chromium Terephthalate (MIL-101) Metal-Organic Frameworks (Mofs)," *International Journal of Technology,* vol. 10, no. 7, p. 1427, 2019, https://doi.org/10.14716/ijtech.v10i7.3706.
[44] W. Tian, Y. Wang, J. Hao, T. Guo, X. Wang, X. Xiang, and Q. Guo, "Amine-Modified Biochar for the Efficient Adsorption of Carbon Dioxide in Flue Gas," *Atmosphere,* vol. 13, no. 4, p. 579, 2022, https://doi.org/10.3390/atmos13040579.
[45] S. Zhu, S. Liang, Y. Tong, X. An, J. Long, X. Fu, and X. Wang, "Photocatalytic Reduction of CO_2 with H_2O to CH_4 on Cu(I) Supported TiO_2 Nanosheets with Defective {001} Facets," *Physical Chemistry Chemical Physics,* vol. 17, no. 15, pp. 9761–9770, 2015, https://doi.org/10.1039/c5cp00647c.
[46] H. Du, C. T. Williams, A. D. Ebner, and J. A. Ritter, "*In Situ* FTIR Spectroscopic Analysis of Carbonate Transformations During Adsorption and Desorption of CO_2 in K-Promoted HTlc," *Chemistry of Materials,* vol. 22, no. 11, pp. 3519–3526, 2010, https://doi.org/10.1021/cm100703e.
[47] R. Shibuya, K. Takeyasu, D. Guo, T. Kondo, and J. Nakamura, "Chemisorption of CO_2 on Nitrogen-Doped Graphitic Carbons," *Langmuir,* vol. 38, no. 47, pp. 14430–14438, 2022.
[48] N. D. McNamara and J. C. Hicks, "CO_2 Capture and Conversion with a Multifunctional Polyethyleneimine-tethered Iminophosphine Iridium Catalyst/adsorbent," *ChemSusChem,* vol. 7, no. 4, pp. 1114–1124, 2014.
[49] S. E. M. Elhenawy, M. Khraisheh, F. AlMomani, and G. Walker, "Metal-organic Frameworks as a Platform for CO_2 Capture and Chemical Processes: Adsorption, Membrane Separation, Catalytic-conversion, and Electrochemical Reduction of CO_2," *Catalysts,* vol. 10, no. 11, p. 1293, 2020.
[50] J. Sreńscek-Nazzal, A. Kamińska, J. Serafin, and B. Michalkiewicz, "Chemical Activation of Banana Peel Waste-Derived Biochar Using KOH and Urea for CO_2 Capture," *Materials,* vol. 17, no. 4, p. 872, 2024.

[51] J. Cai, J. Chen, P. Zeng, Z. Pang, and X. Kong, "Molecular Mechanisms of CO_2 Adsorption in Diamine-cross-linked Graphene Oxide," *Chemistry of Materials,* vol. 31, no. 10, pp. 3729–3735, 2019.
[52] X. Hou, L. Zhuang, B. Ma, S. Chen, H. He, and F. Yin, "Silanol-rich Platelet Silica Modified with Branched Amine for Efficient CO_2 Capture," *Chemical Engineering Science,* vol. 181, pp. 315–325, 2018.
[53] T. D. Pham, M. R. Hudson, C. M. Brown, and R. F. Lobo, "Molecular Basis for the High CO_2 Adsorption Capacity of Chabazite Zeolites," *ChemSusChem,* vol. 7, no. 11, pp. 3031–3038, 2014.
[54] R. S. Bakdash, I. H. Aljundi, C. Basheer, and I. Abdulazeez, "Rice Husk Derived Aminated Silica for the Efficient Adsorption of Different Gases," *Scientific Reports,* vol. 10, no. 1, p. 19526, 2020.
[55] L. Shaojie, S. Jia, T. Nagasaka, H. Bai, and L. Yang, "CO_2 Adsorption Properties of Amine-Modified Zeolites Synthesized Using Different Types of Solid Waste," *Sustainability*, vol. 15, no. 13, p. 10144, 2023, https://doi.org/10.3390/su151310144.
[56] S. Kwon, J. B. Hwang, H. Lee, and W. R. Lee, "Interactive CO_2 Adsorption on the BaO (100) Surface: A Density Functional Theory (DFT) Study," *Bulletin of the Korean Chemical Society,* vol. 31, no. 8, pp. 2219–2222, 2010, https://doi.org/10.5012/bkcs.2010.31.8.2219.
[57] K. Wakuta, N. Tokuda, R. V. Belosludov, S. Ueda, R. Inoue, and T. Ariyama, "Experimental Study and Atomic Level Description of Adsorption Process of CO_2 on Doped Alkaline Earth Oxides," *ISIJ International,* vol. 52, no. 7, pp. 1233–1240, 2012, https://doi.org/10.2355/isijinternational.52.1233.
[58] M. Compagnoni *et al.*, "Surface Probing by Spectroscopy on Titania-Supported Gold Nanoparticles for a Photoreductive Application," *Catalysts,* vol. 8, no. 12, p. 623, 2018, https://doi.org/10.3390/catal8120623.
[59] B. Reif, S. E. Ashbrook, L. Emsley, and M. Hong, "Solid-state NMR Spectroscopy," *Nature Reviews Methods Primers,* vol. 1, no. 1, p. 2, 2021.
[60] M. Sin *et al.*, "In Situ ^{13}C NMR Spectroscopy Study of CO_2/CH_4 Mixture Adsorption by Metal–Organic Frameworks: Does Flexibility Influence Selectivity?" *Langmuir,* vol. 35, no. 8, pp. 3162–3170, 2019, https://doi.org/10.1021/acs.langmuir.8b03554.
[61] J.-H. Du *et al.*, "Identification of CO_2 Adsorption Sites on MgO Nanosheets by Solid-state Nuclear Magnetic Resonance Spectroscopy," *Nature Communications,* vol. 13, no. 1, p. 707, 2022.
[62] A. C. Forse *et al.*, "Elucidating CO_2 Chemisorption in Diamine-appended Metal–organic Frameworks," *Journal of the American Chemical Society,* vol. 140, no. 51, pp. 18016–18031, 2018.
[63] J. Schaber *et al.*, "In Situ Monitoring of Unique Switching Transitions in the Pressure-amplifying Flexible Framework Material DUT-49 by High-pressure 129Xe NMR Spectroscopy," *The Journal of Physical Chemistry C,* vol. 121, no. 9, pp. 5195–5200, 2017.
[64] A. C. Forse *et al.*, "Elucidating CO_2 Chemisorption in Diamine-Appended Metal–Organic Frameworks," *Journal of the American Chemical Society,* vol. 140, no. 51, pp. 18016–18031, 2018, https://doi.org/10.1021/jacs.8b10203.
[65] M. Ilkaeva, R. B. L. Vieira, J. Pereira, M. Sardo, I. Marín-Montesinos, and L. Mafra, "Assessing CO_2 Capture in Porous Sorbents via Solid-State NMR-Assisted Adsorption Techniques," *Journal of the American Chemical Society,* vol. 145, no. 16, pp. 8764–8769, 2023, https://doi.org/10.1021/jacs.3c00281.
[66] Y. Xiao, S. Li, F. Chen, W. Gao, J. Xu, and F. Deng, "Host-Guest Interaction in Ethylene and Ethane Separation on Zeolitic Imidazolate Frameworks as Revealed by Solid-State NMR Spectroscopy," *Chemistry—A European Journal,* vol. 27, no. 44, pp. 11303–11308, 2021, https://doi.org/10.1002/chem.202101779.

[67] M. Inukai, M. Tamura, S. Horike, M. Higuchi, S. Kitagawa, and K. Nakamura, "Storage of CO_2 Into Porous Coordination Polymer Controlled by Molecular Rotor Dynamics," *Angewandte Chemie,* vol. 130, no. 28, pp. 8823–8826, 2018, https://doi.org/10.1002/ange.201805111.

[68] Y. Xiao, S. Li, B. Jiang, X. Liang, Y. Chu, and F. Deng, "Effect of Co-Adsorbed Guest Adsorbates on the Separation of Ethylene/Ethane Mixtures on Metal-Organic Frameworks with Open Metal Sites," *Chemistry–A European Journal,* p. e202401006, 2024.

[69] M. Cavallo *et al.*, "Cooperative CO 2 Adsorption Mechanism in a Perfluorinated Ce IV-based Metal Organic Framework," *Journal of Materials Chemistry A,* vol. 11, no. 11, pp. 5568–5583, 2023.

[70] A. Heydari-Gorji and A. Sayari, "Thermal, Oxidative, and CO_2-induced Degradation of Supported Polyethylenimine Adsorbents," *Industrial & Engineering Chemistry Research,* vol. 51, no. 19, pp. 6887–6894, 2012.

[71] D. V. Quang, T. A. Hatton, and M. R. Abu-Zahra, "Thermally Stable Amine-grafted Adsorbent Prepared by Impregnating 3-aminopropyltriethoxysilane on Mesoporous Silica for CO_2 Capture," *Industrial & Engineering Chemistry Research,* vol. 55, no. 29, pp. 7842–7852, 2016.

[72] M. Wang, L. Yao, J. Wang, Z. Zhang, W. Qiao, D. Long, and L. Ling, "Adsorption and Regeneration Study of Polyethylenimine-impregnated Millimeter-sized Mesoporous Carbon Spheres for Post-combustion CO_2 Capture," *Applied Energy,* vol. 168, pp. 282–290, 2016.

[73] E. Mansfield, K. M. Tyner, C. M. Poling, and J. Blacklock, "Determination of Nanoparticle Surface Coatings and Nanoparticle Purity Using Microscale Thermogravimetric Analysis," *Analytical Chemistry,* vol. 86, no. 3, pp. 1478–1484, 2014, https://doi.org/10.1021/ac402888v.

[74] M. Gorbounov, B. Petrovic, A. Lahiri, and S. M. Soltani, "Application of Nanoporous Carbon, Extracted from Biomass Combustion Ash, in CO_2 Adsorption," In *2021 IEEE 21st International Conference on Nanotechnology (NANO)*, 2021, https://doi.org/10.1109/nano51122.2021.9514288.

[75] K. S. D. Nunes and L. C. Pardini, "Purification and Characterization Methods for Lignin Biomass as a Potential Precursor for Carbon Materials," *Cellulose Chemistry and Technology,* vol. 53, no. 3–4, pp. 227–242, 2019, https://doi.org/10.35812/cellulosechemtechnol.2019.53.23.

[76] B. Velázquez Martí, J. Gaibor-Chavez, I. López Cortés, and L. E. Olivares Aguilar, "Evaluation of the Intermediate Values of the TGA Curves as Indicators of the Proximal Analysis of Biomass," *Agronomy,* vol. 13, no. 10, p. 2552, 2023, https://doi.org/10.3390/agronomy13102552.

[77] K. S. Sultana, D. T. Tran, J. C. Walmsley, M. Rønning, and D. Chen, "CaO Nanoparticles Coated by ZrO_2 Layers for Enhanced CO_2 Capture Stability," *Industrial & Engineering Chemistry Research,* vol. 54, no. 36, pp. 8929–8939, 2015, https://doi.org/10.1021/acs.iecr.5b00423.

[78] J. Samuel and S. Ok, "Petroleum Asphaltene-Based Activated Nanoporous Carbon for CO_2 Capture and H_2 Storage," *Industrial & Engineering Chemistry Research,* vol. 62, no. 26, pp. 9939–9950, 2023, https://doi.org/10.1021/acs.iecr.3c00176.

[79] F. Miccio, A. N. Murri, and E. Landi, "High-Temperature Capture of CO_2 by Strontium Oxide Sorbents," *Industrial & Engineering Chemistry Research,* vol. 55, no. 23, pp. 6696–6707, 2016, https://doi.org/10.1021/acs.iecr.6b00184.

[80] S. B. Jo, L. Cruz, S. Shah, S. Wasantwisut, A. Phan, and K. L. Gilliard-AbdulAziz, "Perspective on Sorption Enhanced Bifunctional Catalysts to Produce Hydrocarbons," *ACS Catalysis,* vol. 12, no. 13, pp. 7486–7510, 2022, https://doi.org/10.1021/acscatal.2c01646.

[81] T. Papalas, A. N. Antzaras, and A. A. Lemonidou, "Evaluation of Calcium-Based Sorbents Derived from Natural Ores and Industrial Wastes for High-Temperature CO_2 Capture," *Industrial & Engineering Chemistry Research,* vol. 59, no. 21, pp. 9926–9938, 2020, https://doi.org/10.1021/acs.iecr.9b06834.

[82] J. M. Valverde, P. Jiménez, and L. A. Pérez-Maqueda, "Calcium-Looping for Post-Combustion CO_2 Capture. On the Adverse Effect of Sorbent Regeneration under CO_2," *Applied Energy,* vol. 126, pp. 161–171, 2014, https://doi.org/10.1016/j.apenergy.2014.03.081.

[83] C. Petit, K. Y. A. Lin, and A. H. A. Park, "Design and Characterization of Liquidlike POSS-Based Hybrid Nanomaterials Synthesized via Ionic Bonding and Their Interactions with CO_2," *Langmuir,* vol. 29, no. 39, pp. 12234–12242, 2013, https://doi.org/10.1021/la4007923.

[84] S. Wu and Y. Zhu, "Behavior of $CaTiO_3$/Nano-CaO as a CO_2 Reactive Adsorbent," *Industrial & Engineering Chemistry Research,* vol. 49, no. 6, pp. 2701–2706, 2010, https://doi.org/10.1021/ie900900r.

[85] A. S. Jalilov *et al.*, "Asphalt-Derived High Surface Area Activated Porous Carbons for Carbon Dioxide Capture," *ACS Applied Materials & Interfaces,* vol. 7, no. 2, pp. 1376–1382, 2015, https://doi.org/10.1021/am508858x.

[86] S. Chen, Y. Chao, J. Wu, H. Ye, X. Luo, and Z. Liang, "A New Solid Adsorbent for CO_2 Capture Based on an Amine Polycarboxylate Ionic Liquid with Multiple Absorption Sites," *Industrial & Engineering Chemistry Research,* vol. 61, no. 32, pp. 11953–11963, 2022, https://doi.org/10.1021/acs.iecr.2c01797.

[87] V. E. Diyuk, A. N. Zaderko, L. M. Grischenko, G. G. Tsapyuk, A. V. Vakaliuk, V. V. Lisnyak, and R. Mariychuk, "CO_2 Adsorption on Pristine, Oxidized, and Diethylamine-Functionalized Activated Carbon Sorbents," *E3s Web of Conferences,* vol. 154, p. 07001, 2020, https://doi.org/10.1051/e3sconf/202015407001.

[88] M. Moreno-González *et al.*, "Sulfuric Acid Electrolyte Impacts Palladium Chemistry at Reductive Potentials," *Chemistry of Materials,* vol. 32, no. 21, pp. 9098–9106, 2020, https://doi.org/10.1021/acs.chemmater.0c01199.

[89] L. Fan, Y. Shen, D. Liu, N. Wahab, and M. Hasan, "Liquid-Phase Hydrogenation of Phenol to Cyclohexanone over Supported Palladium Catalysts," *Bulletin of Chemical Reaction Engineering and Catalysis,* vol. 11, no. 3, pp. 354–362, 2016, https://doi.org/10.9767/bcrec.11.3.575.354–362.

[90] M. Dixit, M. Mishra, P. A. Joshi, and D. O. Shah, "Physico-Chemical and Catalytic Properties of Mg–Al Hydrotalcite and Mg–Al Mixed Oxide Supported Copper Catalysts," *Journal of Industrial and Engineering Chemistry,* vol. 19, no. 2, pp. 458–468, 2013, https://doi.org/10.1016/j.jiec.2012.08.028.

[91] Y. Lei, M. Song, Y. Wei, and J. Xiao, "Combining Carbon Fibers with Ni/Γ–Al_2O_3 Used for Syngas Production: Part A: Preparation and Evaluation of Complex Carrier Catalysts," *Catalysts,* vol. 8, no. 12, p. 658, 2018, https://doi.org/10.3390/catal8120658.

[92] J. Kalyanaraman, Y. Fan, R. P. Lively, W. J. Koros, C. W. Jones, J. Park, and Y. Kawajiri, "Modeling and Experimental Validation of Carbon Dioxide Sorption on Hollow Fibers Loaded with Silica-Supported Poly(ethylenimine)," *Chemical Engineering Journal,* vol. 259, pp. 737–751, 2015, https://doi.org/10.1016/j.cej.2014.08.023.

[93] E. Papadopoulou and T. Ioannides, "Methanol Reforming Over Cobalt Catalysts Prepared from Fumarate Precursors: TPD Investigation," *Catalysts,* vol. 6, no. 3, p. 33, 2016, https://doi.org/10.3390/catal6030033.

[94] H. H. Funke, Y. Luo, M. Z. Chen, G. A. Anderson, J. L. Falconer, and R. D. Noble, "Measuring Mixture Adsorption by Temperature-Programmed Desorption," *Industrial & Engineering Chemistry Research,* vol. 54, no. 18, pp. 5159–5164, 2015, https://doi.org/10.1021/acs.iecr.5b00667.

[95] M. Pachecka, J. M. Sturm, C. Lee, and F. Bijkerk, "Adsorption and Dissociation of CO_2 on Ru(0001)," *The Journal of Physical Chemistry C,* vol. 121, no. 12, pp. 6729–6735, 2017, https://doi.org/10.1021/acs.jpcc.7b00021.

[96] Z. Guo, L. Huang, W. Chu, and S.-Z. Luo, "Effects of Promoter on NiMgAl Catalyst Structure and Performance for Carbon Dioxide Reforming of Methane," *Acta Physico-Chimica Sinica,* vol. 30, no. 4, pp. 723–728, 2014, https://doi.org/10.3866/pku.whxb201402242.

[97] J. He and G. Vidali, "Application of a Diffusion–desorption Rate Equation Model in Astrochemistry," *Faraday Discussions,* vol. 168, pp. 517–532, 2014, https://doi.org/10.1039/c3fd00113j.

[98] W. Liu, V. G. Ruiz, G. Zhang, B. Santra, X. Ren, M. Scheffler, and A. Tkatchenko, "Structure and Energetics of Benzene Adsorbed on Transition-Metal Surfaces: Density-Functional Theory with Van Der Waals Interactions Including Collective Substrate Response," *New Journal of Physics,* vol. 15, no. 5, p. 053046, 2013, https://doi.org/10.1088/1367-2630/15/5/053046.

[99] N. K. Mohammad, A. Ghaemi, and K. Tahvildari, "Hydroxide Modified Activated Alumina as an Adsorbent for CO2 Adsorption: Experimental and Modeling," *International Journal of Greenhouse Gas Control,* vol. 88, pp. 24–37, 2019.

[100] M. Saeidi, A. Ghaemi, K. Tahvildari, and P. Derakhshi, "Exploiting Response Surface Methodology (RSM) as a Novel Approach for the Optimization of Carbon Dioxide Adsorption by Dry Sodium Hydroxide," *Journal of the Chinese Chemical Society,* vol. 65, no. 12, pp. 1465–1475, 2018.

[101] A. Ghaemi, R. Norouzbeigi, and H. Masoumi, *Polymeric Adsorbents: Characterization, Properties, Applications, and Modelling*. Elsevier, 2023.

[102] K. Naderi, F. Bahmanzadegan, B. Parizad, A. Ghaemi, and S. Shahhosseini, "Comprehensive Analysis of O_2 and N_2 Adsorption on Kaolin-Derived Zeolites: Experimental and Computational Insights," *Industrial & Engineering Chemistry Research,* vol. 63, no. 37, pp. 16061–16076, 2024.

4 Synthesis, Preparation, and Pretreatment of CO_2 Adsorbents

Hadiseh Masoumi, Azam Aslani, and Ahad Ghaemi

4.1 INTRODUCTION

CO_2 emission is one of the most critical disasters due to diverse parameters including ignition of fossil sources, reducing the forest, incrementing different industrial sites, and so on [1]. Climate change and global warming are identified as the main effects of CO_2 rise [2]. For inhibiting CO_2 production, other sources are proposed such as wind energy, solar energy, and biomass energy. However, the efficiency of these sources is poor, which is not economical [3]. The other renewable energy is hydrogen energy because it generates no pollution after consumption. This technique has also limitations similar to the other ones especially because more methane gas should be converted to release hydrogen gas, which can produce some toxic gases like CO and CO_2 [4]. Therefore, it is reasonable to reduce the CO_2 amount instead of removing the CO_2 sources [5]. Adsorption and absorption of CO_2 molecules were recognized as the most common and economical methods for mitigating their concentration in the atmosphere [6]. Various solvents containing amine groups such as diethanolamine and monoethanolamine were widely employed in the absorption process. Different adsorbents were utilized in the adsorption process like zeolites [7], carbonaceous materials, metal oxides [8], and porous silica. The latter is preferred between absorption and adsorption procedures because of its low price, excellent yield, easy functionality, and slight energy consumption. Besides, a certain adsorbent is selected in terms of low energy consumption and slight cost of required chemical reagents [9]. The most applicable adsorbents are carbonaceous and organic porous materials. It is indispensable to explain that each of these adsorbents has some drawbacks such as lack of adequate CO_2 selectivity in the presence of other gases, low surface area, and poor uptake capacity in some empirical conditions. In this case, porous aromatic frameworks (PAFs), polymers of intrinsic microporosity (PIMs), polymer-coated polymeric (PCP), and hypercrosslinked polymers (HCPs) exhibit a unique surface area, and biopolymers reveal the slightest. Because the biopolymers cannot adsorb the hazardous moieties as effectively as other organic sorbents, their use may be limited in certain applications requiring high adsorption capacity. However, biopolymers are versatile and of low cost [10]. Therefore, several functional groups have

 DOI: 10.1201/9781003476191-4

been considered for improving the potential of these sorbents, which are explained in detail in this chapter [11]. Sometimes, the incorporation of functional groups can lead to a drop in the CO_2 uptake capacity by occupying the pores, which can prevent the accessibility of CO_2 molecules toward the cavities. Also, the incorporation of multiple functional moieties probably destroys the sorbent pores that mitigate the CO_2 diffusion rate, and this slight kinetic elevates the activation energy, which is not feasible economically [12].

CO_2 molecules can be adsorbed on these adsorbents via physical or chemical interactions. The energy is consumed lower in the physical interaction than in the latter [13]. The adhering of CO_2 molecules on the sorbents obeys one of these mechanisms implied following. These mechanisms are quantum, speed, non-stationary, and vestigial sieve effects [14]. Indeed, the adsorption mechanism depends on the features of the sorbent and the adsorbed species like polarizability, dipole moment, and magnetic sensitivity [15]. In another aspect, there are two main adsorption mechanisms including physical and chemical interactions. In the physical interaction, van der Waals force exists between the adsorbent and the adsorbate. In the case of chemical behavior, hydrogen bonding and acid–base interactions are observed [16]. In terms of the mass transfer concept, four main mechanisms are assumed: bulk, Knudsen, surface penetration, and Poiseuille stream [17]. Poiseuille stream depends on the pressure, while three other mechanisms are the function of temperature. The surface diffusion mechanism is dominant as the surface loading is high, and it will be reduced as the temperature rises. The most uncomplicated mode for interpreting the adsorption behavior is Knudsen diffusion. Knudsen diffusion is proposed when the hole size is lower relative to the average route. In some cases, the adsorption mechanism has only extensive affinity to the features of the adsorbate. Such a condition occurs as the size of the adsorbed particle is close to the diameter of the pores of the sorbent [18]. Previous investigations have revealed that holes lower than five times the molecular size of the adsorbed particles are desired for gas removal at 1 atm. This result was verified experimentally as well as by computational simulation utilizing grand canonical Monte Carlo. Because the molecular diameter of CO_2 is 0.209 nm, only holes lower than 1.0 nm are excellent for CO_2 adsorption at 1 atm. Nevertheless, at elevated pressures, CO_2 is also diffused in the supermicroporosity domain (pore diameters between 0.7 and 2 nm) [19]. Before CO_2 adsorption, the gas stream should be purified from other gases such as NO_x and SO_x. This process is called pretreatment. This process inhibits the filling of other sites of the adsorbents with other gases and also prevents competition of other molecules with CO_2. Besides the pretreatment process, the flue gas should be dehydrated because water molecules can destroy the crystalline architecture of some sorbents. Regardless of the excellent potential of adsorbents in anchoring CO_2 molecules, the adsorbents should be able to be reused. There are two important methods for omitting the adsorbents from CO_2 for the regenerating process, which are pressure swing and temperature swing [20].

The other major topic in the case of carbonaceous and organic adsorbents is their synthesis techniques because many chemical agents in multiple steps of fabrication are applied. Apart from the consumption of chemical reagents, myriad poison gases, especially HCl gases, are released, which generate severe issues for human health

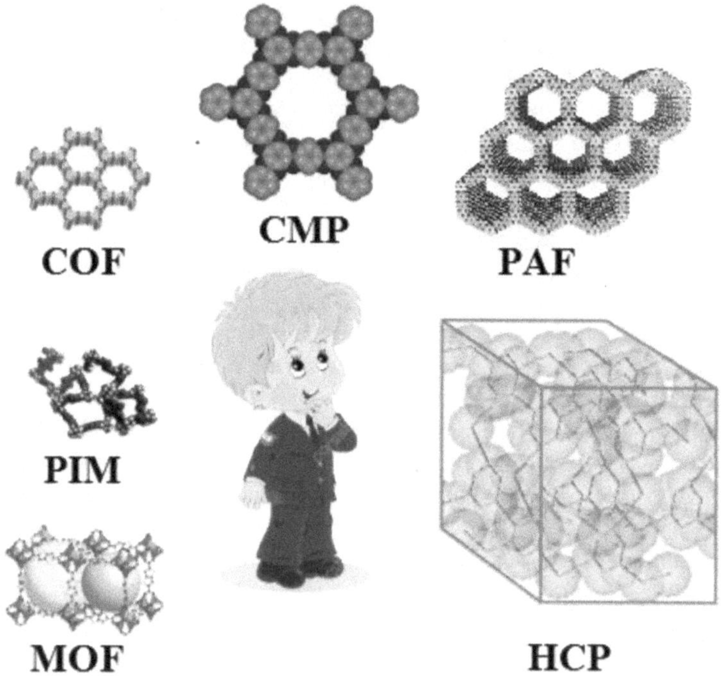

FIGURE 4.1 The schematic of various types of CO_2 adsorbents. Adapted with permission from [21], copyright (2024), Elsevier.

and the environment. Furthermore, most of these synthesis procedures are carried out at high temperatures, which requires strict control on the industrial scale. Hence, it is obligate for researchers who review the diverse synthesis methods to know these methods comprehensively till rectifying these synthesis techniques. It is attempted to explain various synthesis methods in this current chapter. Indeed, synthesis procedures influence directly the hole diameter dispersion and hole type and also can alter the reactivity of the precursors. The use of halogen agents in the skeleton of aromatic substances is a vital technique in organic chemistry because of the presence of halogens, which generate multiple functionalities. This chemical process is mostly employed in diverse applications. For fabricating a wide variety of haloarenes, bromine or chlorine agents are utilized in the traditional electrophilic halogenation. On the other hand, the Lewis basic agents are exploited in the halogenation process in which the nucleophilic oxygen, sulfur, nitrogen, or phosphorus moieties interact with the halogenating agents to generate influential halonium substances [21]. The schematic of well-known polymeric adsorbents is drawn in Figure 4.1.

4.2 CARBONACEOUS MATERIAL

4.2.1 Carbonization

Because of their high carbon and low price, some pristine materials are commonly used to generate activated carbons such as coal, tar, and pitch. From the viewpoint of industrial applications, these materials cannot be employed because they are non-renewable resources. To prevent this challenge, it was recommended to produce materials with a high percentage of carbon via biomass reactions. Carbonization is an inevitable step in a biomass reaction for converting biomass precursors into activated carbon. In this process, two main methods are employed for elevating the content of carbon, which is exposing the produced activated carbon under the inert gas at a certain temperature. Carbonization can be conducted through hydrothermal and pyrolysis reactions [22].

4.2.2 Pyrolysis

Pyrolysis mostly occurs for synthesizing solid carbons like CO, CH_4, and tar. This reaction is feasible in both low and high pressures. It was detected that a lower temperature can create pores with a larger size, which is desired for the adsorption of CO_2 molecules. At low pressure, CO_2 adsorption is conducted dramatically in the micropores than in mesopores, which requires a higher temperature. For instance, the number of micropores in the structure of pyrolyzed date sheets was more at 800°C than at 500°C. Indeed, the pores were in micro and meso sizes at 500°C. By increasing the temperature to 800°C, the range of pore sizes was 0.7–0.9 nm, which were in the category of micropore group. Additionally, the CO_2 adsorption of the mentioned composite has been calculated at 6.4 mmol/g at 1 bar and 0°C. The major technique for enlarging the pore sizes is pyrolyzing the considered substance in the presence of a template. In this case, the melamine was combined with poly(vinylidene chloride) (PVDC). Afterward, the temperature was elevated from 400°C to 800°C. At 400°C, the melem was formed and macropores were generated on the surface of PVDC. By raising the heat degree to 500°C, g-C_3N_4 has been extracted from melem. At 800°C, g-C_3N_4 was decomposed, which caused the enlarging of the pore sizes from micro to meso and macro. This high-temperature-synthesized carbon-based material reveals that CO_2 adsorption was at 2.6 mmol/g in room heat degree and 1 bar. Besides, the influence of temperature on the crystallinity structure and surface functional groups was probed. For this aim, temperatures of 450, 500, 550, 600, and 700°C have been investigated. It was observed that the temperature of 500°C for synthesizing the carbon-based material exhibited higher adsorption capacity, which was 4.17 mmol/g at room temperature and 1 bar [23, 24].

4.2.3 Hydrothermal Treatment

Hydrothermal treatment is known as a chemical reaction. Hydrothermal reaction is employed to complete the carbonization process in an autoclave and an aqueous state. This reaction needs a milder temperature (180–350°C) relative to a pyrolysis reaction, which consumes less energy. The particles with a larger surface area and smaller pore diameter were created during the hydrothermal treatment. For example, palm was carbonized in two synthesis routes including pyrolysis and hydrothermal reactions. The pyrolysis and hydrothermal reactions were carried out at 250°C and 800°C, respectively. It was found that the number of functional moieties in the hydrothermal reaction was higher than those in the pyrolysis one. The CO_2 uptake capacity in the hydrothermal reaction was measured at 5.184 mmol/g in 1 bar and 0°C.

Hydrothermal reaction impacts the functional groups containing oxygen and nitrogen atoms. Generally, the content of oxygen mitigates due to the decomposing of oxygen groups. Also, the effect of hydrothermal reaction on the number of nitrogen atoms is limited because of the Maillard process of the carbonyl and amino moieties [24, 25].

4.2.4 Activation

The uptake capacity of CO_2 molecules relies on diverse variables such as high surface area, porous structure, and surface compositions. To achieve this goal, the activation processes are applied. The activation process is categorized into two main levels: physical and chemical. The CO_2 adsorption exhibited dramatic results at 0°C and both low and high pressure when the carbon-based materials were synthesized via the activation procedure. Several works pointed to the positive impact of the activation technique on progressing CO_2 adsorption [26, 27]. Activation can vary the content of the compositions of the carbonaceous materials. For instance, the content of the oxygen and sulfur atoms is mitigated during the activation process, which is owing to conversion into volatile components. Regarding Figure 4.2, the peak at 1400 1/Cm has a lower intensity for activated carbonaceous materials (the activation agent is KOH) than the non-activated ones because of the degradation of carbonyl groups at a high temperature. The pressure has an influence on the consumed activation agent. At lower pressures like 1 bar, a lower quantity of KOH is required for loading a high amount of oxygen on the carbonaceous surface, which leads to generating more micropores that can elevate the CO_2 adsorption. Besides pressure, temperature has a considerable effect on the activation process. In this case, increasing the temperature from 500°C to 800°C causes

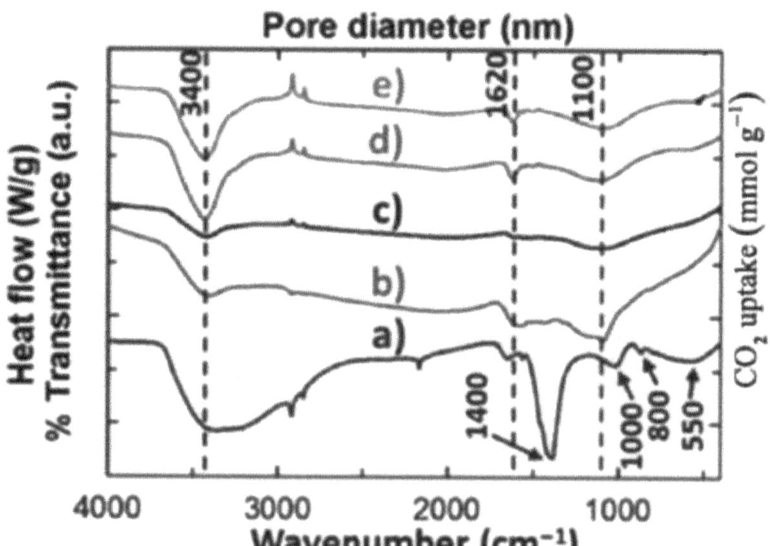

FIGURE 4.2 FTIR pattern of (a) GMB, (b) PAB1, (c) PAB2, (d) PAB3, and (e) PAB4. Adapted with permission from [29], copyright (2020), Elsevier.

diminishing oxygen content from 14.67% to 6.08% because of producing the fugacious substances. Moreover, the quantity of C=O and C–O is promoted, which is due to the decomposition of COOH moieties. In another work, the biomass-derived carbon was boosted at diverse heat degrees, and the activated sorbent at 700°C revealed more CO_2 uptake efficiency of 4.21 mmol/g at room temperature and 1 bar [28].

4.3 ORGANIC POROUS MATERIAL

4.3.1 COVALENT ORGANIC FRAMEWORKS

Covalent organic frameworks (COFs) consist of light elements such as carbon, hydrogen, oxygen, and boron, which are connected extremely via covalent forces. The structure of COFs can be 2D or 3D with a porous crystalline state. The specific surface area of these groups was estimated at 6450 m^2/g or higher than this value. The condensation reactions are mostly conducted for synthesizing COFs. This rigid network was initially fabricated under the names COF-1 and COF-5 through the solvothermal reaction of the dioxane/mesitylene combine. The pores in the COFs are in the category of micro and macro sizes. Furthermore, the topology of this group is determined by the shape of the building units. COFs are widely utilized in multiple cases like catalysts, hydrogen storage, optoelectronics, and CO_2 adsorption. COF groups with the base of boron element disclose a weak selectivity in the adsorption of CO_2 molecules in the presence of other gases because of their high pore diameter. Besides, they don't have adequate resistance to the temperature and moisture of the air, which generates a limit for CO_2 uptake. COFs with the base of imine group is another category of COFs, which was initially synthesized by Yaghi et al. [30] through condensation of aldehydes and amine (Figure 4.3). Banerjee et al. [31]

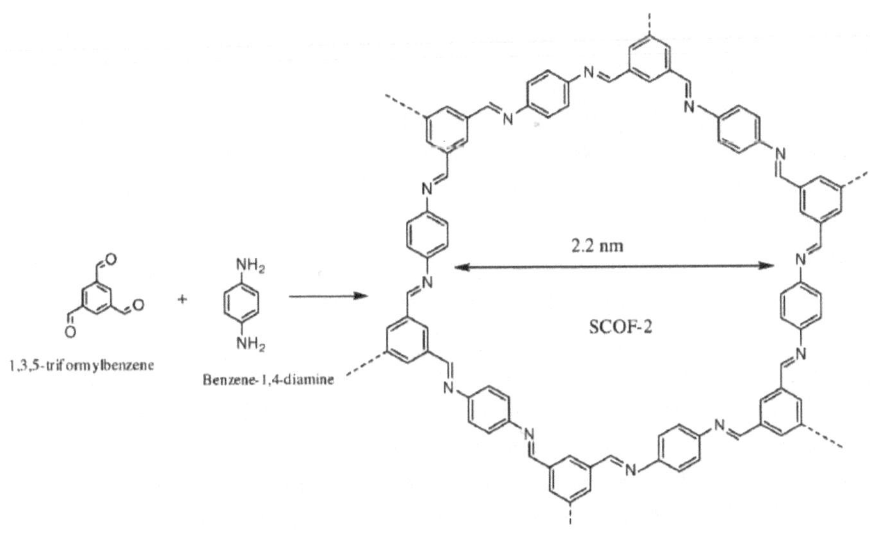

FIGURE 4.3 Synthesis route of imine-based COF. Adapted with permission from [32], copyright (2009), ACS.

synthesized two kinds of 2D-COFs with the base of the imine group for adsorbing CO_2 gases including TpPa-1 and TpPa-2. Tp refers to the 1,3,5-triformylphloroglucinol. These two COFs revealed excellent BET area and CO_2 elimination. The surface area of TpPa-1 and TpPa-2 have been obtained at 535 and 339 m^2/g, respectively. Also, the gas capture of these two composites has been determined at 153.21 and 125.71 mg/g at 0°C and 1 bar, respectively.

4.3.2 Hypercrosslinked Polymers

HCPs have an amorphous structure that contains light elements such as carbon, nitrogen, oxygen, and hydrogen, and polymer chains that have been crosslinked [33, 34]. Also, HCPs with the base of polystyrene were fabricated through the polymerization reaction of vinylbenzyl chloride in the presence of divinylbenzene as a crosslinker that was swelled in a suitable solvent. After swelling the monomer in a certain solvent, the Friedel–Crafts reaction was performed using $FeCl_3$ (Lewis acid) as a catalyst. The surface area of this type of HCP is in the domain of 600–2000 m^2/g. According to the papers, the surface area of HCPs is promoted by increasing the amount of Lewis acid [35, 36]. The novel type of HCPs with the size of nano was synthesized by Germain et al. [37]. In their method, polyaniline was crosslinked with diiodoalkanes or formaldehyde. The benefit of their work is the lack of requirement for the large content of Lewis acid, which inhibits generating a high amount of HCl gas [38]. It was reported that the surface area of hypercrosslinked polyanilines that were synthesized using diiodomethane and formaldehyde has been calculated at 632 and 480 m^2/g, respectively. Besides polyaniline and polystyrene, other chemical reagents have been commonly employed as monomers including dichloroxylene, *bis*(chloromethyl)biphenyl, and *bis*(chloromethyl)anthracene. Luo et al. [39] applied the Friedel–Crafts reaction for crosslinking furan, pyrrole, and thiophene in the presence of formaldehyde dimethyl acetal as a crosslinker agent. In their work, the surface area has been obtained at 437–726 m^2/g with a capacity of 127 mg/g at room temperature and 1 bar. Liu et al. [40] utilized benzene/biphenyl/1,3,5-triphenylbenzene as a monomer and 1,3,5-*tris*(bromomethyl)benzene as a crosslinker and aluminum chloride for eliminating hydrogen gas at 77 K and carbon dioxide at 273 K in the gas pressure of 1 bar. The uptake capacity of hydrogen and carbon dioxide has been determined at 19.1 and 181 mg/g, respectively.

4.3.3 Conjugated Microporous Polymers

Conjugated microporous polymers (CMPs) are categorized as one of the structures of HCPs. However, CMPs pose many differences with HCPs in possessing several C–C links. Chinchilla and Najera [41] synthesized the prime CMP during the Sonogashira–Hagihara in the presence of Pd as a catalyst for joining the aromatic halides to the aromatic alkynes. This constructed CMP was called poly(aryleneethynylene). From the viewpoint of rod length, CMPs are divided into two levels, which are long and short struts. The smaller chain is CMP-0, and the bigger chain is CMP-5. It was observed that the high BET surface area of 1018 m^2/g belongs to the CMPs with the shorter chains. Indeed, the shorter chain length causes the creation of pores

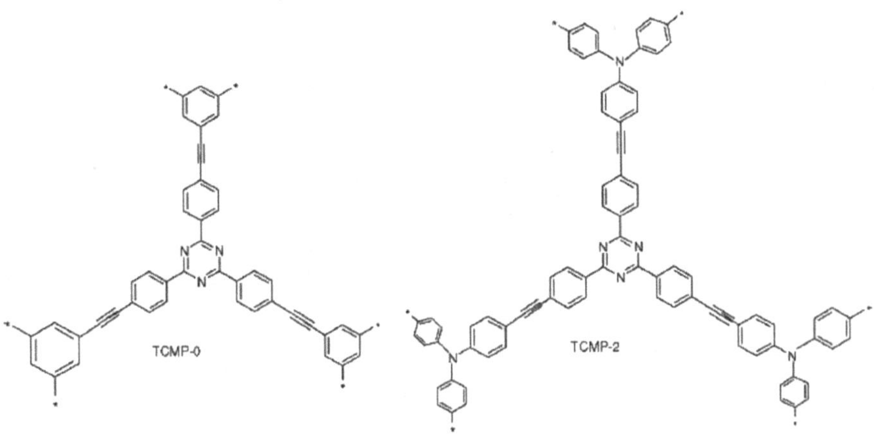

FIGURE 4.4 The chemical structure of CMPs with the base of triazine. Adapted with permission from [43], copyright (2012), RSC.

with small diameters, which increases the number of micropores in their skeleton. However, this case is inverse for the longer chain length. Some researchers such as Jiang et al. [42] and Ren et al. [43] decided to put nitrogen atoms in the structure of CMPs to progress their potential by utilizing monomers with the base of nitrogen. Changing the chemical groups in the knots of the CMPs architecture can impact the CO_2 adsorption. In this case, Ren et al. carried out an experiment that replaced 1,3,5-trisubstituted benzene with 1,3,5-triazine (TCMP) in the knot (Figure 4.4). The results of inquiries disclosed that this substitution can promote thermal resistance and also CO_2 uptake. For instance, TCMP-2 revealed considerable CO_2 uptake (115.28 mg/g) at room temperature and 1 bar relative to the CMP with the base of 1,3,5-trisubstituted benzene (92.4 mg/g). This significant increase is attributed to the stronger interaction of CO_2 molecules with triazine units than those of the phenyl rings. It was reported that employing low-cost precursors instead of expensive ones can also equip the triazine groups in the knots of the CMPs network. Puthiaraj et al. [44] have worked on this topic. They used the Friedel–Crafts process for proceeding with the reaction between the low-cost initiator (cyanuric chloride) and 1,3,5-triphenylbenzene/trans-stilbene in the presence of $AlCl_3$ as a catalyst. Two types of CMPs with the base of triazine were synthesized including MCTP-1 and MCTP-2. MCTP-1 showed more surface area (1452 m^2/g) and uptake capacity (204.3 mg/g) than MCTP-2. Moreover, MCTP-1 revealed higher CO_2 capacity in the presence of CH_4 gas, which was calculated at 494.7 m^2/g at 300 K and 35 bar. The derivatives of a certain material pose a substantial influence on the surface area and uptake capability of CO_2. For example, Rao et al. [45] fabricated polyphenylene with the core of CMPs (poly(phenylene fluorene): PPF) using a Suzuki coupling reaction with two kinds of derivatives, which were p-PPF and m-PPF. The surface area of p-PPF and m-PPF was obtained at 269 and 276 m^2/g, respectively. Moreover, the gas adsorption uptake has been calculated at 113.93 and 80.54 mg/g at 273 K and 30 bar, respectively.

The other functional group containing nitrogen atoms is imine, which is grafted to the CMP network. Neti et al. modified CMP with imine benzothiazole during the Schiff reaction. This reaction was carried out between 1,3,5-*tris*-(4-formylphenyl)-benzene and 2,6-diaminobenzo bisthiazole for 3 days (Figure 4.5). Besides imine and triazine groups, multiple functional groups were incorporated in the CMP skeleton such as carboxylic acids, amines, hydroxyl, and methyl moieties. Each of these functional groups has diverse influences on the CO_2 uptake capacity. For example, the CO_2 adsorption for COOH–CMP was determined at 68.64 mg/g, while the CO_2 adsorption for $(CH_3)_2$-CMP was estimated at 41.36 mg/g. As mentioned, CMPs can be exploited in different cases such as hydrogen storage, nanocomposites, and CO_2 adsorption because of having distinguished features.

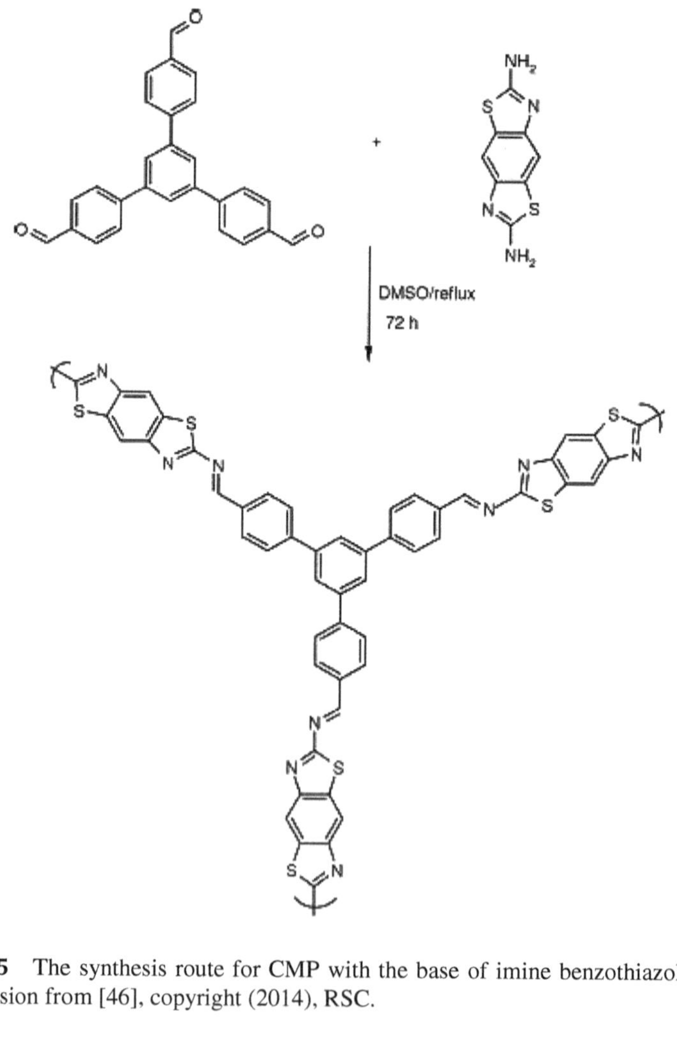

FIGURE 4.5 The synthesis route for CMP with the base of imine benzothiazole. Adapted with permission from [46], copyright (2014), RSC.

4.3.4 Polymers of Intrinsic Microporosity

PIMs consist of light elements such as carbon, hydrogen, oxygen, and nitrogen. The covalent bonds aren't observed in their network. These polymers are widely utilized in diverse fields like gas storage, separation, and catalysts. PIMs can be constructed via a double nucleophilic aromatic replacement process for the tetrahydroxylated precursor and tetrafluorinated precursor (Figure 4.6). McKeown et al. [47] fabricated PIM via several synthesis routes between 4,5-dichlorophthalonitrile and 5,5′,6,6′-tetrahydroxy-3,3,3′-tetramethyl-1,1′-spirobisindane in the presence of metal salt and quinolone at 220°C. Also, the surface area of the produced PIM network has been obtained at 489–895 m^2/g. PIM-1 can dissolve in organic solvents. PIM-1 is frequently employed in sensors and gas separation membranes.

The post-synthesis reaction is applied for modifying PIMs by manipulating the nitrile moieties such as the reaction of the PIM with P_2S_5 to create thioamines. One of the most important functional groups that graft to PIM is imide, called PIM-PI. PIM-PI displays excellent selectivity toward hydrogen/nitrogen (10), carbon dioxide/methane (14.2), and carbon dioxide/nitrogen (23.1). Neti et al. [49] modified porphyrin with imide (PyP). PyP has an amorphous structure with a surface area of 428 m^2/g and a CO_2 capacity of 35 mg/g.

4.3.5 Porous Aromatic Frameworks

PAFs are the same as CMPs with a diamond form and amorphous architecture. The Yamamoto–Ullmann reaction is used for synthesizing them. Also, the tetrahedral materials are applied as precursors. The phenyl groups are connected firmly via releasing bromide. The number of micropores in PAFs is more than that in other organic frameworks because the crosslinking covers the entire site. PAFs possess high thermal resistance, which is owing to the diamond form of these groups and also their slight densities. The first PAFs (PAF-1) were fabricated by Ben et al. In their research, the monomer was tetrakis(4-bromophenyl methane), which was crosslinked through the Yamamoto–Ullmann reaction in the presence of Ni(0) as a catalyst. PAF-1 exhibited an excellent surface area of 5640 m^2/g. Moreover, PAF-1 also shows high adsorption capacity in the elimination of CO_2 (1300 mg/g) and H_2 (107 mg/g) gases. One of the limitations of PAF-1 is its poor interaction with the gas molecules because of its hydrocarbon network. To overcome this challenge, the activation process is suggested, which is the reduction of PAF-1 with Li ions. This

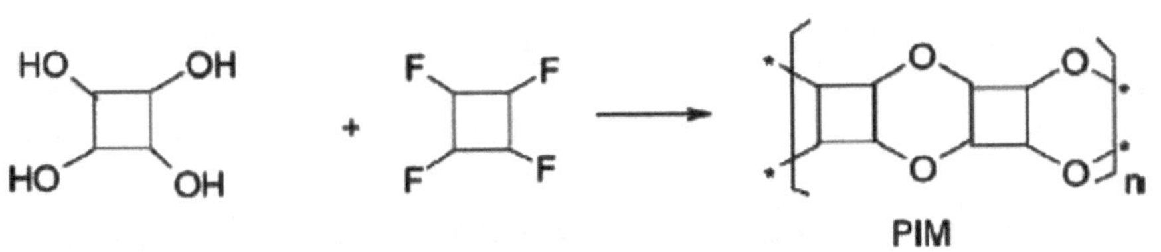

FIGURE 4.6 Synthesis route of PIM. Adapted with permission from [48], copyright (1999). Elsevier.

case mitigates the surface area because Li ions are bound with pores. Besides, this activation promotes the CO_2 uptake capacity to 395.56 mg/g. The other method for elevating the performance of PAF-1 is using monomers with the base of metal in the synthesizing process. There are other chemical reactions for synthesizing PAF networks such as Sonogashira–Hagihara, Suzuki, trimerization reaction of aromatic nitrile compounds, Schiff, and Friedel–Crafts [48].

4.4 CHEMICAL SYNTHESIS METHODS

4.4.1 Chemical Vapor Deposition

The chemical vapor deposition technique is commonly utilized in synthesizing carbon-based nanomaterials. In this method, the chemical reagents react together in the vapor state, producing a narrow layer on the substrate plane. The produced nanomaterials have a 2D network. These chemical reagents should have high purity, suitable volatility, low cost, excellent resistance to evaporation, and high shelf-life. Carbon nanotubes are usually generated through this method. Initially, the substrate is located in the oven for heating at a high temperature. Afterward, the vapor containing carbon is transferred at a low rate in the system as a precursor. The temperature is high in the system, which releases carbon atoms in the form of gas from the decomposition. Then, these gases are combined and carbon nanotubes are produced. In this synthesis procedure, the selection of a catalyst has a considerable impact on the morphology of the produced nanomaterial. For instance, the Ni, Co, or Cu catalysts are employed in the synthesis of graphene. The multilayer graphene is formed as Ni or Co is used. However, the monolayer graphene is created as Cu is employed as a catalyst [50].

4.4.2 Sol–Gel

Sol–gel technique is carried out in a wet media for the preparation of the nanomaterials with the base of the metal oxides. In this method, the liquid agent is incorporated into a sol. Subsequently, the sol is changed to a gel with a certain architecture. The reaction is possibly completed after multiple steps. In the primary synthesis route, the hydrolysis process of the metal oxide is conducted in water or alcohol to generate a sol. In the rest steps, condensation is carried out to create a porous structure by forming hydroxo (M–OH–M) or oxo (M–O–M) bonds between the metals in the aging process. It was found that the aging process declines porosity and the bond length between the colloidal particles is increased. The next step is drying, which eliminates the residual solvents and water molecules. The last step is a calcination reaction, and the ultimate material is generated in this part. The schematic of the sol–gel reaction is brought in Figure 4.7. This reaction has many advantages such as low reaction temperature, homogeneous product, and simple synthesis route [51].

4.4.3 Rigid and No Rigid Constructing Procedures

The no rigid constructing procedure is preferred to produce various 3D porous architectures because of mild empirical conditions, facile synthesis, and different morphologies. Diverse templates are applied in this procedure including block copolymers,

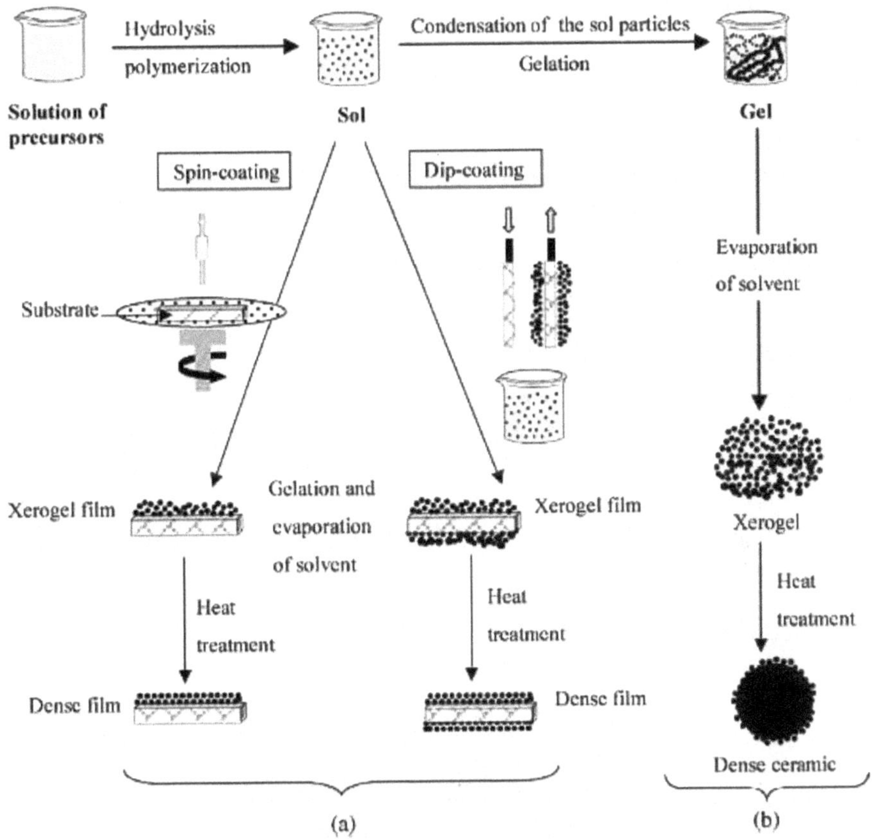

FIGURE 4.7 The schematic of a sol–gel process. Adapted with permission from [52], copyright (2010), Elsevier.

flexible organic molecules, and surfactants. Different forces are created between the soft template and precursors. Multiple parameters impact the quality of the produced material during the soft template reaction such as concentration of precursors and surfactants, surfactant/monomer proportion, experimental conditions, and surfactant network. Single crystal nanorods, porous aluminas, N-doped graphene, and carbonaceous nanospheres are synthesized via the soft templating method [53].

In the hard template method, the solid materials are used as the template and the pores of the solids are occupied by the precursors. One of the benefits of the hard template method is maintaining the mesoporous network during the participation of the precursors in the reaction, and they can also be eliminated from the produced structure without damaging the skeleton. Besides carbonaceous materials, other reagents such as carbon nanotubes are employed as the hard template materials. In this case, three steps should occur. In the first step, a suitable hard template is selected. In the second step, the considered precursor fills the pores of the hard template. In the last step, the template is omitted to obtain the mesoporous structure [54].

4.5 PHYSICAL SYNTHESIS METHODS

4.5.1 Mechanical Milling

Mechanical milling or ball milling is an efficient technique for generating blends of various phases, which is beneficial in the fabrication of nanoadsorbents. It is known as a low-cost procedure for fabricating the material in the domain of nano. This method is extensively utilized for preparing the alloy of the aluminum carbide/oxide, nanocomposite of aluminum/nickel/magnesium/copper, and the rest nanomaterials. The advantages of this technique are energy consumption and environmental compatibility [55].

4.5.2 Sputtering

Sputtering is ascribed to a process in which a high-energy agent including gas or plasma bombards the plane of a solid. The final produced substance is a thin layer in this method. Indeed, the clusters of the tiny atoms were physically ejected in this process (Figure 4.8). The required energy is supplied through various sources like magnetron, radiofrequency diode, and DC diode. This process is conducted in an

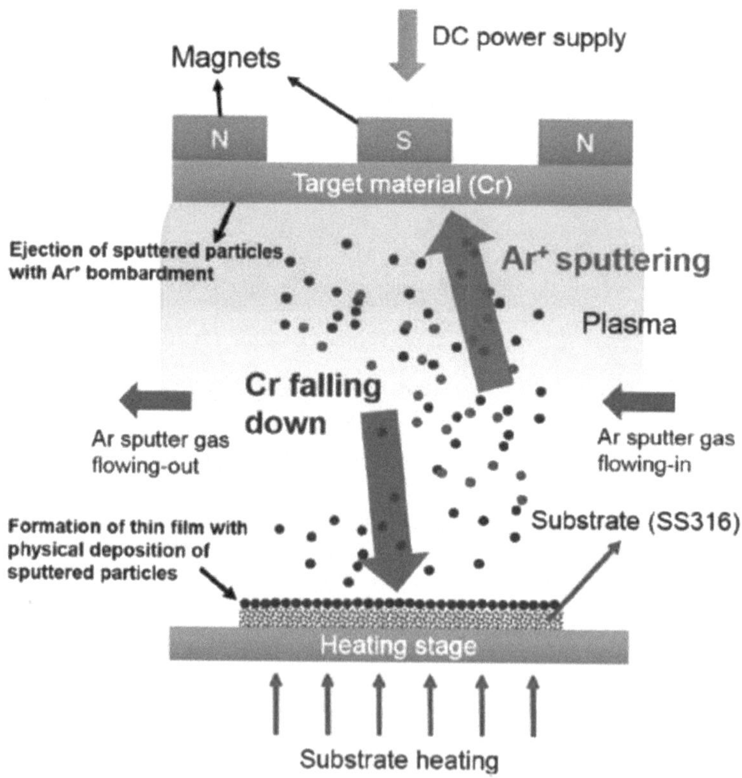

FIGURE 4.8 The schematic of a sputtering process. Adapted with permission from [57], copyright (2017), Elsevier.

evacuated chamber. Subsequently, the electricity current is exerted on the considered agent. The ions with positive charges are extensively excited in the presence of the electricity and transferred to the target substance, which causes the ejection of atoms from the plane of the considered agent. The benefits of this technique are being cost-effective relative to other methods like lithography and lack of creation of a high quantity of impurity [56].

4.5.3 Arc Discharge Method

The arc discharge method is widely exploited in producing carbonaceous materials like fullerenes, carbon nanohorns, carbon nanotubes, and few-layer graphene. In the case of fullerene production, a helium gas should occupy the chamber at a certain pressure and the oxygen and water particles should be depleted because of impeding the fullerene creation. Subsequently, two graphites in rod geometry are incorporated inside the chamber. Ultimately, the arc discharge occurs between the terminal corners of these graphites, leading to evaporation and release of carbon vapors. The experimental conditions pose a substantial significance in the generation of the arc discharge. It is essential to point to this notice that the carbon vapors deposit on the cathode and anode electrodes, and these sediments are collected, which is called the ultimate product. A type of product can reveal diverse morphologies in the presence of various atmospheres. For instance, graphene layers that are fabricated in the presence of hydrogen gases disclose excellent conductivity and also thermal resistance relative to the presence of argon gases [58]. The behavior of the arc discharge technique is shown in Figure 4.9.

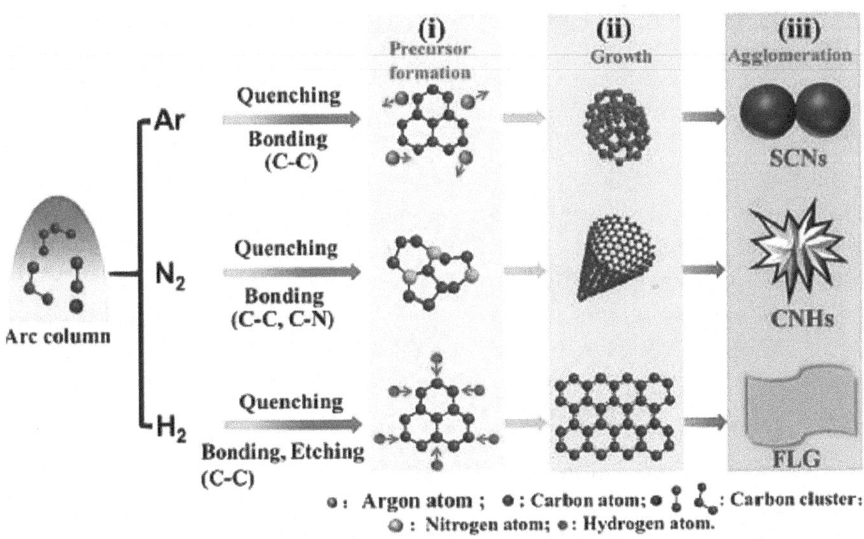

FIGURE 4.9 The behavior of the arc discharge procedure. Adapted with permission from [59], copyright (2018), Elsevier.

4.6 RETREATMENT TECHNIQUES

4.6.1 Carboxylic

The modification of the adsorbents with the carboxylic groups causes dispersion of them in water due to the high density of these functional groups. The carboxylic group on the surface of the sorbents can promote physisorption by enhancing the binding energy. It is attributed to the existence of electrons of the carboxyl group, which can bind to the carbon of CO_2 gas. The impregnation of a carboxylic acid into the network of the adsorbents is associated with raising the enthalpy or heat-releasing. For instance, treatment of metal-organic framework (MOF) with the carboxylic groups can increase the CO_2 capacity from 5.60 to 6.40 mmol/g at 25 bar and 33°C. Besides, the adsorbent sites lacking carboxylic moieties are decomposed by elevating temperature that leads to generating the uncoordinated carbon, which can adsorb CO_2 gas well [60].

4.6.2 Quinone

The treatment of the adsorbent with quinone is proposed by Wang et al. [61]. It was observed that the CO_2 capacity for the modified and unmodified carbonaceous adsorbent is at 3.02 and 2.22 mmol/g at 1 bar and 298 K, respectively. The CO_2 uptake was also increased to 4.78 and 3.02 mmol/g for unmodified and modified carbonaceous adsorbents, respectively, at 273 K. This phenomenon is owing to the slight value of surface area (653 m^2/g) and pore volume (0.38 Cm^3/g) of the modified adsorbent. Moreover, modifying the carbonaceous adsorbent with the quinone group leads to a decrease in the heat release mildly from 23.80 to 23 kJ/mol. The modified adsorbent with quinone improves the selectivity of CO_2 to 4.60 in the blend of CO_2 and CH_4 gases at 298 K and 1 atm. An identical growth was detected for CO_2/N_2 selectivity.

4.6.3 Lactone

Lactone groups are put into the acidic and nonpolar materials. The lactone on the surface can bond with CO_2 molecules through chemical adsorption. It isn't proved that lactone is adverse for adsorbing CO_2 gases. However, other functional moieties are preferred to employ for CO_2 capture. In this case, Bai et al. [62] utilized polar moieties such as hydroxyl and carboxyl groups for modification purposes, which was observed in the uptake capacity and selectivity enhancement.

4.6.4 Carbonyl

The existence of electrons on the oxygen atom of carbonyl groups can be able to promote the adsorption capacity of polarizable particles via the Lewis acid–base interactions. The carbonyl groups are also found in the pyridone moieties [63]. This group is more abundant than the other ones. The bond formation between the CO_2 molecules and carbonyl groups is stronger than that of the other functional groups like ketone and aldehyde. Kim et al. [64] proved the strong interaction between CO_2

and carbonyl groups of cucurbit(6)uril adsorbent. Also, the chemical configuration of this bond is slipped-parallel mode. The studies also exhibited that the existence of both carbonyl and indole groups can increase CO_2 uptake significantly because of the extensive dipole–π interaction. It is essential to note that the carbonyl groups can form a bond with CO_2 particles on the surface of the adsorbent, while the carbonyl–indole groups cannot form a bond with CO_2 gases on the surface of the sorbent. It was detected that the synthesized adsorbent with the indole and carbonyl groups could adsorb CO_2 gases with a capacity of 6.12 mmol/g. Besides, the surface area of this composite was calculated at 1628 m^2/g.

4.6.5 Esters

Esters are almost identical to the ether groups, but the level of polarity is higher than ethers. However, the polarity of ester and ether groups is lower than the alcohols. This case can elevate the CO_2 capacity because of the dipole–quadrupole interaction, which is the chemical adsorption between the esters and CO_2 molecules. Molavi et al. [65] grafted ester groups to the skeleton of MOF to probe their impact on the uptake capability of carbon dioxide. It was derived that the CO_2 uptake was raised from 3.14 to 4.28 mmol/g at 1 bar and 298 K because the pristine MOF has only an amine group while the modified MOF possesses both amine and ester groups. Additionally, the modified composite with ester groups exhibits higher selectivity toward N_2 gas relative to the modified composite with the amine groups.

4.6.6 Imine

The nitrogen of the imine groups exists in the orbital of sp^2 in its chemical configuration. Imine groups are similar to the pyridine ones in terms of the state of nitrogen in the electronic configuration [66]. However, they have a substantial discrepancy, which is the existence of the aromatic ring in the chemical structure of the pyridine, while this case isn't observed in the chemical structure of imine groups. The double link of the carbon and nitrogen in the imine group leads to the generation of an intensive attraction between this double bond and CO_2 molecules. Zeng et al. [67] modified the architecture of COF with imine, triazine, and boron to investigate their impact on CO_2 uptake. According to their research, it was detected that all of these functional groups have a substantial effect on CO_2 uptake because of the moderate sorption enthalpy, high selectivity, and excellent CO_2 adsorption capability. For example, the CO_2 uptake for the imine-modified composite has been determined at 13.38 and 3.11 mmol/g at 273 and 298 K in 5 bar, respectively. The reason for the excellent CO_2 capacity is owing to the large surface area and plenty of nitrogen atoms in the imine structure. Moreover, Gajula et al. [68] implied that the adsorption capacity of CO_2 declines by altering the imine groups (12.80 m^2/g) to the amine ones (5.70 m^2/g).

4.6.7 Pyridone

The nitrogen atom in pyridone has two electrons in the orbital of "p" and also hydrogen links to the surface of the ring [69]. Pyridone and pyrrolic groups cannot be distinguished

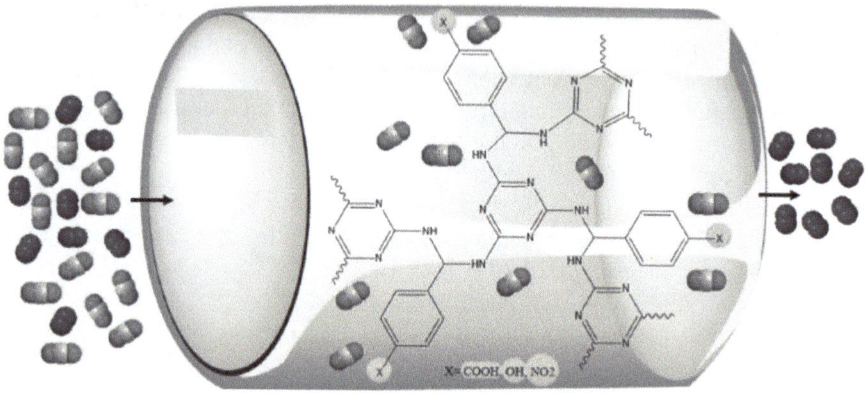

FIGURE 4.10 The selectivity of polyamines toward CO_2 molecules. Adapted with permission from [71], copyright (2021), Elsevier.

by X-ray photoelectron spectroscopy (XPS) analysis; thus, XPS considers these two groups as only one family. Lim et al. [70] measured the bond energy between the modified (pyridone) and unmodified composite with the CO_2 molecules. They concluded that the bond energy between the modified composite and CO_2 molecules (−0.224 eV) is higher than that between the pristine composite and CO_2 gas (−0.218 eV). Indeed, the CO_2 molecules adhere more to the hydroxyl unit of the pyridone, which is due to the strong interaction of the hydrogen binding. Regarding the reports, it was declared that the elevation of the activation temperature and incorporation of more oxidizing agents can mitigate the content of nitrogen atoms in the pyridone groups, which leads to a decrease in CO_2 uptake. For instance, the composite with the pyridone group exhibited CO_2 adsorption capacity of 6.20 mmol/g at 1 atm and 273 K, while this value declined at 4.30 mmol/g with the introduction of the oxygen atoms in the pyridone at the identical experimental conditions. It is noteworthy that the adsorption mechanism between the pyridone group and CO_2 molecules is Lewis and hydrogen bond interactions. Moreover, the effect of the pyridine group on CO_2 adsorption is similar to that of the pyridone group, with only a slight reduction observed. Also, amine groups can adsorb CO_2 gases more selectively than other functional groups (Figure 4.10).

4.6.8 Alkyl

Grafting the alkyl groups into the network of the sorbents promotes the affinity to the CO_2 molecules because of the electron donation property of the alkyl agents [72]. In other words, alkyl agents induct the electron into the aromatic rings of the organic adsorbents, which increases the π–quadrupole interaction. Besides quadrupole mechanism, there is a poor hydrogen link in the interaction of the oxygen of gas and the hydrogen of alkyl moieties [73]. In the research of He et al. [74], alkyl moieties were injected into the architecture of the polymer with the genus of triptycene, which enables it to elevate the adsorption of CO_2 gas to 4.17 mol/g at 1 bar and 273 K. However, it was found that adding the alkyl groups into the composite containing amine groups leads to a decline in the adsorption efficiency of CO_2 gas from 1.04 to 0.98 mmol/g, which is due to the steric hindrance of the electrons in the amine-alkyl composite.

4.6.9 S-Heteroatoms

The functional groups containing sulfur groups like sulfones, sulfoxides, and sulfonic acids can attract CO_2 molecules through polar contacts [75]. Certainly, the incorporation of the sulfur units into the carbonaceous chains of the organic adsorbents causes additional charges on the carbon atoms, which elevates the uptake capacity of CO_2 [76]. From the viewpoint of the chemical concept, the existence of large and polarizable orbitals with the code of "d," which posse electrons leads to forming a bond with the oxygen atoms of CO_2 molecules. Furthermore, sulfur atoms of thiophenic structure induce electrons to the CO_2 particles, which enables to oxidization of the reagents of the atoms on its plane, releasing SO molecules. The other functional group containing sulfur units is the sulfonic group. This group has abundant advantages such as versatility, because of the rapid mobility of the C–S link in the route of the hydroxyl link. This case augments the potential of intermolecular interactions. Biswas et al. [77] treated the surface of the UiO-66 with the sulfonic groups, and the uptake capacity of CO_2 gas was boosted at 5.60 mmol/g in 25 bar and 33°C. In another work, Kronast et al. [78] grafted ethylsulfide into the framework of UiO-66. It was observed that the surface area and CO_2 [70] capacity are improved at 52 m^2/g and 2.40 mmol/g, respectively. Treatment of the adsorbent plane causes an improvement in the growth of the oxygen atoms, which promotes the adsorption potential by elevating the surface polarity, binding energy, and selectivity over nitrogen [79]. The precise probing of the modification process is essential to make sure that those functional groups that are incorporated are useful for gas removal. In the case of functional groups containing nitrogen atoms, the general approach still governs: injecting basic polar units should progress the removal of CO_2 molecules [80].

Typically, functionalization of the reagents with the amine groups will promote CO_2 uptake. Nevertheless, amine moieties pose some defects. Blocking the pores is recognized as a significant problem for high contents of them [81].

4.7 CONCLUSION

There are multiple adsorbents with the origin of carbon atoms that are extensively utilized such as activated carbon, carbon nanotubes, graphene, biomass-derived carbons, and organic frameworks. Biomass-derived carbons have attracted much attention because of their abundance. However, these adsorbents pose several drawbacks especially slight porosity and weak uptake capacity. Thus, the organic frameworks are suggested. The organic adsorbents have different types in terms of porosity, surface area, and uptake capacity. The main species of these materials is the aromatic precursor. The synthesis technique is very crucial in adhering these aromatic rings, which makes these organic composites firm and rigid. Among the diverse synthesis methods, the chemical technique is the most desired one especially sol–gel because of its facile synthesis route and lack of requirement for expensive ingredients. It is noteworthy that CMP groups have revealed a higher CO_2 capacity than the other organic frameworks, which is due to their substantial porosity.

Despite extraordinary features of organic frameworks, they have often exhibited poor performance in CO_2 adsorption. To overcome these challenges, grafting some functional groups to their skeleton is attempted. The most useful functional group is the one with nitrogen atoms such as imine, amine, and pyridine. Furthermore, some other functional groups have no impact on CO_2 adsorption like lactone. Regarding the scientific papers, the main CO_2 adsorption mechanism is the Lewis acid–base interaction between the organic adsorbents and oxygen in the carbon dioxide. The other adsorption mechanisms are hydrogen bonds and pore-filling.

It is proposed to rectify or amend the synthesis methods for employing these organic adsorbents on the commercial scale till the high content of CO_2 is depleted from the atmosphere. The proposed cases are omitting the catalyst, substituting low-cost precursors, ability to accomplish at room temperature, and augmenting the reaction yield. Also, utilizing prediction tools like molecular dynamics (MD) can assist the researchers in detecting the challenges and solving them before conducting the experiments, which reduces the time and price consumption. Besides MD program, using design-expert software can help the researchers probe the effect of two variables on the CO_2 uptake simultaneously, which helps us to adjust the variables logically. Additionally, the design expert can present the most precise equation and the optimum points, which decreases the number of experiments. One of the most appropriate techniques is exploiting the discarded materials with the base of aromatic rings as the monomers including waste polycarbonate and waste polystyrene. Another piece of advice for being economical in the synthesis process is getting least the organic solvent consumption because of releasing the toxic gases and their high cost. In the case of injecting the functional groups, it is more promising to not join nitrogen atoms close to the oxygen atoms because it dramatically mitigates the CO_2 uptake potential. For further inspection of the CO_2 adsorption behavior on the surface of the carbonaceous sorbents, the chemical analysis gives us beneficial information especially the Raman test, zeta potential, and vibrating sample magnetometry (VSM), which were worked rarely. The selectivity of the entire adsorbents whether organic or inorganic toward more than one species of gas molecules is weak. Hence, the researchers can endow their studies on improving the selectivity of sorbents. Treatment of the plane of the adsorbents with amine groups can generate some problems such as corrosion and degradation of the synthesis vessels, which limits its utilization on a larger scale.

REFERENCES

1. Waheed, R., et al., Forest, agriculture, renewable energy, and CO_2 emission. *Journal of Cleaner Production*, 2018. **172**: p. 4231–4238.
2. D'Alessandro, D.M., B. Smit, and J.R. Long, Carbon dioxide capture: Prospects for new materials. *Angewandte Chemie International Edition*, 2010. **49**(35): p. 6058–6082.
3. Gong, J., C. Li, and M.R. Wasielewski, Advances in solar energy conversion. *Chemical Society Reviews*, 2019. **48**(7): p. 1862–1864.
4. Zhao, H., et al., Carbon-based adsorbents for post-combustion capture: A review. *Greenhouse Gases: Science and Technology*, 2018. **8**(1): p. 11–36.
5. Gao, X., J. Ashok, and S. Kawi, A review on roles of pretreatment atmospheres for the preparation of efficient Ni-based catalysts. *Catalysis Today*, 2022. **397**: p. 581–591.

6. Guo, H., et al., Nonaqueous amine-based absorbents for energy efficient CO_2 capture. *Applied Energy*, 2019. **239**: p. 725–734.
7. Chen, S., et al., CO_2 separation from offshore natural gas in quiescent and flowing states using 13X zeolite. *Applied Energy*, 2017. **205**: p. 1435–1446.
8. Jin, S., et al., Fabrication and kinetic study of spherical MgO agglomerates via water-in-oil method for pre-combustion CO_2 capture. *Chemical Engineering Journal*, 2019. **359**: p. 285–297.
9. Liao, Y., et al., Targeted control over the porosities and functionalities of conjugated microporous polycarbazole networks for CO 2-selective capture and H 2 storage. *Polymer Chemistry*, 2017. **8**(46): p. 7240–7247.
10. Ghaemi, A., R. Norouzbeigi, and H. Masoumi, *Polymeric Adsorbents: Characterization, Properties, Applications, and Modelling*. 2023: Elsevier.
11. House, K.Z., et al., The energy penalty of post-combustion CO 2 capture & storage and its implications for retrofitting the US installed base. *Energy & Environmental Science*, 2009. **2**(2): p. 193–205.
12. Sarmah, M., B.P. Baruah, and P. Khare, A comparison between CO_2 capturing capacities of fly ash based composites of MEA/DMA and DEA/DMA. *Fuel Processing Technology*, 2013. **106**: p. 490–497.
13. Drage, T., et al., Comparison of pre and postcombustion CO_2 adsorbent technologies. In *8th International Conference on Greenhouse Gas Control Technologies*. 2006: Citeseer.
14. Abd, A.A., M.R. Othman, and J. Kim, A review on application of activated carbons for carbon dioxide capture: Present performance, preparation, and surface modification for further improvement. *Environmental Science and Pollution Research*, 2021. **28**(32): p. 43329–43364.
15. Mallesh, D., et al., Synthesis, characterization of carbon adsorbents derived from waste biomass and its application to CO_2 capture. *Applied Surface Science*, 2020. **530**: p. 147226.
16. Wang, P., et al., Biomass derived wood ash with amine modification for post-combustion CO_2 capture. *Applied Energy*, 2017. **201**: p. 34–44.
17. Do Duong, D., *Adsorption Analysis: Equilibria and Kinetics*. 1998: Imperial College Press.
18. Suzuki, M. and M. Suzuki, *Adsorption Engineering*. Vol. 14. 1990: Kodansha Tokyo.
19. Martín, C.F., et al., Hypercrosslinked organic polymer networks as potential adsorbents for pre-combustion CO 2 capture. *Journal of Materials Chemistry*, 2011. **21**(14): p. 5475–5483.
20. Pires, J., et al., Recent developments on carbon capture and storage: An overview. *Chemical Engineering Research and Design*, 2011. **89**(9): p. 1446–1460.
21. Masoumi, H., Types of polymeric adsorbents. In *Polymeric Adsorbents*. 2024: Elsevier. p. 1–46.
22. Yu, H., et al., Recent advances of ultrasound-assisted Maillard reaction. *Ultrasonics Sonochemistry*, 2020. **64**: p. 104844.
23. Jürgens, B., et al., Melem (2, 5, 8-triamino-tri-s-triazine), an important intermediate during condensation of melamine rings to graphitic carbon nitride: Synthesis, structure determination by X-ray powder diffractometry, solid-state NMR, and theoretical studies. *Journal of the American Chemical Society*, 2003. **125**(34): p. 10288–10300.
24. Taskin, O.S., et al., One-Pot, One-Step strategy for the preparation of clickable melamine based microporous organic polymer network. *Macromolecular Materials and Engineering*, 2015. **300**(11): p. 1116–1122.
25. Qiu, Y., et al., Biocrude oil production through the Maillard reaction between leucine and glucose during hydrothermal liquefaction. *Energy & Fuels*, 2019. **33**(9): p. 8758–8765.

26. Ello, A.S., et al., Development of microporous carbons for CO_2 capture by KOH activation of African palm shells. *Journal of CO_2 Utilization*, 2013. **2**: p. 35–38.
27. Ma, X., et al., Underlying mechanism of CO_2 uptake onto biomass-based porous carbons: Do adsorbents capture CO_2 chiefly through narrow micropores? *Fuel*, 2020. **282**: p. 118727.
28. Quan, C., R. Su, and N. Gao, Preparation of activated biomass carbon from pine sawdust for supercapacitor and CO_2 capture. *International Journal of Energy Research*, 2020. **44**(6): p. 4335–4351.
29. Ismail, I.S., et al., Oxygen functionalized porous activated biocarbons with high surface area derived from grape marc for enhanced capture of CO_2 at elevated-pressure. *Carbon*, 2020. **160**: p. 113–124.
30. Uribe-Romo, F.J., et al., A crystalline imine-linked 3-D porous covalent organic framework. *Journal of the American Chemical Society*, 2009. **131**(13): p. 4570–4571.
31. Kandambeth, S., et al., Construction of crystalline 2D covalent organic frameworks with remarkable chemical (acid/base) stability via a combined reversible and irreversible route. *Journal of the American Chemical Society*, 2012. **134**(48): p. 19524–19527.
32. Liu, X.-H., et al., On-surface synthesis of single-layered two-dimensional covalent organic frameworks via solid–vapor interface reactions. *Journal of the American Chemical Society*, 2013. **135**(28): p. 10470–10474.
33. Masoumi, H., A. Ghaemi, and H.G. Ghanadzadeh, Elimination of lead from multi-component lead-nickel-cadmium solution using hyper-cross-linked polystyrene: Experimental and RSM modeling. *Journal of Environmental Chemical Engineering*, 2021. **9**(6): p. 106579.
34. Masoumi, H., et al., Benzene-based hypercross-linked polymers as a highly efficient adsorbent for cadmium removal from aqueous solution. *International Journal of Environmental Science and Technology*, 2022: p. 1–16.
35. Masoumi, H., A. Ghaemi, and H.G. Gilani, Synthesis of polystyrene-based hyper-cross-linked polymers for Cd (II) ions removal from aqueous solutions: Experimental and RSM modeling. *Journal of Hazardous Materials*, 2021: p. 125923.
36. Masoumi, H., A. Ghaemi, and H.G. Gilani, Exploiting the performance of Hyper-cross-linked polystyrene for removal of multi-component heavy metal ions from wastewaters. *Journal of Environmental Chemical Engineering*, 2021: p. 105724.
37. Germain, J., J.M. Fréchet, and F. Svec, Hypercrosslinked polyanilines with nanoporous structure and high surface area: Potential adsorbents for hydrogen storage. *Journal of Materials Chemistry*, 2007. **17**(47): p. 4989–4997.
38. Masoumi, H., A. Ghaemi, and H.G. Gilani, Evaluation of hyper-cross-linked polymers performances in the removal of hazardous heavy metal ions: A review. *Separation and Purification Technology*, 2021. **260**: p. 118221.
39. Luo, Y., et al., Hypercrosslinked aromatic heterocyclic microporous polymers: A new class of highly selective CO_2 capturing materials. *Advanced Materials*, 2012. **24**(42): p. 5703–5707.
40. Liu, G., et al., A facile synthesis of microporous organic polymers for efficient gas storage and separation. *Journal of Materials Chemistry A*, 2015. **3**(6): p. 3051–3058.
41. Chinchilla, R. and C. Nájera, The Sonogashira reaction: A booming methodology in synthetic organic chemistry. *Chemical Reviews*, 2007. **107**(3): p. 874–922.
42. Jiang, J.-X., et al., Microporous poly (tri (4-ethynylphenyl) amine) networks: Synthesis, properties, and atomistic simulation. *Macromolecules*, 2009. **42**(7): p. 2658–2666.
43. Ren, S., et al., Functional conjugated microporous polymers: From 1, 3, 5-benzene to 1, 3, 5-triazine. *Polymer Chemistry*, 2012. **3**(4): p. 928–934.
44. Puthiaraj, P., et al., Microporous covalent triazine polymers: Efficient Friedel–Crafts synthesis and adsorption/storage of CO 2 and CH 4. *Journal of Materials Chemistry A*, 2015. **3**(13): p. 6792–6797.

45. Rao, K.V., et al., Extended phenylene based microporous organic polymers with selective carbon dioxide adsorption. *Journal of Materials Chemistry*, 2011. **21**(34): p. 12958–12963.
46. Neti, V.S.P.K., et al., Synthesis of a benzothiazole nanoporous polymer for selective CO 2 adsorption. *RSC Advances*, 2014. **4**(19): p. 9669–9672.
47. McKeown, N.B., Polymers of intrinsic microporosity. *International Scholarly Research Notices*, 2012. **2012**.
48. Olajire, A.A., Synthesis of bare and functionalized porous adsorbent materials for CO_2 capture. *Greenhouse Gases: Science and Technology*, 2017. **7**(3): p. 399–459.
49. Neti, V.S.P.K., et al., Synthesis of a polyimide porous porphyrin polymer for selective CO 2 capture. *Journal of Chemistry*, 2015. **2015**.
50. Jones, A.C. and M.L. Hitchman, Overview of chemical vapour deposition. *Chemical Vapour Deposition: Precursors, Processes and Applications*, 2009. **1**: p. 1–36.
51. Danks, A.E., S.R. Hall, and Z. Schnepp, The evolution of 'sol–gel' chemistry as a technique for materials synthesis. *Materials Horizons*, 2016. **3**(2): p. 91–112.
52. Znaidi, L., Sol–gel-deposited ZnO thin films: A review. *Materials Science and Engineering: B*, 2010. **174**(1–3): p. 18–30.
53. Liu, Y., J. Goebl, and Y. Yin, Themed issue: Chemistry of functional nanomaterials. *Chemical Society Reviews*, 2013. **42**: p. 2610–2653.
54. Yamauchi, Y. and K. Kuroda, Rational design of mesoporous metals and related nanomaterials by a soft-template approach. *Chemistry–An Asian Journal,* 2008. **3**(4): p. 664–676.
55. Yadav, T.P., R.M. Yadav, and D.P. Singh, Mechanical milling: A top down approach for the synthesis of nanomaterials and nanocomposites. *Nanoscience and Nanotechnology*, 2012. **2**(3): p. 22–48.
56. Nie, M., K. Sun, and D.D. Meng, Formation of metal nanoparticles by short-distance sputter deposition in a reactive ion etching chamber. *Journal of Applied Physics*, 2009. **106**(5).
57. Son, H.H., et al., Capillary wicking effect of a Cr-sputtered superhydrophilic surface on enhancement of pool boiling critical heat flux. *International Journal of Heat and Mass Transfer*, 2017. **113**: p. 115–128.
58. Wu, Z.-S., et al., Synthesis of graphene sheets with high electrical conductivity and good thermal stability by hydrogen arc discharge exfoliation. *ACS Nano*, 2009. **3**(2): p. 411–417.
59. Zhang, D., et al., Controllable synthesis of carbon nanomaterials by direct current arc discharge from the inner wall of the chamber. *Carbon*, 2019. **142**: p. 278–284.
60. Kumar, K.V., et al., Effect of nitrogen doping on the CO_2 adsorption behavior in nanoporous carbon structures: A molecular simulation study. *The Journal of Physical Chemistry C*, 2015. **119**(39): p. 22310–22321.
61. Wang, S., et al., A highly permeable graphene oxide membrane with fast and selective transport nanochannels for efficient carbon capture. *Energy & Environmental Science*, 2016. **9**(10): p. 3107–3112.
62. Bai, B.C., et al., Effects of surface chemical properties of activated carbon fibers modified by liquid oxidation for CO_2 adsorption. *Applied Surface Science,* 2015. **353**: p. 158–164.
63. Heydarifard, M., A. Ghaemi, and M. Shirvani, Numerical simulation of CO_2 chemical absorption in a gas-liquid bubble column using the space-time CESE method. *Journal of Environmental Chemical Engineering*, 2020. **8**(5): p. 104111.
64. Kim, H., et al., Highly selective carbon dioxide sorption in an organic molecular porous material. *Journal of the American Chemical Society*, 2010. **132**(35): p. 12200–12202.
65. Molavi, H., et al., Enhancing CO_2/N_2 adsorption selectivity via post-synthetic modification of NH_2-UiO-66 (Zr). *Microporous and Mesoporous Materials*, 2018. **257**: p. 193–201.

66. Behvandi, D., et al., Evaluation of diffusion and Henry's coefficients of CO_2 absorption using Response Surface Methodology and Artificial Neural Network models. *Case Studies in Chemical and Environmental Engineering*, 2024. **9**: p. 100723.
67. Zeng, Y., R. Zou, and Y. Zhao, Covalent organic frameworks for CO_2 capture. *Advanced Materials*, 2016. **28**(15): p. 2855–2873.
68. Gajula, R.K., R. Kishor, and M.J. Prakash, Imine-linked covalent organic cage porous crystals for CO_2 adsorption. *ChemistrySelect,* 2019. **4**(43): p. 12547–12555.
69. Hasanzadeh, A., A. Ghaemi, and S. Shahhosseini, Neural network modeling for development of high-pressure measurement of carbon dioxide solubility in the aqueous AEEA+ sulfolane. *Journal of Chemical & Petroleum Engineering*, 2023. **57**(2).
70. Lim, G., K.B. Lee, and H.C. Ham, Effect of N-containing functional groups on CO_2 adsorption of carbonaceous materials: A density functional theory approach. *The Journal of Physical Chemistry C*, 2016. **120**(15): p. 8087–8095.
71. Sattari, A., et al., The application of polymer containing materials in CO_2 capturing via absorption and adsorption methods. *Journal of CO_2 Utilization*, 2021. **48**: p. 101526.
72. Khosrowshahi, M.S., et al., The role of surface chemistry on CO_2 adsorption in biomass-derived porous carbons by experimental results and molecular dynamics simulations. *Scientific Reports*, 2022. **12**(1): p. 8917.
73. Taheri, F.S., et al., High CO_2 adsorption on amine-functionalized improved mesoporous silica nanotube as an eco-friendly nanocomposite. *Energy & Fuels*, 2019. **33**(6): p. 5384–5397.
74. He, Y., et al., Efficient CO_2 capture by triptycene-based microporous organic polymer with functionalized modification. *Microporous and Mesoporous Materials*, 2015. **214**: p. 181–187.
75. Pashaei, H., et al., Experimental modeling and optimization of CO_2 absorption into piperazine solutions using RSM-CCD methodology. *ACS Omega*, 2020. **5**(15): p. 8432–8448.
76. Wohlgemuth, S.-A., et al., A one-pot hydrothermal synthesis of sulfur and nitrogen doped carbon aerogels with enhanced electrocatalytic activity in the oxygen reduction reaction. *Green Chemistry*, 2012. **14**(5): p. 1515–1523.
77. Biswas, S., et al., Enhanced selectivity of CO 2 over CH 4 in sulphonate-, carboxylate-and iodo-functionalized UiO-66 frameworks. *Dalton Transactions*, 2013. **42**(13): p. 4730–4737.
78. Kronast, A., et al., Gated channels and selectivity tuning of CO_2 over N_2 sorption by post-synthetic modification of a UiO-66-type metal–organic framework. *Chemistry–A European Journal*, 2016. **22**(36): p. 12800–12807.
79. Zafari, P. and A. Ghaemi, Modeling and optimization of CO_2 capture into mixed MEA-PZ amine solutions using machine learning based on ANN and RSM models. *Results in Engineering*, 2023. **19**: p. 101279.
80. Mashhadimoslem, H., et al., Development of predictive models for activated carbon synthesis from different biomass for CO_2 adsorption using artificial neural networks. *Industrial & Engineering Chemistry Research*, 2021. **60**(38): p. 13950–13966.
81. Kazemi, A., et al., Enhanced CO_2 capture potential of UiO-66-NH_2 synthesized by sonochemical method: Experimental findings and performance evaluation. *Scientific Reports*, 2023. **13**(1): p. 19891.

5 CO_2 Nanostructured Adsorbents

Fereshteh Pouresmaeil, Ahad Ghaemi, and Reza Norouzbeigi

5.1 INTRODUCTION

With the world facing serious climate problems, caused by human activities and extensive consumption of nonrenewable fossil fuels, it is crucial to find ways to alleviate this issue. The primary greenhouse gas responsible for climate change is anthropogenic CO_2; thus, it is essential to capture this gas prior to its release into the atmosphere. The goals set out in agreements like the seminal Paris Agreement of 2015 aim to reduce harmful emissions and carbon footprint. Global warming and other impacts of climate change, like making the oceans more acidic, desertification, and causing extreme weather, to some extent, can be controlled by capturing CO_2.

Trapping CO_2 gas, immediately after it is produced, can prevent escaping it into the air. Herein, the nanostructured materials play a pivotal role, as they offer a high pore volume and large surface area, helping trap more CO_2. As a growing research interest, carbon capture and storage (CCS) is tightly associated with developing nanomaterials with optimized properties. The pre-combustion capture from syngas, post-combustion capture from industrial flue gases, and oxy-combustion capture are identified as available approaches for CO_2 capture. The vast majority of CCS research is focused on strategies like absorption, cryogenic separation, membrane separation, and adsorption [1]. The drawbacks of chemical absorption, the most popular technology for CCS, are high regeneration energy, high corrosion rates, large equipment size, expensive operation, and hazardous process [2, 3]. The limitations in physical absorption also make it inappropriate to use for CO_2 capture. The utilization of membranes for CO_2 separation, while benefiting from their permeability and sieving characteristics [4], is hindered by several limitations. These include vulnerability to rare elements such as sulfur compounds, inadequate thermal properties, a reduction in CO_2 absorption in humid environments, and issues related to discontinuity [3]. In this regard, using alternative approaches is of utmost importance.

Adsorption stands out as a major alternative for capturing CO_2 emissions, in terms of high selectivity, energy efficiency, ease of regeneration, less environmental pollution, high performance, and affordable price. CO_2 adsorption on solid surfaces is along with physisorption, which offers weak bonding and low temperature and regeneration energy [2]. Adsorption primarily relies on the van der Waals forces that exist between CO_2 molecules and the adsorbent, with the processes of adsorption and desorption of CO_2 occurring under varying conditions, including temperature and

DOI: 10.1201/9781003476191-5

pressure [4]. To achieve efficient CO_2 adsorption, several critical parameters must be met. These include the maximum achievable CO_2 uptake, high selectivity over other gases, operational capability in humid environments, rapid adsorption kinetics, cost-effectiveness in production, and minimal energy consumption during both adsorption and regeneration processes. Furthermore, structural properties like surface area and porosity, and also molecular characteristics like surface functionalities, are of great importance in synthesizing a suitable adsorbent for CO_2 capture [5]. It is of great importance to choose a proper high-efficient low-cost adsorbent for the industrial CO_2 sequestration process. Accordingly, various porous solid adsorbents such as carbon-based structures [6–9], silicone-based adsorbents [1], clays [10, 11], and polymers [12] have been expanded and are regarded as viable options for sustainable CO_2 capture systems.

Utilizing nanostructured materials as adsorbents can significantly enhance the economic efficiency of the CO_2 capture process [13]. Nanostructured adsorbents exhibit distinctive characteristics that arise from their nanoscale dimensions, including high surface area-to-volume ratios, customizable surface functionalities, and varied pore architectures. These properties endow them with superior CO_2 adsorption capability, enhanced selectivity, and rapid kinetics [14]. The process of CO_2 adsorption can occur either via physisorption through various non-covalent interactions or chemisorption through covalent bonds between the CO_2 molecule and the sorbent [5]. The CO_2 molecule functions as a Lewis acid, which enables it to accept electron pairs from a base. Consequently, an effective CO_2 adsorbent material must possess multiple basic sites on its surface [1]. Furthermore, taking into account the kinetic diameters of different gases, that is, 3.64 Å of N_2, 3.3 Å of CO_2, and 3.8 Å of CH_4, adjusting the pore size and structure of porous nanoadsorbents can significantly enhance the overall capture of CO_2, and it helps design a more selective adsorbent by fine-tuning the pores. Microporous and ultra-microporous structures, characterized by pore sizes less than 0.7 nm, are favored for their strong affinity for CO_2 [5].

This chapter presents an extensive examination of the different categories of nanostructured CO_2 adsorbents, including carbon-based nanoadsorbents, silica-based nanoadsorbents, nanofibers, nanoclays, organic and metal-organic frameworks (MOFs), as well as sol–gel derived mesoporous materials, while detailing their specific characteristics. Different types of nanoadsorbents are presented in Figure 5.1. Furthermore, this chapter explores the latest investigations in each nanostructured adsorbent for CO_2 capture.

5.2 CARBON-BASED NANOADSORBENTS

Carbon-based adsorbents stand out as highly promising candidates for CO_2 capture among various materials, owing to their abundant availability, cost efficiency, substantial surface area, ease of pore structure modification, significant adsorption capacity, low energy requirements for regeneration, hydrophobic properties, and potential for surface functionalization [13]. Pyrogenic carbon materials, such as carbonized biomass, biochar, and charcoal, are particularly effective for carbon capture because of their porous nature. The porosity of pyrogenic carbon depends on pyrolysis or carbonization temperature. Pyrolysis temperature increase leads to

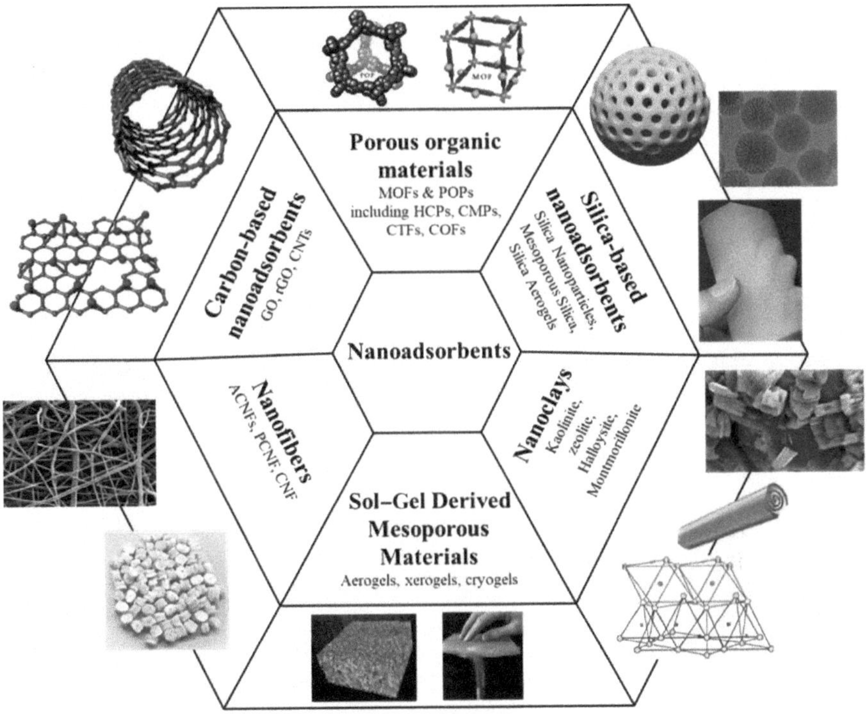

FIGURE 5.1 Different types of nanoadsorbents.

surface area and volume decrease and thus enhanced CO_2 selectivity [15]. Moreover, self-assembly, polymerization, or carbonization can be used for manufacturing carbon nanotubes (CNTs), graphene, nanofibers, fullerenes, and so on [2, 16]. One of the strategies to manufacture porous carbon with controlled porosity is the nanocasting or templating technique, where a porous carbon framework is formed by removing a template. In most cases, enhancing the pore structure necessitates post-synthesis treatments, such as chemical or physical activation, to create an effective porous carbon structure. However, these processes are often energy-intensive and time-consuming. Nanostructured carbon materials were developed through a nanocasting technique, followed by carbonization at 700°C. The temperature played a significant role in determining the textural and elemental characteristics, resulting in a maximum CO_2 uptake of 0.79 mmol/g [13].

To improve the CO_2 interaction with the adsorbent surface, the surface of different adsorbents was modified. One prevalent method of functionalization involves the incorporation of nitrogen functional groups into the carbon matrix. CO_2, as a weak Lewis acid, requires surface modification with amines and metal oxides [2]. Grafting amine groups by impregnation of amine solutions or direct incorporation of nitrogen into the carbon matrix are considered proper approaches for this purpose. The CO_2 uptake is also related to pore volume at different pressure levels. Hierarchical carbon materials were synthesized for CO_2 capture utilizing SBA-15 as a rigid silica template. The resulting specific surface area (SSA) measured 2200 m²/g, accompanied by a total pore volume of 2 cm³/g. The CO_2 adsorption capacity at 308 K and pressures reaching 10 bar attained a peak value of 7 mmol/g, which is attributed to the volumes of both micropores and mesopores. Conversely, at lower pressures, specifically up to 1 bar, the adsorption capacity is predominantly influenced by the micropores [17]. Among different carbon-based materials, graphene and CNTs have shown advantages over other adsorbent materials and are basically used for CO_2 adsorption applications.

5.2.1 Carbon Nanotubes

CNTs have gained recognition as effective nanomaterials for the capture of CO_2, attributed to their distinctive characteristics, including a high surface area, adjustable pore size, and remarkable adsorption capabilities. The various forms of CNTs that are applicable for CO_2 capture include single-walled carbon nanotubes (SWCNTs) and multi-walled carbon nanotubes (MWCNTs). SWCNTs consist of a single layer of graphene rolled into a cylinder, whereas MWCNTs are composed of multiple layers of graphene cylinders nested within one another. This structural variation endows each type of CNT with unique properties that can be harnessed to enhance CO_2 absorption in carbon-based composite materials.

In a study by Cortes-Suarez et al., the SWCNT@HKUST-1 Composite exhibited an impressive CO_2 adsorption capacity of 8.75 mmol/g at 196 K under pressures of up to 100 kPa, highlighting its potential for industrial applications [18]. A comparative analysis of the CO_2 adsorption capacities of SWCNTs and MWCNTs demonstrated that the SWCNTs achieved a CO_2 adsorption capacity of 29.97 g CO_2/kg of adsorbent, representing a 150% enhancement compared to MWCNTs, which had a capacity of 12.09 g CO_2/kg of adsorbent [19].

While MWCNTs may have lower SSA compared to SWCNTs, it is reported in some studies that their hierarchical pore structure enhances CO_2 adsorption capacity by providing additional adsorption sites. Preparing composites from MWCNTs leads to more developed hierarchical porosities. A multi-walled carbon nanotube/carbon foam (MCF) nanocomposite characterized by hierarchical porosity is reported. The material exhibits ultra-microporosity with pore dimensions ranging from 0.50 to 0.80 nm, alongside mesopores averaging approximately 3.70–3.90 nm, which were derived from MWCNTs. The presence of micropores with diameters below 0.80 nm significantly contributes to an enhanced CO_2 adsorption capacity of 4.58 mmol/g. Figure 5.2a depicts the SEM images of hierarchically porous MCF nanocomposites [20]. The hierarchically porous structure provides a combination of macropores, mesopores, and micropores, which can boost gas diffusion and enhance micropores' utilization. Exploiting CNTs in composites can make hierarchically porous structures by building a 3D framework for porous carbon. A porous carbon/CNT composite was doped with N and O, and a material with a porous hierarchical structure and numerous heteroatoms was developed that showed 5.7 and 3.7 mmol/g CO_2 uptake at 1 bar at 273 and 298 K, respectively. The preparation process of this porous CNT composite doped with N and O is schematically presented in Figure 5.2b [21]. In another work, hierarchical Ti_3C_2Tx MXene/CNTs were improved and a promoted CO_2 adsorption capacity was observed [22].

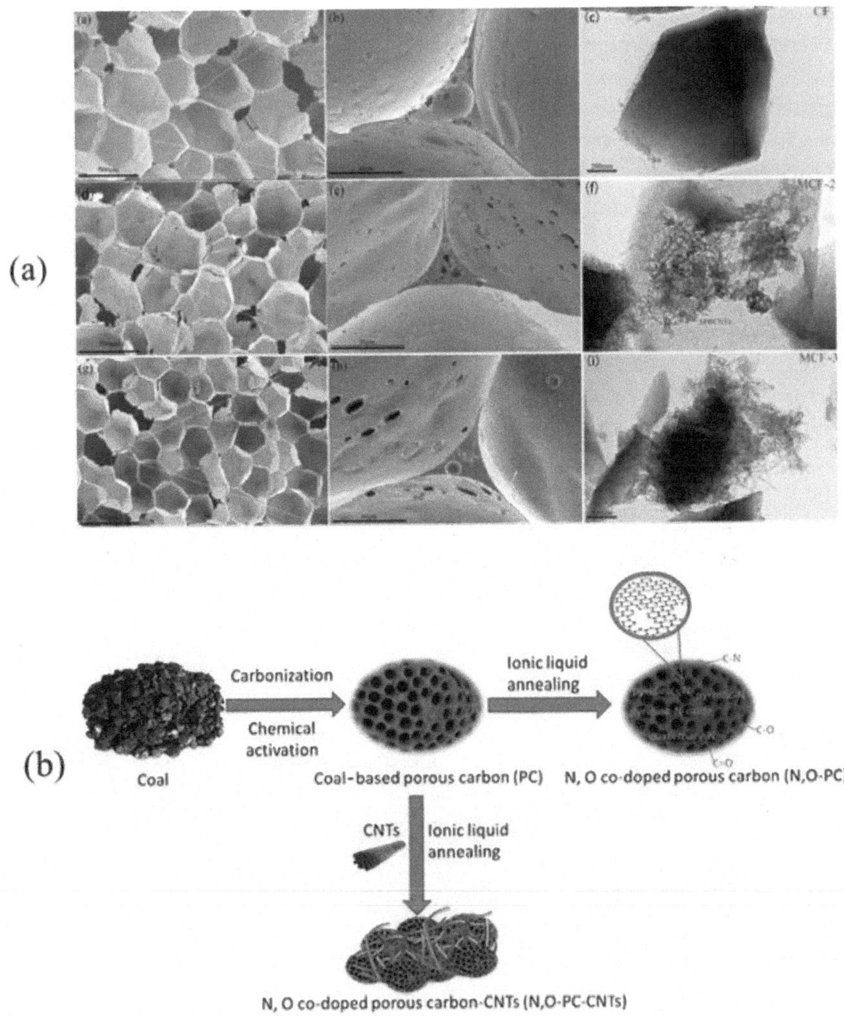

FIGURE 5.2 (a) SEM images of multi-walled carbon nanotube/carbon foam (MCF) nanocomposite with hierarchical porosity (Adapted with permission from [20], copyright (2021), Elsevier). (b) The preparation process of porous carbon/carbon nanotube composite doped with N and O (Adapted with permission from [21], under the Creative Commons Attribution-NonCommercial license (CC BY-NC)).

5.2.2 Graphene

Graphene, a two-dimensional porous carbonaceous nanomaterial, has garnered significant interest for CO_2 capture applications owing to its exceptional mechanical strength, the porosity in its structure, high surface area (theoretical: 2600 m^2/g) [23], tunable surface chemistry, stability, and cost-effectiveness [24]. The graphene-based materials have been examined for adsorption of gas in several researches.

The typical surface area of regular graphene materials is generally below 200 m^2 g^{-1}, which is significantly lower than the theoretical maximum. Graphene oxide (GO) and reduced graphene oxides (rGOs), which are prevalent graphene-based materials, exhibit unique adsorption characteristics for CO_2 capture. Numerous studies have focused on enhancing the surface area of graphene. The development of 3D macro- and microporous graphene, featuring an improved surface area and an optimal pore size of 0.7 nm for CO_2 adsorption, facilitates effective mass transport. Suitable adsorbents for CO_2 removal should have a high density of accessible micropores. 3D graphene with a surface-microporous architecture is applicable for CO_2 capture. This surface-microporous 3D graphene, synthesized directly from CO_2, demonstrated an adsorption capacity of 2.28 mmol g^{-1} at 298 K and 1 bar. The surface micropores play a crucial role in the mass diffusion process and serve as adsorption sites for CO_2. Additionally, KOH activation further enhanced the adsorption capacity by 37.28%, increasing it to 3.13 mmol g^{-1} due to the augmented surface area and the presence of oxygen functional groups. SEM images of 3D graphene before and after activation with KOH are presented in Figure 5.3a and Figure 5.3b, respectively. The significant surface tension of molten KOH results in a mesoporous configuration, which contributes to the disintegration of the large three-dimensional graphene sheets into smaller fragments. The influence of the

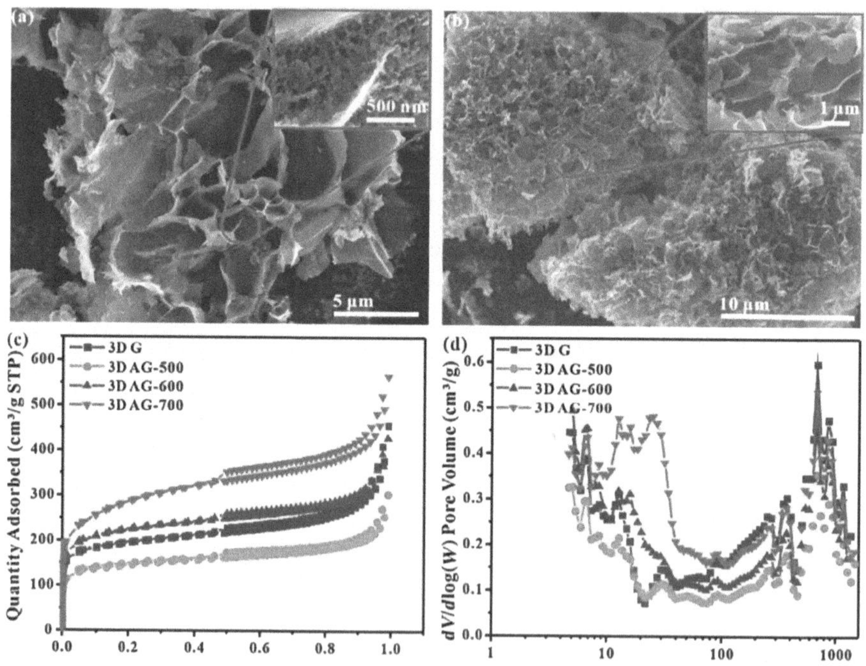

FIGURE 5.3 The SEM images of (a) 3D G and (b) 3D AG-600 (3D G activated by KOH at $T = 600°C$), (c) N_2 adsorption/desorption isotherms (at $T = 77$ K), and (d) curves of the pore size distribution of graphene samples. (Adapted with permission from [23], copyright (2020), Elsevier).

KOH treatment temperature on the surface area of three-dimensional graphene is illustrated through nitrogen adsorption measurements (Figure 5.3c). As shown in Figure 5.3d, the distribution of pore diameters in 3D graphene predominantly falls within the range of 0.5 to 68 nm, thereby ensuring adequate and enhanced accessibility of micropores to CO_2 [23].

5.2.3 Graphene Oxide

GO is a hexagonal structure obtained from graphite by oxidation and subsequent exfoliation, demonstrating a combination of sp^3 and sp^2 hybridization. The rich surface chemistry of GO that entails different oxygen functional groups (OFGs), such as carboxyl, hydroxyl, and epoxy, makes it very appealing to researchers for CO_2 CCS; however, in real-case applications it has commercial limitations due to the highly stacked layers, which can be controlled by increasing the interlayer space (d-spacing) [24]. Thus, as a solution for this drawback, multiple functionalization approaches have been manipulated upon GO for CO_2 adsorption. UV treatment was used as a physical technique to enhance CO_2 capture performance and is considered scalable for industrial applications. The CO_2 adsorption of graphene oxide foam irradiated by UV irradiation (UV-GOF) was studied, and it was found to impact adsorption performance by exhibiting higher CO_2/N_2 selectivity by 30 times and CO_2 capacity of 7 times higher at 100 mbar upon 5-h UV treatment. More CO_2 (40% more) was recovered from the UV-treated GOF at 25°C [24]. The procedure of UV treatment of GO foam and its effect on CO_2 adsorption is depicted in Figure 5.4.

GO can also be utilized as an additive to make composite materials for CO_2 capture. Composites of activated carbon (AC), which was achieved from polymers and metal-MOF with graphene oxide, which contained Cu, were synthesized and at ambient conditions demonstrated selective adsorption of CO_2 over N_2 (9.59 mmol/g) [9]. In another work, a composite of chitosan, graphene oxide, and zinc oxide (CTS/GO/ZnO) was developed as an efficient adsorbent for CO_2, and the optimal values for

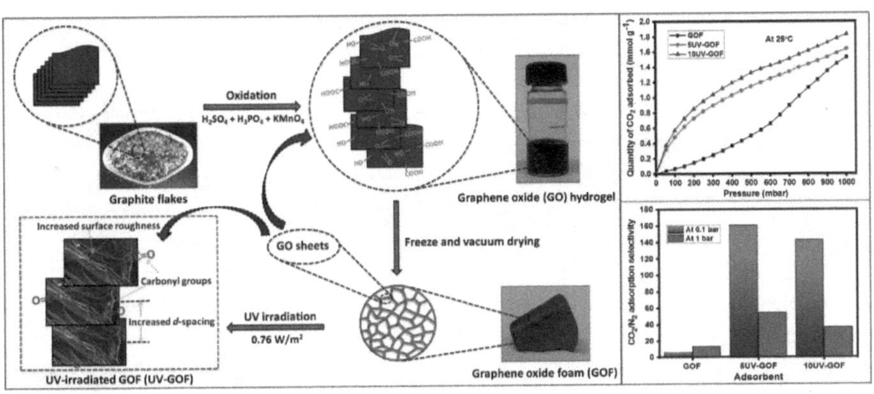

FIGURE 5.4 UV treatment effect of GO on CO_2 adsorption performance (Adapted with permission from [24], copyright (2020), Elsevier).

GO, ZnO, temperature, and pressure were determined to be approximately 23.8 wt%, 18.2 wt%, 30.1°C, and 8.6 bar, respectively. This was achieved through the design of experiment (DOE) utilizing the response surface methodology (RSM)-Box Behnken design (BBD) technique, resulting in a CO_2 uptake capacity of 470.43 mg/g [25]. A magnetic composite from GO was developed in [26], where Fe_3O_4 nanoparticles are integrated into GO to preserve and enhance the surface area of GO. The KOH@GO-Fe_3O_4 composite was evaluated for CO_2 adsorption, and the maximum amount of 6.812 mmol/g was recorded, achieved from 0.5 g of adsorbent at a temperature of 298 K and a pressure of 9 bar.

5.2.4 Reduced Graphene Oxide

Reduced graphene oxide (rGO) is obtained from GO reduction, leading to the restoration of sp^2 carbon bonds and removal of OFGs. While rGO has a high surface area like graphene, it exhibits improved electrical conductivity and enhanced adsorption capacity for CO_2 due to the restoration of the π-electron system. Meanwhile, laminar rGO helps CO_2 separation by providing transport channels [27]. Aerogels of rGO can be utilized also as a porous stabilizing support for sorption applications. The performance of mixed metal oxides of aerogel-supported MgAl (MgAl-MMO/rGO) for CO_2 adsorption was assessed under conditions of elevated temperature and high CO_2 pressure, which are pertinent to pre-combustion carbon capture and hydrogen production technologies. At a CO_2 pressure of 8 bar and a temperature of 300°C, the total CO_2 capacity of the rGO aerogel-supported nanoparticles achieves an adsorption level of 2.36 mmol/g, which is double that of the unsupported nanoparticles [28]. Besides, the incorporation of rGO in a composite material leads to surface and structural modifications such as SSA (SBET) and the total pore and micropore volume (V_{micro} and V_{total}), which shows an influence on the CO_2 adsorption performance. Different concentrations of rGO were used to make a composite of AC nanofibers based on polyacrylonitrile (rGO/ACNF), and, at a pressure of 15 bar, 10% rGO composite (rGO/ACNF0.1) showed 58 mmol/g CO_2 uptake with the highest SBET, V_{micro}, and V_{total} [29]. Furthermore, the synergic effect of exploiting rGO by using heteroatom doping can improve CO_2 adsorption capacity, which was proved in [30], where the porous carbon composites of rGO/N-doped was developed and at 298 K demonstrated a significant CO_2 uptake of 5.77 mmol/g.

A new approach to control CO_2 adsorption is using a switchable electric field, which has been proposed and demonstrated for penta-graphene (PG) nanosheets. CO_2 can be captured or released on PG by activating or deactivating the electric field. In an electric field over 0.025 a.u., the CO_2 physisorption changes to chemisorption [31], and the CO_2 molecule that has been adsorbed can be liberated on the deactivation of the electric field. Figure 5.5 illustrates the electron density distribution for the most stable adsorption configurations of CO_2, H_2, and CH_4 on the PG nanosheet, both in the presence and in the absence of an electric field.

The development of hierarchically porous structures in graphene-based nanoadsorbents, such as CNTs, can enhance CO_2 absorption at ambient temperature. This is achieved by incorporating mesoporosity and minimizing the crystallite size of conventional porous materials to the nanoscale. Graphene-based hierarchical materials,

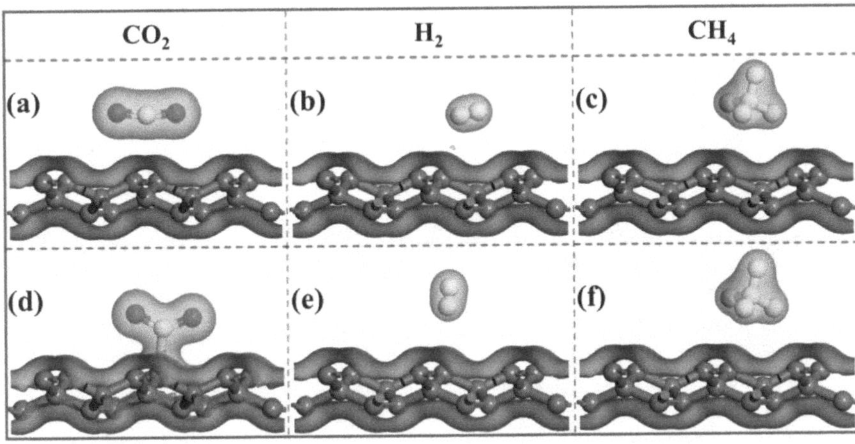

FIGURE 5.5 The distribution of electron density for the most stable adsorption configurations of CO_2, H_2, and CH_4 on the PG nanosheet (a–c) without the application of an electric field and (d–f) under an electric field of 0.030 a.u. (Adapted with permission from [31], copyright (2020), Elsevier.).

achieved through the incorporation of functional heteroatoms such as boron (B), nitrogen (N), phosphorus (P), and sulfur (S) into the graphene lattice, have shown improved affinity for CO_2 adsorption. In contrast, pristine graphene exhibits a limited capacity for CO_2 capture due to its scarcity of micropores and CO_2 trapping sites. The three-dimensional carbon-based material (B/N-CNs), which consists of B and N co-doped graphene-like carbon nanosheets featuring a high density of ultramicropores ranging from 0.47 to 0.9 nm and narrowed mesopores between 2 and 6.5 nm, has demonstrated superior performance in the selective capture of CO_2 [32].

5.3 SILICA-BASED NANOADSORBENTS

Silica-based nanoadsorbents represent a diverse classification of materials with tremendous capability for CO_2 capture applications. Nanoporous silica supports offer a high surface area and significant total pore volume, rendering them suitable for the adsorption of CO_2 gas. Additionally, SiO_2 porous materials facilitate straightforward surface modification, exhibit low energy consumption, possess substantial adsorption capacity, demonstrate good selectivity, and maintain tolerance to moisture [33]. Silica nanoparticles, mesoporous silica materials, and silica aerogels are different types of silica-based nanoadsorbents (Figure 5.6), which are explored in the following:

5.3.1 Silica Nanoparticles

Silica nanoparticles (SiO_2NPs) are nanosized particles with high surface area and adjustable size distribution of pores. These nanoparticles offer excellent adsorption properties and are suitable for various CO_2 capture processes, including

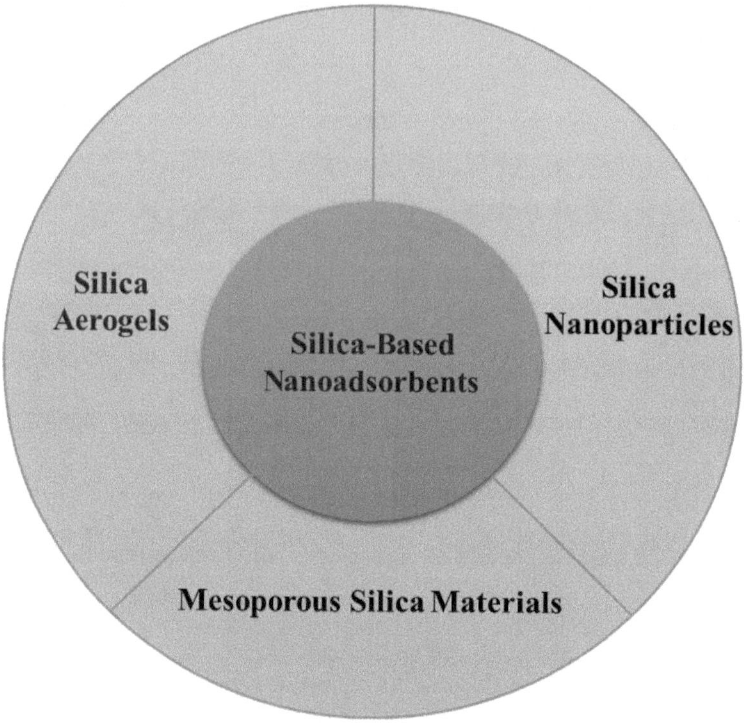

FIGURE 5.6 Schematic diagram of different types of silica-based nanoadsorbents.

post-combustion and pre-combustion capture. Their robust chemical properties and thermal stability of these materials render them suitable for high-temperature applications. Modifying the surface of SiO_2 NPs with functional chemical groups can influence CO_2 sequestration. Hybrid silica nanoparticles, modified with 3-aminopropyl (triethoxysilane) (APTES@SiO_2), demonstrated a significant improvement in CO_2 adsorption when compared to the unmodified material, achieving values of 0.57 and 0.80 mmol/g, respectively [33]. In another study, under both low and high CO_2 gas pressure conditions, standard APTES-functionalized SiO_2 nanoparticles demonstrated superior CO_2 adsorption capacities compared to dendritic and peptide-based nanoparticles that possess amine-polymer functionalities [34]. The amine-functioned mesoporous silica nanoparticles with a hierarchical pore size distribution and surface area of 133 m^2/g demonstrated a significantly improved capacity for CO_2 adsorption, achieving up to 80.5 mg/g at 75°C under conditions of 1 bar and 15 vol% CO_2. This modification also resulted in high selectivity and excellent cycling durability [35]. Silica nanoparticles with a hierarchical dendritic mesoporous structure and three-dimensional center-radial channels are another type of silica nanoparticles, which, owing to the high SSA, large pore volumes, and available internal spaces, have demonstrated that these methods hold high potential for the capture of CO_2. In comparison with conventional mesoporous silica materials, DMSNs have exhibited several superiorities like larger SSA, more open nanochannels, and more accessible internal space [36].

5.3.2 Mesoporous Silica Materials

Mesoporous silica materials, like MCM-41 and SBA-15, feature well-defined mesopores with diameters varying between 2 and 50 nm. These materials offer a large surface area and uniform pore structure, facilitating efficient CO_2 adsorption and desorption kinetics. Adjusting pore size and surface chemistry of mesoporous silica materials can be performed for specific CO_2 capture applications. SBA-15 (Santa Barbara amorphous-15), characterized by its hexagonally ordered structure, exhibited high selective adsorption toward CO_2. The cylindrical capillaries of this material, which are open on both sides, allow the free transit of the adsorptive molecules and the exit of the adsorbate molecules from the structure [1]. Microwave irradiation can facilitate the incorporation of a functionalizing agent for SBA-15, which leads to enhanced CO_2 adsorption. A microwave-assisted detemplating method was utilized for the synthesis of SBA-15 and then functionalized with loadings of polyethyleneimine (PEI), and an SSA of 1271 m^2/g with a pore size of 9.21 nm and pore volume of 2.10 cm^3/g was attained with an optimal loading of 70 wt% PEI. This resulted in a 15% increase in CO_2 uptake, reaching 3.63 mmol CO_2/g (0.16 g/g), compared to the conventional calcined material produced in the furnace, attributed to notable enhancements in structural properties [37]. Not only the synthesis method but also the functionalizing agent can influence CO_2 adsorption capacity. This was evaluated in [38], where two macromolecular and small molecular amines were used for modifying the SBA-15 adsorbent. PEI and diethylenetriamine (DETA) were impregnated into SBA-15, and it was revealed that, in comparison with DETA, the modification of SBA-15 with PEI demonstrated enhanced performance in terms of thermal stability, adsorption capacity, and adsorption kinetics. Specifically, the SBA-15 modified with 50% PEI achieved a CO_2 uptake of 182.11 mg/g. Furthermore, it exhibited exceptional adsorption kinetics, attaining a peak adsorption rate of 350 mg/(g·min) and achieving 90% CO_2 adsorption capacity within 5 min.

Mesoporous silica material structures of MCM, exhibit remarkable characteristics including a high surface area (approximately 900–1300 m^2/g), a uniform hexagonal pore arrangement (with pore sizes ranging from approximately 2 to 4 nm), a substantial total pore volume (around 0.25–0.78 cm^3/g), and an abundance of silicon hydroxyl groups, making them promising candidates for CO_2 adsorption. However, they have low selectivity and adsorption capacity, primarily due to the lack of specific active sites for adsorption. Therefore, modifications are necessary to improve their affinity for CO_2. To this regard, selective MCM-41 (mesoporous silica materials) were fabricated by adding polydopamine to the structure (PDA@MCM), and the total pore volume and pore diameter were increased by 33.7% and 113.4%, showing 0.99 cm^3/g and 7.1 nm, respectively, compared to pristine MCM-41 [3]. In another study, mesoporous MCM-48 modified with cyclic amine (morpholine and 1-methylpiperazine) was developed, and for 1-methylpiperazin-modified MCM-48, high

CO_2 absorption capacity of 4.4 mmol/g was achieved [39]. To reduce the surface area and pore volume, carbon deposition with further acid washing has been exploited for MCM-41 modification, which is due to the mesopores' blockage by carbonaceous species. Aluminum- (Al-MCM-41) and silica-containing (Si-MCM-41) samples were prepared via a hydrothermal method at 150 °C within 24 h. This was followed by a post-synthesis treatment with carbon deposition by using commercial glucose, with the mass ratio of 1 g (support): 2.2 g (glucose): 3.5 g (water), and further 2 h washing with hydrofluoric acid (0.2 M) at 65 °C. The bimodal pore system structure, resulting from the post-synthesis treatment, significantly improves the CO_2 adsorption capacity in comparison to the as-synthesized mesoporous materials, with enhancements of approximately 66% for pure silica and 44% for the Al-containing MCM-41 sample [40]. The adsorption efficiency of amine-functionalized MCM-41 is influenced by the residual amount of surfactant, which can be increased by calcination temperature decrease. The maximum adsorption capacity of 5.495 mmol/g for PEI-MCM-41–100°C was achieved, suggesting that the adsorption capacity may be enhanced through the application of a surfactant [41].

A novel mesoporous CO_2 adsorbent nanocomposite, IMSiNTs-PEI (IMP), was synthesized from natural halloysite nanotubes (HNTs) and by impregnating mesoporous silica nanotubes (IMSiNTs) with PEI. Following the synthesis process, significant improvements were noted, with the SSA increasing approximately 7 times to 340.61 m^2/g and the pore volume rising about 2.5 times to 0.499 m^3/g, while the tubular structure of the halloysite was preserved. The adsorption capacity of IMP-30 PEI reached 7.84 mmol/g at a temperature of 20°C and a pressure of 9 bar, exhibiting excellent thermal stability, favorable reversibility, and consistent performance throughout the adsorption and desorption cycles [42]. In another study, artificial neural networks and RSM were employed for the modeling and optimization of CO_2 adsorption in halloysite adsorbents functionalized with PEI using a five-level, four-factor central composite design (CCD) to optimize the operational conditions for adsorption, specifically targeting a temperature range of 20–50°C, a pressure range of 1–9 bar, and a PEI concentration of 10–40 wt% [43].

5.3.3 Silica Aerogels

Aerogels, classified as synthetic porous materials, are produced by substituting the liquid component of a gel with a gas, all while preserving the original structure. Among various types of porous supports, silica aerogels stand out as highly porous materials characterized by low density, extensive surface area, and comparatively larger pore sizes, which facilitate gas diffusion [44]. A sol–gel method succeeded by supercritical drying is used for their synthesis, resulting in a three-dimensional (3D) and highly porous network. A schematic representation illustrating the process of spherical silica aerogel formation is shown in Figure 5.7. When the aqueous silica sol is introduced into a heated oil bath, it creates a liquid sphere. Due to its immiscibility with oil, the integrity of the spherical droplet is preserved, leading to the formation of spherical silica gel [44]. The various benefits of aerogels, including their low density, high porosity, and substantial surface area, present significant potential for their use in CO_2 capture applications. Silica aerogels show superior adsorption properties and can be modified for CO_2 capture by adjusting pore size and surface functionalization.

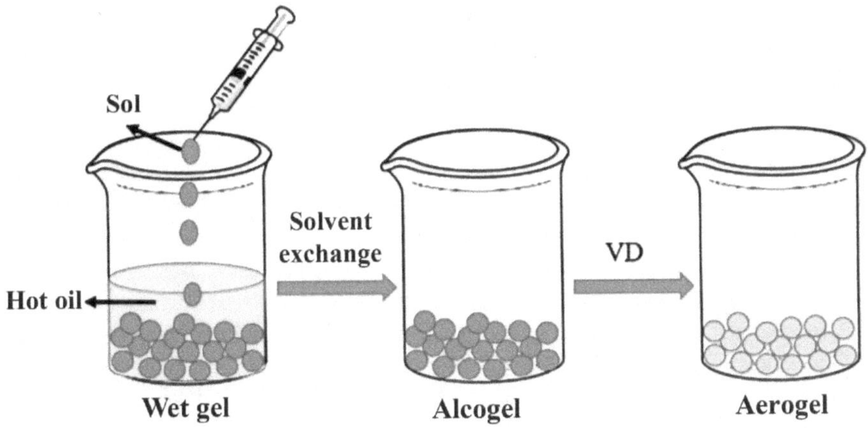

FIGURE 5.7 Schematic diagram of spherical silica aerogel formation. (Adapted with permission from [44], under the Creative Commons Attribution-NonCommercial license (CC BY-NC)).

To improve the capacity and selectivity of aerogels for CO_2 absorption, active materials including amines, potassium carbonate, and ionic liquids have been incorporated onto the surface of the aerogels [45]. Spherical amine-grafted silica aerogels for CO_2 capture were developed and amine-functionalized through the process of grafting APTES onto the silica gel framework. In the case of prolonged grafting duration, there was a reduction in both the SSA and pore volume due to the obstruction of pore spaces. An excessive loading of amine resulted in a diminished capacity for CO_2 adsorption. The maximum CO_2 adsorption capacity recorded was 1.56 mmol/g under dry 1% CO_2 conditions at 35°C. Furthermore, the capacity for CO_2 adsorption was influenced by both the temperature of adsorption and the surface content of amine [44]. The influence of functionalization by amine on CO_2 adsorption capacity was also depicted in [46], where amine-grafted SiO_2 aerogel (AMSA) was formed through an ambient pressure drying process and the effects of pH value and the concentration of 3-(aminopropyl) triethoxysilane (APTES) on the structure and adsorption property were investigated. The synthesized samples exhibited an amorphous nature characterized by a three-dimensional network structure, consisting of nanoparticles that demonstrated a CO_2 adsorption capacity of 3.37 mmol/g at 70°C. Furthermore, these samples maintained a stability of 87.6% after undergoing ten cycles of adsorption and desorption. The kinetic adsorption mechanism of the AMSAs was modeled using a double exponential approach. Additionally, the in-situ AMSA doped with Fe/Ti was developed, showcasing a significantly enhanced CO_2 adsorption capacity of 4.95 mmol/g at 70°C, along with a cyclic stability of 94.3% after ten adsorption–desorption cycles [47].

5.4 NANOFIBERS

Most of the materials used for CO_2 capture own a particle form and thus a difficult recycling process and limitations as gas pipe choking [48]. Nanofibers, as one-dimensional fiber-shaped nanomaterials, hold tremendous promise for CO_2 adsorption due to their high surface area, porosity, enhanced encapsulation efficiency, adjustable morphology, robust chemical and thermal stability, and tunable properties. Moreover, low gas transport resistance and high kinetics make them exclusive materials for CO_2 adsorption. Among different techniques such as template synthesis, heat-induced phase separation, self-assembly, and drawing, for continuous production of highly pure nanofibers with a high surface area, electrospinning is mostly preferred. Carbon nanofibers are suitable candidates for CO_2 adsorption as they exhibit good thermal stability and chemical inertness. Porous-AC nanofibers (ACNFs) were synthesized through facile electrospinning and further pyrolysis process and at 298 K attained CO_2 uptake of 60 cm^3/g. It is demonstrated that by increasing temperature, the CO_2 adsorption decreased [49]. Multi-aperture hierarchical structure of porous carbon nanofibers (PCNF) possessing ~0.71 nm average pore diameter was developed via a three-step procedure of electrospinning, washing, and carbonization. PVP was used as a template for the formation of pores and polyacrylonitrile (PAN) as a carbon nitrogen-rich precursor, the solution of PAN/polyvinylpyrrolidone (PVP) composite was electrospun, and then PVP was selectively removed from the fiber matrix of PAN/PVP, which resulted in highly irregular and significantly porous PAN nanofibers, which were subsequently carbonized to produce multi-aperture PCNF characterized by ultra-small pores, averaging approximately 0.71 nm in diameter. The synthesized PCNF demonstrated a high selectivity for CO_2 gas and achieved a CO_2 adsorption capacity of 3.11 mmol/g, maintaining long-term stability after undergoing 50 cycles of CO_2 adsorption and desorption. A schematic illustration of the CO_2 adsorption process for PCNF is presented in Figure 5.8 [48].

Cellulose nanofibers are another type of nanofibers which hold promise for CO_2 adsorption due to their unique properties like high aspect ratios, large SSA,

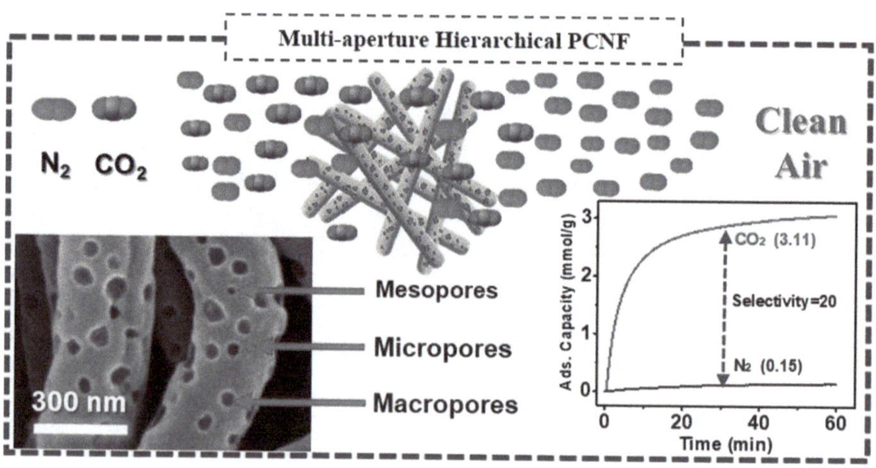

FIGURE 5.8 The schematic representation of CO_2 adsorption of multi-aperture porous hierarchical structure of carbon nanofibers. (Adapted with permission from [48], copyright (2020), Elsevier).

superior biodegradability, exceptional mechanical strength, environmental sustainability, decomposability, non-toxic properties, reduced weight, availability from various biobased sources or residues, and broad possibility of surface modification due to the abundant hydroxyl groups [50, 51]. The intrinsic hydrophilicity of CNF hinders its real application; thus, chemical modification techniques have been adapted to enhance their compatibility. The phthalimide was used as a coupling agent modifier for surface modification of CNF, which offered groups that showed good affinity toward CO_2. Consequently, while the SSA experienced an increase, there was a reduction in overall porosity, pore size, and pore volume [51].

To enhance the performance and properties of cellulose nanofibers, environmentally friendly inorganic particles, such as clays, have been integrated into the CNF matrix. Furthermore, the 2,2,6,6-tetramethylpiperidine-1-oxyl (TEMPO) radical oxidation of cellulose nanofibers can lead to transparent films. The TEMPO-oxidized cellulose nanofibers (TOCNF) were combined with cation- and anion-exchange organoclays, with poly(amido amine) dendrimers incorporated to improve gas adsorption functionality. Due to their distinctive morphology, dendrimers exhibit a significant capacity and selectivity for gas adsorption applications. The adsorption of CO_2 was found to increase with a higher concentration of organoclays in the TOCNF films. Notably, the hydrotalcite organoclay-TOCNF films demonstrated the highest CO_2 adsorption; however, they also exhibited a relatively high rate of CO_2 desorption compared to other systems. The CO_2 adsorption behavior on cationic and anionic dendrimer sites is different; in cation-exchange organoclay-TOCNF films, CO_2 gas is adsorbed and retained on cationic dendrimer sites, whereas in anion-exchange organoclay-TOCNF films, CO_2 gas is adsorbed on cationic clay sites and can be easily desorbed from the films. This difference is demonstrated in Figure 5.9 [52].

Chemical modification of cellulose nanofibril (CNF) films is an effective approach to produce CO_2 adsorbent materials. CNF films derived from different sources (corn husks, oat hulls, and kraft pulp) were obtained and surface-modified by diaminosilane *N*-[3-(trimethoxysilyl)propyl]ethylenediamine. This modification was validated through X-ray photoelectron spectroscopy and Fourier-transform infrared spectroscopy, which displayed a homogeneous coverage of the amino groups. The fabricated CNF films of corn husks, oat hulls, and kraft pulp with surface modification exhibited CO_2 adsorption of 0.90, 1.27, and 2.11 mmol CO_2 g^{-1}, respectively [53]. Bioaerogels can be synthesized from cellulose nanofibers as bio-CO_2 adsorbents. An efficient amination process was employed in a suspension titration method to create ultra-lightweight, spherical cellulose nanofiber aerogels with superior porosity exceeding 96.54% and a three-dimensional network structure achieved through freeze-drying of the spherical cellulose nanofiber hydrogel. The maximum CO_2 adsorption capacity recorded was 1.78 mmol/g, with a regeneration rate of more than 10 cycles [54].

5.5 NANOCLAYS

Clay minerals are available low-cost raw materials that are versatile for adsorption applications. However, carbon-based adsorbents, MOFs, silicas, and zeolites exhibit significantly higher sorption capacities, but, regarding operating costs, especially for

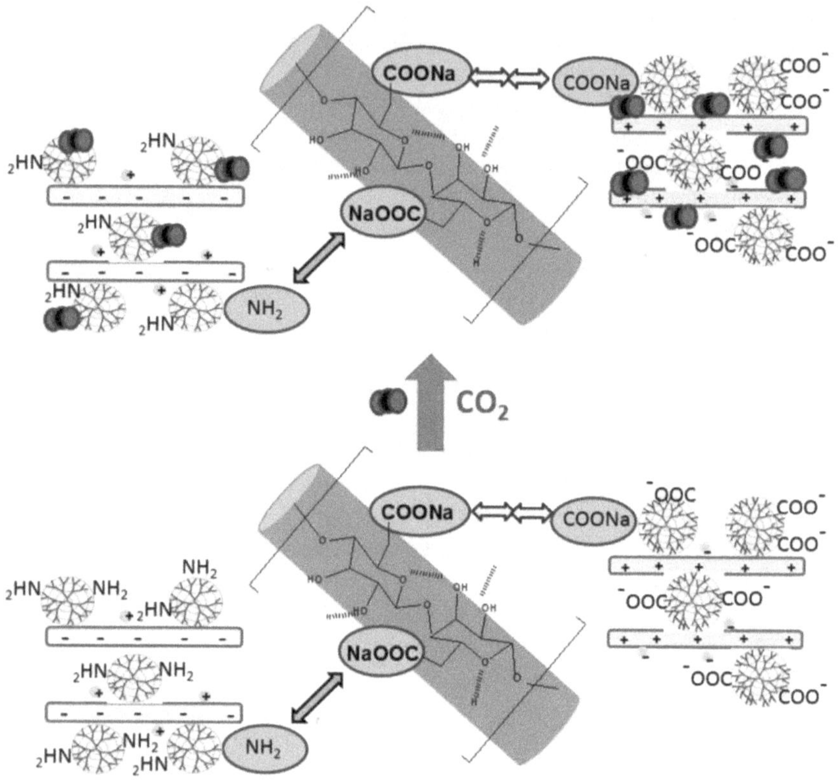

FIGURE 5.9 Schematic diagram of CO_2 and NH_3 gas adsorption on organoclay-TOCNF films (Adapted with permission from [52], copyright (2016), American Chemical Society).

applications where large amounts of adsorbents are used, clay minerals have shown to have 100 times lower costs [55]. Clay minerals consisted of tetrahedral and octahedral sheets, and, based on the ratio of these layers, they are classified as 1:1 or 2:1 type. In the 1:1 type (Kaolinite, etc.), one tetrahedral sheet is combined with one octahedral sheet, while in the 2:1 structure (Montmorillonite, etc.) an octahedral sheet is positioned between two tetrahedral sheets. The composition and arrangement of tetrahedral and octahedral sheets make up the chemical and physical characteristics of various clays, highlighting their intrinsic differences. CO_2 adsorption can take place on the outer surface (physical adsorption) or interlayer space of clay minerals (chemical adsorption), and the total surface area of the clay mineral can influence CO_2 adsorption capacity. The presence of hydrogenous species within the nanospace of a clay mineral creates sites for CO_2 adsorption, leading to the formation of HCO_3- through the reaction of CO_2 with hydroxyl groups on the clay surface, and thus further physical and chemical adsorption process reactions occur on the high-energy adsorption sites [4].

5.5.1 Kaolinite and Zeolite

Kaolinite, as a 1:1 clay mineral, a cheap abundant natural source and layered aluminosilicate clay mineral with the chemical formula of $Al_4Si_4O_{10}(OH)_8$, owing to its stability and environmental friendliness, is used in environmental applications [56, 57]; however, the use of kaolinite for CO_2 sequestration is constrained by its minimal SSA and the absence of interlayer domains [4]. Kaolinite showed a low affinity of ~0.106 mmol/g toward CO_2 [58]. Although kaolinite shows less CO_2 affinity in contrast to AC, as investigated in [59], it is still considered promising and has significantly influenced the amount of CO_2 adsorbed by revealing 12.98 mg CO_2/g capacity.

Calculations of density functional theory were also used to investigate the effects of element doping on the electronic structures and CO_2 capture performance of kaolinite and revealed that the CO_2 adsorption is influenced by the presence of –OH groups on the surfaces of kaolinite as well as the atomic radius of the dopants [57].

Kaolinite serves as an excellent precursor for zeolite production, and due to its favorable adsorption kinetics and substantial adsorption capacity, zeolites are regarded as effective adsorbents for CO_2 capture from post-combustion emissions [60]. The conditions under which zeolite is synthesized from kaolinite can significantly impact its CO_2 adsorption capacity. A type A zeolite was synthesized in alkaline solution through a hydrothermal method using metakaolinite, which is produced by calcining kaolinite at 600°C for 2 hours. The optimal synthesis conditions were determined to be 100°C for a duration of 48 hours. CO_2 adsorption tests conducted at 25°C and 1 bar revealed an adsorption capacity of 2.5 mmol/g for the synthesized zeolite [61].

4A-zeolite obtained from kaolin with surface modifications through impregnation with tetraethylenepentamine and diethanolamine were optimized for CO_2 adsorption capacity via RSM, resulting in a maximum adsorption capacity of 579.468 mg/g at a temperature of 25.270°C, a pressure of 8.870 bar, and an amine concentration of 11.112 wt% [62]. In another work, modification of zeolite 13X by a piperazine solution increased its adsorption capacity from 193.844 to 242.084 mg g^{-1} at 25°C and 8 bar, with a piperazine concentration of 2 wt%, an adsorbent amount of 1 g, and a particle size of 200 μm [63].

5.5.2 Halloysite

Among various clay minerals, halloysite is considered a material with remarkable sorption properties, and several studies are dedicated to investigating the CO_2 adsorption of halloysite nanotubes [64]. Halloysite, with a similar chemical composition to kaolinite, is a more suitable clay mineral for CO_2 adsorption applications in contrast to kaolinite, owing to the higher surface area and more hydroxyl groups on the surface, which is due to the intrinsic tubular morphology and the presence of bound interlayer water. The CO_2 sorption characteristics of halloysite have been very attractive when there is high amount required of adsorbents. Due to the availability and abundance of halloysite and the lack of synthesis challenges, compared to synthetic nanoadsorbents like zeolites and MOFs, halloysite is deliberated as a competitive natural nanoadsorbent. Figure 5.10 depicts the internal and external surface of hallysite and its potential to be used as a gas adsorbent.

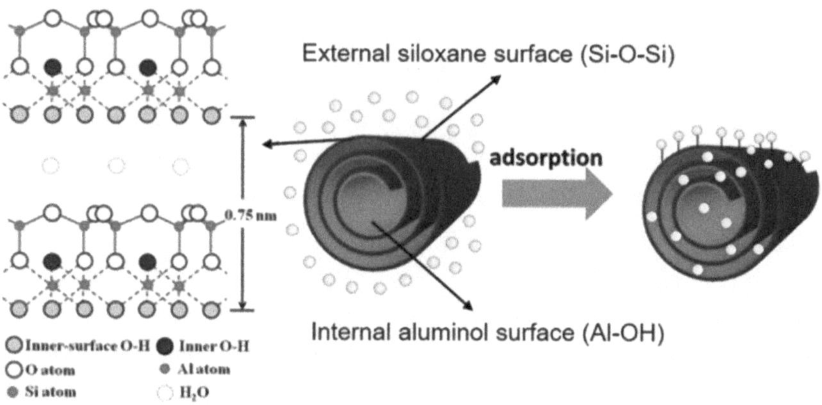

FIGURE 5.10 The halloysite surface structure and the schematic adsorption of a gas on its surface (Adapted with permission from [4], copyright (2022), Elsevier).

Calcination can influence the pore structure and sorption characteristics of halloysite. The pore structure and sorption properties of raw and calcined halloysite were evaluated. It is demonstrated that the surface area of halloysite mesopores and macropores, porosity, total pore volume within these pore types, and CO_2 sorption capacities were increased in halloysite calcined at 873 K. CO_2 sorption capacities for raw and calcined halloysite were measured at 0.54 and 0.63 mmol/g, respectively [55]. Likewise, the functionalization of HNTs increases CO_2 adsorption capacity. Vacuum loading and solvothermal reactions were utilized for the fabrication of HKUST-1@HNT composites. Etching the inner layers led to HNT modification to improve the loading efficiency and also amine functionalizing the outer surface for increasing the CO_2 adsorption capacity [65].

Ramadass et al. [66] synthesized nanoporous carbon flakes from halloysite nanotubes and investigated their CO_2 capture capacity. A comparison among pristine halloysite nanoclay, the halloysite nanoporous carbon (HNC), and the zinc chloride ($ZnCl_2$) activated halloysite nanoporous carbon (AHNC) revealed that HNC and AHNC were able to adsorb CO_2, but AHNC had a higher capacity (25.7 mmol/g at 0°C), nearly double that of HNC. The study also revealed that increasing temperature negatively affected CO_2 capture, leading to a decrease in adsorption capacity.

Additionally, halloysite serves as a cost-effective and abundant raw material for the large-scale production of X zeolites. Zeolite-X with a high CO_2 adsorption rate and adsorption selectivity was synthesized from natural mineral halloysite nanotubes through a process involving purification, calcination, and hydrothermal synthesis. This process was influenced by factors such as alkali concentration, calcination duration, and crystallization time and temperature. The optimal X zeolite was produced under conditions of high-temperature calcination at 650°C for 4 hours, followed by a reaction with 3 mol/L alkali at 88°C, resulting in a surface area of 440 m^2/g and a total pore volume of 0.26 cm^3/g. This zeolite demonstrated a significant capacity for CO_2 adsorption, measuring 4.3 mmol/g at a pressure of 113 kPa and a temperature of 303 K [67].

5.5.3 Montmorillonite

Montmorillonite, a natural and abundant 2:1 clay mineral, exhibits potential for application as a CO_2 adsorbent [68]. While natural montmorillonite demonstrates slightly superior performance compared to kaolinite, its overall adsorption capacity for CO_2 remains relatively low, approximately 0.114 mmol/g, which may be attributed to the interlayer capture mechanism [4].

The study of Roth et al. [69] confirmed this fact, when the untreated montmorillonite nanoclay did not show a significant CO_2 adsorption amount, and, on the contrary, the treated montmorillonite with aminopropyltrimethoxysilane and polyethylenimine was able to adsorb CO_2 by providing CO_2 capture sites. The pressure and temperature were found to be positively effective on CO_2 capture amount when removal efficiency was increased from 8% at 690 kPa to 11.5% at 2070 kPa and the highest removal efficiency of 7.5% was at 85°C for amine-modified montmorillonite nanoclay.

Acid treatment can also enhance the affinity of montmorillonite toward CO_2 capture due to the modified textural characteristics, increasing its surface area, and facilitating the formation of an amorphous phase. Montmorillonite was subjected to acid treatment for 3 h in an HCl solution to improve its physicochemical properties. Consequently, the SSA of the clay rose from 39 to 202 m^2/g, while the pore volume expanded from 0.05 to 0.31 cm^3/g, showcasing that the acid-treated clays had more efficiency for CO_2 adsorption than raw materials. The effect of pressure and temperature was evaluated and revealed that at low pressures both before and after acid activation, montmorillonite had low CO_2 affinity, and only with an increase in CO_2 pressure, the adsorption was enhanced. With the temperature increase, the CO_2 uptake decreases, especially at high pressures, proving that CO_2 adsorption is a physical process. Therefore, at 0°C the maximum adsorption was achieved, where 24 h-treated montmorillonite samples reached 33.2 and 67.4 mg/g at 1 and 5 bar, respectively. The treatment time is also effective, and it was shown that 24 h- and 8 h-treated montmorillonite had better affinity than 3 h-treated samples, despite their lower surface area [70].

The effect of montmorillonite nanoclay surface activation with HCl and modification with strontium hydroxide (SH) on CO_2 adsorption capacity was investigated, and it was revealed that mesopores–micropores during the surface modification with SH decrease gradually. By surface activation, the specifc surface area reached the highest amount of 144.35 m^2/g and the pore volume of 0.18 cm^3/g. At $T = 25°C$, $P = 9$ bar, the acid concentration of 4.03 mol/L, and SH concentration of 16.25 wt.%, an optimal adsorption capacity of 102.21 mg/g was obtained [71].

A green chemistry approach was implemented to functionalize montmorillonite using choline chloride: urea as a deep eutectic solvent, and the adsorption of CO_2, O_2, and N_2 was analyzed for comparison. The individual uptake of each gas revealed the highest amount for CO_2 with 208.6 mg/g and an exothermic ($\Delta H < 0$) adsorption nature with − 14.49 Kj/mol value, showing a proper selectivity over N_2 and O_2. The smaller kinetic diameter of CO_2 and the variations in critical temperature and quadrupole moment values facilitate a more efficient separation of CO_2 from porous media compared to N_2 and O_2. The adsorbent weight can also affect the adsorption capacity, and increasing the weight could decrease capacity, with a maximum adsorption value of 217 mg/g achieved at 35°C, 5 bar, and 0.5 g adsorbent [72].

The optimization of different parameters in the CO_2 adsorption process, including operational conditions, reactor temperature and pressure, acid concentration for surface activation, and the weight percentage of the modifying agent, is of great importance [71, 73]. The optimization of CO_2 adsorption using sodium-montmorillonite as an adsorbent was conducted through RSM in conjunction with CCD to study their interactions, thereby addressing the limitations of traditional methods that require independent investigation of each parameter through the study. The range of 25–65°C for temperature and 1–9 bar for pressure were the operating parameters for CO_2 adsorption. It was revealed that the variables influencing CO_2 adsorption capacity are temperature and pressure, with optimal conditions identified as 25°C and 9 bar, respectively. Under these conditions, an adsorption capacity of 100.67 mg/g was observed, which correlates well with the experimental value of 104.701 mg/g [73].

In another work, the same investigation was carried out to figure out the operational conditions and properties of modified montmorillonite nanoclay (MMT) treated with NaOH, aimed at enhancing the CO_2 adsorption capacity. The analysis of variance revealed that the weight percentage of NaOH, along with pressure and temperature, were the key parameters influencing CO_2 affinity. The CO_2 adsorption capacity of 105.55 mg/g was achieved under the optimum temperature of 65°C, pressure of 1.00 bar, HCl concentration of 5.99 M, and NaOH concentration of 39.76% [74].

5.6 POROUS ORGANIC MATERIALS

Porous organic materials, such as MOFs and porous organic polymers (POPs) including hypercrosslinked polymers (HCPs), conjugated microporous polymers (CMPs), covalent triazine-based frameworks (CTFs), and covalent organic frameworks (COFs), have garnered significant interest among researchers for applications in CO_2 adsorption. This interest is attributed to their inherent porous structures and strong affinity for CO_2. The characteristics of the pore structure and the functional properties of the pores in both organic and metal-organic systems are influenced by their constituent building blocks. The potential of different POPs and MOFs for CO_2 capture is further discussed. The porous organic materials are classified in Figure 5.11.

5.6.1 Porous Organic Polymers

POPs due to their significant SSA, uniform pore size distribution, structural tenability and diversity, chemical stability, and relatively low cost have been competitive candidates for CO_2 capture [5, 75]. Despite the amourphous structure of POPs, resulting from kinetic conditions, POPs have the advantage of simple and convenient synthesis strategies, leading to low production costs. Adding heteroatoms to the structure, pore engineering, and surface functionalization are the strategies to enhance the CO_2 affinity, resulting in highly microporous POPs with rich active sites for CO_2 capture. A wide variety of POPs can be synthesized using various precursors, such as aldehydes, amines, carboxylic acids, triazines, and melamine, through various synthesis methods, namely, solvothermal, polymerization, and condensation reactions.

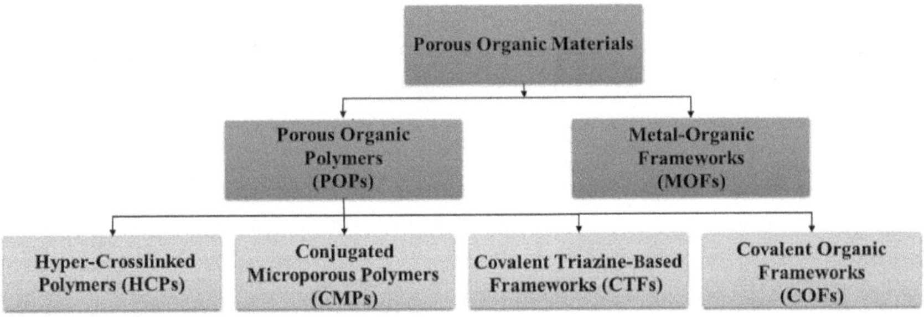

FIGURE 5.11 Classification of porous organic materials as adsorbents.

Heteroatoms such as nitrogen, oxygen, phosphorus, sulfur, or fluorine in porous materials, including POPs, enhance their surface properties, improving CO_2 affinity by dipole–quadrupole interactions, as CO_2 is a symmetric molecule with quadrupole moment [5].

POPs have been utilized to create porous carbon materials exhibiting enhanced porosity of 3367 m^2/g and pore volumes reaching 1.224 cm^3/g, with a CO_2 adsorption capacity of 7.78 mmol/g at 273 K and 1 bar, alongside a high CO_2/N_2 selectivity [76]. Melamine-based POPs have been specifically developed for CO_2 capture due to the abundant nitrogen content in melamine, which facilitates interactions with CO_2 through acid–base interactions and chelating coordination. Melamine has been used as a rigid crosslinker and then POPs with an SSA (SBET) of 645 m^2/g and a total pore volume (V_{total}) of 0.84 cm^3/g, characterized by hierarchical porosity, were synthesized based on Schiff-base chemistry, which showed an efficient CO_2 capture of 153 mg/g [75]. In another study, amino-functionalized porphyrin-based POPs were synthesized through Friedel–Crafts acylation and Schiff-base reactions, utilizing melamine as both a rigid cross-linking agent and an amino-functionalized component. These polymers exhibited an SBET of 587 m^2/g, a pore volume of 0.46 cm^3/g, and a CO_2 uptake of 172 mg/g at 273 K and 1.0 bar [77].

Different chemical routes and precursors for POP synthesis have terminated to different new structures including HCPs, CMPs, CTFs, and COFs.

5.6.1.1 Hypercrosslinked Polymers

HCPs were first identified in 1970 by Davankov as the initial microporous polymer materials characterized by nanosized pores. Although they are not a novel material, the increasing utilization of HCPs in CO_2 sequestration has rendered this porous organic substance particularly appealing to researchers in recent years. This interest is attributed to their diverse synthesis methods, high SSA, proper physical, chemical, and thermal stability, small pore size, low density, and favorable cost-effectiveness for optimal operational conditions. The Friedel–Crafts alkylation reaction is the most common method used for forming a highly crosslinked network, by providing fast kinetics and porous structure. This approach offers a cost-efficient, one-step method for the large-scale synthesis of HCPs, which is essential for gas storage applications [78, 79]. The main advantage of HCPs for capturing CO_2 is their high surface area, which is achieved through the interconnection of aromatic monomers and forms the polymer network. Moreover, the simple preparation method and the low-cost reagents with simple reaction conditions make HCPs competitive alternatives for traditional microporous adsorbents, like AC. The main proposed techniques for HCP synthesis are post-crosslinking of polymer precursors, one-step polycondensation of monomers, and knitting building blocks by external crosslinkers [78].

Nitrogen-rich functionalized HCPs have demonstrated enhanced adsorption capacity and better selectivity toward CO_2. Physical impregnation, chemical grafting, and direct synthesis of amine species in a single pot are followed by researchers. A highly effective carbazole-based HCP adsorbent was developed via a Friedel–Crafts reaction, followed by amine functionalization. Although the SSA decreased from 705.35 to 471.21 m^2/g due to the filling of pores with amines, this modification did not diminish the adsorption capacity. In fact, the CO_2 adsorption capacity increased from 182.13 to 236.37 mg/g at a temperature of 25°C and a pressure of 5 bar [79].

The application of DOEs and RSM in modelling the adsorption process and evaluating the interconnection of different parameters on CO_2 affinity is a hotspot in this field. The influence of factors such as pressure, temperature, synthesis duration, and the ratio of precursors on the CO_2 adsorption efficiency of benzene-based HCPs has been analyzed using RSM to enhance optimization outcomes. The optimal conditions identified were a pressure of 7.8 bar, a temperature of 294 K, a crosslinker to benzene ratio of 2.75, and a synthesis time of 13.6 hours, leading to an adsorption capacity of 262 mg/g [80]. In another work [81], an HCP adsorbent was developed through a Friedel–Crafts reaction utilizing carbazole as a monomer for CO_2 capture. RSM, employing a CCD, was utilized to assess the impacts of the crosslinker-to-monomer ratio, time, and catalyst type. It was depicted that the optimum adsorption capacity of 191.2 mg/g was achieved with a crosslinker to carbazole ratio of 2.7, a synthesis time of 11.2 hours, and $FeCl_3$ as the catalyst. Moreover, the optimization of synthesis parameters for maximizing the CO_2 adsorption capacity of an HCP derived from polystyrene was explored using DOE in conjunction with RSM. The optimal conditions for maximizing capacity were determined to be a crosslinker amount of 17.7 mmol, a synthesis duration of 14.6 hours, and the use of $FeCl_3$ as the catalyst [82].

5.6.1.2 Conjugated Microporous Polymers

CMPs represent a novel category of porous polymers with enormous potential for gas adsorption applications, providing high surface area and porosity, chemically reactive groups, chemical stability, and electrical conductivity. In addition, the structural diversity of π systems is another attractive advantage of CMPs [78, 83]. A wide range of CMPs have been developed by different chemical reactions, various building blocks, and synthesis methods. The porous structure of CMPs gives accessible sites for guest molecules and metal ions, to construct organic–inorganic hybrids. In addition, CMPs possess a conjugated framework, wherein the building components are covalently linked through a p-conjugated bond during the synthesis process [78].

Two distinct types of two-dimensional CMPs, namely Py-BSU and TBN-BSU CMPs, were synthesized using the Sonogashira cross-coupling reaction involving BSU-Br_2 (2,8-dibromothianthrene-5,5′,10,10′-tetraoxide) with Py-T (1,3,6,8-tetraethynylpyrene) and TBN-T (2,7,10,15-tetraethynyldibenzo[g,p]chrysene), respectively.

The TBN-BSU CMP demonstrated a CO_2 uptake capacity of 1.60 mmol/g, while the Py-BSU CMP exhibited a capacity of 1.45 mmol/g at 298 K and 1 bar [83]. In another work, two CMPs, SNCMP-1 and SNCMP-2, featuring multiple active sites, were synthesized through a palladium(II)-catalyzed coupling reaction. These polymers displayed remarkable CO_2 adsorption capacities, achieving uptakes of 3894 and 3878 mg/g under conditions of 70 bar and 323 K, respectively [84].

5.6.1.3 Covalent Triazine-Based Frameworks

The high nitrogen content, chemical stability, tunable pore structure, large surface area, rich nitrogen content, rapid modification, and abundant micropore structure of CTFs contribute to their increasing interest in a variety of applications, particularly in CO_2 capture. CTFs have a crystalline structure with nanoscale-ordered pores, and it helps to better accept guest molecules on accessible active sites than amorphous porous structures [78]. The SSA of CTFs typically remains below approximately 3000 m^2/g, which limits their broader application in CO_2 capture. Deng et al. introduced a novel 7,7,8,8-tetracyanoquinodimethane-derived covalent triazine framework exhibiting an SSA of up to 4000 m^2/g, achieving a notable CO_2 adsorption capacity of 5.99 mmol/g at 273 K under ambient pressure, in contrast with other reported results for CTFs. Also, by theoretical calculations, it was demonstrated that SSA and surface functionalization with nitrogen are effective for CO_2 affinity, but SSA contributes more CO_2 uptake [85].

The formation of the Lewis acid–base pairs in CTFs has a synergistic effect on improving the reactivity of CO_2 with epoxy compounds. Pyridine-rich CTFs were synthesized using ionic 1,3,5-*tris*(4-cyanopyridinium-1-ylmethyl)-benzene tribromide (TPM) as the monomer through an ionothermal trimerization reaction employing $ZnCl_2$ as both the catalyst and reaction medium. The resulting CTF-TPMs are characterized by a rich presence of acid–base ion pairs, which are advantageous for CO_2 capture and conversion. The porosity and ionic pore characteristics of CTFs were adjusted by controlling the polymerization temperature. CTF-TPMs exhibited strong electrostatic interactions with CO_2 attributed to their high SSA, abundant pyridine cation sites, and the presence of free Cl– ions [86]. A nitrogen-rich amine-functionalized CTF was synthesized from cyanuric chloride and 1,4-phenylenediamine through catalyst-free condensation polymerization for CO_2/CH_4 adsorption. The amine functionalization resulted in a 41.9% increase in CO_2 adsorption capacity and a 3.43% improvement in CO_2/CH_4 selectivity [87].

5.6.1.4 Covalent Organic Frameworks

COFs represent a class of porous, large crystalline structures composed of organic building blocks and light elements (O, B, N, H, C) with strong covalent bonds. They hold tremendous potential for the adsorption of carbon dioxide (CO_2) in substantial amounts with excellent selectivity, which is owing to their stable porous structures, large surface areas of up to 6450 m^2/g, and high thermal stability (up to 600°C). They have the lowest density among other organic materials. The ability to engineer their pore structures through various synthesis methods and a diverse range of building blocks makes COFs exceptional materials for removing different target contaminants by pre-designing their structure [78]. COFs are recognized as promising candidates for gas adsorption, as they make a highly porous solid and consist of lightweight components bound together by covalent bonds. The molecular components in COFs are arranged into periodic networks via covalent bonds, and therefore their structure is predictable.

The COFs' adsorption capacity is affected by different factors like the molecular building components. Interactions with amines can enhance the selective adsorbance of CO_2, facilitating its capture from exhaust emissions. A novel COF, designated as $Me_3TFB\text{-}(NH_2)_2BD$, was synthesized by dynamic linker exchange. Despite the linker exchange, the porosity of the COF was largely preserved, yielding a high Brunauer–Emmett–Teller surface area of 1624 ± 89 m²/g. The CO_2 adsorption at 273 and 295 K were measured at 1.12 ± 0.26 and 0.72 ± 0.07 mmol/g, respectively, at a pressure of 1 bar. The imine bonds with primary amine substituents or benzimidazole bonds can be formed in theory [88] (Figure 5.12).

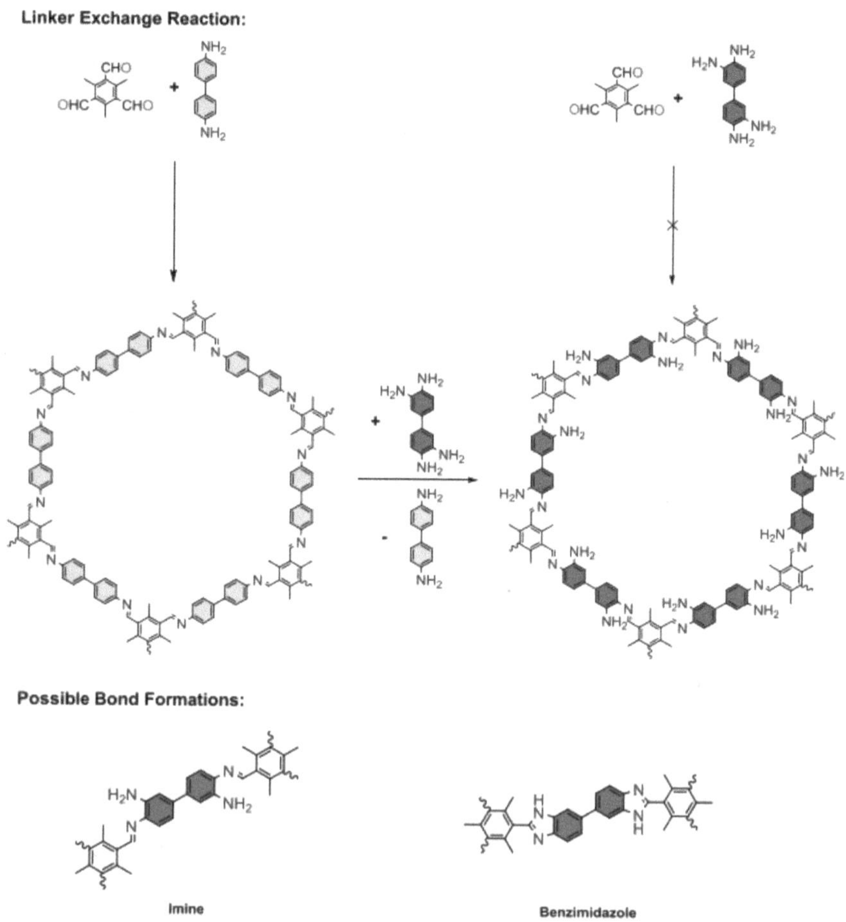

FIGURE 5.12 Schematic overview of the linker exchange with (NH_2)2BD to form Me_3TFB-(NH_2)2BD from Me_3TFB-BD COF (Adapted with permission from [88], under the Creative Commons Attribution (CC BY 4.0)).

Reactive aliphatic amine species were incorporated into COFs through a systematic process. Initially, an imine-linked COF, designated as COF-609-Im, was crystallized. This was followed by the transformation of its imine linkage and the subsequent covalent incorporation of tris(3-aminopropyl)amine into the framework. The resulting COF-609 demonstrated a remarkable 1360-fold increase in CO_2 adsorption capacity, along with a 29% improvement in performance under humid conditions, in comparison to the original framework [89]. Additionally, amino-functionalized ionic-liquid-grafted COFs were synthesized, and it was shown that, by the polar groups (amino groups) and its microporous nature, the obtained composite exhibited higher CO_2 capture activity (117.4 mg/g) than non-modified (62.6 mg/g) [90]. The pore size of COFs is also a very decisive factor in CO_2 capture ability. COFs with narrow pores (the largest cavity diameter < 15 Å) and low porosities ($\phi < 0.75$) are more effective adsorbents for the selective separation of CO_2 from H_2, based on the performance analysis of the structural properties [91].

5.6.2 Metal-Organic Frameworks

MOFs are characterized by their three-dimensional structure, uniform pore distribution, and a network of channels, consisting of two primary components: metal ion clusters and organic molecular linkers. This configuration allows MOFs to reversibly adsorb guest molecules [92]. With their excellent structural stability and high porosity, evidenced by surface areas reaching approximately 2900 $m^2 \cdot g^{-1}$ and total pore volumes ranging from 1 to 2 $cm^3 \cdot g^{-1}$, along with adjustable pore structures and multifunctional capabilities, MOFs have shown enormous potential, and sometimes they are often regarded as highly promising adsorbents for gas adsorption, particularly in CO_2 capture applications [93, 94]. However, they have some drawbacks such as high synthesis cost, low selectivity for CO_2—especially in the presence of other gases, and the adsorption capacity decrease under humid conditions [1]. MOF's surface chemistry, pore size, acidity, and surface modifications can significantly enhance their capacity for CO_2 adsorption. One effective strategy for improving CO_2 uptake is the functionalization of MOFs. A series of MOF-808 structures were synthesized using 11 different amino acids for this purpose, with glycine and dl-lysine-functionalized MOF-808 (designated as MOF-808-Gly and dl-Lys) demonstrating the highest adsorption capacities Additionally, a novel adsorbent for CO_2 capture was developed, featuring a greater SSA and pore volume than the unmodified MOF-808, while maintaining similar mesoporous morphology and exhibited optimal operating conditions at 25°C, 9 bar, and 20 wt% NH_2 incorporated into the MOF structure, resulting in a maximum equilibrium adsorption capacity of 369.11 mg/g. The molecular structure of the MOF-808/NH_2 nanocomposite and its CO_2 adsorption mechanism are illustrated in Figure 5.13 [95].

The synthesis method can influence on properties of the achieved MOFs, especially in terms of size distribution, affecting the CO_2 capture capacity. Sonochemical synthesis of MOFs due to the fast nucleation and growth of MOFs, providing narrow size distribution and offering highly pure crystallines with minimal impurities, has been an effective approach. This method utilizes high-frequency ultrasonic waves to induce cavitation, creating bubbles that contain metal and organic precursors.

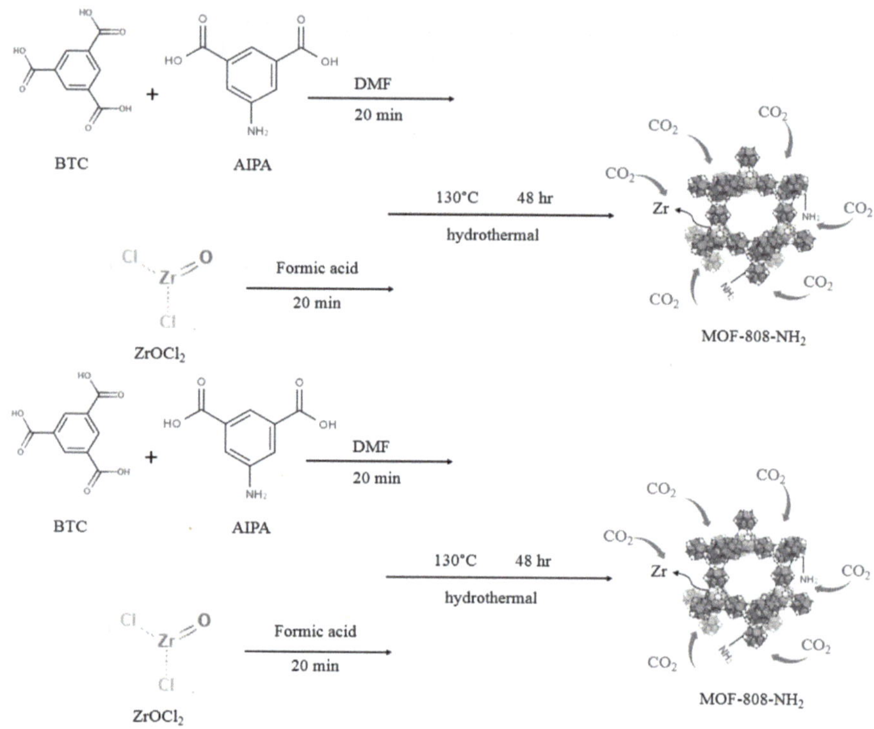

FIGURE 5.13 The schematic molecular structure of MOF-808/NH_2 nanocomposite and its CO_2 adsorption mechanism (Adapted with permission from [95], under the Creative Commons Attribution (CC BY 4.0)).

By collapsing these bubbles, high pressure and temperature are generated for the formation of crystalline MOFs. Kazemi et al. used a sonochemical method for the synthesis of UiO-66-NH_2 in 1 hour, achieving a notable increase in CO_2 adsorption capacity (3.2 mmol/g), in comparison with the solvothermal technique, which yielded 2.3 mmol/g at 25°C and 1 bar [96].

The CO_2 adsorption capacity of MOFs has been also modeled by studying four different types: MOF-2 (Zn_2 $(BDC)_2$), MOF-74 (Zn_2 (DHBDC)), IRMOF-1 (Zn_4O $(BDC)_3$), and MOF-177 (Zn_4O $(BTB)_2$), showing different performances toward gas adsorption. The models used for this study were in good agreement with experimental data, and it was revealed that MOF-74 and MOF-2 were suitable for CO_2 adsorption [97].

5.7 SOL–GEL-DERIVED MESOPOROUS MATERIALS

Sol–gel-derived mesoporous materials such as aerogels, xerogels, and cryogels with remarkable characteristics such as low density, high surface area, high porosity, and adjustable surface chemistry are promising candidates for environmental remediation applications including CO_2 adsorption. Unlike aerogels, which are obtained by supercritical drying, xerogels are dried in ambient pressure, resulting in a denser structure with lower porosity, while cryogels are fabricated at subzero temperatures. Chitosan hydrogel microspheres were used as templates for the growth of MOFs, and as a result the MOF-type HKUST-1@chitosan aerogels, cryogels, and xerogels were fabricated through various drying techniques (supercritical CO_2, freeze-drying, and vacuum drying) and were tested for gas adsorption. The maximum CO_2 adsorption capacity recorded was 1.98 mmol·g^{-1} at 298 K and 1 bar, demonstrating favorable selectivity for CO_2 over CH_4 and Xe over Kr gases [98]. The CO_2 capacity of xerogels has a linear correlation with both micropore volume and SSA. Microporous carbon xerogels (CX-HMTAs) were fabricated with diverse morphologies and well-defined porosity. The CO_2 adsorption capacity of 4.21 mmol/g at 273 K and 1 bar along with a selectivity of 70 and high stability (99.4% after 1000 cycles) was obtained benefiting from reasonable pore size distribution [99].

Cellulose-based carbon aerogels (CAs) possess remarkable characteristics, that is, biocompatibility, biodegradability, non-toxicity, cost-effectiveness, thermal stability, and excellent sorption capabilities. However, in terms of CO_2 capture, the lack of active adsorption sites entails modifying the structure by functionalizing agents, especially N-containing functional groups, which provide an enhanced interaction between the adsorptive material and CO_2, regulate the microstructure preserving the fundamental characteristics of carbon materials, and elevate the adsorption efficiency of carbon-based porous materials. The cellulose-derived nitrogen-doped carbon aerogel was synthesized from cellulose by dissolving it in a mixed solution of urea and NaOH, which influenced the formation and microstructure of CAs. Subsequent processes, including hydrothermal polymerization, freeze-drying, carbonization, and activation, resulted in a well-developed porous structure with a rich presence of nitrogen functional groups. The adsorption capacity reached 3.65 mmol CO_2/g with a selectivity of 19.69 for simulated flue gas comprising 15 vol% CO_2 and 85 vol% N_2 [100].

Urea plays a dual role in carbon cryogel porous framework, that is, as a porogen and as a heteroatom dopant. The introduction of heteroatoms significantly influences the textural characteristics and CO_2 adsorption performance of the carbon cryogels. Heteroatom (N)-doped porous carbon cryogels, characterized by a high density of fine micropores, an SSA of 2084 m^2/g, and a substantial micropore volume of 1.1806 cm^3/g were fabricated by modification with urea and CO_2 adsorption of 280.57 mg/g at 273 K and 1 bar with CO_2/N_2 selectivity (up to 115) was achieved [101].

5.8 FUTURE DIRECTIONS AND CHALLENGES

While nanostructured materials show great promise for CO_2 capture, challenges such as industrial scalability, chemical and physical stability, and cost-effectiveness of adsorbents still need to be addressed. Future research efforts may focus on developing scalable synthesis methods, enhancing stability under harsh operating conditions, and optimizing the cost-effectiveness of nanostructured-based CO_2 capture adsorbents. In the following, there are presented some directions that need to be followed by researchers in the future:

1. Future investigation efforts should be focused on improving the performance of nanostructured CO_2 adsorbents by optimizing their structural and chemical properties. This includes exploring novel synthesis methods, surface functionalization techniques, and pore engineering strategies to further increase CO_2 adsorption capacity, selectivity, and stability.
2. Addressing the cost-effectiveness of nanostructured CO_2 adsorbents is an important issue for their industrial-scale CO_2 capture applications. Research in this field should aim to develop synthesis methods that are scalable and require cost-effective precursors by maintaining their high performance.
3. Regeneration of nanostructured CO_2 adsorbents can reduce operational costs and also environmental impact due to reusing the adsorbents. Improving the reusability and long-term stability of adsorbent materials, through investigating regeneration methods can be an important direction for future research.
4. Commercialization of nanostructured CO_2 capture technologies necessitates proper scaling up from laboratory scale to industrial scale. This important issue can be effectively addressed by process engineering, equipment design, and integration with existing infrastructure.
5. Considering the environmental impact of nanostructured CO_2 adsorbents is very crucial, as in nanoscale many materials show toxicity and can affect the lives of humans and other living beings. Research can be focused on potential safe methods for the synthesis of nanostructured adsorbents like green approaches. Reducing energy consumption and material toxicity and waste management are other challenges to be addressed in this field.
6. Integrating additional functionalities in nanostructured CO_2 adsorbents, like catalytic activity and electrochemical or photothermal properties, can be an interesting direction for future research in this field. Multifunctionality could introduce new advanced materials for various applications with improved performance and synergistic effects.

In summary, although nanostructured CO_2 adsorbents hold great promise for mitigating climate change and reducing greenhouse gas emissions, numerous challenges need to be addressed to achieve their full potential such as improving adsorption selectivity, enhancing stability under harsh operating conditions, and reducing energy consumption during regeneration. By addressing these challenges and proceeding with research in this area, the nanostructured CO_2 adsorption technologies can play a vital role in a more sustainable future. Future research efforts may focus on developing advanced CO_2 nanoadsorbents with tailored properties and improved performance for CO_2 capture applications.

5.9 CONCLUSIONS

In conclusion, in the quest for mitigating climate change by capturing anthropogenic CO_2 emissions, the development of nanostructured CO_2 adsorbents plays a pivotal role. Through superior properties such as adjustable surface functionalities, tunable pore structures, and high surface area-to-volume ratios, nanostructured materials offer superior CO_2 adsorption capabilities, enhanced selectivity, and rapid kinetics compared to traditional adsorbents. Carbon-based nanoadsorbents, silica-based nanoadsorbents, nanofibers, nanoclays, porous organic materials like MOFs and POPs, and sol–gel derived mesoporous materials have all demonstrated great potential for CO_2 capture applications. Nanostructured CO_2 adsorbents represent a versatile and promising class of materials for CO_2 adsorption, with each type offering unique properties and advantages. By exploring their characteristics and latest investigations, this chapter provides valuable insights into the diverse range of nanostructured CO_2 adsorbents and their suitability for sustainable CO_2 capture systems. However, further studies and analyses are essential to enhance the efficiency and economic viability of the nanostructured CO_2 adsorbents. For the development of nanostructured porous materials and improving the uptake capacity, similar efforts have been implemented for different adsorbents. Heteroatom doping, surface functionalization, optimization of operating conditions experimentally or by computational modelling, and using various synthetic procedures are the most common strategies for boosting CO_2 adsorption capacity in different nanostructured adsorbents. Moreover, many investigations are dedicated to the selective separation of CO_2 over other industrial flue gases, as in real-case applications there is mostly a combination of gases, which need to be separated. Furthermore, to overcome the limitations associated with each nanomaterial, it is essential to explore novel structures, introduce new functional groups, clarify the adsorption mechanisms, and develop hybrid systems and technologies.

REFERENCES

1. Medina-Juárez, O., et al., Optimal surface amino-functionalization following thermo-alkaline treatment of nanostructured silica adsorbents for enhanced CO_2 adsorption. *Materials*, 2016. 9(11): p. 898.
2. Kanjilal, B., et al., Chapter 8—CO_2 adsorption by carbonaceous materials and nanomaterials, in *Advances in Carbon Capture*, M.R. Rahimpour, M. Farsi, and M.A. Makarem, Editors. 2020, Woodhead Publishing. p. 173–192.
3. Hossein Tehrani, N.H.M., et al., Polydopamine-modified mesoporous silica materials as a novel adsorbent for superior CO_2 adsorption: Experimental and DFT study. *Journal of Environmental Chemical Engineering*, 2023. 11(5): p. 110451.
4. Tao, H., et al., Research progress of clay minerals in carbon dioxide capture. *Renewable and Sustainable Energy Reviews*, 2022. 164: p. 112536.
5. Song, K.S., P.W. Fritz, and A. Coskun, Porous organic polymers for CO_2 capture, separation and conversion. *Chemical Society Reviews*, 2022. 51(23): p. 9831–9852.
6. Creamer, A.E. and B. Gao, Carbon-based adsorbents for postcombustion CO_2 capture: A critical review. *Environmental Science & Technology*, 2016. 50(14): p. 7276–7289.
7. Zaker, A., et al., Carbon-based materials for CO_2 capture: Their production, modification and performance. *Journal of Environmental Chemical Engineering*, 2023. 11(3): p. 109741.
8. Wu, F., et al., Assessing the environmental impact and payback of carbon nanotube supported CO_2 capture technologies using LCA methodology. *Journal of Cleaner Production*, 2020. 270: p. 122465.

9. Szczęśniak, B. and J. Choma, Graphene-containing microporous composites for selective CO_2 adsorption. *Microporous and Mesoporous Materials*, 2020. 292: p. 109761.
10. Ramezanipour Penchah, H., A. Ghaemi, and H. Godarziani, Eco-friendly CO_2 adsorbent by impregnation of diethanolamine in nanoclay montmorillonite. *Environmental Science and Pollution Research*, 2021. 28(39): p. 55754–55770.
11. Xu, M., et al., Evaluation and optimization of VPSA processes with nanostructured zeolite NaX for post-combustion CO_2 capture. *Chemical Engineering Journal*, 2019. 371: p. 693–705.
12. Gupta, S., et al., 1—CO_2-philic adsorbents: An overview, in *CO_2-philic Polymers, Nanocomposites and Chemical Solvents*, A. Kumar Nadda, S. Sharma, and S. Kalia, Editors. 2023, Elsevier. p. 1–15.
13. Goel, C., H. Bhunia, and P.K. Bajpai, Development of nitrogen enriched nanostructured carbon adsorbents for CO_2 capture. *Journal of Environmental Management*, 2015. 162: p. 20–29.
14. Firdaus, R.M., et al., Progress in adsorption capacity of nanomaterials for carbon dioxide capture: A comparative study. *Journal of Cleaner Production*, 2021. 328: p. 129553.
15. Tan, J.S. and F.N. Ani, Carbon molecular sieves produced from oil palm shell for air separation. *Separation and Purification Technology*, 2004. 35(1): p. 47–54.
16. Lu, A.-H., G.-P. Hao, and X.-Q. Zhang, Porous carbons for carbon dioxide capture, in *Porous Materials for Carbon Dioxide Capture*, A.-H. Lu and S. Dai, Editors. 2014, Springer Berlin Heidelberg: Berlin, Heidelberg. p. 15–77.
17. Montiel-Centeno, K., et al., Hierarchical nanostructured carbons as CO_2 adsorbents. *Adsorption*, 2019. 25(7): p. 1287–1297.
18. Cortés-Súarez, J., et al., Synthesis and Characterization of an SWCNT@HKUST-1 Composite: Enhancing the CO_2 Adsorption Properties of HKUST–1. *ACS Omega,* 2019. 4(3): p. 5275–5282.
19. Osler, K., et al., Synthesis and evaluation of carbon nanotubes composite adsorbent for CO_2 capture: A comparative study of CO_2 adsorption capacity of single-walled and multi-walled carbon nanotubes. *International Journal of Coal Science & Technology*, 2017. 4(1): p. 41–49.
20. Zhang, Y., et al., Multi-walled carbon nanotubes/carbon foam nanocomposites derived from biomass for CO_2 capture and supercapacitor applications. *Fuel*, 2021. 305: p. 121622.
21. Hao, J., et al., Hierarchical structure N, O-co-doped porous carbon/carbon nanotube composite derived from coal for supercapacitors and CO_2 capture. *Nanoscale Advances*, 2020. 2(2): p. 878–887.
22. Hu, Z., et al., Hierarchical Ti_3C_2Tx MXene/carbon nanotubes for low overpotential and long-life Li-CO_2 batteries. *ACS Nano,* 2021. 15(5): p. 8407–8417.
23. Tian, Y., et al., Surface-microporous graphene for CO_2 adsorption. *Catalysis Today*, 2020. 356: p. 514–518.
24. Varghese, A.M., et al., Performance enhancement of CO_2 capture adsorbents by UV treatment: The case of self-supported graphene oxide foam. *Chemical Engineering Journal,* 2020. 386: p. 124022.
25. Fathalian, F., et al., Efficient CO_2 adsorption using chitosan, graphene oxide, and zinc oxide composite. *Scientific Reports*, 2024. 14(1): p. 3186.
26. Helmi, M., et al., Synthesis and characterization of KOH@Graphene oxide-Fe_3O_4 as a magnetic composite adsorbent for CO_2 capture. *Journal of Physics and Chemistry of Solids*, 2023. 178: p. 111338.
27. Wang, Y., et al., N-doped porous carbon derived from rGO-Incorporated polyphenylenediamine composites for CO_2 adsorption and supercapacitors. *Journal of Power Sources*, 2020. 472: p. 228610.
28. Xia, D., et al., Electrically heatable graphene aerogels as nanoparticle supports in adsorptive desulfurization and high-pressure CO_2 capture. *Advanced Functional Materials*, 2020. 30(40): p. 2002788.

29. Che Othman, F.E., et al., Electrospun composites made of reduced graphene oxide and polyacrylonitrile-based activated carbon nanofibers (rGO/ACNF) for enhanced CO_2 adsorption. *Polymers*, 2020. 12(9): p. 2117.
30. Xiao, J., et al., rGO/N-porous carbon composites for enhanced CO_2 capture and energy storage performances. *Journal of Alloys and Compounds*, 2021. 857: p. 157534.
31. Wang, M., et al., Penta-graphene as a promising controllable CO_2 capture and separation material in an electric field. *Applied Surface Science*, 2020. 502: p. 144067.
32. Shen, Z., et al., Construction of hierarchically porous 3D graphene-like carbon material by B, N co-doping for enhanced CO_2 capture. *Microporous and Mesoporous Materials*, 2021. 322: p. 111158.
33. Cueto-Díaz, E.J., et al., APTES-based silica nanoparticles as a potential modifier for the selective sequestration of CO_2 gas molecules. *Nanomaterials*, 2021. 11(11): p. 2893.
34. Cueto-Díaz, E.J., et al., CO_2 adsorption capacities of amine-functionalized microporous silica nanoparticles. *Reactive and Functional Polymers*, 2022. 170: p. 105100.
35. Song, T., et al., Facile assembly of mesoporous silica nanoparticles with hierarchical pore structure for CO_2 capture. *Chinese Chemical Letters*, 2019. 30(12): p. 2347–2350.
36. Wang, Y., et al., The capture and catalytic conversion of CO_2 by dendritic mesoporous silica-based nanoparticles. *Energy & Environmental Materials*, 2024. 7(2): p. e12593.
37. Dey, R. and A. Samanta, Microwave-synthesized high-performance mesoporous SBA-15 silica materials for CO_2 capture. *Korean Journal of Chemical Engineering*, 2020. 37(11): p. 1951–1962.
38. Lin, L., et al., Comparison of characteristics and performance between PEI and DETA impregnated on SBA-15 for CO_2 capture. *Separation and Purification Technology*, 2023. 322: p. 124346.
39. Ravutsov, M., et al., CO_2 adsorption on modified mesoporous silicas: The role of the adsorption sites. *Nanomaterials*, 2021. 11(11): p. 2831.
40. Silva, D.P.S., et al., Modification of MCM-41 type structures by carbon deposition and acid washing for CO_2 adsorption. *Journal of Sol-Gel Science and Technology*, 2021. 97(2): p. 382–392.
41. Huang, W., et al., Influence of surfactant on CO_2 adsorption of amine-functionalized MCM–41. *Environmental Technology*, 2022. 43(28): p. 4545–4553.
42. Taheri, F.S., et al., High CO_2 adsorption on amine-functionalized improved mesoporous silica nanotube as an eco-friendly nanocomposite. *Energy & Fuels*, 2019. 33(6): p. 5384–5397.
43. Khoshraftar, Z., A. Ghaemi, and F.S. Taheri, Polyethylenimine-functionalized halloysite nanotube as an adsorbent for CO_2 capture: RSM and ANN methodology. *Current Research in Green and Sustainable Chemistry*, 2023. 7: p. 100389.
44. Jiang, X., et al., Spherical amine grafted silica aerogels for CO_2 capture. *RSC Advances*, 2020. 10(43): p. 25911–25917.
45. Keshavarz, L., M.R. Ghaani, and N.J. English, The importance of precursors and modification groups of aerogels in CO_2 capture. *Molecules*, 2021. 26(16): p. 5023.
46. Feng, J., et al., An efficient amine-modified silica aerogel sorbent for CO_2 capture enhancement: Facile synthesis, adsorption mechanism and kinetics. *Colloids and Surfaces A: Physicochemical and Engineering Aspects*, 2023. 656: p. 130510.
47. Fan, L., et al., In-situ Fe/Ti doped amine-grafted silica aerogel from fly ash for efficient CO_2 capture: Facile synthesis and super adsorption performance. *Chemical Engineering Journal,* 2023. 452: p. 138945.
48. Zainab, G., et al., Electrospun carbon nanofibers with multi-aperture/opening porous hierarchical structure for efficient CO_2 adsorption. *Journal of Colloid and Interface Science*, 2020. 561: p. 659–667.
49. Othman, F.E.C., et al., Activated carbon nanofibers incorporated metal oxides for CO_2 adsorption: Effects of different type of metal oxides. *Journal of CO_2 Utilization*, 2021. 45: p. 101434.

50. Mahfoudhi, N. and S. Boufi, Nanocellulose as a novel nanostructured adsorbent for environmental remediation: A review. *Cellulose*, 2017. 24(3): p. 1171–1197.
51. Sepahvand, S., et al., A promising process to modify cellulose nanofibers for carbon dioxide (CO_2) adsorption. *Carbohydrate Polymers*, 2020. 230: p. 115571.
52. Shah, K.J. and T. Imae, Selective gas capture ability of gas-adsorbent-incorporated cellulose nanofiber films. *Biomacromolecules*, 2016. 17(5): p. 1653–1661.
53. Valdebenito, F., et al., CO_2 adsorption of surface-modified cellulose nanofibril films derived from agricultural wastes. *ACS Sustainable Chemistry & Engineering*, 2018. 6(10): p. 12603–12612.
54. Liu, S., et al., High CO_2 adsorption by amino-modified bio-spherical cellulose nanofibres aerogels. *Environmental Chemistry Letters*, 2018. 16(2): p. 605–614.
55. Pajdak, A., et al., Sorption of CO_2 and CH_4 on raw and calcined halloysite—structural and pore characterization study. *Materials*, 2020. 13(4): p. 917.
56. Chen, M., et al., The application of mineral kaolinite for environment decontamination: A review. *Catalysts*, 2023. 13(1): p. 123.
57. Hou, J., et al., Regulating the effect of element doping on the CO_2 capture performance of kaolinite: A density functional theory study. *Applied Surface Science*, 2020. 512: p. 145642.
58. Quiroz-Estrada, K., et al., N_2 and CO_2 adsorption by soils with high kaolinite content from San Juan Amecac, Puebla, México. *Minerals*, 2016. 6(3): p. 73.
59. Akpasi, S.O. and Y.M. Isa, Effect of operating variables on CO_2 adsorption capacity of activated carbon, kaolinite, and activated carbon—kaolinite composite adsorbent. *Water-Energy Nexus*, 2022. 5: p. 21–28.
60. Bahmanzadegan, F. and A. Ghaemi, Modification and functionalization of zeolites to improve the efficiency of CO_2 adsorption: A review. *Case Studies in Chemical and Environmental Engineering*, 2024. 9: p. 100564.
61. Cecilia, J.A., et al., Kaolinite-based zeolites synthesis and their application in CO_2 capture processes. *Fuel*, 2022. 320: p. 123953.
62. Bahmanzadegan, F., M.A. Pordsari, and A. Ghaemi, Improving the efficiency of 4A-zeolite synthesized from kaolin by amine functionalization for CO_2 capture. *Scientific Reports*, 2023. 13(1): p. 12533.
63. Fashi, F., A. Ghaemi, and A.H. Behroozi, Piperazine impregnation on Zeolite 13X as a novel adsorbent for CO_2 capture: Experimental and modeling. *Chemical Engineering Communications*, 2021. 208(8): p. 1104–1120.
64. Lutyński, M., P. Sakiewicz, and M.G. Gonzalez, Halloysite as mineral adsorbent of CO_2–kinetics and adsorption capacity. *Inżynieria Mineralna*, 2014. 15.
65. Park, S., et al., Halloysite nanotubes loaded with HKUST-1 for CO_2 adsorption. *Colloids and Surfaces A: Physicochemical and Engineering Aspects*, 2022. 651: p. 129750.
66. Ramadass, K., et al., Carbon nanoflakes and nanotubes from halloysite nanoclays and their superior performance in CO_2 capture and energy storage. *ACS Applied Materials & Interfaces*, 2020. 12(10): p. 11922–11933.
67. Lu, X., et al., Zeolite-X synthesized from halloysite nanotubes and its application in CO_2 capture. *Journal of the Taiwan Institute of Chemical Engineers*, 2022. 133: p. 104281.
68. Godarziani, H., H. Ramezanipour Penchah, and A. Ghaemi, Triethanolamine-modified montmorillonite clay as a new adsorbent for CO capture: Characterization, adsorption, and RSM modeling. *Journal of the Chinese Chemical Society*, 2022. 69(12): p. 1981–1996.
69. Roth, E.A., S. Agarwal, and R.K. Gupta, Nanoclay-based solid sorbents for CO_2 capture. *Energy & Fuels*, 2013. 27(8): p. 4129–4136.
70. Horri, N., et al., Effect of acid activation on the CO_2 adsorption capacity of montmorillonite. *Adsorption*, 2020. 26(5): p. 793–811.

71. Khajeh, M. and A. Ghaemi, Strontium hydroxide-modified nanoclay montmorillonite for CO_2 capture: Response surface methodology and adsorption mechanism. *International Journal of Environmental Analytical Chemistry*, 2023. 103(17): p. 5311–5336.
72. Ansari, A., A. Ghaemi, and S. Shahhosseini, Impregnation of nanoclay montmorillonite by choline chloride-urea for selective CO_2, N_2, and O_2 adsorption: Experimental and modeling. *Arabian Journal of Chemistry*, 2024. 17(4): p. 105688.
73. Khajeh, M. and A. Ghaemi, Nanoclay montmorillonite as an adsorbent for CO_2 capture: Experimental and modeling. *Journal of the Chinese Chemical Society*, 2020. 67(2): p. 253–266.
74. Khajeh, M. and A. Ghaemi, Exploiting response surface methodology for experimental modeling and optimization of CO_2 adsorption onto NaOH-modified nanoclay montmorillonite. *Journal of Environmental Chemical Engineering*, 2020. 8(2): p. 103663.
75. Wang, L., et al., Melamine-supported porous organic polymers for efficient CO_2 capture and Hg^{2+} removal. *Chemical Engineering Journal*, 2020. 387: p. 124070.
76. Chen, J., et al., Constructing highly porous carbon materials from porous organic polymers for superior CO_2 adsorption and separation. *Journal of Colloid and Interface Science*, 2022. 609: p. 775–784.
77. Jiangfei, G., et al., Amino-functionalized porphyrin-based porous organic polymers for CO_2 capture and Hg^{2+} removal. *Energy & Fuels*, 2020. 34(8): p. 9771–9778.
78. Khakbaz, M., A. Ghaemi, and G. Mir Mohamad Sadeghi, Synthesis methods of microporous organic polymeric adsorbents: A review. *Polymer Chemistry*, 2021. 12(48): p. 6962–6997.
79. Torkashvand, A., M.R. Moradi, and A. Ghaemi, Amine grafting of carbazole-based hypercrosslinked polymer as an adsorbent to enhance CO_2 capture. *Case Studies in Chemical and Environmental Engineering*, 2023. 8: p. 100472.
80. Ramezanipour Penchah, H., A. Ghaemi, and H. Ganadzadeh Gilani, Benzene-based hyper-cross-linked polymer with enhanced adsorption capacity for CO_2 capture. *Energy & Fuels*, 2019. 33(12): p. 12578–12586.
81. Ramezanipour Penchah, H., et al., Characterization of hypercrosslinked polymer adsorbent based on carbazole to achieve higher CO_2 capture. *Environmental Progress & Sustainable Energy*, 2021. 40(4): p. e13586.
82. Ramezanipour Penchah, H., A. Ghaemi, and H. Ghanadzadeh Gilani, Efficiency increase in hypercrosslinked polymer based on polystyrene in CO_2 adsorption process. *Polymer Bulletin*, 2022. 79(6): p. 3681–3702.
83. Mohamed, M.G., et al., Design and synthesis of bisulfone-linked two-dimensional conjugated microporous polymers for CO_2 adsorption and energy storage. *Molecules*, 2023. 28(7): p. 3234.
84. Wang, L., et al., A series of thiophene- and nitrogen-rich conjugated microporous polymers for efficient iodine and carbon dioxide capture. *New Journal of Chemistry*, 2021. 45(42): p. 19636–19640.
85. Deng, Q.W., et al., Hydrogen and CO_2 storage in high surface area covalent triazine–based frameworks. *Materials Today Energy*, 2020. 18: p. 100506.
86. Zhao, Y., et al., Design and synthesis of novel pyridine-rich cationic covalent triazine framework for CO_2 capture and conversion. *Microporous and Mesoporous Materials*, 2022. 329: p. 111526.
87. Mukhtar, A., et al., Impact of amine functionality on the selective CO_2/CH_4 adsorption behavior of porous covalent triazine adsorbent. *Journal of Natural Gas Science and Engineering*, 2020. 83: p. 103582.
88. Dautzenberg, E., G. Li, and L.C.P.M. de Smet, Aromatic amine-functionalized covalent organic frameworks (COFs) for CO_2/N_2 separation. *ACS Applied Materials & Interfaces*, 2023. 15(4): p. 5118–5127.

89. Lyu, H., et al., Covalent organic frameworks for carbon dioxide capture from air. *Journal of the American Chemical Society*, 2022. 144(28): p. 12989–12995.
90. Yin, M., L. Wang, and S. Tang, Amino-functionalized ionic-liquid-grafted covalent organic frameworks for high-efficiency CO_2 capture and conversion. *ACS Applied Materials & Interfaces*, 2022. 14(50): p. 55674–55685.
91. Aksu, G.O., et al., Computational selection of high-performing covalent organic frameworks for adsorption and membrane-based CO_2/H_2 separation. *The Journal of Physical Chemistry C*, 2020. 124(41): p. 22577–22590.
92. Dashti, A., et al., Estimation of CO_2 adsorption in high capacity metal–organic frameworks: Applications to greenhouse gas control. *Journal of CO_2 Utilization*, 2020. 41: p. 101256.
93. Ghanbari, T., F. Abnisa, and W.M.A. Wan Daud, A review on production of metal organic frameworks (MOF) for CO_2 adsorption. *Science of the Total Environment*, 2020. 707: p. 135090.
94. Li, Z., et al., Porous metal–organic frameworks for carbon dioxide adsorption and separation at low pressure. *ACS Sustainable Chemistry & Engineering*, 2020. 8(41): p. 15378–15404.
95. Esfahani, H.J., S. Shahhosseini, and A. Ghaemi, Improved structure of Zr-BTC metal organic framework using NH_2 to enhance CO_2 adsorption performance. *Scientific Reports*, 2023. 13(1): p. 17700.
96. Kazemi, A., et al., Enhanced CO_2 capture potential of UiO-66-NH_2 synthesized by sonochemical method: Experimental findings and performance evaluation. *Scientific Reports*, 2023. 13(1): p. 19891.
97. Ghaemi, A. and A. Hemmati, Modeling of two parameters isotherm of CO_2 and H_2S adsorption using metal organic frameworks (MOFs). *Iranian Chemical Engineering Journal*, 2021. 20(114): p. 22–36.
98. Hammi, N., et al., Enhanced gas adsorption in HKUST-1@chitosan aerogels, cryogels, and xerogels: An evaluation study. *ACS Applied Materials & Interfaces*, 2023. 15(46): p. 53395–53404.
99. Wang, S., et al., Hierarchical porous N-doped carbon xerogels for high performance CO_2 capture and supercapacitor. *Colloids and Surfaces A: Physicochemical and Engineering Aspects*, 2021. 616: p. 126285.
100. Li, T., et al., One-step synthesis of cellulosed-based nitrogen-doped carbon aerogel and its CO_2 adsorption performance. *Energy & Fuels*, 2024. 38(1): p. 555–564.
101. Nazir, G., A. Rehman, and S.-J. Park, Self-activated, urea modified microporous carbon cryogels for high-performance CO_2 capture and separation. *Carbon*, 2022. 192: p. 14–29.

6 Organic and Inorganic CO_2 Adsorbents

Zohreh Khoshraftar and Ahad Ghaemi

6.1 INTRODUCTION OF ORGANIC AND INORGANIC ADSORBENTS

In the realm of combating climate change, the quest for efficient carbon capture and storage (CCS) technologies has intensified [1–6]. Organic CO_2 adsorbents, emerging as a promising solution, represent a significant breakthrough in this field. Unlike traditional inorganic adsorbents, organic CO_2 adsorbents offer distinct advantages, including tunable properties, high selectivity, and ease of synthesis. Organic CO_2 adsorbents function by selectively capturing CO_2 molecules from gas streams, such as those emitted from industrial processes or power plants, thus mitigating their release into the atmosphere. These adsorbents typically consist of porous materials with tailored chemical functionalities, allowing for strong interactions with CO_2 molecules while minimizing the adsorption of other gases. The design and synthesis of organic CO_2 adsorbents involve a multidisciplinary approach, integrating principles from chemistry, materials science, and engineering. Researchers employ various techniques, such as molecular modeling, surface modification, and structure–property relationship studies, to optimize the adsorption capacity and kinetics of these materials. One of the key advantages of organic CO_2 adsorbents lies in their versatility and customizability. By adjusting parameters such as pore size, surface area (SA), and functional groups, researchers can tailor adsorbents to specific applications and operating conditions. This flexibility opens up possibilities for their use in diverse settings, ranging from large-scale industrial processes to portable carbon capture devices [7, 8].

Furthermore, organic CO_2 adsorbents hold the potential to address challenges associated with traditional CCS technologies, such as energy-intensive regeneration processes and limited scalability [9, 10]. Some organic adsorbents exhibit reversible adsorption behavior, enabling facile regeneration under mild conditions, which can reduce energy consumption and operational costs. As research in this area continues to advance, organic CO_2 adsorbents are poised to play a crucial role in achieving global carbon reduction targets. Their implementation could facilitate the transition to a more sustainable energy future by enabling efficient capture and utilization of CO_2 emissions [11–19].

This section explores the distinct attributes and possible uses of various organic CO_2 adsorbents in CCS. These organic alternatives present a favorable option compared to conventional inorganic adsorbents due to their economical, abundant, and

DOI: 10.1201/9781003476191-6

eco-friendly nature [20]. Initially, we scrutinize natural polymers such as chitosan, cellulose, and lignin. Chitosan, sourced from crustaceans' exoskeletons, exhibits considerable potential for CO_2 capture due to its large SA and chemical reactivity. Similarly, cellulose, a plant cell wall polysaccharide, and lignin, a complex polymer in plant tissues, have been investigated for their CO_2 adsorption capabilities. Another compelling organic CO_2 adsorbent we explore is fly ash, a by-product of coal combustion renowned for its elevated carbon content. Studies indicate that fly ash effectively captures CO_2 due to its porous structure and surface functional groups. Finally, we delve into the potential of carbon nanotubes (CNTs) as organic CO_2 adsorbents. Comprising carbon atoms, CNTs boast unique properties like a high SA and remarkable mechanical strength, making them highly appealing for CO_2 capture applications. In conclusion, the adoption of organic CO_2 adsorbents presents a distinctive and promising approach to mitigating greenhouse gas emissions. Through the examination of natural polymers, fly ash, and CNTs, we aspire to contribute to the advancement of efficient and sustainable CO_2 capture technologies.

In our study of inorganic adsorbents for CO_2 adsorption in another section, we explored a variety of materials, including clays such as kaolinite, halloysite, montmorillonite, and bentonite, as well as zeolites and dolomite. These natural materials exhibit unique properties that enhance their ability to capture carbon dioxide effectively. Additionally, we investigated activated carbon (AC)-based adsorbents derived from wood biochar, rice husk, and sawdust, which are not only sustainable but also demonstrate significant adsorption capacities. The combination of these diverse adsorbents highlights the potential for utilizing both natural and bio-based materials in the quest for efficient CO_2 capture solutions, contributing to environmental sustainability and climate change mitigation efforts. The classifications of absorbents examined in this chapter are depicted in Figure 6.1.

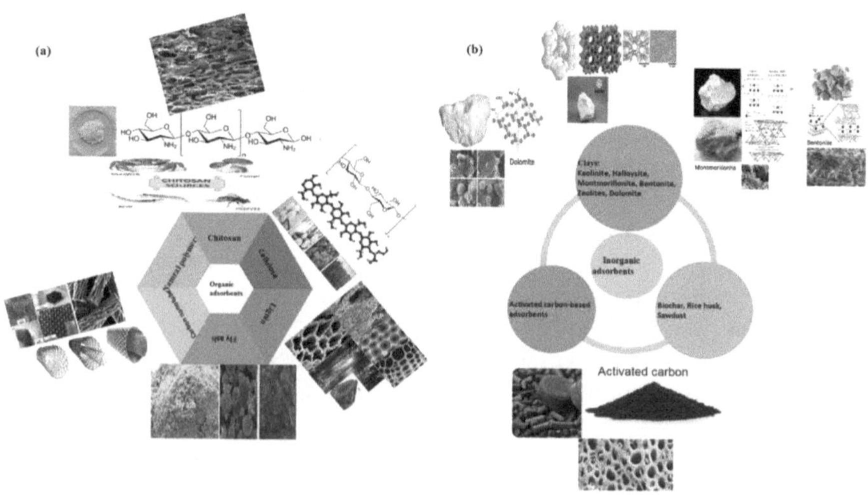

FIGURE 6.1 (a) Organic adsorbents and (b) inorganic adsorbents.

6.1.1 Natural Polymers

6.1.1.1 Chitosan

Chitosan, a biopolymer derived from chitin, has gained significant attention for its potential application as a CO_2 adsorbent due to its unique properties [21]. It is known for its high nitrogen content, which enables its adsorption capacity for various gases, including carbon dioxide. Chitosan also offers several advantages over other common CO_2 adsorbents, such as zeolites or AC, including its abundance, low cost, and biodegradability. One of the key factors that contributes to chitosan's CO_2 adsorption capabilities is its porous structure, which provides a large SA for gas adsorption. Chitosan can be chemically modified to enhance the specific surface area (SSA), porosity, and surface functional groups, which enhance its CO_2 adsorption performance.

Chitosan is a distinctive biodegradable non-toxic polymer primarily derived from chitin, a natural and abundant material found in the exoskeletons of arthropods like crabs, shrimps, and lobsters. These exoskeletons, which would typically be considered waste, can now be utilized in CO_2 capture, thereby helping combat climate change and reducing surface waste pollution. Chitin ranks second in abundance among polymers, following cellulose, and its production is inexhaustible. In the animal kingdom, arthropods composed of chitin represent 106 species out of the total 1.2 million species. Chitin has a white appearance and is characterized by its hardness and lack of elasticity. It is considered a nitrogenous polysaccharide polymer, consisting of a β (1→4) linkage of 2-acetamido 2-deoxy-β-D-glucose, with the hydroxyl C-2 modified with an acetamido group. Despite the presence of nitrogen in its chain, this polymer is commonly referred to as "cellulose" due to the substitution of the hydroxyl group. Chitosan is a nitrogenous polysaccharide that is composed of N-acetyl glucosamine and glucosamine units linked together in a linear (1→4) structure. It shares similar physical properties with chitin, such as its white appearance and hard, inelastic nature [22, 23].

The relationship between DDA and the CO_2 adsorption capacity of chitosan is unique and important when synthesizing chitosan for CO_2 capture applications. The DDA process involves the removal of the acetyl groups from chitin, resulting in a higher number of amino groups (NH_2) on the chitosan molecule. These amino groups are responsible for the interaction and adsorption of CO_2 molecules. The increase in the number of NH_2 groups during the DDA process enhances the adsorption capacity of chitosan for CO_2. This is because CO_2 molecules can form strong chemical bonds with the NH_2 groups through a process known as chemisorption. The more NH_2 groups available, the more CO_2 molecules can be adsorbed. Furthermore, the removal of impurities, such as proteins, minerals, and pigments, in the preceding steps (deproteinization, demineralization, and decolorization: DDA) also contributes to the improved adsorption capacity of chitosan. These impurities can potentially block or hinder the adsorption sites on the chitosan molecule, reducing the overall adsorption capacity. Therefore, the DDA process is crucial in maximizing the CO_2 capture efficiency of chitosan. By increasing the number of available NH_2 groups and removing impurities, chitosan with higher adsorption capacity can be synthesized. This is important for CO_2 capture, as it enables more efficient extraction of CO_2 from gas streams, thereby aiding in the reduction of greenhouse gas emissions and tackling climate change [22].

Chitosan has a high SA and porosity, allowing for efficient adsorption of CO_2 molecules. It also has a high affinity for polar molecules, including CO_2, due to its positive charge and hydrogen bonding capabilities. The chemisorption of CO_2 onto chitosan occurs through a reaction between the amine groups on chitosan and the CO_2 molecules. This results in the formation of a stable and reversible carbamate complex. The adsorption capacity of chitosan for CO_2 can be enhanced by modifying its surface properties through various methods, such as crosslinking, functionalization, or incorporation of additives like surfactants or nanoparticles. Chitosan's unique properties make it a promising adsorbent for CO_2 capture. It offers advantages over traditional adsorbents like AC, zeolites, or amine-based adsorbents, as it is renewable, biodegradable, and non-toxic. Additionally, chitosan can be sourced from waste materials, such as crustacean shells, making it a cost-effective and sustainable option. Further research is still needed to optimize the CO_2 capture performance of chitosan and understand its adsorption mechanisms in greater detail. However, its potential applications in CO_2 capture and storage, as well as in other fields, make it an interesting material for future studies [22].

According to Sneddon et al. (2015) [24], they developed CO_2 adsorbents by applying a coating of chitosan onto mesoporous silicas. These chitosan-based adsorbents demonstrated a range of CO_2 adsorption capacities, specifically between 12.76 to 43.12 gCO_2 per kg of adsorbent. The experiments were conducted using 100% CO_2 as the gas, a pressure of 1 atmosphere, and a temperature of 20°C. The study conducted by Fujiki and Yogo (2016) [25] presented a remarkable CO_2 adsorption capacity of chitosan-derived AC adsorbents at 70.4 gCO_2/kg adsorbent. It is important to note that these experiments were carried out at 25°C and 15 kPa, which is significantly higher than the typical pressure of flue gas, which usually operates at atmospheric pressure. Operating at such high pressures may incur substantial costs. In order to overcome this challenge and minimize expenses, it becomes essential to comprehend the CO_2 adsorption capacity of chitosan at pressures closer to atmospheric. However, there is limited information available in the public domain regarding the CO_2 adsorption capacity of pure chitosan at lower partial pressures. Further research in this aspect would be beneficial to explore the potential of using chitosan for CO_2 capture under more practical operating conditions. In conclusion, chitosan presents a unique opportunity as a CO_2 adsorbent, owing to its abundant availability, low cost, and biodegradability. Its porous structure and chemical modifications enable improved CO_2 adsorption capacity, making it a promising material for various carbon capture applications. Figure 6.2 presents a schematic illustration of chitosan extraction methods, including chemical extraction, biological extraction, microwave-assisted extraction, and extraction using green solvents.

The adsorption mechanism of CO_2 on chitosan-based adsorbents involves a combination of physical adsorption and chemical reactions leading to carbamate formation, hydrogen bonding, and potentially electrostatic interactions, making it a multifaceted process (Figure 6.3). It is important to refer to Yang et al. [27] and Hsan et al. [28] for specific figures and experimental details that visualize and elucidate these mechanisms further. While carbonization has long been regarded as the most traditional and effective method for creating chitosan-based adsorbents with desirable morphology, hydrothermal carbonization (HTC) has recently gained attention. HTC operates at a

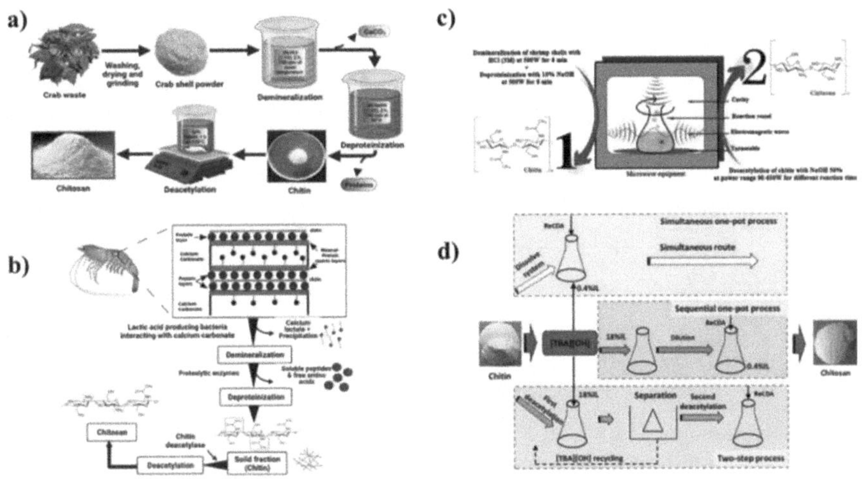

FIGURE 6.2 Chitosan extraction methods, including (a) chemical extraction, (b) biological extraction, (c) microwave-assisted extraction, and (d) extraction using green solvents [26]. Reproduced with permission from Elsevier, license number, 5874321254875.

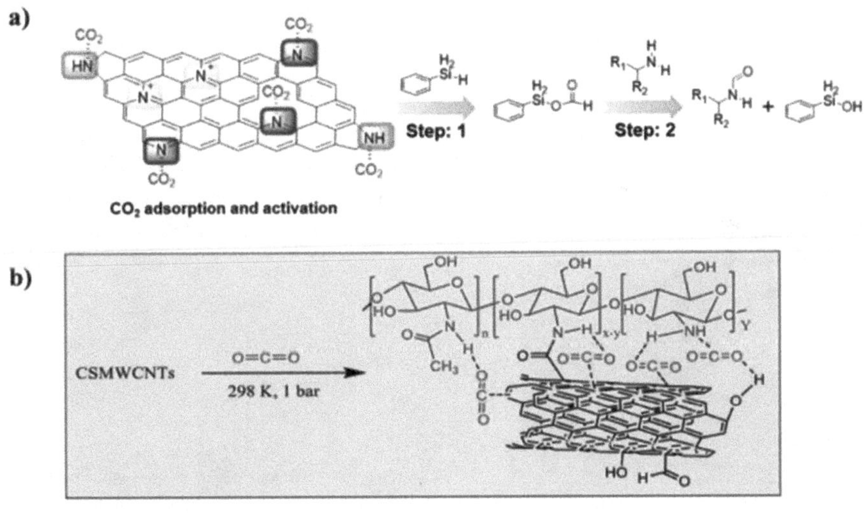

FIGURE 6.3 (a) and (b) Possible mechanisms for the adsorption of CO_2 onto chitosan-based adsorbents [26]. Reproduced with permission from Elsevier, license number, 5874321254875.

relatively low temperature of around 200°C, which is believed to reduce production costs and eliminate the need for organic solvents, catalysts, or surfactants. The chitosan treated through HTC produces hydrochar with advantageous properties, including a rougher surface texture and a more irregular, less folded morphology. This is attributed to the more favorable interactions between nitrogen atoms and CO_2 gas [26].

6.1.1.2 Cellulose

Cellulose, as a renewable and abundant natural polymer, possesses unique properties that make it an attractive material for carbon capture technologies. First, cellulose is biodegradable and non-toxic, making it environmentally friendly and safe to use. This is crucial in carbon capture processes where the materials used should not contribute to pollution or harm ecosystems. Additionally, cellulose has a high adsorption capacity due to its porous structure [29]. It can adsorb and retain a broad spectrum of gases, including methane (CH_4), carbon dioxide (CO_2), and nitrogen oxides (NO_x). This adsorption capacity can be further enhanced by converting cellulose into cellulose derivatives such as cellulose nanofibrils (CNFs) and cellulose nanocrystals (CNCs). CNFs and CNCs are nanoscale materials derived from cellulose, and they possess several unique characteristics. CNFs have a high aspect ratio, large SA, and strong mechanical properties, which make them excellent candidates for gas adsorption. CNCs, on the other hand, have a rod-like shape and high crystallinity, providing them with superior adsorption properties compared to CNFs (Figure 6.4). These nanomaterials can be tailored to have specific pore sizes and surface chemistries, enabling selective adsorption of different gases. They can be used to develop adsorbent materials for capturing greenhouse gases from industrial flue gas streams or exhaust emissions. The adsorbed gases can then be released, separated, and stored or utilized for various purposes, such as enhanced oil recovery or fertilizer production. Additionally, cellulose-based materials can be used in gas separation membranes, where they selectively allow certain gases to pass through while blocking others [29–31].

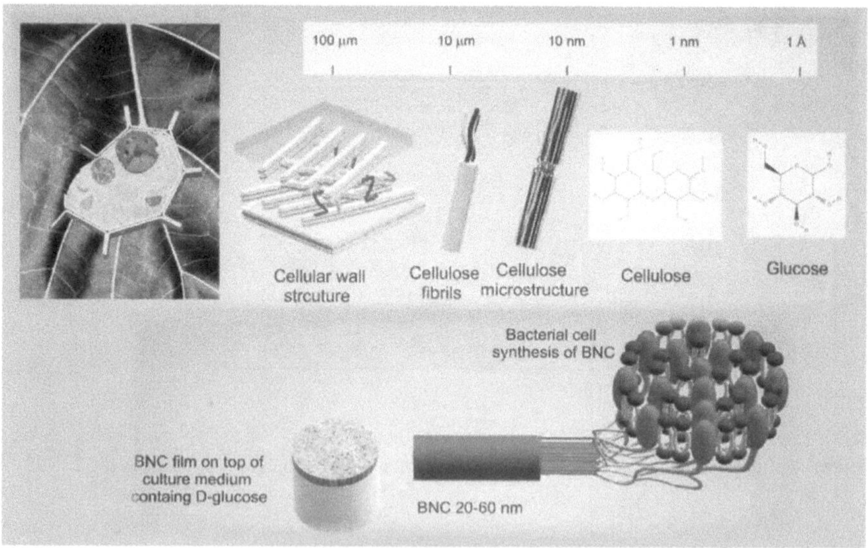

FIGURE 6.4 The process of isolating cellulose into nanocellulose at the top, and the cultivation of bacteria to produce bacterial nanocellulose (BNC) at the bottom [30]. Reproduced with permission from Elsevier, license number, 5746890883577.

These materials have the potential to contribute to the reduction of greenhouse gas emissions and are crucial in mitigating climate change. Further research and development in this field are needed to explore their full potential and optimize their performance in carbon capture applications. One unique aspect of bacterial nanocellulose (BNC) is its production method. Unlike other types of cellulose nanofibers or CNCs that are derived from plant sources, BNC is generated by specific bacteria from sources like glucose and glycerol. This bacterial synthesis of cellulose results in a unique twisted polymer structure. Another unique characteristic of BNC is its purity. When produced under controlled conditions, BNC can have very minimal impurities compared to CNFs or CNCs. This purity is advantageous in that a significant level of cleanliness is required, such as in medical or pharmaceutical settings. In terms of morphology, BNC typically has a diameter within the range of 20–100 nm and can reach lengths of several micrometers. This elongated structure contributes to its high aspect ratio, which can be beneficial in reinforcing composite materials and improving their strength and modulus. However, a limitation of BNC is that its production at a large scale is challenging. The cultivation and fermentation process required for BNC production is time-consuming and requires carefully controlled conditions. This makes large-scale production difficult and expensive compared to other cellulose nanomaterials. Overall, the unique properties of BNC, including its bacterial synthesis, high purity, elongated structure, and potential for reinforcement in composites, make it a promising material for various applications [30].

Nanocellulose, priced as low as USD$2 per kilogram, has emerged as a promising material for carbon capture applications due to its potential as an effective adsorbent [32]. Several studies have explored the use of cellulosic materials as CO_2 adsorbents, and these strategies are summarized in Table 6.1. The key strategies utilized include the chemical alteration of nanocellulose, the addition of inorganic particles to nanocellulose, and the modification of cellulose itself. In contrast to other adsorbents like AC, zeolites, and metal-organic frameworks (MOFs), which demonstrate moderate adsorption capacities of up to 6 mmol/g at room temperature, there is a limited number of cellulose-based adsorbents that display compelling adsorption capabilities.

Overall, the availability and cost-effectiveness of nanocellulose as an adsorbent, along with the potential for high CO_2 adsorption capacities, make it an attractive candidate for further research and implementation in carbon capture technologies.

6.1.1.3 Lignin

Lignin, a complex polymer found in the cell walls of plants, has gained significant attention as a potential feedstock for the development of environmentally friendly and sustainable materials. One particularly promising application is its utilization in the production of adsorbents for capturing carbon dioxide (CO_2) from industrial emissions and other sources [16]. CO_2 capture is an essential step in mitigating climate change, as it reduces the amount of this greenhouse gas released into the atmosphere. While various technologies have been developed for CO_2 capture, many of them rely on energy-intensive processes or require the use of expensive materials. Lignin-based absorbents offer a potential solution to this problem due to their abundant availability, low cost, and unique characteristics. Lignin is typically obtained

TABLE 6.1
The Process of Adsorbing CO_2 Using Cellulosic Materials [30]

Cellulosic Materials	Functionalization	CO_2 Adsorption Capacity (mmol/g)	Comment	Reference
CNF aerogel	3-Aminopropylmethyldiethoxysilane	2.26	Adsorption capacity increased slightly with increasing humidity.	[33]
CNF aerogel	*N*-(2-aminoethyl)-3-aminopropylmethyldimethoxysilane and acetic acid	1.91	Surface area reduced after modification.	[34]
CNF aerogel	*N*-(2-aminoethyl)-3-aminopropylmethyldimethoxysilane	1.78	Surface area reduced after modification.	[35]
CNF aerogel	*N*-(2-aminoethyl)-3-aminopropylmethyldimethoxysilane	1.01	Surface area reduced after modification.	[36]
CNC aerogel	*N*-(2-aminoethyl)-3-aminopropylmethyldimethoxysilane	1.68	Surface area reduced after modification.	[37]
CNC aerogel	*N*-(2-aminoethyl)-3-aminopropylmethyldimethoxysilane	1.59	Surface area reduced after modification.	[38]
CNF aerogel	Phthalimide (1,3-dihydro-1,3-dioxoisoindole)	5.3	BET surface area was improved to 335 m^2/g.	[39]
Cellulose suspension	(3-Chloro-2 hydroxypropyl) trimethylammonium chloride	0.14	Adsorption capacity increased with humidity raise.	[40]
Cellulose aerogel	Acrylamide grafting	1.07		[41]

Source: Reproduced with permission from Elsevier, license number, 5746890883577.

as a by-product from the paper and biofuel industries, making it an attractive sustainable alternative to petroleum-based materials. The structure of lignin makes it an excellent absorbent material. It consists of a three-dimensional matrix with numerous hydroxyl and phenolic functional groups, which provide binding sites for CO_2 molecules. Additionally, lignin has a high porosity and SA, enabling efficient absorption and retention of CO_2.

The production of lignin-based absorbents involves several steps. First, lignin is extracted from plant biomass or industrial waste using various extraction methods. The lignin is then chemically modified to enhance its CO_2 capture properties, such as by incorporating amine groups that can react with CO_2 molecules. This modification can be achieved through functionalization or grafting techniques. The resulting lignin-based adsorbent materials have shown promising results in CO_2 capture experiments. They have demonstrated high CO_2 absorption capacities, good selectivity, and stability over multiple absorption-desorption cycles. Furthermore, the renewable nature of lignin ensures the potential for large-scale production without depleting valuable resources. In addition to CO_2 capture, lignin-based absorbents have the potential for other applications in environmental remediation. Lignin's unique properties can be harnessed for the removal of other pollutants, such as volatile organic compounds and heavy metals, from air and water.

One unique aspect of lignin is its three-dimensional and partly random amorphous structure. This means that lignin does not have a uniform or predictable arrangement of its constituent molecules. This randomness and complexity contribute to the diverse properties and functions of lignin in the cell walls of plants. Another unique feature of lignin is its synthesis through the oxidative coupling of monolignol precursors. These monolignol precursors, such as coniferyl alcohol, sinapyl alcohol, and p-coumaryl alcohol units, are enzymatically polymerized to form the lignin polymer network. The different proportions of these precursors in biomass lead to variations in the composition and structure of lignin. For example, softwood lignin is primarily made up of coniferyl alcohol, while hardwood lignin contains a similar amount of coniferyl alcohol and sinapyl alcohol units. This difference in monolignol composition results in distinct structural characteristics, such as higher branching in softwood lignin compared to hardwood lignin. The unique properties of lignin, including its complexity, randomness, and variation in monolignol composition, contribute to its diverse functions in plant cell walls. Lignin provides mechanical support, enhances water impermeability, and contributes to the protection against pathogens and environmental stresses. These unique characteristics make lignin a key component of plant biomass and a valuable material for various applications, such as biofuels, chemicals, and materials [42].

The unique properties of carbon materials derived from lignin make them highly suitable for adsorption applications. First, lignin-derived ACs have a large internal SA, which is crucial for adsorption as it provides a large contact area for the adsorbate molecules. This high SA can range from 500 to 2000 m^2/g, allowing for efficient adsorption of target molecules. Second, the starting material lignin influences the properties of the resulting carbon materials. Lignin is a renewable resource, which means that carbon materials derived from lignin are sustainable and environmentally friendly. Additionally, lignin has a high carbon content, making it an excellent precursor for carbon materials. The phenolic structure of lignin also provides functional groups on the carbon materials, which can enhance their adsorption capacity. Furthermore, the porosity and morphology of carbon materials from lignin are important factors that contribute to their adsorption capabilities. The porous nature of ACs allows for the easy diffusion of adsorbate molecules into the material, increasing the efficiency of adsorption. The morphology of the carbon materials can also influence their adsorption capacity, as different morphologies can provide different SAs and pore structures. Overall, carbon materials derived from lignin possess a unique combination of properties, including a large SA, renewable nature, high carbon content, and functionalized phenolic structure. These properties make them highly effective adsorbents for a wide range of applications [42, 43].

Gong and Bao in 2023 [44] investigated the use of dealkalized lignin as a precursor for the production of N,O-codoped porous carbon materials. This was achieved through a method that included precarbonization and chemical activation, as shown in Figure 6.5. They found that the temperature during precarbonization significantly influences the chemical makeup, SA, and porosity of the final carbon product. To optimize the adsorption performance, the researchers examined the effects of various

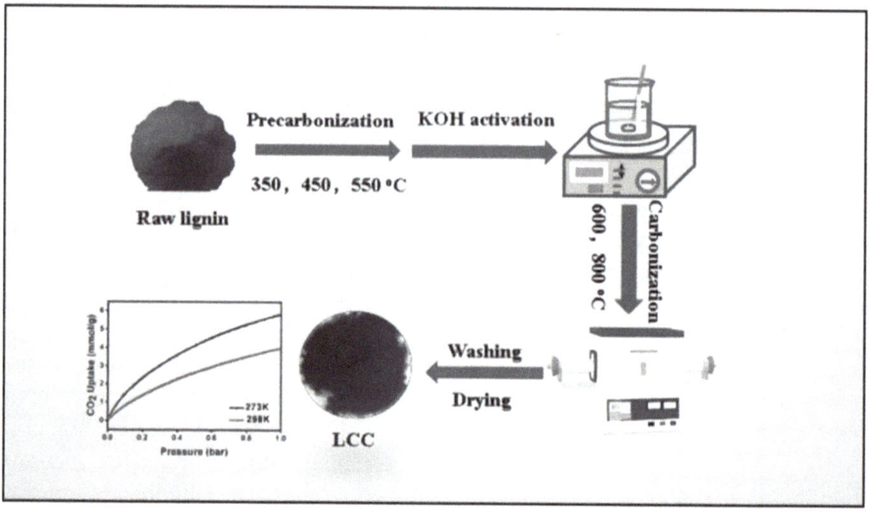

FIGURE 6.5 The preparation process for lignin-derived porous carbon (LCC) [44].

factors, including pretreatment temperature, the ratio of KOH used for activation, and the carbonization temperature. They selected KOH as the activating agent due to its superior ability to enhance the SA and develop a more pronounced microporous structure compared to other activators like NaOH. The results indicated that the resulting porous carbon materials exhibited a substantial micropore volume and SA, enriched with N,O-containing functional groups, which contributed to their impressive CO_2 adsorption capabilities. Specifically, the gas adsorption capacities of the LCC-450-2 K-600 material at 1 bar were measured at 5.82 mmol/g at 273 K and 3.98 mmol/g at 298 K.

Overall, lignin-derived absorbents offer a promising avenue for sustainable CO_2 capture solutions. Utilizing lignin as a feedstock not only reduces environmental impact but also contributes to the development of a circular economy by valorizing an underutilized waste product. With further research and development, lignin-based absorbents have the potential to play a vital role in combating climate change and creating a cleaner future. Lignin and its application in adsorbing CO_2 gas are indicated in Table 6.2.

6.1.2 Fly Ash

Fly ash, a significant by-product generated from the combustion of coal, constitutes a major portion of the overall waste derived from coal-fired power plants, accounting for approximately 60–88% of the total residues produced. Despite some utilization of fly ash in the production of bricks, as well as in the construction of roads, and dam building, a considerable amount still ends up being disposed of in landfills or ash ponds, leading to severe environmental consequences. Environmental threats from fly ash disposal occur because particles are released into the air and toxic components may leach into soil or water bodies [50].

TABLE 6.2
Lignin and Their Application in Adsorbing CO_2 Gas [42]

Material	Activator	Activator to Biomass Ratio	Surface Area (m^2/g)	Total Pore Volume (cm^3/g)	Adsorption Capacity (mmol/g)	Ref.
De-alkaline lignin	KOH	3/1 activator to biochar	2254	1.02	7.24 (1 bar, 298 K) 9.4 (1 bar, 273 K)	[45]
Lignin waste	KOH	2/1 activator to hydrochar	1924	0.95	4.0 (1 bar, 298 K) 17.3 (20 bar, 298 K) 6.5 (1 bar, 273 K)	[46]
Eucalyptus lignin/Norway spruce lignin	KOH	Approx. 3.3/1 activator to biochar	1647	0.69	6.0 (1 bar, 273 K) 1.8 (0.15 bar, 273 K)	[47]
De-alkaline lignin	KOH and NH_3 treatment	1/1 activator to biochar	2922	1.36	5.12 (1 bar, 298 K) 8.6 (1 bar, 273 K)	[48]
Corn straw lignin	H_3PO_4	2/1 activator to biomass at 500°C	820	0.8	0.4 (273 K)	[49]

Source: Reproduced with permission from Elsevier, license number, 501890471.

One unique aspect of using FA for CO_2 emission reduction is its reliance on mineral carbonation technology. This technology takes advantage of the high content of alkali components in FA, particularly the presence of CaO and MgO. By using FA as a caustic material, the pH of acidic brine solutions generated from oil and gas production can be increased, leading to CO_2 emission reduction. Another unique advantage of using FA is its on-site application in coal-fired power plants. This eliminates the need for costly transfer, treatment, and disposal of FA, resulting in significant cost savings for CCS efforts. Additionally, the carbonated FA products that result from this process can be reused as construction materials or additives. Their physical structure is altered, enhancing their mechanical strength and leaching resistance, making them valuable resources for various applications [51].

Fly ash can serve as a direct solid sorbent, but its transformation into more porous materials significantly enhances its adsorption capabilities. Among the various derivatives of fly ash, ACs, mesoporous silica, alkali silicate, and zeolites stand out as the primary categories of porous materials. The choice of which porous material to synthesize is heavily influenced by the specific composition of the fly ash in question. Each type of porous material offers distinct advantages and unique properties tailored to various applications. By converting fly ash into these advanced materials, researchers aim to maximize its utility while simultaneously addressing environmental concerns. Fly ashes that contain a substantial amount of unburned carbon are typically employed in the production of AC, which possesses a larger SA compared to the original fly ash. On the other hand, fly ashes without unburned carbon can be utilized in the synthesis of zeolites or mesoporous aluminosilicate materials. When fly ash has a significant amount of aluminosilicate glass, it is more effective for synthesizing mesoporous silica or alkali silicates, as these silicates can be readily extracted with mild processing conditions. Furthermore, fly ashes that are high in CaO and MgO can be directly utilized for simultaneous capture and mineralization [50].

In their study, Lee et al. [52] focused on synthesizing CO_2 sorbents using a mixture of fly ash, CaO, and NaOH. They found that the inclusion of fly ash in the sorbent resulted in a 9% increase in CO_2 capacity compared to the sorbent without fly ash. Moreover, the sorbent with fly ash showed better desorption characteristics. The researchers attributed this improvement to the dispersing ability of the fly ash, which allowed for a larger SA available for CO_2 binding. This dispersion of CaO and NaOH increased the sorbent's efficiency in capturing CO_2.

However, the recyclability of the fly ash-based sorbent posed a challenge. These new solid phases lacked the ability to adsorb CO_2 effectively.

Overall, the study highlights the potential benefits of incorporating fly ash into CO_2 sorbents, such as increased capacity and easier desorption. However, further research is needed to address the issue of recyclability and develop strategies to enhance the sorbent's performance in subsequent cycles. Table 6.3 outlines the various uses of raw fly ash in the field of CO_2 capture.

The study conducted by Yahia et al. [56] took a different approach to activate the fly ash for carbon dioxide adsorption. Instead of concentrating the carbon through acid digestion like Maroto-Valer et al., Yahia et al. chemically activated the fly ash carbon directly without any prior concentration step. Their results showed that the maximum CO_2 capacity achieved using this method was 26 mg/g of AC at 40°C. This capacity is lower than the maximum capacity observed in the study by Maroto-Valer et al. [57], which reached 68.6 mg/g at 30°C for the MEA-impregnated sorbent. It is important to note that these experiments were conducted under different conditions and with different sorbents. The impregnated sorbents used by Maroto-Valer et al. involved adding MEA, DEA, and MDEA to the activated fly ash carbon, while Yahia et al. used chemically activated fly ash carbon without any additional impregnation. Overall, the two studies highlight different approaches to utilizing fly ash carbon for CO_2 capture. While the impregnated sorbents showed higher capacities in Maroto-Valer et al.'s study, Yahia et al. demonstrated that chemically activating fly ash carbon directly can also result in significant CO_2 capture [50].

TABLE 6.3
Utilizations of Untreated Fly Ash in the Domain of Carbon Dioxide (CO_2) Capture [50]

Support	Additives	Capacity (mg/g)	T_{ads} (°C)	T_{reg} (°C)	P_{CO2} (bar)	CO_2%	Ref.
Fly ash	CaO, NaOH	9% increase due to fly ash	25–800	25–800 (pure N_2)	1	15	[52]
Fly ash	CaO-MgO-FA (50:10:40)	396	650	950 (pure N_2)	1	100	[53]
Fly ash	Ca oxalate-FA (90:10)	380	700–800	900 (pure N_2)	1	100	[54]
Fly ash	MEA/DMA DEA/DMA	303	30 60	—	1	100	[55]

Source: Reproduced with permission from Elsevier, license number, 5747071304247.

6.1.3 Carbon Nanotube

The process of carbon dioxide adsorption by CNTs is gaining attention in both experimental and theoretical studies. This is due to the similarities between CNTs and ACs, as well as their advantageous surface properties, hollow tube structures, high mechanical strength, low weight, thermal and chemical stability, and beneficial pore size distribution. CNTs can have cross-sectional diameters in the range of tens of nanometers, while their length can reach thousands of times that. The internal structure of CNTs consists of rolled-up graphene layers, forming a uniform cylinder with an empty interior. This tubular structure enables the regulation of gas distribution and residence time. The quantity of CO_2 captured by CNTs depends on factors such as gas pressure, temperature, the diameter of the nanotubes throughout the substance, as well as the space between neighboring graphene layers [8].

The unique properties of each type of CNT make them effective adsorbent materials for specific gaseous media. Single-walled carbon nanotubes (SWCNTs) have a large SSA, allowing them to efficiently absorb gases. On the other hand, multi-walled carbon nanotubes (MWCNTs) are considered to be superior in capturing carbon dioxide (CO_2). This is due to their structure, which consists of multiple graphene layers. The presence of multiple layers provides more binding sites for CO_2 molecules, enhancing their adsorption capacity. In addition to their adsorption properties, the number of graphene layers in CNTs affects their mechanical and thermal parameters. SWCNTs are known for their exceptional tensile strength, making them highly durable materials. The mechanical strength is directly related to the number of graphene layers, with the SWCNTs being the strongest. Furthermore, the thermal properties of CNTs, such as thermal conductivity, are influenced by the number of graphene layers. In conclusion, the systematic division of CNTs according to the quantity of graphene layers allows for the identification of unique properties and applications. SWCNTs are characterized by a large SSA and high tensile strength, while MWCNTs are particularly effective in capturing CO_2. These distinctions make CNTs versatile materials for various industries, including gas adsorption, energy storage, and catalysis. In Figure 6.6, you can observe spatial representations of CNTs. Panel (a) displays an SWCNT, while panel (b) showcases a DWCNT. Finally, panel (c) portrays a multi-walled CNT.

Based on numerous studies, various modification strategies have been explored to improve the CO_2 adsorption capacity by CNTs (Table 6.4). One commonly studied

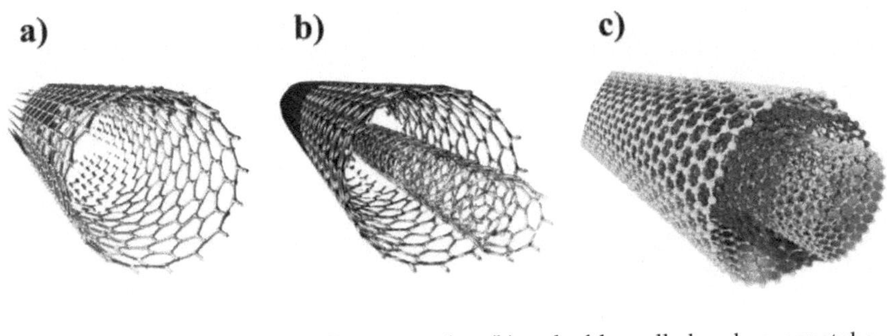

FIGURE 6.6 A single-walled carbon nanotube, (b) a double-walled carbon nanotube, and (c) a multi-walled carbon nanotube [8].

TABLE 6.4

Effect of Modification Techniques on the Textural Characteristics and Capacity for CO_2 Adsorption of Carbon Nanotube [8]

Carbon Nanotube-Based Adsorbent	BET Surface Area (m^2/g)	Total Pore Volume (cm^3/g)	Micropore Volume (cm^3/g)	Average Pore Diameter (nm)	CO_2 Selectivity at 1 bar	CO_2 Uptake (mmol/g)	Adsorption Temperature (°C)	Adsorption Pressure (bar)	Reported Treatment Sources/Chemicals	Purpose of Modification Method	Reference
TEPA-CNT	2.20	0.03	e	e	e	5.00	60	e	Amine functionalization (Tetraethylene-pentamine); Chemical activation (KOH)	Increasing basicity, affinity between CO_2 molecules–adsorbent surface and level-up	[59]
K_2CO_3-APF-CNT	1005	0.61	0.45	0.92	8.80	4.50	0	1	Amine functionalization (3-aminophenol/formaldehyde resin); Chemical activation (K_2CO_3)	N-contents; Increasing surface area and porosities	[60]
TEPA-CNT	8.94	0.0565	e	19.04	e	3.56	40	e	Amine functionalization (Tetraethylenepentamine)	Increasing basicity, affinity between CO_2 molecules–	[61]
PEI-MWCNT	34.7	23.4	e	e	196.00	2.11	25	1	amine functionalization (Polyethylenimine)	adsorbent surface and level-up N-contents	[58]
PEI-MWCNT	174	1.646	0.015	e	e	2.14	25	0.15	HNO_3/H_2SO_4 pre-treatment; amine functionalization (Polyethylenimine)	Carboxylation, oxidation of MCNT; Increasing basicity affinity between CO_2 molecules–adsorbent surface and level-up N-contents	[62]
BPEI-CNT	14.1	0.10	e	e	e	2.43	70	e	Amine functionalization (Polyethylenimine)	Increasing basicity, affinity between CO_2 molecules–adsorbent surface and level-up N-contents	[63]

PEI-purine-CNT	e	e	e	e	e	3.875	50	1	Amine functionalization (Polyethylenimine); Purine	Increasing specific area of CNTs, basicity, affinity toward CO_2 and level-up N-contents	[64]
APS-CNT	198	0.63	e	12.2	e	2.19	25	1	Amine functionalization (3-aminopropyltrie-thoxysilane)	Increasing basicity and affinity between CO_2 molecules–adsorbent surface and level-up N-contents	[65]
PAA-MWCNT	60.44	0.40	e	13.76	e	1.59	25	1	Amine functionalization (Polyaspartamide)	Building bonding between amino-containing compound and substrate material to enhance the surface area and pore volume	[66]
HNO_3/H_2SO_4-DAP-MCNT	112.8	0.65	e	11.6	e	2.11	30	17.3	HNO_3/H_2SO_4 pre-treatment; amine functionalization (1,3-diaminopropane)	Carboxylation, oxidation of MCNT	[67]
HNO_3/H_2SO_4-PDA	90.7	0.53	e	23.80	e	0.54	25	2	HNO_3/H_2SO_4 pre-treatment; diamine	Increasing basicity and affinity between CO_2 molecules-adsorbent surface and level-up N-contents	[68]

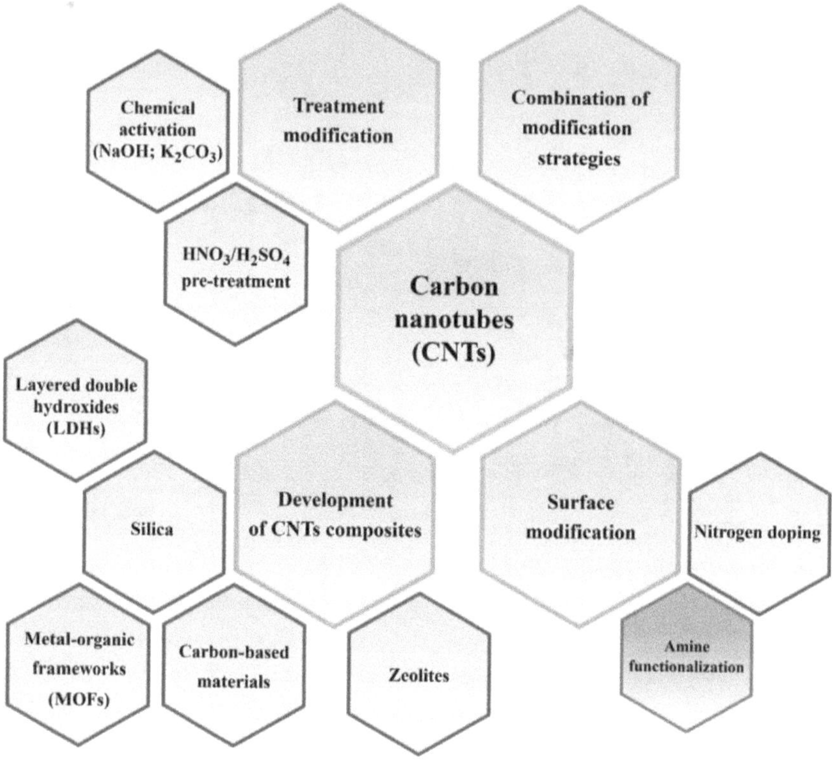

FIGURE 6.7 The modification strategies and research directions of carbon nanotubes [8].

approach is nitrogen doping through the use of amines on the surface of CNTs. Additionally, treatment methods involving chemical agents have been investigated, as well as the synthesis of hybrid materials with MOFs, silica, layered double hydroxides, and carbon foams. Some studies have also combined multiple approaches to further enhance CO_2 adsorption. Figure 6.7 illustrates the generic modification strategies and research paths for CNTs. Based on the reviewed literature, amine-functionalized CNTs are proposed as highly suitable and non-toxic materials for the further development of CO_2 capture processes from flue gas streams [8]. Keller et al. [58] studied the functionalization of highly porous multi-walled CNT-based microtubes using polyethyleneimine (PEI). The study revealed that utilizing a low concentration of PEI in the impregnating solution resulted in carbon materials demonstrating a significant SSA and porosity within microtubes. Consequently, the uptake of carbon dioxide (CO_2) was found to rise with an increasing amount of PEI. The researchers recorded a peak CO_2 uptake of 2.10 mmol/g when the PEI content reached 20%. However, an increase in PEI concentration beyond 20 wt% led to a decline in CO_2 adsorption efficiency.

6.2 INORGANIC ADSORBENTS

6.2.1 Clays

In the 1:1 type structure, each tetrahedral sheet is directly bonded to an octahedral sheet [69]. This type of clay mineral, such as kaolinite, has a single layer of tetrahedral and octahedral sheets alternately stacked. This structure gives 1:1 clay minerals their characteristic low cation exchange capacity and relatively low swelling behavior. In contrast, the 2:1 type structure consists of two tetrahedral sheets sandwiched between two octahedral sheets. This type of clay mineral, including smectite and illite, has a more complex layering arrangement. The octahedral sheet is positioned between two tetrahedral sheets, creating a three-layer configuration. The unique properties of clay minerals arise from their layered structure and ability to undergo ion exchange and water adsorption. The tetrahedral and octahedral sheets provide reactive sites for chemical reactions and interactions with other substances. These minerals play crucial roles in various industries, such as agriculture, geology, and materials science, due to their adsorption, catalytic, and mechanical properties. It is important to note that clay minerals exhibit a wide range of compositions, with different types of cations occupying the octahedral and interlayer positions. This variability in their chemical composition contributes to their diverse physical and chemical properties. Additionally, the arrangement and stacking of layers can vary within a clay mineral group, leading to variations in their behavior and reactivity. Figure 6.8 illustrates various layer structures found in phyllosilicates [70].

There are several factors that can influence the performance of natural clay minerals in adsorbing CO_2 (Figure 6.9). These include:

a. The presence of bound water and the tubular morphology of halloysite: Halloysite is a type of clay mineral that has a unique tubular structure. This structure provides a larger SA for CO_2 adsorption, and the presence of bound water molecules can also enhance the adsorption process.
b. Higher lower layer charge: Clay minerals with higher lower layer charge, such as montmorillonite, tend to have increased CO_2 adsorption abilities. This is because the charged particles can attract and hold CO_2 molecules more effectively.
c. Influence of clay surface: The surface of clay minerals is critical for CO_2 adsorption. Different surface domains or areas of the clay can have different affinities for CO_2 molecules, leading to variations in adsorption performance.
d. Swelling of K- and Ca-exchanged montmorillonites: Montmorillonite is a type of clay mineral that can undergo ion exchange, where cations such as potassium (K) or calcium (Ca) can replace the original interlayer cations. When these exchanged montmorillonites interact with CO_2, they can swell, further increasing the SA available for adsorption.

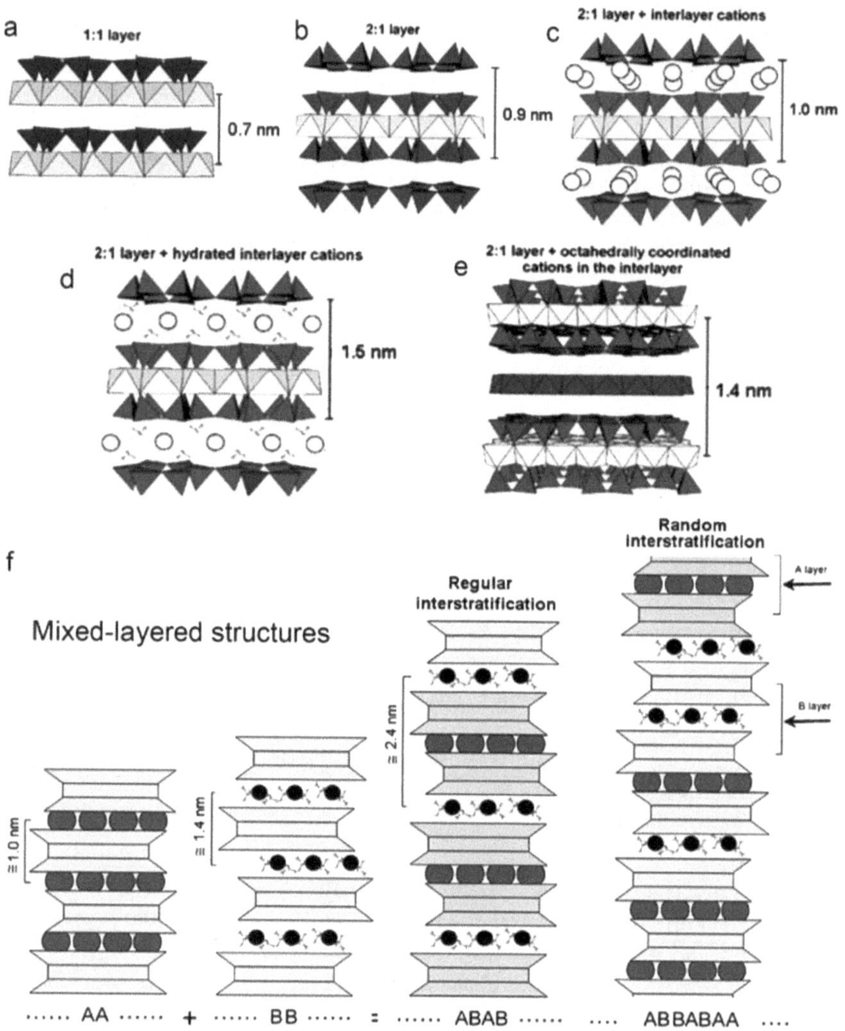

FIGURE 6.8 (a) The 1:1 layer consists of alternating layers of kaolinite- and serpentine-like layers. (b) The 2:1 layer consists of pyrophyllite- and talc-like layers. (c) The 2:1 layer with anhydrous interlayer cations resembles the mica-like layer. (d) The 2:1 layer with hydrated interlayer cations resembles the smectite- and vermiculite-like layer. (e) The 2:1 layer with octahedrally coordinated interlayer cations resembles the chlorite-like layer. (f) There are regularly and randomly interstratified phyllosilicates, with A and B layers exhibiting different periodicities along the c direction [70]. Reproduced with permission from Elsevier, license number, 5746480005339.

Modified clay minerals have demonstrated varying degrees of improvement in CO_2 adsorption performance when compared to their natural counterparts (Table 6.5) [70]. Clay materials, such as montmorillonite, kaolinite, and bentonite, have been suggested as suitable supports for commercial CO_2 capture due to their affordable cost

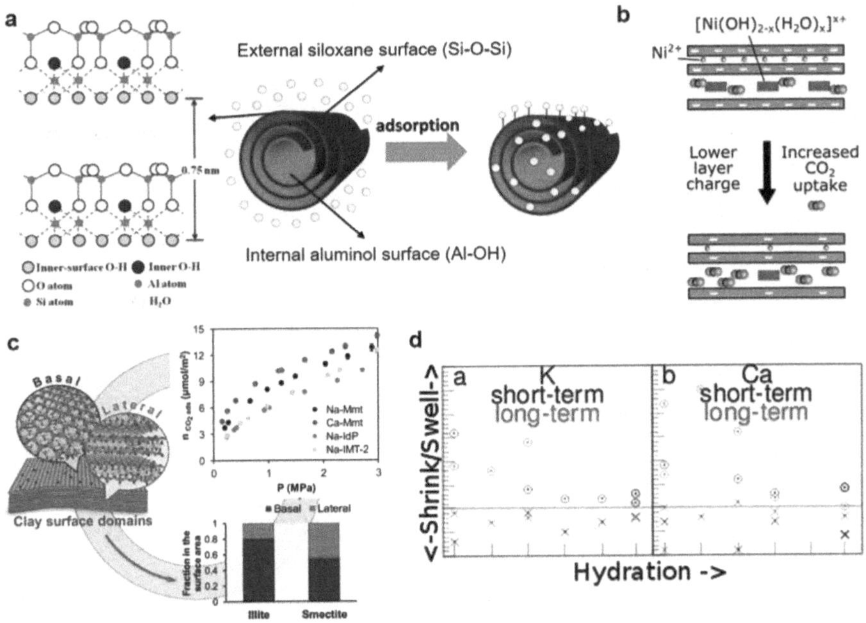

FIGURE 6.9 Several factors that can influence the performance of natural clay minerals in adsorbing CO_2 [70]. Reproduced with permission from Elsevier, license number, 5746480005339.

and moderate porosity. Clays can contain a variety of different cations within their structure, which can influence their properties significantly. To improve their textural characteristics and increase both pore volumes and SAs, clay materials often undergo acid treatment prior to the incorporation of amines. For example, the CO_2 adsorption capacities of untreated kaolinite, bentonite, montmorillonite, and smectite at 25°C and 1 bar varied from 0 to 9.40 mg of CO_2 per gram of adsorbent. In contrast, following acid treatment, the CO_2 uptake was enhanced, rising to values between 3.40 and 15.70 mg CO_2/g ads [71]. Figure 6.10 shows the general schematic of the mineral adsorbents that we are investigating.

6.2.1.1 Kaolinite

One of the most plentiful natural materials is clay mineral, making it readily accessible at a reduced expense. To illustrate, kaolinite, a type of clay mineral, comprises tetrahedral silicate sheets and octahedral hydroxide sheets in a balanced 1:1 proportion. An acid treatment was used to modify kaolinite, an inexpensive and easily accessible clay mineral. The acid-treated kaolinite sample exhibited significantly higher SA and porosity compared to the untreated sample, indicating that acid treatment is an effective method to enhance the textural properties of kaolinite. After subjecting the kaolinite sample to a 10-hour reaction with 3M H_2SO_4, it was discovered that the treated sample displayed the highest Brunauer–Emmett–Teller (BET) SA (ranging from 18.9 to 74.3 m^2/g) and pore volume (0.11 to 0.31 cm^3/g).

TABLE 6.5
The Capacity of Clay Minerals Functionalized with Amines to Adsorb CO_2 [70]

Clay Minerals	Modifier	Loading Factor (wt%)	Method	Pretreatment	Conditions	Adsorption Capacity (mmol CO_2/g)	Reference
Kaolinite	MEA	20	Impregnation	/	298 K, 1 bar	1.36	[108]
	EDA	33.3	Impregnation	/	298 K, 1 bar	2.94	[108]
	4MEA-1EDA	50	Impregnation	/	298 K, 1 bar	3.39	[108]
Halloysite	APTES	20 vol %	Grafting	Acid treatment	760 mmHg	0.113	[109]
Bentonite	AP	2.8 mmol N g^{-1}	Grafting	/	318 K, 1 bar	0.98	[71]
	AP	2.3 wt% N	Grafting	Acid treatment	303 k	0.66	[77]
	DT	1.5 mmol N g^{-1}	Grafting	/	318 K, 1 bar	0.73	[71]
	DT	2.3 wt % N	Grafting	Acid treatment	303 k	0.29	[77]
	APTES	20	Grafting	Pillaring	298 K, 1 bar	1.143	[110]
	TEPA	60	Impregnation	Pillaring	298 K, 1 bar	1.645	[110]
	PEI	60	Impregnation	Pillaring	298 K, 1 bar	1.455	[110]
	PEI	37	Impregnation	/	318 K, 1 bar	1.03	[71]
	PEI	30	Impregnation	/	348 K, 1 bar	1.068	[111]
Montmorillonite	DT	3.4 mmol N g^{-1}	Grafting	/	318 K, 1 bar	1.14	[71]
	PEI	37	Impregnation	/	318 K, 1 bar	1.37	[71]
	AP-PEI	/	Double functionalization	/	318 K, 1 bar	0.3	[71]
	AEAPTS	10.92	Grafting	CTAB intercalate	298 K	2.4	[112]
	AP	2.7 mmol N g^{-1}	Grafting	/	318 K, 1 bar	0.75	[71]
	AEAPTS	10.84	Grafting	/	298 K	1.6	[112]
	CTAB	10.48	Intercalate	/	298 K	0.2	[112]
	APTES	20 vol %	Grafting	Pillaring	298 K, 760 mmHg	1.01	[113]
	PEI	50	Impregnation	Pillaring	348 K, 1 bar	3.6	[113]
	Octadecylamine	20–30	Impregnation	/	298 K, 50 bar	7.16	[114]

Source: Reproduced with permission from Elsevier, license number, 5746480005339.

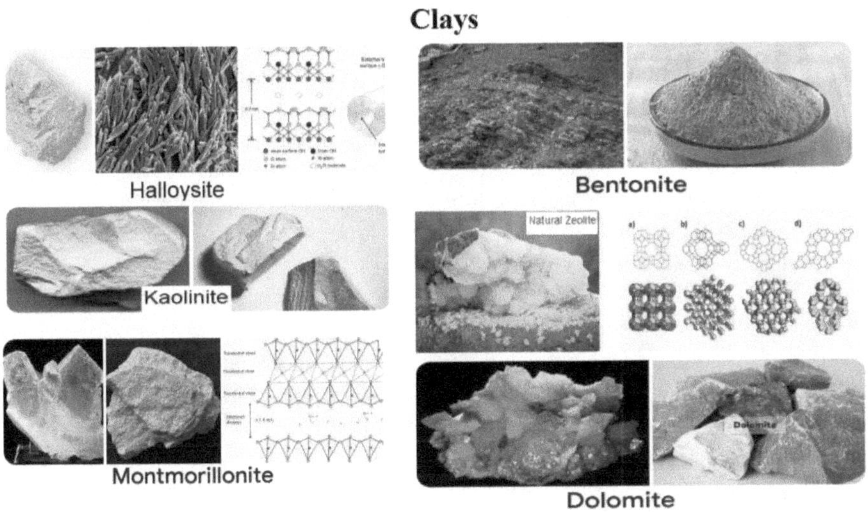

FIGURE 6.10 Mineral adsorbents.

The original CO_2 adsorption capacity of kaolinite at 25°C, which was approximately 0 mg-CO_2/g-sorbent, experienced a significant increase to 3.4 mg-CO_2/g-sorbent following the acid treatment. Additionally, the research revealed that the adsorption capacity of kaolinite for CO_2 was higher at room temperature compared to elevated temperatures [72].

6.2.1.2 Halloysite

Halloysites are a distinct and sustainable type of clay mineral that originates from the weathering of aluminosilicate minerals. These minerals have a distinct structure consisting of an aluminum-based octahedral plane and a silicon-based tetrahedral plane, with no substitution of aluminum or silicon. Halloysites tubes have a double-layered aluminosilicate composition and are similar chemically to kaolinite, with a hollow tubular structure measuring around micrometers. A common method for modifying the structure of halloysite is through acid-base conditions, which lead to structural and morphological changes. Halloysite is readily available, cost-effective, and experiences minimal surface changes compared to CNTs and montmorillonites. As a result, halloysites are commonly used as nanofillers in the production of nanocomposites [3, 6].

The exploration of naturally available HNTs from various locations in South Australia involved a series of steps carried out by Ramadass et al. [73]. These steps included crushing, blunging, reblunging, sedimentation, and filtration aimed at purifying and processing the samples to different degrees. Microscopic and adsorption analyses unveiled that the samples possessed a highly porous structure and displayed a tubular morphology with a notable SSA. Notably, the halloysite content varied considerably among the samples from different locations, although the processing steps introduced minor variations in the final halloysite content. Among the characterized materials, HNT1, derived from Camel Lake, boasted the highest halloysite content (88%) and showcased an SSA of approximately 50.8 m^2/g. HNT1 had a CO_2 adsorption capacity of 3.4 mmol/g at 25°C and increased to 6.1 mmol/g at 0°C. The examined HNT1 has a CO_2 adsorption rate of 120 μmol/m^2, surpassing that of current halloysites and well-known commercial materials such as AC, mesoporous silica, and carbon nitrides. This improved adsorption capability of HNT is chiefly attributed to its exceptional surface properties, tubular morphology, small pores, and the presence of free O–H on both the inner and outer surfaces of HNT, which significantly contribute to the CO_2 adsorption process [73].

6.2.1.3 Montmorillonite

The montmorillonite nanoclay is a commonly used material in the polymer nanocomposite and drilling fluids industry [74]. Its widespread availability is attributed to its distinctive characteristics, including its ability to expand, large SA, and strong surface reactivity [74]. This clay is classified as a cationic smectite and comprises two tetrahedral silicate layers with a central alumina or magnesia octahedral layer [75]. Roth et al. [75] devised an innovative approach to enhance montmorillonite by first grafting it with aminopropyl (AP) groups. They then further improved the material by impregnating it with PEI. Additionally, they explored a method of impregnating montmorillonite that had already been grafted with AP, combining both AP and PEI in the process. This multifaceted modification strategy aimed to optimize the material's properties for enhanced performance in various applications. They found that an optimal temperature of approximately 85°C resulted in high selectivity to CO_2 for all three materials. At this temperature, the clays that were treated with a single amine compound attained a gas uptake of 60 mg CO_2 per gram of adsorbent when exposed to a 10 vol. % CO_2 concentration at 1 bar. However, the double-functionalized sample displayed an even higher gas uptake of 75 mg CO_2/g ads.

A study aimed to develop an affordable and easily accessible solid sorbent for capturing CO_2, using montmorillonite nanoclay as the base material. APTMS and PEI were added to introduce amine functionality to the nanoclay [76]. The experiments showed that the treated nanoclay had fast CO_2 sorption kinetics and high capture capacities, while also being easily regenerated. In tests conducted with pure CO_2 and a 10% CO_2 gas stream diluted with nitrogen, the nanoclay treated with a single amine compound exhibited a capture capacity of around 6 wt%. However, when both amines were used, the capture capacity increased to about 7.5 wt% at 85°C and atmospheric pressure. The sorbent could be regenerated using nitrogen at 100°C over multiple cycles of sorption and desorption. Interestingly, regeneration could also be achieved using dry or humid CO_2 at 155°C, as well as through vacuum desorption. The latter two methods may have more practical applications in industry. Under high pressures of 2.07 MPa, the sorbent demonstrated capture capacities reaching 17 wt% at 85°C, and vacuum desorption was also effective for regeneration in this case. Overall, this research demonstrated that a solid sorbent with high capacity could be created using a cost-effective nanoclay material, and it could perform effectively under a broad spectrum of sorption conditions while being easily regenerated under practical conditions [76].

6.2.1.4 Bentonite

Bentonite mineral is a versatile natural adsorbent that is widely used in various industries and applications. It is composed of the clay mineral montmorillonite, which is formed from volcanic ash and sediment over millions of years. A recent study revealed that untreated bentonite exhibited a modest CO_2 adsorption capacity of just 6.2 mg/g, primarily due to its limited SSA and pore volume. However, following acid modification, significant enhancements in both SSA and pore volume were noted. This improvement was linked to the leaching of octahedral cations, which facilitated the development of an amorphous silica phase, ultimately boosting the material's ability to capture CO_2 effectively. Despite the slight improvement in CO_2 adsorption after clay structure degradation, it was not significant. To improve the CO_2 adsorption capacity, both raw and acid-treated clay minerals were modified by grafting with two distinct amine compounds: aminopropyl-trimethoxysilane (AP) and diethylene-triamine-trimethoxysilane (DT). The addition of AP groups resulted in a more robust chemical interaction with CO_2 molecules compared to the DT groups. This modification led to a remarkable increase in CO_2 adsorption capacity, reaching 29.2 mg/g at 30°C for the sample subjected to an extended acid treatment. Furthermore, the samples functionalized with AP demonstrated excellent stability and reusability, successfully undergoing eight cycles of adsorption and desorption without significant loss of performance [77].

6.2.1.5 Zeolites

Zeolites are microporous aluminosilicate minerals characterized by a specific chemical composition and distinctive porous structure [78]. Zeolites find extensive applications across multiple sectors, including purification processes, adsorption techniques, separation methods, catalysis, and drug delivery systems [12]. These versatile materials are known for their diverse physicochemical properties, which include impressive thermal stability and effective ion exchange capabilities. Their unique structure allows them to perform efficiently in various industrial applications, making them invaluable in enhancing performance and efficiency in these fields [79]. Recent research has concentrated on evaluating how zeolites and mixed matrix membranes that incorporate zeolites perform in gas separation applications, with a particular emphasis on their responsiveness to moisture levels. Furthermore, investigations have delved into the distinctive characteristics of zeolites that are infused with metal nanoparticles, revealing their potential advantages in various applications. Table 6.6 presents a summary of the CO_2 capture capabilities of various zeolite-based adsorbents across different temperatures and pressures [80].

6.2.1.6 Dolomite

Dolomite is mainly composed of calcium carbonate ($CaCO_3$) and magnesium carbonate ($MgCO_3$) and is a common mineral found in sedimentary beds. It is widely used in various industries such as construction, glass production, cement manufacturing, rubber industries, agriculture (fertilizers), gas purification (ammonia removal), and wastewater treatment for heavy metal removal. Dolomite is not actively formed on the earth's surface today; instead, it exists in substantial layers within ancient rocks. These samples exhibit different structures due to extraction from various mines. An interesting characteristic of dolomite is its resistance to mineral alteration after sediment formation. It is

TABLE 6.6
The Effect of Modification Techniques on the Textural Characteristics and CO_2 Adsorption Capabilities of Zeolites [8]

Area (m^2/g)		Volume (cm^3/g)	Volume (cm^3/g)	Volume (cm^3/g)	At 1 bar		mmol/g	Temperature, bar	°C		Method	
Zeolite 4A	e	e	e	e	0.74	e	4.8	25	20.7	e	e	[82]
Zeolite 13X	e	e	e	e	1.08	e	5.2	25	20.7	e	e	[82]
Zeolite 13X	164.3	0.21	e	e	e	e	4.54	25	30	e	e	[83]
Zeolite 5A	e	e	e	e	e	e	4.5	25	1	e	e	[84]
NaY	723	e	0.35	e	e	e	4.8	30	1	e	e	[85]
Nanc-zeolite	698.19	0.369	e	e	e	18.65	4.81	20	1	Novel synthetic methodologies of synthesis	Enhancement of CO_2 capture	[86]
Silicalite-1 (CON)	390.14	0.24	e	e	e	~2.75	2.36	30	0.8			[87]
Silicalite-1 (SON)	382.83	0.31	e	e	e	~3.20	2.12	30	0.8			[87]
A5 zeolite	179.44	e	e	e	11.9	e	5.2	0	30			[88]
Z4A	39	0.14	e	0.13	e	e	3.77	0	1	Novel synthetic methodologies of synthesis; Textural properties adjustment; Moisture effect	Implementation of mesoporosity into zeolite structure	[89]
							3.39	25	1			
							3.00	40	1			
Humid-Z4A	e	e	e	e	e		1.56	40	1			
HZ4A-1–3	126	0.45	e	0.44	e	e	3.41	0	1			[89]
							3.09	25	1			
							2.86	40	1			

Humid-HZ4A-1–3	e	e	e	e	e	e	1.36	40	1			[89]
HZ4A-1–1	112	0.42	e	0.41	e	e	2.72	0	1			[89]
							2.38	25	1			
							2.05	40	1			
HZ4A-3–1	65	0.53	e	0.52	e	e	0.47	0	1			[89]
							0.45	25	1			
							0.38	40	1			
ZY	662	0.30	0.28	0.02	2.2	e	4.5	25	1	Novel synthetic methodologies of synthesis; Textural properties adjustment	Improving structural properties (increasing the pore volume and surface area)	[90]
PDY-3	699	0.36	0.27	0.09	2.3	e	5.1	25	1			[90]
PDY-6	705	0.38	0.26	0.12	2.1	~100.00	5.2	25	1			[90]
PDY-7	773	0.39	0.26	0.13	2.0	160.00	5.9	0	1			[90]
							5.4	25	1			
ZSM-5	383	0.21	0.19	0.02	25	~72.50	2.37	0	1	Novel synthetic methodologies of synthesis; Textural properties adjustment; Moisture effect	Improving structural properties (combining micro- and mesoporous structures)	[91]
							1.81	25	1			
Cl-HP-ZSM-5	438	0.29	0.17	0.12	26	~67.50	2.60	0	1			[91]
							1.95	25	1			
Humid-Cl-HP-ZSM-5	e	e	e	e	e	e	0.26	50	1			[91]
N-HP-ZSM-5	405	0.25	0.18	0.07	25.5	~65.00	2.53	0	1			[91]
							1.91	25	1			

(Continued)

TABLE 6.6 *(Continued)*

The Effect of Modification Techniques on the Textural Characteristics and CO_2 Adsorption Capabilities of Zeolites [8]

Area (m^2/g)		Volume (cm^3/g)	Volume (cm^3/g)	Volume (cm^3/g)	At 1 bar		mmol/g	Temperature, bar °C			Method	
LTA-000	656	0.35	0.29	e	e	e	e	e	e	Textural properties adjustment; Amine functionalization (3-aminopropyl) trimethoxysilane)	Implementation of mesoporosity into zeolite structure; Increasing basicity, affinity between CO_2 molecules-adsorbent surface and level-up N-contents	[92]
APTMS-LTA-050	708	0.64	0.23	e	e	e	1.4	60	0.15			[92]
APTMS-LTA-100	708	0.58	0.14	e	e	e	1.9	60	0.15			[92]
APTMS-LTA-125	499	1.02	0.11	e	e	e	2.3	60	0.15			[92]
TEPA-ZSM-5	19.045	e	0.049	e	e	e	1.80	100	1	Amine functionalization (Tetraethylenepentamine) Amine functionalization (Polyethylenimine)	Increasing basicity, affinity	[93]
Meso-13X	728	0.46	0.25	0.21	e	e	e	100	1		between CO_2 molecules-adsorbent surface and level-up N-contents	[94]
PEI-Meso-13X	1	0.001	e	e	e	e	1.82	100	1			[94]
Ag(I)-SSZ-13	770.88	0.33	0.26	e	11.9	18.80	3.14	25	1	Cation exchange (Ag^+, Co^{2+}, Ni^+)	Increasing basicity, affinity between CO_2 molecules-adsorbent surface	[95]
Co(II)-SSZ-13	786.75	0.44	0.23	e	9.92	52.55	4.49	25	1			[95]
Ni(I)-SSZ-13	836.01	0.44	0.23	e	10.04	42.61	4.45	25	1			[95]
Li-LSX	662	0.32	0.31	e	1.00	128.00	400	60	0.15	Cation exchange (Li^+)		[96]

LiX	560.25	0.30	e	e	e	103.00	e	e	e	Cation exchange (Li^+, Ag^+, Pd^{2+})		[97]
LiPdAgX	549.19	0.29	e	e	e	128.00	e	e	e			[97]
Clino	37	0.14	0.005	e	4.79	e	2.25	10	1	Cation exchange (Na^+);	Increasing basicity, affinity	[98]
							2.16	30		Natural materials as zeolite precursor	between CO_2 molecules-adsorbent surface; Reduction of production costs	
Na-Clino	32	0.14	0.004	e	5.75	e	2.17	10	1			[98]
							2.00	30				
Na-Zeolite b	508	0.52	e	e	7.4	16.80	3.45	0	1	Cation exchange (Na^+, K^+, Li^+)	Increasing basicity, affinity between CO_2 molecules-adsorbent surface	[99]
							2.96	25				
K-Zeolite b	446	0.41	e	e	7.4	14.70	3.57	0	1			[99]
							2.70	25				
Li-Zeolite b	474	0.45	e	e	7.4	20.00	3.08	0	1			[99]
							2.39	25				
Z-Y-3	819	0.33	0.32	e	2.25	e	2.59	30	1	Solid wastes as zeolite precursor (gasified rice husk)	Reduction of production costs	[100]
Coal fly ash zeolites (CFAZ)	486	0.31	0.13	0.17	e	e	3.02	0	1	Solid wastes as zeolite precursor		[101]
Zeolite 13X	643	e	0.35	e	3.0e5.0	e	5.07	25	1	Solid wastes as zeolite precursor (fly ash)		[102]
Fly ash zeolite (F)	414	0.251	0.188	e	e	e	4.16	50	e	Solid wastes as zeolite precursor;	Reduction of production costs; Enhancement of CO_2 capture	[103]
Fly ash zeolite (H)	39	0.138	0.016	e	e	e	1.2	50	e			[103]

(Continued)

TABLE 6.6 *(Continued)*

The Effect of Modification Techniques on the Textural Characteristics and CO_2 Adsorption Capabilities of Zeolites [8]

Area (m²/g)		Volume (cm³/g)	Volume (cm³/g)	Volume (cm³/g)	At 1 bar		mmol/g	Temperature,	bar °C		Method	
Fly ash zeolite (TS)	34	0.069	0.013	e	e	e	0.24	50	e	Novel synthetic methodologies of synthesis		[103]
Zeolite 13X (13X-C)	588	e	0.240	e	e	e	6.9	25	1	Natural materials as zeolite precursor (kaolin, bentonite, feldspar)	Reduction of production costs	[104]
							6.3	35				
Zeolite 13X (13X-K)	591	e	0.250	e	e	6.88	5.5	25	1			[104]
							6.2	35				
Zeolite 13X (13X-B)	505	e	0.160	e	e	e	4.9	25	1			[104]
							4.2	35				
Zeolite 13X (13X-F)	472	e	0.140	e	e	e	3.9	25	1			[104]
							3.2	35				
C-13X	876	0.35	0.35	e	e	26.00	6.18	25	1	Natural materials as zeolite		[105]
B-13X	688	0.38	0.30	e	e	37.00	4.80	25	1	precursor (bentonite)		[105]
Silicalite-1/ TOCNF/gelatin	~272	~0.22	e	e	e	e	~1.24	25	1	Materials based on the gelatin/nanocellulose	Composite/hybrid materials formation	[106]
Zeolite-Y/ Chitosan	795	e	e	e	e	e	1.64	25	1	Chitosan biopolymer		[107]
ZSM-5/Chitosan	444	e	e	e	e	e	2.39	25	1			[107]

found in several regions worldwide. Dolomite's elevated SSA, along with the textural characteristics and pH-sensitive charge on its surface, renders it an outstanding adsorbent for the extraction of metal ions from water sources. The unique physicochemical properties and non-toxic nature of natural clays and rocks, such as dolomite, make them ideal for use as adsorbents [4, 5].

A recent study highlights that natural dolomite may serve as a superior alternative to limestone for developing sorbent precursors in post-combustion CO_2 capture using the calcium looping (CaL) technology. Thermogravimetric analysis (TGA) conducted under conditions mimicking actual sorbent regeneration indicates that the capture capacity of CaO obtained from limestone is considerably influenced by the precalcination parameters and the presence of an intermediate recarbonation phase. In contrast, dolomite-derived sorbents demonstrate a robust capture capacity that remains largely unaffected by these varying conditions. This predictable performance of dolomite, regardless of the precalcination and recarbonation methods used, presents a significant advantage over limestone, which shows a marked sensitivity to these factors. Furthermore, SEM analysis reveals that dolomite samples exhibit improved stability with segregated MgO grains and a CaO layer, which improves the conversion of multicyclic CaO and increases the sorbent's capture capacity. Moreover, dolomite may offer the benefit of quicker decomposition in the presence of CO_2 compared to limestone. This can lead to a significant decrease in the calciner temperature, which is the primary energy-intensive aspect of the process [81].

6.2.2 AC-Based Adsorbents

The utilization of AC in reducing CO_2 emissions is widespread due to its excellent adsorption properties. The adsorption capabilities of AC rely primarily on its internal SA, typically ranging from 1500–500 m^2/g for industrial varieties. AC is favored as a cost-effective absorber for several reasons, including its low sensitivity to moisture, high capacity at normal pressure, and ease of regeneration [115].

The preparation of AC usually involves two processes: carbonization and activation. Carbonization entails the removal of volatile compounds at high temperatures, while activation involves forming intricate pores through the reaction of the remaining material with an active agent. The main constituents for AC production can include fossil fuels such as coal, biomass like wood, almond, and hazelnut skins, as well as resins or other carbon-rich materials. Many studies focus on AC derived from non-renewable fossil fuels, as they offer high CO_2 adsorption capacities [116].

The carbonization process involves heating stagnant air to break down non-carbon elements (hydrogen, oxygen, nitrogen) and obtain carbonaceous matter. Activation, either physically or chemically, creates porosity and active sites, increasing the SA. Physical activation can be achieved using carbon dioxide, steam, or air at temperatures of 1250–1100 K. Chemical activation involves using agents like potassium hydroxide (KOH)[2,50], phosphoric acid (H_3PO_4) [117, 118], or zinc chloride ($ZnCl_2$) [119–121]. Ding and Liu studied the production of ACs using different biomass types (*Sargassum* and *Enteromorpha*) with KOH as the activating agent. They obtained AC with varying surface characteristics by maintaining the activation conditions unchanged [122]. Other studies have also explored different impregnation ratios and activation techniques to produce ACs with the highest SA [123]. Research has shown that increasing the temperature (above 830) results in the formation of more micropores [124]. A notable outcome of using $ZnCl_2$ was the activation of Arundo Biomass from Donax using the dry blending method [125].

The increasing global population and rapid urbanization have led to a surge in CO_2 emissions and solid waste generation [126]. In this context, employing biomass and biochar as low-cost precursors for CO_2 adsorbents presents a dual solution: facilitating effective carbon capture while simultaneously addressing the challenge of waste management. Biomass, derived from organic materials such as agricultural residues, forestry waste, and municipal solid waste, is both abundant and renewable. When processed into biochar through pyrolysis, it transforms into a porous carbon-rich material with unique properties that enhance its effectiveness as a CO_2 adsorbent. Biochar exhibits a hydrophobic character, which significantly increases its affinity for gaseous CO_2, allowing for more effective adsorption compared to traditional materials. The cyclic adsorption process leverages the unique properties of biochar to create an efficient and sustainable method for CO_2 capture (Figure 6.11). This process involves alternating between periods of adsorption and regeneration, thereby maximizing the utilization of biochar as an adsorbent over repeated cycles. Employing biomass and biochar as low-cost CO_2 adsorbents through a cyclic adsorption process not only addresses the urgent challenge of rising greenhouse gas emissions but also provides a sustainable strategy for solid waste management. This intertwined approach enhances resource efficiency, contributes to environmental sustainability, and aligns with the global agenda of mitigating climate change while

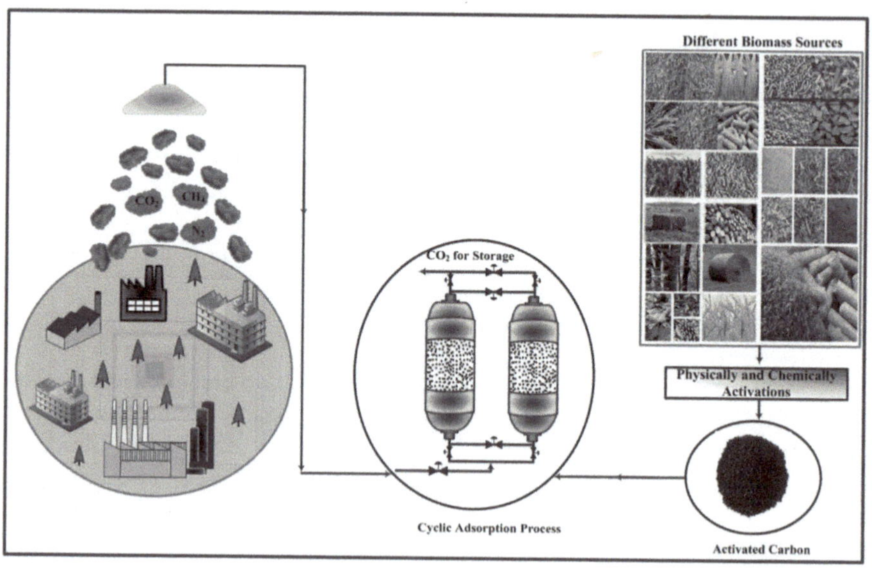

FIGURE 6.11 Activated carbon for CO_2 capture [127]. Reproduced with permission from Elsevier, license number, 5874090689297.

promoting circular economies. The unique properties of biochar, combined with the dynamic nature of cyclic adsorption, pave the way for innovative advancements in carbon capture technologies [127].

6.2.2.1 Solid Wastes

Enormous quantities of solid waste are generated across various sectors such as coal engineering, agricultural manufacturing, water management, and households. According to the United Nations Environment Programme, the global annual production of biomass from agriculture alone is estimated at around 140 billion metric tons. Unfortunately, conventional waste treatment methods like landfills and incineration contribute to climate change and pollution of water, air, and soil. To address these environmental challenges, extensive research is underway to devise innovative strategies for waste management.

One crucial focus is the development of high-value products derived from abundant biomass waste. This is particularly significant in the context of mitigating the cost of carbon dioxide (CO_2) capture and minimizing the energy penalty associated with the capture process. Effectively reducing CO_2 emissions from flue gases poses a significant challenge, and creating valuable products from biomass waste can play a pivotal role in overcoming this challenge. Additionally, the diverse compositions of biowastes have led to the proposal of solid adsorbents with varied physicochemical properties for efficient CO_2 capture within specific temperature ranges.

Lignocellulosic biomasses, including wood and agricultural or forestry by-products, represent the most prevalent type of non-edible biomass. These materials consist mainly of cellulose, hemicellulose, lignin, and a minor proportion of extractives. Transforming lignocellulosic biomass into functional materials, such as carbon-based CO_2 adsorbents, presents a promising strategy to not only repurpose biomass waste but also mitigate environmental pollution resulting from landfilling. The creation of AC using agricultural waste by-products has garnered interest for its potential to repurpose these materials while simultaneously lowering production costs when compared to conventional sources like coal. The processes of recycling and reusing waste are not only energy-efficient but have also gained popularity due to their benefits for the environment and their ability to reduce expenses [128].

The olive stones were subjected to carbonization at 600°C under a N_2 flow, followed by activation at 800°C. The resultant biochar adsorbents displayed a notably elevated pH of 8.7 due to the presence of solid Lewis base sites. The SA and basicity of biochar carbons changed based on the methods used for activation and amine grafting. Notably, biochar adsorbents made from olive stones demonstrated a greater capacity for CO_2 adsorption, reaching 107 mg/g, compared to 86 mg/g for those created from amine-impregnated olive stone carbon at 25°C. Biochars produced from olive stones underwent an NH_3 treatment to improve their CO_2 capture capabilities. The process involved exposing the adsorbents to ammonia at a temperature of 800°C for two hours, which resulted in a CO_2 adsorption capacity of 110 mg g^{-1} surpassing the 88 mg g^{-1} capacity of the unmodified stone-derived AC when tested in flue gas at 30°C. This enhancement can be attributed to the reduction in pore size within the adsorbents, achieved by using higher concentrations of ammonia. Additionally, palm kernel shells were utilized as raw materials following a similar preparation method.

6.2.2.2 Biochar

The high SSA and highly porous structure of biochar are a reason for its polar and hydrophilic nature and possess excellent characteristics for capturing carbon dioxide (CO_2). However, scientists are currently focused on enhancing its CO_2 adsorption capacity through different surface properties can be achieved through the modification of biochar with novel structures. Various methods can be employed to modify biochar, including the utilization of a variety of activation conditions, precursors, and additives. These modifications can be implemented either prior to or following the pyrolysis process, based on the intended changes to the characteristics of the biochar. Chemical modification involves treating biochar with different chemicals to introduce specific functional groups that can enhance CO_2 adsorption. Physical modification, on the other hand, involves altering the structure and surface properties of biochar through processes like grinding or milling. Impregnation with elements involves incorporating metals or metal oxides into the biochar matrix to enhance its sorption capacity. The process of grafting is the addition of specific molecules to the surface of biochar.

Recent studies have concentrated on investigating the application of modified biochar for carbon dioxide adsorption, and several key findings have emerged. These studies have demonstrated that modifying biochar can significantly enhance its CO_2 adsorption capacity. For example, chemical modification with nitrogen-containing compounds has been found to improve the biochar's affinity for CO_2. Similarly, physical modification techniques like thermal treatment or high-temperature carbonization have been shown to enhance the microporosity and biochar's SSA, leading to improved CO_2 adsorption. The impregnation of biochar with metals, such as iron or copper, has also been found to enhance CO_2 adsorption. The metal particles serve as active sites for the adsorption of CO_2, enhancing the total adsorption capability of the modified biochar. Grafting techniques have also been explored, with researchers attaching specific molecules, such as amine groups, onto the surface of biochar to enhance its CO_2-capturing capabilities. In conclusion, modifying biochar through various methods can significantly enhance its CO_2 adsorption capacity. Chemical, physical, impregnation, and grafting techniques can all be employed to tailor the surface properties of biochar and improve its performance as a CO_2 adsorbent. These modifications offer the potential for the development of engineered or designer biochar with enhanced CO_2 capturing capabilities, contributing to initiatives aimed at combating climate change and lowering emissions of greenhouse gases.

Li et al. [129] provide a method that utilizes biochar and porous carbon to synthesize high-pressure CO_2 capture in one step in their study. They specifically focus on the use of biochar derived from mesquite wood as a starting material. The researchers found that by activating potassium hydroxide (KOH) with the mesquite wood biochar, they were able to produce CO_2 sorbents that have a high SA. The SA achieved during optimal activation conditions was 3167 m^2/g. In addition, the researchers recorded a CO_2 absorption rate of 26 mmol/g at a pressure of 30 bar and a temperature of 25°C. To determine the effectiveness of biochar-derived porous carbon for capturing CO_2, they examined its cycling stability and CO_2/CH_4 selectivity and conducted various activity screenings. Their findings indicated that wood waste pyrolyzed at 450°C stood out as the most effective biochar source for generating high-pressure CO_2 sorbents. Overall, this research highlights a promising strategy for developing efficient

CO_2 capture materials derived from biochar. The use of mesquite wood biochar and activation with KOH allowed for the generation of sorbents with high SAs and significant CO_2 uptake capacity. These results aid in the creation of effective and sustainable techniques for capturing CO_2 and could significantly impact efforts to reduce greenhouse gas emissions. In a study by Mazdaki et al. (2016) [130], a comparison was made between the characteristics of raw sawdust biochar and sawdust biochar that had been treated with amine. According to the researchers, gasification of raw sawdust biochar at 850°C was successful, and it resulted in the highest CO_2/kg biochar adsorption. Specifically, at a temperature of 30°C, the raw sawdust biochar was able to adsorb 0.47 kg of carbon dioxide per kg of biochar. Upon amine treatment, the pore structures of the sawdust biochar remained relatively unchanged. The porous nature of the material was not significantly affected by the treatment. Additionally, the surfaces of both the raw and amine-treated sawdust biochar were found to be porous and fragmented, indicating a high SA for potential adsorption.

6.2.2.3 Wood Biochar

Wood biochar is an effective substance for adsorbing CO_2 because it contains a large amount of carbon (ranging from 82.14% to 84.70%) and low ash concentration (3.28% to 4.61%) [131]. The amount of CO_2 adsorbed by wood biochar (78 mg/g) is higher than that of PEI-impregnated silicon carbide fiber (8.3 mg/g) [132] and TiO_2 (27.3 mg/g) [133]. The efficacy of wood biochar in CO_2 adsorption is dependent on its porosity, alkaline nature, and SA [134]. Biochar pores develop due to the carbonization and development of pores in the woody feedstock with high levels of lignin, hemicellulose, and cellulose [135]. Moreover, key functional groups present on the surface of wood biochar, including phenols, C–OH, –OH, C–H, and C=O, enhance its ability to adsorb CO_2 [136]. Research on the CO_2 adsorption properties of biochar produced from wood pyrolyzed at 500°C in a fixed-bed reactor revealed that KOH-activated white wood demonstrated the highest adsorption capacity, reaching 78 mg/g. This significant adsorption capacity can be ascribed to its larger SSA of 1400 m^2/g and mesopore volume of 0.62 cm^3/g. In contrast, another type of biochar, referred to as WCMK char, exhibited a maximum CO_2 adsorption capacity of 2.92 mol/kg, outperforming that of wood chips. This phenomenon can be attributed to its remarkable SSA of 1408 m^2/g, a micropore volume (PV) of 0.36 cm^3/g, and a pore area measuring 690.18 m^2/g. The activation of biochar can lead to further enhancement of the adsorption capacity as discovered. Moreover, combining wood with chicken manure proved to be an effective way to obtain an excellent adsorbent with enhanced adsorption capabilities [137].

The adsorption capacity of wood biochar, which stands at 78 mg/g, is affected by several variables including SA, heating rate, residence time, adsorption temperature, pyrolysis pressure, porosity, and material composition. A significantly larger SA, such as 1408.8 m^2/g, offers more active sites for adsorption, leading to enhancements in capacity. However, a high ash content in biochar can decrease its SSA by clogging pores, which negatively impacts its ability to adsorb. On the other hand, using feedstock rich in carbon can enhance the CO_2 adsorption capacity of biochar [134, 138].

The reactivity of biochar is greatly affected by the heating rate. Higher heating rates lead to the development of more macropores, which contribute to increased adsorption capacity. Pyrolysis temperature is another critical factor. Elevating the temperature from 250°C to 700°C creates internal pores, significantly impacting the SA. Micropores have a greater influence on adsorption than other pores, and a higher number of micropores results in a larger SA. Biochar that is produced through high-pressure pyrolysis has a higher carbon content and a longer residence time, which is a factor in its ability to absorb substances. Increasing the duration of residence between 30 and 90 minutes further enhances the SA. The development of porous structures in biochar can be enhanced by discharging organic and volatile molecules during the pyrolysis process. However, an increase in adsorption temperature from 0°C to 65°C can decrease the adsorption capacity of biochar. At higher temperatures, the Brownian motion of CO_2 molecules promotes desorption as a result. Table 6.7 showcases a variety of factors that exert influence on the adsorption of CO_2 on wood biochar [134].

TABLE 6.7
Several Factors Can Influence the Adsorption of CO_2 on Wood-Based Biochars [134]

Biomass	Process, *T* (°C), *t* (h), and FR (L/S)	CO_2 Adsorption Capacity (mg/g)	SSA (m^2/g)	O/C	PV (cm^3/g)	Reference
Food and wood	Gasification, 500°C	0.088	294	—	0.1	[139]
White wood	Fast pyrolysis, *T*—25–65°C	78	1400	0.21	0.62	[137]
Wood chip + chicken manure	Gasification, *t*—3 h, gas FR—4	0.128	1408.8	0.36	0.85	[140]
Douglas fir wood	Pyrolysis, *T*—20°C	—	277	0.36	0.43	[141]
Hybrid poplar wood	N_2 FR—15×10^{-3}		257	0.32	0.35	
Hickory wood	Slow pyrolysis, *T*—600°C, *t*—3 h	41.23–45.85	347.96	0.10	0.471	[136]
Pinewood			356.87	0.14	0.464	
HW 300	Slow pyrolysis (300°C, 450°C, 600°C)	11.15–73.55	0.10	0.35	—	[142]
HW 450			12.90	0.14		
HW 600			401	0.17		
Pure sewage sludge and Leucaena wood	Microwave CO—pyrolysis	61.97	—	—	—	[143]
	N_2 FR—41.6×10^{-5}					
Maple wood chips	Pyrolysis, 500°C	1.63–3.26	-	—	—	[144]
Wood chip—W	Pyrolysis, 500°C	0.0005–0.0018	196	—	—	[145]
Rubber-wood-sawdust	Pyrolysis, HR—10°C/min	18	10–200	0.130	—	[146]

6.3 RICE HUSK

Rice husk, an agricultural waste generated during rice milling, has received considerable attention as a possible adsorbent for CO_2. The high carbon content in rice husk, along with its porous structure, is a material that has the potential to be used for CO_2 capture and storage applications. The CO_2 molecules and the surface of the rice husk have physical and chemical interactions that cause the adsorption process to take place. The abundant silicates and cellulose present in rice husk provide a large SA for CO_2 adsorption. The porosity of rice husk allows for easy diffusion of CO_2 molecules, enhancing the adsorption capacity. Additionally, the presence of alkali and alkaline earth metals in rice husk, such as potassium and magnesium, can also improve its CO_2 adsorption capacity. These metal ions can react with CO_2, forming carbonates and bicarbonates, which further increase the adsorption efficiency. Research studies have explored different methods to improve the capacity of CO_2 sorption in rice husks. Surface modification techniques, such as impregnation with amine-based compounds or alkali activation, have been employed to increase the quantity of active sites accessible for CO_2 adsorption. Moreover, the utilization of rice husk in composite materials, such as mixing it with polymers or activating it with other carbon-based materials, has shown improved CO_2 adsorption performance [147–151].

AC is a versatile material that is widely used for adsorption processes. Its adsorption capabilities are greatly influenced by its acidic and basic surface groups. In the case of CO_2 adsorption, the basicity of AC is particularly important. To enhance the fundamentality of AC for CO_2 adsorption, it can be subjected to high temperatures. This process involves heating the material, which leads to the decomposition of acidic surface groups. By removing these acidic groups, the basicity of the AC is increased, making it more effective at adsorbing CO_2 [2, 15, 152–156].

To synthesize AC, researchers have explored various methods, including chemical activation. AC from rice husk was synthesized using the chemical activation method by Muniandy et al. and Guo et al. In their research, they compared the use of activating agents including NaOH and KOH. According to their findings, KOH was found to be superior in terms of its activation capabilities. This suggests that using KOH as an activating agent can lead to the production of AC with enhanced basicity for CO_2 adsorption [157, 158].

The study was focused on developing adsorption compounds from rice hulls to obtain carbon dioxide (CO_2) from a gas stream in an industrial setting [159]. To achieve this, various pretreatment processes, including desilicalization, chemical activation, and K_2CO_3 impregnation, were applied to prepare the adsorbents. The adsorbents were then subjected to detailed physicochemical characterization. A fixed-bed reactor was employed to evaluate the CO_2 capture capabilities of the newly developed adsorbents. The research focused on examining how various process parameters, including adsorption temperature, moisture levels, and the amount of active component loaded, affected the CO_2 adsorption capacity. The outcomes were contrasted with those acquired from a commercial AC. Remarkably, the 20K-ARH adsorbent exhibited a maximum adsorption capacity of 67 g of CO_2 per kilogram of adsorbent. The research highlighted the critical role of water in CO_2 adsorption, demonstrating that the presence of moisture facilitated the formation of $KHCO_3$ during the adsorption process. Furthermore, investigations pinpointed the ideal parameters for maximizing CO_2 removal from a model flue gas, which included temperatures between 30°C and 40°C, relative humidity levels between 80 and 90%, and a K_2CO_3 concentration of roughly 20 wt %. Notably, the findings indicated that about 94% of the captured CO_2 could be effectively released at a temperature of 180°C, allowing for the adsorbent to be utilized in several cycles while retaining a substantial capture capability. In summary, this work successfully prepared adsorbents from rice husks for carbon dioxide capture applications. Through comprehensive physicochemical characterization and examination of various process parameters, the researchers gained valuable insight into the adsorption capacity and behavior of the adsorbents. These findings contribute to the development of effective and sustainable CO_2 capture technologies [159].

The FESEM images of rice husk biochar (RHBC) and KOH-activated porous carbons (PAC1 and PAC2) are presented in Figure 6.11a–c. The morphological examination reveals that RHBC is characterized by a lack of pores, while the application of KOH activation results in the formation of a densely porous structure within the ACs, with visible micron-sized pores on the surfaces of PAC1 and PAC2. The KOH impregnation ratio increased significantly, causing a noticeable rise in pore formation. Additionally, following chemical activation, the previously smooth and even surface of RHBC transformed into a network of pores, with PAC2 exhibiting a markedly uniform and highly porous architecture compared to PAC1. The activated surface is marked by larger pores measuring approximately 2–5 μm, alongside micropores and mesopores that are embedded within the structure. SA analysis of PAC2 indicates that about 76.34% of its total SA consists of micropores. KOH plays a crucial role in facilitating the oxidation of rice husk; during the activation process, it is believed that the interaction between KOH and carbon generates gases that promote pore formation, particularly contributing to the development of small mesopores and micropores. CO_2 adsorption was assessed at three temperatures: 0°C, 25°C, and 50°C, across various pressure conditions. Figure 6.12 illustrates the comparative CO_2 adsorption

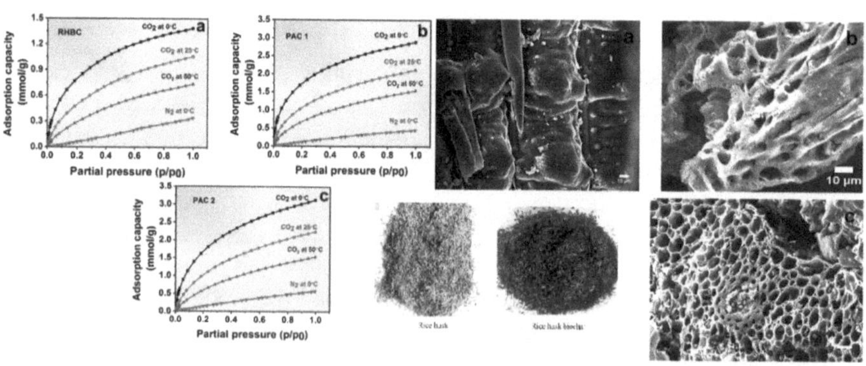

FIGURE 6.12 Adsorption isotherms of CO_2 and FESEM for rice husk-derived carbons: (a) RHBC, (b) PAC1, and (c) PAC2 [160].

capacities of different samples under these experimental conditions. Notably, PAC2 achieved the peak adsorption rate of 3.13 mmol/g at 0°C and 1 bar pressure. Even at a lower pressure of 0.1 bar, the samples demonstrated favorable performance, likely attributed to the existence of narrower micropores ($V < 0.7$ nm) [160].

6.4 SAWDUST

Sawdust wastes were utilized as a cost-effective source to produce porous carbons through the activation process using KOH. The researchers conducted an orthogonal experiment to optimize the preparation parameters. The study focused on four key activation parameters: sawdust kind, KOH to sawdust ratio, activation temperature, and activation period. Findings indicated that both the activation temperature and the KOH to sawdust ratio exerted a substantial influence on the porous texture characteristics, specifically the SSA and micropore volume of the porous carbons produced from sawdust waste. The best preparation conditions produced porous carbon from sawdust waste, which exhibited an SSA of 1601 m^2/g and a micropore volume of 0.75 cm^3/g. This carbon material had a suitable CO_2 adsorption capacity of 4.25 mmol/g at 0°C and ambient pressure. In comparison, the extrudated form of the sawdust wastes-derived porous carbon showed a lower CO_2 adsorption capacity, ranging from 2.85 to 3.03 mmol/g. The reduction in adsorption capacity was mainly due to the extrusion process and the incorporation of binders, which led to the collapse and obstruction of the original porous structure. As a result, these changes heightened the resistance to CO_2 diffusion and adsorption. Nevertheless, the extruded porous carbon derived from sawdust waste offered several practical benefits. It demonstrated no pressure build-up, enhanced bulk density, and impressive compressive strength, measuring between 7.3 and 12.3 MPa. These advantages make the extrudated porous carbon a favorable choice for real-world applications [161]. Comparing the carbon dioxide uptake capacity, various biomass-derived carbon materials and commercial adsorbents are used as adsorbents indicated in Table 6.8.

6.5 THE IMPACT OF BIOCHAR MODIFICATION ON ITS PROPERTIES AND ABILITY TO ADSORB CO_2

Biochar, a substance resembling charcoal, is created through the pyrolysis process of biomass. It has garnered considerable interest in recent years for its potential benefits in carbon storage and enhancing soil quality. One environmentally friendly material that can be used for capturing CO_2 is biochar. Biochar can be generated from readily available biomass and waste materials, including agricultural residues, wood by-products, animal excrement, food scraps, municipal solid waste, and sewage sludge. Utilizing biochar produced from waste for carbon dioxide sequestration not only helps capture CO_2 but also facilitates sustainable waste management [129, 193–195]. However, the properties of biochar can vary depending on the feedstock used and the pyrolysis conditions employed. To enhance the performance of biochar and expand its applications, researchers have been exploring various modification techniques. These modifications can involve physical, chemical, or biological treatments that alter the biochar's surface structure and chemical composition (Figure 6.13a). One effect of biochar modification

TABLE 6.8
Comparing the Carbon Dioxide Uptake Capacity, Various Biomass-Derived Carbon Materials and Commercial Adsorbents Are Used as Adsorbents [19]

Precursor	Adsorption Conditions	Preparation Method	BET Surface Area (m^2/g)	Total Pore Volume (cm^3/g)	CO_2 (mmol/g)	Ref
Celtuce leaves	25 °C, 1 bar, pure CO_2	Carbonization-activation (KOH)	3404	1.88	4.40	[162]
Palm kernel shell	30°C, 4 bar, pure CO_2	HTC-calcination at 700 °C	167.08	0.089	7.32	[163]
Almond shell	25°C, 1 bar, pure CO_2	Carbonization-activation (CO_2)	831	0.37	2.60	[164]
Walnut shell	25°C, 1 bar	Mg loading	292	0.157	1.86	[165]
Sawdust	70°C, 1 bar	Treatment with monoethanolamine (MEA)	0.15	—	0.51	[130]
White wood	25°C, 1 bar	Steam activation	840	0.55	1.34	[137]
White wood	25°C, 1 bar	Carbonization-activation (KOH)	1400	0.62	1.80	[137]
Olive stone	25°C, 1 bar, pure CO_2	Carbonization-activation (CO_2)	1079	0.502	2.40	[166]
Olive stone	25°C, 1 bar	Agent: O_2 (3%) at 650°C	697	0.27	2.00	[167]
Wood ashes	60°C, 1 bar, 5% CO_2	Tetraethylenepentamine (TEPA)	0.71	0.00076	2.02	[168]
Mesua ferra L. Seed cake	25°C, 1 bar, 10% CO_2	HTC-calcination at 650 °C	184.1	0.13	2.60	[169]
Coffee grounds	25 °C and 15 vol% of CO_2	Functionalization by deep eutectic solvent	1033	0.52	5.50	[170]
Palm shell activated carbon	25°C and 1 bar	Choline hydroxide and urea	682.94	0.31	0.90	[171]
Coconut shell-based AC	25°C and 1 bar	Char + urea and KOH	1535	0.56	4.80	[172]
Coconut shell	25°C and 1 bar	Carbonization-activation (CO_2)	370.72	0.15	1.80	[173]
Bamboo-AC	25°C and 1 bar	Carbonization-activation (CO_2)	787	—	2.52	[174]
Bamboo-AC	25°C and 1 bar	Modification of AC with air oxidation	811	—	2.26	[174]
Waste wool-derived N-doped hierarchical porous carbon	25°C and 1 bar	KOH chemical activation	1352	0.78	2.78	[175]
Pine cones	25°C and 1 bar	HTC-carbonization at 220°C	688	0.49	4.06	[176]
Walnut shell-based carbons	25°C and 1 bar	Carbonization with chemical activation using KOH	1868	1.06	5.17	[177]

Carbon spheres	0°C and 20 bar	Synthesized from $KHCO_3$ activation of glucose derived hydrochar	3576	1.91	35.57	[178]
Pistachio shells	25°C and 5.6 bar	Physical activation by steam	715.34	0.34	3.29	[155]
Entada Africana Guill. & Perr	25°C and 1 bar	A single-step KOH activation	2556	0.78	4.34	[15]
Spent coffee grounds	25°C and 1 bar	Modified with K_2CO_3 at 700°C	2337	1.15	4.54	[179]
Hazelnut Shell	25°C and 1 bar	KOH activation	1099	0.41	4.32	[180]
Cerastoderma glaucum (CG)	25°C and 4.5 bar	Calcined it for 30 min at 850°C	1.9312	0.0055	0.48	[156]
Sludge	25°C and 1 bar	Modified with sodium hydroxide (NaOH)	178.9	0.25	1.30	[181]
Nutshell	25°C and 1 bar	Agent: CO_2 at 950°C	573	0.17	3.48	[182]
Pine wood pellets	30°C and 1 bar	Agent: CO_2 at 900°C	561	0.22	6.66	[183]
Argan fruits shells	25°C and 1 bar	Modified with NaOH	1889	0.87	5.63	[184]
Pecan nutshell-based biochar	0°C and 1 bar	Prepared using microwave pyrolysis	187	0.075	2.50	[185]
Activated mesoporous carbons	25°C and 1 bar	Activated at 850°C with a KOH	1586	1.12	2.48	[186]
Coal-AC	25°C, 2.12 bar	—	915	—	0.48	[187]
Coal-AC	25°C, 2.12 bar	Modified with sodium hydroxide	860	—	1.16	[188]
Fly ash and poly(vinyl chloride)	55°C	10 wt % PVC_{62} ($M_w = 62{,}000$), 17 wt % FA	—	—	1.33	[188]
Commercial AC	40°C, 0.15 bar	—	—	—	0.30	[189]
Commercial activated carbon	0°C and 1 bar	KOH was used as modification agents	2406	0.46	5.80	[190]
Commercial activated carbon	25°C and 1 bar	KOH was used as modification agents	2406	0.46	3.67	[190]
Commercial granular AC: GAC-10–500	25°C and 6 bar	Modified with $ZnCl_2$	719.13	0.31	4.00	[191]
GAC-pure	25°C and 2 bar	—	921.09	0.41	1.80	[191]
Commercially available porous Norit RB-2 activated carbon: Norit-RB2-AC	25°C and 1.25 bar	—	816.31	0.44	0.99	[192]
Date stones-based biomass: AC-DS	25°C and 1.25 bar	Physical activation	821.69	0.45	1.66	9
β-Zeolite	40°C and 1 bar	—	—	—	1.48	44

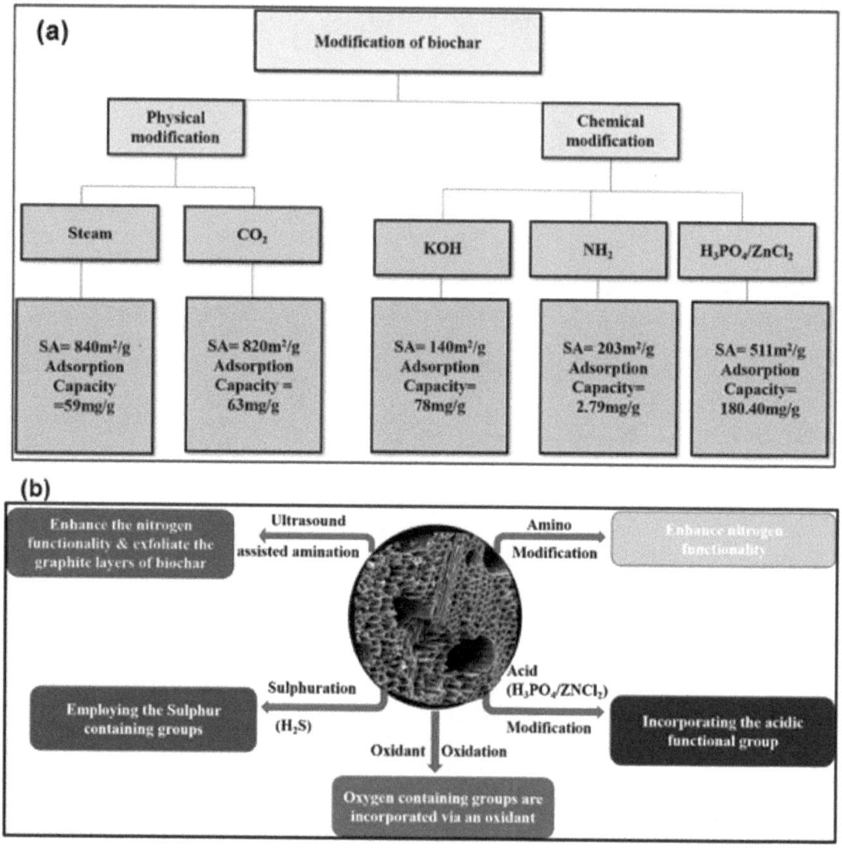

FIGURE 6.13 (a) Various variations of biochar, specifically those with different surface areas measured as SA, are available as modifications [134]. (b) Various techniques were utilized to modify the surface of biochar [134].

is an improvement in its SA and porosity. By increasing the SA, the active sites available for gas adsorption, including CO_2, can be significantly enhanced. This can result in a greater CO_2 adsorption capacity, making modified biochar a potential material for carbon capture applications.

Surface-functionalized biochar has demonstrated significant efficiency as an adsorbent for carbon dioxide, attributed to its enhanced SA (greater than 203 m^2/g) and the provision of additional active sites conducive to CO_2 adsorption. A range of surface modifications including amino functionalization, acid treatment, sulfuration, and ultrasound-assisted amination are utilized to introduce targeted functional groups onto the biochar, either prior to or following the pyrolysis process. These functional groups help enhance the structure and improve the functionality of biochar for different applications. Compared to raw biochar, functionalized biochar offers a higher SA. Engineered biochar can be characterized by specific surface functional groups, such as hydroxyl (O–H), carbonyl (C=O), and ether (C–O) groups. These functional groups are categorized into various types, which include nitrogen-containing groups (such as amine-N, graphitic-N, and pyrrolic-N), oxygen-containing groups (including carboxylic and phenolic groups), and sulfur-containing groups (like sulfonic groups). These functional groups contribute to the enhanced CO_2 adsorption capabilities of functionalized biochar (Figure 6.13b) [134].

Moreover, modification techniques can also enhance the functional groups found on the surface of biochar. Certain chemical treatments, such as oxidation or functionalization with nitrogen-containing compounds, can introduce additional oxygen or nitrogen groups. These functional groups can offer additional adsorption sites for CO_2 molecules, thereby improving the adsorption capacity of the modified biochar. Additionally, modification techniques can also improve the thermal stability of biochar. The pyrolysis process can result in the formation of volatile compounds that can be released upon heating, potentially reducing the biochar's stability. However, by modifying the biochar, these volatile compounds can be removed or transformed into more stable forms, making the modified biochar more resistant to thermal degradation. Furthermore, biochar modification can also influence its water-holding capacity and nutrient retention abilities. Through physical or chemical modifications, the hydrophilic character of biochar can be altered, leading to enhanced water retention capacity. This can be beneficial for agricultural applications as it helps to improve soil moisture levels and reduce irrigation needs. In summary, biochar modification techniques can have profound effects regarding its characteristics and ability to adsorb CO_2 (Table 6.9). By increasing the SA, modifying functional groups, enhancing thermal stability, and improving water-holding capacity, modified biochar exhibits enhanced performance for CCS, soil improvement, and agricultural applications. These modifications offer great potential for expanding the applications of biochar as a sustainable and effective tool for mitigating climate change.

6.6 PROCESS OF CARBON DIOXIDE CAPTURE ON BIOCHAR

Physisorption and chemisorption are two mechanisms of adsorption that have been extensively studied. Physisorption is influenced by temperature, with increased CO_2 adsorption occurring at lower temperatures. The linear configuration and polar characteristics of CO_2 facilitate its interaction with the active sites on the adsorbent's surface. The forces that draw CO_2 to the adsorbent are based on intermolecular interactions, and this process does not require any activation energy. In contrast, chemisorption involves a covalent interaction between an acidified carbon dioxide molecule and the active site on the adsorbent's surface. Different adsorbents display a range of chemisorption processes, and compounds containing nitrogen, such as amines, can be used to incorporate functional groups into the substrate, resulting in chemisorption. The reaction between CO_2 and amines can lead to the formation of carbamate, where the carbon atom in CO_2 reacts with the amine group. Additionally, through a base-catalyzed hydration process, carbon dioxide can be converted into bicarbonate ions, which subsequently interact with tertiary amines to enhance adsorption. It has been found that raising the temperature can promote chemisorption, as it frequently requires activation energy. For instance, ammoniated biochar, created using ammonia, is capable of chemically adsorbing CO_2 at a temperature of 120°C. The process

TABLE 6.9
Impact of Biochar Modification on Its Characteristics and CO_2 Adsorption Potential [202]

Feedstock	Pyrolysis Temperature (°C)	Modification Method	BET Surface Area (m^2/g)	Surface Area of Micropores (m^2/g)	Total Pore Volume (cm^3/g)	Micropore Volume (cm^3/g)	Adsorption Temperature (°C)	CO_2 Adsorption Capacity (mg/g)	Reference
Whitewood	500	Steam activation	840	N/A	0.55	N/A	25	59	[137]
Whitewood	500	CO_2 activation	820	N/A	0.45	N/A	25	63	[137]
Whitewood	500	KOH activation	1400	N/A	0.62	N/A	25	78	[137]
Soybean straw	500	Raw biochar without activation	0.04	250	N/A	0.1	30	45 (approx.)	[203]
Soybean straw	500	CO_2 activation at 500°C	5.5	300	N/A	0.12	30	46 (approx.)	[203]
Soybean straw	500	CO_2 activation at 600°C	2.6	342	N/A	0.14	30	58 (approx.)	[203]
Soybean straw	500	CO_2 activation at 700°C	22	398	N/A	0.16	30	60 (approx.)	[203]
Soybean straw	500	CO_2 activation at 800°C	346	473	N/A	0.19	30	76 (approx.)	[203]
Soybean straw	500	CO_2 activation at 900°C	397	445	N/A	0.18	30	66 (approx.)	[203]
Soybean straw	500	Ammonification with NH_3 at 500°C	1.5	311	N/A	0.13	30	48 (approx.)	[203]
Soybean straw	500	Ammonification with NH_3 at 600°C	5.8	339	N/A	0.14	30	57 (approx.)	[203]
Soybean straw	500	Ammonification with NH_3 at 700°C	221	433	N/A	0.17	30	62 (approx.)	[203]
Soybean straw	500	Ammonification with NH_3 at 800°C	365	479	N/A	0.19	30	79 (approx.)	[203]
Soybean straw	500	Ammonification with NH_3 at 900°C	469	461	N/A	0.19	30	74 (approx.)	[203]
Soybean straw	500	Treatment with CO_2–NH_3 mixture at 500°C	2	318	N/A	0.13	30	55 (approx.)	[203]
Soybean straw	500	Treatment with CO_2–NH_3 mixture at 600°C	1.2	370	N/A	0.15	30	60 (approx.)	[203]
Soybean straw	500	Treatment with CO_2–NH_3 mixture at 700°C	41	439	N/A	0.18	30	64 (approx.)	[203]
Soybean straw	500	Treatment with CO_2–NH_3 mixture at 800°C	491	534	N/A	0.21	30	89 (approx.)	[203]

Soybean straw	500	Treatment with CO_2–NH_3 mixture at 900°C	764	489	N/A	0.2	30	82 (approx.)	[203]
Cotton stalk	600	Unmodified biochar	N/A	224	N/A	0.07	20	38 (approx.)	[204]
Cotton stalk	600	Modified with CO_2 at 500°C	N/A	289	N/A	0.12	20	53 (approx.)	[204]
Cotton stalk	600	Modified with CO_2 at 600°C	N/A	351	N/A	0.13	20	64 (approx.)	[204]
Cotton stalk	600	Modified with CO_2 at 700°C	N/A	372	N/A	0.14	20	66 (approx.)	[204]
Cotton stalk	600	Modified with CO_2 at 800°C	N/A	610	N/A	0.24	20	99.42	[204]
Cotton stalk	600	Modified with CO_2 at 900°C	N/A	556	N/A	0.21	N/A	96 (approx.)	[204]
Cotton stalk	600	Modified with NH_3 500°C	N/A	161	N/A	0.06	N/A	26 (approx.)	[204]
Cotton stalk	600	Modified with NH_3 600°C	N/A	252	N/A	0.1	N/A	52 (approx.)	[204]
Cotton stalk	600	Modified with NH_3 700°C	N/A	255	N/A	0.1	N/A	50 (approx.)	[204]
Cotton stalk	600	Modified with NH_3 800°C	N/A	349	N/A	0.14	N/A	75 (approx.)	[204]
Cotton stalk	600	Modified with NH_3 900°C	N/A	435	N/A	0.17	N/A	78 (approx.)	[204]
Cotton stalk	600	Modified with CO_2 and NH_3 mixture 500°C	N/A	95	N/A	0.04	N/A	15 (approx.)	[204]
Cotton stalk	600	Modified with CO_2 and NH_3 mixture 600°C	N/A	297	N/A	0.12	120	52 (approx.)	[204]
Cotton stalk	600	Modified with CO_2 and NH_3 mixture 700°C	N/A	336	N/A	0.13	N/A	65 (approx.)	[204]
Cotton stalk	600	Modified with CO_2 and NH_3 mixture 800°C	N/A	627	N/A	0.25	N/A	95 (approx.)	[204]
Cotton stalk	600	Modified with CO_2 and NH_3 mixture 900°C	N/A	469	N/A	0.19	N/A	90 (approx.)	[204]

(Continued)

TABLE 6.9 *(Continued)*
Impact of Biochar Modification on Its Characteristics and CO_2 Adsorption Potential [202]

Feedstock	Pyrolysis Temperature (°C)	Modification Method	BET Surface Area (m^2/g)	Surface Area of Micropores (m^2/g)	Total Pore Volume (cm^3/g)	Micropore Volume (cm^3/g)	Adsorption Temperature (°C)	CO_2 Adsorption Capacity (mg/g)	Reference
Cotton stalk	600	Unmodified biochar	224.12	N/A	N/A	0.07	20	58 (approx.)	[205]
							120	14 (approx.)	
Cotton stalk	600	Modified with NH_3 at 500°C	N/A	160.89	N/A	0.06	20	46 (approx.)	[205]
							120	36 (approx.)	
Cotton stalk	600	Modified with NH_3 at 600°C	N/A	251.91	N/A	0.08	20	50 (approx.)	[205]
							120	35 (approx.)	
Cotton stalk	600	Modified with NH_3 at 700°C	N/A	254.97	N/A	0.14	20	60 (approx.)	[205]
							120	28 (approx.)	
Cotton stalk	600	Modified with NH_3 at 800°C	N/A	348.56	N/A	0.17	20	72 (approx.)	[205]
							120	13 (approx.)	
Cotton stalk	600	Modified with NH_3 at 900°C	N/A	434.92	N/A	0.19	20	78 (approx.)	[205]
							120	10 (approx.)	
Cotton stalk	600	Modified with CO_2 at 500°C	N/A	289.07	N/A	0.12	20	64 (approx.)	[205]
							120	10 (approx.)	
Cotton stalk	600	Modified with CO_2 at 600°C	N/A	351.49	N/A	0.13	20	54 (approx.)	[205]
							120	12 (approx.)	
Cotton stalk	600	Modified with CO_2 at 700°C	N/A	371.65	N/A	0.14	20	72 (approx.)	[205]
							120	13 (approx.)	
Cotton stalk	800	Modified with CO_2 at 800°C	N/A	610.04	N/A	0.24	20	96 (approx.)	[205]
							120	20 (approx.)	
Cotton stalk	600	Modified with CO_2 at 900°C	N/A	556.35	N/A	0.21	20	80 (approx.)	[205]
							120	16 (approx.)	

Sawdust	450	Unmodified biochar	8.76	N/A	N/A	N/A	30	19.7	[130]
Sawdust	450	Unmodified biochar	8.76	N/A	N/A	N/A	70	13.5	[130]
Sawdust	450	Treatment with monoethanolamine	0.61	N/A	N/A	N/A	30	19.1	[130]
Sawdust	450	Treatment with monoethanolamine	0.61	N/A	N/A	N/A	70	12.1	[130]
Sawdust	450	Treatment with monoethanolamine	0.61	N/A	N/A	N/A	70	12.1	[130]
Sawdust	750	Unmodified biochar	1.36	N/A	N/A	N/A	30	45.2	[130]
Sawdust	750	Unmodified biochar	1.36	N/A	N/A	N/A	70	25.4	[130]
Sawdust	750	Treatment with monoethanolamine	0.15	N/A	N/A	N/A	30	39.7	[130]
Sawdust	750	Treatment with monoethanolamine	0.15	N/A	N/A	N/A	70	22.6	[130]
Sawdust	850	Unmodified biochar	182.04	N/A	N/A	N/A	30	47.5	[130]
Sawdust	850	Unmodified biochar	182.04	N/A	N/A	N/A	70	28.8	[130]
Sawdust	850	Treatment with monoethanolamine	3.17	N/A	N/A	N/A	30	44.8	[130]
Sawdust	850	Treatment with monoethanolamine	3.17	N/A	N/A	N/A	70	25.2	[130]
Walnut shell	500	Unmodified biochar	94.509	N/A	0.054	0.021	N/A	N/A	[165]
Walnut shell	900	Unmodified biochar	397.015	N/A	0.198	0.159	25	72.6	[165]
							70	30.07	
Walnut shell	900	Mg loaded	292.002	N/A	0.157	0.118	25	82.04	[165]
							70	43.76	
Cottonwood	600	Unmodified biochar (CW)	99	N/A	0.01	N/A	25	57.96	[206]
Cottonwood	600	Mg:CW = 0.01	275	N/A	0.01	N/A	25	63.69	[206]
Cottonwood	600	Mg:CW = 0.25	244	N/A	0.03	N/A	25	47.69	[206]
Cottonwood	600	Mg:CW = 1	184	N/A	0.1	N/A	25	35.35	[206]
Cottonwood	600	Mg:CW = 3	228	N/A	0.12	N/A	25	33.83	[206]
Cottonwood	600	Mg:CW = 6	197	N/A	0.29	N/A	25	27.79	[206]

(Continued)

TABLE 6.9 *(Continued)*

Impact of Biochar Modification on Its Characteristics and CO_2 Adsorption Potential [202]

Feedstock	Pyrolysis Temperature (°C)	Modification Method	BET Surface Area (m^2/g)	Surface Area of Micropores (m^2/g)	Total Pore Volume (cm^3/g)	Micropore Volume (cm^3/g)	Adsorption Temperature (°C)	CO_2 Adsorption Capacity (mg/g)	Reference
Cottonwood	600	Mg:CW = 20	289	N/A	0.25	N/A	25	35.05	[206]
Cottonwood	600	Mg:CW = 40	262	N/A	0.27	N/A	25	32.33	[206]
Cottonwood	600	Al:CW = 0.025	256	N/A	0.01	N/A	25	63.87	[206]
Cottonwood	600	Al:CW = 0.25	206	N/A	0.03	N/A	25	62.98	[206]
Cottonwood	600	Al:CW = 2.5	331	N/A	0.3	N/A	25	69.3	[206]
Cottonwood	600	Al:CW = 1	263	N/A	0.25	N/A	25	64.63	[206]
Cottonwood	600	Al:CW = 3	370	N/A	0.39	N/A	25	69.49	[206]
Cottonwood	600	Al:CW = 4	367	N/A	0.37	N/A	25	71.05	[206]
Cottonwood	600	Fe:CW = 0.01	302	N/A	0.01	N/A	25	64.3	[206]
Cottonwood	600	Fe:CW = 0.05	NA	N/A	NA	N/A	25	55.61	[206]
Cottonwood	600	Fe:CW = 0.1	458	N/A	0.04	N/A	25	66.57	[206]
Cottonwood	600	Fe:CW = 5	665	N/A	0.59	N/A	25	60.68	[206]
Cottonwood	600	Fe:CW = 6	654	N/A	0.19	N/A	25	65.26	[206]
Cottonwood	600	Fe:CW = 10	749	N/A	0.33	N/A	25	53.79	[206]

Source: Reproduced with permission from Elsevier, license number, 5747091207739.

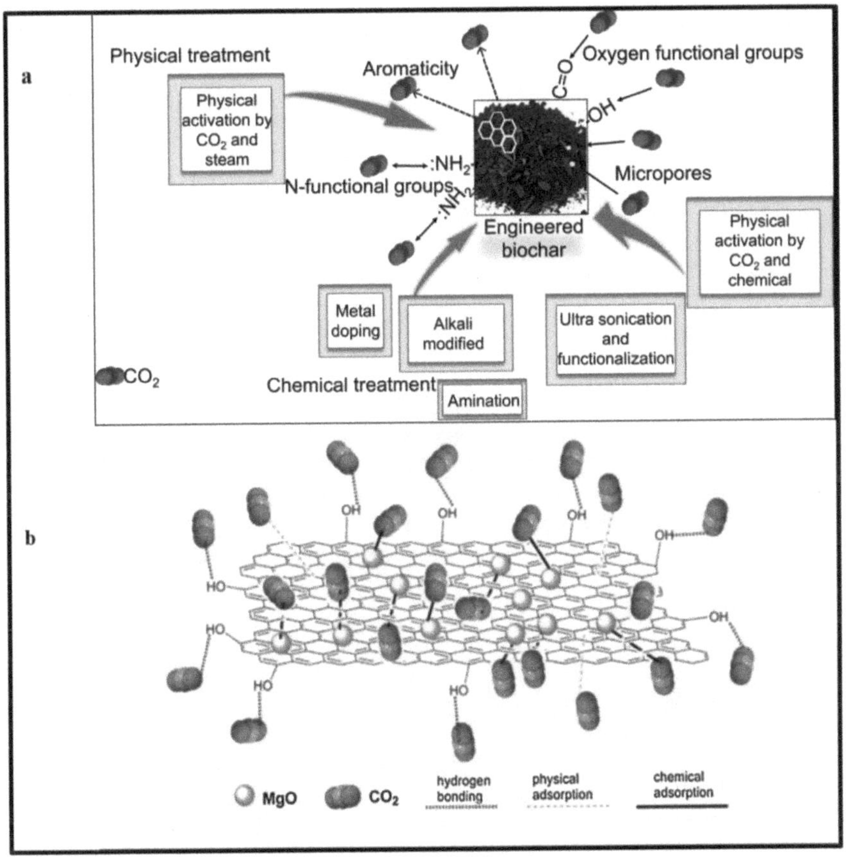

FIGURE 6.14 (a) Understanding the mechanism of CO_2 adsorption on biochar. (b) Investigating the adsorption of CO_2 on sawdust biochar doped with MgO [134].

by which CO_2 is adsorbed on engineered biochar is depicted in Figure 6.14a. It's important to highlight that, as the temperature rises from 20°C to 120°C, the adsorption of CO_2 on biochar shifts from physisorption to chemisorption. In conclusion, physisorption and chemisorption are two important mechanisms for the adsorption of CO_2. Physisorption is temperature-dependent and relies on intermolecular forces, while chemisorption involves covalent bonding between CO_2 and the active site on the adsorbent surface. The use of amines and ammonia can enhance chemisorption, and raising the temperature can facilitate this process [129, 134, 196–199]. In a groundbreaking study led by Liu et al. [197, 200, 201], they sought to investigate the CO_2 adsorption capabilities of activated biochars. Surprisingly, they found that the adsorption capacity of these biochars changed based on the temperature. Initially, when mesoporous-activated biochar infused with magnesium oxide (MgO) was tested, at 50°C, the material showed a CO_2 adsorption capability of 2.7 mmol/g. This capacity significantly increased to 5.45 mmol/g at 80°C but then dropped to 1.3 mmol/g when the temperature reached 400°C. The notable increase in CO_2 adsorption at 80°C was ascribed to a chemical reaction between CO_2 and magnesium oxide, along with the establishment of a hydrogen bond between CO_2 molecules and the hydroxyl (OH) groups found on the biochar. This unique interaction greatly improved the adsorption capability. The mechanism illustrating the CO_2 adsorption process on the MgO-doped sawdust biochar is depicted in Figure 6.14b. It is worth highlighting that while the CO_2 adsorption capacity increased with temperature at first, it eventually declined at higher temperatures. This observation indicates that there are optimal conditions for effective CO_2 adsorption on activated biochars, beyond which the capacity for adsorption decreases [134].

6.7 CONCLUSION

In conclusion, organic and inorganic adsorbents have shown promise in the removal of carbon dioxide. Natural polymers, AC, CNT, and fly ash have been investigated and found to be effective in capturing CO_2. Additionally, clay minerals such as zeolite and dolomite have proven to be viable mineral adsorbents. The use of these adsorbents is appealing due to their low cost and abundance. The capacity for CO_2 adsorption and the structural characteristics of the adsorbents have been extensively studied. Moving forward, further research should focus on developing adsorbents that are both effective and economically feasible. The development of adsorbents from natural plant-based materials, such as biomass and biochar, shows promise. These materials have been proven to have a lesser environmental impact compared to other absorbent materials. Agricultural solid waste adsorbents are particularly attractive as they are readily available and affordable. Future research should also explore the optimization of CO_2 adsorption processes using these natural adsorbents. This includes studying factors such as adsorption kinetics, temperature, pressure, and regeneration techniques. Additionally, the scalability and commercial viability of these adsorbents should be assessed to determine their potential for large-scale CO_2 capture. In summary, the research on natural adsorbents for CO_2 capture has shown great potential. Further studies should focus on developing cost-effective and environmentally friendly adsorbents, optimizing adsorption processes, and assessing the scalability of these materials for practical applications.

REFERENCES

1. Z. Khoshraftar, A. Ghaemi, Prediction of CO_2 solubility in water at high pressure and temperature via deep learning and response surface methodology, *Case Stud. Chem. Environ. Eng.* (2023) 100338.
2. Z. Khoshraftar, A. Ghaemi, A.H.M. Sigaroodi, The effect of solid adsorbents in Triethanolamine (TEA) solution for enhanced CO_2 absorption rate, *Res. Chem. Intermed.* (2021) 1–20.
3. Z. Khoshraftar, A. Ghaemi, F.S. Taheri, Polyethylenimine-functionalized halloysite nanotube as an adsorbent for CO_2 capture: RSM and ANN methodology, *Curr. Res. Green Sustain. Chem.* (2023) 100389.
4. Z. Khoshraftar, H. Masoumi, A. Ghaemi, An insight into the potential of dolomite powder as a sorbent in the elimination of heavy metals: A review, *Case Stud. Chem. Environ. Eng.* 7 (2023) 100276.

5. Z. Khoshraftar, H. Masoumi, A. Ghaemi, Experimental, response surface methodology (RSM) and mass transfer modeling of heavy metals elimination using dolomite powder as an economical adsorbent, *Case Stud. Chem. Environ. Eng.* (2023) 100329.
6. Z. Khoshraftar, F.S. Taheri, S. Nezami, A. Ghaemi, Using halloysite nanotubes modified by tetraethylenepentamine for advanced carbon capture: Experimental and modeling via RSM and ANNs, *Chem. Eng. J. Adv.* (2023) 100543.
7. B. Wang, A.P. Côté, H. Furukawa, M. O'Keeffe, O.M. Yaghi, Colossal cages in zeolitic imidazolate frameworks as selective carbon dioxide reservoirs, *Nature.* 453 (2008) 207–211.
8. B. Dziejarski, J. Serafin, K. Andersson, R. Krzyżyńska, CO_2 capture materials: A review of current trends and future challenges, *Mater. Today Sustain.* (2023) 100483.
9. Z. Khoshraftar, A. Ghaemi, Modeling and prediction of CO_2 partial pressure in methanol solution using artificial neural networks, *Curr. Res. Green Sustain. Chem.* (2023) 100364.
10. Z. Khoshraftar, A. Ghaemi, Modeling of CO_2 solubility in piperazine (PZ) and diethanolamine (DEA) solution via machine learning approach and response surface methodology, *Case Stud. Chem. Environ. Eng.* (2023) 100457.
11. A. Ghaemi, M.K. Dehnavi, Z. Khoshraftar, Exploring artificial neural network approach and RSM modeling in the prediction of CO_2 capture using carbon molecular sieves, *Case Stud. Chem. Environ. Eng.* (2023) 100310.
12. F. Bahmanzadegan, M.A. Pordsari, A. Ghaemi, Improving the efficiency of 4A-zeolite synthesized from kaolin by amine functionalization for CO_2 capture, *Sci. Rep.* 13 (2023) 12533.
13. F. Fashi, A. Ghaemi, P. Moradi, Piperazine-modified activated alumina as a novel promising candidate for CO_2 capture: Experimental and modeling, *Greenh. Gases Sci. Technol.* 9 (2019) 37–51.
14. A. Kazemi, F. Moghadaskhou, M.A. Pordsari, F. Manteghi, A. Tadjarodi, A. Ghaemi, Enhanced CO_2 capture potential of UiO-66-NH2 synthesized by sonochemical method: Experimental findings and performance evaluation, *Sci. Rep.* 13 (2023) 19891.
15. Z. Khoshraftar, A. Ghaemi, Preparation of activated carbon from Entada Africana Guill. & Perr for CO_2 capture: Artificial neural network and isotherm modeling, *J. Chem. Pet. Eng.* 56(1) (2022) 165–180.
16. Z. Khoshraftar, A. Ghaemi, Enhanced carbon dioxide adsorption using lignin-derived and nitrogen-doped porous carbons: A machine learning approaches, RSM and isotherm modeling, *Case Stud. Chem. Environ. Eng.* (2024) 100668.
17. Z. Khoshraftar, A. Ghaemi, A. Hemmati, Comprehensive investigation of isotherm, RSM, and ANN modeling of CO_2 capture by multi-walled carbon nanotube, *Sci. Rep.* 14 (2024) 5130.
18. Z. Khoshraftar, A. Ghaemi, H. Mashhadimoslem, Evaluation of the silica gel adsorbent potential for carbon dioxide capture: Experimental and modeling, *Iran. J. Chem. Eng.* 18 (2021) 64–80.
19. Z. Khoshraftar, H. Masoumi, A. Ghaemi, Characterization and evaluation of low-cost biomass-based-AC for CO_2 capture: A review, *Case Stud. Chem. Environ. Eng.* (2023) 100373.
20. H.J. Esfahani, S. Shahhosseini, A. Ghaemi, Improved structure of Zr-BTC metal organic framework using NH2 to enhance CO_2 adsorption performance, *Sci. Rep.* 13 (2023) 17700.
21. M. Helmi, F. Moazami, A. Ghaemi, A. Hemmati, Synthesis, characterization and performance evaluation of NaOH@ Chitosan-Fe3O4 as an adsorbent for CO_2 capture, *Fuel.* 338 (2023) 127300.
22. K.C. Osler, Synthesis and performance evaluation of Chitosan based adsorbent for CO_2 capture, University of the Witwatersrand, faculty of engineering and the built~. . . (2017). https://hdl.handle.net/10539/24850.

23. K. Divya, S. Rebello, M.S. Jisha, A simple and effective method for extraction of high purity chitosan from shrimp shell waste, *Proc. Int. Conf. Adv. Appl. Sci. Environ. Eng.* (2014) 141–145.
24. G. Sneddon, A.Y. Ganin, H.H.P. Yiu, Sustainable CO_2 adsorbents prepared by coating chitosan onto mesoporous silicas for large-scale carbon capture technology, *Energy Technol.* 3 (2015) 249–258.
25. J. Fujiki, K. Yogo, The increased CO_2 adsorption performance of chitosan-derived activated carbons with nitrogen-doping, *Chem. Commun.* 52 (2016) 186–189.
26. S.Y. Foong, others, Sustainable CO_2 capture via adsorption by chitosan-based functional biomaterial: A review on recent advances, challenges, and future directions, *Renew. Sustain. Energy Rev.* 181 (2023) 113342.
27. C. Yang, T. Zhao, H. Pan, F. Liu, J. Cao, Q. Lin, Facile preparation of N-doped porous carbon from chitosan and NaNH2 for CO_2 adsorption and conversion, *Chem. Eng. J.* 432 (2022) 134347.
28. N. Hsan, P.K. Dutta, S. Kumar, N. Das, J. Koh, Capture and chemical fixation of carbon dioxide by chitosan grafted multi-walled carbon nanotubes, *J. CO_2 Util.* 41 (2020) 101237.
29. P. Panchal, E. Ogunsona, T. Mekonnen, Trends in advanced functional material applications of nanocellulose, *Processes.* 7 (2018) 10.
30. N.A.D. Ho, C.P. Leo, A review on the emerging applications of cellulose, cellulose derivatives and nanocellulose in carbon capture, *Environ. Res.* 197 (2021) 111100.
31. H. V Lee, others, Conversion of lignocellulosic biomass to nanocellulose: structure and chemical process, *Sci. World J.* 2014 (2014).
32. K.L. Spence, R.A. Venditti, O.J. Rojas, Y. Habibi, J.J. Pawlak, A comparative study of energy consumption and physical properties of microfibrillated cellulose produced by different processing methods, *Cellulose.* 18 (2011) 1097–1111.
33. C. Gebald, J.A. Wurzbacher, A. Borgschulte, T. Zimmermann, A. Steinfeld, Single-component and binary CO_2 and H_2O adsorption of amine-functionalized cellulose, *Environ. Sci. Technol.* 48 (2014) 2497–2504.
34. Y. Wu, Y. Zhang, N. Chen, S. Dai, H. Jiang, S. Wang, Effects of amine loading on the properties of cellulose nanofibrils aerogel and its CO_2 capturing performance, *Carbohydr. Polym.* 194 (2018) 252–259.
35. S. Liu, Y. Zhang, H. Jiang, X. Wang, T. Zhang, Y. Yao, High CO_2 adsorption by amino-modified bio-spherical cellulose nanofibres aerogels, *Environ. Chem. Lett.* 16 (2018) 605–614.
36. Y. Li, P. Jia, J. Xu, Y. Wu, H. Jiang, Z. Li, The aminosilane functionalization of cellulose nanofibrils and the mechanical and CO_2 adsorption characteristics of their aerogel, *Ind. Eng. Chem. Res.* 59 (2020) 2874–2882.
37. T. Zhang, Y. Zhang, H. Jiang, X. Wang, Aminosilane-grafted spherical cellulose nanocrystal aerogel with high CO_2 adsorption capacity, *Environ. Sci. Pollut. Res.* 26 (2019) 16716–16726.
38. T. Zhang, W. Zhang, Y. Zhang, M. Shen, J. Zhang, Gas phase synthesis of aminated nanocellulose aerogel for carbon dioxide adsorption, *Cellulose.* 27 (2020) 2953–2958.
39. S. Sepahvand, M. Jonoobi, A. Ashori, F. Gauvin, H.J.H. Brouwers, K. Oksman, Q. Yu, A promising process to modify cellulose nanofibers for carbon dioxide (CO_2) adsorption, *Carbohydr. Polym.* 230 (2020) 115571.
40. C. Hou, Y. Wu, T. Wang, X. Wang, X. Gao, Preparation of quaternized bamboo cellulose and its implication in direct air capture of CO_2, *Energy Fuels.* 33 (2018) 1745–1752.
41. Y. Tang, S. Tang, T. Zhang, Homogeneous preparation of aerocellulose grafted acrylamide and its CO_2 adsorption properties, *Cellulose.* 27 (2020) 3263–3275.
42. N. Supanchaiyamat, K. Jetsrisuparb, J.T.N. Knijnenburg, D.C.W. Tsang, A.J. Hunt, Lignin materials for adsorption: Current trend, perspectives and opportunities, *Bioresour. Technol.* 272 (2019) 570–581.

43. C.R. Correa, T. Otto, A. Kruse, Influence of the biomass components on the pore formation of activated carbon, *Biomass and Bioenergy*. 97 (2017) 53–64.
44. L. Gong, A. Bao, High-value utilization of lignin to prepare N, O-codoped porous carbon as a high-performance adsorbent for carbon dioxide capture, *J. CO_2 Util.* 68 (2023) 102374.
45. D. Saha, B. Taylor, N. Alexander, D.F. Joyce, G.I. Faux, Y. Lin, V. Shteyn, G. Orkoulas, One-step conversion of agro-wastes to nanoporous carbons: Role in separation of greenhouse gases, *Bioresour. Technol.* 256 (2018) 232–240.
46. W. Sangchoom, R. Mokaya, Valorization of lignin waste: Carbons from hydrothermal carbonization of renewable lignin as superior sorbents for CO_2 and hydrogen storage, *ACS Sustain. Chem. Eng.* 3 (2015) 1658–1667.
47. W. Hao, F. Björnerbäck, Y. Trushkina, M. Oregui Bengoechea, G. Salazar-Alvarez, T. Barth, N. Hedin, High-performance magnetic activated carbon from solid waste from lignin conversion processes. 1. Their use as adsorbents for CO_2, *ACS Sustain. Chem. Eng.* 5 (2017) 3087–3095.
48. D. Saha, S.E. Van Bramer, G. Orkoulas, H.-C. Ho, J. Chen, D.K. Henley, CO_2 capture in lignin-derived and nitrogen-doped hierarchical porous carbons, *Carbon N. Y.* 121 (2017) 257–266.
49. Y. Sun, G. Yang, J. Zhang, Y. Wang, M. Yao, Activated carbon preparation from lignin by H_3PO_4 activation and its application to gas separation, *Chem. Eng. & Technol.* 35 (2012) 309–316.
50. A. Dindi, D.V. Quang, L.F. Vega, E. Nashef, M.R.M. Abu-Zahra, Applications of fly ash for CO_2 capture, utilization, and storage, *J. CO_2 Util.* 29 (2019) 82–102.
51. J.-H. Wee, A review on carbon dioxide capture and storage technology using coal fly ash, *Appl. Energy*. 106 (2013) 143–151.
52. J. Lee, S.-J. Han, J.-H. Wee, Synthesis of dry sorbents for carbon dioxide capture using coal fly ash and its performance, *Appl. Energy*. 131 (2014) 40–47.
53. B. Sreenivasulu, I. Sreedhar, B.M. Reddy, K. V Raghavan, Stability and carbon capture enhancement by coal-fly-ash-doped sorbents at a high temperature, *Energy Fuels*. 31 (2017) 785–794.
54. F. Yan, J. Jiang, K. Li, S. Tian, M. Zhao, X. Chen, Performance of coal fly ash stabilized, CaO-based sorbents under different carbonation—calcination conditions, *ACS Sustain. Chem. Eng.* 3 (2015) 2092–2099.
55. M. Sarmah, B.P. Baruah, P. Khare, A comparison between CO_2 capturing capacities of fly ash based composites of MEA/DMA and DEA/DMA, *Fuel Process. Technol.* 106 (2013) 490–497.
56. Y.A. Alhamed, S.U. Rather, A.H. El-Shazly, S.F. Zaman, M.A. Daous, A.A. Al-Zahrani, Preparation of activated carbon from fly ash and its application for CO_2 capture, *Korean J. Chem. Eng.* 32 (2015) 723–730.
57. M.M. Maroto-Valer, Z. Lu, Y. Zhang, Z. Tang, Sorbents for CO_2 capture from high carbon fly ashes, *Waste Manag.* 28 (2008) 2320–2328.
58. L. Keller, B. Ohs, J. Lenhart, L. Abduly, P. Blanke, M. Wessling, High capacity polyethylenimine impregnated microtubes made of carbon nanotubes for CO_2 capture, *Carbon N. Y.* 126 (2018) 338–345.
59. M. Irani, A.T. Jacobson, K.A.M. Gasem, M. Fan, Modified carbon nanotubes/tetraethylenepentamine for CO_2 capture, *Fuel*. 206 (2017) 10–18.
60. H. Wang, C. Xu, Y. Zhou, W. Zhao, J. Zhong, W. Huang, R. Chen, Fabrication of hierarchical N-doped carbon nanotubes for CO_2 adsorption, *Nano*. 14 (2019) 1950072.
61. Q. Ye, J. Jiang, C. Wang, Y. Liu, H. Pan, Y. Shi, Adsorption of low-concentration carbon dioxide on amine-modified carbon nanotubes at ambient temperature, *Energy Fuels*. 26 (2012) 2497–2504.

62. M.-S. Lee, S.-Y. Lee, S.-J. Park, Preparation and characterization of multi-walled carbon nanotubes impregnated with polyethyleneimine for carbon dioxide capture, *Int. J. Hydrogen Energy*. 40 (2015) 3415–3421.
63. Z. Zhou, C.M. Anderson, S.K. Butler, S.K. Thompson, K.J. Whitty, T.-C. Shen, K.J. Stowers, Stability and efficiency of CO_2 capture using linear amine polymer modified carbon nanotubes, *J. Mater. Chem. A*. 5 (2017) 10486–10494.
64. M. Deng, H.G. Park, Spacer-assisted amine-coiled carbon nanotubes for CO_2 capture, *Langmuir*. 35 (2019) 4453–4459.
65. C. Lu, H. Bai, B. Wu, F. Su, J.F. Hwang, Comparative study of CO_2 capture by carbon nanotubes, activated carbons, and zeolites, *Energy Fuels*. 22 (2008) 3050–3056.
66. J.M. Ngoy, N. Wagner, L. Riboldi, O. Bolland, A CO_2 capture technology using multi-walled carbon nanotubes with polyaspartamide surfactant, *Energy Procedia*. 63 (2014) 2230–2248.
67. K. Rahimi, S. Riahi, M. Abbasi, Z. Fakhroueian, Modification of multi-walled carbon nanotubes by 1, 3-diaminopropane to increase CO_2 adsorption capacity, *J. Environ. Manage*. 242 (2019) 81–89.
68. H. Hu, T. Zhang, S. Yuan, S. Tang, Functionalization of multi-walled carbon nanotubes with phenylenediamine for enhanced CO_2 adsorption, *Adsorption*. 23 (2017) 73–85.
69. M. Khajeh, A. Ghaemi, Exploiting response surface methodology for experimental modeling and optimization of CO_2 adsorption onto NaOH-modified nanoclay montmorillonite, *J. Environ. Chem. Eng*. 8 (2020) 103663.
70. H. Tao, X. Qian, Y. Zhou, H. Cheng, Research progress of clay minerals in carbon dioxide capture, *Renew. Sustain. Energy Rev*. 164 (2022) 112536.
71. G. Gómez-Pozuelo, E.S. Sanz-Pérez, A. Arencibia, P. Pizarro, R. Sanz, D.P. Serrano, CO_2 adsorption on amine-functionalized clays, *Microporous Mesoporous Mater*. 282 (2019) 38–47.
72. Y.-H. Chen, D.-L. Lu, CO_2 capture by kaolinite and its adsorption mechanism, *Appl. Clay Sci*. 104 (2015) 221–228.
73. K. Ramadass, G. Singh, K.S. Lakhi, M.R. Benzigar, J.-H. Yang, S. Kim, A.M. Almajid, T. Belperio, A. Vinu, Halloysite nanotubes: Novel and eco-friendly adsorbents for high-pressure CO_2 capture, *Microporous Mesoporous Mater*. 277 (2019) 229–236.
74. M. Khajeh, A. Ghaemi, Nanoclay montmorillonite as an adsorbent for CO_2 capture: Experimental and modeling, *J. Chinese Chem. Soc*. 67 (2020) 253–266.
75. E.A. Roth, S. Agarwal, R.K. Gupta, Nanoclay-based solid sorbents for CO_2 capture, *Energy Fuels*. 27 (2013) 4129–4136.
76. A. Sushant, others, Nanoclay-based solid sorbents for CO_2 capture, *Energy & Fuels*. 27(8) (2013) 4129–4136. https://doi.org/10.1021/ef302017m.
77. N. Horri, E.S. Sanz-Pérez, A. Arencibia, R. Sanz, N. Frini-Srasra, E. Srasra, Amine grafting of acid-activated bentonite for carbon dioxide capture, *Appl. Clay Sci*. 180 (2019) 105195.
78. F. Bahmanzadegan, A. Ghaemi, Modification and functionalization of zeolites to improve the efficiency of CO_2 adsorption: A review, *Case Stud. Chem. Environ. Eng*. (2023) 100564.
79. F. Fashi, A. Ghaemi, A.H. Behroozi, Piperazine impregnation on Zeolite 13X as a novel adsorbent for CO_2 capture: Experimental and modeling, *Chem. Eng. Commun*. 208 (2021) 1104–1120.
80. S. Kumar, R. Srivastava, J. Koh, Utilization of zeolites as CO_2 capturing agents: Advances and future perspectives, *J. CO_2 Util*. 41 (2020) 101251.
81. J.M. Valverde, P.E. Sanchez-Jimenez, L.A. Pérez-Maqueda, Ca-looping for postcombustion CO_2 capture: A comparative analysis on the performances of dolomite and limestone, *Appl. Energy*. 138 (2015) 202–215.
82. R.V. Siriwardane, M.-S. Shen, E.P. Fisher, J.A. Poston, Adsorption of CO_2 on molecular sieves and activated carbon, *Energy Fuels*. 15 (2001) 279–284.

83. Z. Zhang, W. Zhang, X. Chen, Q. Xia, Z. Li, Adsorption of CO_2 on zeolite 13X and activated carbon with higher surface area, *Sep. Sci. Technol.* 45 (2010) 710–719.
84. I. Noshadi, T. Jafari, B. Kanjilal, E. Moharreri, N. Khakpash, A. Masoumi, F. Liu, S.L. Suib, Materials today energy, *Mater. Today.* 4 (2017) 81e88.
85. W. Shao, L. Zhang, L. Li, R.L. Lee, Adsorption of CO_2 and N 2 on synthesized NaY zeolite at high temperatures, *Adsorption.* 15 (2009) 497–505.
86. T.-H. Pham, B.-K. Lee, J. Kim, C.-H. Lee, Enhancement of CO_2 capture by using synthesized nano-zeolite, *J. Taiwan Inst. Chem. Eng.* 64 (2016) 220–226.
87. M. Razavian, S. Fatemi, M. Masoudi-Nejad, A comparative study of CO_2 and CH4 adsorption on silicalite-1 fabricated by sonication and conventional method, *Adsorpt. Sci. Technol.* 32 (2014) 73–87.
88. S.K. Wahono, J. Stalin, J. Addai-Mensah, W. Skinner, A. Vinu, K. Vasilev, Physico-chemical modification of natural mordenite-clinoptilolite zeolites and their enhanced CO_2 adsorption capacity, *Microporous Mesoporous Mater.* 294 (2020) 109871.
89. D. Panda, E.A. Kumar, S.K. Singh, Introducing mesoporosity in zeolite 4A bodies for rapid CO_2 capture, *J. CO_2 Util.* 40 (2020) 101223.
90. A.A. Dabbawala, I. Ismail, B.V. Vaithilingam, K. Polychronopoulou, G. Singaravel, S. Morin, M. Berthod, Y. Al Wahedi, Synthesis of hierarchical porous zeolite-Y for enhanced CO_2 capture, *Microporous Mesoporous Mater.* 303 (2020) 110261.
91. Q. Liu, P. He, X. Qian, Z. Fei, Z. Zhang, X. Chen, J. Tang, M. Cui, X. Qiao, Y. Shi, Enhanced CO_2 adsorption performance on hierarchical porous ZSM-5 zeolite, *Energy Fuels.* 31 (2017) 13933–13941.
92. T.H. Nguyen, S. Kim, M. Yoon, T.-H. Bae, Hierarchical zeolites with amine-functionalized mesoporous domains for carbon dioxide capture, *ChemSusChem.* 9 (2016) 455–461.
93. Y. Wang, T. Du, Y. Song, S. Che, X. Fang, L. Zhou, Amine-functionalized mesoporous ZSM-5 zeolite adsorbents for carbon dioxide capture, *Solid State Sci.* 73 (2017) 27–35.
94. C. Chen, S.-S. Kim, W.-S. Cho, W.-S. Ahn, Polyethylenimine-incorporated zeolite 13X with mesoporosity for post-combustion CO_2 capture, *Appl. Surf. Sci.* 332 (2015) 167–171.
95. M. Sun, Q. Gu, A. Hanif, T. Wang, J. Shang, Transition metal cation-exchanged SSZ-13 zeolites for CO_2 capture and separation from N2, *Chem. Eng. J.* 370 (2019) 1450–1458.
96. R. Kodasma, J. Fermoso, A. Sanna, Li-LSX-zeolite evaluation for post-combustion CO_2 capture, *Chem. Eng. J.* 358 (2019) 1351–1362.
97. S.J. Chen, M. Zhu, Y. Fu, Y.X. Huang, Z.C. Tao, W.L. Li, Using 13X, LiX, and LiPdAgX zeolites for CO_2 capture from post-combustion flue gas, *Appl. Energy.* 191 (2017) 87–98.
98. E. Davarpanah, M. Armandi, S. Hernández, D. Fino, R. Arletti, S. Bensaid, M. Piumetti, CO_2 capture on natural zeolite clinoptilolite: Effect of temperature and role of the adsorption sites, *J. Environ. Manage.* 275 (2020) 111229.
99. S.-T. Yang, J. Kim, W.-S. Ahn, CO_2 adsorption over ion-exchanged zeolite beta with alkali and alkaline earth metal ions, *Microporous Mesoporous Mater.* 135 (2010) 90–94.
100. P. Murge, S. Dinda, S. Roy, Zeolite-based sorbent for CO_2 capture: Preparation and performance evaluation, *Langmuir.* 35 (2019) 14751–14760.
101. S. Boycheva, I. Marinov, D. Zgureva-Filipova, Studies on the CO_2 capture by coal fly ash zeolites: Process design and simulation, *Energies.* 14 (2021) 8279.
102. Z. Zhang, Y. Xiao, B. Wang, Q. Sun, H. Liu, Waste is a misplayed resource: Synthesis of zeolites from fly ash for CO_2 capture, *Energy Procedia.* 114 (2017) 2537–2544.
103. N. Czuma, I. Casanova, J. Baran Pawełand Szczurowski, K. Zarębska, CO_2 sorption and regeneration properties of fly ash zeolites synthesized with the use of differentiated methods, *Sci. Rep.* 10 (2020) 1825.
104. V. Garshasbi, M. Jahangiri, M. Anbia, Equilibrium CO_2 adsorption on zeolite 13X prepared from natural clays, *Appl. Surf. Sci.* 393 (2017) 225–233.

105. C. Chen, D.-W. Park, W.-S. Ahn, CO_2 capture using zeolite 13X prepared from bentonite, *Appl. Surf. Sci.* 292 (2014) 63–67.
106. L. Valencia, W. Rosas, A. Aguilar-Sanchez, A.P. Mathew, A.E.C. Palmqvist, Bio-based micro-/meso-/macroporous hybrid foams with ultrahigh zeolite loadings for selective capture of carbon dioxide, *ACS Appl. Mater. Interfaces.* 11 (2019) 40424–40431.
107. S. Kumar, R. Bera, N. Das, J. Koh, Chitosan-based zeolite-Y and ZSM-5 porous biocomposites for H_2 and CO_2 storage, *Carbohydr. Polym.* 232 (2020) 115808.
108. Y.-H. Chen, D.-L. Lu, Amine modification on kaolinites to enhance CO_2 adsorption, *J. Colloid Interface Sci.* 436 (2014) 47–51.
109. J.A. Cecilia, E. Vilarrasa-García, C.L. Cavalcante Jr, D.C.S. Azevedo, F. Franco, E. Rodríguez-Castellón, Evaluation of two fibrous clay minerals (sepiolite and palygorskite) for CO_2 capture, *J. Environ. Chem. Eng.* 6 (2018) 4573–4587.
110. E. Vilarrasa-García, J.A. Cecilia, D.C.S. Azevedo, C.L. Cavalcante Jr, E. Rodríguez-Castellón, Evaluation of porous clay heterostructures modified with amine species as adsorbent for the CO_2 capture, *Microporous Mesoporous Mater.* 249 (2017) 25–33.
111. C. Chen, D.-W. Park, W.-S. Ahn, Surface modification of a low cost bentonite for post-combustion CO_2 capture, *Appl. Surf. Sci.* 283 (2013) 699–704.
112. L. Stevens, K. Williams, W.Y. Han, T. Drage, C. Snape, J. Wood, J. Wang, Preparation and CO_2 adsorption of diamine modified montmorillonite via exfoliation grafting route, *Chem. Eng. J.* 215 (2013) 699–708.
113. E. Vilarrasa-García, J.A. Cecilia, E. Rodríguez Aguado, J. Jiménez-Jiménez, C.L. Cavalcante, D.C.S. Azevedo, E. Rodríguez-Castellón, Amino-modified pillared adsorbent from water-treatment solid wastes applied to CO_2/N2 separation, *Adsorption.* 23 (2017) 405–421.
114. M. Atilhan, S. Atilhan, R. Ullah, B. Anaya, T. Cagin, C.T. Yavuz, S. Aparicio, High-pressure methane, carbon dioxide, and nitrogen adsorption on amine-impregnated porous montmorillonite nanoclays, *J. Chem. Eng. Data.* 61 (2016) 2749–2760.
115. K. Sumida, D.L. Rogow, J.A. Mason, T.M. McDonald, E.D. Bloch, Z.R. Herm, T.-H. Bae, J.R. Long, Carbon dioxide capture in metal—organic frameworks, *Chem. Rev.* 112 (2012) 724–781.
116. M. Younas, M. Sohail, L.K. Leong, M.J.K. Bashir, S. Sumathi, Feasibility of CO_2 adsorption by solid adsorbents: A review on low-temperature systems, *Int. J. Environ. Sci. Technol.* 13 (2016) 1839–1860.
117. G. Dobele, T. Dizhbite, G. Rossinskaja, G. Telysheva, D. Meier, S. Radtke, O. Faix, Pre-treatment of biomass with phosphoric acid prior to fast pyrolysis: A promising method for obtaining 1, 6-anhydrosaccharides in high yields, *J. Anal. Appl. Pyrolysis.* 68 (2003) 197–211.
118. F. Quesada-Plata, others, Activated carbons prepared through H_3PO_4-assisted hydrothermal carbonisation from biomass wastes: Porous texture and electrochemical performance, 81(12) (2016) 1349–1359. https://doi.org/10.1002/cplu.201600412.
119. A. Kumar, H.M. Jena, High surface area microporous activated carbons prepared from Fox nut (Euryale ferox) shell by zinc chloride activation, *Appl. Surf. Sci.* 356 (2015) 753–761.
120. Q. Lu, Z. Wang, C. Dong, Z. Zhang, Y. Zhang, Y. Yang, X. Zhu, Selective fast pyrolysis of biomass impregnated with ZnCl2: Furfural production together with acetic acid and activated carbon as by-products, *J. Anal. Appl. Pyrolysis.* 91 (2011) 273–279.
121. I. Ozdemir, M. Şahin, R. Orhan, M. Erdem, Preparation and characterization of activated carbon from grape stalk by zinc chloride activation, *Fuel Process. Technol.* 125 (2014) 200–206.
122. S. Ding, Y. Liu, Adsorption of CO_2 from flue gas by novel seaweed-based KOH-activated porous biochars, *Fuel.* 260 (2020) 116382.

123. A. Hai, G. Bharath, K.R. Babu, H. Taher, M. Naushad, F. Banat, Date seeds biomass-derived activated carbon for efficient removal of NaCl from saline solution, *Process Saf. Environ. Prot.* 129 (2019) 103–111.
124. C. Zhang, W. Song, Q. Ma, L. Xie, X. Zhang, H. Guo, Enhancement of CO_2 capture on biomass-based carbon from black locust by KOH activation and ammonia modification, *Energy Fuels.* 30 (2016) 4181–4190.
125. G. Singh, K.S. Lakhi, I.Y. Kim, S. Kim, P. Srivastava, R. Naidu, A. Vinu, Highly efficient method for the synthesis of activated mesoporous biocarbons with extremely high surface area for high-pressure CO_2 adsorption, *ACS Appl. Mater. Interfaces.* 9 (2017) 29782–29793.
126. M. Ahmadi, F. Bahmanzadegan, M. Qasemnazhand, A. Ghaemi, H. Ramezanipour Penchah, Experimental, RSM modelling, and DFT simulation of CO_2 adsorption on Modified activated carbon with LiOH, *Sci. Rep.* 14 (2024) 13595.
127. M. Karimi, M. Shirzad, J.A.C. Silva, A.E. Rodrigues, Biomass/Biochar carbon materials for CO_2 capture and sequestration by cyclic adsorption processes: A review and prospects for future directions, *J. CO_2 Util.* 57 (2022) 101890.
128. Z. Rouzitalab, D.M. Maklavany, S. Jafarinejad, A. Rashidi, Lignocellulose-based adsorbents: A spotlight review of the effective parameters on carbon dioxide capture process, *Chemosphere.* 246 (2020) 125756.
129. Y. Li, G. Ruan, A.S. Jalilov, Y.R. Tarkunde, H. Fei, J.M. Tour, Biochar as a renewable source for high-performance CO_2 sorbent, *Carbon N. Y.* 107 (2016) 344–351.
130. H. Madzaki, others, Carbon dioxide adsorption on sawdust biochar, *Procedia Eng.* 148 (2016) 718–725.
131. C.A. Peterson, R.C. Brown, Oxidation kinetics of biochar from woody and herbaceous biomass, *Chem. Eng. J.* 401 (2020) 126043.
132. O.D. Nartey, others, Biochar preparation, characterization, and adsorptive capacity and its effect on bioavailability of contaminants: An overview, *Adv. Mater. Sci. Eng.* 2014 (2014).
133. S. Campbell, F.L. Bernard, D.M. Rodrigues, M.F. Rojas, L.Á. Carreño, V. V Chaban, S. Einloft, Performance of metal-functionalized rice husk cellulose for CO_2 sorption and CO_2/N2 separation, *Fuel.* 239 (2019) 737–746.
134. J.C. Francis, A. Nighojkar, B. Kandasubramanian, Relevance of wood biochar on CO_2 adsorption: A review, *Hybrid Adv.* (2023) 100056.
135. S. Wang, B. Gao, A.R. Zimmerman, Y. Li, L. Ma, W.G. Harris, K.W. Migliaccio, Physicochemical and sorptive properties of biochars derived from woody and herbaceous biomass, *Chemosphere.* 134 (2015) 257–262.
136. L. Cao, X. Zhang, Y. Xu, W. Xiang, R. Wang, F. Ding, P. Hong, B. Gao, Straw and wood based biochar for CO_2 capture: Adsorption performance and governing mechanisms, *Sep. Purif. Technol.* 287 (2022) 120592.
137. S. Shahkarami, R. Azargohar, A.K. Dalai, J. Soltan, Breakthrough CO_2 adsorption in bio-based activated carbons, *J. Environ. Sci.* 34 (2015) 68–76.
138. C. Goel, S. Mohan, P. Dinesha, CO_2 capture by adsorption on biomass-derived activated char: A review, *Sci. Total Environ.* 798 (2021) 149296.
139. A.D. Igalavithana, S.W. Choi, P.D. Dissanayake, J. Shang, C.-H. Wang, X. Yang, S. Kim, D.C.W. Tsang, K.B. Lee, Y.S. Ok, Gasification biochar from biowaste (food waste and wood waste) for effective CO_2 adsorption, *J. Hazard. Mater.* 391 (2020) 121147.
140. P.D. Dissanayake, S.W. Choi, A.D. Igalavithana, X. Yang, D.C.W. Tsang, C.-H. Wang, H.W. Kua, K.B. Lee, Y.S. Ok, Sustainable gasification biochar as a high efficiency adsorbent for CO_2 capture: A facile method to designer biochar fabrication, *Renew. Sustain. Energy Rev.* 124 (2020) 109785.
141. W. Suliman, J.B. Harsh, N.I. Abu-Lail, A.-M. Fortuna, I. Dallmeyer, M. Garcia-Perez, Modification of biochar surface by air oxidation: Role of pyrolysis temperature, *Biomass Bioenergy.* 85 (2016) 1–11.

142. A.E. Creamer, B. Gao, M. Zhang, Carbon dioxide capture using biochar produced from sugarcane bagasse and hickory wood, *Chem. Eng. J.* 249 (2014) 174–179.
143. Y.-F. Huang, P.-T. Chiueh, S.-L. Lo, Carbon capture of biochar produced by microwave co-pyrolysis: Adsorption capacity, kinetics, and benefits, *Environ. Sci. Pollut. Res.* 30 (2023) 22211–22221.
144. L. Krounbi, A. Enders, C.R. Anderton, M.H. Engelhard, R. Hestrin, D. Torres-Rojas, J.J. Dynes, J. Lehmann, Sequential ammonia and carbon dioxide adsorption on pyrolyzed biomass to recover waste stream nutrients, *ACS Sustain. Chem. Eng.* 8 (2020) 7121–7131.
145. C.F. Chiu, Z.D. Huang, Microbial methane oxidation and gas adsorption capacities of biochar-modified soils, *Int. J. Geosynth. Gr. Eng.* 6 (2020) 1–14.
146. B. Shanmughan, A. Nighojkar, B. Kandasubramanian, Exploring the future of 2D catalysts for clean and sustainable hydrogen production, *Int. J. Hydrogen Energy.* (2023).
147. A.L. Yaumi, M.Z.A. Bakar, B.H. Hameed, Melamine-nitrogenated mesoporous activated carbon derived from rice husk for carbon dioxide adsorption in fixed-bed, *Energy.* 155 (2018) 46–55.
148. N.A. Rashidi, S. Yusup, B.H. Hameed, Kinetic studies on carbon dioxide capture using lignocellulosic based activated carbon, *Energy.* 61 (2013) 440–446.
149. E. Papa, E. Landi, V. Medri, Adsorption of gaseous pollutants by alkali-activated materials, *Alkali-Activated Mater. Environ. Technol. Appl.* (2022) 243–275, Elsevier.
150. Y. Chen, Y. Zhu, Z. Wang, Y. Li, L. Wang, L. Ding, X. Gao, Y. Ma, Y. Guo, Application studies of activated carbon derived from rice husks produced by chemical-thermal process—A review, *Adv. Colloid Interface Sci.* 163 (2011) 39–52.
151. A. Boonpoke, S. Chiarakorn, N. Laosiripojana, A. Chidthaisong, Enhancement of carbon dioxide capture by amine-modified rice husk mesoporous material, *Environ. Prog. Sustain. Energy.* 35 (2016) 1716–1723.
152. Z. Khoshraftar, A. Ghaemi, Presence of activated carbon particles from waste walnut shell as a biosorbent in monoethanolamine (MEA) solution to enhance carbon dioxide absorption, *Heliyon.* 8 (2022) e08689.
153. R.C. Bansal, M. Goyal, *Activated carbon adsorption*, CRC Press, 2005.
154. M.S. Shafeeyan, W.M.A.W. Daud, A. Houshmand, A. Shamiri, A review on surface modification of activated carbon for carbon dioxide adsorption, *J. Anal. Appl. Pyrolysis.* 89 (2010) 143–151.
155. Z. Khoshraftar, A. Ghaemi, Evaluation of pistachio shells as solid wastes to produce activated carbon for CO_2 capture: Isotherm, response surface methodology (RSM) and artificial neural network (ANN) modeling, *Curr. Res. Green Sustain. Chem.* 5 (2022) 100342. https://doi.org/10.1016/j.crgsc.2022.100342.
156. Z. Khoshraftar, A. Ghaemi, Evaluation of CaO derived from Cerastoderma glaucum of Caspian beach as a natural sorbent for CO_2 capture, *Curr. Res. Green Sustain. Chem.* 6 (2023) 100360.
157. Y. Guo, S. Yang, K. Yu, J. Zhao, Z. Wang, H. Xu, The preparation and mechanism studies of rice husk based porous carbon, *Mater. Chem. Phys.* 74 (2002) 320–323.
158. L. Muniandy, F. Adam, A.R. Mohamed, E.-P. Ng, The synthesis and characterization of high purity mixed microporous/mesoporous activated carbon from rice husk using chemical activation with NaOH and KOH, *Microporous Mesoporous Mater.* 197 (2014) 316–323.
159. P. Murge, S. Dinda, S. Roy, Adsorbent from rice husk for CO_2 capture: Synthesis, characterization, and optimization of parameters, *Energy Fuels.* 32 (2018) 10786–10795.
160. R. Nandi, M.K. Jha, S.K. Guchhait, D. Sutradhar, S. Yadav, Impact of KOH activation on rice husk derived porous activated carbon for carbon capture at flue gas alike temperatures with high CO_2/N2 selectivity, *ACS Omega.* 8 (2023) 4802–4812.

161. C. Jin, J. Sun, Y. Chen, Y. Guo, D. Han, R. Wang, C. Zhao, Sawdust wastes-derived porous carbons for CO_2 adsorption. Part 1. Optimization preparation via orthogonal experiment, *Sep. Purif. Technol.* 276 (2021) 119270. https://doi.org/https://doi.org/10.1016/j.seppur.2021.119270.
162. R. Wang, P. Wang, X. Yan, J. Lang, C. Peng, Q. Xue, Promising porous carbon derived from celtuce leaves with outstanding supercapacitance and CO_2 capture performance, *ACS Appl. Mater. Interfaces*. 4 (2012) 5800–5806.
163. N.S. Nasri, U.D. Hamza, S.N. Ismail, M.M. Ahmed, R. Mohsin, Assessment of porous carbons derived from sustainable palm solid waste for carbon dioxide capture, *J. Clean. Prod.* 71 (2014) 148–157.
164. M.G. Plaza, C. Pevida, C.F. Martín, J. Fermoso, J.J. Pis, F. Rubiera, Developing almond shell-derived activated carbons as CO_2 adsorbents, *Sep. Purif. Technol.* 71 (2010) 102–106.
165. P. Lahijani, M. Mohammadi, A.R. Mohamed, Metal incorporated biochar as a potential adsorbent for high capacity CO_2 capture at ambient condition, *J. CO_2 Util.* 26 (2018) 281–293.
166. M.G. Plaza, C. Pevida, B. Arias, J. Fermoso, M.D. Casal, C.F. Martín, F. Rubiera, J.J. Pis, Development of low-cost biomass-based adsorbents for postcombustion CO_2 capture, *Fuel*. 88 (2009) 2442–2447.
167. M.G. Plaza, A.S. González, J.J. Pis, F. Rubiera, C. Pevida, Production of microporous biochars by single-step oxidation: Effect of activation conditions on CO_2 capture, *Appl. Energy*. 114 (2014) 551–562.
168. P. Wang, Y. Guo, C. Zhao, J. Yan, P. Lu, Biomass derived wood ash with amine modification for post-combustion CO_2 capture, *Appl. Energy*. 201 (2017) 34–44.
169. L.K.G. Bhatta, S. Subramanyam, M.D. Chengala, U.M. Bhatta, N. Pandit, K. Venkatesh, Investigation of CO_2 adsorption on carbon material derived from Mesua ferrea L. seed cake, *J. Environ. Chem. Eng.* 3 (2015) 2957–2965.
170. A. Mukherjee, B. Saha, C. Niu, A.K. Dalai, Preparation of activated carbon from spent coffee grounds and functionalization by deep eutectic solvent: Effect of textural properties and surface chemistry on CO_2 capture performance, *J. Environ. Chem. Eng.* 10 (2022) 108815.
171. F. Hussin, M.K. Aroua, R. Yusoff, Adsorption of CO_2 on palm shell based activated carbon modified by deep eutectic solvent: Breakthrough adsorption study, *J. Environ. Chem. Eng.* 9 (2021) 105333.
172. J. Chen, J. Yang, G. Hu, X. Hu, Z. Li, S. Shen, M. Radosz, M. Fan, Enhanced CO_2 capture capacity of nitrogen-doped biomass-derived porous carbons, *ACS Sustain. Chem. Eng.* 4 (2016) 1439–1445.
173. N.A. Rashidi, S. Yusup, A. Borhan, L.H. Loong, Experimental and modelling studies of carbon dioxide adsorption by porous biomass derived activated carbon, *Clean Technol. Environ. Policy*. 16 (2014) 1353–1361.
174. W. Dilokekunakul, P. Teerachawanwong, N. Klomkliang, S. Supasitmongkol, S. Chaemchuen, Effects of nitrogen and oxygen functional groups and pore width of activated carbon on carbon dioxide capture: Temperature dependence, *Chem. Eng. J.* 389 (2020) 124413.
175. Y. Li, R. Xu, X. Wang, B. Wang, J. Cao, J. Yang, J. Wei, Waste wool derived nitrogen-doped hierarchical porous carbon for selective CO_2 capture, *RSC Adv.* 8 (2018) 19818–19826.
176. K.K. Kishibayev, J. Serafin, R.R. Tokpayev, T.N. Khavaza, A.A. Atchabarova, D.A. Abduakhytova, Z.T. Ibraimov, J. Sreńscek-Nazzal, Physical and chemical properties of activated carbon synthesized from plant wastes and shungite for CO_2 capture, *J. Environ. Chem. Eng.* 9 (2021) 106798.

177. J. Serafin, B. Dziejarski, O.F.C. Junior, J. Sreńscek-Nazzal, Design of highly microporous activated carbons based on walnut shell biomass for H_2 and CO_2 storage, *Carbon N. Y.* 201 (2023) 633–647.
178. J. Shi, H. Cui, J. Xu, N. Yan, Carbon spheres synthesized from $KHCO_3$ activation of glucose derived hydrochar with excellent CO_2 capture capabilities at both low and high pressures, *Sep. Purif. Technol.* 294 (2022) 121193.
179. M.-J. Kim, S.W. Choi, H. Kim, S. Mun, K.B. Lee, Simple synthesis of spent coffee ground-based microporous carbons using K2CO3 as an activation agent and their application to CO_2 capture, *Chem. Eng. J.* 397 (2020) 125404.
180. S. Liu, R. Ma, X. Hu, L. Wang, X. Wang, M. Radosz, M. Fan, CO_2 adsorption on hazelnut-shell-derived nitrogen-doped porous carbons synthesized by single-step sodium amide activation, *Ind. Eng. Chem. Res.* 59 (2019) 7046–7053.
181. J.M. de Andrés, L. Orjales, A. Narros, M.D. de la Fuente, M.E. Rodríguez, Carbon dioxide adsorption in chemically activated carbon from sewage sludge, *J. Air Waste Manag. Assoc.* 63 (2013) 557–564.
182. J.-S. Bae, S. Su, Macadamia nut shell-derived carbon composites for post combustion CO_2 capture, *Int. J. Greenh. Gas Control.* 19 (2013) 174–182.
183. J.F. Vivo-Vilches, A.F. Pérez-Cadenas, F.J. Maldonado-Hódar, F. Carrasco-Marin, R.P. V Faria, A.M. Ribeiro, A.F.P. Ferreira, A.E. Rodrigues, Biogas upgrading by selective adsorption onto CO_2 activated carbon from wood pellets, *J. Environ. Chem. Eng.* 5 (2017) 1386–1393.
184. O. Boujibar, A. Souikny, F. Ghamouss, O. Achak, M. Dahbi, T. Chafik, CO_2 capture using N-containing nanoporous activated carbon obtained from argan fruit shells, *J. Environ. Chem. Eng.* 6 (2018) 1995–2002.
185. G. Durán-Jiménez, E.T. Kostas, L.A. Stevens, W. Meredith, M. Erans, V. Hernández-Montoya, A. Buttress, C.N. Uguna, E. Binner, Green and simple approach for low-cost bioproducts preparation and CO_2 capture, *Chemosphere.* 279 (2021) 130512.
186. E. García-Díez, A. Castro-Muñiz, J.I. Paredes, M.M. Maroto-Valer, F. Suárez-García, S. García, CO_2 capture by novel hierarchical activated ordered micro-mesoporous carbons derived from low value coal tar products, *Microporous Mesoporous Mater.* 318 (2021) 110986.
187. S.S.R. Dehkordi, Q. Delavar, H.A. Ebrahim, S.S. Partash, CO_2 adsorption by coal-based activated carbon modified with sodium hydroxide, *Mater. Today Commun.* 33 (2022) 104776.
188. W.C. Wilfong, M.L. Gray, B.W. Kail, B.H. Howard, Pelletization of immobilized amine carbon dioxide sorbents with fly ash and poly (vinyl chloride), *Energy Technol.* 4 (2016) 610–619.
189. P. Ammendola, F. Raganati, R. Chirone, CO_2 adsorption on a fine activated carbon in a sound assisted fluidized bed: Thermodynamics and kinetics, *Chem. Eng. J.* 322 (2017) 302–313.
190. J. Sreńscek-Nazzal, K. Kiełbasa, Advances in modification of commercial activated carbon for enhancement of CO_2 capture, *Appl. Surf. Sci.* 494 (2019) 137–151.
191. H. Mashhadimoslem, M. Safarzadeh Khosrowshahi, M. Jafari, A. Ghaemi, A. Maleki, Adsorption equilibrium, thermodynamic, and kinetic study of O2/N2/CO_2 on functionalized granular activated carbon, *ACS Omega.* 7(22) (2022) 18409–18426. https://doi.org/10.1021/acsomega.2c00673.
192. M. Danish, P. Vijay, M.K. Almesfer, Evaluating CO_2 capture performance by adsorption using synthesized and commercial activated carbon, *Sep. Sci. Technol.* 57 (2022) 1392–1407.
193. X. Yang, A. Tsibart, H. Nam, J. Hur, A. El-Naggar, F.M.G. Tack, C.-H. Wang, Y.H. Lee, D.C.W. Tsang, Y.S. Ok, Effect of gasification biochar application on soil quality: Trace metal behavior, microbial community, and soil dissolved organic matter, *J. Hazard. Mater.* 365 (2019) 684–694.

194. A.U. Rajapaksha, Y.S. Ok, A. El-Naggar, H. Kim, F. Song, S. Kang, Y.F. Tsang, Dissolved organic matter characterization of biochars produced from different feedstock materials, *J. Environ. Manage.* 233 (2019) 393–399.
195. X. He, Z. Liu, W. Niu, L. Yang, T. Zhou, D. Qin, Z. Niu, Q. Yuan, Effects of pyrolysis temperature on the physicochemical properties of gas and biochar obtained from pyrolysis of crop residues, *Energy*. 143 (2018) 746–756.
196. Y.G. Ko, S.S. Shin, U.S. Choi, Primary, secondary, and tertiary amines for CO_2 capture: Designing for mesoporous CO_2 adsorbents, *J. Colloid Interface Sci.* 361 (2011) 594–602.
197. S. Jung, Y.-K. Park, E.E. Kwon, Strategic use of biochar for CO_2 capture and sequestration, *J. CO_2 Util.* 32 (2019) 128–139.
198. D.H. Everett, Manual of symbols and terminology for physicochemical quantities and units, appendix II: Definitions, terminology and symbols in colloid and surface chemistry, *Pure Appl. Chem.* 31 (1972) 577–638.
199. R.T. Yang, *Adsorbents: Fundamentals and applications*, John Wiley & Sons, 2003.
200. X. Yu, others, Highly disordered cobalt oxide nanostructure induced by sulfur incorporation for efficient overall water splitting, *Nano Energy*. 71 (2020) 104652.
201. Z. Shen, J. Zhang, D. Hou, D.C.W. Tsang, Y.S. Ok, D.S. Alessi, Synthesis of MgO-coated corncob biochar and its application in lead stabilization in a soil washing residue, *Environ. Int.* 122 (2019) 357–362.
202. P.D. Dissanayake, others, Biochar-based adsorbents for carbon dioxide capture: A critical review, *Renew. Sustain. Energy Rev.* 119 (2020) 109582.
203. X. Zhang, J. Wu, H. Yang, J. Shao, X. Wang, Y. Chen, S. Zhang, H. Chen, Preparation of nitrogen-doped microporous modified biochar by high temperature CO_2-NH3 treatment for CO_2 adsorption: Effects of temperature, *RSC Adv.* 6 (2016) 98157–98166.
204. X. Zhang, S. Zhang, H. Yang, Y. Feng, Y. Chen, X. Wang, H. Chen, Nitrogen enriched biochar modified by high temperature CO_2–ammonia treatment: Characterization and adsorption of CO_2, *Chem. Eng. J.* 257 (2014) 20–27.
205. Z. Xiong, Z. Shihong, Y. Haiping, S. Tao, C. Yingquan, C. Hanping, Influence of NH3/CO_2 modification on the characteristic of biochar and the CO_2 capture, *BioEnergy Res.* 6 (2013) 1147–1153.
206. A.E. Creamer, B. Gao, S. Wang, Carbon dioxide capture using various metal oxyhydroxide—biochar composites, *Chem. Eng. J.* 283 (2016) 826–832.

7 Polymeric CO_2 Adsorbents

Azam Aslani, Hadiseh Masoumi, and Ahad Ghaemi

7.1 INTRODUCTION

In the last decade, global warming has emerged as a critical issue in the current global scenario, leading to climate change despite uncertainty about its primary source. Environmental effects, mainly due to the increase in industrialization and the increase in the world population, lead to the utilization of fossil fuels such as coal and oil. Among the sources of various human and natural greenhouse gas emissions, we can mention those caused by fossil fuel power stations, industrial operations, transportation, jungle fires, and volcanic explosions. As a result, the scientific community became increasingly interested in studying the implications of these emissions and their effects on the environment. CO_2 emissions have been identified as the primary greenhouse gas among the key components of combustion flue gases, such as CO_2, SO_2, N_2O, N_2, and unburned fuel. These emissions are increasing alarmingly and have reached their highest atmospheric levels. According to the registered reports, the total emission of CO_2 fossil gases globally for 2011 was 351,015 gr. According to investigations conducted by the Goddard Institute for Space Studies, the earth's temperature has been rising by an average of 1°C each year since 1951 due to global warming. This trend is expected to continue in the 21st century, as CO_2 emissions increase as a significant greenhouse gas. The implications of this trend are important and require immediate attention from policymakers and stakeholders across the globe. Research by governmental panels has established that redundant CO_2 issuances are significantly associated with universal climate variation and can affect global temperatures. Presently, the level of CO_2 in the atmosphere stands at approximately 401 ppm, equivalent to 0.04% by volume of the earth's atmosphere; this is 70% higher than the pre-industrial era of 280 ppm and is projected to increase to approximately 550 ppm by 2050. Recent reports by NOAA show that atmospheric CO_2 levels on July 29, 2020, were 413.72 ppm, which is an increase compared to July 29, 2019 (408.96 ppm). The presence of CO_2 in the atmosphere as one of the most significant greenhouse gases causes global warming, sea level lift, and variable pH of the oceans, which may lead to floods, droughts, ecosystem destruction, species overthrow, and crisis in the food supply. These effects could have long-term impacts, including restrictions on human operations and harmful effects on future productions [1, 2]. Given the significant impact of increased CO_2 concentration on human activities, developing strategies to reduce CO_2 emissions is imperative. Capture and

DOI: 10.1201/9781003476191-7

individuation are currently the most outstanding options for reducing atmospheric CO_2 accumulation and its negative efficacies and often offer a noteworthy blend of efficiency and expense. Recently, global efforts have been directed toward the enlargement of various technologies toward CO_2 capture and storage (CCS), which comprise three stages: pre-combustion, intra-combustion, and post-combustion. The first stage involves the adsorption of CO_2 from imperfect combustion synthesis gas until the CO_2 concentration is less than 50%. However, one of the challenges of this approach is the lack of sufficient commercial availability. In the intermediate stage of CO_2 adsorption combustion, two main technologies include oxy-fuel combustion and carbonate-calcining oxy-fuel processes, where combustion is performed with net oxygen instead of air. Although the CO_2 concentration is simultaneously enhanced to 90%, this method is not very popular due to the vast cost of air segregation required to acquire net oxygen [3]. In the final phase, CO_2 adsorption from flue gas with a CO_2 concentration of around 5–15% occurs after the complete combustion of the fuel. This approach is proficient in retrofitting extant power plants and can be utilized sustainably in full-scale commercial plants. While the CO_2 concentration may differ, both systems can be extended successfully with energy and cost-effectively. The separation of CO_2 at an economical cost is obligatory, and traditional CO_2 separation technologies include adsorption, chemical ring combustion, hydrate-based separation, and cryogenic distillation. Due to the deficiency of developed technological options for the widespread use of renewable energy sources, fossil fuels remain the primary sources of energy. As a result, extensive research efforts at the international level are being made to develop new technologies for continuous and efficient performance, given the impossibility of the immediate cessation of CO_2 emissions. Various classical and modern technological options have been presented for CO_2 separation, including adsorption, membrane, surface adsorption, cryogenic, chemical ring combustion, and hydrate-based separation. Carbon CCS has been offered to reduce atmospheric CO_2 concentration. Gas sweetening, one of the most mature and common CO_2 separation technologies, uses amines such as monoethanolamine (MEA), diethanolamine (DEA), and methyldiethanolamine (MDEA) as adsorbents for CO_2 adsorption. However, this technology has various inherent problems, including equipment erosion, solvent boiling, energy-consuming, costly solvent regeneration, excessive use of amines, toxic unpleasant odor, formation of harmful by-products in nature, and excessive consumption of water that leads to an increase in CO_2 adsorption costs in power plants [4, 5]. An alternative and delicate approach to overcoming these issues is the adsorption of target molecules on porous solids, which has been feasible and proven to be a versatile and economical approach. Porous solids have many advantages over conventional amine solvents, which can be attributed to non-corrosiveness, lower specific heat capacity, selectivity to CH_4, high CO_2 adsorption capacity, easy handling, high specific surface area, reusability of adsorbent, and adjustable pore chemistry. Also, these adsorbents have chemical stability to H_2O, O_2, and NO_x. The advantages mentioned and the potential for tailoring adsorbent properties and adsorption process design position porous solids as an up-and-coming option for efficient CO_2 separation. The chemistry related to linking building units through covalent bonds, along with the development and synthesis of nanoporous materials containing light elements such as C, H, N, and O, leads to the development of design

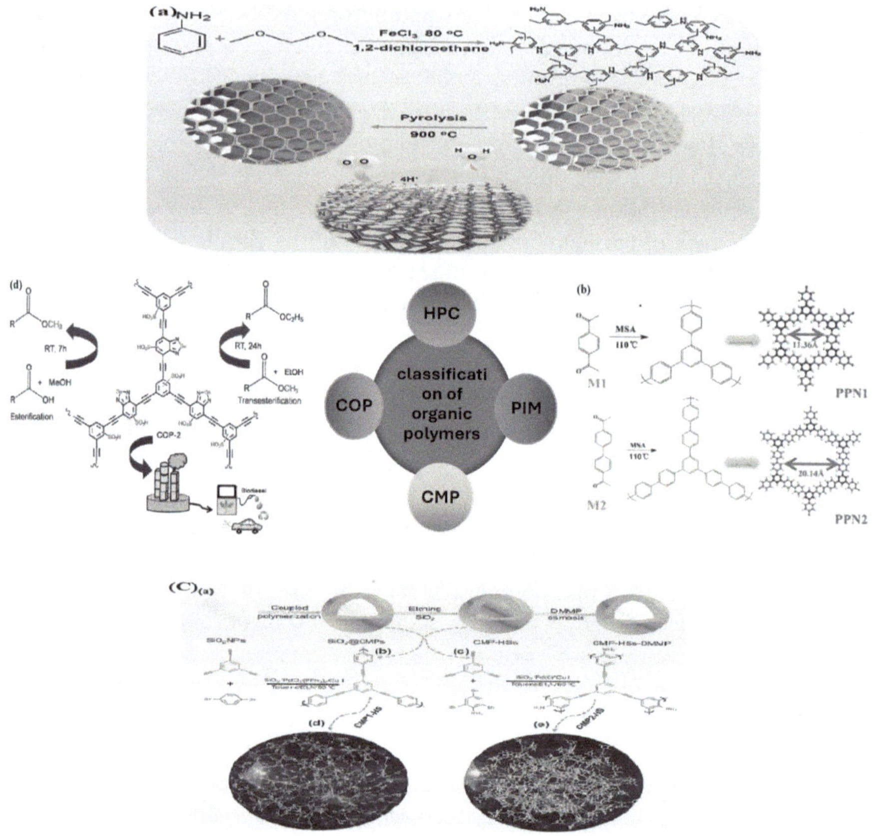

FIGURE 7.1 Schematic of classification organic polymers.

methods and synthetic polymers. These polymer adsorbents with pore sizes ranging from 1 to 100 nm have diverse properties that make them essential for industrial and refining purposes. Within the realm of polymer adsorbents utilized for CO_2 gas adsorption, various types such as microporous organic polymers (MOPs), porous organic polymers (POPs), covalent organic frameworks (COFs), polymers of intrinsic microporosity (PIMs), hypercrosslinked polymers (HCPs), covalent triazine-based frameworks (CTFs), conjugated microporous polymers (CMPs), tetrazine-based organic frameworks (TZFs), and covalent organic polymers (COPs) have been discussed in detail within this chapter [6, 7]. The schematics of organic and inorganic polymers are shown in Figure 7.1 [8–11] and Figure 7.2 [12–14], respectively.

7.2 CLASSIFICATION OF POLYMERIC ADSORBENT

7.2.1 Organic Polymer

MOPs and POPs represent a burgeoning category of porous substances, existing in both amorphous and crystalline forms, that offer excellent CO_2 selectivity, substantial storage capacity, and significant surface at a cost-effective rate. These materials have

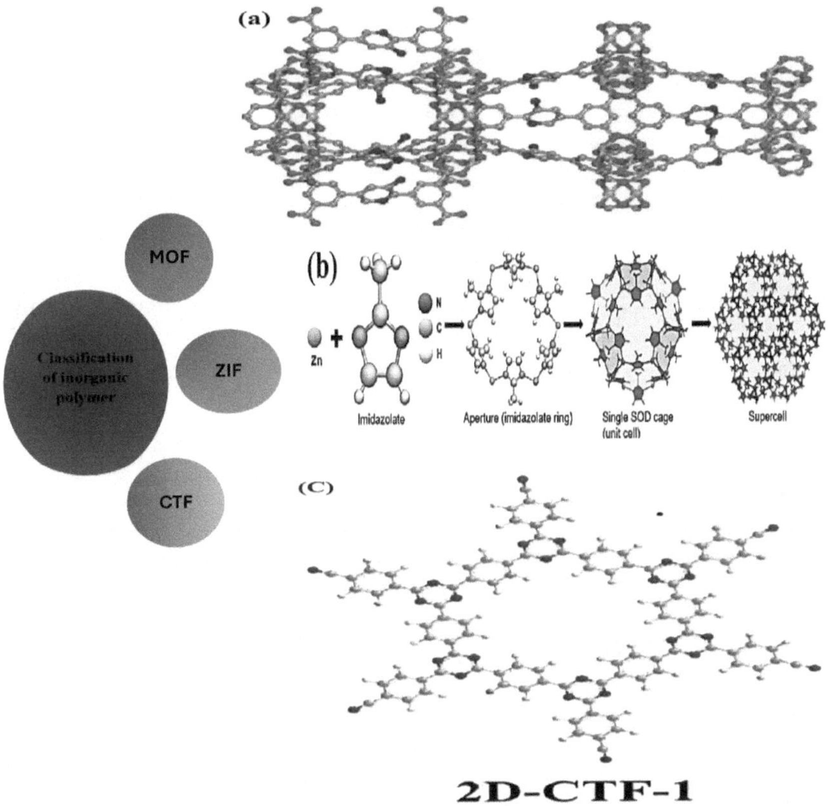

FIGURE 7.2 Schematic of inorganic polymers [12–14].

shown promise in various applications including adsorption, catalysis, gas separation, and storage. Furthermore, their potential in addressing the pressing global need to mitigate CO_2 emissions has garnered substantial attention. Recent research efforts have been carried out toward synthesizing various organic polymers, particularly polymers of MOPs to CO_2 adsorption, owing to their superior adsorption capacity and physicochemical perpetuity. MOPs, which encompass various polymeric networks such as PIMs, COFs, HCPs, CMPs, and CTFs, offer disparate advantages. Their exceptional thermal and chemical resistance under harsh conditions distinguishes them from contrary porous solids, such as MOFs [15].

7.2.1.1 Hypercrosslinked Polymers

HCPs, mainly prepared by the Friedel–Kroft alkylation reaction, are cost-effective robust microporous organic materials. The permanent porosity in HCPs results from extensive crosslinking reactions that exclude the polymer chains from collapsing into a dense, non-porous state. The highly crosslinked character of the materials renders them extremely thermal resistant not normally envisaged for organic polymers. These materials can have a surface of up to 2090 m^2/g. Typically, most pores in such a substance are about 1.8 nm. The high surface and little pore extent make these materials excellent gas adsorbers for neat energy and environmental usage [16, 17]. During research, Li et al. have considered the capacity of CO_2 adsorption to hypercrosslinked polymers. Poly(divinylbenzene (DVB)-co-vinyl benzyl chloride (VBC)) with various amounts of DVB as a precursor, a chain of crosslinked HCP–DVB–VBC materials was polymerized. As the DVB amount changes from 0% to 10%, the pore size of HCP–DVB–VBC decreases from macropore to micropore, the hole scale distribution narrows, and the micropore volume enhances from 6.82% to 61.90%. CO_2 adsorption studies showed that polymers with small vent sizes and huge microporous volumes indicated higher CO_2 adsorption capacity. For instance, DVB–VBC with 10% DVB indicates a net microporous building and a large micropore volume; this may adsorb the maximum CO_2 gas (2.82 mmol/g, 12.41 wt%) at 1 bar and 273.15 K. The preparation of crosslinked mesopore organic polymers is a topic of immense investigation interest. It involves various procedures, such as polyaniline, thiophene oxidation polymerization crosslinking of polypyrrole, organolithium compounds, Lewis acids catalyzed by $FeCl_3$ or $AlCl_3$, and Friedel–Kroft alkylation polymerization, which provides the alkylation polymerization of aromatic loops [18]. Crosslinked polymers form extended networks in rigid and poriferous crystalline frameworks, mesopore organic polymers, and zeolites, percolated by rather polymer solvents, and are considered thermodynamically bad solvents. These crosslinked mesopore organic polymers are permanent in organic and aqueous solvents and enable fast-scale provision. Over the years, significant advances have been made in preparing crosslinked mesopore organic polymers. For instance, researchers have expended phosphonium salt containing HCPs and benzene-relying HCPs, fused aromatic monomers containing hydroxystyrene group, tetraphenylmethane, triazine, tetraphenylethylene, and tetraphenylbiphenyldiamine for synthesis. However, the governed procurement of silicon and germanium-based HCPs is still challenging, as the process involves expensive catalysts, high temperatures, and lengthy reaction times. Hence, further research is required to develop cost-effective and efficient methods that enable the controlled procurement of these HCPs. Puthiaraj and Ahn have sketched an ordinary and inexpensive strategy to acquire HCPs containing tetrahedral units combined with germanium and silicone with a higher potential for CO_2 uptake [19]. Constructing porous polymer networks from multifunctional monomers or crosslinks is an area of study that has gained significant interest in recent years. In this regard, a feasible mechanism has been offered to show the constitution of vents in the incorporated polymer, which sheds light on the underlying reaction pathways. Production of benzyl carbocation at the elementary step in the polycondensation reaction leads to the formation of a dimer by attacking another monomer molecule. The dimer, in turn, grows by attacking other molecules with the hydroxyethyl group of the benzyl carbonium ion and via benzyl alcohol (BA) attack. The resulting much-replaced loops and little branches lead to porosity from incompetent packing of extremely branched polymer loops. Importantly, the gas adsorption valence of HCP–BA is observed to be 8.46 wt% of CO_2 at 1 bar and 273 K, and the isosteric therm of adsorption on CO_2 is evaluated to be 27.4–24.1 kJ/mol—this is a promising outcome. However, despite significant progress in this field, there is a need to optimize the synthetic diversity and porous attributes of HCPs to increase gas storage performance. Recently, researchers have explored new approaches to achieve this target.

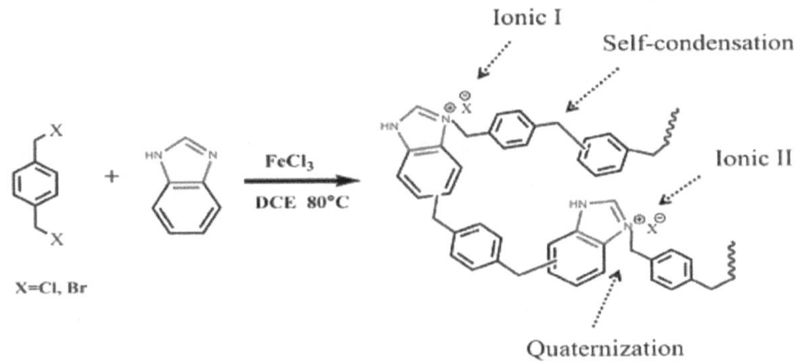

FIGURE 7.3 Schematic method toward synthesizing the benzimidazole-relying hyper crosslinked polyionic liquids [21].

For example, formaldehyde dimethyl ether has been used as a crosslinking factor, and $FeCl_3$ is used as a catalyst to develop an inverse tiling strategy to prepare super-crosslinked MOPs. This approach has shown promising results, as specific HCPs with a BET surface of up to 1391 m^2/g have been produced, by a CO_2 adsorption of 13.5 wt% at 1 bar and 273 K. Furthermore, by crosslinking the tetrahedral monomer, triphenylmethane, CO_2 uptake at 298 K and 1 bar was also measured to be 1.77 $mmol^{-1}$, among the highest capacities stated for MOPs. Importantly, this HCP synthesis method has also shown tolerance to different functional groups, as evidenced by the successful polymerization of phenol. The formed network had a surface area of nearly 400 m^2/g and could adsorb 9.4 wt% of CO_2 at 1 bar and 273.15 K [20]. In 2019, Song and Huang synthesized a type of benzimidazole-seated hypercrosslinked poly(ionic liquids) through a hypercrosslinking strategy; the prepared adsorbents have abundant mesoporosity and ionic active sites with high SBET of 485–780 m^2/g^2. Hypercrosslinked poly(ionic liquids) hyper-branched polyionic liquids (HPILs) showed significant CO_2 uptake of 79–143 mg/g at 273 K and 1.0 bar, average isosteric heat of 32–45 kJ/mol, and CO_2/N_2 selectivity. Interestingly, HPILs-Cl-2 exhibited excellent catalytic activity for CO_2 cycloaddition, and 99% of propylene oxide was successfully converted to cyclocarbonates under the mild conditions of 70°C, 0.1 MPa, and 9 h CO_2. This synthetic tactic is of major significance for functionalized hypercrosslinked polymers, and the amalgamated HPILs are promising for efficient CO_2 capture and alteration under gentle situations. Figure 7.3 shows that a minimum of three chemical reactions co-occur for benzimidazole (BMZ) and benzyl halides in the present study. First the $FeCl_3$-catalyzed Friedel–Crafts alkylation reaction of benzyl chloride or benzyl bromide provides polymers by vast SBET and plentiful micropores. Then, the nucleophilic replacement of Bmz with benzyl halides should be the initial reaction and provide the polymers with several active imidazolium salts (ionic I and ionic II). Finally, another intermolecular Friedel–Crafts alkylation reaction with $FeCl_3$ between Bmz and benzyl halides elevates HPIL with the benefits of HCPs and multifold ionic active sites. CO_2 adsorption in these polymers at 273 K and 1.0 bar follows the arrangement of HCP-Cl (148 mg/g) > HPILs-Cl-1 (143 mg/g) > HPILs-Br 1 (101 mg/g) > HCP-Br (98 mg/g) > HPILs-Br-2 (83 mg/g) > HPILs-Cl-2 (79 mg/g). At 298 K and 1.0 bar, CO_2 capture decreased considerably, indicating an exothermic method. Interestingly, at 273 K and 0.15 bar HPILs-Cl-1(53 mg/g) had partly superior CO_2 capture than HCP-Cl (45 mg/g), and the equal phenomenon became visible for the Br-replaced monomers (HPILs-Br-1: 40 mg/g; HCP-Br: 33 mg/g). The presented Bmz as the positive or negative burdens in polymer frames should render a positive portion. The high solution N1s determined that the imidazoline and imidazolinium cations were the original configurations for the polymers, which can increase CO_2 caught by the electrostatic strength. Additionally, it indicated that the CO_2 captures approximately had no damage after five adsorption–desorption periods, offering supreme reusability [21].

Maliki et al. have recently utilized styrofoam waste as a precursor for synthesizing a crosslinked polymeric adsorbent (HCP) applying the Friedel–Crafts method for CO_2 adsorption. The super crosslinked adsorbent was schemed to employ formaldehyde dimethylacetal as a binder. The adsorbent synthesis method involved the establishment of a complex betwixt a Lewis acid catalyst and a crosslinking molecule, weakening the bond between the methoxyl group and the carbon atom and generating numerous average carbocations in the methylene chloride solvent. As the reaction proceeded, carbocations combined with phenyl rings and added polymethoxymethyl groups to aromatic molecules. Subsequently, the highly reactive methoxymethyl groups engaged with extra phenyl loops, converting to methylene bonds and resulting in extremely rigid network structures with high surface area and significant microporosity. This study thoroughly examined the adsorption process, utilizing isotherm, kinetic, and thermodynamic models. The discoveries demonstrated that the Sips isotherm and Elwich kinetic models outperformed other models in terms of accuracy. Thermodynamic analysis revealed negative values for the parameters ΔH^0, ΔG^0, and ΔS^0, suggesting an exothermic and spontaneous reaction accompanied by a reduced disorder during absorption. Moreover, the adsorbent demonstrated an impressive adsorption efficiency of up to 96.1% after 10 cycles. The hypercrosslinked adsorbent (WSHC) exhibited a notable surface area of 802.84 m^2/g and an average pore diameter of 2.869 m^3/g. Additionally, the WSHC adsorbent displayed a maximum CO_2 absorption of 11.053 mmol/g at 298 K and 10 bar [22]. Moradi et al. synthesized a carbazole-based crosslinked polymer through an alkaline Friedel–Crafts reaction, employing $FeCl_3$ and $AlCl_3$ as catalysts. This polymer was designed to adsorb CO_2. The surface adsorption process was analyzed using artificial neural network modeling and the response surface method (RSM). RSM is employed to assess the impact of various effects and their combined effect on the system answer, employing a blend of mathematical and statistical approaches. In this perusal, the efficacy of parameters such as adsorption time, crosslinking relation, and synthesis time, and their efficacies were investigated. The central composite design was exploited in the response level method, and the correlation factor for the response was considered based on the quadratic model. In ANN, two algorithms, MLP and RBF, were used for modeling, and both prepared acceptable prediction methods. According to synthesizing this adsorbent, a $FeCl_3$ catalyst is better than $AlCl_3$. When $FeCl_3$ was applied as a catalyst for the synthesized adsorbent, the correlation coefficient for the response in RSM was 0.9432, and in MLP and RBF the maximum correlation coefficients were 0.9960 and 0.9957, respectively. The optimum adsorption capacity values for $AlCl_3$ and $FeCl_3$ were 172.68 and 174.59 mg/g, respectively [23].

7.2.1.2 Polymers of Intrinsic Microporosity

PIMs, an innovative class of macromolecules, have emerged as a subject of extensive research in the field of polymer chemistry since 2003 when Mckeown and Budd discovered them. These polymers possess an amorphous structure, and their pores arise from the inefficient packing of non-network polymer chains, which are hard and constricted. A crucial aspect of designing and synthesizing PIMs is ensuring that the polymer chains are adequately rigid and constricted to prevent rotation along the chain. Typically, these chains comprise aromatic heterocyclic ladder polymers. If the polymer chains lack a non-network structure, the arrangement of the chains can be rearranged, leading to the crumbling of the poriferous structure. The initial PIMs developed were created by linking phthalocyanines or porphyrins with tough spirobisindane groups. These substances represented a microporous structure with a BET surface area of 895 and 910 m^2/g for phthalocyanine- and porphyrin-relying lattices. Liebl and Senker synthesized a class of triazine-based polyamide (TPI) porous polymer lattices with a great grade of functionalization. The TPI polymer network displayed a relatively supreme heat of CO_2 adsorption between 29 and 34 kJ/mol. TPI-1, with a BET of 809 m^2/g, can adsorb 2.45 $mmol^{-1}$ CO_2 at 273 K and 1 bar, while TPI-7 exhibits excellent binding selectivity (56) for CO_2 upon N_2 [24]. Farha et al. also synthesized lithium-doped polyamide by chemically detracting naphthalene diimide employing lithium metal as a reducing agent. The porosity manufactured in lithium-doped polyamide illustrates a sharp increase in selectivity (38) for CO_2 upon CH_4. Recently, polyamide-containing triazine loops have been related to handling melamine and inexpensive commercial anhydrides as monomers. Due to the large percentage of existent nitrogen, it can adsorb 7.3% by weight of CO_2 at 273 K and 1.13 bar despite the average level [25]. In 2019, Akbarzadeh et al. synthesized a novel series of PIMs-1–9 polyamines that incorporated ortho-connected thiazole units and pliable thioether bonds derived by diamine monomers (DA-1–3) and several commercially accessible aromatic dialdehydes (terephthalaldehyde, isophthalaldehyde, and 2,5-pyrazinedialdehyde). The solubility of the PIMs was investigated, given its importance as a significant index for determining the polymer's morphology, processability, and efficiency in gas separation. The properties of the solvent, including polarity, vaporizability, and particular interplay characteristics with the polymer substance, play a fundamental pattern in the degree of polymerization, level, porosity, CO_2 adsorption capacity, and selectivity of the polymer. To this end, the solvent template's roles that direct the polymers' pore value with its close dimensions were examined. The experiments indicated that the polymers have supreme solubility in polar solvents such as NMP, DMF, DMAc, DMSO, and pyridine at environment temperature and medium solubility in acetone and THF. These explorations are consistent with the synergistic factors of pliable thioether units with nonplanar structure and massive pendant segments, phenyl and naphthyl substituents, weakening H-bonding, and reduced polymer packing. Notably, solvents with great ebullition points and polarity, such as DMSO and DMF, can significantly impact the polymerization process, forming highly porous and polymerized materials. This is due to the reduction in entanglement and the consequent improvement in the free volume during synthesis. However, it is important to mention that the effect of such solvents is more pronounced in highly porous polymers with a great surface area. On the

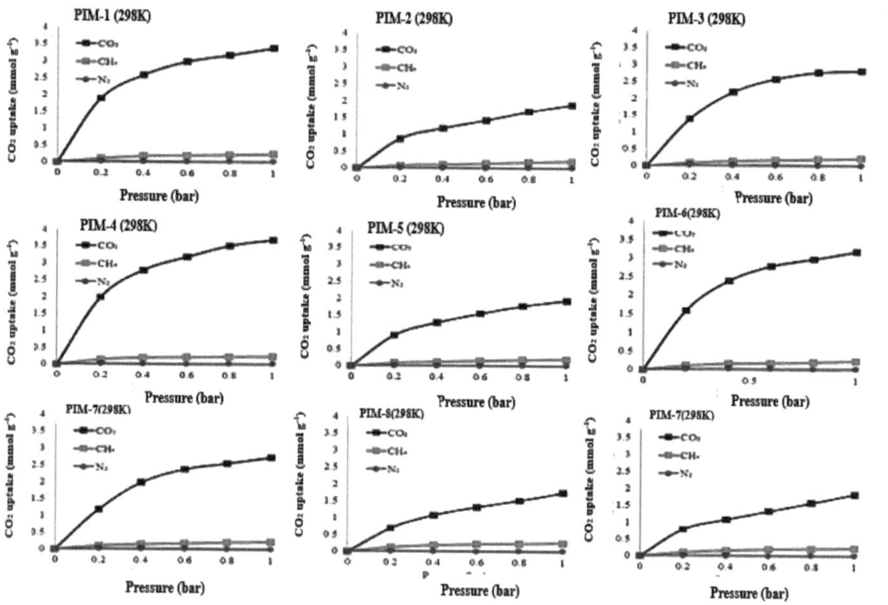

FIGURE 7.4 Selective adsorption of CO_2, CH_4, and N_2 gases for different PIMs at 298 [26].

other hand, the solubility of polymers primarily depends on the sort of aldehydes applied in polymerization operations. For instance, 2,5-thiophendicarbaldehyde, as a primary heteroaromatic dialdehyde in the synthesis of PIM-3 and PIM-6, renders these polymers insoluble in THF solvent. Unlike another aldehyde, this compound fully corresponds with the polar structure of the corresponding dialdehyde. As for naphthyl-dependent polymers, such as PIM-7, PIM-8, and PIM-9, exhibit relatively better solubility in less polar solvents, such as acetone and chloroform. This can be attributed to the inherent outcome of the non-polar pendant naphthyl group and its bulkiness, which increases the separation of polymer strings and pursuantly disrupts the order and efficiency of compact loop packing. According to Figure 7.4, PIM-4 and PIM-1, respectively, show better ideal selectivity ratios due to linear packing among other polymers. A structure that leads to a greater preference for CO_2 (CO_2/N_2: 77.3, 70.2 and CO_2/CH_4: 13.7, 12.5). In addition, PIM-8 and PIM-9, which have naphthalene groups, show the lowest gas selectivity ratio, which is CO_2/N_2: 55, 57 and CO_2/CH_4: 6.2, 7, respectively. This is due to the increase in the interval of the polymer chains with the tendency of bulky naphthyl groups along with the development of steric effects, which overtakes a reduction in the interaction between gas and polymer, and, as a consequence, the selectivity of these polymers decreases [26].

7.2.1.3 Conjugated Microporous Polymers

CMPs are a subclass of highly conjugated polymers. They are determined by a porous structure and an extensive conjugated network, formed by multiple C–C bonds or aromatic rings. For instance, porous poly (arylene ethynylene) can be cited as one of the earliest instances of CMPs, synthesized by Cooper et al. in 2007 through Pd-catalyzed Sonogashira–Hagihara crosscoupling betwixt arylethynylenes and aryl halides. Despite their planar nature, the easy rotation of the monomer around the alkyne bonds allows them to form three-dimensional networks. Furthermore, adjustments in the length of the monomer column enable precise control over the surface area, mean pore amount, and pore volume [27]. Ben et al. reported synthesizing the porous aromatic framework (PAF-1) via Yamamoto homocoupling of tetrakis(4-bromophenyl)methane. This polymer adsorbent has good solubility and thermal stability with a record surface area of 5640 m^2/g. In addition, PAF-1 can adsorb CO_2 up to 1300 mg/g at 298 K/40 bar, making it a suitable option for CO_2 storage [28]. It is worth noting that Cooper's group and Zhou's group have recently synthesized microporous synthetic polymers with remarkable BET surface areas by substituting tetrakis(4-bromophenyl) methane with 1,3,5,7-tetrakis(4-bromophenyl) adamantine (TBPA). Specifically, Network 3 and PPN-3 exhibited BET surface areas of 3180 and 2840 m^2/g, respectively. In 2010, the Cooper group reported the synthesis of PAF containing tetravalent Si (lattice 2) with a BET surface area of 1102 m^2/g. This structure was subsequently resynthesized by the group of Zhou (PPN-4) and Qiu (PAF-3) exerting an optimized Yamamoto pair. Notably, PPN-4 exhibited the highest BET surface area reported for any porous material at around 6461 m^2, and it could adsorb 212 wt% of CO_2 at 295 K/50 bar with an extremely promising selectivity for the adsorption of CO_2 over N_2. Similarly, PAF-3 illustrated a BET surface area of 2932 m^2/g and an excellent selectivity for the adsorption of CO_2/N_2 in ambient conditions. These findings significantly advance the development of porous substances with unique properties for various usages. PAFs are fine-applied candidates for CO_2 capture and sequestration but have a relatively small isosteric heat of CO_2 adsorption (Qst) (15–19 kJ/mol), which limits the application of gas storage at upper temperatures [27, 29]. Babarao et al. have designed modern PAFs by presenting polar organic groups to the biphenyl unit using Monte Carlo simulation for CO_2 separation. The isosteric heat of CO_2 capture for functionalized PAF-1 is about 21.05 (NH2-PAF-1), 33.68 (MO-PAF-1), and 30.19 kJ/mol (DHF-PAF-1). All these amounts are much preferable to pure PAF-1 (12.76 kJ/mol) [28]. Lu et al. have synthesized sulfonate-linked PPN-6, sulfonic acid-linked PPN-6-SO_3H, and lithium sulfonate-linked PPN-6-SO_3Li. The surface area of PPN-6-SO_3H and PPN-6-SO_3Li decreased from 4023 to 1254 m^2/g and 1186 m^2/g compared to PPN–6. However, the isosteric therm of CO_2 capture augmented from 17 to 30.4 and 35.7 kJ/mol, respectively. This indicates powerful interplays betwixt the sulfonate-linked PPN-6 samples and CO_2, significantly enhancing CO_2 adsorption capacity. For example, uncrosslinked PPN-6 has a CO_2 adsorption of 5.1 wt% at 295 K and 1 bar, while PPN-6-SO_3H and PPN-6-SO_3Li can capture 13.1 wt% of CO_2 under similar conditions. Additionally, the selectivity over N_2 of PPN-6-SO_3Li and the high adsorption capacity of CO_2 can be mostly attributed to the potent electrostatic interactions among CO_2. The sulfonate-linked PPN-6 networks also exhibit extremely high CO_2/N_2 selectivities with amounts of 155 and 414 for PPN-6-SO_3H and PPN-6-SO_3Li, respectively, as observed from the results of IAST computing using single-component isotherm information and CO_2/N_2 ratio 85/15 at 295 K and 1 bar [30]. The synthesis of a novel sort of difunctional CMPs including sulton and hydroxyl groups by Zhang et al. was achieved through the Pd-catalyzed Sunogashira–Hagihara crosscoupling reaction of bromophenol blue with 1,4-diethynylbenzene or 1,3,5-triethynylbenzene. The polymerization reactions were conducted at a molar concentration of 300 mmol/L at the temperature of 100°C and a reaction time of 36 hours. The molar ratio of ethynyl to bromine functionalities in the monomer feed was optimized to 1.5:1 to maximize the surface area in this class of CMP networks. The BET apparent surface area is 1316 m^2/g and 1470 m^2/g for BFCMP-1 and BFCMP-2, respectively, notably higher than other reported CMPs. However, BFCMP-2 exhibited a smaller surface area, attributed to the higher crosslinking of 1,3,5-triethyl benzene linkage with more linkages than 1,4-diethylbenzene. Structural analysis of the polymer adsorbents revealed that BFCMP-2 has a lower surface area and micropore volume than BFCMP-1, consistent with nitrogen adsorption isotherms. Furthermore, BFCMP-2 demonstrated a hydrogen adsorption capacity of 156 cm^3/g (~1.39 wt%) at 77 K/1.13 bar and a CO_2 adsorption capacity of 2.77 mmol/g at 273 K and 1.13 bar. Compared to other reported non-functionalized CMPs, both functionalized polymer networks show relatively high isosteric heats of CO_2 adsorption due to the introduction of polar sulton and hydroxy functional groups that increase the binding affinity between the polymer networks and CO_2 molecules. CMP-1 networks functionalized with different polar groups such as eCOOH, eNH_2, $e(CH_3)_2$, and eOH increase the heat of CO_2 adsorption. The conclusions show that the presentation of powerful polar groups into a polymer skeleton is effective for producing CO_2-philic MOPs with increased binding vicinity to CO_2 molecules. Both polymer lattices showed nearly identical thermal attributes because they have similar functional categories and major chain structures. Bifunctionalized polymer nets investigated by thermogravimetric analysis (TGA) analysis have shown lower thermal stability (250°C) in nitrogen atmosphere than non-functionalized CMPs reported with numerous thermal resistance up to 400°C. The reason can be attributed to sulton and hydroxyl groups in BFCMP. Though the presentation of polar functional groups post-modification leads to an increase in the binding dependency betwixt adsorbent molecules and CO_2, the level of improved MOPs is a lot lower than the original substances because the introduced functional groups engross the pores that reduce CO_2 adsorption capacity in some cases [31]. Subsequently, Xia et al. synthesized conjugated (1,3,5-triethynylbenzene) microporous polymeric adsorbents using Co and AL metals as modifiers. These polymers are a new class of polymeric adsorbents with excellent CO_2 absorption at atmospheric pressure and room temperature. These polymers can reserve CO_2 by adsorption valences comparable to metal-organic frameworks (MOFs). Coordinated CMPs can altogether act as heterogeneous catalysts for CO_2 and propylene oxide reactions at atmospheric pressure and room temperature. Combining the subordinates of gas repositories and catalysts, this strategy prepares a path to affordable CO_2 devaluation processes. Nitrogen adsorption analyses investigated the porous properties of polymers at 77.3 K and CO_2 adsorption analyses at 298 K. All polymers illustrated type I nitrogen gas adsorption isotherms matching the IUPAC classification, denoting that the metallopolymers are composed of micropores and mesopores. Significant change in pore size distribution was not observed between Co-CMP and Al-CMP. Corresponding to the size dispensation curves computed by NLDFT, the average pore size value in polymers is 0.5 nm. Other porosity attributes such as CO_2 adsorption capacity, Brunauer–Emmett–Teller (BET) surface area, and pore volume were also

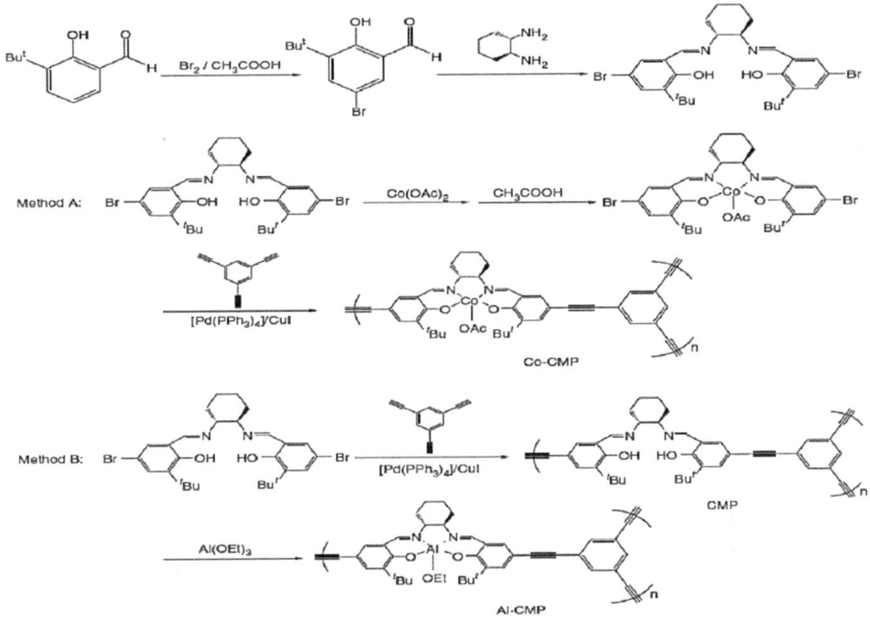

FIGURE 7.5 Synthesis method of Co-CMP and Al-CMP [32].

investigated. Figure 7.5 shows the synthesis method of Co-CMP and Al-CMP [32]. According to the analysis results, it was found that Co-CMP has a much higher BET level than Al-CMP and CMP, which are 965, 798, and 772 cm^2/g, respectively. Also, the total pore volume of Co-CMP is significantly larger than several well-known MOFs with high pore volume, including UMCM-2, MOF-177, and NOTT-112, which are 2.81, 2.31, 1.89, and 1.62 cm^2/g, respectively [32].

7.2.1.4 Covalent Organic Polymers

COFs have become an important research topic in CO_2 sequestration. Owing to their high porosity, controllable structural appearances, high surface area, and diverse chemical performance have been widely studied for their potential in this area. Among these frameworks, a category of high-capacity ultrastable porous substances made of rather stable C–C, C–H, and C–N covalent bonds has attracted increasing attention for CO_2 adsorption. The presence of relatively abiding covalent bonds ensures their hydrothermal stability, an important factor for large-scale applications as water or steam is inevitably present during the adsorption of CO_2 [33]. In this regard, Xiang et al. have prepared a series of COPs with high hydrothermal constancy, which show promising potential for CO_2 adsorption. These polymers have been synthesized through self-polymerization of 1,3,5-*tris*((4-bromophenyl)-ethynyl) benzene, *tris*(4-bromophenyl)amine, 1,3,5-*tris*(4-bromophenyl)benzene monomers, and 2,4,6-*tris*-(4-bromophenyl)-[1,3,5]triazine by the nickel(0)-catalyzed Yamumoto reaction. Furthermore, the efficiency of COP-1, COP-2, COP-3, and COP-4 has been investigated for adsorbing post-combustion and oxy-fuel. This study's results highlight the potential of COPs for CO_2 CCS and pave the way for further research in this area [34]. A recent study conducted by Ullah et al. has appraised the potential of two porous polymers, COP-9 and COP-10, for the adsorption of CO_2 at different pressure ranges. COFs of amine and amide have been used as CO_2 absorbers in different isotherms and wide-pressure conditions. An accurate DFT analysis was also performed for the COP-9 and COP-10 structures for their interaction with CO_2 and N_2 alone and CO_2/N_2 present together about the COP structures. In this study, BET, TGA, and Fourier-transform infrared (FTIR) spectroscopy analyses were conducted to assess the physical properties of the porous polymers. The TGA analysis results indicated that COP-9 exhibits greater stability than COP-10, although its gradual degradation is noticeable after 400°C Conversely, severe degradation was observed in COP-10. The BET analysis revealed that the surface area of COP-9 is significantly greater than that of COP-10, with values of 146.4 and 84.5 cm^2/g, respectively. Moreover, the FTIR analysis confirmed the formation of amine and amide COPs. These findings suggest that COP-9 has a higher potential for CO_2 adsorption, which could have significant implications for developing sustainable and cost-effective carbon capture technologies [35]. Troger-based nanoporous COPs exhibit three-dimensional architectural structures and high surface area. Byun et al. have synthesized polymers using a simple one-pot reaction under relatively mild conditions through the one-pot polymerization of tetraanilmethane or tetraaniladamantane with dimethoxymethane in the presence of triureoacetic acid (Figure 7.2). This process was carried out in a nitrogen atmosphere at room temperature. The establishment of formaldehyde moieties enabled subsequent alkylation and cyclization of amino groups, leading to the formation of crosslinked amorphous polymers bridged by the Troger base. It was observed that TB-COP-1 tolerates intermolecular cyclization, while TB-COP-2 favors intramolecular as well as intermolecular cyclization due to the vicinity of the amino moieties in the tetraanillmethane monomer. This suggests that the co-stiffness of the monomer and the interval amid amino groups play a crucial inscription in the synthesis of covalently linked organic polymers with the Troger base. Furthermore, the surface area, CO_2 adsorption capacity, and CO_2/N_2 selectivity of nanoporous COPs with Troger base (TB-COPs) were interrogated. The formation of nanoporous polymers from two tetrahedral monomers, TB-COP-1 and TB-COP-2, resulted in a notable contrast in surface area. TB-COP-1 exhibited a high surface area of 1340 m^2/g, while TB-COP-2 demonstrated a considerably lower surface area of 0.094 m^2/g. This disparity can be attributed to the spatial vicinity of the amino groups in the monomer unit. The concise interlude between the amino groups facilitates intramolecular and intermolecular cyclization, leading to reduced porosity in TB-COP-2. TB-COP-1 displayed substantial CO_2 capture capacities of 5.19 and 3.16 mmol/g at 273 and 298 K, respectively, under ambient pressure. Additionally, it exhibited CO_2/N_2 selectivities of 79.2 and 68.9 at 273 and 298 K, respectively, at 1 bar for a gas mixture with a CO_2:N_2 ratio of 0.15:0.85. Notably, TB-COP-1 retained significant selectivity even as the temperature increased from 273 to 298 K. Using the T-plot method, microporosity TB-COP-1 was determined to be 0.58 cm^3/g, which indicates that the majority of pores are concentrated in the microporous area. Furthermore, the NLDFT pore size distribution analysis revealed a narrow repartition with a mediocre pore size of

1.58 nm. These porous properties position TB-COP-1 in alignment with well-known high-surface area microporous materials such as MOFs and zeolites [36].

Table 7.1 presents an exact comparison of COPs that employ cinnauric chloride and 1-3-5-benzenetricarbonyl trichloride as polymer cores and comprise amine- and amide-based grafts. Among the COPs that employ amine-based grafts, COP-1 synthesized from piperazine shows superior CO_2 adsorption efficiency at high pressures compared to COP-9. COP-9 has shown CO_2 adsorption performance almost twice as well as COP-2 at high pressures, which uses bipiperidine as a linking element. Although COP-1 has a large volume of 1.36 mmol/g sample, in contrast to the efficiency evaluation at vast pressures, COP-2 shows premiere performance (0.93 mmol CO_2/g sample) compared to COP-9 (0.60 mmol/g). Furthermore, the less CO_2 uptake capacity of COP-2 notwithstanding the relatively higher measured surface area can be illustrated by the major percentage of mesopores in COP-2, which leads to increased surface area but is not necessarily beneficial for the CO_2 uptake capacity performance. These conclusions display that the CO_2 affinity and gas adsorption performance can be equiponderant by manipulating the type of linker, which steers to modify the pore size and pore volume of COPs and has a crucial role in the adsorption capacity over the surface area. COPs rely on CO_2 adsorption performance, the following order is observed: COP-33 > COP-32 > COP-34 > COP-10 based on presented adsorption amounts at 200 bar and in the 308 K isotherm.

Graphene oxide (GO) with the atomic plane structure of graphite, which consists of sp^2 and sp^3 hybridized carbons with various surface oxygen functionalities such as epoxy, hydroxyl, carbonyl, and carboxyl, is one of the main products of graphene.

TABLE 7.1
Evaluation of COPs That Are Incorporated by Applying Amine and Amide-Based Linkers [35]

				CO_2 Sorption (mmol/g)		N_2 Sorption (mmol/g)		BET	
								Surface Area	Pore Volume
COP NO.	Core	Linker Group	Linker Specific	200 bar 308K	1 bar 298K	200 bar 308K	1 bar 298K	(m^2/g)	(m^3/g)
COP-1	Cynanuric chloride	Amine	Piperazine	128	1.36	—	0.086	168	0.25
COP-9			Trans-2,6-dimethyl-piperazine	79	0.60	21	0.6	146	0.24
COP-2	1-3-5 benzene-tricarbonyl trichloride	Amide	bipiperadine	47	0.93	—	0.186	158	0.67
COP-33			1,4-phenylenediamine	133	0.6	30	0.02	53	0.2
COP-32			4-aminobenzylamine	128	0.44	31	0.02	46	0.14
COP-34			1,3-phenylenediamine	63	0.47	15	0.025	33	0.1
COP-10			2-imidazolidinone	52	0.7	12	0.12	85	0.63

The porosity of this material is based on the multilayered, interwoven, and layered forms as well as the crosslinking of graphene sheets after development. GO can be easily customized with various functional materials for targeted absorption applications [37]. The preparation of CO_2 adsorbent based on GO impregnated with polyethyleneimine (PEI) was reported by Shin et al. The authors performed acidic pretreatment of GO handling a combination of HNO_3 and H_2SO_4 to introduce more carboxyl and hydroxyl moieties and thus increase PEI loading. The effect of PEI on GO resulted in superior CO_2 valency and selectivity due to the interaction of amine moieties with CO_2. Specifically, the adsorption capacity of the studied models was appended by increasing the amount of PEI up to 60% by weight [38].

7.3 INORGANIC POLYMER

The issue of CO_2 capture has been the issue of immense research for multiple decades. Inorganic and organic polymers have been developed as potential solutions for this problem. Among them, porous inorganic polymers such as MOFs and zeolitic imidazolate frameworks (ZIFs) have gained significant attention because of their flexible structure and impressive gas adsorption properties. Due to their unique properties, CTFs have emerged as a promising new class of high-performance porous organic materials (POPs). CTFs, frequently classified as CMPs, are a group of amorphous porous polymers that combine π-conjugated skeletons with constant nanopores. These CTFs as a viable solution for CO_2 adsorption and could play a significant role in addressing the issue of carbon emissions. Skeleton chemistry theory suggests that reversible covalent bond formation is necessary for crystallization, but it frequently limits stability. Various strategies for synthesizing porous CTFs have been reported in recent literature. These strategies include utilizing thermal ion conditions, strong acidic Brønsted conditions, and the Friedel–Kroft reaction. The ionothermal reaction with $ZnCl_2$ produces defined CTFs and porous carbon materials with surface areas of up to 1152 m^2/g. Furthermore, applying strong Brønsted acid (CF_3SO_3H) under room temperature and microwave situations has facilitated the synthesis of CTFs with wide surface areas. While acidic Brønsted conditions yield fewer surface areas and effectively prevent disintegration and condensation reactions, such as C–H bond failure and carbonization, resulting in fewer structural imperfections. The Friedel–Crafts reactions catalyzed with anhydrous aluminum chloride have also been employed to synthesize specific CTF substances with surface areas of up to 1452 m^2/g. These methods offer easy synthesis and yield high-performance materials with significant surface areas [39]. To overcome this issue, Yuan et al. synthesized a series of triazine covalent framework materials relying on phthalazinone with triphenylamine (PHCTF-8) with a high special surface area. This material was successfully obtained from a novel tricyano precursor, symmetric and wheel-like center (TPAHPZ-TN), through thermal ion polymerization for the surface capture of CO_2. The synthesis process involves the preparation of tricyano building blocks by an easy and inexpensive aromatic nucleophilic subrogation reaction, in which the rigid and twisted part of phthalazinone supports the porous structure. The three functional groups increase the crosslinking density, provide thermal stability of polymer networks, and cause effective space division. Additionally, the high specific surface area, high porosity with dominant microporous disseminations, and electron-rich N-heterocycle skeleton feature lead to a high CO_2 storage capacity of PHCTF-8 up to 17.8 wt% at 273 K and 1 bar. The high competitive capacity of 11.2 wt% (2.55 mmol/g) at 298 K and 1 bar and high valence at low pressures (4.3–5.7 wt%, 0.15 bar at 273 K) are noteworthy. High reaction temperatures frequently result in a higher fraction of mesopores and a broad pore size dissemination, along with a higher surface area. As the final synthesis temperature increases from 400 to 700, the adsorbent surface area and pore volume increment from 1188 and 0.7 cm^3/g to 1665 and 0.96 cm^3/g, respectively. In general, the micropore area contributes to supreme adsorption and gas separation properties, especially for little-sized CO_2 molecules [40]. In other studies, Liu et al. designed a hydrazine-bridged triazine covalent polymer (HB-CTP) by a plain nucleophilic substitution reaction of 2,4,6-trihydrazinyl-1,3,5-triazine, with cyanuric chloride, which was synthesized by sodium carbonate at 110°C. One of the significant advantages of this adsorbent used for CO_2 and N_2 adsorption is the presence of multiple triazine units to comfort the reversible capture of CO_2 main nitrogen sites that dress up the pores of the substance, as well as forming hydrogen bonds. Triazines activate large hydrazine moieties. Indeed, HB-CTP is a stable CO_2 scavenger and an active and recyclable catalyst for CO_2 cycloaddition with different epoxides to produce high-efficiency cyclic carbonates. The polymer adsorbent was analyzed by FTIR, scanning electron microscopy (SEM), TEM, BET, and XRD. The morphology of HB-CTP shows that material mostly exists in the altogether form. Also, the BET surface area was 51.2 m^2/g, and the total pore volume was 0.28 cm^3/g. The XRD pattern has shown an amorphous and non-absorbent structure [41]. Day et al. studied the synthesis of CTF-triptycene (TPC) and CTF-FL porous frameworks through Friedel–Crafts reactions betwixt TPC and fluorene as a linker with cyanuric chloride as a node in the attendance of anhydrous $AlCl_3$ as a catalyst in dichloromethane (Figure 7.6a). The XRD analysis indicated the amorphous nature of the adsorbent. TGA results showed that CTF-TPC and CTF-FL adsorbents beginning to break down in the O_2 atmosphere at temperatures of 380 and 360°C, respectively. BET surface area and pore volume for CTF-TPC and CTF-FL were measured at 1668 and 773 m^2/g and 0.65 and 0.31 cm^3/g, respectively. Also, SEM determines the association of spherical particles (Figure 7.6b and c). The non-uni-thermal CTF-TPC and CTF-FL displayed great CO_2 adsorption capacity, up to 4.24 mmol/g at 273K and 1 bar. Furthermore, BET evaluations of the further two batches confirmed the reproducibility of the results. It is evident from the research results that the CTF level of TPC, synthesized by $AlCl_3$, surpassed all the quantities reported for similar lattices. Additionally, the surface of CTF-FL was minor compared to that of formerly announced fluorene-based CTF [42].

Inorganic polymer frameworks, especially CTFs, have consideration by researchers due to their synthetic modularity, chemical constancy, high surface area, and suitable surface performance for CO_2 adsorption and conversion [43]. In a recent experimental study, Song et al. utilized $ZnCl_2$, a reactive molten salt, as a soft pattern and a catalyst to prepare CTFs. The method of thermal ion synthesis, involving molten $ZnCl_2$ salts and a $ZnCl_2$/KCl/NaCl eutectic salt mixture, has been employed to form polytriazine networks that are reversible at 400°C but irreversible at high temperatures. The $ZnCl_2$/KCl/NaCl eutectic salt mixture facilitates the polymerization

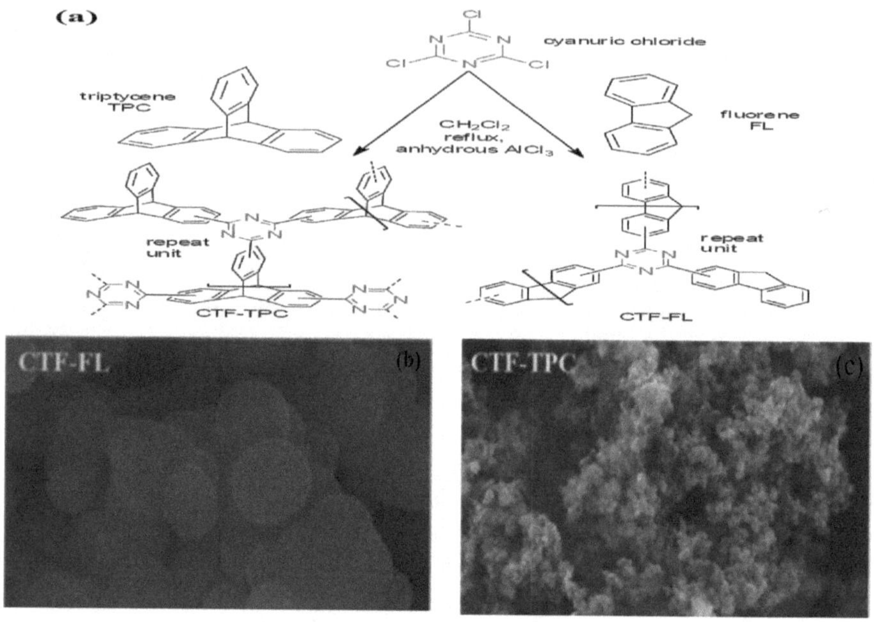

FIGURE 7.6 Schematic of synthesis paths for CTF-TPC and CTF-FL [42].

reaction at lower temperatures. Furthermore, 1,4-phenylene diisocyanate has excellent solubility in these salt melts. Cyclic imide-containing covalent frameworks (CICFs) were synthesized by heating 1,4-phenylene diisocyanate in molten $ZnCl_2$ salts and its eutectic mixture $ZnCl_2$/NaCl/KCl at 400°C and 500°C for 24 h in a sealed ampoule. In this context, the trimerization of 1,4-phenylene diisocyanate under isothermal situations at various reaction temperatures using $ZnCl_2$ (CICF) and eutectic salt mixture KCl/NaCl/$ZnCl_2$ (CICF-KCl/NaCl) as reactive solvents through cyclotimerization reaction of triisocyanate molecules. Remarkable differences in the morphology of the two polymers were observed, with CICF displaying disordered micrometer-sized particles and CICF-KCl/NaCl exhibiting a film-like morphology. This study investigates the influence of ion–dipole interplays betwixt alkali metal cations and monomer oxygen atoms on the two-dimensional development and fully microporous framework formation of POPs. The CICF-KCl/NaCl exhibited nearly theoretical preservation of nitrogen quantity at 500°C due to favorable ion–dipole interactions, resulting in a remarkable increase in the BET surface (590 m^2/g) and CO_2 adsorption valency (5.9 $mmol^{-1}$ at 273 K and 1.1 bar). This research highlights the significance of interactions between salts and monomers in thermal ion synthesis, which can be utilized to control the morphology, porosity, and gas adsorption properties of POPs. Furthermore, the trimerization reactions of various aromatic nitriles, ethynyls, ortho-aminonitriles, isocyanates, and boronic acids are significant for preparing POPs. For instance, the trimerization reaction of aromatic nitriles under an ionothermal situation, using $ZnCl_2$ as a reactive solvent, led to the progress of CTFs. This study emphasizes the potential of thermal ion synthesis in fabricating microporous POPs with desirable properties for diverse usages. The reaction temperature and the quantity of $ZnCl_2$ heavily influence the formation of COFs. $ZnCl_2$ acts as a Lewis acid in the trimerization of aromatic nitriles, where the nitrogen atom interacts with the Zn^{2+} cation through ion–dipole interactions. Such noncovalent interactions lead to a soft template effect that facilitates the production of ultrafine porous CTFs at low temperatures. Enhancing the reaction temperature results in a substantial augment in surface area and conductivity, but it also comes with a decline in crystallinity and a significant reduction in nitrogen quantity. At higher temperatures, mesopores form due to partial carbon, which indicates the loss of the mold effect. The CTF formation approach has the main advantage of lowering the melting point of $ZnCl_2$, making CTF synthesis possible at lower temperatures. It's worth noting that the attendance of alkali metal salts at high temperatures results in the formation of significant mesopores, which can be explained by the low affinity of nitrogen atoms to alkali metal cations. Polyisocyanurates have various applications in insulation and flame retardant materials due to their high thermal constancy and in the form of polymer adsorbents for CO_2 capture [44]. SBA-15 is a siliceous substance that was identified by Zhao et al. SBA-15 is a widely used material whose structure is regular hexagonal pores in the range of 4.6–30 nm [45]. Sanz-Perez et al. have investigated the reusability and recycling of silica amine materials as CO_2 absorbers. Adsorbents were obtained by grafting diethylenetriamine or impregnating PEI on silica SBA-15 [46]. Chang et al. in 2017 researched the effect of silica base and PEG additive on the adsorption efficiency of molecular adsorbent (MBS) for CO_2 adsorption consisting of PEI and one of the following bases: SBA-15 (2D structure), TUD-1 (3D sponge-like structure), and silica fume HS-5 (3D disordered structure). The results showed that the addition of PEG that the support pore characteristics containing 3D pore structure, pore size, and pore volume may not be directly related to the CO_2 adsorption performance of the MBS sample. In contrast, these characteristics play a more significant role in CO_2 capture after PEI loading. MBS capacity with 3D pore structure indicated better CO_2 adsorption valence and amine performance than 2D structure support. Among the perusal adsorbents, fumed silica-based MBS (HS-5) demonstrated a superior CO_2 adsorption temperature range of 30–95°C, perchance due to its inimitable interstitial pores loaded with SiO_2 particle aggregation. Adding PEG can substantially increase the CO_2 adsorption valence and ameliorate the amine performance of MBS by reducing the propagation obstacle in PEI bulk layers via the intermolecular interaction between PEI and PEG [47]. MOFs are also a novel category of crystalline materials that consist of metal ions or organic ligand-bound clusters ordered in an open lattice with extensive porosity. The CO_2 adsorption performance of MOF adsorbents can be easily adjusted by controlling the chemical potential of the surface and modifying the size and morphology of the pores through changes to the organic ligand. Zhu et al. synthesized novel MOF-polymer composites grafted with PEI@PGD-H through HIPE template polymerization of DVB and glycidyl methacrylate in the presence of hydrophobic rectified CuO nanoparticles pursued by in-situ production of the HKUST type. MOF via reaction of CuO with H_3BTC followed by functionalization of PEI via epoxy-amine reaction and interaction of amine-metal sites. The products have been evaluated as CO_2 absorbers. PEI70@PGD-H illustrates a CO_2/N_2 separation coefficient of up to 76. After 20 cycles of CO_2 adsorption and desorption, the adsorbent still maintained a superb and stable CO_2 adsorption valency of more than 2.8 mmol/g CO_2 for the flue gas. The heat of desorption is 48 kJ/mol of CO_2. These properties make PEI@PGD-H extremely effective, strong, and low-energy CO_2 adsorbents for feasible usage in CO_2 adsorption [48]. During their research, Ding et al. synthesized a ZIF-L with a hierarchical morphology using a hydrothermal manner to enhance CO_2 adsorption. The resulting hierarchical product exhibited unique physical characteristics, such as ZIF-L sheets measuring several micrometers in length, 12 μm in width, and 300 nm in thickness. The hydrothermal temperature was observed to play a crucial role in forming this hierarchical nanostructure. Time-dependent experiments revealed the formation mechanism to be a two-step crystal development process. A comparison of the adsorption capacity of the hierarchical ZIF-L with 2D samples of ZIF-L and ZIF-8 indicated significant differences. Despite ZIF-8 possessing a larger surface area, hierarchical ZIF-L exposed superior adsorption valency owing to the unique cushion-shaped cavities of ZIF-L and the powerful interaction between CO_2 and 2-methylimidazole molecules. Furthermore, the higher CO_2 adsorption efficiency of hierarchical ZIF-L units, compared to 2D ZIF-L obtained at room temperature, can be attributed to the superior surface area and highly efficient hierarchical nanostructures. The surface area and adsorption capacity values for hierarchical ZIF-L, 2D ZIF-L, and ZIF-8 were found to be 304 m^2/g and 1.56 mmol/g, 161 m^2/g and 0.94 mmol/g, and 1450 m^2/g and 0.68 mmol/g, respectively. In conclusion, the study demonstrates the potential of hierarchical ZIF-L as an effective CO_2 adsorbent, offering insights into the impact of morphological characteristics and surface interactions on adsorption performance [49].

7.4 INTRODUCTION OF FUNCTIONALIZED CO_2 ADSORBENTS

7.4.1 Nitrogen-Containing Functional Groups

N-functional solid adsorbents (NFSAs) are a group of adsorbents with stronger CO_2 dependence than hollow and porous adsorbents and better humidity endurance. Hence, NFSAs have been identified as one of the most promising post-combustion adsorbents for adsorbing CO_2 from the chimney gas mixture. NFSAs are manufactured electronegative N sources into solid adsorbents, which utilize both the pore structure of porous solid adsorbents and the acidic gas affinity of N compounds. In the following, the modified polymer adsorbents with nitrogen-containing groups will be explained [50].

7.4.1.1 Amine-Functionalized Polymer Adsorbent

Functionalizing polymer adsorbents using amines as a nitrogen source has become increasingly popular recently. This modification increases the adsorbents' performance and ability to adsorb CO_2 through chemisorption. Therefore, there is a growing interest in developing amine/polymeric adsorbent material composites to improve the efficiency of solid adsorbents. Two primary methods are conjugated to enhance the performance of solid adsorbents: chemical modification and physical impregnation. While chemical modification is generally considered simpler than physical impregnation, adsorbents functionalized through chemical modification at higher temperatures exhibit better chemical stability than those prepared using physical impregnation. Thus, the election of the modification method is a significant consideration in tailoring the properties of solid adsorbents for specific applications. Fayemiwo et al. have synthesized a series of nitrogen-rich crosslinked poly[methacrylamide-co (ethylene glycol dimethacrylate (EGDMA))] (HCP MAAMs) polymers convenient for bulk polymerization and CO_2 adsorption (Figure 7.7). The polymers were intrinsically amine-functionalized and exhibited great CO_2/N_2 selectivity at low CO_2 pressures. The physicochemical properties of the polymer particles were characterized, and their efficiency under typical situations was evaluated. Polymers with a molar ratio of MAAM to EGDMA) ranging from 0.3 to 0.9 were made rich with nitrogen and demonstrated a high association for elective CO_2 capture at low pressures. The attendance and density of NH_2 fragments in the polymer lattice were specified exerting FTIR spectroscopy and X-ray photoelectron spectroscopy (XPS). TGA illustrated that the polymers were thermally stable up to 515–532 K. The peak CO_2 adsorption capacity at 273 K was 1.56 mmol/g, and the isosteric heat of adsorption was 28–35 kJ/mol. An increment in the density of amide groups in the polymer lattice led to a higher affinity for CO_2 at low pressure. At a CO_2:N_2 ratio of 15:85, the CO_2/N_2 selectivity was 52 at 1 bar at 273 K and reached 104 at a very low CO_2 partial pressure. It is noteworthy that the economic efficiency of the HCP-MAAM strengthening scenario is competitive with chemical solvent leaching scenarios. Additionally, amine-modified porous polymers have been drafted to enhance CO_2 adsorption efficiency. A modern category of triazine-based microporous frameworks with high thermal and chemical solidity has been created through the Sonogashira–Hagihara reaction using Pd-based catalysts to rectify the CO_2 adsorption of polymeric substances [51]. In 2017, Jafari et al. reported the microwave-assisted synthesis of extremely hydrophobic POPs from

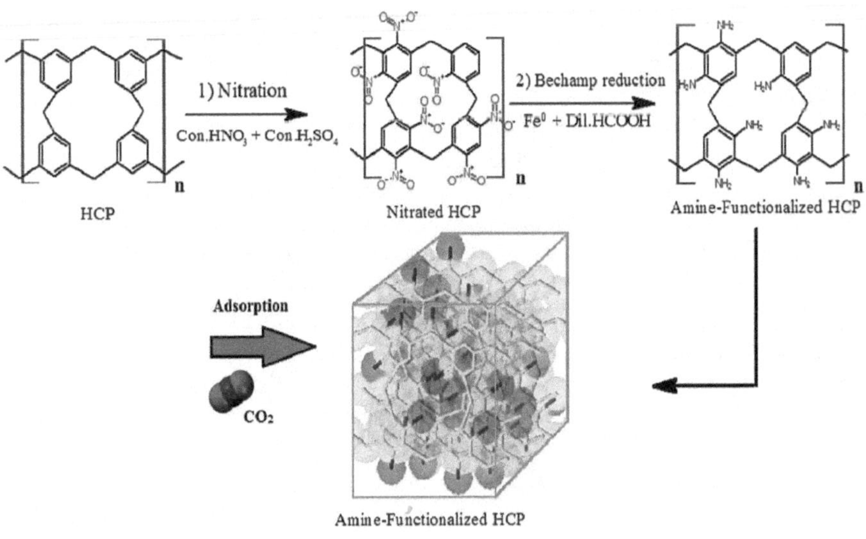

FIGURE 7.7 Amine functionalization method of the benzene-based HCP [54].

polyvinyl benzene (PDVB) modified by amine-rich monomers vinyl imidazole (VI) and vinyl triazole (VT). The PDVB-VI and PDVB-VT adsorbers exhibited an optimal amount of homomonomer and initiator incorporated, leading to a high adsorption activity toward CO_2. Adding amine moieties enhanced atmospheric CO_2 uptake while maintaining optimal surface and pore volume. A certain amount of initiator resulted in better incorporating the VT monomer while keeping the surface and pores accessible. The maximum CO_2 adsorption of 2.65 $mmol^{-1}$ at 273 K/1 bar was caught for the triazole-based adsorbent (PDVB-VT) with 0.7 g of VT and 0.07 g of starter. The adsorption efficiency increased more than twofold compared to a non-functionalized substance (PDVB) with 1.2 $mmol^{-1}$ CO_2 adsorption, and the adsorbent retained its performance for up to seven cycles. Theoretical modeling corroborates that the active position is the nitrogen on the imidazole/triazole hoop, and incorporation of VT into polymer lattice enhances adsorption properties more than VI. The N_2 adsorption–desorption isotherms of different PDVB porous substances were synthesized below various co-monomer situations with several nitrogen densities, co-monomer amounts, and initiator amounts. A higher level of co-monomer incorporation in PDVB resulted in lower surface area and pore volume. The surface area of unfunctionalized PDVB (without co-monomer) decreased from 789 to 500 m^2/g for the maximum value of VT (PDVB-VT0.9) [52].

Among the diverse approaches for synthesizing MOPs, chemical amination is recognized as a powerful implement to control their morphology and surface polarity. This chemistry provides the evolution of MOPs from nanospheres to nanorods with uniform channel surface polarity. The selection of amination effects authorizes good verification of the diameter and length of nanorods. Incorporating CO_2-biophilic features, such as π-rich moieties, acidic groups, or basic units onto the backbone of porous solids, is a reputable method to ameliorate CO_2 uptake. Tao et al. have recently demonstrated a one-step modification of carboxyl-rich MOPs (PTPMA) synthesized by Friedel–Crafts' acylation reaction of tryptase (TPC) and pyrolytic dihydride consuming hydrazine hydrate and ethylenediamine. This modification resulted in CO_2 adsorption, making it a low-cost, straight CO_2 absorption technology. These detections provide new perspectives for developing highly efficient and cost-effective gas storage and separation materials. Polymer-based materials have inimitable properties that make them glamorous for diverse usages, including gas separation. In this regard, PTPMA has been found to have an excellent adsorption capacity for CO_2 due to the presence of a carboxyl group. To further enhance the functionality of PTPMA, amino-terminal attachment to the backbone was successfully achieved through chemical amidation coupling between the carboxyl and hydrazine or ethylenediamine, leading to the formation of PTPNHs. The selection of hydrazine and ethylenediamine as simultaneous reactions was based on structural similarity and difference in chain length, facilitating the study of their effect on the products after modification. Compared to known post-reformation methods such as click reaction or metal catalyst cycling, this catalyst-free method is easier and more impressive. Notably, this modification of auxiliary groups also enables the external evolution of metal-organic polymers, which is unprecedented. Therefore, the modification of PTPMA with hydrazine and ethylenediamine represents a promising approach to increase its performance and opens new opportunities for the development of MOPs. FTIR spectroscopic analysis was employed to describe the chemical structures of poly(4-vinyl pyridine) (PTPMA), poly(4-vinylpyridine-co-styrene) modified with hydrazine (PTPNH-1), and poly(4-vinylpyridine-co-styrene) modified with ethylenediamine (PTPNH-2) to determine the successful synthesis of polymers. The infrared spectra of PTPNH-1 and PTPNH-2 acquired by hydrazine or ethylenediamine modification revealed that the carboxyl group stretching vibration peaks at 2638 and 1714 cm^{-1} disappeared, indicating the successful conversion of the carboxyl group. Additionally, BET analysis demonstrated that the surface area of the adsorbent after modification decreased from 738 m^2/g for the PTPMA adsorbent to 554 m^2/g (PTPNH-1) and 441 m^2/g (PTPNH-2). The comparison of CO_2/N_2 selectivity of the functionalized polymer adsorbent at 273 K and 1 bar showed that CO_2 capture of these two substances is much higher than N_2 adsorption [53]. Moradi et al. (2021) have conducted a study on the synthesis and modification of a benzene-based crosslinked polymer (HCP) adsorbent for the capture of CO_2 from industrial flue gas streams. The HCP adsorbent was synthesized using the Knitting method presented by Li et al. (2018) (Figure 7.7). The synthesis process involved mixing benzene,1,2 dichloroethane and dimethylacetal formaldehyde in a three-necked flask at room temperature under a nitrogen atmosphere, followed by the addition of iron (III) chloride to the mixture at 40°C. The contents of the flask were mixed under nitrogen atmosphere and reflux conditions at 80°C. The resulting polymer lattice was then purified with deionized water and ethanol using a Soxhlet extractor. To improve the adsorption efficiency and electrical properties of the HCP adsorbent, a chemical modification was performed by attaching an amine group to the HCP lattice. This was achieved through a crosslinking reaction in which three of the six carbons in the benzene ring were inserted into abutting benzene molecules by the crosslinking molecule. The modification significantly increased the CO_2 capture valency of the HCP adsorbent through amine linkages. The study demonstrated that the benzene-based crosslinked polymer (HCP) adsorbent modified with amine linkages can effectively capture CO_2 from industrial flue gas streams. This research has promising implications for the development of more efficient adsorbents for the reduction of CO_2 emissions in industrial processes. Polymer adsorbents were analyzed by FTIR, XPS, EDS, and BET analyses to inquire about the morphological characteristics of the adsorbent surface. The results of BET analysis indicate a decrease in the specific surface area of the HCP sample from 806 to 453 m^2/g after surface modification, and CO_2 adsorption examinations illustrated that the amine bond of the HCP sample enhanced the CO_2 adsorption capacity from 301.67 to 414.41 mg/g. FTIR and XPS analysis results show the amine group's successful bonding to the HCP sample's skeleton and the increase of the nitrogen quantity from 0% to 5.17% after the amine modification. The thermodynamic verification of CO_2/N_2 adsorption evidences the spontaneous and exothermic nature of CO_2/N_2 adsorption by both types of models. An evaluation of the capture selectivity between the HCP sample and the amine-linked HCP adsorbent shows a higher selectivity for the adsorption of CO_2 over N_2 at the specific CO_2/N_2 combination of 15:85 [54].

7.4.1.2 Azo-Functionalized Polymer Adsorbent

Azo-functional groups have garnered remarkable regard due to their selectivity and sensibility to varied metal ions, making them more appealing than other nitrogen functional groups. The reactive character of azobenzene and its derivatives renders azo-functionalized porous materials responsive to vigorous adsorbents, deploying them as potential nominations for stimuli-responsive applications. Lee et al. have reviewed the studies of POPs decorated with azobenzene groups, like UCBZ and Azo-COP-2. In the case of UCBZ, CO_2 uptake enhanced following UV irradiation, attributed to the isomerization of azobenzene from trans to cis, resulting in the inauguration of more holes and a simultaneous increase in gas absorption. Similarly, Azo-COP-2 represented effectual CO_2 photo-switch control, enabling the immediate extrication of adsorbed CO_2 of up to 25% and superb CO_2/N_2 dissociation post-UV irradiation. While photoresponsive POPs have been less prospected compared to photoresponsive MOFs, they indicate superior performance in terms of lightweight properties, physicochemical stability, and structural tunability [55]. Giao et al. synthesized extremely stable HCPs with beneficent CO_2 uptake and CO_2/N_2 selectivity in 2021, under an easy synthesis situation using amine and azo-functional groups derived from azobenzene. Functionalized HCPs were gathered through the Friedel–Crafts alkylation reaction of 4-amino-*p*-terphenyl with dichloromethane (P1) as the solvent and crosslinker, in the attendance of an $AlCl_3$ catalyst. Then, azo-functionalized HCPs were provided by polymerization of P1 and nitrosobenzene because nitrosobenzene and aniline can form azobenzene(P2). In addition, P0, without any functional group, was incorporated from p-terphenyl and DCM under the same conditions, with the same monomer ratio, for comparison. Azo functionals are first reported and used as CO_2 adsorbents (3.02 mmol/g) with good CO_2/N_2 selectivity. Nitrogen elements on the polymer adsorbent were surveyed with different analyses. The appearance of peaks at 1305 cm^{-1} to the N=N stretching band in P2 by FTIR analysis demonstrates that the azobenzene group is stable and preserved in the frameworks. In addition, the nitrogen content was assayed by elemental analysis. Due to the azo group's N=N bond, P2 has the highest amount of N, containing 5.88% nitrogen, which is higher than P0 and P1, which are <0.3% and 3.89%, respectively. BET surface area and pore volume P0 are the largest with 1062 and 0.69 cm^3/g, respectively. The cause that the surface area and pore volume of P1 and P2 (P1: 447 m^2/g, 0.21 cm^3/g, P2: 242 m^2/g, 0.12 cm^3/g) is less than that of P0 is probably the occupation of the pore by the amine group. The synthesis of azo-functionalized HCPs caught up with the polymerization of P1 and nitrosobenzene to produce azobenzene. For comparison, P0, lacking functional groups, was synthesized from p-terphenyl and DCM under the same situations and monomer ratio. Azo functionalities were initially employed as CO_2 adsorbents, performing a CO_2 sorption capacity of 3.02 mmol/g at 273.15 K and a favorable CO_2/N_2 selectivity of 34.52 at 298.15 K. The attendance of nitrogen in the polymer adsorbent was authenticated through multiple analyses. FTIR analysis revealed the attendance of peaks at 1305 cm^{-1}, corresponding to the N–N stretching band in P2, indicating the stability and presence of the azobenzene group within the frameworks. Furthermore, the elemental analysis demonstrated varying nitrogen content, with P2 exhibiting the highest nitrogen amount at 5.88%, surpassing P0 and P1, which illustrated nitrogen contents of <0.3% and 3.89%, respectively [56]. The work by Tao et al. describes the development of a series of azo-bridged, aromatic, nitrogen-rich, and microporous covalent porphyrin polymers (Azo-CPPs) via a metal-catalyzed direct coupling of a tetranitro building block, $4NO_2TPP$, and various diamine compounds or tetraamine. The azo-CPPs synthesized under these conditions were found to be amorphous with a surface area ranging from 197 to 591 m^2/g. The CO_2 adsorption capacity of the azo-CPPs was investigated, revealing values between 7.36% and 9.22% at 273 K and 1.10 bar. Furthermore, the selectivity of CO_2/N_2 was measured to be between 27.4 and 53.9 at 273 K and 31.2 ~ 107.8 at 303 K, indicating an increase in selectivity with rising temperature. As the length of the linker increased gradually in the order of p-PDA < DABP < DAAzo < $4NH_2TPP$, the BET surface areas decreased gradually in the order of Azo-CPP-1 >Azo-CPP-3 > Azo-CPP-6 > Azo-CPP-7, under otherwise same conditions. For example, when the linker changed from p-PDA to DABP, the BET surface area showed a loss of 13% (from 591 to 514 m^2/g). This phenomenon corresponds well with the "base length" effects [57]. Panic et al. synthesized novel 2D POPs based on azo-bridged benzene and triazine. Different synthetic approaches to the azo-bridged porous polymers were tested, including reductive homocoupling of nitroaromatic monomers, oxidative homocoupling of aromatic amino monomers, and heterocoupling of nitroaromatic monomers and a series of aromatic diamines with different lengths and strengths (Figure 7.8). The results showed that the synthetic methods and building blocks significantly affect the final porosity. Reductive and oxidative homocoupling of nitro and amino aromatic building blocks lead to two-dimensional azo-bridged polymers with significantly higher porosity than polymers produced by heterocoupling reactions. In particular, the highest BET surface areas of 351 and 50.8 m^2/g were observed for azo-bridged polymers with triazine central units, AZO-T-P2 and AZO-T-P3, prepared by NaBH4-mediated reductive homocoupling of the nitro monomer [58].

7.4.1.3 Amid-Functionalized Polymer Adsorbent

The synthesis of amide-functionalized microporous organic polymer (Am-MOP) from trimesic acid, *p*-phenylenediamine, and thionyl chloride as reagents has been documented by Suresh et al. In this polymer, polar amide functional groups (–CONH–) act as a connecting unit between the loop and the spacer, forming the pore wall of the unremitting polymer lattice. The amide groups, characterized by a large and basic quadrupole moment, exhibit significant interaction with CO_2 molecules. This interaction is attributed to the high thermal and chemical stability the amide linkages provide, which results from strong covalent bonds between trimesic acid and *p*-phenylenediamine. The extremely polar hole surface of Am-MOP authorizes selective adsorption of CO_2 at 195 K over other gases such as N_2, Ar, and O_2. Through DFT computation, it has been indicated that the CO_2 molecule interacts with the amide functional groups through Lewis acid–base type interactions. It is important to note that the CO_2 absorption capacity is not solely dependent on the surface other functional groups such as $-NH_2$, –C=N, –OH, and F on the porous surface action a role in adjusting the isosteric heat of absorption.

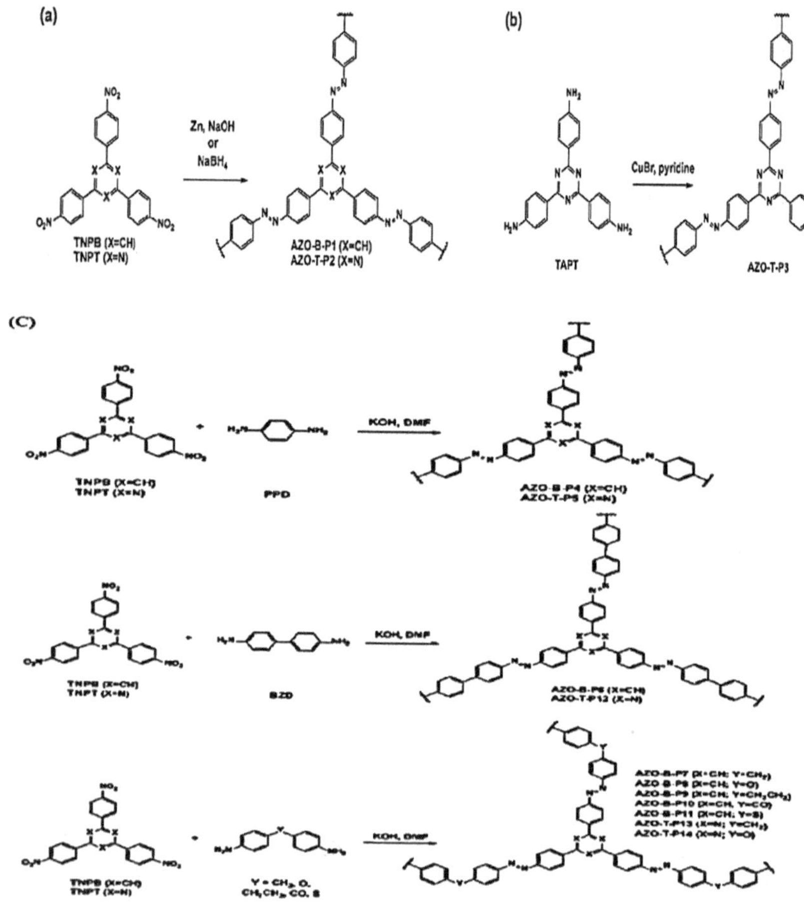

FIGURE 7.8 (a and b) Synthesis method of a series of benzene-based (AZO-B-P) and triazine-based (AZO-T-P) azo-bridged polymers with zinc and copper reductants, respectively. (c) Heterocoupling of nitroaromatic monomers (TNPB and TNPT) and diverse aromatic diamines [58].

This adjustment is crucial for achieving higher capture and selectivity in organic porous polymers [59].

In the realm of CO_2 capture, polymeric adsorbents have emerged as valuable tools. Upon analyzing the data presented in Table 7.2, it becomes evident that HCP adsorbents exhibit a notably higher specific surface area and CO_2 uptake than other polymeric adsorbents. Notably, CTFs demonstrate superior CO_2 absorption, and CMP, a subset of HCPs, boasts a larger pore volume than other polymers. Furthermore, the findings in Table 7.2 indicate that elevated temperatures favorably influence the synthesis of polymer adsorbents. As temperatures rise, the surface area of the polymer increases, consequently leading to heightened CO_2 uptake.

TABLE 7.2
A Summary of the Comparison of Physical Properties and CO_2 Uptake of Different Polymer Adsorbents

No	Polymeric Adsorbent	Surface Area (m^2/g)	Pore Volume (cm^3/g) p/p_0=0.995	Co_2 Uptake (mmol/g) (Wt%) at 1 bar and 273.15K	Ref.
1	HCP–VBC	1420	1.29[a]	2.26	[18]
2	HCP-(0.5%) DVB–VBC	1790	1.86[a]	2.27	[18]
3	HCP-(1%) DVB–VBC	1860	1.26[a]	2.27	[18]
4	HCAP-1	1137	0.81[a]	(114.4)	[19]
5	HCAP-2	1059	0.73[a]	(104.3)	[19]
6	HCP-BDM	847	0.55[a]	—	[20]
7	HCP–BA	742	0.51[a]	—	[20]
8	WSHC	802.84	0.57	11.05	[22]
9	TPI-1	809	0.45[b]	2.45	[24]
10	TPI-2	706	0.4[b]	2.45	[24]
11	TPI-3	40	—	0.68	[24]
12	TPI-4	245	0.22[b]	1.85	[24]
13	TPI-5	201	0.17[b]	1.57	[24]
14	BFCMP-1	1316	1.2[a]	2.45	[31]
15	BFCMP-2	1470	1.35[a]	2.77	[31]
16	CMP	772[c]	1.21[a]	(71)	[32]
17	CO-CMP	965[c]	2.81[a]	(79.3)	[32]
18	AL-CMP	798[c]	1.41[a]	(76.5)	[32]
19	PHCTF-8(400)	1188[d]	0.70[e]	12.8	[40]
20	PHCTF-8(500)	1197[d]	0.71[e]	12.9	[40]
21	PHCTF-8(550)	1340[d]	0.78[e]	15.2	[40]
22	PHCTF-8(600)	1400[d]	0.81[e]	15.3	[40]
23	PHCTF-8(650)	1471[d]	0.82[e]	17.6	[40]
24	PHCTF-8(700)	1665[d]	0.96[e]	17.8	[40]
25	CIFT-400	657[f]	0.32[a]	—	[44]
26	CIFT-500	1674[f]	0.95[a]	3.95	[44]
27	CIFT-KCL/NaCl-500	590[f]	0.27[a]	5.90	[44]
27	SBA-PEI (10)	399[g]	0.79[h]	—	[46]
28	SBA-PEI (30)	153[g]	0.27[h]	—	[46]
29	SBA-PEI(50)	49[g]	0.09[h]	—	[46]
30	HMC-1	855	—	10.5	[60]
31	HMC-2	425	—	13.7	[60]
32	HMC-3	566	—	14.2	[60]
33	HCP-MAAM-1	298	0.47	1.56	[51]
34	HCP-MAAM-2	142	0.87	1.45	[51]
35	HCP-MAAM-3	83	0.24	1.28	[51]

(Continued)

TABLE 7.2 *(Continued)*
A Summary of the Comparison of Physical Properties and CO_2 Uptake of Different Polymer Adsorbents

No	Polymeric Adsorbent	Surface Area (m^2/g)	Pore Volume (cm^3/g) p/p_0=0.995	Co_2 Uptake (mmol/g) (Wt%) at 1 bar and 273.15K	Ref.
36	P_0	1062	0.69[a]	3.79	[56]
37	P1	447	0.21[a]	4.24	[56]
38	P_2	242	0.12[a]	3.02	[56]
39	DA-CMP1	662[g]	0.84[i]	1.35	[61]
40	DA-CMP1	603[g]	0.44[i]	0.95	[61]
50	AZO-CMP1	1146[g]	1.46[i]	2.15	[61]
51	AZO-CMP2	898[g]	0.59[i]	1.96	[61]

[a] (P/P0)= 0.995.
[b] (P/P0)= 0.95.
[c] (P/P0)= 0.05–0.3.
[d] (P/P0)= 0.05–0.1.
[e] (P/P0)= 0.05–0.1.
[f] (P/P0)= 0.01–0.11.
[g] (P/P0) = 0.05–0.2.
[h] (P/P0)= 0.96.
[i] (P/P0)= 0.98.

7.5 CONCLUSION

The pressing topic of global climate alter emissions affected by the continuous consumption of fossil fuels has become the primary concern of environmentalists and climate experts worldwide. The advancement of extremely selective and pliable technologies for sustainable large-scale capture and conversion of CO_2 is now considered a significant imperative, as it provides multiple opportunities for CO_2 research and investigation. Mesoporous organic polymers are the most promising areas in current materials science. CO_2 adsorbent should exhibit excellent capacity, selectivity, fast kinetics, low energy expenditure, chemical and thermal stability, low manufacturing cost, large-scale mechanical strength, and stable performance for multiple cycles. Synthetic porous polymers have the potential to exhibit good physicochemical stability, making them a promising candidate for addressing this pressing global issue. Incorporating different functional groups into MOPs can be beneficial for tuning their level interactions with various molecules. In the last decade, significant progress has been made in this scope, leading to the development of a large number of MOPs, including HCPs readied via Friedel–Crafts paths, crystalline COFs, soluble and crosslinked PIMs, and CMPs prepared via various chemistries. This chapter provides an overview of the state-of-the-art progress on the design, synthesis, and CO_2 adsorption properties of MOPs, covering most of the reported work on this subject. CO_2 adsorption typically carried out at 0°C or 25°C is a crucial process in reducing greenhouse gas emissions However, the actual flue gas temperature is usually much higher, around 90°C, which presents a challenge in CO_2 adsorption under operational conditions. One approach to address this problem is to cool the flue gas, which requires significant energy. Additionally, conventional coal-fired flue gas contains a small percentage of CO_2, about 1–14% by volume, which makes CO_2 adsorption in real operating conditions more challenging. Despite substantial research efforts, the volume of theoretical and computational studies focused on the comprehensive analysis of polymer-based CO_2 adsorption mechanisms, particularly throughout the adsorption process, remains relatively limited. So, using mathematical modeling and simulation to explore the causes of adsorbent inconstancy and less adsorption factor as a guide for designing new renewable and cost-effective polymer-based adsorbents is a research area that needs attention. Most MOPs are inherently insoluble and amorphous, which makes them relatively difficult to characterize at the molecular level compared to soluble or crystalline polymers. Nonetheless, recent advances in characterization techniques, such as X-ray diffraction, solid-state nuclear magnetic resonance, and electron microscopy, have greatly improved the understanding of the structural properties of MOPs. A comprehensive knowledge of MOP structures is vital to synthesizing MOP materials more judiciously. Furthermore, the insolubility of MOPs in organic solvents poses a significant challenge in the material processing stage. Consequently, it is imperative to design and synthesize soluble MOPs with high surface area or to explore innovative and efficient recognition technologies in future research. In practical applications, cost estimation is a crucial factor that requires careful consideration. The cost is influenced not only by the initial materials and the synthesis method but also by the lifespan of the resulting material and the number of operational cycles it can endure. Therefore, it is crucial to minimize material costs while maximizing their durability.

REFERENCES

1. Xu, C., et al., Microporous organic polymers as CO_2 adsorbents: Advances and challenges. *Materials Today Advances*, 2020. 6: p. 100052.
2. Kupgan, G., et al., Modeling amorphous microporous polymers for CO_2 capture and separations. *Chemical Reviews*, 2018. 118(11): p. 5488–5538.
3. Bolourchian Tabrizi, S.Z., S. Shahhosseini, and A. Ghaemi, Insights into the mass transfer mechanisms of nanofluids: A CO_2 absorption study. *Energy & Fuels*, 2021. 35(24): p. 20172–20184.
4. Yuan, Z., M.R. Eden, and R. Gani, Toward the development and deployment of large-scale carbon dioxide capture and conversion processes. *Industrial & Engineering Chemistry Research*, 2016. 55(12): p. 3383–3419.
5. Mukhtar, A., et al., A review on CO_2 capture via nitrogen-doped porous polymers and catalytic conversion as a feedstock for fuels. *Journal of Cleaner Production*, 2020. 277: p. 123999.
6. Lee, S.-Y. and S.-J. Park, A review on solid adsorbents for carbon dioxide capture. *Journal of Industrial and Engineering Chemistry*, 2015. 23: p. 1–11.

7. Masoumi, H., A. Ghaemi, and H.G. Gilani, Evaluation of hyper-cross-linked polymers performances in the removal of hazardous heavy metal ions: A review. *Separation and Purification Technology*, 2021. 260: p. 118221.
8. Han, W., et al., Post-modification of PIM-1 and simultaneously in situ synthesis of porous polymer networks into PIM-1 matrix to enhance CO_2 separation performance. *Journal of Membrane Science*, 2021. 636: p. 119544.
9. Xu, J., et al., Hierarchical micro/mesoporous nitrogen-doped carbons derived from hypercrosslinked polymers for highly efficient oxygen reduction reaction. *Carbon*, 2018. 138: p. 348–356.
10. Wu, S., et al., Facile preparation of composite flame retardantbased on conjugated microporous polymer hollow spheres. *Journal of Colloid and Interface Science*, 2021. 586: p. 152–162.
11. Maiti, S., A.R. Chowdhury, and A.K. Das, Benzoselenadiazole-based nanoporous covalent organic polymer (COP) as efficient room temperature heterogeneous catalyst for biodiesel production. *Microporous and Mesoporous Materials*, 2019. 283: p. 39–47.
12. Younas, M., et al., Recent progress and remaining challenges in post-combustion CO_2 capture using metal-organic frameworks (MOFs). *Progress in Energy and Combustion Science*, 2020. 80: p. 100849.
13. Yu, T., et al., Mechanisms behind high CO_2/CH_4 selectivity using ZIF-8 metal organic frameworks with encapsulated ionic liquids: A computational study. *Chemical Engineering Journal*, 2021. 419: p. 129638.
14. Li, G., et al., Covalent triazine frameworks membrane with highly ordered skeleton nanopores for robust and precise molecule/ion separation. *Journal of Membrane Science*, 2020. 595: p. 117525.
15. Gao, H., Q. Li, and S. Ren, Progress on CO_2 capture by porous organic polymers. *Current Opinion in Green and Sustainable Chemistry*, 2019. 16: p. 33–38.
16. Huang, J. and S.R. Turner, Hypercrosslinked polymers: A review. *Polymer Reviews*, 2018. 58(1): p. 1–41.
17. Najafi, P., H.R. Penchah, and A. Ghaemi, Synthesis and characterization of Benzyl chloride-based hypercrosslinked polymers and its amine-modification as an adsorbent for CO_2 capture. *Environmental Technology & Innovation*, 2021. 23: p. 101746.
18. Li, B., et al., Tailoring the pore size of hypercrosslinked polymers. *Soft Matter*, 2011. 7(22): p. 10910–10916.
19. Puthiaraj, P. and W.-S. Ahn, CO_2 capture by porous hyper-cross-linked aromatic polymers synthesized using tetrahedral precursors. *Industrial & Engineering Chemistry Research*, 2016. 55(29): p. 7917–7923.
20. Luo, Y., et al., Microporous organic polymers synthesized by self-condensation of aromatic hydroxymethyl monomers. *Polymer Chemistry*, 2013. 4(4): p. 1126–1131.
21. Sang, Y. and J. Huang, Benzimidazole-based hyper-cross-linked poly (ionic liquid) s for efficient CO_2 capture and conversion. *Chemical Engineering Journal*, 2020. 385: p. 123973.
22. Maleki, F., A. Ghaemi, and G. Mir Mohamad Sadeghi, Synthesis and characterization of waste Styrofoam hypercrosslinked polymer as an adsorbent for CO_2 capture. *Environmental Progress & Sustainable Energy*, 2023. 42(1): p. e13954.
23. Moradi, M.R., H. Ramezanipour Penchah, and A. Ghaemi, CO_2 capture by benzene-based hypercrosslinked polymer adsorbent: Artificial neural network and response surface methodology. *The Canadian Journal of Chemical Engineering*, 2023. 101(10): p. 5621–5642.
24. Liebl, M.R. and J.R. Senker, Microporous functionalized triazine-based polyimides with high CO_2 capture capacity. *Chemistry of Materials*, 2013. 25(6): p. 970–980.
25. Farha, O.K., et al., Chemical reduction of a diimide based porous polymer for selective uptake of carbon dioxide versus methane. *Chemical Communications*, 2010. 46(7): p. 1056–1058.

26. Akbarzadeh, E., A. Shockravi, and V. Vatanpour, Efficient thiazole-based polyimines as selective and reversible chemical absorbents for CO2 capture and separation: Synthesis, characterization and application. *Polymer*, 2019. 182: p. 121840.
27. Lu, W., et al., Porous polymer networks: Synthesis, porosity, and applications in gas storage/separation. *Chemistry of Materials*, 2010. 22(21): p. 5964–5972.
29. Holst, J.R., et al., High surface area networks from tetrahedral monomers: Metal-catalyzed coupling, thermal polymerization, and "click" chemistry. *Macromolecules*, 2010. 43(20): p. 8531–8538.
28. Babarao, R., S. Dai, and D.-E. Jiang, Fucd selective CO_2 separation: A molecular simulation study. *Langmuir*, 2011. 27(7): p. 3451–3460.
30. Lu, W., et al., Sulfonate-grafted porous polymer networks for preferential CO_2 adsorption at low pressure. *Journal of the American Chemical Society*, 2011. 133(45): p. 18126–18129.
31. Zhang, C., et al., Bifunctionalized conjugated microporous polymers for carbon dioxide capture. *Polymer*, 2015. 61: p. 36–41.
32. Xie, Y., et al., Capture and conversion of CO_2 at ambient conditions by a conjugated microporous polymer. *Nature Communications*, 2013. 4(1): p. 1960.
33. Ding, S.-Y. and W. Wang, Covalent organic frameworks (COFs): From design to applications. *Chemical Society Reviews*, 2013. 42(2): p. 548–568.
34. Xiang, Z., et al., Covalent-organic polymers for carbon dioxide capture. *Journal of Materials Chemistry*, 2012. 22(42): p. 22663–22669.
35. Ullah, R., et al., A combined experimental and theoretical study on gas adsorption performance of amine and amide porous polymers. *Microporous and Mesoporous Materials*, 2019. 279: p. 61–72.
36. Byun, J., et al., Nanoporous covalent organic polymers incorporating Tröger's base functionalities for enhanced CO2 capture. *Journal of Materials Chemistry A*, 2014. 2(31): p. 12507–12512.
37. Varghese, A.M., et al., Performance enhancement of CO2 capture adsorbents by UV treatment: The case of self-supported graphene oxide foam. *Chemical Engineering Journal*, 2020. 386: p. 124022.
38. Shin, G.-J., K. Rhee, and S.-J. Park, Improvement of CO2 capture by graphite oxide in presence of polyethylenimine. *International Journal of Hydrogen Energy*, 2016. 41(32): p. 14351–14359.
39. Hijikata, Y., et al., Relationship between channel and sorption properties in coordination polymers with interdigitated structures. *Chemistry–A European Journal*, 2011. 17(18): p. 5138 5144.
40. Yuan, K., et al., Construction of triphenylamine functional phthalazinone-based covalent triazine frameworks for effective CO_2 capture. *Polymer*, 2018. 151: p. 65–74.
41. Liu, A., J. Zhang, and X. Lv, Novel hydrazine-bridged covalent triazine polymer for CO_2 capture and catalytic conversion. *Chinese Journal of Catalysis*, 2018. 39(8): p. 1320–1328.
42. Dey, S., et al., Covalent triazine-based frameworks (CTFs) from triptycene and fluorene motifs for CO_2 adsorption. *Journal of Materials Chemistry A*, 2016. 4(17): p. 6259–6263.
43. Wang, H., et al., Covalent triazine frameworks for carbon dioxide capture. *Journal of Materials Chemistry A*, 2019. 7(40): p. 22848–22870.
44. Song, K.S., et al., Molten salt templated synthesis of covalent isocyanurate frameworks with tunable morphology and high CO_2 uptake capacity. *ACS Applied Materials & Interfaces*, 2021. 13(22): p. 26102–26108.
45. Hakiki, A., et al., Synthesis and characterization of mesoporous silica SBA-15 functionalized by mono-, di-, and tri-amine and its catalytic behavior towards Michael addition. *Materials Chemistry and Physics*, 2018. 212: p. 415–425.

46. Sanz-Pérez, E.S., et al., CO_2 adsorption performance of amino-functionalized SBA-15 under post-combustion conditions. *International Journal of Greenhouse Gas Control*, 2013. 17: p. 366–375.
47. Zhang, L., et al., CO_2 capture over molecular basket sorbents: Effects of SiO2 supports and PEG additive. *Journal of Energy Chemistry*, 2017. 26(5): p. 1030–1038.
48. Zhu, J., et al., Polyethylenimine-grafted HKUST-type MOF/PolyHIPE porous composites (PEI@ PGD-H) as highly efficient CO_2 adsorbents. *Industrial & Engineering Chemistry Research*, 2019. 58(10): p. 4257–4266.
49. Ding, B., et al., Hydrothermal preparation of hierarchical ZIF-L nanostructures for enhanced CO_2 capture. *Journal of Colloid and Interface Science*, 2018. 519: p. 38–43.
50. Sekizkardes, A.K., et al., Amine-functionalized porous organic polymers for carbon dioxide capture. *Materials Advances*, 2022. 3(17): p. 6668–6686.
51. Fayemiwo, K.A., et al., Nitrogen-rich hyper-crosslinked polymers for low-pressure CO_2 capture. *Chemical Engineering Journal*, 2018. 334: p. 2004–2013.
52. Jafari, T., et al., Microwave-assisted synthesis of amine functionalized mesoporous polydivinylbenzene for CO_2 adsorption. *Journal of CO_2 Utilization*, 2017. 19: p. 79–90.
53. Tao, J., et al., Engineering pore surface and morphology of microporous organic polymers for improved affinity towards CO_2. *Chemical Engineering Journal*, 2019. 373: p. 338–344.
54. Moradi, M.R., et al., Amine functionalized benzene based hypercrosslinked polymer as an adsorbent for CO_2/N_2 adsorption. *Scientific Reports*, 2023. 13(1): p. 9214.
55. Li, S., N. Prasetya, and B.P. Ladewig, Investigation of Azo-COP-2 as a photoresponsive low-energy CO_2 adsorbent and porous filler in mixed matrix membranes for CO_2/N_2 separation. *Industrial & Engineering Chemistry Research*, 2019. 58(23): p. 9959–9969.
56. Qiao, Y., et al., Amine or Azo functionalized hypercrosslinked polymers for highly efficient CO_2 capture and selective CO_2 capture. *Materials Today Communications*, 2021. 27: p. 102338.
57. Tao, L., et al., Azo-bridged covalent porphyrinic polymers (Azo-CPPs): Synthesis and CO_2 capture properties. *RSC Advances*, 2015. 5(117): p. 96871–96878.
58. Panić, B., et al., Synthesis and characterization of benzene-and triazine-based azo-bridged porous organic polymers. *Polymers*, 2023. 15(1): p. 229.
59. Suresh, V.M., et al., Amide functionalized microporous organic polymer (Am-MOP) for selective CO_2 sorption and catalysis. *ACS Applied Materials & Interfaces*, 2014. 6(7): p. 4630–4637.
60. Kundu, S.K. and A. Bhaumik, Novel nitrogen and sulfur rich hyper-cross-linked microporous poly-triazine-thiophene copolymer for superior CO_2 capture. *ACS Sustainable Chemistry & Engineering*, 2016. 4(7): p. 3697–3703.
61. Wang, X., et al., Nitrogen-rich conjugated microporous polymers: Impact of building blocks on porosity and gas adsorption. *Journal of Materials Chemistry A*, 2015. 3(42): p. 21185–21193.

8 Metallic and MOF Adsorbents for CO_2 Capture

Amir Kazemi, Faranak Manteghi, and Ahad Ghaemi

8.1 INTRODUCTION

In recent years, the pressing demand to reduce greenhouse gas emissions and address climate change has driven significant research into efficient CO_2 capture methods. Among the various approaches explored, metallic adsorbents have gained significant attention as a promising low-cost and efficient method for CO_2 capture. Solid adsorbents play a crucial role in various industrial and environmental applications, including gas separation, purification, and moisture control. The effectiveness of these materials is largely determined by their structural properties and chemical characteristics. This diagram categorizes solid adsorbents into four main groups: metal oxides and hydroxides, zeolites, silica gels, and metal-organic frameworks (MOFs). Each category exhibits unique properties that make them suitable for various applications. Metal oxides and hydroxides are known for their high stability and effectiveness in adsorption processes, particularly in environmental remediation. Zeolites are characterized by their porous structure and ion-exchange capabilities, making them ideal for gas separation and purification. Silica gels are widely recognized for their excellent moisture control in packaging and storage, while MOFs offer tunable pore sizes and functionalization potential, enhancing their versatility in gas storage and drug delivery systems. This classification underscores the diverse functionalities and advantages of solid adsorbents in various industrial and environmental applications. As illustrated in Figure 8.1, solid adsorbents are categorized into four main groups: metal oxides and hydroxides, zeolites, silica gels, and MOFs. Each category exhibits unique properties that make them suitable for various applications. Metal oxides and hydroxides are known for their high stability and effectiveness in adsorption processes, particularly in environmental remediation. Zeolites are characterized by their porous structure and ion-exchange capabilities, making them ideal for gas separation and purification. Silica gels are widely recognized for their excellent moisture control in packaging and storage, while MOFs offer tunable pore sizes and functionalization potential, enhancing their versatility in gas storage and drug delivery systems. This classification underscores the diverse functionalities and advantages of solid adsorbents in various industrial and environmental applications [1–3].

DOI: 10.1201/9781003476191-8

FIGURE 8.1 Diagram of popular and widely used solid adsorbents, highlighting their advantages and applications.

This chapter aims to delve into the properties of metallic adsorbents, with a particular focus on MOFs, and their role in capturing CO_2. Metallic adsorbents present various benefits compared to conventional methods, including amine-based solvents and physical absorption techniques. They possess high surface areas, tunable pore sizes, and adjustable functionalities, enabling them to exhibit exceptional adsorption capacities and selectivities toward CO_2. These unique characteristics make them attractive candidates for CO_2 capture applications. The discussion in this chapter revolves around the design of metallic adsorbent structures and the underlying adsorption mechanisms that contribute to their CO_2 capture properties. In particular, the focus is on MOFs, which are a category of crystalline materials formed by metal ions or clusters linked to organic ligands. MOFs have emerged as a fascinating area of research due to their exceptional porosity, structural diversity, and customizable properties. Understanding the synthesis of MOFs tailored specifically for CO_2 capture is crucial for harnessing their full potential. The chapter explores various synthesis techniques employed to fabricate MOFs with optimized properties for efficient CO_2 capture. These methods encompass a range of approaches, including solvothermal, microwave-assisted, and template-directed synthesis, among others. By analyzing the synthesis strategies, researchers can obtain valuable insights into managing the physicochemical characteristics of MOFs to improve their performance in CO_2 adsorption. Overall, this chapter provides a comprehensive overview of metallic adsorbents, with a focus on MOFs, as effective solutions for CO_2 capture. By investigating the design principles, adsorption mechanisms, and synthesis techniques of these materials, researchers can gain a deeper understanding of their potential applications in mitigating CO_2 emissions and combating climate change [4–8].

The pressing necessity to tackle rising CO_2 emissions and fight climate change has led to significant research efforts aimed at developing effective CO_2 capture methods. Among the various approaches evaluated, metallic adsorbents have emerged as a low-cost and highly efficient method for CO_2 capture. This chapter aims to explore the CO_2 capture properties of metallic adsorbents, with a particular focus on MOFs, by examining the design of adsorbent structures and the underlying adsorption mechanisms. Furthermore, the synthesis of MOFs for CO_2 capture will be investigated, along with an exploration of mineral carbonation using metal oxide, metal hydroxide, metal silicates, and silica gel. Metallic adsorbents offer distinct advantages as an attractive solution for CO_2 capture. They provide a cost-effective alternative compared to traditional techniques, and their superior efficiency ensures a high degree of CO_2 adsorption.

In this chapter, we investigate the properties and attributes of metallic adsorbents, specifically focusing on the importance of MOFs. MOFs represent a class of crystalline materials formed from metal ions or clusters coordinated with organic ligands. They possess exceptional porosity, adjustable pore sizes, and customizable properties, which contribute to their outstanding CO_2 capture capabilities. Understanding the design of adsorbent structures and the underlying adsorption mechanisms is crucial for optimizing the CO_2 capture performance of metallic adsorbents. The chapter explores the principles of designing metallic adsorbents, including MOFs, with a focus on tailoring their structures to enhance CO_2 adsorption. The discussion also encompasses the exploration of various adsorption mechanisms involved in CO_2 capture, such as physisorption and chemisorption, shedding light on the interactions between the adsorbent and CO_2 molecules. Furthermore, the chapter investigates the synthesis of MOFs specifically designed for CO_2 capture [9]. Researchers can explore various synthesis techniques, such as solvothermal synthesis, microwave-assisted synthesis, and template-directed synthesis, to fabricate MOFs with tailored properties for efficient CO_2 adsorption. The chapter also delves into mineral carbonation, a process where CO_2 reacts with metal oxides, metal hydroxides, and silica gel to produce stable carbonate minerals. This comprehensive approach includes investigating metallic adsorbents, their CO_2 capture capabilities, MOF synthesis, and mineral carbonation

TABLE 8.1
Benefits and Drawbacks of Capturing CO_2 with Solid Adsorbents

Adsorbent Type	Benefits	Drawbacks
Metal oxide	- Plentiful and easily obtainable [14–16] - Non-toxic [14, 17, 18] - **Economical material** and resources with favorable cost performance [14, 19, 20] - Broad temperature and pressure operability for adsorption [15, 21, 22]	- **Unfavorable regeneration thermodynamics** [23, 24]
Zeolite	- High durability under repeated loading/unloading [25, 26] - Exceptional resistance to high temperatures [27–29] - Highly porous and extensive surface area [30–32]	- **Limited preference for CO_2 over other gases** [27, 32, 33] - **Low CO_2 storage potential** - **High operating costs due to energy demands for regeneration** [28, 31] - **Economically disadvantageous due to high material cost** [27, 29, 33]
MOFs	- Highly porous and extensive surface area [11, 34] - Tunable pore architecture and tailorable porosity [35, 36] - Excellent chemical and physical stability [37, 38] - Modifiable surface properties and tunable functional groups [36, 39]	- **Limited preference for CO_2 over other gases** [40, 41] - **Economically disadvantageous due to high material cost** - **Loss of integrity during solvent evaporation and structural collapse upon solvent removal** [37, 42, 43]
Silica	- Highly porous and extensive surface area [44–46] - Controlled pore size distribution [47, 48] - Excellent resistance to heat and deformation [46, 49, 50]	- **Limited CO_2 uptake without additional energy input** [45, 51] - **High operating costs due to energy demands for regeneration** [41, 44, 46]

to understand their potential in mitigating CO_2 emissions and advancing CO_2 capture technologies. By concentrating on design principles, adsorption mechanisms, and synthesis strategies, researchers can advance the creation of affordable and efficient CO_2 capture technologies, facilitating the transition to a more sustainable future. Mineral carbonation, which includes metal oxides, metal hydroxides, metal silicates, and silica gel, is essential for CO_2 capture and storage, as it involves the chemical transformation of CO_2 into stable carbonate minerals (Table 8.1) [10–13].

8.2 MINERAL CARBONATION: METAL OXIDE

Fighting climate change requires capturing excess CO_2 from the atmosphere. Metal oxides are emerging as a promising group of materials for this task due to several advantages. They are abundant and inexpensive, making them suitable for large-scale CO_2 capture. Additionally, some metal oxides, like magnesium oxide (MgO) and calcium oxide (CaO), have a strong attraction for CO_2 molecules, leading to high capture capacity [15, 52]. They can also be regenerated through heat or chemicals, allowing for multiple capture and release cycles. This reusability makes them an environmentally friendly option. However, there are some hurdles to overcome before widespread adoption. Metal oxides might sometimes capture other unwanted gases along with CO_2, requiring extra purification steps. The speed of CO_2 capture and release can also be slow for some oxides, impacting their efficiency. In certain cases, the regeneration process might require high temperatures, increasing the overall energy consumption of CO_2 capture [53, 54].

8.2.1 Magnesium Oxide

Magnesium oxide (MgO) adsorbents have emerged as a frontrunner in the pursuit of CO_2 capture technologies. Their porous structures hold immense promise, characterized by high surface area, large pore size, and substantial pore volume. These attributes provide an abundance of adsorption sites, significantly enhancing the material's capacity to capture CO_2 molecules. In the fight against climate change, scientists are exploring MgO as a potential game-changer for capturing CO_2 directly from the atmosphere [55, 56]. In the fight against rising temperatures, MgO emerges as a potential hero. This innovative material can pull CO_2 directly from the air, like a giant sponge, and then securely store it away underground, preventing it from contributing to climate change. This method offers several advantages. First, MgO is naturally abundant, making it a scalable solution compared to other capture methods. Another key strength is its reusability. MgO can react with CO_2 in the air, essentially pulling it out. This captured CO_2 can then be separated and stored underground, preventing its release back into the environment [57, 58]. After capturing CO_2, MgO can be regenerated through a heating process, releasing the CO_2 in a concentrated form for storage and allowing the MgO itself to be reused for further capture cycles. Magnesium oxide (MgO) has become a promising option for CO_2 capture applications, thanks to its distinctive physicochemical properties. One notable characteristic is its favorable surface basicity, which promotes the formation of oxygen vacancies. These vacancies are thought to contribute to improving the material's CO_2 uptake performance. MgO offers a distinct advantage over other metal oxide adsorbents, such as those based on lithium (Li) and calcium (Ca). Notably, MgO requires a significantly lower regeneration temperature (below 773 K), making it a more energy-efficient option. Additionally, MgO demonstrates effectiveness in the intermediate temperature range of 473–673 K, offering flexibility in its application. Theoretical studies have shown a high CO_2 uptake capacity for MgO, reaching up to 24.8 mmol CO_2 per gram of adsorbent. The CO_2 capture process in MgO involves a chemical reaction between the acidic CO_2 gas and the basic O_2^--Mg^{2+} sites within the material. This reaction leads to the formation of various carbonate species, including bicarbonates, unidentate carbonates, and bidentate carbonates. Despite its promising characteristics, pure MgO faces some limitations that hinder its widespread industrial application. These challenges encompass a low surface area and sluggish adsorption kinetics, which hinder CO_2 molecules' access to the internal cavities of the adsorbent, ultimately resulting in reduced CO_2 uptake capacity. Additionally, MgO-based adsorbents necessitate relatively high temperatures during the desorption phase, which can pose thermal stability concerns. Elevated regeneration temperatures may lead to sintering, a process that diminishes the material's surface area. This reduction in surface area negatively impacts both CO_2 capture performance and the recyclability of the adsorbent. As a result of these limitations, MgO-based adsorbents are currently less economically viable for large-scale industrial applications. However, due to MgO's potential for CO_2 capture, extensive research efforts are underway to improve its stability and enhance its CO_2 uptake performance. These advancements aim to make MgO a more viable and cost-effective solution for combating climate change through CO_2 capture technologies. Despite its promising attributes, widespread adoption of MgO for CO_2 capture necessitates addressing certain limitations. The CO_2 capture process exhibits slow adsorption kinetics, particularly at ambient temperatures. This translates to a longer time required for capturing substantial quantities of CO_2, potentially hindering practical applications. Furthermore, the regeneration process, which often involves thermal desorption, can be energy-intensive. If the energy source for regeneration isn't clean (e.g., reliant on fossil fuels), it could diminish the overall environmental benefits of CO_2 capture. Recognizing the limitations of MgO for widespread adoption in CO_2 capture applications, the scientific community has embarked on a concerted research effort to enhance its capabilities. One key strategy involves developing MgO materials with a higher surface area. This increased surface area provides more adsorption sites for CO_2 molecules, thereby enhancing the capture capacity. Additionally, researchers are optimizing reaction conditions, such as temperature and pressure, to accelerate the CO_2 capture process. Some promising advancements involve doping MgO with specific elements or creating composites with other materials to improve its overall performance.

A notable example by Hanif et al. demonstrated the potential for creating improved MgO-based CO_2 capture materials. Their research centered on a MgO adsorbent developed through a straightforward wet impregnation-calcination process. This method utilizes non-porous, fibrous silica nanospheres, often referred to as "flower-like KCC-1," as a support structure for MgO nanocrystallites. The study successfully deposited small MgO nanocrystallites on the external surface of the petal-like silica support. This approach led to a significant enhancement in the adsorbent's surface area and pore volume. These improved characteristics translate to a demonstrably improved CO_2 uptake performance. The resulting MgO-based adsorbent exhibits a high level of CO_2 adsorption, particularly at a relatively low temperature (473 K), when compared to other unsupported MgO materials. This finding highlights the potential of utilizing tailored materials design to address limitations associated with conventional MgO adsorbents for CO_2 capture [59]. While MgO shows promise for capturing CO_2, a major drawback is its relatively low CO_2 uptake capacity. This means a single unit of MgO can only capture a limited amount of CO_2 molecules. Imagine it like a sponge with limited soaking ability. Because of this, large quantities of MgO might be needed to capture significant amounts of CO_2, potentially making the process expensive and impractical. Researchers are actively devising techniques to enhance the CO_2 adsorption capacity of MgO, leading toward a more effective approach for CO_2 capture. Gao et al. [60] proposed a groundbreaking approach to improve MgO adsorbents for CO_2 capture. Instead of focusing solely on optimizing adsorption sites, they investigated the use of molten salt modifiers. Their research revealed a dramatic increase in CO_2 uptake (15.7 mmol/g) at 350°C with the addition of 10 mol % sodium nitrite ($NaNO_2$) to MgO, compared to negligible uptake with unmodified MgO. Utilizing transient kinetic experiments, they identified a three-step process for CO_2 capture with modified MgO at 300°C: fast surface adsorption, chemical reaction forming magnesium carbonate, and a slow diffusion step as the rate-limiting factor. The validity of these findings was confirmed through a combination of techniques, including transient kinetic studies, mass spectrometry, and TGA measurements. This research provides a deeper understanding of the enhanced CO_2 capture performance achieved and holds significant promise for the design of practical pre-combustion CO_2 capture technologies. In another study, Gong et al. [61] reported that their study aimed to enhance rapid and efficient CO_2 capture, which is vital for CO_2 capture and utilization (CCU) techniques designed to combat global warming. They focused on alkali molten salt (AMS)-promoted MgO sorbents, which are known for their high capacity but face limitations due to slow adsorption kinetics and poor stability. To address these issues, they developed MgO sorbents combined with AMS and Fe_2O_3 (Mg100-*x*Fex-AMS10) using an improved methanol evaporation-induced surface precipitation method. Characterization revealed that Mg98Fe_2-AMS10 had a loose flake structure, offering a large surface area and abundant active sites. This structure enabled Mg98Fe_2-AMS10 to demonstrate a robust CO_2 capacity of 14.86 mmol CO_2/g sorbent at 300°C along with good cyclic durability. Through exploration of three kinetic models, the Avrami model accurately predicted the CO_2 adsorption behavior, highlighting the rapid adsorption kinetics of Mg98Fe_2-AMS10. The researchers also examined the adsorption mechanism through density functional theory calculations. They found that there was an increase in the charge transfer and bonding strengths of surface oxygen atoms on MgO, which, in the presence of Fe catalysis, preferentially formed bonds with activated CO_2 (Figure 8.2). This study provides valuable insights into enhancing CO_2 capture and kinetics of magnesium oxide (MgO) through the integration of 3d transition metals.

8.2.2 Calcium Oxide

Calcium oxide (CaO) is a versatile compound with numerous applications, including its role in carbon capture processes. When calcium oxide reacts with CO_2, it undergoes a chemical reaction to form calcium carbonate ($CaCO_3$). This reaction is a crucial step in carbon capture and storage (CCS) technologies, where CO_2 emissions are captured from industrial processes or power plants to mitigate their impact on the environment. The process of capturing CO_2 using calcium oxide involves the absorption of CO_2 into a slurry of calcium oxide in water, forming calcium carbonate. This reaction not only helps in reducing CO_2 emissions but also produces a valuable byproduct in the form of calcium carbonate, which can be used in various industries such as construction, agriculture, and pharmaceuticals [62, 63]. Overall, the use of calcium oxide in CO_2 capture plays a significant role in addressing climate change by reducing greenhouse gas emissions and promoting sustainable practices in various sectors of the economy. CaO-based sorbents are invariably exposed to water vapor and various other components, which have been shown to affect their performance. This review highlights the role of steam/H_2O in the CO_2 capture process involving CaO-based sorbents. It discusses the impact of steam on sorbent

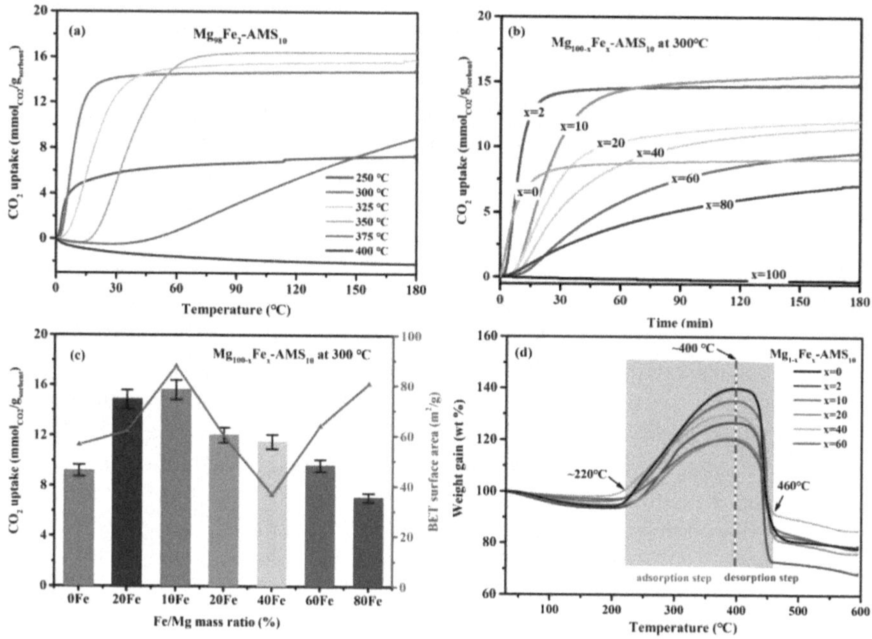

FIGURE 8.2 (a) Profiles of CO_2 adsorption for Mg98Fe_2-AMS10 at temperatures spanning 250 to 400°C; (b) CO_2 uptake capacities for all Mg100-*x*Fex-AMS10 samples (where $0 \leq x \leq 100$) measured at 300°C; (c) the relationship between CO_2 uptake capacity and specific surface area (SBET); (d) dynamic thermograms for Mg100-*x*Fex-AMS10 as it undergoes heating from 30 to 600°C. Reproduced from Ref. [61].

reactivation, its effects on carbonation and calcination, and how its presence influences both the carbonation and calcination stages simultaneously [64]. Traditionally, CaO-based CO_2 adsorbents may have limitations in terms of environmental compatibility, CO_2 absorption capacity, or stability in multiple uses. Yasipourtehrani et al. [65] investigated the potential of using blast furnace slag (BFS) as a sustainable feedstock for CaO-based CO_2 sorbents. Their research highlights two key advantages of this approach. First, they found that the chemical composition of the slag significantly impacts the efficiency of extracting key elements like calcium (Ca) from the slag. This underlines the importance of understanding the slag's composition for optimal sorbent production. Second, the study reports a significant breakthrough in developing sorbents with a balanced performance. Traditionally, there might be a trade-off between high CO_2 capture capacity and the sorbent's ability to maintain its effectiveness over multiple cycles (cyclic stability). However, this study revealed that the efficient separation of calcium and silicon (Si) during the acid extraction process greatly enhances the performance of the sorbent. Consequently, this led to the creation of a sorbent exhibiting an exceptional CO_2 capture capacity of 0.37 g CO_2 per gram of sorbent, along with an impressively low cyclic deactivation rate of just 2.3%. This balanced performance surpasses most other CaO based sorbents derived from waste materials. Overall, the study paves the way for utilizing

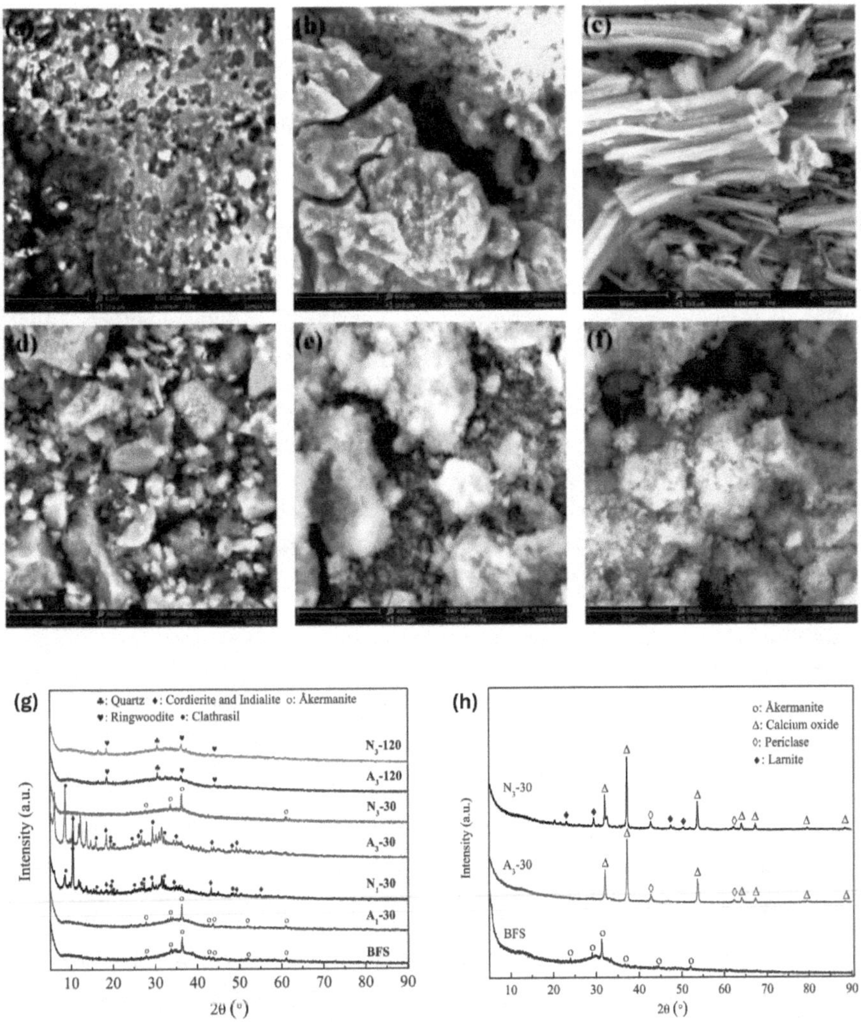

FIGURE 8.3 SEM images of BFS residues obtained after acid extraction under various leaching conditions: 30 minutes in 1 mol/L of (a) acetic acid and (d) nitric acid, 30 minutes in 3 mol/L of (b) acetic acid and (e) nitric acid, and 120 minutes in 3 mol/L of (c) acetic acid and (f) nitric acid. Additionally, XRD patterns are presented for (g) the raw BFS and its corresponding residues related to the synthesized CO_2 sorbents post-acid extraction under different conditions, as well as (h) the selected BFS-derived, CaO-based CO_2 sorbents A3–30 and N3–30, with the raw BFS pattern included for comparative analysis. Reproduced from Ref. [65].

BFS as a sustainable and effective alternative to traditional sorbents for CO_2 capture (Figure 8.3). Their findings suggest that controlling the extraction process to optimize the calcium-to-silicon ratio is crucial for achieving superior CO_2 capture performance and cyclic stability in BFS-derived sorbents. This research holds promise for the industrial utilization of CaO-based CO_2 capture technologies.

8.3 METAL HYDROXIDE

Capturing excess CO_2 to combat climate change requires efficient and sustainable technologies. Metal hydroxides are emerging as a promising option due to their unique properties. Some, like lithium and potassium hydroxide, have a strong chemical bond with CO_2 molecules, allowing for efficient capture. Additionally, they can be regenerated through processes like temperature swing adsorption or chemical conversion, enabling multiple capture cycles. This reusability makes them an attractive and sustainable solution. However, there are some limitations to address before widespread use. Similar to metal oxides, metal hydroxides might sometimes capture other unwanted gases alongside CO_2, requiring extra purification steps. Another challenge is their sensitivity to water. Some hydroxides lose effectiveness or degrade in humid environments, limiting their real-world application. Finally, the regeneration process for certain hydroxides might require significant energy input, impacting the overall efficiency of the CO_2 capture system. Ongoing research is focused on overcoming these limitations. This includes developing composite materials that combine metal hydroxides with other materials for improved CO_2 selectivity and water resistance. Additionally, exploring alternative regeneration methods with lower energy demands is crucial for making metal hydroxides a viable and sustainable technology for CO_2 capture.

Alkali hydroxides are emerging as attractive materials for capturing CO_2. These materials have a strong affinity for CO_2 molecules, allowing them to capture significant amounts of CO_2. This high capacity makes them suitable for large-scale CO_2 capture applications, potentially contributing to reducing greenhouse gas emissions. Unlike some CO_2 capture methods requiring high temperatures, alkali hydroxides can react with CO_2 at moderate temperatures or even room temperature. This eliminates the need for energy-intensive preheating steps, improving the overall efficiency of the capture process. Some alkali hydroxides are well-suited for capturing CO_2 from environments with high moisture content. This makes them potentially useful for capturing CO_2 from exhaust gases or directly from ambient air, which naturally contains moisture. After capturing CO_2, the alkali hydroxide needs to be regenerated for reuse. This regeneration process can require additional energy input, impacting the overall efficiency of the CO_2 capture system. Depending on the specific alkali hydroxide chosen and the regeneration method employed, utilizing these materials for large-scale CO_2 capture might be cost-prohibitive. Overall, alkali hydroxides are promising candidates for CO_2 capture due to their high capacity, favorable reaction conditions, and compatibility with moist environments. However, research and development are needed to address regeneration challenges, optimize costs for large-scale applications, and develop strategies for specific alkali hydroxides [66].

8.3.1 Sodium Hydroxide

One promising approach involves the use of sodium hydroxide (NaOH) aqueous solutions as adsorbents for CO_2 capture. NaOH aqueous solutions have garnered attention as potential adsorbents for CO_2 capture due to several advantages [67, 68]. First, NaOH is a relatively inexpensive and widely available chemical. Additionally, the

CO_2 absorption capacity of NaOH solutions is generally higher than that of other commonly used adsorbents, such as monoethanolamine (MEA). Furthermore, the CO_2 capture process using NaOH solutions can be designed to operate at relatively low temperatures, potentially reducing the energy requirements compared to other CO_2 capture methods [69, 70]. This solid precipitation is the key to the CO_2 capture capacity of the NaOH solution, as it removes the CO_2 from the gas phase and traps it in the solid form. Yoo et al. discovered that NaOH solutions exhibit a significantly higher capacity for absorbing CO_2. This suggests that NaOH could be a more efficient material for capturing CO_2 from various sources. The study also shed light on the chemical reactions that take place during CO_2 capture with NaOH. As CO_2 interacts with the NaOH solution, it primarily converts into sodium carbonate (Na_2CO_3) and sodium bicarbonate ($NaHCO_3$). Understanding these reactions is crucial for optimizing the CO_2 capture process. This research highlights the potential of NaOH for CO_2 capture, particularly due to its high initial absorption capacity and the clear reaction mechanisms involved. However, they also identified the challenge of limited long-term CO_2 capture as the solution becomes saturated with reaction products. Further research is needed to address this limitation and explore potential regeneration methods for NaOH solutions to make them a more viable option for CO_2 capture technologies. [71]. NaOH has several important advantages as an adsorbent, including a high CO_2 absorption rate and low vapor pressure. Absorbed CO_2 is not simply stored but converted into useful products through a conversion process. A key aspect to encourage the use of this adsorbent is the regeneration of the spent NaOH solution that is used for resorption. This regeneration process is critical to ensure the continued use of NaOH to absorb CO_2 without causing significant additional carbon emissions. Of course, the overall efficiency of the proposed cycle is very important [72].

8.3.2 Calcium Hydroxide

Calcium hydroxide ($Ca(OH)_2$), also known as slaked lime, is emerging as a promising contender in this fight due to its unique properties. One of the key advantages of $Ca(OH)_2$ is its abundance and affordability. This readily available and inexpensive material makes it a cost-effective solution for large-scale CO_2 capture, particularly attractive for industrial applications. Furthermore, $Ca(OH)_2$ boasts a high CO_2 capture capacity [73]. It exhibits a strong chemical attraction to CO_2 molecules, readily capturing them through a reaction that forms calcium carbonate ($CaCO_3$). This capture process can be represented by the following equation:

$$Ca(OH)_2 + CO_2 \text{->} CaCO_3 + H_2O$$

Another significant benefit of $Ca(OH)_2$ is its reusability. The captured CO_2 can be released from $CaCO_3$ through a heating process called calcination. This calcination process regenerates calcium oxide (CaO), which can then react with CO_2 again, allowing for multiple capture and release cycles. This cyclical nature makes $Ca(OH)_2$ an attractive option for sustainable CO_2 capture technologies. However, there are still some hurdles to overcome before the widespread adoption of $Ca(OH)_2$. Selectivity is one such challenge. While $Ca(OH)_2$ has a strong affinity for CO_2, it might also capture other unwanted gases present in industrial flue gas streams. This can necessitate additional purification steps to isolate the captured CO_2, increasing the complexity and potentially the cost of the process. Energy consumption during the regeneration process is another concern. The calcination process for regenerating CaO requires significant heat input. Optimizing this process to reduce energy consumption is crucial. Lowering the energy demands would improve the overall efficiency and environmental impact of CO_2 capture using $Ca(OH)_2$. Finally, the reaction rate between $Ca(OH)_2$ and CO_2 can be slower compared to some other CO_2 capture materials. This slower reaction rate can limit the capture efficiency of $Ca(OH)_2$. Research efforts are actively exploring methods for accelerating the reaction rate. These methods might involve modifications of the $Ca(OH)_2$ surface to enhance its reactivity or the use of additives that can act as catalysts to speed up the CO_2 capture reaction. Despite these challenges, calcium hydroxide presents a promising avenue for CO_2 capture due to its abundance, low cost, high CO_2 capture capacity, and regenerability. Ongoing research is focused on addressing the limitations related to selectivity, energy consumption, and reaction kinetics. By addressing these challenges, $Ca(OH)^{2-}$ $OH)^{2}$-based technologies hold promise as a feasible and sustainable solution for CO_2 capture and storage, making a substantial contribution to efforts aimed at mitigating climate change. Han et al. investigated the effectiveness of calcium hydroxide ($Ca(OH)_2$) aqueous solutions for capturing CO_2 from gas mixtures with a high CO_2 concentration (around 30%). While $Ca(OH)_2$ solutions are known CO_2 absorbents, their performance with such concentrated CO_2 streams had not been extensively studied. Their research revealed a key finding: the concentration of $Ca(OH)_2$ in the solution significantly impacts its CO_2 capture performance. They observed that higher $Ca(OH)_2$ concentrations might not always be beneficial. The simultaneous processes of $Ca(OH)_2$ dissolving and calcium carbonate ($CaCO_3$) production within the solution can hinder the interaction between calcium ions (Ca^{2+}) and carbonate ions (CO_3^{2-}), which are crucial for capturing CO_2. This resulted in a reduced CO_2 absorption capacity and the formation of large, poorly crystallized $CaCO_3$ particles at higher $Ca(OH)_2$ concentrations. Interestingly, the study found that a $Ca(OH)_2$ saturated aqueous solution, where all the $Ca(OH)_2$ has dissolved, exhibited the best CO_2 capture performance among the solutions tested. This saturated solution had the highest absorption rate and capture ratio. The determined CO_2 absorption capacity of this solution was threefold that of a 1% $Ca(OH)_2$ suspension. This observation highlights that a completely dissolved $Ca(OH)_2$ solution is the most effective absorbent for CO_2 capture in an aqueous $Ca(OH)_2$ solution system [74].

8.3.3 Lithium Hydroxide

Capturing excess CO_2 from the atmosphere is crucial for mitigating climate change. Lithium hydroxide (LiOH) is gaining recognition as a promising material for this application because of its distinctive properties. LiOH exhibits a remarkable capacity for capturing CO_2 molecules. This high capacity makes LiOH a promising candidate for large-scale CO_2 capture applications. Ahmadi et al. [75] investigated the potential of lithium hydroxide (LiOH) as a high-capacity adsorbent for capturing

CO_2. Their research focused on understanding the effectiveness of LiOH for CO_2 capture and identifying factors that influence its performance. This study confirms that LiOH exhibits a remarkable capacity for capturing CO_2 molecules. This high capacity makes LiOH a promising candidate for large-scale CO_2 capture applications. The research goes beyond simply demonstrating the high capacity. Studies explored how various factors can influence the CO_2 capture process with LiOH. These factors include temperature, pressure, the size of the LiOH particles, and the amount of LiOH used (LiOH loading). By optimizing these parameters, researchers can significantly improve the CO_2 capture efficiency of LiOH [75]. Unlike some CO_2 absorbents that require high temperatures, LiOH reacts irreversibly with CO_2 at room temperature. This eliminates the need for energy-consuming preheating steps. Also, due to its low molar mass, LiOH can absorb a significant amount of CO_2 molecules per unit mass of LiOH used. This makes it a potentially efficient material for CO_2 absorption. A key advantage for internal combustion engines is the ability of LiOH to perform well in environments with high moisture content. Unlike some materials that lose their effectiveness in humid conditions, studies have shown that LiOH exhibits maximum CO_2 absorption at about 65–70% relative humidity, which is typical for engine exhaust [76].

8.3.4 Potassium Hydroxide

Potassium hydroxide (KOH) exhibits a strong affinity for binding CO_2 molecules. Research has demonstrated its effectiveness in capturing CO_2 from ambient air. KOH adsorbents are materials that can capture CO_2 from gas streams. When CO_2 interacts with KOH, a chemical reaction takes place, resulting in the absorption of CO_2 by potassium hydroxide. This process helps in capturing and removing CO_2 from industrial emissions or other sources, contributing to CO_2 capture and mitigation efforts [77, 78]. Rastegar et al. investigated the effectiveness of capturing CO_2 using a potassium hydroxide (KOH) aqueous solution. Their study aimed to find the optimal operating conditions for maximizing CO_2 capture efficiency. They utilized response surface methodology grounded in central composite design to structure experiments and evaluate the impacts of different variables on CO_2 capture. This study identified the opposite effects of temperature, pressure, and KOH concentration on CO_2 uptake, such that higher temperatures and KOH concentrations resulted in decreased CO_2 loading, while increased pressure resulted in a beneficial increase in CO_2 uptake in the system [79, 80]. Potassium hydroxide (KOH) adsorbents are used for CO_2 capture because of their high reactivity with CO_2. When CO_2 interacts with KOH, it forms potassium carbonate through a chemical reaction. This absorption mechanism enables the removal of CO_2 from gas streams effectively. Additionally, KOH can be regenerated by releasing the captured CO_2 under specific conditions, making it a reusable and cost-effective option for CO_2 capture technologies.

8.4 METAL SILICATES

Metal silicates, such as lithium silicate or calcium silicate, can act as adsorbents for CO_2 capture. These materials feature a high surface area and are capable of chemically adsorbing CO_2 through processes such as chemisorption and physisorption. Metal silicates offer the potential for reversible CO_2 capture, making them promising candidates for applications in CCS technologies. In addition to lithium and calcium silicates, other metal silicates like magnesium silicate and aluminum silicate are also studied for CO_2 capture due to their tunable properties and high CO_2 adsorption capacities. Metal silicates can be tailored for specific CO_2 capture applications by modifying their surface chemistry or structure, making them versatile adsorbents in the realm of carbon capture technologies [81].

8.4.1 Lithium Silicates

Lithium silicates are a specific type of metal silicate that shows promise for capturing CO_2. Lithium silicates boast a strong affinity for CO_2 molecules, offering a high theoretical CO_2 capture capacity. This translates to potentially capturing a significant amount of CO_2 per unit of material used, making them attractive for large-scale CO_2 capture systems. Similar to other adsorbents, lithium silicates can be regenerated after capturing CO_2. This regeneration allows for the reuse of the material, reducing waste and promoting a more sustainable CO_2 capture solution. Unlike some CO_2 capture methods requiring high temperatures, lithium silicates can react with CO_2 at room temperature. This eliminates the need for preheating steps, reducing the energy consumption of the capture process. Lithium silicates exhibit good thermal stability, meaning they can withstand high temperatures that might be encountered during regeneration cycles without significant degradation. The effectiveness of lithium silicates for CO_2 capture can be influenced by moisture content in the gas stream. High humidity might reduce their CO_2 capture capacity, requiring strategies to manage moisture levels in the capture environment. Overall, lithium silicates present a promising avenue for CO_2 capture due to their high capacity, regeneration potential, and favorable reaction conditions. However, addressing moisture sensitivity, optimizing regeneration methods, and reducing costs are key areas for further development. Lithium consumption is rising due to the widespread use of lithium-ion batteries. This has led to increased lithium prices. $LONS_2$ is a solid sodium-based lithium silicate adsorbent that uses less lithium for CO_2 absorption compared to traditional Li_4SiO_4 adsorbents. $LONS_2$ is prepared by a simple method consisting of mixing LiOH with sodium silicate solution in a specific ratio (2:1 M) [82]. This adsorbent consistently maintains its ability to adsorb CO_2 over multiple cycles, indicating good reusability. $LONS_2$ absorbs CO_2 faster than traditional Li_4SiO_4 adsorbents. In another example, Zhang et al. [83] identified limitations of Li_4SiO_4 for practical CO_2 capture and proposed strategies to address them. They achieved improvements in cycling stability, capture rate, and effectiveness at lower temperatures and CO_2 concentrations through modifications with zeolite precursors, metal doping, and alkali carbonate coating. The study synthesized Li_4SiO_4 using four different zeolite precursors (MCM-41, MCM-48, TS-1, and ZSM-5). They evaluated the CO_2 uptake, cycling stability, and optimal conditions for CO_2 capture. The researchers further investigated improving the CO_2 capture rate (kinetics) by modifying the MCM-41 precursor with different metals (Al, Ti, Ca, and Na). They found that Na-MCM-41-Li_4SiO_4 achieved the fastest CO_2 capture rate, reaching its maximum capacity (close to the theoretical value of 36.7 wt%) within 20 minutes. At 550°C and 20 vol% CO_2, coated sorbents (15 wt% K_2CO_3 and 10 wt% Na_2CO_3) achieved significantly higher CO_2 uptake (32.2 and 34.7 wt%) compared to uncoated MCM-41-Li_4SiO_4 (11.8 wt%).

8.5 METAL-ORGANIC FRAMEWORK

MOFs, also referred to as coordination polymers or networks, constitute an innovative category of functional materials specifically engineered for gas adsorption. These hybrid materials are formed by integrating metal ions or clusters with organic ligands via coordination bonds. MOFs offer numerous advantages, including robust three-dimensional structures, exceptionally high surface areas, customizable pore architectures, and tunable pore surface properties [84, 85]. These attributes make MOFs highly effective for gas adsorption, especially for CO_2 capture and sequestration. Beyond CO_2 adsorption, MOFs find applications in areas such as sensing [86, 87], drug delivery [88, 89], and catalysis [90, 91]. While extensive reviews have explored MOFs across various domains, this paper focuses on the latest advancements in MOFs specifically for CO_2 adsorption and separation. It covers synthesis methods, CO_2 uptake and selectivity, strategies to enhance CO_2 adsorption, and the water stability of MOFs in CO_2 capture. Over the past two decades, MOFs have emerged as a significant area of research due to their novel properties. MOFs are hybrid porous materials composed of organic ligands and metal-containing clusters or nodes (Figure 8.4).

The versatility of MOF synthesis allows for a wide range of metal and organic combinations, resulting in diverse MOF structures and properties. The main attributes of MOFs encompass their stable three-dimensional frameworks, inherent porosity, and modular characteristics. These features ensure that MOFs retain their structural integrity and function effectively even when guest molecules are removed from their open channels and pores. Consequently, MOFs are highly stable and adaptable for various applications, including CO_2 adsorption. Additional notable properties of MOFs include their extraordinary surface areas, ultrahigh porosity, tunable porous structures, and customizable surface functionalities [92].

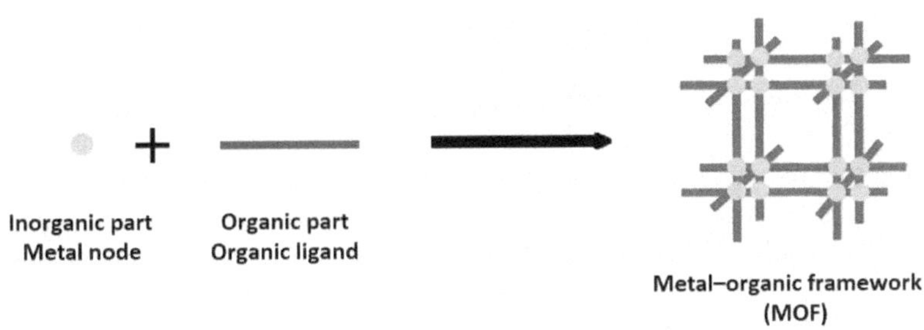

FIGURE 8.4 Schematic illustrates that metal-organic frameworks (MOFs), also known as coordination polymers or networks, are hybrid porous materials composed of organic ligands and metal-containing clusters or nodes.

8.5.1 Selective CO_2 Adsorption

Gas adsorption is critically dependent on the selectivity of the target gas over other components in a mixture. Four primary mechanisms facilitate gas separation in porous materials: (a) molecular screening (size and shape exclusion), (b) thermodynamic equilibrium influences, (c) kinetic factors, and (D) quantum screening [93, 94]. Molecular sieving depends on the size and shape of adsorbate molecules, with the kinetic or collision diameter, the minimum distance at which two molecules can collide without any kinetic energy, being a key factor. This mechanism is also prevalent in zeolites and molecular sieves, although it can be temperature-sensitive, affecting sieving efficiency [95]. Equilibrium separation is effective when the pore size of MOFs accommodates all gas components. In this context, the relationship between the surface of the MOF and the adsorbate is essential for determining selectivity. This interaction is influenced by the properties of both the adsorbate and the MOF, such as polarizability and dipole moments. When equilibrium separation is impractical, kinetic separation is employed, necessitating precise control over the pore size of the MOF relative to the kinetic diameters of the molecules being separated [96]. The selection of an MOF is critical for effective adsorptive separation, with the ideal MOF exhibiting both high selectivity and capacity for the target molecules. Key factors include the nature of the MOF, particularly its pore characteristics and the specifics of the adsorption process. In bulk gas separation, ease of desorption is also an important consideration. For general applications, MOF selection is based on the properties of the adsorbate molecules, including their size, shape, dipole moment, quadrupole moment, and polarizability. For instance, molecules with high dipole moments are best adsorbed by MOFs with highly polarized surfaces [97]. Molecules with high quadrupole moments, like CO_2, require adsorbents with surfaces that have strong electric field gradients. For non-polar molecules with high polarizability, adsorbents with a large surface area are preferred. Additionally, separations and purifications often rely on interactions such as π–π stacking and hydrogen bonding between the adsorbate and the adsorbent. Ultimately, the selection process should also consider equilibrium isotherms and diffusivity of the adsorbates [94]. Salehi et al. [98] described the synthesis of a nanocomposite composed of multi-walled carbon nanotubes (MWCNTs) and the MOF-199, referred to as CNT@MOF-199. This composite was formed by integrating MWCNTs into MOF-199, resulting in increased micropore volume, enhanced gas storage capacity, and improved stability compared to MOF-199 alone. The study indicated that the adsorption capacities of gas molecules in both MOF-199 and CNT@MOF-199 decreased with increasing temperature, suggesting that the adsorption process is exothermic. Additionally, CNT@MOF-199/piperazine (PZ) was created by impregnating various amounts (10, 20, and 30 wt%) of PZ onto synthetic CNT@MOF-199. Although modifications led to reductions in surface area, pore size, and pore volume, CNT@MOF-199/PZ exhibited superior selective CO_2 adsorption compared to CNT@MOF-199. This improvement is attributed to the strong interaction between CO_2 molecules, which behave as a Lewis acid, and the basic amine sites present on CNT@MOF-199/PZ. In another study, Li et al. [99] reported the development of stable Al-based MOFs with multiple adsorption sites for efficient and selective CO_2 capture and separation. Their study focused

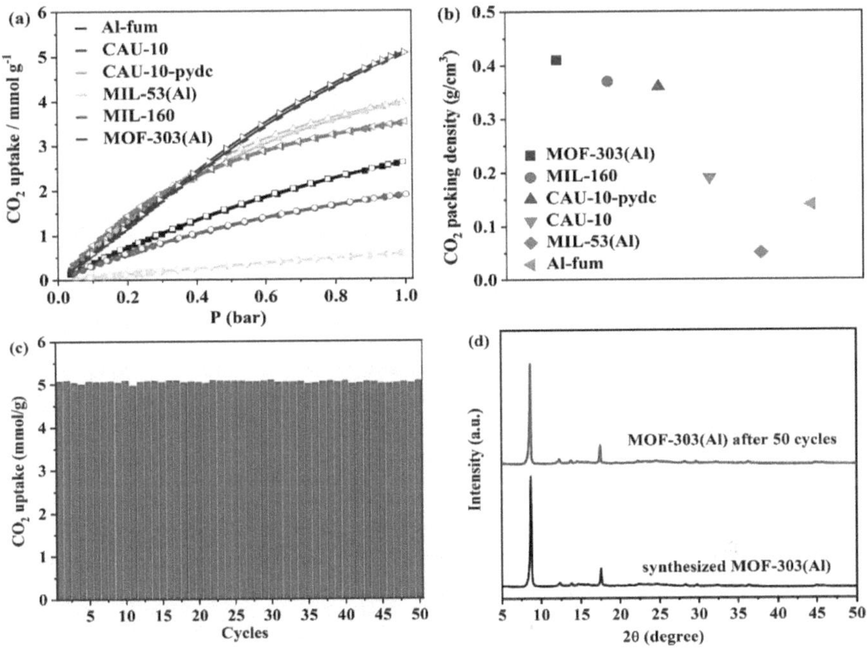

FIGURE 8.5 (a) CO_2 adsorption isotherms for the Al-based MOFs at 25.0°C; (b) CO_2 packing densities of the Al-based MOFs at 25.0°C; (c) CO_2 uptake of MOF-303(Al) over 50 cycles, with CO_2 absorption occurring at 25.0°C and MOF regeneration taking place at 60°C under vacuum; (d) PXRD patterns of MOF-303(Al) before and after 50 cycles. Reproduced from Ref. [99].

on MOF-303(Al), which features various active sites, including μ-OH, N–H, N–N, and H–C. This framework exhibited a high CO_2 adsorption capacity of 5.1 mmol/g and demonstrated the ability to be recycled over 50 times. Importantly, MOF-303(Al) achieved excellent selective CO_2 adsorption from CO_2/CH_4 and CO_2/N_2 mixtures, positioning it as a promising material for natural gas purification (Figure 8.5). The mechanism underlying this performance was attributed to the multiple adsorption sites, with μ-OH and N–H sites contributing more significantly to CO_2 adsorption compared to N–N and H–C sites.

8.5.2 Capacity of CO_2 Adsorption

A key aspect of adsorption technology is the capacity of the adsorbate, which largely relies on the surface area of the adsorbent. MOFs, characterized by their high and ultrahigh surface areas, exhibit enhanced CO_2 uptake capacities in comparison to other adsorbents, such as zeolites and activated carbons. Beyond surface area, factors like adsorbate–adsorbent interactions, pressure, and adsorption temperature also play a significant role in determining CO_2 uptake capacity. Considerable research has concentrated on the gravimetric capacity of adsorbents, which assesses the weight percentage of the adsorbed gas in relation to the total weight of the system [100, 101]. A notable study by Millward and Yaghi systematically investigated the correlation between the surface area of MOFs and their CO_2 adsorption capacity. This study evaluated nine MOFs with various characteristics and geometries, including MOF-2, MOF-505, $Cu_3(BTC)_2$, MOF-74, IRMOF-11, IRMOFs-3, IRMOF-6, IRMOF-1, and MOF-177, under consistent conditions (298 K and 35 bar). The results indicated that MOF-177, which had the highest surface area of 4,508 m²/g, exhibited the highest CO_2 uptake capacity of 60.0 wt% [102]. At high pressure, CO_2 adsorption capacity is closely linked to the surface area and pore volume of MOFs. Therefore, enhancing these properties can significantly improve CO_2 uptake capacity. This can be achieved by using larger organic ligands, such as BTB and BTC, or by extending the MOF structure during synthesis. The adsorption heat, or isosteric heat (Q_{st}), is a crucial parameter in evaluating gas adsorption on MOFs and other solid sorbents. This value represents the affinity of the MOFs for CO_2, highlighting the strength of the interaction between the adsorbent and the adsorbate molecules. The magnitude of Q_{st} correlates with the binding strength and also signifies the energy required for the regeneration process. A high Q_{st} suggests that the adsorbent has a strong affinity for CO_2 but also implies that substantial energy is needed for regeneration. Therefore, an optimal MOF for CO_2 uptake should have a balanced enthalpy of adsorption, measured in kJ mol^{-1}, to ensure effective performance and manageable regeneration energy requirements [96, 103]. Qasem et al. [104] investigated the CO_2 storage efficiency of MOF-5 and MOF-177. Their study concentrated on CO_2 adsorption at moderate storage pressures between 5 and 50 bar. They created an experimentally validated model using a User Defined Function (written in C) integrated with the ANSYS Fluent program, which effectively simulated gas adsorption systems. The outcomes from their multidimensional modeling indicated that MOF-5 is suitable for CO_2 storage at pressures below 5 bar, whereas MOF-177 demonstrates superior performance at higher pressures (≥10 bar). The ideal storage pressure for both adsorbents was determined to be 30 bar, with MOF-5 exhibiting a CO_2 uptake of 21.07 mmol/g (0.93 kg CO_2/kg MOF) and an energy requirement of 231 kJ/ton CO_2. Conversely, MOF-177 demonstrated a higher CO_2 uptake of 32.5 mmol/g (1.43 kg CO_2/kg MOF) and a marginally increased energy consumption of 233 kJ/ton CO_2. The study pointed out that CO_2 adsorptive storage using MOF-177 represents a more efficient strategy with lower energy demands compared to traditional storage systems, such as gas cylinders. MOFs can significantly enhance their CO_2 adsorption capacity, particularly at low pressures. This improvement is due to the modification's effect on the interaction between the MOF and CO_2, which boosts the framework's ability to capture the gas. Crucially, this heightened adsorption capacity proves effective even at high temperatures, indicating that the alterations enhance both the performance and stability of MOFs across diverse conditions. This capability is essential for applications such as CCS, where efficient gas adsorption is needed in varying operational environments. In another study, Gaikwad et al. [105] reported that modifying adsorbents with amine groups significantly enhances their CO_2 adsorption capacity, particularly at low pressures. Their study demonstrated that the amine modified adsorbents, especially with tetraethylenepentamine (TEPA), exhibited a remarkable 4.8-fold increase in adsorption capacity compared to the

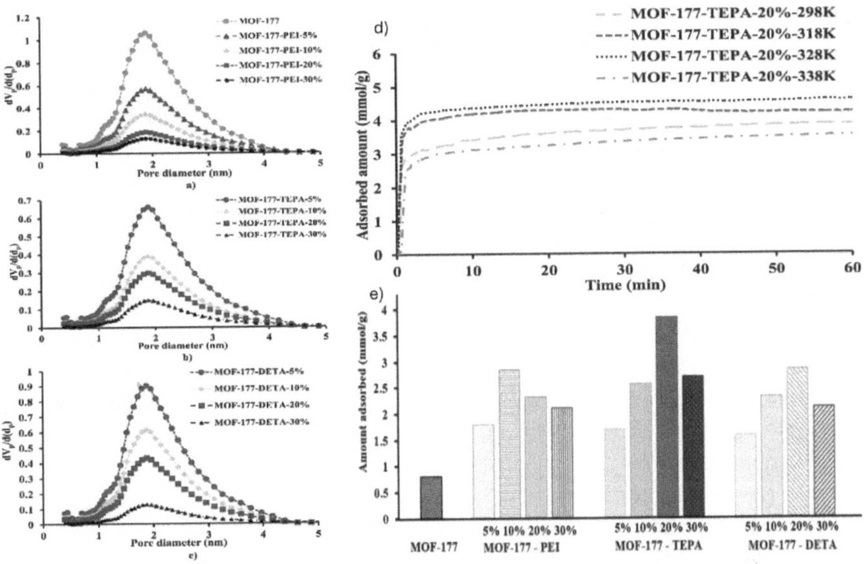

FIGURE 8.6 Pore size distribution graphs for (a) MOF-177-PEI, (b) MOF-177-TEPA, and (c) MOF-177-DETA at varying loading percentages. (d) CO_2 adsorption curves for MOF-177-TEPA-20% across different temperatures within the 298–338 K range, and (e) CO_2 adsorption capacities of MOF-177, MOF-177-PEI, MOF-177-TEPA, and MOF-177-DETA powder samples at 298 K and 1 bar. Reproduced from Ref. [105].

unmodified adsorbent at 298 K. The modified adsorbents also showed increased uptake with rising temperatures, reaching 4.6 mmol/g at 328 K, surpassing previously reported amine-modified adsorbents (Figure 8.6).

Additionally, the modified adsorbents maintained consistent capture performance through multiple adsorption–desorption cycles, highlighting their effectiveness and stability for selective adsorption.

8.5.3 Synthesis of MOFs

MOFs are formed by metal (inorganic) ions or clusters connected to organic ligands via coordination bonds. Transition metals are frequently utilized in the construction of MOFs due to their incomplete d-block, which provides high reactivity and adaptability. These metals can display a variety of coordination numbers and oxidation states, allowing for the creation of MOFs with different geometries, such as square-planar, tetrahedral, octahedral, trigonal, and pentagonal configurations. The geometric structure of the MOFs is influenced by factors including the metal sites, their coordination numbers, ligands, solvents, and even counter-ions. Coordination reactions typically take place between metal salts and ligands in the presence of organic solvents like dimethylformamide, ethanol, or methanol, leading to MOFs with diverse geometric shapes based on the reaction conditions [106, 107]. The organic part of MOFs consists of ligands that feature Lewis base-binding atoms or functional groups, including halides, nitriles, cyanides, anionic organic molecules (such as benzenedicarboxylic acid), and neutral organic molecules (like 4,4′-bipyridine). The tunability of MOFs is largely attributed to the ligands' ability to incorporate different functional groups under various conditions. The capacity for gas storage is significantly influenced by the length of the ligand, with extended organic linkers typically resulting in more substantial storage volumes due to the increased number of functional groups they provide. However, excessively long ligands may lead to interpenetrating structures and chain intertwining. Overall, the ligands play a critical role in enhancing the gas adsorption capacity of MOFs through their functionalization. Recently, there has been significant emphasis on the synthesis of MOFs, driven by the ability to design their structures for specific applications. Various synthesis methods, including conventional, microwave-assisted, sono-chemical, mechanochemical, and electrochemical techniques, have been employed (Figure 8.7). The main difference between these methods is the type of energy input employed, which will be elaborated upon in the subsequent sections [108, 109]. The most common approach for synthesizing MOFs is the conventional method, which relies on traditional heating sources to generate the necessary energy. Temperature is a pivotal element in these chemical processes, differentiating this method from both solvothermal (or hydrothermal) and

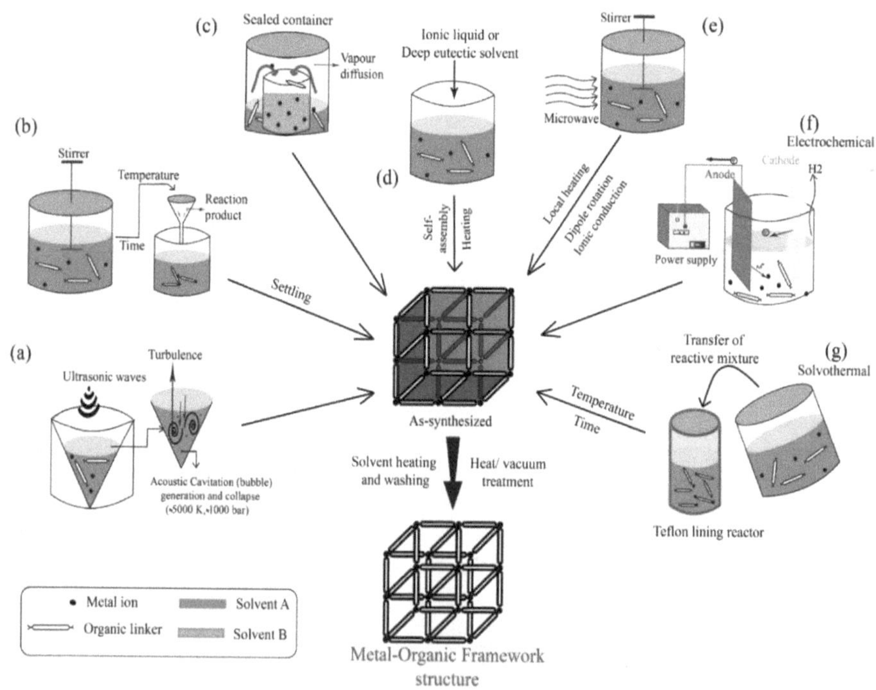

FIGURE 8.7 Various methods for synthesizing MOFs include (a) sono-chemical synthesis, (b) conventional solution synthesis, (c) diffusion synthesis, (d) iono-thermal processing, (e) microwave synthesis, (f) electrochemical synthesis, and (g) solvothermal synthesis. Reproduced from Ref. [112].

non-solvothermal techniques. In solvothermal reactions, a sealed reactor, often made from Teflon, is used, allowing for autogenous pressure and temperatures that surpass the boiling point of the solvent. On the other hand, non-solvothermal reactions are conducted at ambient pressure and at or below the solvent's boiling point. Typically, MOF synthesis temperatures range from room temperature to approximately 250°C, with solvothermal reactions being the preferred method due to their favorable synthesis conditions [110, 111].

Fractals in the synthesis of MOFs are a fascinating and emerging topic that significantly enhances gas adsorption and framework stability. Fractal science studies complex geometric shapes that exhibit self-similarity across different scales, meaning their patterns repeat regardless of magnification. These self-similar structures create hierarchical pore designs that increase surface area and accessibility for gas molecules. The varied pore sizes mimic fractal patterns, improving gas separation by selectively adsorbing different-sized molecules. Understanding fractal geometry is essential for optimizing MOF design and synthesis, ultimately boosting their performance in gas storage and separation applications [113–115].

8.5.4 Methods to Improve CO_2 Adsorption

To increase the capacity and selectivity for CO_2 adsorption in MOFs, multiple approaches have been proposed. These approaches involve the incorporation of open metal sites (OMSs), the utilization of modifications before and after synthesis, and the optimization of pore size.

8.5.4.1 Open Metal Sites

Metal sites play a crucial role in influencing the CO_2 adsorption capacity and selectivity of MOFs. During the synthesis process, certain metal sites may interact with solvent molecules, which can diminish the framework's ability to capture guest molecules. To mitigate this problem, an activation step is often employed, typically involving heat treatment and/or vacuum application. This activation removes excess solvent from the metal sites, thereby generating OMSs. The presence of OMS significantly boosts the MOF's CO_2 adsorption capacity by increasing the number of active sites available for binding CO_2. These metal centers enhance the interaction energy between the MOF surface and CO_2 molecules, promoting adsorption through dipole-quadrupole interactions that effectively capture CO_2. Almost all transition metals in the periodic table can be utilized in MOF construction. These metals interact with other atoms and molecules via their d atomic orbitals, making the study of their bonding mechanisms both fascinating and essential. The interactions between MOF frameworks and small guest molecules can be further elucidated through the d band center theory, which clarifies how OMS engages with guest molecules. Understanding the nature of these interactions is critical for the development of MOFs with improved adsorption and separation properties. Additionally, it is important to consider the variations among transition metal cations. The binding between OMS and adsorbates results from a complex interplay of molecular orbital hybridization, electrostatic interactions, van der Waals forces, and Pauli repulsion. Therefore, the selection of transition metal cations is key to determining the adsorption characteristics of MOFs

[116, 117]. Chowdhury et al. [118] conducted a comparative study on the adsorption of CO, CO_2, and CH_4 using two adsorbents: Cu-BTC (HKUST-1) and Cr-BDC (MIL-101). Their results indicated that at low loading levels, the enthalpy of adsorption was greater for MIL-101 compared to Cu-BTC, implying the presence of OMSs in MIL-101. Furthermore, the study showed that MIL-101 experienced a pronounced reduction in the enthalpy of adsorption as pressure increased, whereas Cu-BTC exhibited only a slight decrease. Additionally, MIL-101 had a higher Henry's constant for CO, while Cu-BTC exhibited greater CO_2 capacity across all pressures. The analysis highlighted the critical role of OMS in enhancing gas adsorption performance and provided insights into the influence of adsorbate polarity and temperature on electrostatic interactions.

8.5.4.2 Amine Functionalization in Pre-Synthetic Procedure

Ligand functionalization greatly enhances MOFs for CO_2 adsorption and separation. The process is effective due to its ease and the range of functional groups available. The CO_2 quadrupole moment is enhanced by polar groups within the MOF structure, enhancing its selectivity over non-polar gases such as CH_4, N_2, and H_2. Functionalization can be performed pre- or post-synthesis using various methods, including amine, phosphonate, sulfonate, and multifunctional groups. Amine-functionalized ligands are particularly effective for CO_2 adsorption due to their high affinity for CO_2. This is because amine groups act as Lewis bases, which readily interact with CO_2, a Lewis acid. The interaction between the Lewis basic amines and the acidic CO_2 enhances the adsorption and selectivity of CO_2 over other gases. This strong chemical interaction is a key reason why amine functionalization is a popular method for improving CO_2 capture in MOFs [119, 120]. Shin et al. [121] reported that the selective removal of CO_2 from industrial offgases is crucial, and MOFs have been studied for this purpose. They introduced a novel approach by loading a basic polymer, polyvinylamine (PVAm), into the highly porous MOF MIL-101 using the ship-in-a-bottle technique. This amine functionalization, performed during the pre-synthetic procedure, significantly enhanced the CO_2 adsorption capacity and selectivity of MIL-101. The PVAm@MIL-101 exhibited approximately 11 times higher CO_2/N_2 selectivity and 2.5 times greater CO_2 adsorption capacity compared to the pristine MIL-101 at 298 K. Additionally, the modified MOF demonstrated good recyclability and an optimal isosteric heat of adsorption for CO_2, making it a promising candidate for efficient CO_2 capture from offgases (Figure 8.8).

8.5.4.3 Nitrogen, Phosphonate, and Sulfonate Functionalization in Pre-Synthetic Procedure

The inclusion of uncoordinated nitrogen sites in MOF frameworks typically improves CO_2 adsorption capabilities. For example, triazoles and tetrazoles serve as prominent illustrations of this effect. CPF-6, a MOF characterized by its three-dimensional structure and a high density of uncoordinated nitrogen sites derived from tetrazole ligands (specifically, 1,3,5-*tris*(2H-tetrazol-5-yl)benzene), demonstrates significant CO_2 adsorption and selectivity over N_2, even in the absence of OMSs. This remarkable performance is attributed to the abundant open nitrogen sites located on the pore surfaces. Nonetheless, although open nitrogen sites

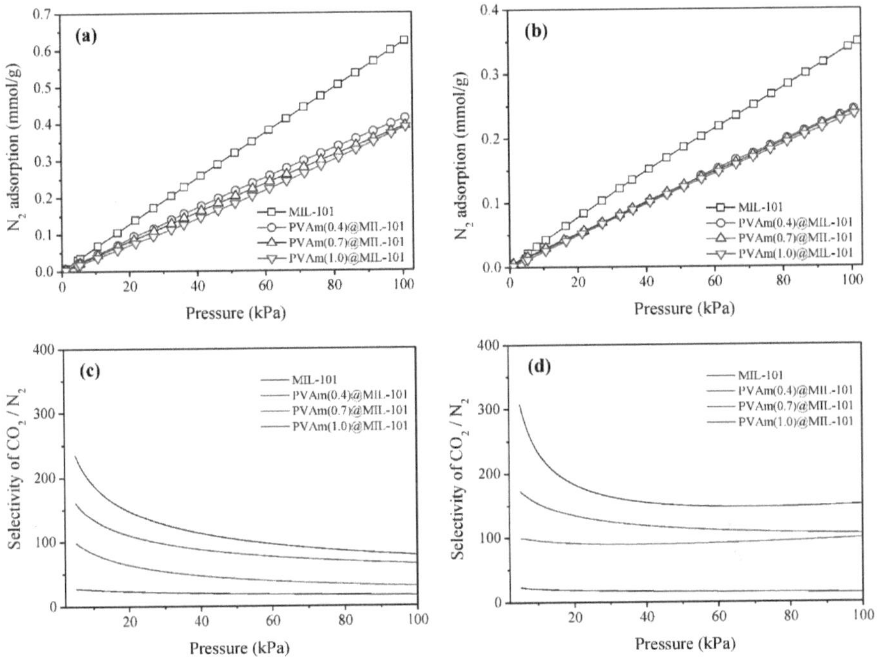

FIGURE 8.8 (a) and (b) The N_2 adsorption isotherms for MIL-101 and several selected PVAm@MIL-101 samples at temperatures of 273 and 298 K, respectively. (c) and (d) The IAST selectivities (CO_2/N_2 = 0.15/0.75) for both MIL-101 and the selected PVAm@MIL-101 samples at the same temperatures of 273 and 298 K, respectively. Reproduced from Ref. [121].

substantially enhance CO_2 adsorption, they may not be strictly necessary for achieving optimal CO_2 capture efficiency [122, 123]. Adenine groups are ideal for ligand functionalization in MOFs due to their uncoordinated nitrogen sites. They offer several advantages: the different nitrogen positions allow for diverse MOF structures, the rigid nature of adenine contributes to the formation of robust frameworks, and the coplanarity of adenine atoms facilitates π–π interactions between ligands during MOF assembly. These properties make adenine a valuable component in designing MOFs with enhanced structural integrity and functional performance. These notable properties make adenine an effective building block for constructing MOFs aimed at CO_2 adsorption. A computational study has demonstrated that the interaction energy between adenine and CO_2 molecules is superior compared to other MOFs with nitrogen linkers [124]. Ligand functionalization significantly influences CO_2 adsorption and selectivity, as well as improves framework stability. In particular, phosphonate and sulfonate functional groups enhance the water stability of MOFs, which is crucial given that flue gases often contain water vapor that can affect CO_2 adsorption. Improving water stability is therefore essential for optimizing gas adsorption technologies [125].

8.5.4.4 Mixed-Ligand-Functionalization in Pre-Synthetic Procedure

The concept of incorporating mixed ligands into the structure of MOFs presents an intriguing approach, particularly for modifying frameworks and enhancing adsorption characteristics. The integration of various functional groups within the framework not only increases CO_2 capacity and selectivity but also enhances the stability of the MOFs. As a result, the presence of multiple functional sites within the framework contributes to an improved CO_2 uptake capability of the MOF. In recent years, these functionalization techniques have garnered significant attention due to the impressive outcomes they can achieve [122, 126]. Park et al. [127] reported a successful synthesis and operational evaluation of MIL-101 (Cr)-NH_2-F0.5 (CrA-F0.5) by utilizing a mixed-ligand strategy for the enhancement of CO_2 adsorption in humid conditions (Figure 8.9). This material had 2.9% nitrogen and 0.5% fluorine and displayed a calculated CO_2/N_2 selectivity of 108 (CO_2/N_2 = 15/85) at 20°C and 1 bar. This selectivity value represents a significant improvement compared to both MIL-101 (Cr) and MIL-101 (Cr)-NH_2: 3.7 times and 2.1 times, respectively. The results suggest that the amine and fluorine groups within CrA-F0.5 function as CO_2 adsorption sites. The CO_2 adsorption isosteric heat of CrA-F0.5 was moderately estimated at 45 kJ mol-1, indicating preferential CO_2 desorption characteristics. By performing a hydrophobicity index test, CrA-F0.5 showed approximately 2.5 times the hydrophobicity relative to MIL-101 (Cr)-NH_2. In the breakthrough test of CO_2 and N_2 in a humid environment, CO_2 breakthrough time was considerably decreased at 30°C 1 bar by 40% comparing to MIL-101 (Cr)-NH_2 and 10% by CrA-F0.5 at 60% RH, this result corresponds to hydrophobicity index, showing that CrA-F0.5 has excellent water repellency. Consequently, CrA-F0.5 was highlighted as a candidate for CO_2 adsorption in the actual post-combustion process.

8.5.4.5 Post-Synthetic Procedure

Post-synthesis modification (PSM) is another effective method for tuning MOF structures to enhance CO_2 adsorption performance. Unlike pre-synthesis modification, where functional groups are introduced during MOF synthesis, PSM involves adding functional groups to preexisting MOFs. This approach seeks to enhance CO_2 capacity and selectivity while minimizing energy requirements for regeneration.

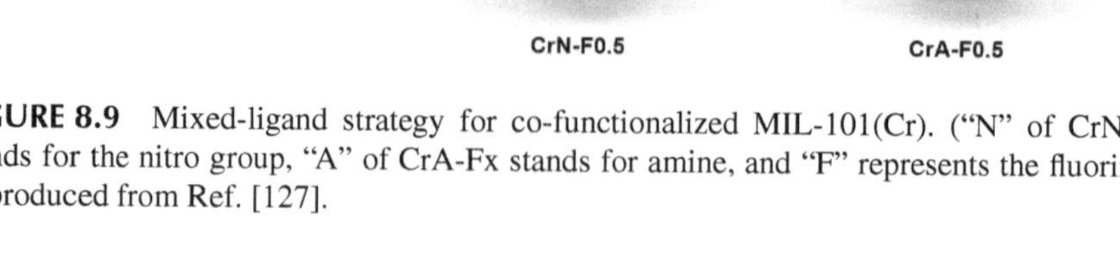

FIGURE 8.9 Mixed-ligand strategy for co-functionalized MIL-101(Cr). ("N" of CrN-Fx stands for the nitro group, "A" of CrA-Fx stands for amine, and "F" represents the fluorine). Reproduced from Ref. [127].

PSM presents benefits over pre-synthesis techniques, notably by preventing possible changes in the crystal structure that may arise from the interaction of functional groups with metal sites during the initial synthesis phase. Additionally, PSM can circumvent issues related to the instability of functional groups under synthesis conditions and the potential formation of side-products. Overall, PSM is a valuable approach for optimizing CO_2 adsorption properties. As previously mentioned, amine groups exhibit a strong affinity for CO_2 molecules. Among them, ethylenediamine (en) is the most commonly used functional group in PSM for enhancing CO_2 adsorption [128, 129]. Lin et al. [130] investigated a series of MIL-101 adsorbents incorporated with polyethyleneimine (PEI) at various loadings. Despite significant reductions in surface area and pore volume following PEI incorporation, the composites showed a substantial increase in CO_2 adsorption capacity at low pressures. At a 100 wt% PEI loading, the CO_2 adsorption capacity reached 4.2 mmol/g at 0.15 bar and 25°C, and 3.4 mmol/g at 50°C. Furthermore, the adsorbents exhibited fast adsorption kinetics and exceptionally high selectivity for CO_2 over N_2, with selectivity values reaching 770 at 25°C and 1200 at 50°C. The findings suggest that PEI-functionalized MOFs hold significant potential for efficient CO_2 capture.

8.5.4.6 Tuning Pore Size

As previously discussed, the pore dimensions of MOFs play a crucial role in their adsorption characteristics, especially regarding selectivity. This behavior is referred to as the size-exclusion effect, where smaller molecules can traverse the MOF pores while larger ones are blocked. For effective selective adsorption of CO_2 in comparison to N_2 and CH_4, optimal pore diameters are typically between 3.3 and 3.6 Å. This approach is particularly advantageous for the selective adsorption of small gas molecules like H_2, N_2, CO_2, and CH_4. The desired pore size can be achieved through methods such as ligand or metal exchange. Literature suggests that employing bulky linkers, utilizing short ligands, creating interpenetrated networks, and opting for smaller metal ions can yield effective results. Additionally, the post-synthetic exchange process allows for precise tuning of pore sizes [131].

8.5.5 Challenges and Disadvantages

MOFs have gained significant attention for their applications in gas separations and storage, attributed to their high surface areas, tunable pore sizes, and diverse chemical functionalities. However, despite these advantages, there are notable limitations to their practical application. Challenges in large-scale production and commercialization arise from the complexities involved in their development under inert conditions, along with the scarcity of single-crystalline MOFs. These factors hinder the broader adoption and utilization of MOFs in real-world scenarios. Several obstacles remain in maximizing the function of MOF materials, most notably in the design of pillar ligands and the development of post-synthetic modification to MOFs. Understanding MOF behavior in hydrocarbon separations is critical to broadening the use of MOFs, as there is less data about hydrocarbon separation performance compared to other gaseous separations. Future progress in expanding the available library of MOFs for specific adsorption and separation tasks could further realize MOFs potential for contemporary challenges observed in separations like chemical processing. By focusing development on increasing the adsorption capacity and purity, researchers are working toward overcoming limitations to improve gas separation efficiency and sustainability.

8.6 SILICA GEL

Silica gel is a form of silicon dioxide that is porous and amorphous, and it is often used as a drying agent to take up water. In the context of CO_2 adsorption, silica gel can function as a sorbent material with its porosity and high surface area. Silica gel can be effective in adsorbing CO_2 molecules when structured or modified appropriately. Silica gel, characterized by its porous structure, offers a substantial surface area that facilitates the absorption and adherence of CO_2 molecules. The adsorption mechanism primarily involves physical interactions, including van der Waals forces and surface adsorption. By modifying properties such as pore size distribution and surface functionalization, the affinity of silica gel for CO_2 can be enhanced. The specific adsorption of CO_2 is influenced by various factors, including temperature, pressure, and the presence of other gases. Furthermore, effective regeneration of silica gel post-CO_2 desorption is crucial for its reutilization. Techniques such as heating or purging with a different gas are employed to release the adsorbed CO_2 molecules, and thus the silica gel can be used for multiple adsorption–desorption cycles. In summary, silica gel is an inexpensive and widely available material for CO_2 adsorption, especially when a simple and highly efficient adsorption process is required. Its versatility and tunability make it an attractive candidate for a variety of CO_2 capture and separation processes. Jahanbakhshi et al. [3] reported on their investigation into enhancing CO_2 capture performance by impregnating porous silica gel with a eutectic choline chloride-monoethanolamine solvent (ChCl-MEA). They varied the weight percentages of ChCl-MEA from 10% to 60% at 25°C to assess the influence of ChCl-MEA loading on the structural characteristics of the DES-modified silica samples, employing BET, FTIR, and TGA analyses. Their findings indicated that the modified silica gel with 50% ChCl-MEA (silica-CM50) achieved the highest CO_2 capture capacity, recorded at 89.32 mg/g. In terms of kinetic modeling, the fractional order model provided the best correlation with the experimental data, yielding a correlation coefficient of 0.998. The adsorption behavior of silica-CM50 was effectively represented by the dual-site Langmuir isotherm model, which suggests the existence of two distinct adsorption sites, supported by a correlation coefficient of 0.999. Additionally, the researchers calculated thermodynamic parameters at 25°C, finding the enthalpy at −2.770, the entropy at −0.005, and the Gibbs free energy at −1.162. These thermodynamic values indicate that the adsorption process of CO_2 on the modified silica gel is exothermic, spontaneous, and feasible.

8.7 CONCLUSION

In this chapter, we have examined the potential of metallic adsorbents and MOFs as innovative approaches for CO_2 capture. The pressing need to combat greenhouse gas emissions has spurred the development of efficient and cost-effective CO_2 capture methods, with metallic adsorbents and MOFs emerging as promising candidates. Metallic adsorbents, which include metal oxides, hydroxides, and silicates, present significant advantages due to their high reactivity and capability for mineral carbonation. This process captures CO_2 and converts it into stable, solid carbonates, thereby contributing to long-term storage solutions. Their versatility, combined with abundance and low cost, positions these materials as attractive options for large-scale CO_2 capture applications. In contrast, MOFs are distinguished by their exceptional structural attributes, including high porosity, tunable pore sizes, and the capacity for selective CO_2 adsorption. The various synthesis methods explored, such as solvothermal, microwave-assisted, and template-directed approaches, enable precise control over the properties of MOFs, facilitating enhancements in both CO_2 adsorption capacity and selectivity. Additionally, the incorporation of functional groups and the adjustment of pore sizes further optimize their performance, establishing MOFs as powerful tools in the battle against climate change. Despite their promising attributes, challenges persist in optimizing these materials for practical applications. Key issues, such as stability, cost, and scalability, must be addressed to fully harness their potential for CO_2 capture and storage. Nonetheless, advancements in the design and synthesis of metallic adsorbents and MOFs mark significant progress in the development of effective CO_2 capture technologies. Both metallic adsorbents and MOFs offer a promising pathway toward reducing CO_2 emissions and mitigating climate change. By continuing to refine these materials and tackle existing challenges, we can advance toward achieving sustainable and efficient CO_2 capture solutions.

REFERENCES

1. Bahmanzadegan, F. and A. Ghaemi, Exploring the effect of zeolite's structural parameters on the CO_2 capture efficiency using RSM and ANN methodologies. *Case Studies in Chemical and Environmental Engineering*, 2024. 9: p. 100595.
2. Esfahani, H.J., S. Shahhosseini, and A. Ghaemi, Improved structure of Zr-BTC metal organic framework using NH2 to enhance CO_2 adsorption performance. *Scientific Reports*, 2023. 13(1): p. 17700.
3. Jahanbakhshi, M., A. Ghaemi, and M. Helmi, Impregnation of silica gel with choline chloride-MEA as an eco-friendly adsorbent for CO_2 capture. *Scientific Reports*, 2024. 14(1): p. 15208.
4. Khan, U., et al., Assessing absorption-based CO_2 capture: Research progress and techno-economic assessment overview. *Carbon Capture Science & Technology*, 2023. 8: p. 100125.
5. Yang, M., et al., Circular economy strategies for combating climate change and other environmental issues. *Environmental Chemistry Letters*, 2023. 21(1): p. 55–80.
6. Ghamami, S., A. Kazemi Korayem, and N. Baqeri, Production and purification of titanium dioxide with titanium tetrachloride nanoparticles from eliminate concentrate of Kahnooj mine in Kerman. *Applied Chemistry Today*, 2020. 15(55): p. 189–206.
7. Pordsari, M.A., M. Moradi, and A. Ghaemi, Fabrication of a cost-effective metal oxide-based adsorbent from industrial waste slag for efficient CO_2 separation under flue gas conditions. *Journal of CO2 Utilization*, 2024. 88: p. 102930.
8. Bahmei, F., et al., Improving hypercrosslinked polymer CO_2/N_2 selective separation through tuning polymer's porous properties: Optimization using RSM-BBD. *Journal of CO2 Utilization*, 2024. 88: p. 102926.

9. Aldy, J.E. and Z.M. Halem, The evolving role of greenhouse gas emission offsets in combating climate change. *Review of Environmental Economics and Policy*, 2024. 18(2): p. 000–000.
10. Dziejarski, B., et al., CO_2 capture materials: A review of current trends and future challenges. *Materials Today Sustainability*, 2023: p. 100483.
11. Singh, G., et al., Emerging trends in porous materials for CO_2 capture and conversion. *Chemical Society Reviews*, 2020. 49(13): p. 4360–4404.
12. Kazemi, A., F. Manteghi, and Z. Tehrani, Metal electrocatalysts for hydrogen production in water splitting. *ACS Omega*, 2024. 9(7): p. 7310–7335.
13. Torabi, E., et al., Sacrificial MOF-derived MnNi hydroxide for high energy storage supercapacitor electrodes via DFT-based quantum capacitance study. *Heliyon*, 2025. 11(1): p. 41261.
14. Pardakhti, M., et al., Trends in solid adsorbent materials development for CO_2 capture. *ACS Applied Materials & Interfaces*, 2019. 11(38): p. 34533–34559.
15. Dunstan, M.T., et al., CO_2 capture at medium to high temperature using solid oxide-based sorbents: Fundamental aspects, mechanistic insights, and recent advances. *Chemical Reviews*, 2021. 121(20): p. 12681–12745.
16. Gao, W., et al., Industrial carbon dioxide capture and utilization: State of the art and future challenges. *Chemical Society Reviews*, 2020. 49(23): p. 8584–8686.
17. Sharma, H. and A. Dhir, Capture of carbon dioxide using solid carbonaceous and non-carbonaceous adsorbents: A review. *Environmental Chemistry Letters*, 2021. 19(2): p. 851–873.
18. Danish, M.S.S., et al., A systematic review of metal oxide applications for energy and environmental sustainability. *Metals*, 2020. 10(12): p. 1604.
19. Khdary, N.H., et al., Metal oxides as catalyst/supporter for CO_2 capture and conversion, review. *Catalysts*, 2022. 12(3): p. 300.
20. Omodolor, I.S., et al., Dual-function materials for CO_2 capture and conversion: A review. *Industrial & Engineering Chemistry Research*, 2020. 59(40): p. 17612–17631.
21. Lai, J.Y., L.H. Ngu, and S.S. Hashim, A review of CO_2 adsorbents performance for different carbon capture technology processes conditions. *Greenhouse Gases: Science and Technology*, 2021. 11(5): p. 1076–1117.
22. Ruhaimi, A., M. Aziz, and A. Jalil, Magnesium oxide-based adsorbents for carbon dioxide capture: Current progress and future opportunities. *Journal of CO_2 Utilization*, 2021. 43: p. 101357.
23. Alivand, M.S., et al., Catalytic solvent regeneration for energy-efficient CO_2 capture. *ACS Sustainable Chemistry & Engineering*, 2020. 8(51): p. 18755–18788.
24. Zhang, X., et al., Amine-based CO_2 capture aided by acid-basic bifunctional catalyst: Advancement of amine regeneration using metal modified MCM-41. *Chemical Engineering Journal*, 2020. 383: p. 123077.
25. Luzzi, E., et al., Mechanically coherent zeolite 13X/chitosan aerogel beads for effective CO_2 capture. *ACS Applied Materials & Interfaces*, 2021. 13(17): p. 20728–20734.
26. Liu, L., et al., Alkyl-functionalization of (3-Aminopropyl) triethoxysilane-grafted zeolite beta for carbon dioxide capture in temperature swing adsorption. *Chemical Engineering Journal*, 2020. 382: p. 122834.
27. Fu, D. and M.E. Davis, Carbon dioxide capture with zeotype materials. *Chemical Society Reviews*, 2022. 51(22): p. 9340–9370.
28. Panda, D., E.A. Kumar, and S.K. Singh, Introducing mesoporosity in zeolite 4A bodies for Rapid CO_2 capture. *Journal of CO2 Utilization*, 2020. 40: p. 101223.
29. Debost, M., et al., Synthesis of discrete CHA zeolite nanocrystals without organic templates for selective CO_2 capture. *Angewandte Chemie International Edition*, 2020. 59(52): p. 23491–23495.

30. Dabbawala, A.A., et al., Synthesis of hierarchical porous Zeolite-Y for enhanced CO_2 capture. *Microporous and Mesoporous Materials*, 2020. 303: p. 110261.
31. Wu, T., et al., Porous carbon frameworks with high CO_2 capture capacity derived from hierarchical polyimide/zeolitic imidazolate frameworks composite aerogels. *Chemical Engineering Journal*, 2020. 395: p. 124927.
32. Rahmah, W., et al., Small-pore zeolite and zeotype membranes for CO_2 capture and sequestration–a review. *Journal of Environmental Chemical Engineering*, 2022. 10(6): p. 108707.
33. Boer, D.G., J. Langerak, and P.P. Pescarmona, Zeolites as selective adsorbents for CO_2 separation. *ACS Applied Energy Materials*, 2023. 6(5): p. 2634–2656.
34. Ma, X., et al., Activated porous carbon with an ultrahigh surface area derived from waste biomass for acetone adsorption, CO_2 capture, and light hydrocarbon separation. *ACS Sustainable Chemistry & Engineering*, 2020. 8(31): p. 11721–11728.
35. Zhang, W., et al., Regulation of porosity in MOFs: A review on tunable scaffolds and related effects and advances in different applications. *Journal of Environmental Chemical Engineering*, 2022. 10(6): p. 108836.
36. Kazemi, A., et al., Enhanced CO_2 capture potential of UiO-66-NH2 synthesized by sonochemical method: Experimental findings and performance evaluation. *Scientific Reports*, 2023. 13(1): p. 19891.
37. Younas, M., et al., Recent progress and remaining challenges in post-combustion CO_2 capture using metal-organic frameworks (MOFs). *Progress in Energy and Combustion Science*, 2020. 80: p. 100849.
38. Usman, M., et al., Advanced strategies in metal-organic frameworks for CO_2 capture and separation. *The Chemical Record*, 2022. 22(7): p. e202100230.
39. Gu, C., et al., Effects of functional groups for CO_2 capture using metal organic frameworks. *Frontiers of Chemical Science and Engineering*, 2021. 15: p. 437–449.
40. Azmi, A. and M. Aziz, Mesoporous adsorbent for CO_2 capture application under mild condition: A review. *Journal of Environmental Chemical Engineering*, 2019. 7(2): p. 103022.
41. Yu, C.-H., C.-H. Huang, and C.-S. Tan, A review of CO_2 capture by absorption and adsorption. *Aerosol and Air Quality Research*, 2012. 12(5): p. 745–769.
42. Cotlame-Salinas, V.D.C., et al., CO_2 capture enhancement in MOFs via the confinement of molecules. *Reaction Chemistry & Engineering*, 2021. 6(3): p. 441–453.
43. Ghanbari, T., F. Abnisa, and W.M.A.W. Daud, A review on production of metal organic frameworks (MOF) for CO_2 adsorption. *Science of the Total Environment*, 2020. 707: p. 135090.
44. Shen, Z., et al., Facile synthesis of silica nanosheets with hierarchical pore structure and their amine-functionalized composite for enhanced CO_2 capture. *Chemical Engineering Science*, 2020. 217: p. 115528.
45. Zhang, G., P. Zhao, and Y. Xu, Development of amine-functionalized hierarchically porous silica for CO_2 capture. *Journal of Industrial and Engineering Chemistry*, 2017. 54: p. 59–68.
46. Cherevotan, A., J. Raj, and S.C. Peter, An overview of porous silica immobilized amines for direct air CO_2 capture. *Journal of Materials Chemistry A*, 2021. 9(48): p. 27271–27303.
47. Zhang, G., et al., Structure property–CO_2 capture performance relations of amine-functionalized porous silica composite adsorbents. *ACS Applied Materials & Interfaces*, 2018. 10(40): p. 34340–34354.
48. Wang, H.C., et al., Pilot-scale production of mesoporous silica-based adsorbent for CO_2 capture. *Applied Surface Science*, 2012. 258(18): p. 6943–6951.
49. Quan, C., et al., Amine-modified silica zeolite from coal gangue for CO_2 capture. *Fuel*, 2022. 322: p. 124184.

50. Niu, M., et al., Amine-impregnated mesoporous silica nanotube as an emerging nanocomposite for CO_2 capture. *ACS Applied Materials & Interfaces*, 2016. 8(27): p. 17312–17320.
51. Fan, Y. and X. Jia, Progress in amine-functionalized silica for CO2 capture: Important roles of support and amine structure. *Energy & Fuels*, 2022. 36(3): p. 1252–1270.
52. Zhu, X., et al., Efficient CO_2 capture from ambient air with amine-functionalized Mg-Al mixed metal oxides. *Journal of Materials Chemistry A*, 2020. 8(32): p. 16421–16428.
53. Alkadhem, A.M., M.A. Elgzoly, and S.A. Onaizi, Novel amine-functionalized magnesium oxide adsorbents for CO_2 capture at ambient conditions. *Journal of Environmental Chemical Engineering*, 2020. 8(4): p. 103968.
54. Rajamathi, R., N. Bhojaraj, and C. Nethravathi, Porous CaO-MgO nanostructures for CO_2 capture. *ACS Applied Nano Materials*, 2021. 4(10): p. 10969–10975.
55. Chammingkwan, P., T. Ikeda, and P. Mohan, Nanostructured magnesium oxide microspheres for efficient carbon dioxide capture. *Journal of CO_2 Utilization*, 2021. 51: p. 101652.
56. Tan, C., et al., Structurally improved MgO adsorbents derived from magnesium oxalate precursor for enhanced CO_2 capture. *Fuel*, 2020. 278: p. 118379.
57. Wang, L., et al., Mesoporous MgO enriched in Lewis base sites as effective catalysts for efficient CO_2 capture. *Journal of Environmental Management*, 2023. 332: p. 117398.
58. Yu, H., et al., Explainable molecular simulation and machine learning for carbon dioxide adsorption on magnesium oxide. *Fuel*, 2024. 357: p. 129725.
59. Hanif, A., et al., Silica supported MgO as an adsorbent for precombustion CO_2 capture. *ACS Applied Nano Materials*, 2019. 2(10): p. 6565–6574.
60. Gao, W., et al., Molten salt-promoted MgO adsorbents for CO_2 capture: Transient kinetic studies. *Environmental Science & Technology*, 2021. 55(8): p. 4513–4521.
61. Gong, F., et al., MgO promoted by Fe2O3 and nitrate molten salt for fast and enhanced CO2 capture: Experimental and DFT investigation. *Separation and Purification Technology*, 2023. 307: p. 122766.
62. Liu, W., et al., Performance enhancement of calcium oxide sorbents for cyclic CO_2 capture A review. *Energy & Fuels*, 2012. 26(5): p. 2751–2767.
63. Sun, H., et al., Progress in the development and application of CaO-based adsorbents for CO_2 capture—a review. *Materials Today Sustainability*, 2018. 1: p. 1–27.
64. Zhang, L., et al., The role of water on the performance of calcium oxide-based sorbents for carbon dioxide capture: A review. *Energy Technology*, 2015. 3(1): p. 10–19.
65. Yasipourtehrani, S., et al., Development of robust CaO-based sorbents from blast furnace slag for calcium looping CO_2 capture. *Chemical Engineering Journal*, 2020. 387: p. 124140.
66. Sen, R., et al., Hydroxide based integrated CO_2 capture from air and conversion to methanol. *Journal of the American Chemical Society*, 2020. 142(10): p. 4544–4549.
67. Helmi, M., et al., Synthesis, characterization and performance evaluation of NaOH@ Chitosan-Fe3O4 as an adsorbent for CO_2 capture. *Fuel*, 2023. 338: p. 127300.
68. Khajeh, M. and A. Ghaemi, Exploiting response surface methodology for experimental modeling and optimization of CO_2 adsorption onto NaOH-modified nanoclay montmorillonite. *Journal of Environmental Chemical Engineering*, 2020. 8(2): p. 103663.
69. Mahmoudkhani, M. and D.W. Keith, Low-energy sodium hydroxide recovery for CO_2 capture from atmospheric air—thermodynamic analysis. *International Journal of Greenhouse Gas Control*, 2009. 3(4): p. 376–384.
70. Stolaroff, J.K., D.W. Keith, and G.V. Lowry, Carbon dioxide capture from atmospheric air using sodium hydroxide spray. *Environmental Science & Technology*, 2008. 42(8): p. 2728–2735.
71. Yoo, M., S.-J. Han, and J.-H. Wee, Carbon dioxide capture capacity of sodium hydroxide aqueous solution. *Journal of Environmental Management*, 2013. 114: p. 512–519.

72. Kar, S., et al., A carbon-neutral CO_2 capture, conversion, and utilization cycle with low-temperature regeneration of sodium hydroxide. *Journal of the American Chemical Society*, 2018. 140(49): p. 16873–16876.
73. Ghaemi, A. and A.H. Behroozi, Comparison of hydroxide-based adsorbents of Mg (OH) 2 and Ca (OH) 2 for CO_2 capture: Utilization of response surface methodology, kinetic, and isotherm modeling. *Greenhouse Gases: Science and Technology*, 2020. 10(5): p. 948–964.
74. Han, S.-J., et al., Carbon dioxide capture using calcium hydroxide aqueous solution as the absorbent. *Energy & Fuels*, 2011. 25(8): p. 3825–3834.
75. Ahmadi, M., A. Ghaemi, and M. Qasemnazhand, Lithium hydroxide as a high capacity adsorbent for CO_2 capture: Experimental, modeling and DFT simulation. *Scientific Reports*, 2023. 13(1): p. 7150.
76. Costagliola, M.A., M.V. Prati, and G. Perretta, Post combustion CO_2 capture with calcium and lithium hydroxide. *Scientific Reports*, 2022. 12(1): p. 10518.
77. Mourad, A.A.H., et al., Effects of potassium hydroxide and aluminum oxide on the performance of a modified solvay process for CO_2 capture: A comparative study. *International Journal of Energy Research*, 2021. 45(9): p. 13952–13964.
78. Mourad, A.A., et al., A process for CO_2 capture and brine salinity reduction through reaction with potassium hydroxide: A multi-stage evaluation. *Journal of Natural Gas Science and Engineering*, 2022. 106: p. 104756.
79. Rastegar, Z. and A. Ghaemi, CO_2 absorption into potassium hydroxide aqueous solution: Experimental and modeling. *Heat and Mass Transfer*, 2022. 58(3): p. 365–381.
80. Yang, F., et al., Polyvinyl chloride (PVC) derived microporous carbons prepared via hydrothermal dechlorination and potassium hydroxide activation for efficient CO_2 capture. *Renewable and Sustainable Energy Reviews*, 2023. 180: p. 113279.
81. Cho, M.S., et al., Characterization of new potassium-based solid sorbents prepared using metal silicates for post-combustion CO_2 capture. *Process Safety and Environmental Protection*, 2018. 117: p. 296–306.
82. Kwon, Y.M., et al., Regenerable sodium-based lithium silicate sorbents with a new mechanism for CO_2 capture at high temperature. *Renewable Energy*, 2019. 144: p. 180–187.
83. Zhang, Y., et al., Fabrication of lithium silicates from zeolite for CO_2 capture at high temperatures. *Journal of energy chemistry*, 2019. 33: p. 81–89.
84. Millward, A.R. and O.M. Yaghi, Metal– organic frameworks with exceptionally high capacity for storage of carbon dioxide at room temperature. *Journal of the American Chemical Society*, 2005. 127(51): p. 17998–17999.
85. Kazemi, A., et al., Polydopamine-coated Zn-MOF-74 nanocarriers: Versatile drug delivery systems with enhanced biocompatibility and cancer therapeutic efficacy. *Journal of Inorganic and Organometallic Polymers and Materials*, 2024: p. 1–14.
86. Sohrabi, H., et al., Metal-organic frameworks (MOF)-based sensors for detection of toxic gases: A review of current status and future prospects. *Materials Chemistry and Physics*, 2023. 299: p. 127512.
87. Wu, W., et al., Metal-organic framework (MOF)-based sensors for exogenous contaminants in food: Mechanisms, advances, and prospects. *Trends in Food Science & Technology*, 2023. 138: p. 238–271.
88. Kazemi, A., et al., Tunable Zn-MOF-74 nanocarriers coated with sodium alginate as versatile drug carriers. *Polymer Bulletin*, 2024: p. 1–17.
89. Rabiee, N., Sustainable metal-organic frameworks (MOFs) for drug delivery systems. *Materials Today Communications*, 2023. 35: p. 106244.
90. Konnerth, H., et al., Metal-organic framework (MOF)-derived catalysts for fine chemical production. *Coordination Chemistry Reviews*, 2020. 416: p. 213319.
91. Rao, R., et al., Recent advances of metal-organic framework-based and derivative materials in the heterogeneous catalytic removal of volatile organic compounds. *Journal of Colloid and Interface Science*, 2023. 636: p. 55–72.

92. Kazemi, A., et al., Unveiling the power of defect engineering in MOF-808 to enhance efficient carbon dioxide adsorption and separation by harnessing the potential of DFT analysis. *Chemical Engineering Journal*, 2024. 494: p. 153049.
93. Yang, R.T., *Adsorbents: Fundamentals and applications*. 2003, John Wiley & Sons.
94. Li, J.-R., R.J. Kuppler, and H.-C. Zhou, Selective gas adsorption and separation in metal–organic frameworks. *Chemical Society Reviews*, 2009. 38(5): p. 1477–1504.
95. Smit, B. and T.L. Maesen, Molecular simulations of zeolites: adsorption, diffusion, and shape selectivity. *Chemical Reviews*, 2008. 108(10): p. 4125–4184.
96. Sumida, K., et al., Carbon dioxide capture in metal–organic frameworks. *Chemical reviews*, 2012. 112(2): p. 724–781.
97. Liu, J., et al., Progress in adsorption-based CO_2 capture by metal–organic frameworks. *Chemical Society Reviews*, 2012. 41(6): p. 2308–2322.
98. Samira, S. and A. Mansoor, High CO_2 adsorption capacity and CO2/CH4 selectivity by nanocomposites of MOF–199. *Energy & Fuels*, 2017. 31(5): p. 5376–5384.
99. Li, Z., et al., Constructing multiple sites of metal-organic frameworks for efficient adsorption and selective separation of CO2. *Separation and Purification Technology*, 2023. 307: p. 122725.
100. Singh, M. and S. Neogi, Selective and multicyclic CO_2 adsorption with visible light-driven photodegradation of organic dyes in a robust metal–organic framework embracing heteroatom-affixed pores. *Inorganic Chemistry*, 2022. 61(28): p. 10731–10742.
101. Zhang, R., et al., Three metal–organic framework isomers of different pore sizes for selective CO_2 adsorption and isomerization studies. *Dalton Transactions*, 2020. 49(17): p. 5618–5624.
102. Yaghi, O.M. and A.R. Millward, *Metal-organic frameworks with exceptionally high capacity for storage of carbon dioxide at room-temperature*. 2010, Google Patents.
103. Zeman, F., Energy and material balance of CO_2 capture from ambient air. *Environmental Science & Technology*, 2007. 41(21): p. 7558–7563.
104. Qasem, N.A., R. Ben-Mansour, and M.A. Habib, An efficient CO_2 adsorptive storage using MOF-5 and MOF-177. *Applied energy*, 2018. 210: p. 317–326.
105. Gaikwad, S., et al., Enhanced CO_2 capture capacity of amine-functionalized MOF-177 metal organic framework. *Journal of Environmental Chemical Engineering*, 2021. 9(4): p. 105523.
106. Chen, X.-M. and M.-L. Tong, Solvothermal in situ metal/ligand reactions: A new bridge between coordination chemistry and organic synthetic chemistry. *Accounts of Chemical Research*, 2007. 40(2): p. 162–170.
107. Park, J., J.Y. Koo, and H.C. Choi, Solvent-effected coordination variation of flexible ligands to Cu (II) for the formation of 1D and 2D secondary building units for metal–organic frameworks. *Inorganic Chemistry*, 2021. 60(7): p. 5376–5382.
108. Sud, D. and G. Kaur, A comprehensive review on synthetic approaches for metal-organic frameworks: From traditional solvothermal to greener protocols. *Polyhedron*, 2021. 193: p. 114897.
109. Mahreni, M. and Y. Ristiyaningsih. *Review on metal-organic framework (MOF): Synthesis and solid catalyst applications*. 2020, LPPM UPN "Veteran" Yogyakarta Conference Series Proceeding on Engineering . . . .
110. Al Obeidli, A., et al., Recent advancements in MOFs synthesis and their green applications. *International Journal of Hydrogen Energy*, 2022. 47(4): p. 2561–2593.
111. Lee, Y.-R., J. Kim, and W.-S. Ahn, Synthesis of metal-organic frameworks: A mini review. *Korean Journal of Chemical Engineering*, 2013. 30: p. 1667–1680.
112. Sharanyakanth, P. and M. Radhakrishnan, Synthesis of metal-organic frameworks (MOFs) and its application in food packaging: A critical review. *Trends in Food Science & Technology*, 2020. 104: p. 102–116.
113. Kazemi Korayem, A., S. Ghamami, and Z. Bahrami, Fractal properties and morphological investigation of nano-amiodarone using image processing. *Signal, Image and Video Processing*, 2019. 13: p. 281–287.

114. Kang, Y., et al., Controlled architectures in supported photocatalysis: Exploring the potential of cold plasma discharge, additive manufacturing, and fractal geometry to form hierarchically immobilized Ni-MOF/BiOI/AgVO3. *ACS Applied Materials & Interfaces*, 2023. 15(26): p. 31849–31866.
115. Kazemi Korayem, A., S. Ghamami, and Z. Bahrami, Fractal properties and morphological investigation of Nano hydrochlorothiazide is used to treat hypertension. *BMC Pharmacology and Toxicology*, 2018. 19: p. 1–9.
116. Liu, Y., J. Liu, and Y. Lin, Strong binding site molarity of MOFs and its effect on CO_2 adsorption. *Microporous and Mesoporous Materials*, 2015. 214: p. 242–245.
117. Nandi, S., et al., Preferential adsorption of CO_2 in an ultramicroporous MOF with cavities lined by basic groups and open-metal sites. *Inorganic Chemistry*, 2018. 57(9): p. 5267–5272.
118. Chowdhury, P., et al., Adsorption of CO, CO_2 and CH4 on Cu-BTC and MIL-101 metal organic frameworks: Effect of open metal sites and adsorbate polarity. *Microporous and Mesoporous Materials*, 2012. 152: p. 246–252.
119. Jo, H., et al., Fine-tuning of the carbon dioxide capture capability of diamine-grafted metal–organic framework adsorbents through amine functionalization. *ChemSusChem*, 2017. 10(3): p. 541–550.
120. Lawson, S., et al., Amine-functionalized MIL-101 monoliths for CO_2 removal from enclosed environments. *Energy & Fuels*, 2019. 33(3): p. 2399–2407.
121. Shin, S., et al., Polyvinylamine-loaded metal–organic framework MIL-101 for effective and selective CO_2 adsorption under atmospheric or lower pressure. *Chemical Engineering Journal*, 2020. 389: p. 123429.
122. Kazemi, S. and V. Safarifard, Carbon dioxide capture in MOFs: The effect of ligand functionalization. *Polyhedron*, 2018. 154: p. 236–251.
123. Gupta, V. and S.K. Mandal, A highly stable triazole-functionalized metal–organic framework integrated with exposed metal sites for selective CO_2 capture and conversion. *Chemistry–A European Journal*, 2020. 26(12): p. 2658–2665.
124. Song, Y., et al., Metal–organic frameworks constructed from mixed infinite inorganic units and adenine. *CrystEngComm*, 2014. 16(15): p. 3082–3085.
125. Taddei, M., et al., Synthesis, breathing, and gas sorption study of the first isoreticular mixed-linker phosphonate based metal–organic frameworks. *Chemical Communications*, 2013. 49(13): p. 1315–1317.
126. Chen, D.-M., et al., Functionalization of metal–organic framework via mixed-ligand strategy for selective CO_2 sorption at ambient conditions. *Crystal Growth & Design*, 2015. 15(2): p. 961–965.
127. Park, J.M., et al., Amine and fluorine co-functionalized MIL-101 (Cr) synthesized via a mixed-ligand strategy for CO_2 capture under humid conditions. *Chemical Engineering Journal*, 2022. 444: p. 136476.
128. Al-Janabi, N., et al., A facile post-synthetic modification method to improve hydrothermal stability and CO_2 selectivity of CuBTC metal–organic framework. *Industrial & Engineering Chemistry Research*, 2016. 55(29): p. 7941–7949.
129. He, W., et al., Post-synthesis modification of metal–organic frameworks: Synthesis, characteristics, and applications. *Journal of Materials Chemistry A*, 2023. 11(45): p. 24519–24550.
130. Lin, Y., et al., Polyethyleneimine incorporated metal-organic frameworks adsorbent for highly selective CO_2 capture. *Scientific Reports*, 2013. 3(1): p. 1859.
131. Kim, M., et al., Postsynthetic ligand and cation exchange in robust metal–organic frameworks. *Journal of the American Chemical Society*, 2012. 134(43): p. 18082–18088.

9 Innovations and Emerging Trends in CO_2 Adsorbents

Fatemeh Bahmanzadegan and Ahad Ghaemi

9.1 INTRODUCTION

In response to the pressing global issue of climate change, significant efforts have been directed toward developing effective strategies for mitigating carbon dioxide (CO_2) emissions [1]. This chapter delves into recent advancements and emerging paradigms in CO_2 adsorbents, shedding light on breakthroughs in materials science, process engineering, and innovative methodologies poised to reshape CO_2 capture technologies. At the core of this exploration lies the concept of tailored sorbent materials meticulously engineered to exhibit heightened affinity for CO_2 molecules. This tailored approach represents a paradigm shift in the pursuit of efficient CO_2 capture, aiming to maximize adsorption efficiency while minimizing energy input and operational costs. Among these strategies, adsorption has emerged as a promising avenue for capturing CO_2 from industrial emissions due to its versatility and efficiency [2]. In this context, modifying adsorbents is crucial in improving their performance and selectivity for CO_2 capture [3]. The modification of adsorbents' structure is fundamental for developing materials that are optimized for CO_2 capture [4]. Researchers aim to enhance the adsorption capacity and kinetics of adsorbents by altering their porosity and surface characteristics. Additionally, exploring novel materials such as metal-organic frameworks (MOFs) [5, 6], zeolites [7], and activated carbons allows for tailoring pore sizes and surface functionalities to optimize CO_2 adsorption. Chemical modification involves functionalizing adsorbent surfaces with specific chemical groups to enhance their affinity for CO_2 [8]. This strategy typically involves attaching amine groups or other functional ligands to the adsorbent matrix, facilitating stronger interactions with CO_2 molecules and improving selectivity. Moreover, surface modification techniques like plasma treatment and chemical vapor deposition offer precise control over surface properties, thereby promoting enhanced CO_2 uptake and stability. Thermal modification presents another avenue for enhancing adsorbent performance through controlled heat treatment processes. By subjecting adsorbents to high temperatures, structural rearrangements and surface activation can be induced, leading to improved CO_2 adsorption capacities and thermal stability. Additionally, thermal treatments facilitate the regeneration of spent adsorbents, prolonging their operational lifespan and reducing overall process costs. Hybrid approaches, which integrate multiple modification techniques, have gained

DOI: 10.1201/9781003476191-9

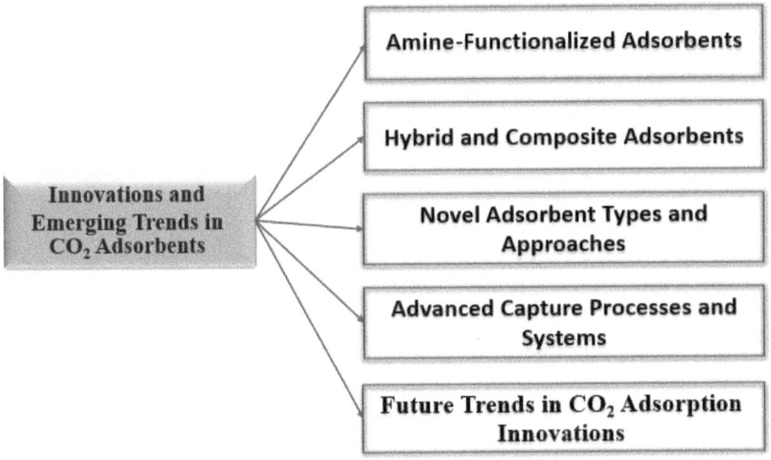

FIGURE 9.1 Innovation in CO_2 adsorbents.

attention for their synergistic effects on enhancing CO_2 capture performance [9]. By combining structural, chemical, and thermal modifications, researchers aim to design adsorbents with tailored properties optimized for CO_2 adsorption under various conditions. These hybrid materials offer enhanced adsorption capacities, selectivity, and regeneration capabilities, addressing critical challenges in CO_2 capture processes. Furthermore, repairing structural defects and incorporating actuating mechanisms represent innovative strategies for improving adsorbent performance. Repairing pore defects and optimizing material morphology help minimize diffusion limitations and enhance CO_2 adsorption kinetics. Additionally, integrating actuating elements enables dynamic control over adsorbent properties, facilitating on-demand CO_2 capture and release. Researchers endeavor to optimize parameters such as pore structure, surface area, and chemical composition to fine-tune sorbent performance, with the goal of meeting diverse industrial requirements, from power generation to carbon capture and utilization initiatives. This chapter comprehensively evaluates recent progress in adsorbent modification techniques for CO_2 capture, highlighting key strategies, challenges, and future directions. Through a systematic analysis of structural, chemical, thermal, and hybrid modification approaches, this chapter aims to provide valuable insights for designing and optimizing next-generation adsorbents for sustainable CO_2 capture applications. Innovation and emerging adsorbents discussed in this chapter are schematically illustrated in Figure 9.1.

9.2 AMINO-FUNCTIONALIZED ADSORBENTS

The adsorbent surface becomes more effective in adsorbing CO_2 when amine groups are present on it [10]. This method can activate sites, increase surface area, reduce pore size, enhance porosity volume, and improve stability and regeneration properties. Monoethanolamine (MEA) [11, 12], tetraethylenepentamine (TEPA) [13, 14], and diethanolamine (DEA) [15] have been used in this process. Eqs. (9.1) and (9.2) illustrate the reaction between CO_2 and primary and secondary amines, respectively.

$$CO_2 + 2RNH_2 \Leftrightarrow RNH_3^+ + RNHCOO^- \tag{9-1}$$

$$CO_2 + 2R_1R_2NH \Leftrightarrow R_1R_2NH_2^+ + R_1R_2NCOO^- \tag{9-2}$$

The TEPA molecule has two functional groups: primary amine (R_1NH_2) and secondary amine (R_1R_2NH). Both of these functional groups can react with CO_2 to produce a carbamate ion. In contrast, the MEA and isopropanol amine (IPA) molecules have only one functional group (–NH_2), which is a primary amine and responsible for the chemical reaction with CO_2 [16]. The process entails the establishment of covalent bonds between the acidic CO_2 molecules and the active surface sites of the adsorbents [17]. There are two approaches to incorporating amine onto zeolite: impregnation and grafting [18]. Amines that are mostly utilized in CO_2 capture are explained in Table 9.1.

Amino-functionalized adsorbents (Figure 9.2), such as amine-modified silica or polymer resins, exhibit strong interactions with CO_2 molecules through chemisorption mechanisms. The presence of amine groups facilitates the reversible formation of stable carbamate species, enabling efficient CO_2 capture under ambient or mild conditions. Recent advancements in ligand design and immobilization techniques have led to the development of highly stable and regenerable amino-functionalized adsorbents with improved selectivity and contaminant resistance. Amino-functionalized CO_2 adsorbents have been developed to enhance CO_2 uptake capability and selectivity.

Wu et al. immobilized an amino-functionalized ionic liquid (IL) on a chromatographic column filler to create a novel supported IL with high CO_2 adsorption capacity and stability. The supported IL with 60 wt% IL content had the best adsorption performance at 40°C. The adsorption capacity of the supported IL was 1.29 mmol/g at 0.1 MPa, six times greater than that of the pure carrier [20]. Yin et al. have developed a method to enhance CO_2 capture and generate cyclic carbonate more efficiently. They achieved this by chemically bonding an amino-functionalized imidazole IL onto the channel walls of mesoporous covalent organic framework materials. The resulting material, named [AeImBr]83%-TAPT-COF, has a CO_2 capture capacity of 117.4 mg/g and a yield of 99.1% for cyclochloroallyl carbonate [21].

Tan et al. investigated TEPA as an adsorbent promoter in CO_2 capture (Figure 9.3), highlighting its advantages, including high amine density, low viscosity, low cost, and low toxicity properties [13].

Chabon et al. have explored the potential of amino-functionalized graphene as a material for capturing CO_2. They used density functional theory to investigate the capturing ability of extensively amino-functionalized graphene. The study found that amino-functionalized graphene can chemically capture CO_2 by forming carbamic acid. Each benzene ring can accommodate up to six amino groups [22]. Ramenzanipour et al. modified nanoclay montmorillonite with DEA, and the maximum CO_2 adsorption capacity was 219.9 mg/g [23]. Kazemi et al. conducted a study investigating the CO_2 adsorption capabilities of sonochemically synthesized UiO-66-NH_2, comparing it with its solvothermal counterpart. They observed a significant improvement in CO_2

TABLE 9.1
Types of Amines Utilized in CO_2 Capture

Amine Type	Abbreviation	Structure
Monoethanolamine	MEA	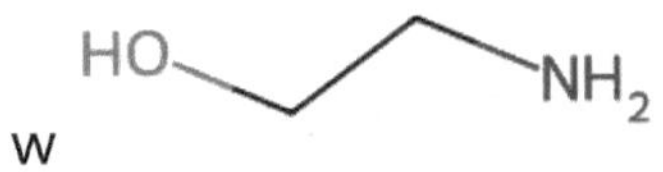
Diethanolamine	DEA	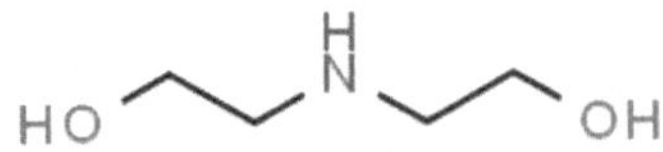
Ethylenediamine	EDA	
Diethylenetriamine	DETA	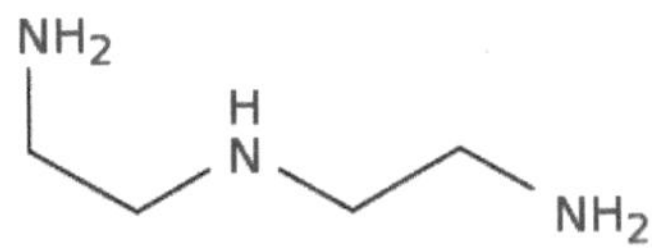
Tetraethylenepentamine	TEPA	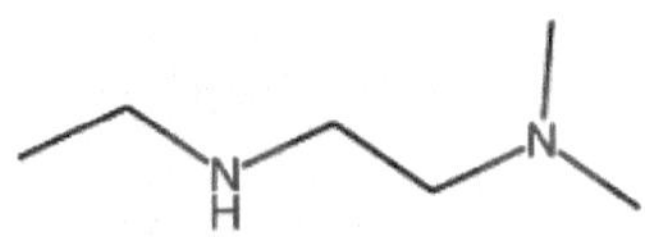
Polyethyleneimine	PEI	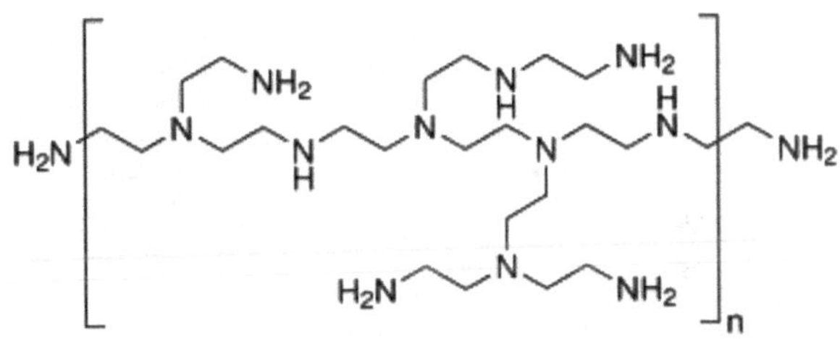
(3-Aminopropyl) trimethoxysilane	APTMS	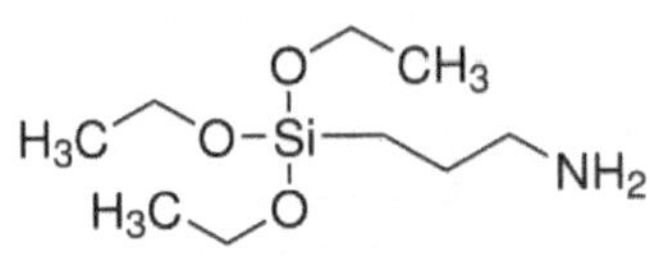

adsorption with sonochemical synthesis, attributed to the material's increased surface area and smaller particle sizes. Isotherm modeling revealed the Langmuir isotherm as the best fit, indicating monolayer adsorption behavior and emphasizing chemisorption phenomena. Kinetic modeling using the Elovich model further highlighted the importance of chemisorption. Isosteric heat analysis revealed significant energies associated with the interaction of CO_2 molecules with the adsorbent surface, supporting the occurrence of chemical reactions with amine groups [5]. Khoshraftar et al. explored the capability of halloysite nanotubes to serve as a natural adsorbent for CO_2. They prepared sorbents by impregnating TEPA onto halloysite nanotube support at varying

FIGURE 9.2 Amino-functionalized CO_2 adsorbents [19], copyright (2021), Elsevier.

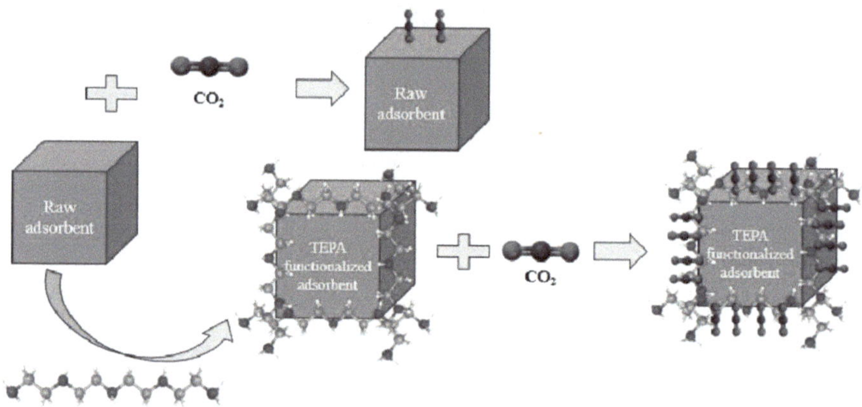

FIGURE 9.3 Proposed mechanism of CO_2 adsorption over a raw adsorbent and a TEPA-functionalized adsorbent [13], copyright ©2022.

concentrations (10% to 40% by weight). Using artificial neural network modeling and response surface methodology (RSM), they analyzed the effects of parameters on CO_2 adsorption. Optimization via the surface response method yielded optimal operating conditions. The adsorption capacity was evaluated across temperatures (20°C to 50°C) and pressures (1 to 9 bar) using TEPA-functionalized halloysite nanotubes. Their artificial neural network models, MLP and RBF, predicted the adsorption process. Response surfaces, generated using central composite design, demonstrated strong agreement between experimental and simulated CO_2 capture results. The study achieved a maximum CO_2 adsorption capacity of 9.3041 mmol/g at 9 bar pressure and 20°C temperature [24]. Girimonte et al. examined the capability of mesoporous silica functionalized with 20% APTES concentration which exhibited high efficacy for CO_2 capture [25]. The study conducted by Gu et al. found that the TEPA-SG-20 adsorbent with 20 wt% TEPA loaded on a silica gel support had the highest CO_2 adsorption capacity of 1.90 mmol/g. Furthermore, the TEPA-SG-20 adsorbent was found to be efficiently regenerable with a regeneration efficacy of 97%. It exhibits fast adsorption kinetics, good regeneration efficacy, and working stability. Overall, the TEPA-SG-20 adsorbent shows promise for low-concentration CO_2 removal in environmental control and life support systems [26]. Table 9.2 outlines a comparison of utilizing various amines for modifying zeolite, considering factors such as adsorption capacity and operational conditions.

TABLE 9.2
Comparison of Utilizing Various Amines for Modifying Zeolite

Zeolite	Amine Type	CO_2 Adsorption Capacity(mmol/g)	*P* (bar)	*T* (°C)	Ref
Clinoptilolite	MEA	3.58	5	25	[27]
Clinoptilolite	TEA	5	5	25	[27]
LTA	APTMS	2.3	1	60	[28]
4A	IBA	2.56	1	25	[29]
HZSM-5	AEEA	4.44	1	48	[30]
Zeolite 13X	MEA	0.78	1	30	[11]
NaY	TEPA	2.11	1	75	[31]
Zeolite 13X	Piperazine	4.54	5	25	[32]
Zeolite Y	TEPA	2.6	1	30	[33]
Beta	Melamine	1.8	1	25	[34]
4A	IPA (0.3)	2.3	1	20	[35]
4A	APTMS	1.05	1	20	[35]
4A	TEPA	9.2	5	25	[7]
4A	DEA	9.4	5	25	[7]
Beta	MEA	0.77	1	30	[12]
Y	MEA	1.97	1	25	[16]
Beta-25	APTES	4.7	1	35	[36]
Beta-25	TEPA	2.56	1	35	[36]
Zeolite Y	APTES	5.15	1	35	[37]
Zeolite Y	APMDES	4.36	1	35	[38]
Beta	TEPA	2.07	1	55	[39]
13X	DTEA	1.054	1	75	[40]
13X	TEA	1.69	1	75	[40]
13X	0.2% MEA	3.1	1	25	[41]
13X	50% MEA	1.45	1	25	[41]

9.3 HYBRID AND COMPOSITE ADSORBENTS

Composite adsorbents involve the incorporation of functional additives or modifiers into a porous matrix to augment CO_2 adsorption performance. For instance, graphene-based composites have garnered attention due to their exceptional mechanical strength and high surface area. By incorporating nanoparticles or metal nanoparticles into graphene matrices, researchers have demonstrated significant enhancements in CO_2 adsorption capacity and regeneration stability. Composite adsorbents present a versatile platform for tailoring the physicochemical properties of adsorbents to meet specific application requirements.

9.3.1 Carbon-Based Hybrid Adsorbents

Carbon-based hybrid adsorbents are innovative materials utilized for CO_2 adsorption due to their exceptional properties (Figure 9.4). These adsorbents combine the advantages of carbonaceous materials, such as high surface area and tunable surface chemistry, with other components like MOFs or metal oxides [42]. The synergistic effects of these combinations enhance CO_2 capture efficiency by providing additional active sites and improving selectivity. Carbon-based hybrid adsorbents offer promise in addressing the challenge of CO_2 capture from various emission sources, including

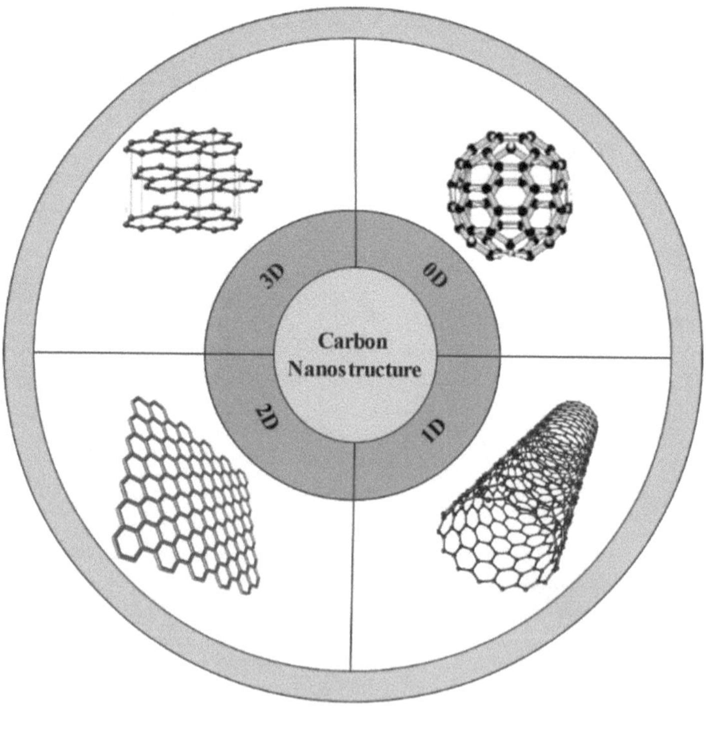

FIGURE 9.4 Types of carbon materials adsorbent for CO_2.

industrial processes and power plants, contributing to mitigating climate change [43]. For instance, graphene, a single layer of carbon atoms arranged in a two-dimensional honeycomb lattice, exhibits exceptional mechanical strength, high thermal conductivity, and large surface area. Functionalized graphene sheets or graphene oxide (GO) have been explored for CO_2 capture due to their high surface area and the potential for chemical modification to enhance CO_2 adsorption capacity. Studies have shown that oxygen-containing functional groups on graphene surfaces can significantly improve CO_2 adsorption properties [44]. Fathalian et al. developed a novel composite of CTS/GO/ZnO to enhance CO_2 adsorption efficiency. Employing a design of experiment methodology utilizing the RSM-BBD technique, they synthesized samples with diverse loadings of GO and zinc oxide (ZnO) nanoparticles. They constructed a quadratic model to forecast adsorption capacity within the design parameters, identifying optimized GO, ZnO, temperature, and pressure values, achieving a CO_2 uptake capacity of 470.43 mg/g, the maximum observed in their study [45]. Ghaemi et al. used NiO and MgO to modify activated carbon for enhanced CO_2 adsorption capacity. The optimum adsorption capacities were obtained at 121.35 and 105.17 mg/g for AC/NiO-3 and AC/MgO-3, respectively [46]. Helmi et al. focused on developing and utilizing Persian gum (Pg)@GO as a novel adsorbent for capturing CO_2. They optimized operational parameters, including temperature, pressure, and adsorbent weight, employing RSM coupled with Box-Behnken design (BBD). The study found that the highest CO_2 adsorption capacity of 4.80 mmol/g was achieved under 300 K and 7.8 bar pressure. Various isotherm and kinetic models were utilized, yielding correlation coefficients (R^2) of 0.955 and 0.986, respectively, indicating heterogeneous CO_2 adsorption behavior. Thermodynamic assessment indicated that the CO_2 adsorption process is both exothermic and spontaneous, with a ΔG° value of −8.169 at 300 K [9]. Khoshraftar et al. conducted a study wherein chemical vapor deposition was employed to produce multi-walled carbon nanotubes (MWCNTs), subsequently modified by Fe–Ni/AC catalysts to augment CO_2 adsorption. While the surface area of MWCNT was initially measured at 240 m^2/g, modification resulted in a reduction to 11 m^2/g. Nevertheless, the modified MWCNT exhibited heightened adsorption capacity due to introducing new adsorption sites and favorable interactions at lower temperatures. Notably, at 25°C and 10 bar, it demonstrated a maximum adsorption capacity of 424.08 mg/g [47].

Pełech et al. conducted a comprehensive investigation into the characteristics of composites comprising carbon spheres in conjunction with either titanium dioxide or zinc oxide. Their analysis encompassed thermal oxidative assessment and Fourier-transform infrared spectroscopy (FTIR), affirming the presence of diverse carbon functional groups within the samples, primarily including unsaturated C=C from amorphous carbon and C–O(H), alongside additional groups in the form of C–H. The study elucidated that those heightened levels of spherical carbon within the composites expanded the combustion temperature range while augmenting metal oxide content constricted this range. Structural examinations revealed elevated total pore volume in materials containing titanium dioxide, attributed to a more significant contribution of mesopores; however, this phenomenon did not significantly influence carbon dioxide adsorption. Conversely, a reduction in CO_2 adsorption was observed in CS/TiO_2 materials featuring pronounced declines in mesopore volume. CS/ZnO samples displayed marginally diminished CO_2 adsorption compared to CS/TiO_2 due to a slightly lower micropore content. Nevertheless, beyond a carbon content threshold of 1 g, both composite types exhibited uniform CO_2 adsorption regardless of oxide variation. The integration of heightened carbon content facilitated ultra-microporosity development, with pores measuring below 0.6 nm demonstrating efficacy in CO_2 adsorption. Furthermore, the synthesized composite materials showcased commendable stability and regenerability across ten consecutive CO_2 adsorption–desorption cycles. Notably, the CO_2/N_2 selectivity analysis, a novel endeavor for spherical carbon-based composites, underscored a pronounced preference for CO_2 to interact with carbon pores over N_2 [48]. Helmi et al. investigated the incorporation of Fe_3O_4 nanoparticles onto GO layers to preserve the surface area of GO while maintaining its nanostructure. To enhance the alkaline nature of GO-Fe_3O_4, they employed an impregnation method with a 0.1 N KOH solution. Brunauer-Emmett-Teller analysis revealed a mesoporous structure with a pore diameter of 15.05 nm in the resulting adsorbent. The efficacy of KOH@GO-Fe_3O_4 for CO_2 adsorption was evaluated over a range of temperatures from 298 to 338 K and pressures from 1 to 9 bar. The highest CO_2 adsorption capacity of 6.812 mmol/g was achieved using 0.5 g of adsorbent at 298 K and 9 bar pressures. Thermodynamic analysis indicated that CO_2 capture reached its maximum capacity at lower temperatures compared to other GO-based adsorbents, with confirmation of exothermic and physical CO_2 adsorption. Additionally, thermogravimetric analysis and regeneration studies demonstrated the high thermal stability and regenerability of KOH@GO-Fe_3O_4 [49]. Figure 9.5 illustrates the preparation process of nitrogen-doped cellulose-amine-activated porous carbons. The diagram outlines the steps involved, starting from the selection of cellulose as the precursor material. It includes the introduction of nitrogen doping through chemical treatments with amine groups, followed by activation procedures that create a porous structure. The resulting material exhibits enhanced properties for CO_2 adsorption due to the combined effects of nitrogen doping and increased porosity [50].

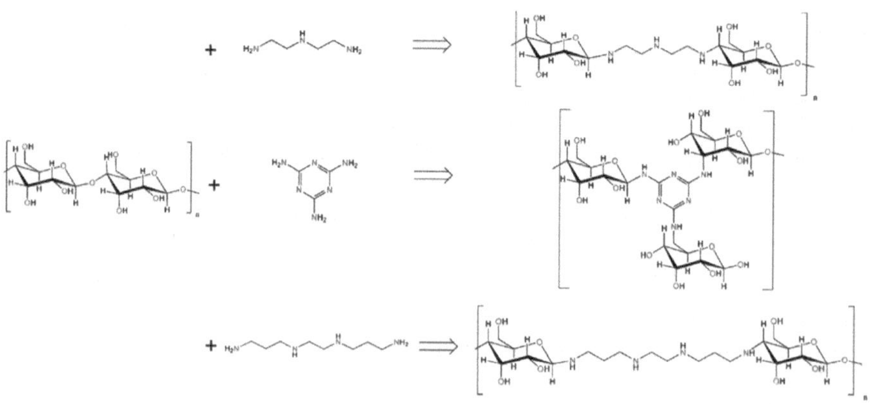

FIGURE 9.5 Schematic for preparation of nitrogen-doped cellulose–amine activated porous carbons [50], copyright (2024), MDPI.

TABLE 9.3
Compression of CO_2 Adsorption Capacity of Carbon-Based Materials

References	Adsorption Capacity (mmol/g)	T (k)	P (bar)	BET Surface Area, m²/g	Material
[51]	7.19	273	1	—	Black locust-KOH
[52]	5.60	273	1	—	Coconut shell-CO_2
[53]	5.50	273	1	—	Sugarcane bagasse-LiOH
[54]	5.31	303	0.05	656	AC/zeolite 13X
[55]	1.14	298	1	182	AC/Na-A zeolite
[56]	12.24	298	32	1115	Pd-AC/MOF-74(Ni)
[56]	11.42	298	32	1088	Pd-AC/MOF-74(Co)
[57]	2.26	303	1	599.41	AC/CueZn
[58]	2.08	298	1	413	AC nanofibers/MgO
[59]	6.78	303	1	1954	AC/CuO
[59]	6.48	303	1	1954	AC/NiO
[60]	1.14	473	3	441	AC/hydrotalcite
[61]	3.578	298	1	79.83	AC/phenolic resin/carbon fibers

Table 9.3 provides a comparative analysis of the CO_2 adsorption capacities of various carbon-based materials.

9.3.2 Zeolite-Based Hybrid Adsorbents

Tailored hybrid or composite materials based on zeolites for CO_2 adsorption offer a promising method for efficient carbon capture technologies. Zeolites, known for their well-defined pore structures and high surface areas, provide an excellent platform for designing materials that selectively capture CO_2 from flue gases and industrial emissions. These tailored hybrid or composite materials enhance CO_2 adsorption capacity, selectivity, and stability through precise engineering and incorporation of complementary components. One critical approach involves customizing the properties of zeolites through the creation of hybrid materials by combining them with other functional additives. Here are some strategies used to tailor hybrid or composite materials based on zeolites for CO_2 adsorption:

- **Modification of Zeolite Surface Chemistry**
  Zeolite surfaces can be modified by functionalizing specific chemical groups to enhance their affinity for CO_2 molecules. Functionalization alters the surface properties of zeolites, creating additional adsorption sites and enhancing their interaction with CO_2, thereby augmenting adsorption capacity and selectivity. Kongnoo et al. explore the CO_2 adsorption capacity of zeolite 13X derived from fly ash sourced from both inactive and activated plants, as depicted in Figure 9.6. Their investigation focuses on an acidic setting.

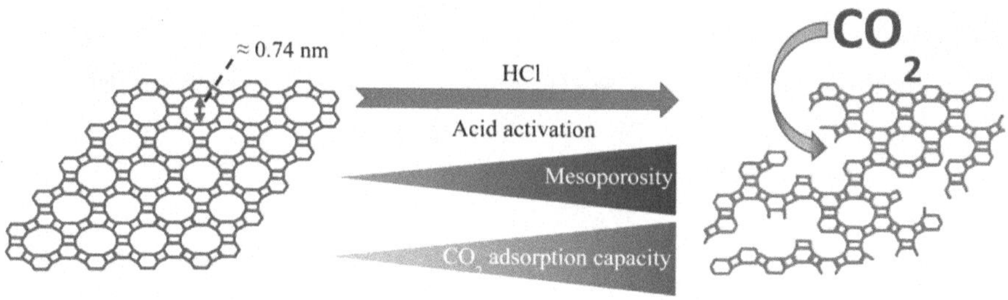

FIGURE 9.6 Surface modification process of zeolite employing acidic treatment [62], copyright (2017), Elsevier.

Laboratory experiments revealed that exposing zeolite 13X to an acidic environment led to a rise in CO_2 adsorption, reaching 6.42 mmol/g [62].

- **Incorporation of Metal Oxides or Organic Polymers**
  Zeolites can be combined with metal oxides or organic polymers to form hybrid materials with synergistic properties for CO_2 capture. Metal oxides provide additional adsorption sites and enhance the stability of zeolites, while organic polymers can improve the mechanical strength and flexibility of the composite material.
- **Hybridization with Activated Carbon**
  Combining zeolites with activated carbon creates hybrid materials with complementary pore structures and enhanced CO_2 adsorption capacities. Activated carbon presents significant surface area and pore volume, while zeolites provide selective adsorption sites, improving overall CO_2 capture performance [63].
- **Design of Hierarchical Structures**
  Tailored hybrid materials based on zeolites can be designed with hierarchical pore structures, combining micropores, mesopores, and macropores to maximize CO_2 adsorption efficiency. Hierarchical structures offer enhanced mass transfer kinetics and accessibility of adsorption sites, leading to improved CO_2 capture performance.
- **Optimization of Synthesis Parameters**
  Precise control over synthesis parameters, such as temperature, pressure, and composition, is crucial for tailoring the properties of hybrid materials based on zeolites. Optimization of synthesis conditions enables the fabrication of materials with tailored pore structures, surface chemistries, and mechanical properties optimized for CO_2 adsorption.

Qian et al. developed a series of alkali metal cations (Na^+, K^+, Rb^+, Cs^+)-exchanged ERI zeolites for CO_2 separation from gas mixtures. The K-ERI zeolite exhibited superior performance with an adsorption volume of 70.84 cm^3/g and an IAST selectivity of CO_2/CH_4: 6254, CO_2/N_2: 2872 at 298 K, surpassing previously reported metal cation exchanged zeolites. A combination of powder X-ray diffraction crystallography and DFT calculations identified that a small portion of

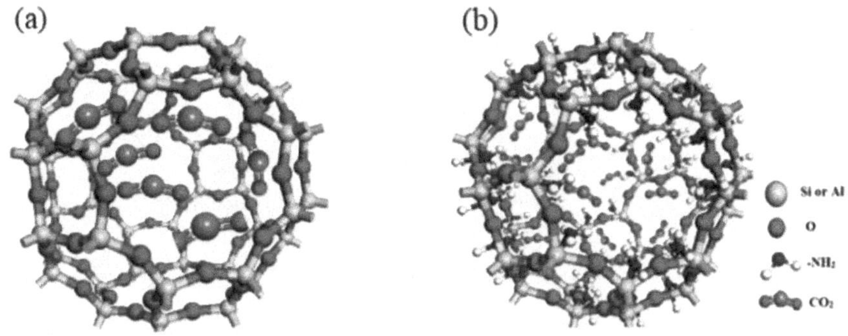

FIGURE 9.7 CO_2 adsorption on (a) unmodified zeolite, and (b) modified zeolite [7], copyright (2023), Nature.

exchanged extra-framework alkali metal cations modified the 8-MR aperture, playing a crucial role in the gas separation process. The adsorption and passage through 8-MR capacities of ion-exchanged zeolites were found to be influenced by the cation-type dominated zeolite openings and cation–gas molecule interactions, with Na-ERI exhibiting strong affinity between the gas molecule and zeolite framework [64]. The study by Bahmanzadegan et al. focused on augmenting the adsorption efficacy of zeolite through alterations involving DEA and TEPA. The results elucidated that the 4A-DEA adsorbent attained a noteworthy optimal adsorption capacity of 579.468 mg/g as shown in Figure 9.7. Furthermore, the investigation delineated operational variables crucial for optimizing adsorption effectiveness, encompassing temperature, pressure, and amine concentration [7].

9.3.3 MOF-Based Hybrid Adsorbents

MOFs are porous materials of metal ions or clusters coordinated to organic ligands, forming highly ordered and tunable structures with large surface areas (Figure 9.8). These properties make MOFs ideal candidates for gas adsorption applications, including CO_2 capture (Figure 9.9). Recently, there has been a lot of interest in hybrid or composite materials based on MOFs for CO_2 adsorption. They have great potential to help with environmental issues related to carbon capture and storage. One prominent strategy involves customizing the composition and structure of hybrid or composite materials by integrating MOFs with other functional components.

This tailored approach aims to synergistically enhance the adsorption performance of MOFs for CO_2 molecules through several vital mechanisms:

- **Fine-tuning Pore Structures**
  MOFs possess intrinsic porosity, which can be further tailored to optimize CO_2 adsorption. By carefully selecting MOF structures and controlling pore sizes, researchers can increase the accessibility of adsorption sites and improve the uptake capacity for CO_2 molecules [65].

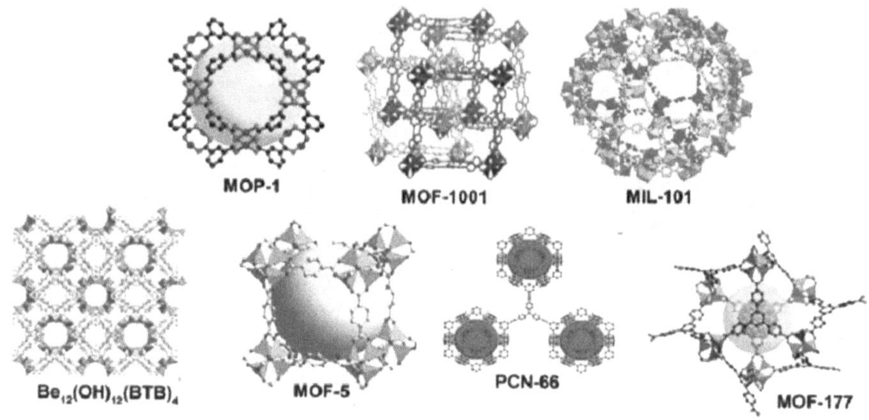

FIGURE 9.8 Types of MOF utilized in CO_2 capture.

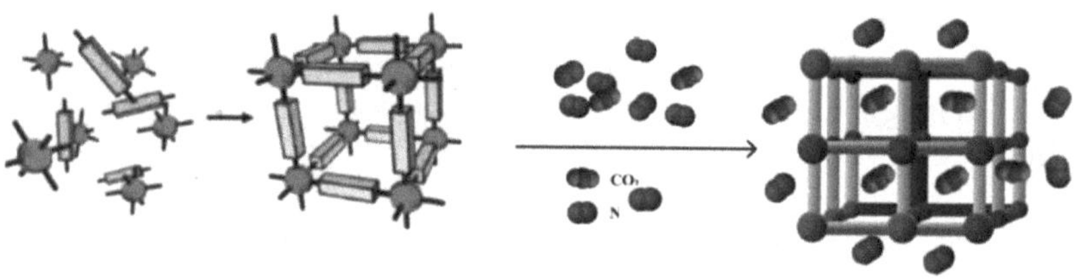

FIGURE 9.9 CO_2 adsorption mechanism on MOF.

- **Functionalization for Enhanced Affinity**
  Functionalizing MOFs with specific chemical groups or metal centers enables the creation of adsorption sites with a high affinity for CO_2. This tailored functionalization enhances the interaction between MOFs and CO_2 molecules, increasing adsorption selectivity and efficiency [65].
- **Hybridization with Synergistic Materials**
  Integrating MOFs with complementary materials, such as activated carbon, polymers, or inorganic matrices, allows for the development of hybrid or composite materials with enhanced adsorption properties. These synergistic combinations leverage the strengths of each component, such as the high surface area of activated carbon or the tunable properties of polymers, to achieve superior CO_2 capture performance [66].

Huang et al. modified a MOF with aminosilane to synthesize a new MOF adsorbent (SAP-MIL-125) with improved CO_2 adsorption capacity [67]. Jo et al. synthesized a cuprous-based ultramicroporous MOF with amino groups, which exhibited high CO_2 adsorption capacity and retention even in moisture [68]. Joodaki et al. conducted a study focusing on CALF-20, a Zn-triazole-based MOF renowned for its potential in

CO_2 capture. However, its performance is hindered by competitive water adsorption, particularly above 40% relative humidity (RH), which limits its applicability in scenarios like post-combustion flue gas where CO_2 is mixed with moisture. To address this, the researchers introduced three novel MOFs, named NU-220, CALF-20M-w, and CALF-20M-e, featuring hydrophobic methyltriazolate linkers. Adding methyl groups aimed to enhance CO_2 uptake in the presence of water. CALF-20M-w and CALF-20M-e exhibited remarkable retention of CO_2 capture efficiency, surpassing 20% at 70% RH, where CALF-20's CO_2 uptake is minimal. Grand canonical Monte Carlo simulations demonstrated that the presence of methyl groups hinders the formation of a water network within the pores of CALF-20M-w and CALF-20M-e, thereby enhancing their selectivity for CO_2 over N_2 under high moisture conditions. Additionally, calculations of radial distribution functions revealed that the insertion of methyl groups into the triazolate linker increases the distance between water molecules and Zn coordination bonds, explaining the enhanced moisture stability observed in CALF-20M-w and CALF-20M-e compared to CALF–20. This design approach offers more robust sorbents capable of efficiently capturing CO_2 in practical industrial environments, presenting a promising solution to the challenge of moisture interference in CO_2 capture processes [69]. Esfahani et al. investigated the synthesis and enhancement of Zr-BTC MOF-808 with NH_2 ligand, aiming to develop a novel mixed ligand MOF-based adsorbent for CO_2 capture. Their study aimed to improve the modified adsorbent's specific surface area and pore volume compared to pure MOF-808 while preserving its mesoporous morphology. The research focused on optimizing three key factors—pressure, temperature, and the amount of loaded NH_2—to maximize CO_2 adsorption capacity. Their investigation unveiled that optimal operating parameters for achieving the highest equilibrium adsorption capacity of 369.11 mg/g were found to be at 25°C, under 9 bar pressure, with a 20 wt% NH_2 loading incorporated into the MOF structure. The investigation also highlighted the direct correlation between adsorption capacity and pressure and NH_2 amount, while temperature exhibited an inverse relationship. Kinetic modelling using the Ritchie second-order model and isotherm modelling with the Sips model provided the best fit for experimental data, demonstrating high accuracy in describing CO_2 adsorption behavior. Furthermore, the thermodynamic analysis indicated an exothermic, spontaneous reaction mechanism with reduced irregularity [6]. Zeolitic imidazolate frameworks (ZIFs) belong to the MOF family and share a structural similarity with zeolites. ZIFs consist of transition metal ions (such as Fe, Co, Cu, and Zn) linked by organic imidazole connectors in tetrahedral coordination. Due to the resemblance of the metal–imidazole–metal angle to the Si–O–Si angle in zeolites (approximately 145°), ZIFs exhibit zeolite-like topologies. ZIFs are known for their robust thermal and moisture resistance, surpassing many other MOF types. Like zeolites and other porous materials, ZIF membranes offer significant potential for gas separation applications. This is due to their highly porous nature, substantial accessible pore volume with fully exposed organic link edges and surfaces, pore sizes matching the kinetic diameter of various gas molecules, and a notable capacity for CO_2 adsorption (Table 9.4). Using MOF for CO_2 adsorption capacity compares MOF-based adsorbent materials for CO_2 adsorption.

TABLE 9.4
Using MOF for CO_2 Adsorption Capacity

Material	Surface Area (m^2/g)	T (K)	P (bar)	q (mmol/g)	Ref.
2D MOF	340.8	298	1	1.9	[70]
Li-HKUST-1	985.8	298	18	7.89	[71]
MOF-177	6240	298	50	35.23	[72]
MOF-210	12	298	80	65.23	[72]
pt-UiO-66(Zr)(OH)2	1230	298	1	5.63	[73]
MIL-101(Cr)/activated carbon (type Maxsorb-III)	2670	300	1	7.14	[74]
PCN-124	1372	295	1	5.01	[75]
MIL-101(Cr)/MIL-53(Al)	1746	298	40	5.7	[76]
MIL-101(Cr)/MCM-41	2843	298	1	2.09	[77]

9.3.4 Polymer-Based Hybrid Adsorbents

Using polymer-based hybrid adsorbents, as shown in Figure 9.10, is a novel approach to CO_2 capture technology, combining the advantages of polymers with other materials to enhance CO_2 adsorption efficiency. These hybrids offer tunable properties,

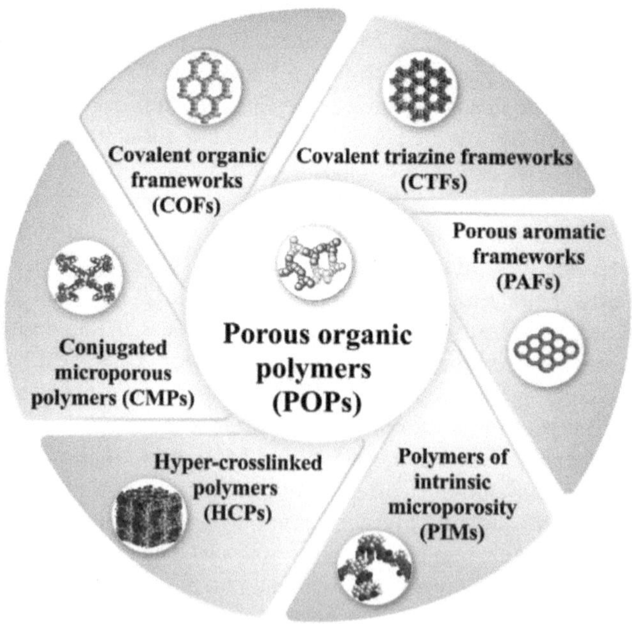

FIGURE 9.10 Types of polymers utilized for CO_2 capture [80], copyright (2021), Elsevier.

high selectivity, and improved stability, making them promising candidates for mitigating carbon emissions [78]. Polymer-based hybrid adsorbents, designed through innovative concepts like the "molecular basket," demonstrate high adsorption capacity and selectivity for CO_2, showcasing the potential of polymer-based materials in carbon capture applications [79].

Moradi et al. modified a benzene-based hypercrosslinked polymer (HCP) with amine groups, enhancing CO_2 adsorption capacity and selectivity. The adsorbent shows higher selectivity for CO_2 over N_2. The selectivity of the amine-modified HCP for CO_2/N_2 adsorption was improved by 43% [3]. Ramenzanipour et al. investigated the CO_2 adsorption capacity of activated carbon modified with piperazine; the surface modification of activated carbon improved the CO_2 adsorption capacity more than the unmodified adsorbent, and the highest CO_2 adsorption was obtained at 203.842 mg/g at 25°C and 8 bar [81]. Torkashvand et al. studied synthesizing a polymeric solid sorbent, specifically an HCP derived from carbazole monomer via the Friedel–Crafts reaction—the research aimed to investigate the impact of amination on the CO_2 uptake capability of the resulting sample. Chemical modification was employed to introduce additional groups to the porous polymer skeleton. Furthermore, the study utilized a batch reactor to analyze the adsorption process across temperatures ranging from 298 to 328 K and pressures from 1 to 9 bar. The investigation also assessed the influence of temperature and pressure on adsorption capacity. Adding amine groups to the porous polymer network significantly enhanced the adsorption capacity of the adsorbent, increasing from 182.13 to 236.37 mg/g at 298 K and 5 bar. Various isotherm, kinetic, and thermodynamic models were employed to explore adsorption. Among these models, the Hill and fractional order models exhibited excellent fitting with the experimental data for isotherm and kinetic analyses, respectively. Additionally, thermodynamic modeling provided valuable insights into the enthalpy, entropy, and Gibbs free energy values for the carbazole-based HCP adsorbent [8]. Helmi et al. investigated the adsorption of CO_2 experimentally using NaOH@Chitosan-Fe_3O_4 as a magnetic biodegradable adsorbent. Fe_3O_4 nanoparticles were used to magnetize chitosan and to enhance the alkaline nature of Chitosan-Fe_3O_4, a 4 M NaOH solution was immobilized on it (Figure 9.11). The findings indicated that as pressure increased from 1 to 9 bar and temperature decreased from

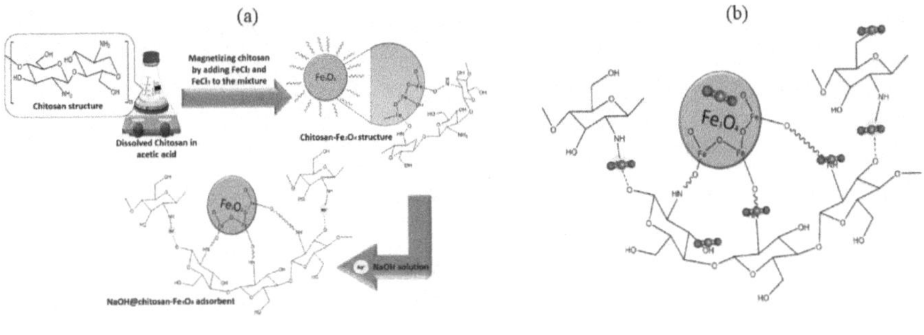

FIGURE 9.11 (a) Preparation of composite and (b) CO_2 adsorption mechanism on the composite [82], copyright (2023), Elsevier.

TABLE 9.5
Polymer Utilized in CO_2 Adsorption

Material	BET Surface Area (m^2/g)	*T* (k)	*P* (bar)	Adsorption Capacity (mmol/g)	Ref.
HCPs-5%	1165	273	1	2.25	480
InCz-HCP2	1421	273	1.13	3.58	475
BAHCP	1101	273	1	3.03	482
NPC-700-ZnC12	1105	298	1	4.02	479
NDPC-10%	1226	273	1	6.11	478
HCP	992	273	1	3.55	471

25°C to 65°C, the adsorption capacity showed improvement. The highest adsorption capacity of 3.21 mmol/g was achieved at 25°C and 9 bar pressure [82]. Table 9.5 reviewed polymeric adsorbents for CO_2 adsorption capacity.

9.4 NOVEL ADSORBENT TYPES AND APPROACHES

ILs constitute a category of molten salts distinguished by their distinctive properties, encompassing negligible vapor pressure and customizable chemical functionalities. These attributes make ILs promising candidates for CO_2 capture applications. Researchers can design ILs with high CO_2 solubility and selectivity by selecting appropriate cation and anion combinations. Moreover, immobilizing ILs onto solid supports or incorporation into porous materials enhances their processability and recyclability, paving the way for scalable CO_2 capture processes. Zhang et al. synthesized a Li_4SiO_4 sorbent (BR-Li_4SiO_4) for CO_2 capture, utilizing spent lithium cobalt oxide ($LiCoO_2$) cathodes and rice husk ash. They also prepared R-Li_4SiO_4/P-Li_4SiO_4 using commercial Li_2CO_3, rice husk ash, or commercial crystalline SiO_2 for comparison purposes. The results indicate that BR-Li4SiO4 exhibited remarkable CO_2 capture capacity and cycling stability at 675°C, maintaining a consistent capacity of 0.28 g/g under 100% CO_2 after 30 cycles, while also offering reduced costs and superior microstructure compared to R-Li_4SiO_4 and P-Li_4SiO_4. Moreover, the impact of simulated flue gas containing 15% CO_2 and 10% H_2O on Li_4SiO_4 adsorption was investigated, revealing a decrease in CO_2 adsorption at low concentrations but a slight improvement with the addition of 10% H_2O. When exposed to a mixture of 15% CO_2 + 10% H_2O balanced with N_2, BR-Li_4SiO_4 demonstrated a maximum adsorption capacity of 0.17 g/g, with its optimal adsorption temperature decreasing by 50°C to 625°C compared to conditions with 15% CO_2 alone [83]. Ansari et al. conducted a study introducing a novel approach to compare the adsorption of CO_2 adsorption compared to N_2 and O_2. This methodology opens avenues for future investigations. MMT100-M–DESn materials were synthesized through improved HCl activation of montmorillonite (MMT) nano clay and impregnation with Choline Chloride: Urea (ChCl: U) as a deep eutectic solvent. Acid surface activation significantly increased CO_2 adsorption while decreasing O_2 adsorption due to cationic ion leaching, with minimal effect on N_2 adsorption.

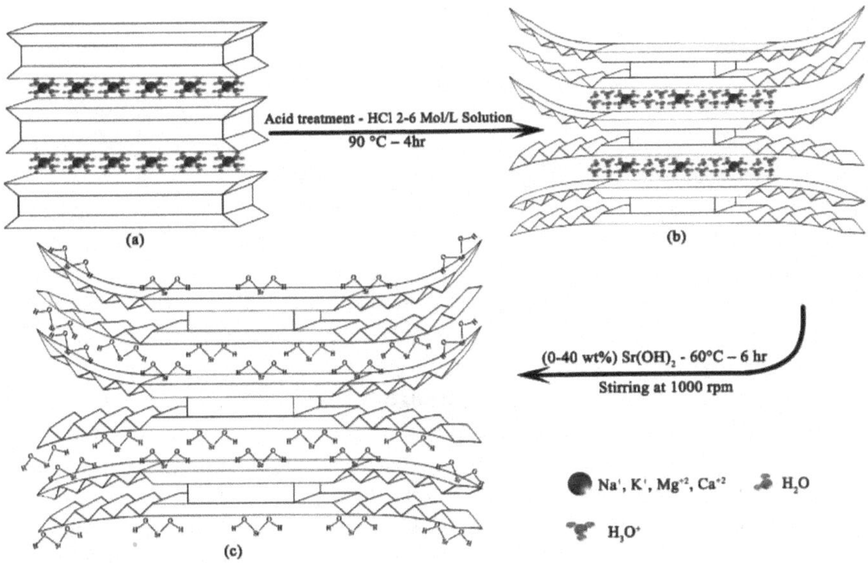

FIGURE 9.12 Surface activation of nano clay [85], copyright (2021).

This enhanced active groups for acidic gas molecules, boosting CO_2 adsorption by approximately 19%. The CO_2 isotherm model fits well with the heterogeneous multilayer Hill isotherm model, indicating higher CO_2 uptake adsorption than O_2 and N_2, which showed higher regression than the Langmuir monolayer isotherm model. These findings underscore the potential application of the developed adsorbent in green industrial CO_2 capture processes [84]. Khaje et al. conducted a thorough investigation to evaluate the influence of surface activation and modification on the CO_2 adsorption capacity of nanoclay adsorbents. Their research employed SEM analysis, which unveiled the maintenance of the nanoclay layer structure following activation, suggesting the preservation of crystallinity. Furthermore, FTIR analysis highlighted structural alterations even in the initial stages of surface treatment with HCl. BET analysis underscored the significant increase in specific surface area, with the MMT-4 sample exhibiting the highest value of 144.35 m^2/g, alongside a notable pore volume of 0.18 cm^3/g. Furthermore, the mesoporous-microporous content showed a gradual decline during the surface modification with strontium hydroxide (SH), as depicted in Figure 9.12. Through rigorous optimization employing RSM, the study systematically explored a range of operational parameters, including temperature, pressure, acid concentration for adsorbent surface activation, and the weight percentage of SH. Remarkably, an optimal adsorption capacity of 102.21 mg/g was attained under specific conditions: temperature of 25°C, pressure of 9 bar, acid concentration of 4.03 mol/L, and SH content of 16.25 wt.%[85].

Noorani et al. investigated the synthesis and characterization of aluminum terephthalate MOFs (AlTp) infused with 1-butyl-4-methyl pyridinium and 1-butyl-3-methylimidazolium–based ILs featuring various anions, including tetrafluoroborate

($[BF4]^-$), thiocyanate ($[SCN]^-$), chloride ($[Cl]^-$), and bromide ($[Br]^-$). The study analyzed CO_2 adsorption in AlTp infused with ILs at 298.15 K and pressures up to 5 bar using QCM. A comparative analysis of CO_2 adsorption in ILs/AlTp with different anions ($[Cl]^-$, $[Br]^-$, $[SCN]^-$, $[BF_4]^-$) underscored the significant role of anions in CO_2 adsorption. Results indicated that, among the systems investigated, [BMPyr][Cl]/AlTp demonstrated the highest CO_2 adsorption capacity. Moreover, while the regeneration efficiency marginally decreased after several adsorption–desorption cycles, indicating the reusability of these systems, they remain viable for gas separation applications [86]. Hoque et al. have conducted a rigorous investigation into the imperative role of CO_2 capture and separation within sustainable energy production, particularly emphasizing the attainment of net zero energy sources. They have delved into membrane-based gas separation technologies, notably mixed matrix membranes (MMMs), recognized for their potential scalability, cost-effectiveness, and minimal environmental footprint. To enhance MMM performance, the researchers meticulously selected COF-316, a 2D nanofiller distinguished by pendant nitrile (–CN) groups, to augment CO_2 permeability and CO_2/N_2 selectivity. The findings unveiled the successful integration of COF-316 into the PEBAX-1657 matrix, resulting in expanded membrane-free volume and heightened gas permeability. Furthermore, –CN groups facilitated dipole–quadrupole interactions with CO_2, further enhancing CO_2/N_2 selectivity. Optimization endeavors concerning nanofiller ratio and feed gas pressure unveiled the 0.5-MMM, harboring 0.5 wt% COF-316, as the frontrunner, showcasing superior performance metrics with significantly elevated CO_2/N_2 selectivity (from 29.85 to 73.4) and CO_2 permeability (from 89.5 to 330 barrier) in comparison to pristine PEBAX-1657 membranes. Notably, the performance of the 0.5-MMM approached the theoretical limit delineated by Robeson, underscoring its substantial promise for real-world gas separation applications [87]. Asim et al. conducted an investigation focusing on the development of a novel 2D g-PTAP filler, which was subsequently incorporated into Pebax polymer matrices to fabricate MMMs with varying concentrations (0–10 wt.%). The study evaluated the transport properties of these MMMs for CO_2 separation from N_2, revealing that the inclusion of g-PTAP led to an enhancement in CO_2 permeability. This improvement was attributed to the formation of H-bonding interactions between Pebax and g-PTAP and intra-molecular H-bonding within Pebax's PE and PA segments. The fractional free volume initially increased with the rising content of g-PTAP in the membranes (up to 5.0 wt.%), followed by a subsequent decrease, impacting the permeability and selectivity of gas molecules. Particularly noteworthy was the superior gas performance of PX2.5 ($pCO_2 = 154.6$ Barrer and $\alpha CO_2/N_2 = 83.5$) compared to the neat Pebax membrane ($pCO_2 = 82.3$ Barrer, $\alpha CO_2/N_2 = 49.5$), supported by the filler enhancement index (Findex). These findings underscore the potential of g-PTAP characteristics to significantly contribute to gas separation processes, suggesting its viability as a 2D material for various high-performance polymer applications [88].

9.5 ADVANCED CAPTURE PROCESSES AND SYSTEMS

Advanced capture processes encompass innovative techniques for enhancing the efficiency, selectivity, and scalability of CO_2 capture systems:

> Breakthroughs in adsorption/desorption kinetics facilitate faster and more efficient CO_2 capture, contributing to improved overall system performance and cost-effectiveness [89].

Process intensification strategies optimize CO_2 capture processes by maximizing resource utilization and minimizing energy consumption, leading to enhanced operational efficiency [90].

Integrating renewable energy sources, such as solar or wind power, offers sustainable solutions for powering CO_2 capture processes, reducing operational costs and environmental impact [91].

Pressure swing adsorption (PSA) and temperature swing adsorption (TSA) processes have been particularly instrumental in advancing CO_2 capture technology:

> Novel cycle designs and advanced sorbent materials have optimized PSA and TSA processes, resulting in higher purity and recovery rates of captured CO_2 [92].

By leveraging PSA and TSA processes, researchers have significantly improved CO_2 capture efficiency and sustainability, marking important strides toward mitigating climate change [93].

9.6 FUTURE TRENDS IN CO_2 ADSORPTION INNOVATIONS

Several key trends are poised to shape the future of CO_2 adsorption technologies.

9.6.1 Multifunctional Adsorbents

Future innovations in CO_2 adsorption aim to create adsorbents that not only capture CO_2 efficiently but also convert it into valuable products like formic acid. This multifunctionality enhances the economic viability of CO_2 capture processes by producing useful chemicals alongside carbon capture, thus maximizing the utilization of captured CO_2 [94].

9.6.2 Novel Materials Inspired by Nature

Researchers are expected to draw inspiration from natural processes to develop CO_2 capture materials. Bio-inspired polymers and MOFs offer the potential for creating more efficient and sustainable adsorbents. Mimicking natural systems can lead to the discovery of materials with enhanced CO_2 capture capabilities and improved environmental sustainability [95].

9.6.3 Integration of Artificial Neural Network

The application of artificial neural network (ANN) in materials discovery and process optimization is anticipated to revolutionize CO_2 capture technologies. By analyzing vast datasets and identifying complex patterns, ANN can expedite the development of novel materials and optimize operational parameters, leading to faster innovation and more efficient CO_2 capture processes [96]. Ghaemi et al. used RSM and ANN to model CO_2 adsorption and reduction by carbon molecular sieves, with the MLP network showing good agreement between experimental and predicted data [97].

9.6.4 Interdisciplinary Collaboration

Future advancements in CO_2 adsorption will rely on interdisciplinary collaboration among academia, industry, and government. This collaboration fosters knowledge exchange, resource sharing, and expertise integration, driving innovation toward sustainable CO_2 capture solutions. Collaborative efforts can accelerate developing and deploying effective CO_2 capture technologies by leveraging diverse perspectives and resources [95]. By embracing these future trends, the field of CO_2 adsorption is positioned for continuous improvement and plays a crucial role in global efforts to combat climate change.

9.7 CONCLUSION

Materials science plays a central role in designing and optimizing CO_2 adsorbents. Future advancements in this field are expected to focus on developing materials with enhanced adsorption capacities, selectivity, and stabilities. Researchers are exploring novel synthesis techniques, such as bottom-up assembly and templating methods, to tailor the structure and properties of adsorbents at the molecular level. Additionally, advancements in computational modeling and high-throughput screening are accelerating the discovery of new materials with superior CO_2 capture performance. This chapter underscores notable progressions and emerging patterns in CO_2 capture materials research and alterations, as indicated by the considerable volume of literature from 2010 to 2024. Researchers have innovated a range of CO_2 adsorbents, encompassing carbon-based materials, amine-based materials, MOFs, silica, zeolites, alumina, and polymers, each demonstrating varying degrees of advancement. The activation process is pivotal for augmenting specific surface area and porosity. Functionalizing amines, a widely adopted technique, significantly improves CO_2 uptake, offering comparable performance to more complex and expensive synthesis methods. Additionally, creating hybrid materials that combine diverse adsorbents results in unique properties for CO_2 adsorption from flue gases. The exploration of CO_2 adsorbents has revealed significant advancements across various materials, each exhibiting unique structural characteristics that contribute to their effectiveness in CO_2 capture. Polymers, particularly those modified with amine groups, demonstrate high adsorption capacities due to their enhanced interaction with CO_2 molecules. MOFs, characterized by their customizable pore architectures and extensive surface areas, present considerable promise for selective CO_2 adsorption. Zeolites, known for their microporous nature and thermal stability, continue to be effective in CO_2 capture applications. Carbon-based materials, including activated carbons and carbon nanotubes, also exhibit remarkable adsorption properties due to their high surface areas and pore volumes. Overall, the advancements in CO_2 adsorbent materials underscore ongoing innovation and emerging trends in this field. The development of new materials and modifications to existing ones are paving the way for more efficient and effective CO_2 capture technologies. Integrating these various methodologies is crucial for addressing the challenges associated with CO_2 emissions and contributes to the broader efforts in mitigating climate change.

REFERENCES

1. F. Bahmanzadegan and A. Ghaemi, "Modification and functionalization of zeolites to improve the efficiency of CO_2 adsorption: A review," *Case Studies in Chemical and Environmental Engineering,* p. 100564, 2023.
2. F. Maleki, A. Ghaemi, and G. Mir Mohamad Sadeghi, "Synthesis and characterization of waste styrofoam hypercrosslinked polymer as an adsorbent for CO_2 capture," *Environmental Progress & Sustainable Energy,* vol. 42, no. 1, p. e13954, 2023.
3. M. R. Moradi, A. Torkashvand, H. Ramezanipour Penchah, and A. Ghaemi, "Amine functionalized benzene based hypercrosslinked polymer as an adsorbent for CO2/N2 adsorption," *Scientific Reports,* vol. 13, no. 1, p. 9214, 2023.
4. F. S. Taheri, A. Ghaemi, A. Maleki, and S. Shahhosseini, "High CO_2 adsorption on amine-functionalized improved mesoporous silica nanotube as an eco-friendly nanocomposite," *Energy & Fuels,* vol. 33, no. 6, pp. 5384–5397, 2019.
5. A. Kazemi, F. Moghadaskhou, M. A. Pordsari, F. Manteghi, A. Tadjarodi, and A. Ghaemi, "Enhanced CO_2 capture potential of UiO-66-NH2 synthesized by sonochemical method: experimental findings and performance evaluation," *Scientific Reports,* vol. 13, no. 1, p. 19891, 2023.
6. H. J. Esfahani, S. Shahhosseini, and A. Ghaemi, "Improved structure of Zr-BTC metal organic framework using NH2 to enhance CO_2 adsorption performance," *Scientific Reports,* vol. 13, no. 1, p. 17700, 2023.
7. F. Bahmanzadegan, M. A. Pordsari, and A. Ghaemi, "Improving the efficiency of 4A-zeolite synthesized from kaolin by amine functionalization for CO_2 capture," *Scientific Reports,* vol. 13, no. 1, p. 12533, 2023.
8. A. Torkashvand, M. R. Moradi, and A. Ghaemi, "Amine grafting of carbazole-based hypercrosslinked polymer as an adsorbent to enhance CO_2 capture," *Case Studies in Chemical and Environmental Engineering,* vol. 8, p. 100472, 2023.
9. M. Helmi, Z. Khoshdouni Farahani, A. Hemmati, and A. Ghaemi, "Facile synthesis of Persian gum–graphene oxide composite as a novel adsorbent for CO_2 capture: Characterization and optimization," *Scientific Reports,* vol. 14, no. 1, p. 5511, 2024.
10. J. Hack, N. Maeda, and D. M. Meier, "Review on CO_2 capture using amine-functionalized materials," *ACS Omega,* vol. 7, no. 44, pp. 39520–39530, 2022.
11. P. D. Jadhav, R. V. Chatti, R. B. Biniwale, N. K. Labhsetwar, S. Devotta, and S. S. Rayalu, "Monoethanol amine modified zeolite 13X for CO_2 adsorption at different temperatures," *Energy & Fuels*, vol. 21, no. 8, pp. 3555–3559, 2007, doi: https://doi.org/10.1021/ef070038y
12. X. Xu, X. Zhao, L. Sun, and X. Liu, "Adsorption separation of carbon dioxide, methane and nitrogen on monoethanol amine modified β-zeolite," *Journal of Natural Gas Chemistr,* 2009, doi: https://doi.org/10.1016/S1003-9953(08)60098-5
13. K. Tan, A. Ruhaimi, C. Hitam, and M. Taib, "A brief on tetraethylenepentamine (TEPA) functionalized-adsorbents in CO_2 capture application," *Journal of Physics: Conference Series*, vol. 2259, no. 1, p. 012002, 2022, IOP Publishing.
14. Y.-K. Kim, Y.-H. Mo, J. Lee, H.-S. You, C.-K. Yi, Y. C. Park, and S.-E. Park, "Tetraethylenepentamine embedded zeolite A for carbon dioxide adsorption," *Journal of Nanoscience and Nanotechnology,* vol. 13, no. 4, pp. 2703–2707, 2013.

15. B. Ozturk, S. Demiral, and H. A. Ozen, "Adsorptive capture of CO_2 by amine-impregnated activated carbon, pumice, and zeolite 4A," *Greenhouse Gases: Science and Technology,* vol. 12, no. 5, pp. 602–615, 2022.
16. S. C. Lee, C. C. Hsieh, C. H. Chen, and Y. S. Chen, "CO2 adsorption by Y-type zeolite impregnated with amines in indoor air," *Aerosol and Air Quality Research,* 2013, doi: http://dx.doi.org/10.4209/aaqr.2012.05.0134
17. D. Bonenfant, M. Kharoune, P. Niquette, M. Mimeault, and R. Hausler, "Advances in principal factors influencing carbon dioxide adsorption on zeolites," *Science and Technology of Advanced Materials,* vol. 9, no. 1, p. 013007, 2008.
18. I. W. Zapelini and D. Cardoso, "Amine-grafted Na-LTA zeolite precursors as basic catalysts for Knoevenagel condensation," *Microporous and Mesoporous Materials,* vol. 324, p. 111270, 2021.
19. P. Zhao, G. Zhang, H. Yan, and Y. Zhao, "The latest development on amine functionalized solid adsorbents for post-combustion CO_2 capture: Analysis review," *Chinese Journal of Chemical Engineering,* vol. 35, pp. 17–43, 2021.
20. J. Wu, Z. Yang, J. Xie, P. Zhu, J. Wei, R. Jin, and H. Yang, "Porous polymer supported amino functionalized ionic liquid for effective CO_2 capture," *Langmuir,* vol. 39, no. 7, pp. 2729–2738, 2023.
21. M. Yin, L. Wang, and S. Tang, "Amino-functionalized ionic-liquid-grafted covalent organic frameworks for high-efficiency CO_2 capture and conversion," *ACS Applied Materials & Interfaces,* vol. 14, no. 50, pp. 55674–55685, 2022.
22. V. V. Chaban and N. A. Andreeva, "Extensively amino-functionalized graphene captures carbon dioxide," *Physical Chemistry Chemical Physics,* vol. 24, no. 42, pp. 25801–25815, 2022.
23. H. Ramezanipour Penchah, A. Ghaemi, and H. Godarziani, "Eco-friendly CO_2 adsorbent by impregnation of diethanolamine in nanoclay montmorillonite," *Environmental Science and Pollution Research,* vol. 28, no. 39, pp. 55754–55770, 2021.
24. Z. Khoshraftar, F. S. Taheri, S. Nezami, and A. Ghaemi, "Using halloysite nanotubes modified by tetraethylenepentamine for advanced carbon capture: Experimental and modeling via RSM and ANNs," *Chemical Engineering Journal Advances,* vol. 16, p. 100543, 2023.
25. R. Girimonte, F. Testa, M. Turano, G. Leone, M. Gallo, and G. Golemme, "Amine-functionalized mesoporous silica adsorbent for CO_2 capture in confined-fluidized bed: Study of the breakthrough adsorption curves as a function of several operating variables," *Processes,* vol. 10, no. 2, p. 422, 2022.
26. Z. Gu, et al., "Experimental and kinetics investigations of low-concentration CO_2 adsorption on several amine-functionalized adsorbents," *Process Safety and Environmental Protection,* vol. 160, pp. 573–583, 2022.
27. N. Mortazavi, et al., "Enhancement of CO_2 adsorption on natural zeolite, modified clinoptilolite with cations, amines and ionic liquids," *Sustainable Chemistry and Pharmacy,* 2021, doi: https://doi.org/10.1016/j.scp.2021.100495
28. T. H. Nguyen, et al., "Hierarchical zeolites with amine-functionalized mesoporous domains for carbon dioxide capture," *ChemSusChem,* 2016, doi: https://doi.org/10.1002/cssc.201600004
29. D. Panda, E. A. Kumar, and S. K. Singh, "Amine modification of binder-containing zeolite 4A bodies for post-combustion CO_2 capture," *Industrial & Engineering Chemistry Research,* 2019, doi: https://doi.org/10.1021/acs.iecr.8b03958
30. H. Cheng, H. Song, S. Toan, B. Wang, K. A. Gasem, M. Fan, and F. Cheng, "Experimental investigation of CO_2 adsorption and desorption on multi-type amines loaded HZSM-5 zeolites," *Chemical Engineering Journal,* 2021, doi: https://doi.org/10.1016/j.cej.2020.126882

31. M. Banaei, M. Anbia, and M. Kazemipour, "Enhancement of CO2/CH4 adsorptive selectivity by functionalized nano zeolite," *Journal of Ultrafine Grained and Nanostructured Materials,* 2018, doi: https://doi.org/10.22059/jufgnsm.2018.02.10
32. F. Fashi, et al., "Piperazine impregnation on zeolite 13X as a novel adsorbent for CO_2 capture: Experimental and modeling," *Chemical Engineering Communications,* 2021, doi: https://doi.org/10.1080/00986445.2020.1746657
33. P. Murge, et al., "Zeolite-based sorbent for CO_2 capture: Preparation and performance evaluation," *Langmuir,* 2019, doi: https://doi.org/10.1021/acs.langmuir.9b02259
34. K. Ahmad, et al., "A melamine-modified β-zeolite with enhanced CO_2 capture properties," *Energy Technology,* 2013, doi: https://doi.org/10.1002/ente.201300027
35. D. Panda, et al., "A comparative study of CO_2 capture by amine grafted vs amine impregnated zeolite 4A," *IOP Conference Series: Materials Science and Engineering,* 2018, doi: https://doi.org/10.1088/1757-899X/377/1/012148
36. D. Madden and T. Curtin, "Carbon dioxide capture with amino-functionalised zeolite-β: A temperature programmed desorption study under dry and humid conditions," *Microporous and Mesoporous Materials,* 2016, doi: https://doi.org/10.1016/j.micromeso.2016.03.041
37. O. G. Nik, et al., "Aminosilanes grafting on FAU/EMT zeolite: Effect on CO_2 adsorptive properties," *Microporous and Mesoporous Materials,* 2011, doi: https://doi.org/10.1016/j.micromeso.2011.03.002
38. O. G. Nik, et al., "Amine-functionalized zeolite FAU/EMT-polyimide mixed matrix membranes for CO2/CH4 separation," *Journal of Membrane Science,* 2011, doi: https://doi.org/10.1016/j.memsci.2011.06.019
39. J. C. Fisher, et al., "Oxide-supported tetraethylenepentamine for CO_2 capture," *Environmental Progress & Sustainable Energy,* 2009, doi: https://doi.org/10.1002/ep.10363
40. V. Tejavath, et al., "Technoeconomic investigation of amine-grafted zeolites and their kinetics for CO2 capture," *ACS Omega,* 2021, doi: https://doi.org/10.1021/acsomega.0c05397
41. D. P. Bezerra, et al., "Adsorption of CO_2 on nitrogen-enriched activated carbon and zeolite 13X," *Adsorption*, 2011, doi: http://dx.doi.org/10.1007/s10450-011-9320-z
42. S. K. Gebremariam, L. F. Dumée, P. L. Llewellyn, Y. F. AlWahedi, and G. N. Karanikolos, "Metal-organic framework hybrid adsorbents for carbon capture–a review," *Journal of Environmental Chemical Engineering,* vol. 11, no. 2, p. 109291, 2023.
43. T.-D. Hoang, S. A. Bandh, F. A. Malla, I. Qayoom, S. Bashir, S. B. Peer, and A. Halog, "Carbon-based synthesized materials for CO_2 adsorption and conversion: Its potential for carbon recycling," *Recycling,* vol. 8, no. 4, p. 53, 2023.
44. R. Balasubramanian and S. Chowdhury, "Recent advances and progress in the development of graphene-based adsorbents for CO_2 capture," *Journal of Materials Chemistry A,* vol. 3, no. 44, pp. 21968–21989, 2015.
45. F. Fathalian, H. Moghadamzadeh, A. Hemmati, and A. Ghaemi, "Efficient CO_2 adsorption using chitosan, graphene oxide, and zinc oxide composite," *Scientific Reports,* vol. 14, no. 1, p. 3186, 2024.
46. A. Ghaemi, H. Mashhadimoslem, and P. Zohourian Izadpanah, "NiO and MgO/activated carbon as an efficient CO_2 adsorbent: Characterization, modeling, and optimization," *International Journal of Environmental Science and Technology,* vol. 19, no. 2, pp. 727–746, 2022.
47. Z. Khoshraftar, A. Ghaemi, and A. Hemmati, "Comprehensive investigation of isotherm, RSM, and ANN modeling of CO_2 capture by multi-walled carbon nanotube," *Scientific Reports,* vol. 14, no. 1, p. 5130, 2024.
48. I. Pełech, et al., "CO_2 sorbents based on spherical carbon and photoactive metal oxides: Insight into adsorption capacity, selectivity and regenerability," *Molecules,* vol. 27, no. 20, p. 6802, 2022.

49. M. Helmi, F. Moazami, A. Hemmati, and A. Ghaemi, "Synthesis and characterization of KOH@ graphene oxide-Fe3O4 as a magnetic composite adsorbent for CO_2 capture," *Journal of Physics and Chemistry of Solids,* vol. 178, p. 111338, 2023.
50. S. Krupšová and M. Almáši, "Cellulose–amine porous materials: The effect of activation method on structure, textural properties, CO_2 capture, and recyclability," *Molecules,* vol. 29, no. 5, p. 1158, 2024.
51. C. Zhang, W. Song, Q. Ma, L. Xie, X. Zhang, and H. Guo, "Enhancement of CO_2 capture on biomass-based carbon from black locust by KOH activation and ammonia modification," *Energy & Fuels,* vol. 30, no. 5, pp. 4181–4190, 2016.
52. A. S. Ello, L. K. de Souza, A. Trokourey, and M. Jaroniec, "Coconut shell-based microporous carbons for CO_2 capture," *Microporous and Mesoporous Materials,* vol. 180, pp. 280–283, 2013.
53. Y. Guo, C. Tan, J. Sun, W. Li, J. Zhang, and C. Zhao, "Porous activated carbons derived from waste sugarcane bagasse for CO_2 adsorption," *Chemical Engineering Journal,* vol. 381, p. 122736, 2020.
54. M. Rostami, M. Mofarahi, R. Karimzadeh, and D. Abedi, "Preparation and characterization of activated carbon–zeolite composite for gas adsorption separation of CO_2/N_2 system," *Journal of Chemical & Engineering Data,* vol. 61, no. 7, pp. 2638–2646, 2016.
55. T. Okutani, T. Utsumi, and M. Ohnishi, "Synthesis of conjunctive zeolite-activated carbon composite adsorbent from rice hulls for simultaneous adsorption of CO_2 and H_2O," *42nd International Conference on Environmental Systems*, p. 3429, 2012.
56. A. K. Adhikari and K.-S. Lin, "Improving CO_2 adsorption capacities and CO_2/N_2 separation efficiencies of MOF-74 (Ni, Co) by doping palladium-containing activated carbon," *Chemical Engineering Journal,* vol. 284, pp. 1348–1360, 2016.
57. S. Hosseini, I. Bayesti, E. Marahel, F. E. Babadi, L. C. Abdullah, and T. S. Choong, "Adsorption of carbon dioxide using activated carbon impregnated with Cu promoted by zinc," *Journal of the Taiwan Institute of Chemical Engineers,* vol. 52, pp. 109–117, 2015.
58. F. E. C. Othman, et al., "Activated carbon nanofibers incorporated metal oxides for CO_2 adsorption: Effects of different type of metal oxides," *Journal of CO_2 Utilization,* vol. 45, p. 101434, 2021.
59. M. Nowrouzi, H. Younesi, and N. Bahramifar, "Superior CO_2 capture performance on biomass-derived carbon/metal oxides nanocomposites from Persian ironwood by H_3PO_4 activation," *Fuel,* vol. 223, pp. 99–114, 2018.
60. X. Zhu, Q. Wang, Y. Shi, and N. Cai, "Layered double oxide/activated carbon-based composite adsorbent for elevated temperature H_2/CO_2 separation," *International Journal of Hydrogen Energy,* vol. 40, no. 30, pp. 9244–9253, 2015.
61. J.-S. Bae and S. Su, "Macadamia nut shell-derived carbon composites for post combustion CO_2 capture," *International Journal of Greenhouse Gas Control,* vol. 19, pp. 174–182, 2013.
62. A. Kongnoo, et al., "Surface characteristics and CO_2 adsorption capacities of acid-activated zeolite 13X prepared from palm oil mill fly ash," *Fuel,* 2017, doi: https://doi.org/10.1016/j.fuel.2016.12.087
63. H. Wu, Y. Chen, Q. Chen, Y. Ding, X. Zhou, and H. Gao, "Synthesis of flexible aerogel composites reinforced with electrospun nanofibers and microparticles for thermal insulation," *Journal of Nanomaterials,* vol. 2013, pp. 10–10, 2013.
64. Y. Guo, et al., "Tailoring the performance of Ni-CaO dual function materials for integrated CO_2 capture and conversion by doping transition metal oxides," *Separation and Purification Technology,* vol. 305, p. 122455, 2023.
65. T. Ghanbari, F. Abnisa, and W. M. A. W. Daud, "A review on production of metal organic frameworks (MOF) for CO_2 adsorption," *Science of the Total Environment,* vol. 707, p. 135090, 2020.

66. M. Kalaj, K. C. Bentz, S. Ayala Jr, J. M. Palomba, K. S. Barcus, Y. Katayama, and S. M. Cohen, "MOF-polymer hybrid materials: From simple composites to tailored architectures," *Chemical Reviews,* vol. 120, no. 16, pp. 8267–8302, 2020.
67. Z. Huang, L. Ying, F. Gong, J. Lu, W. Wang, J. Ding, and J. Yan, "Aminosilane-functionalized Ti-based metal–organic framework for efficient and selective CO_2 adsorption," *Journal of Environmental Chemical Engineering,* vol. 11, no. 3, p. 109739, 2023.
68. D. Jo, S.-K. Lee, K. H. Cho, J. W. Yoon, and U.-H. Lee, "An amine-functionalized ultramicroporous metal–organic framework for postcombustion CO_2 capture," *ACS Applied Materials & Interfaces,* vol. 14, no. 51, pp. 56707–56714, 2022.
69. X. Wang, et al., "Tailoring hydrophobicity and pore environment in physisorbents for improved carbon dioxide capture under high humidity," *Journal of the American Chemical Society,* vol. 146, no. 6, pp. 3943–3954, 2024.
70. Q. Yan, et al., "Designed synthesis of functionalized two-dimensional metaleorganic frameworks with preferential CO_2 capture," *ChemPlusChem,* vol. 78, no. 1, p. 86e91, 2013.
71. Y. Cao, Y. Zhao, F. Song, and Q. Zhong, "Alkali metal cation doping of metal-organic framework for enhancing carbon dioxide adsorption capacity," *Journal of Energy Chemistry,* vol. 23, no. 4, pp. 468–474, 2014.
72. H. Furukawa, et al., "Ultrahigh porosity in metal-organic frameworks," *Science,* vol. 329, no. 5990, pp. 424–428, 2010.
73. Z. Hu, Y. Wang, S. Farooq, and D. Zhao, "A highly stable metal-organic framework with optimum aperture size for CO_2 capture," *AIChE Journal,* vol. 63, no. 9, pp. 4103–4114, 2017.
74. S. Kayal and A. Chakraborty, "Activated carbon (type Maxsorb-III) and MIL-101 (Cr) metal organic framework based composite adsorbent for higher CH4 storage and CO_2 capture," *Chemical Engineering Journal,* vol. 334, pp. 780–788, 2018.
75. J. Park, et al., "A versatile metal–organic framework for carbon dioxide capture and cooperative catalysis," *Chemical Communications,* vol. 48, no. 80, pp. 9995–9997, 2012.
76. A. Taheri, E. G. Babakhani, J. T. Darian, and S. Pakseresht, "Hybrid MIL-101 (Cr)@ MIL-53 (Al) composite for carbon dioxide capture from biogas," *RSC Advances,* vol. 9, no. 26, pp. 15141–15150, 2019.
77. C. Chen, et al., "Template-directed fabrication of MIL-101 (Cr)/mesoporous silica composite: Layer-packed structure and enhanced performance for CO_2 capture," *Journal of Colloid and Interface Science,* vol. 513, pp. 891–902, 2018.
78. V. Indira and K. Abhitha, "A review on polymer based adsorbents for CO_2 capture," *IOP Conference Series: Materials Science and Engineering*, vol. 1114, no. 1, p. 012081, 2021, IOP Publishing.
79. H. R. Penchah and F. Maleki, "Polymeric adsorbents for gas adsorption," in *Polymeric adsorbents*: Elsevier, 2024, pp. 205–258.
80. B. Dziejarski, J. Serafin, K. Andersson, and R. Krzyżyńska, "CO_2 capture materials: A review of current trends and future challenges," *Materials Today Sustainability,* p. 100483, 2023.
81. H. Ramezanipour Penchah, A. Ghaemi, and F. Jafari, "Piperazine-modified activated carbon as a novel adsorbent for CO_2 capture: Modeling and characterization," *Environmental Science and Pollution Research,* pp. 1–10, 2021.
82. M. Helmi, F. Moazami, A. Ghaemi, and A. Hemmati, "Synthesis, characterization and performance evaluation of NaOH@ Chitosan-Fe3O4 as an adsorbent for CO_2 capture," *Fuel,* vol. 338, p. 127300, 2023.
83. C. Zhang, C. Li, W. Li, and X. Guo, "High-performance Li4SiO4 sorbents prepared from solid waste for CO_2 capture," *Separation and Purification Technology,* vol. 326, p. 124730, 2023.
84. A. Ansari, A. Ghaemi, and S. Shahhosseini, "Impregnation of nanoclay montmorillonite by choline chloride-urea for selective CO2, N2, and O2 adsorption: Experimental and modeling," *Arabian Journal of Chemistry,* vol. 17, no. 4, p. 105688, 2024.

85. M. Khajeh and A. Ghaemi, "Strontium hydroxide-modified nanoclay montmorillonite for CO_2 capture: Response surface methodology and adsorption mechanism," *International Journal of Environmental Analytical Chemistry,* vol. 103, no. 17, pp. 5311–5336, 2023.
86. N. Noorani and A. Mehrdad, "Improving performance of mesoporous MOF AlTp impregnated with ionic liquids for CO_2 adsorption," *Scientific Reports,* vol. 13, no. 1, p. 3227, 2023.
87. B. Hoque, et al., "Tailored design of CO2-selective mixed-matrix membranes using nitrile-functionalized COFs as 2D nanofillers," *Journal of Environmental Chemical Engineering,* p. 112695, 2024.
88. M. Asim, A. Khan, A. Helal, W. Alshitari, U. A. Akbar, and M. Y. Khan, "A 2D graphitic-polytriaminopyrimidine (g-PTAP)/poly (ether-block-amide) mixed matrix membrane for CO_2 separation," *Chemistry–An Asian Journal,* vol. 16, no. 13, pp. 1839–1848, 2021.
89. A. Das, et al., "Advancements in adsorption based carbon dioxide capture technologies-a comprehensive review," *Heliyon,* vol. 9, no. 12, p. e22341, 2023.
90. J. Buckingham, T. R. Reina, and M. S. Duyar, "Recent advances in carbon dioxide capture for process intensification," *Carbon Capture Science & Technology,* vol. 2, p. 100031, 2022.
91. L. Riboldi and O. Bolland, "Overview on pressure swing adsorption (PSA) as CO_2 capture technology: State-of-the-art, limits and potentials," *Energy Procedia,* vol. 114, pp. 2390–2400, 2017.
92. A. M. James, J. Reynolds, D. G. Reed, P. Styring, and R. Dawson, "A pressure swing approach to selective CO_2 sequestration using functionalized hypercrosslinked polymers," *Materials,* vol. 14, no. 7, p. 1605, 2021.
93. R. Ben-Mansour and N. A. Qasem, "An efficient temperature swing adsorption (TSA) process for separating CO_2 from CO2/N2 mixture using Mg-MOF-74," *Energy Conversion and Management,* vol. 156, pp. 10–24, 2018.
94. F. Hussin and M. K. Aroua, "Recent trends in the development of adsorption technologies for carbon dioxide capture: A brief literature and patent reviews (2014–2018)," *Journal of Cleaner Production,* vol. 253, p. 119707, 2020.
95. B. Dziejarski, R. Krzyżyńska, and K. Andersson, "Current status of carbon capture, utilization, and storage technologies in the global economy: A survey of technical assessment," *Fuel,* vol. 342, p. 127776, 2023.
96. X. Yu, C. O. Catanescu, R. E. Bird, S. Satagopan, Z. J. Baum, L. M. Lotti Diaz, and Q. A. Zhou, "Trends in research and development for CO_2 capture and sequestration," *ACS Omega,* vol. 8, no. 13, pp. 11643–11664, 2023.
97. A. Ghaemi, M. K. Dehnavi, and Z. Khoshraftar, "Exploring artificial neural network approach and RSM modeling in the prediction of CO_2 capture using carbon molecular sieves," *Case Studies in Chemical and Environmental Engineering,* vol. 7, p. 100310, 2023.

10 RSM and ANN Modelling of CO_2 Adsorbents

Zohreh Khoshraftar, Fatemeh Bahmanzadegan, and Ahad Ghaemi

10.1 INTRODUCTION

Response surface methodology (RSM) is particularly useful in optimizing processes, improving product quality, and reducing costs by identifying the optimal operating conditions. By using statistical techniques and experimental design, RSM helps researchers explore the relationships between variables and predict the optimal response values within the experimental region. Overall, RSM is a powerful tool for improving process efficiency and achieving desired outcomes in research and industry [1–3]. The development of the RSM model for CO_2 adsorption has revolutionized the field of carbon capture and storage (CCS). Its unique features, such as determining optimal conditions, capturing interaction effects, and facilitating optimization, have made it an invaluable tool for researchers in the quest for effective CO_2 adsorption materials and processes. With further advancements and improvements in this modeling approach, RSM has the potential to contribute significantly to the development of sustainable and energy-efficient solutions for addressing climate change [4, 5].

ANN is a valuable tool for analyzing complex patterns and making predictions, making it a highly sought-after topic in research [4–8]. Using machine learning algorithms, such as ANN, for predicting adsorbent properties for CO_2 capture offers a unique advantage in handling complex and nonlinear interactions between input variables and output qualities. Traditional methods rely on empirical correlations or simple mathematical models, which may not fully capture the complexity of the system. By training an ANN with a large dataset of experimental data on adsorbent properties and performance, the algorithm can accurately predict the optimal adsorbent for CO_2 capture based on factors like pore size, surface area, and functional groups. This approach not only simplifies the selection process but also allows for exploring a broader range of potential materials that traditional methods may overlook. Moreover, ANN algorithms can optimize the synthesis conditions of adsorbents for CO_2 capture by identifying key variables that affect adsorption performance and recommending adjustments to improve desired properties. This can result in more efficient and cost-effective production of adsorbents with enhanced performance for CO_2 capture applications [9].

ANNs are machine learning algorithms inspired by the human brain, used for pattern recognition and predictive modeling. RSM relies on statistical principles and

DOI: 10.1201/9781003476191-10

regression analysis to model relationships, while neural networks use interconnected nodes to learn from data. RSM is better suited for simple systems with defined inputs and outputs, while neural networks excel with nonlinear relationships and large datasets. By understanding the strengths and weaknesses of each method, researchers can choose the most appropriate approach for their specific needs [1, 10]. This chapter proposes to offer researchers and practitioners a detailed understanding of how RSM and ANN techniques can be effectively used to optimize CO_2 adsorbents, ultimately contributing to the improvement of more efficient and low-cost solutions for mitigating climate change. Through the exploration of these modelling approaches, this chapter book aims to enhance the knowledge and capabilities of those working in the field of CO_2 adsorption.

10.2 RESPONSE SURFACE METHODOLOGY

RSM is an effective tool used for assessing, modeling, and analyzing the effects of different operational factors and their interactions. It involves using experimental models to fit data sets obtained from experimental tests. These models incorporate mathematical relationships and statistical methods within the RSM framework to provide insights into the impact of variables and their interdependencies. In many studies on CO_2 adsorption using adsorbents, the influence of operational variables has been researched independently, which means that the impact of one variable on another cannot be investigated [6, 11]. This limitation makes it difficult to optimize processes, especially in industrial settings where extensive testing and time are required. To overcome these challenges, it is necessary to apply RSM for optimization. RSM helps researchers determine the best combination of operational parameters to improve the adsorption efficiency of CO_2 [2]. This technique including RSM based on central composite design (CCD) provides statistical models that help understand how the optimized parameters interact with each other. By utilizing RSM, the operating conditions can be balanced effectively. The experimental design helps in minimizing the total number of trials, thereby reducing both time and cost in the modification process. This efficient approach enables researchers to efficiently investigate various parameters and their effects on the adsorbent modification process. Moreover, RSM aids in process modeling and optimization [8, 12].

10.2.1 Optimization Algorithm

RSM is a statistical technique widely employed in optimization processes to maximize or minimize a response of interest. The optimization in RSM typically involves a series of steps, and Figure 10.1 illustrates the algorithm used in this particular case, which can be apportioned into six distinct steps:

1. **Selection of Possible Factors and Responses**
   This initial step involves identifying the factors that may influence the response variable and selecting the response variable itself. These factors are often referred to as independent variables or factors, while the response variable is the dependent variable [13].

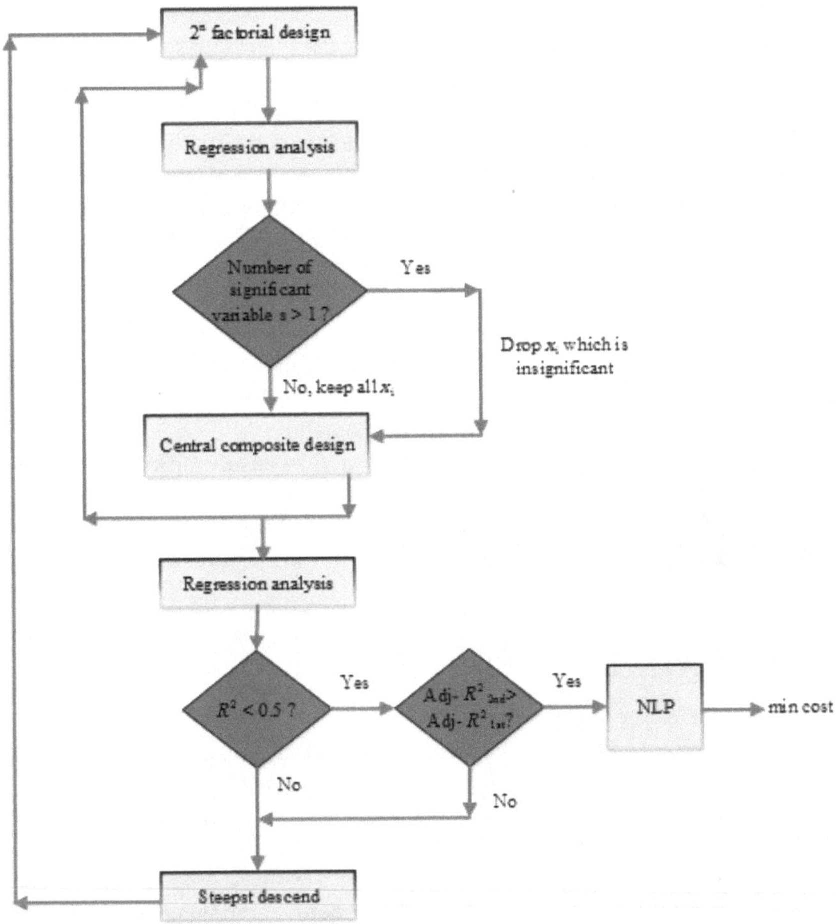

FIGURE 10.1 The scheme of study for response surface methodology [17].

2. **Choice of Experimental Design Strategy**
   Once the factors and responses are identified, the next step is to choose a suitable experimental design strategy. This could involve techniques such as factorial designs, CCDs, or Box–Behnken designs, among others. The goal is to efficiently explore the factor space with a limited number of experiments [14].
3. **Performance of Experiments and Obtaining Results**
   In this step, the designed experiments are conducted, and data on the responses are collected. The experiments are carried out systematically based on the chosen design strategy, and the results are recorded for each combination of factor levels [15].
4. **Model Fit with the Experimental Data**
   Subsequently, a mathematical model is developed to explain how the factors are related to the response variable. The model is fitted to the experimental

data, often using regression analysis. This step aids in comprehending the numerical relationship between the factors and the response [16].

5. **ANOVA**

   Once the model is established, response diagrams are created to visualize the impact of individual factors and their interactions on the response variable. Analysis of variance (ANOVA) is often performed to assess the significance of each factor and identify the most influential ones. Confirmation from the model ensures that the model adequately represents the system under consideration.

6. **Obtain the Optimal Conditions**

Optimization algorithms are applied to identify the factor settings that maximize or minimize the response variable, based on the established model. This step guides the decision-making process for achieving the best outcome.

In summary, the optimization algorithm in this case employs a systematic approach, from the initial selection of factors and responses to the final determination of optimal conditions. Each step contributes to a comprehensive understanding of the system and aids in making informed decisions for process optimization [17].

10.3 PRINCIPLES OF RESPONSE SURFACE METHODOLOGY

- **Factorial Design Integration**

  RSM often incorporates factorial experimental designs to analyze multiple factors simultaneously. By varying input factors at different levels, researchers can systematically explore the effects and interactions among these factors, improving experimentation efficiency.

- **CCD**

  CCD is a crucial element of RSM, utilizing both factorial and axial points at the center of the experimental space. This design facilitates the estimation of linear, quadratic, and interaction effects, making it effective in capturing the curvature in response surfaces. This section is more useful for adsorbing carbon dioxide, and its details are given in Section 10.5.

- **Response Surface Modeling**

  RSM utilizes mathematical models derived from regression analysis to depict the connection between input variables and the system's response. The visualization of the response surface, achieved through contour plots or 3D graphs, aids in comprehending the optimal regions conducive to the desired response. The primary goal of RSM is to identify optimal process conditions that either maximize or minimize a specific response.

- **Robustness and Sensitivity Analysis**

  RSM allows for assessing the robustness of optimized conditions. Sensitivity analysis identifies significant factors influencing the response, enabling researchers to focus on controlling these factors for consistent and reliable outcomes.

- **Iterative Nature**
  RSM follows an iterative process where the initial experimental design is refined based on obtained results. By progressively narrowing down the experimental space, researchers converge toward optimal conditions, enhancing the efficiency of the optimization process.

In conclusion, RSM offers a systematic and efficient approach to comprehending complex relationships between input variables and system responses. Through the integration of factorial designs, utilization of CCDs, and application of response surface modeling, RSM empowers researchers to optimize processes in diverse fields, contributing to advancements in both research and industrial applications.

10.4 STATISTICAL ANALYSIS AND INTERPRETATION OF THE RSM RESULTS

RSM utilizes statistical techniques to optimize responses by adjusting variables in an experimental design. By fitting a mathematical model to experimental data, RSM allows for the prediction and optimization of response variables. Statistical analysis is crucial in interpreting the relationship between variables and making informed decisions. The first step in statistical analysis involves fitting a response surface model to the data by estimating regression coefficients and determining the significance of each term. ANOVA helps identify influential factors and their interactions [18]. Interpreting the coefficients helps determine the direction and magnitude of main effects and interactions. Understanding the optimal region for response optimization is important. Contour plots or three-dimensional surface plots visually depict the relationship between variables and the response, aiding in identifying the optimal combination of factors. Confidence intervals and hypothesis testing assist in interpreting the results. Confidence intervals estimate the true value of a parameter, while hypothesis testing assesses significant differences between variable levels, validating the model and quantifying uncertainty.

ANOVA in the RSM model is used to assess the quality of the selected model. This method involves choosing separate inputs and decreasing the amount of data, which makes the processing step easier. The analysis of responses is done without making assumptions. Encoding the X inputs in the design stage is beneficial for preprocessing. These inputs can be easily converted to real values, while the outputs consist of observed data. The importance of the models can be assessed by examining the P and F values. A higher F value signifies the credibility of the model, which is determined by Eqs. (10.1) and (10.2).

$$F_0 = \frac{MS_R}{MS_E} \tag{10.1}$$

$$MS_R = \frac{SS_R}{P-1} \tag{10.2}$$

$$MS_E = \frac{SS_E}{n - P^{\cdot}} \tag{10.3}$$

which is the following equation that estimates SSR and SSR, where SSR stands for the sum of squared errors and SSR stands for the sum of regression squares [19].

$$SS_R = \sum(\hat{Y}_i - \overline{Y}_i)^2 \tag{10.4}$$

$$SS_E = \sum(Y_i - \hat{Y}_i)^2 \tag{10.5}$$

In this scenario, the variables $\bar{Y}_i$ and Y_i represent the average and actual value of a target, respectively. Additionally, $\hat{Y}_i$ represents the predicted value of the response obtained from the regression model. Put differently, $\bar{Y}_i$ and Y_i represent the mean and actual value of a response, while $\hat{Y}_i$ is the value predicted by the regression model [10]. The Adeq precision measure serves as a vital statistical tool within the framework of RSM, gauging the alignment between a model and the data. By contrasting the average prediction error against the standard deviation of the response variable, this metric provides insight into the model's accuracy within the defined range. A heightened Adeq precision measure reflects an enhanced fit, affirming the model's proficiency in forecasting the response variable [6].

In the realm of regression analysis, the adjusted R^2 emerges as a refined iteration of the conventional coefficient of determination (R^2). Its distinctiveness lies in its consideration of the number of predictors embedded in the model, offering a more reliable assessment of the model's goodness of fit. Unlike R^2, adjusted R^2 penalizes the inclusion of extraneous predictors that fail to significantly bolster the model's predictive capacity. Elevated values of adjusted R^2 not only signify robust fit but also underscore the model's simplicity by discouraging the inclusion of superfluous variables.

Predicted R^2 assumes a crucial role in evaluating the predictive prowess of a regression model, particularly in the context of RSM. This metric estimates the fraction of variance in the response variable that can be elucidated by predictors beyond the observed input space, often derived through extrapolation. A higher predicted R^2 underscores the model's adeptness in forecasting beyond the confines of the experimental region, thereby scrutinizing the reliability of predictions in uncharted territories [11]. In the context of RSM, the coefficient of variation (CV%) is a critical metric that provides valuable information about the accuracy and relative variability of the response variable. It is computed by representing the standard deviation of the response variable as a percentage of its mean. A reduced CV% indicates a more exact and stable response, while an increased CV% implies heightened variability, necessitating a careful examination of the stability and dependability of experimental outcomes. This metric holds significant importance in the determination of optimal operational conditions, as it assists in evaluating the stability of the response [10].

The effectiveness of the polynomial model was measured using specifications coefficients, average deviation (AAD), and R^2 and adj-R^2 values, as outlined in Eqs. (10.6) to (10.8). AAD should be minimized to indicate a close match between predicted and observed values, while adj-R^2 should approach a value of 1.0. AAD demonstrates the differences between experimental and calculated quantities, while adj-R^2 indicates the proportion of data variables accounted for by the model [20].

$$AAD\% = 100\left[\sum_{i=1}^{n}\left(\left|y_{i,\exp} - y_{i,cal}\right| / y_{i,\exp}\right)\right] / n \tag{10.6}$$

In cases where n represents the number of experiments conducted and $y_{i,cal}$ and $y_{i,exp}$ refer to the predicted and observed responses, respectively, the statement can be paraphrased as:

"When conducting multiple experiments, the values $y_{i,cal}$ and $y_{i,exp}$ represent the calculated and experimental responses, respectively."

$$R^2 = 1 - \frac{SS_{res}}{SS_{\mathrm{mod}\,el} + SS_{res}} \tag{10.7}$$

$$Adj - R^2 = 1 - \frac{SS_{res} / d.f._{\cdot res}}{(SS_{\mathrm{mod}\,el} + SS_{res}) / (d.f._{\cdot \mathrm{mod}\,el} + d.f._{\cdot res})} \tag{10.8}$$

Table 10.1 displays equations that utilize parameters to affect diagnostics such as SS and DF.

TABLE 10.1
Various Parameters and Their Utilization in the RSM-CCD Model [17]

Definition	Equation	Objective of Study
Small subset	$\hat{\hat{Y}} = X\hat{\beta} \cdot e = Y - \hat{\hat{Y}}$	• The regression yields an asymmetrical outcome where the predicting variable is more influenced by the impression subset rather than the majority information.
Leverage	$H = X(X^T X)^{-1} X^T - [h_{ij}]$	• Specifies the proportion of variability in the error that is explained by a particular factor or design point, along with an estimated value. • Indicates the extent to which a design point accurately represents the fit coefficients of the model. • A value ranging from 0 to 1 that signifies the influence of a design point on the values of the model. • When $H = 1$, it suggests that the experiments and predicted values are nearly identical, resulting in a residual of zero: rephrase all the information.
Internally Studentized Residual	$r_i = \frac{e_i}{\hat{\sigma}\sqrt{1 - h_{ii}}}$	• Differentiating between estimated and real values.

(Continued)

TABLE 10.1 *(Continued)*

Various Parameters and Their Utilization in the RSM-CCD Model [17]

Definition	Equation	Objective of Study
Externally Studentized Residual	$\hat{\sigma}^2_{(-i)} = \frac{(n-p).\sigma^2 - \frac{e_i^2}{(1-h_{ii})}}{\sigma\sqrt{1-h_{ii}}}$	• After excluding each individual run from the survey, predict the responses based on the remaining runs.
t-Value	$t_i = \frac{e_i}{\sqrt{\hat{\sigma}^2_{(-i)}(1-h_{ii})}}$	• Measures the standard deviation difference between predicted and actual response values. • Quantifies the number of standard deviations by which the parameter estimate deviates from zero.
DFFITS	$DFFITS = \frac{\hat{Y} - \hat{Y}_{(-i)}}{\sqrt{\hat{\sigma}^2_{(-i)}.h}}$ $\hat{Y}_{(-i)} = Y + \frac{e}{1-h}$	• The studentized discrepancy between the predicted and observed value for i.
DFBETAS	$DFBETAS_{j,(-i)} = \frac{\hat{\beta}_j - \hat{\beta}_{j,(-i)}}{\sqrt{\hat{\sigma}^2_{(-i),i}.(X^TX)^{-1}_{jj}}}$	• Demonstrates the impact of the ith observation on each regression coefficient. • Represents the square result of a uniform leverage function and the internally studentized residual for the ith observation.
Cook's distance	$D_i = \frac{r_i^2}{p+1}(\frac{h}{1-h})$	• Demonstrates the model's suitability for representing the ith observation, taking into account the corresponding outcome (y_i), and incorporates a component that gauges the proximity of that data point to the overall dataset.

In the context of the equation, e represents the error, y_i stands for the measured response data, and $\hat{y}$ denotes the predicted value derived from the model. X and H are matrices for the model, with X having n rows and p columns, where n is the total number of runs minus one, and p is the count of terms in the model, encompassing the intercept.

10.5 CENTRAL COMPOSITE DESIGN

The CCD was initially introduced informally in Example 1 as a way to fit a second-order model, and it has since gained popularity and had significant theoretical and practical developments. Its effectiveness in sequential experimentation has been a major motivation for its use. A CCD consists of strategically distributed experiments that capture important information about the system being studied. Its versatility and efficiency in fitting second-order models make it a valuable tool. The layout of a CCD allows researchers to thoroughly explore the response surface, which helps identify optimal conditions and understand complex system interactions. The CCD design, proposed by Box and Wilson, has become the preferred choice for fitting second-order models due to its strong theoretical foundation and practical usefulness. Its systematic arrangement of experimental points facilitates a comprehensive examination of the factors influencing the response, making it an important tool in experimental design and optimization. The significant attention given to CCD in both theory and practice highlights its importance in the field of experimental design and statistical modeling [21].

1. A two-level factorial design requires conducting *nF* experiments, which are the only ones that provide valuable information for estimating the two-factor interactions. To achieve optimal results, it is recommended to use either a complete 2^k design or a fractional 2^{k-r} design with resolution *V*.
2. The "star" or "axial" design consists of conducting 2^k experiments with specific coordinates (±α, 0, . . ., 0), (0, ±α, . . ., 0), (0, 0, . . ., ±α). While the axial points are not useful for estimating interaction terms, the other points contribute to this estimation.
3. The n_C center points, which are conducted at the origin (0, 0, . . ., 0), are particularly valuable for estimating the quadratic terms. Additionally, they provide an estimate of MS_{PE} (mean square prediction error) and aid in the estimation process.

This study employs a CCD, where factors are tested at five levels (−α, 1,0, 1, and α), excluding the α = 1 level. The design, as depicted in Figure 10.2, is conducted sequentially. The factorial points in the design are optimal for fitting a first-degree model or a model with two-factor interactions. In summary, the design used in this study is distinctive and offers valuable insights for modeling the system.

The use of CCDs offers flexibility in their application, particularly in selecting the axial distance (α) and the number of center points (n_C). The choice of these two

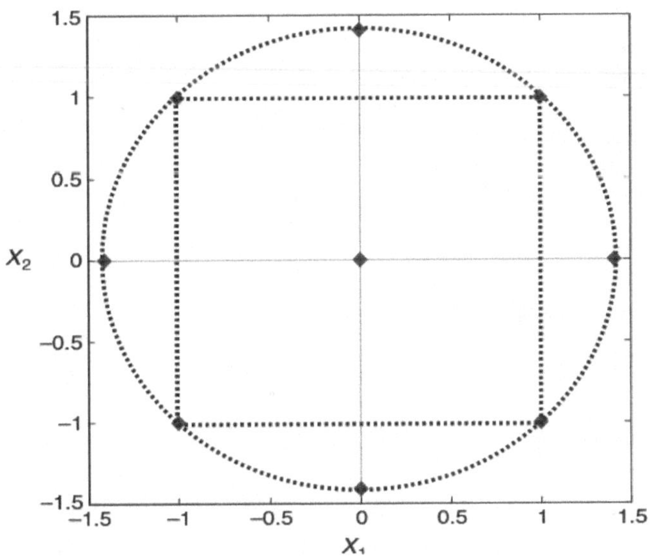

FIGURE 10.2 The CCD for two factors with $\alpha = \sqrt{2}$ consists of solid points [21]. Reproduced with permission from Elsevier, license number, 5699631323253.

parameters is crucial as it can significantly impact the optimality of the resulting design. The value of α is influenced by the dimensions and configuration of the experimental area. To ensure that the CCD remains rotatable, it is important to select appropriate values of a. More specifically, the design will be rotatable if certain conditions relating to the chosen value of α are met [21].

$$\alpha = n_F^{\ 1/4} \tag{10.9}$$

The value of n_C greatly affects the shape and magnitude of the variance function $d(x)$ in the experimental domain. This determines the level of uniformity in the variance function. It is not possible to have a constant variance function, but it can vary minimally and consistently. To achieve uniform precision, it is necessary for the design to be rotatable. When the design is rotatable, it has been proven that the number of center points needed (n_C) to ensure uniformity.

$$n_C = K(\sqrt{n_F} + 2)^2 - n_F - 2K \tag{10.10}$$

$$K = \left((K+3) + (9K^2 + 14K - 7)^{\frac{1}{2}} \right) / (4(K+2) \tag{10.11}$$

Although CCD cannot be orthogonal, the correlation between the estimated coefficients can be minimized if the relation α:

$$\alpha = \left(n_F(\sqrt{N} - \sqrt{n_F})^2 / 4 \right)^{1/4} \tag{10.12}$$

It is not possible to simultaneously have both near orthogonality and uniform precision. The main rationale behind choosing uniform precision is that each approach relies on a distinct number of center points and achieving near orthogonality is generally more challenging than obtaining uniform precision. As a result, it is common practice to prioritize maintaining uniform precision. When dealing with more than four variables, the number of runs required for a complete factorial design becomes very large. As a result, it is common to use fractional designs instead. Table 10.2

TABLE 10.2
Most Used CCD up to Six Factors [21]

	2	3	4	5	6
Factorial design	22	23	24	25–1 I =12345	26–1 I = 123456
n_F	4	8	16	16	32
2	1.414	1.682	2	2	2.378
nC, up	5	6	7	5	9
nC, min	2–3	3	3–4	5	5
p	6	10	15	21	28

presents several CCDs, some of which are fractional and accompanied by their respective generator. This option is highly valuable in practical applications, as it allows for the use of alternative fractions of $2k$, as long as they maintain a V resolution.

In the context of rotatable designs, Table 10.2 provides the values of nC that ensure consistent precision according to Eq. (10.9), as well as the minimum value required for a satisfactory design. With the availability of software, it is now feasible to explore how both parameters interact in solving the problem at hand. For instance, when dealing with a problem involving two variables, it may be necessary to expand a factorial design to accommodate a second-order model. In other words, more factors and levels need to be incorporated to adequately represent the relationships between variables and capture the complexity of the problem.

Figure 10.3 displays the level curves for various cases when the variable "α" takes values of 1, 1.414, or 2, and there are either one or five planned center points. The shape of the experimental domain differs depending on the value of "α": for α value of 1, the domain is a square with sides parallel to the axes (Figure 10.3a and b); for α value of 1.414, the domain is a circle (Figure 10.3c and d); and for α value of 2, the domain is a square with its vertices on the axes (Figures 10.3e and f). Additionally, the surfaces of these domains differ significantly, with surface areas of 4, 6.28, and 8, respectively. Figure 10.3 demonstrates that in the experimental range, various acceptable values for the parameter $d_{\max}(x)$ are observed. These values are 1, 1, and 0.8 for ±=1, 1.41, and 2, respectively. However, when $n_C = 1$, the surface d(x) shows a maximum point near the center with a value that depends on the parameter a. Specifically, when α =1.414 (as shown in Figure 10.3c), the maximum point is global and equals 1.0. On the other hand, in the other two cases, the maximum point is local and equals 0.5. The minimum values of 0.4, 0.5, and 0.4 are also observed in the same manner. Furthermore, these minimum values are achieved at distinct points and the function $d(x)$ exhibits significantly different behavior in the vicinity of each of these points, as depicted in Figure 10.3a, c, and e, respectively [21].

In recent years, the RSM has become a popular approach among researchers. This method helps in understanding the impact of various input variables on the performance and characteristics of a process. It employs mathematical and statistical techniques to analyze the interactive relationships between the operating variables of a system and its responses. By doing so, it can identify the optimal conditions for achieving the desired response. To accurately describe a process and potential changes in operating conditions, it is necessary to create a model using a quadratic equation or a higher degree nonlinear second-order model. The equation for the quadratic model regression is represented as Eq. (10.13) [5].

$$Y = \beta_0 + \sum_{i=1} \beta_i X_i + \sum_{i=1} \beta_{ii} X_i^2 + \sum_{i=1} \sum_{j=i+1} \beta_{ij} X_i X_j + \varepsilon$$
$$x_i \text{ For } i = 1, 2, \ldots, n \text{ factors} \tag{10.13}$$

The nonlinear programming (*NLP*) model:

$$Y = \beta_0 + \sum_{i=1} \beta_i X_i + \sum_{i=1} \beta_{ii} X_i^2 + \sum_{i=1} \sum_{j=i+1} \beta_{ij} X_i X_j$$
$$-1 \le x_i \le 1 \text{ For } i = 1, 2, \ldots, n \text{ factors} \tag{10.14}$$

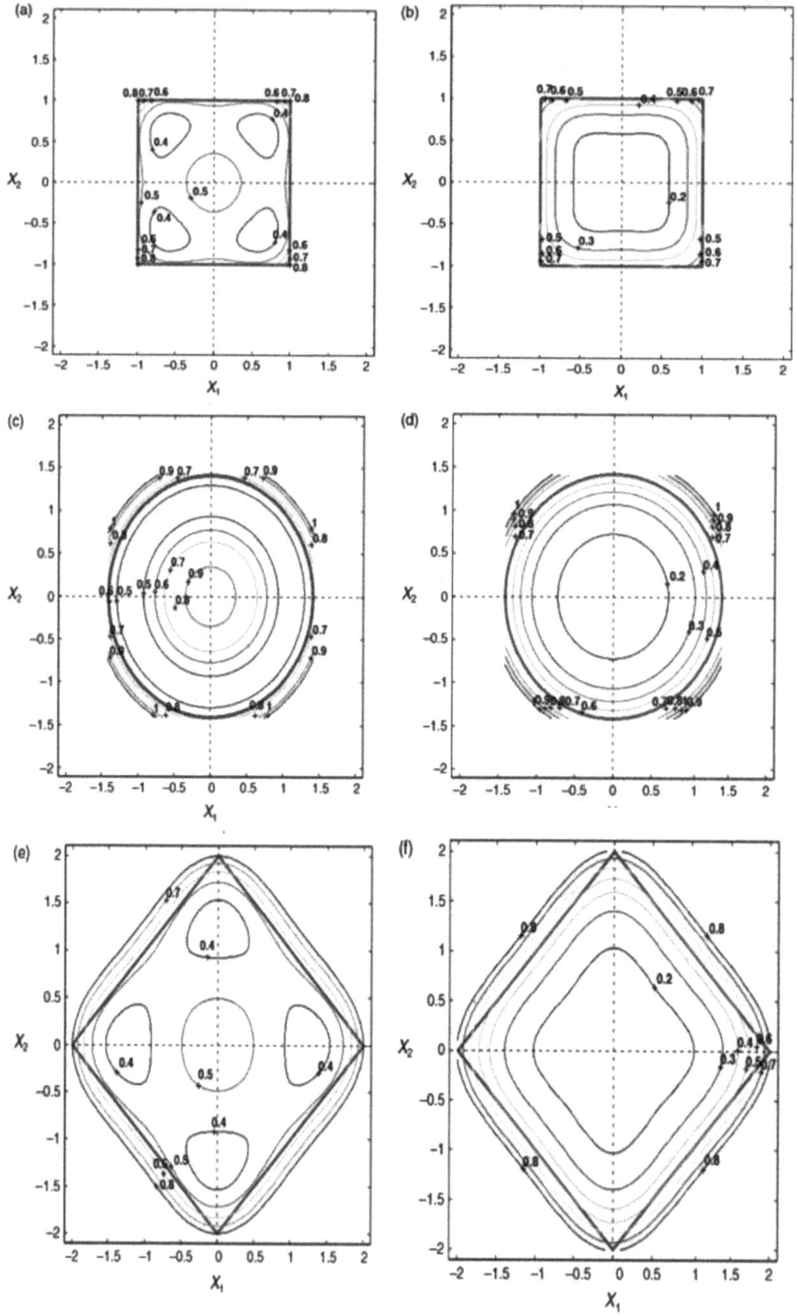

s

FIGURE 10.3 The variance function, represented as $d(x)$, for a two-factor second-order central composite design. Variance values vary based on axial distance and number of center points [21]. Reproduced with permission from Elsevier, license number, 5699631323253.

In this context, Y denotes a function linked to the anticipated response. X_i and X_j represent the model-derived values for variables i and j, respectively. β_0 serves as the baseline, while β_i and β_{ii} correspond to the coefficients associated with linear and quadratic parameters, respectively, reflecting the impact of interaction. ε stands for the residual connected to the experimental outcomes [22].

The purpose of using the method of least squares (MLS) in this study was to minimize the residual value of the coefficient "*b*." This method involved using multiple regression analysis on experimental data to fit (Eq. 10.13). The resulting model was used to describe the behavior of the system response within the experimental range, with the independent variable being the model itself. The steps of the study were designed in accordance with the CCD model, where the reaction time differed from the maximum adsorption time. The optimization process aimed to find the best values for all variables, including temperature, pressure, and amine loading concentration, to achieve the optimal response [10].

The decision variables optimized through the response optimizer have been established. Following the acquisition of data pertaining to each practical level within the chosen scheme, it becomes imperative to scrutinize the patterns exhibited by points sharing similar responses and subsequently derive a mathematical equation to characterize them. This involves approximating the parameter "b" from Eq. (10.13). Therefore, when annotating the matrix, Eq. (10.13) can be expressed as a rephrased statement [17]:

$$y_m x_i = X_m x_n b_n x_1 + e_m x_1 \tag{10.15}$$

In the given context, b represents the model parameter vector, while y denotes the response parameter vector. The variables m and n correspond to the numbers of rows and columns in the matrices, respectively. X represents the experimental design matrix with a selective nature, and e signifies the residual. Eq. (10.15) is resolved through a statistical technique known as the MLS [20]. The computation of the vector b can be expressed as per Eq. (10.16) [23]:

$$b_{n-1} = (X^T_{n-m} y_{m-i})(X^T_{n-m} X_{m-n})^{-1} \tag{10.16}$$

The suitability and validity of the model were further validated through the application of ANOVA in the Design Expert software. ANOVA operates on the principle that every observed value is influenced by three main components: the overall mean value (α), the impact of the factors being measured on the system response (β), and the residual error (ε) [8].

$$Y_i = \alpha_i + \beta_i + \varepsilon_i \tag{10.17}$$

The sum of squares for the residual errors [8]:

$$RSS = \sum (Y_i - \alpha_i - \beta_i)^2 \tag{10.18}$$

In ANOVA, changes in a dataset are assessed by examining their variability. This variability is measured by calculating the deviation (d_i) of each observation (y_i) or its iteration (y_{ij}) from the mean ($\overline{y}$). More specifically, the squared value of this deviation is represented in Eq. (10.19):

$$d_i^2 = (y_{ij} - \overline{y})^2 \tag{10.19}$$

The total sum of squares (SS_{tot}) is the aggregate of the squared differences of each observation from the average. It can be further divided into two components: the sum of squares due to regression (SS_{reg}) and the sum of squares due to residuals generated by the model (SS_{res}). This partitioning allows us to assess the proportion of variability in the data that can be accounted for by the regression model and the proportion that remains unexplained [17].

$$SS_{tot} = SS_{res} + SS_{reg} \tag{10.20}$$

When the center point is replicated in an experiment, we can estimate the pure error that occurs during repetition. This estimation helps us separate the sum of squares for residuals into two components: the sum of squares for pure error (SS_{pe}) and the sum of squares for lack of fit (SS_{lof}). This division enables us to distinguish between the variability caused by pure error and the variability caused by the model's lack of fit [17].

$$SS_{res} = SS_{pe} + SS_{lof} \tag{10.21}$$

10.6 DEVELOPMENT OF THE RSM MODEL FOR CO_2 ADSORPTION

The development of the RSM model for CO_2 adsorption has been a noteworthy breakthrough in the field of CCS technologies. This model focuses on understanding and optimizing the adsorption process of CO_2 onto various solid materials, such as porous materials, zeolites, and activated carbon (AC) [4]. The RSM approach integrates statistical experimental design and mathematical regression analysis to create a predictive model that can effectively represent the intricate CO_2 adsorption behavior. One unique aspect of the RSM model is its ability to determine the optimal conditions for CO_2 adsorption. By utilizing factorial designs and response surface analysis, the model can identify the critical parameters that influence the adsorption process. These parameters can include factors such as temperature, pressure, gas composition, and the characteristics of the adsorbent material. By systematically varying these factors within a designed parameter space, the RSM model generates a response surface that predicts the CO_2 adsorption capacity under different conditions. This enables researchers to determine the best combination of parameters that maximizes the adsorption efficiency and minimizes the cost of the process [2, 8].

Another distinct feature of the RSM model is its capability to capture the interaction effects between different parameters. In many cases, the influence of one parameter on the CO_2 adsorption capacity is not independent of other parameters.

For example, the temperature may have a different effect on adsorption efficiency at different pressure levels. This enables a more accurate prediction of CO_2 adsorption performance, as it reflects the complex nature of the adsorption process. Furthermore, the RSM model can be used for optimization and process intensification. Once the response surface is developed, optimization techniques, such as desirability functions or numerical optimization algorithms, can be employed to find the global maximum or minimum of the response surface. This allows researchers to identify the optimal operating conditions that maximize the adsorption capacity of CO_2. By optimizing the process, the RSM model can help in the development of more cost-effective and efficient CCS technologies, which are crucial for mitigating greenhouse gas emissions [4, 5].

Previous studies on CO_2 adsorption and modeling using RSM approaches have investigated various materials and conditions to develop effective and sustainable strategies for CCS.

These studies (Table 10.3) demonstrate the significant contribution of RSM approaches in understanding and optimizing CO_2 adsorption processes. By employing RSM, researchers can comprehensively explore the effects of various parameters and develop reliable models for predicting and optimizing CO_2 adsorption capacity. This enables the development of efficient and sustainable CCS techniques, thus contributing toward mitigating climate change. Several research studies have been conducted to evaluate the efficiency of carbon dioxide (CO_2) capture using RSM techniques. These investigations have focused on assessing the performance of various methods for capturing CO_2 emissions. By utilizing RSM, researchers aim to optimize the capture process by considering multiple factors that influence CO_2 capture efficiency. These studies have provided valuable insights into the effects of different parameters, such as temperature, pressure, solvent type, and flow rate, on the overall performance of CO_2 capture systems (Table 10.4). The application of RSM in these studies has enabled researchers to develop mathematical models that can predict and optimize the CO_2 capture process, leading to more sustainable and effective CO_2 mitigation strategies.

One notable study on CO_2 adsorption and modeling using RSM was conducted by Zhou et al. (2017) [46]. The researchers explored the use of carbon nanotube-supported amine adsorbents for CO_2 capture and investigated the effects of various operating parameters such as pressure, temperature, and amine loading on the capture efficiency. RSM was employed to establish a mathematical model that could predict the CO_2 adsorption performance based on the input variables. The model provided insights into the interactions among the variables and allowed for the optimization of the process to achieve maximum CO_2 capture efficiency.

One unique study by Khoshraftar et al. (2023) [5] focused on the design and optimization of CO_2 adsorption onto tetraethylenepentamine-functionalized halloysite using RSM. The researchers employed RSM to determine the optimum operating parameters including temperature, pressure, and concentration of the amine functional groups. They found that the amine-functionalized exhibited high CO_2 adsorption capacities and that the RSM model accurately predicted the adsorption performance.

TABLE 10.3

Review Studies on CO_2 Adsorption and Modeling Using RSM

Adsorbents	Design	Inputs	Values	Optimum Parameters	Q (mg/g) Output	Ref
Triethanolamine-modified montmorillonite clay	CCD	• Temperature • Pressure • Percentage triethanolamine	30–70°C 1–9 bar 10–30% wt	Temperature of 40°C, pressure of 7 bar, and percentage triethanolamine of 18.36%w	219.86	[24]
Activated alumina (AA) was modified with sodium hydroxide	CCD	• Temperature • Pressure	20–80°C 2–10 bar	Temperature of 20 °C, pressure of 6 bar and 30% NaOH	146.70	[25]
Zeolite 4A from kaolin was modified with tetraethylenepentamine (TEPA) and diethanolamine (DEA)	CCD	• Temperature • Pressure • Wt% amine • Amine Type	25–65°C 1–9 bar 5–25% DEA-TEPA	Temperature of 25 °C, pressure of 9 bar and 15% TEPA, 10% DEA	579.468	[26]
Activated alumina (AA) was modified with potassium hydroxid	CCD	• Temperature • Pressure	20–80°C 2–10 bar	Temperature of 20 °C, pressure of 6 bar, and 30% KOH	130.84	[25]
NaOH-modified nanoclay montmorillonite	CCD	• Temperature • Pressure • Acid concentration • wt% NaOH	25–65°C 1–9 bar 2–6 mol/L 0–40 %w/w	39.76 wt% of NaOH, pressure of 1.00 bar, temperature of 65°C, and acid concentration of 5.99 M	105.55	[27]
Carbazole-based hypercrosslinked polymeric	CCD	• Adsorption time • Ratio • Synthesis time	0–3600 0.378–4.621 5.930–20.070	Adsorption time of 810 s, crosslinker ratio of 1.54, and synthesis time of 8.8 h at constant temperature and pressure of 25°C and 5 bar	174.59	[28]
Strontium hydroxide-modified nanoclay	CCD	• Temperature • Pressure • Acid concentration • wt% of strontium hydroxide	25–65°C 1–9 bar 2–6 mol/L 0–40 wt%	Temperature of 25°C, pressure of 9 bar, the acid concentration of 4.03 mol/L, and the weight percent of strontium hydroxide (16.25 wt.%).	102.21	[29]

Benzene-based hypercrosslinked polymer	CCD	• Synthesis time, crosslinker • Ratio (formaldehyde dimethyl acetal [FDA]/benzene) adsorption time, • Pressure • Temperature	1–12 h 0–1 mol/L 1–9 bar 55–88°C	The optimized values of synthesis time (10.11 h), crosslinker ratio (formaldehyde dimethyl acetal [FDA]/benzene) of 1, adsorption time of 220 s, pressure of 9 bar, and temperature of 55°C	167	[30]
AC/NiO-3	CCD	• Pressure • Temperature • Absorbent type	1–9 bar 25–80°C	3 wt% of optimal concentration, pressure of 7.8 bar, and temperature of 62.7°C	121.35	[31]
AC/MgO-3	CCD	• Pressure • Temperature • Absorbent type	1–9 bar 25–80°C	3 wt% of optimal concentration, pressure of 7.8 bar, and temperature of 57.1°C	105.17	[31]
Amination of Commercial palm-shell-based granular activated carbon (GAC)	CCD	• Amination temperature (°C) • Amination time (h)	200–1000°C 1–3	The most effective outcome as determined by numerical optimization was achieved by aminating the pre-oxidized sorbent at a temperature of 425°C for a duration of 2.1 hours.	26.47	[32]
Regenerable alkali metal carbonate: K_2CO_3/Al_2O_3 adsorbent	Box–Behnken design (BBD)	• Temperature • H_2O/CO_2 • Vapor pretreatment Time	50–88°C 0.5–1.5 3–9	The numerical optimization yielded the best values for the process variables as follows: 61.3°C for the adsorption temperature, 1 for the mole ratio of H_2O/CO_2, and 9 minutes for the vapor pretreatment time.	87.71	**[16]**
Blended amines into mesoporous silica	CCD	• TEPA • DEA • MeOH	31.59–48.41 wt% 13.18-46-82 wt% 50–117.05 g	The optimum amounts of TEPA, DEA, and MeOH were 1.33 g (7.03 mmol), 0.85 g (8.08 mmol), and 58.2 g for the preparation of 3 g of sorbent at 40°C and 1 bar	301.84	[33]

TABLE 10.4
Summary Works about CO_2 Capture Using RSM Techniques with Models [10]

Sorbent	Variables	Response	Method	Model	R^2, Adj-R^2	Ref.
DEA, TEPA, MSU-F	C_{DEA}, C_{TEPA}, C_{MECH}	q_{CO_2}	Full factorial, CCD	$Y = 5.62192 + 0.0830368X_1 - 0.0345798X_2\ 0.0208987X_3 - 0.295X_1X_2 - 0.215X_1X_3 - 0.1375X_2X_3 - 0.156328X_1^2 - 0.619369X_2^2 - 0.140422X_3^2$	0.953	[33]
Amine	T, P, Q_g, and P_{CO_2}	T_{rise}	CCD	$T_{rise} = -29.789 + 0.081T + 0.93P + 1.95P_{CO_2} + 0.37Q_g - 0.00493P^2 - 0.015P_{CO_2}^2 + 0.00503TP + 0.1TP_{CO_2} - 0.00813TQ_g - 0.0604PP_{CO_2} - 0.00249PQ_g$	0.989, n.d.	[34]
NaCl	T, Q_g, C_B, and S	$\%R$	CCD	$\%R = 60.55 - 0.328X_1 + 3.118X_2 + 0.0833X_3 + 33.44X_4 - 0.01844X_1^2 - 0.1002X_2^2 - 0.0003X_3^2 - 20.55X_4^2 + 0.03632X_1X_2 + 0.307X_1X_4$	0.988, n.d.	[35]
NaCl	T, Q_g, and M_r	$\%R$	CCD	$\%R = 97.9 - 1.36T - 49.3Q_g - 10.5M_r - 0.0153T^2 - 8.65Q_g^2 - 0.89M_r^2 + 0.127TQ_g - 0.278TM_r + 21.47Q_gM_r$	0.994, close to 1	[36]
Water	T, C_{CO2}, gas/liquid ratio, P_{CO_2}	$\%R$	CCD	$q \times 10^4 = -8.71 - 30.71X_1 - 0.63X_2 + 11.16X_3 + 1.47X_4 + 32.82X_1^2 + 0.01X_2^2 - 4.88X_3^2 + 1.61X_1X_2 + 24.34X_1X_3 - 1.65X_1X_4 + 0.45X_2X_3 - 0.02X_2X_4 - 0.74X_3X_4$	0.982, 0.972	[37]
Activated carbon	T, P_{CO_2}	q_{CO_2}, t_b	Full factorial	$Q = 2.146 - 0.048T + 2.128P_{CO_2} - 0.106P_{CO_2}{}^2 + 0.0003T^2 - 0.0109T \times P_{CO_2}$ $t_b = 5.394 - 0.056T + 0.0003T^2 - 0.017T \times P_{CO_2} + 0.981P_{CO_2} + 0.516P_{CO_2}{}^2$	0.999, 0.998 0.990, 0.984	[38]
MEA	T, Q_l, Q_{reb}, solvent loading	$\%R$	CCD	$\%R = 84.77 - 4.64A4.63B - 3.28C3.9D - 3.28E + 0.73AB - 0.5AC + 0.59AD - 0.56AE + 0.81BC - 0.74BD + 0.53BE + 0.5CD - 0.54CE + 0.5DE - 0.6B^2 - 0.3C^2 - 0.53D^2 - 0.37E^2$	0.974, 0.989	[39]

MEA	Q_l, C_{MEA}, and P	CO_2 purity	n.d.	$Y \times 10^4 = 9900 - 7.282A - 2.166B - 5.130C - 1.433D + 20.253E$ $+0.0.04409AD + 3.540AE - 3.049BE + 2.044CE - 2.650DE$	n.d., n.d.	[40]
HMPD, AEEA	P_{CO_2}, C_{HMPD}, C_{AEEA}	α_{CO_2}, absorption rate	n.d.	$(CO_2 loading)^{1.78} = 9.5268 - 0.6534A - 0.15606B$ $-0.042156C + 0.04238A^2 + 5.04679 \times 10^{-5}C^2$ $-2.88344 \times 10^{-7}D^2 + 1.72214 \times 10^3 D + 0.035597A \times B$ $+7.89928 \times 10^{-4} A \times C - 2.30338 \times 10^6 C \times D$ $Ln(R_A) = -5.7302 + 0.6334A + 1.8444B + 5.1935 \times 10^{-3} C$ $+5.547 \times 10^{-4} D - 0.1278A \times B + 5.2492 \times 10^{-5} A \times D$ $-0.16196A^2 - 0.53106B^2 - 1.62695 \times 10^{-7} D^2$	0.962, 0.949 0.969, 0.967	[41]
MEA, Glycerol	C_{MEA}, $C_{glycerol}$ T, Q_g	α_{CO_2}	CCD	CO_2 Solubility$= -56.737 + 15.15341A + 12.246B + 0.376C - 0.0665D$ $-1.743A \times B + 7.691 \times 10^{-3} B \times D + 2.2110A^2 + 1.196B^2$	0.953, 0.935	[42]
DGA, DEPG	P, T, P_{CO_2}	q_{CO_2}, Q_{reb}, $Q_{cooling}$	CCD	$Power\ requirement = +76783 + 49159A + 33010B + 25703AB$ $Cooling\ duty = +80575.1 + 50957.3A + 32256.2B - 4236.4C + 25294.1AB$ $Recovered\ CH_4 = +97.4 - 3.9B + 2.3C - 0.4D + 2.1BC + 0.6CI$ $Captured\ CO_2 = +89.6 + 13.9A + 27.1B - 9.1C - 23.3B^2$	0.999, 0.998 0.997, 0.995 0.977, 0.944 0.961, 0.907	[43]
MEA, DGA DEA, MDEA, TEA	$n_{s,\,absorber}$, $n_{s,stripper}$, C_{amine}, Q_{reb}, $T_{stripper}$	CO_2 capture	Full factorial, Box–Behnken	n.d.	n.d.	[44]
Aminated activated carbons	$T_{amination}$	q_{CO_2}, $q_{CO_2,\,des}$	CCD	$Y_S = 27.121 + 2.962x_1 + 0.569x_2 + 0.787x_3 - 1.237x_1^2$ $-1.486x_2^2 - 0.292x_1x_2 + 0.83x_1x_3$ $Y_D = 93.70 - 2.63x_1 - 0.478x_2 - 0.253x_3 + 1.357x_1^2$ $+0.859x_2^2 + 0.192x_1x_2 - 0.328x_1x_3$	0.976, 0.964 0.988, 0.982	[32]

(Continued)

TABLE 10.4 *(Continued)*

Summary Works about CO_2 Capture Using RSM Techniques with Models [10]

Sorbent	Variables	Response	Method	Model	R^2, Adj-R^2	Ref.
Activated carbon	P_{des}, T_{des}, P/F ratiodes	q_{CO_2}	CCD	$q_{CO_2} = 2.647 + 0.001T_{des} + 0.268P/F\ ratio - 0.139(P/F\ ratio)^2$	0.879, 0.746	[45]
				$CO_2\ recovery = 93.377 + 0.029T_{des} + 9.447P/F\ ratio - 4.891(P/F\ ratio)^2$	0.880, 0.746	
				$Productivity = 7.942 + 0.003T_{des} + 0.804P/F\ ratio - 0.416(P/F\ ratio)^2$	0.880, 0.746	
				$H_2\ purity = 97.2066 + 0.0123T_{des} + 4.0052P/F\ ratio - 2.0771(P/F\ ratio)^2$	0.878, 0.742	
				$CO_2\ purity = 90.7025 - 0.0949T_{des} + 1.4473P_{des} - 12.9794P/F\ ratio$	0.867, 0.720	
				$r_{max\ des} = 29.100 + 0.208T_{des} - 3.436P_{des} - 11.930P/F\ ratio$	0.804, 0.586	

n.d.= not defined.

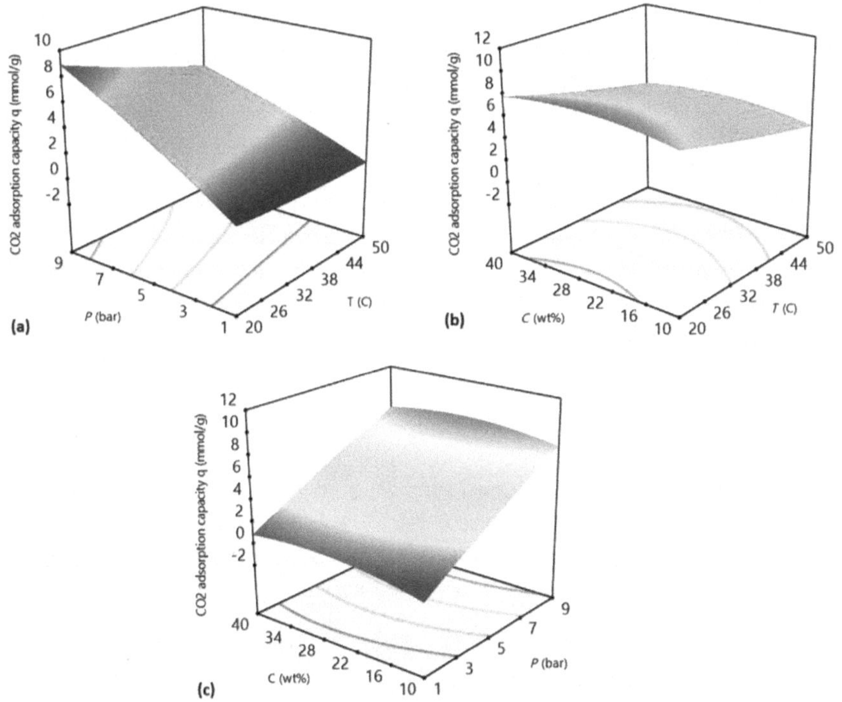

FIGURE 10.4 The response surface graphs depicting the CO_2 uptake capacity in relation to (a) the concentration of amine in weight percent, (b) pressure, and (c) temperature are illustrated [5], copyright (2023), Elsevier.

The results shown in Figure 10.4 highlighted that the influence of pressure and temperature on the capacity for CO_2 adsorption was more substantial than the impact of TEPA loading. These crucial factors in the CO_2 capture process are consistent with the results from the ANOVA analysis. Additionally, the relationships between CO_2 adsorption efficiency and the variables of pressure, temperature, and amine loading were illustrated in Figure 10.4 as the output. Overall, this study demonstrated the complex interplay between pressure, temperature, and amine concentration on CO_2 adsorption capacity. Increasing pressure and decreasing temperature were found to enhance CO_2 capture, while increasing amine concentration beyond a certain point led to decreased adsorption due to pore blockage and decreased physisorption. The findings suggest that a careful balance between these variables is necessary for optimal CO_2 capture efficiency. The ANOVA also highlighted the significance of the pressure-temperature interaction in influencing adsorption trends. These results emphasize the importance of considering multiple variables and their interactions when designing adsorption systems for CO_2 capture and suggest that model optimization may be more effective than factor optimization in achieving optimal performance (Figure 10.4a–c).

In another noteworthy study by Khoshraftar et al. (2023) [10], RSM was employed to model and optimize the adsorption of CO_2 using polyethylenimine (PEI)-functionalized halloysite adsorbents. The study utilized a five-level four-factor CCD to optimize the operational conditions for adsorption, including temperature (ranging from 20°C to 50°C), pressure (ranging from 1 to 9 bar), and PEI concentration (ranging from 10 to 40 wt%). The optimal values for temperature, pressure, and PEI concentration were determined to be 20°C, 9.00 bar, and 29.49 wt%, respectively. The adsorption capacity was found to be 8 mmol/g. The statistical analysis revealed a high correlation coefficient ($R^2 = 0.9868$) and adjusted-R^2 value (0.9846) for the models [10].

A research focused on enhancing the adsorption capacity of the nano clay montmorillonite adsorbent by modifying it with diethanolamine and optimizing the operating conditions for CO_2 adsorption [47]. The presented results demonstrate that the adsorption capacity reaches its maximum when operating conditions include high pressure and low temperature. This observation can be attributed to the physisorption mechanism of adsorption [47]

Physisorption refers to the attraction between gas molecules and the surface of the adsorbent, typically driven by weak intermolecular forces such as van der Waals interactions. At high pressure, the gas molecules are more likely to come into contact with the adsorbent surface, resulting in a higher adsorption capacity. Furthermore, the effect of temperature on adsorption capacity is seen to decrease with an increase in the concentration of diethanolamine. This can be explained by considering the chemical properties of diethanolamine. Diethanolamine contains nitrogen atoms which possess unpaired electrons. These unpaired electrons can form additional bonds with gas molecules, enhancing the adsorption process. As a result, the impact of temperature on adsorption is attenuated by the increased presence of unpaired electrons from diethanolamine. This aligns with the principles behind physisorption, as higher pressures promote a greater number of gas molecules coming into contact with the adsorbent surface, thereby enhancing adsorption capacity [47].

The primary objective of a different research study is to investigate methods for enhancing the CO_2 adsorption process on AC derived from olive waste through chemical modifications. To achieve this, the researchers used the RSM to investigate how different variables and their interactions affect the effectiveness of the process to determine the ideal conditions. Through the analysis of the data, it was determined that pressure and temperature have the greatest impact on the capacity of the AC to adsorb CO_2. The parameters identified through this model were statistically significant with a value of $p < 0.05$. Furthermore, the researchers developed a semiempirical correlation based on the optimal operating conditions of 25°C and 9 bar. The results demonstrated that RSM models are a valuable tool for evaluating CO_2 adsorption. These findings suggest that AC has a strong potential as an adsorbent for capturing CO_2. Additionally, the model's significance has been evaluated using a correlation coefficient (R^2) and adjusted coefficients ($R_{adj}{}^2$). However, a smaller p-value indicates a larger F value, indicating the significance of the model. Moreover, p-values below 0.04 indicate a strong statistical fit for the model. A strong indication of the reliability of this model is the high correlation value of 0.998 between the experimental values and calculated results. This suggests that the model is accurate in predicting the experimental data and can be effectively used in the analysis of CO_2 adsorption. To optimize the reaction conditions, it is important to analyze the relationship between pressure, temperature, and adsorption ability. It is clear that pressure plays a crucial role in determining the adsorption capacity [48].

Bahmanzadegan et al. gathered data on five synthesized zeolite sorbents to develop machine learning-based and response surface models for predicting CO_2 adsorption capacity. The graph in Figure 10.5a shows the relationship between predicted CO_2 uptake and actual CO_2 uptake via zeolites from kaolin [49]. The data appear to come from an average population, and the residuals (the difference between predicted and actual values) are randomly spread around a straight line. The adj-R^2 and pred-R^2 values for the quadratic model are 0.9254 and 0.9231, respectively, which are both above the acceptable threshold of 0.2. The accuracy of the model was also assessed using other parameters, and it was found to be satisfactory. The signal-to-noise ratio is 114.5087, indicating a good signal. The model is considered reliable and fit. The actual and predicted values show good agreement, and the standard plot of residuals confirms that the residuals follow a normal distribution. The expression perturbation

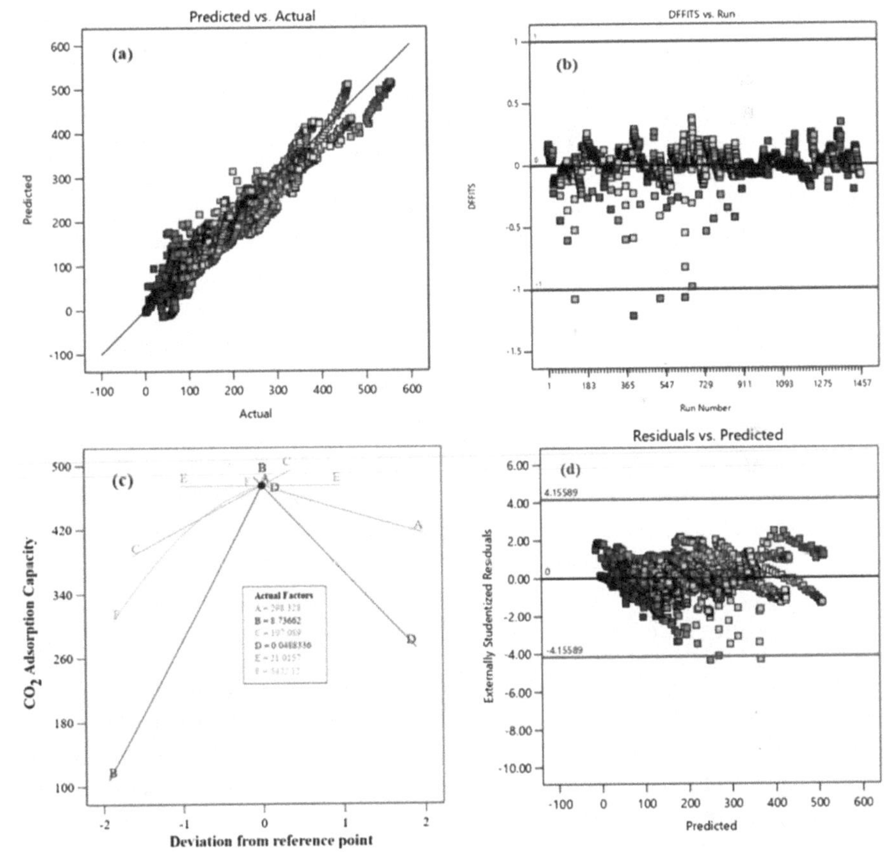

FIGURE 10.5 (a) The predicted adsorption capacity based on the CCD model; actual adsorption versus predicted, (b) predicted versus externally studentized residuals. (b) Assessing model fit: DFFTIS versus Run analysis, (c) a diagram showing the q values in relation to the deviation from the reference point (Perturbation diagram), (d) a normal representation of the residue [49], copyright (2023), Elsevier.

diagram in Figure 10.5c shows the deviations from the reference point, with pressure (P) having the highest deviation and diameter (D_p), having the lowest. Figure 10.5d depicts residual plots showing random variations without identifiable systematic trends. Overall, the model appears to be the most appropriate for predicting the CO_2 uptake in response to the given parameters.

Through the use of RSM graphics, specifically 3D surface plots, valuable insights were gained into the impact of different operating parameters on system conditions (Figure 10.6a–f). As evidenced by figures, the augmentation of pressure correlates with heightened CO_2 adsorption capacity within zeolites. This phenomenon arises from the intensified force exerted on CO_2 molecules, rendering them more susceptible to adhesion onto the zeolite surface. Consequently, elevated pressure engenders a greater adsorption capacity by accommodating more CO_2 molecules. In Figure 10.6d, the escalation in pore volume further facilitates the availability of active sites for CO_2 adsorption. A larger pore volume translates to increased spatial accommodation within the zeolite structure, enhancing the entrapment of CO_2 molecules and thereby amplifying adsorption capacity. Furthermore, an increase in the size of the pores in the zeolite material leads to an increase in the number of active sites available for the adsorption of CO_2 (Figure 10.6e). The presence of larger pores furnishes additional surface area for CO_2 molecules to interact with, thereby intensifying adsorption capacity. It was discovered that the highest level of CO_2 adsorption via lignin occurred when pressure was increased and temperature was lowered (Figure 10.6g and h) [9].

10.7 OPTIMIZATION OF CO_2 APPLICATION

10.7.1 Application of RSM Model for Optimization Purposes

RSM is a powerful statistical technique used for optimizing and improving the performance of processes. When applied to the field of CO_2 adsorbents, RSM becomes a valuable tool for designing experiments, analyzing the effects of various factors, and optimizing the conditions for maximum CO_2 adsorption efficiency. Comparing multiple approaches to enhance the efficiency of CO_2 capture constitutes a key focus of this research. The optimization efforts encompassed the refinement of all elements and factors associated with the adsorption process. This involved concurrent enhancements in the adsorbent structure, alongside the optimization of variables and operational conditions, ultimately resulting in the attainment of maximum adsorption capacity. Through an evaluation of variable efficiency, quadratic polynomial models were employed to identify the ideal conditions for temperature, pressure, and amine concentration. The TEPA-modified halloysite adsorbent demonstrated an optimal adsorption capacity of 9.3041 mmol/g (Table 10.5) [5].

The primary objective of a different study was to identify the most effective conditions for achieving the highest CO_2 adsorption capacity using a modified nano-halloysite adsorbent via polyethylenimine (PEI). To do this, the researchers analyzed various input and output variables and used numerical optimization to select the optimal values. They developed second order polynomial models to determine the optimal response conditions within the range of input variables. By studying the

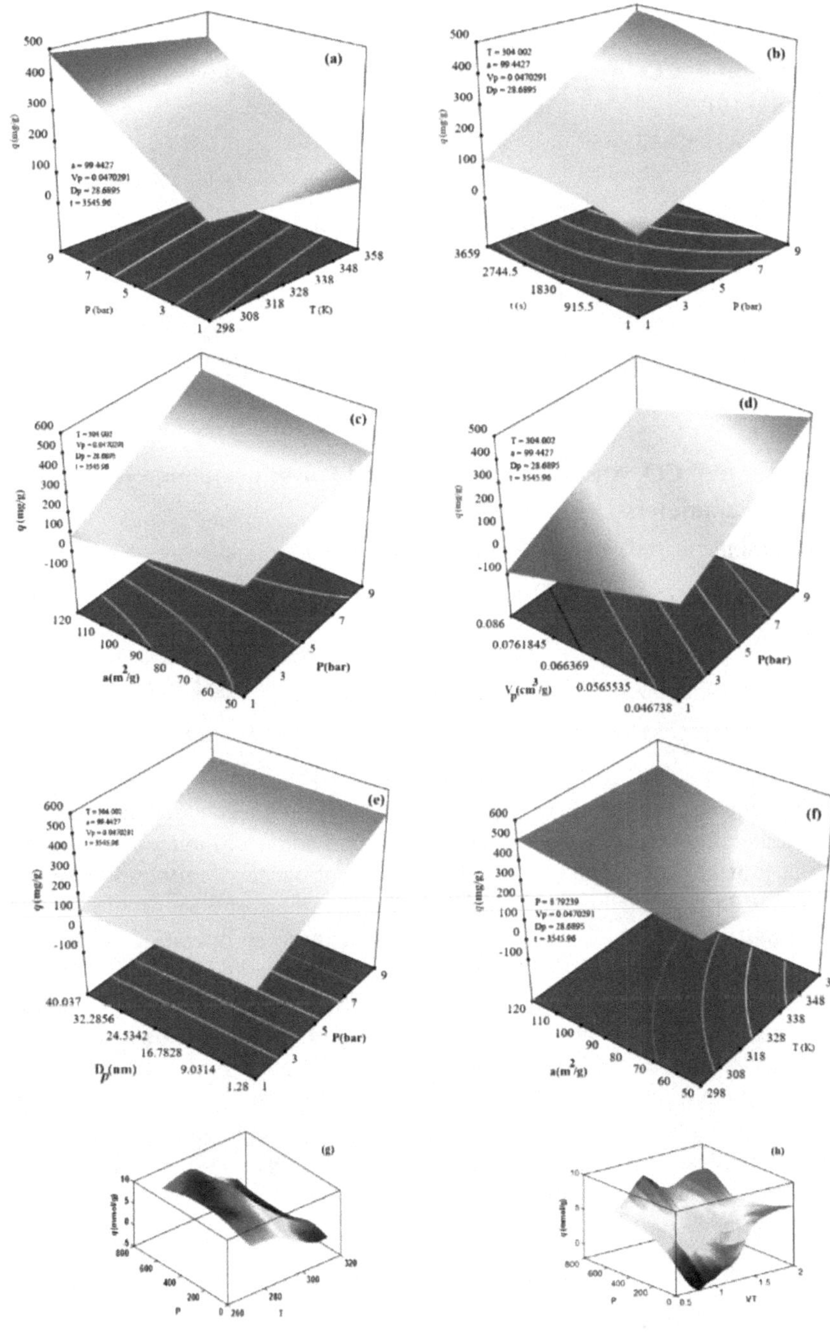

FIGURE 10.6 (a–f) RSM plots of carbon dioxide capture capacity influence of (a) T and P, (b) P and t, (c) P and a, (d) P and V_p, (e) P and D_p, and (f) T and a [49]. (g) The effects of pressure and temperature on CO_2 adsorption are depicted in one graph and (h) the effect of pressure and total pore volume [9], copyright (2024), Elsevier.

TABLE 10.5
Optimization of CO_2 Adsorption on TEPA-Modified Halloysite Parameters Using RSM [5]

Parameters	Constrain	Low	High	Lower Weight	Upper Weight	Importance
Temperature T (°C)	Is in range	20	50	1	1	3
Pressure P (bar)	Is in range	1	9	1	1	3
Concentration C (wt%)	Is in range	10	40	1	1	3
CO_2 adsorption capacity q (mmol/g)	Maximize	0.54	9.30	1	1	5

TABLE 10.6
Optimization of CO_2 Adsorption on Modified Polyethylenimine (PEI)-Nano-Halloysite Parameters Using RSM [10]

Parameters Units	Constrain	Low	High			
Temperature (°C)	In range	20	50	1	1	3
Pressure (bar)	In range	1	9	1	1	3
Concentration (wt%(	In range	10	40	1	1	3
CO_2 adsorption capacity (mmol/g)	Maximize	0.5536	7.8408	1	1	5

pressure, temperature, and loaded amine concentration, the researchers were able to achieve a maximum adsorption capacity of 7.598 mmol/g for the modified nano-halloysite adsorbent. The optimal values for this performance were found to be a concentration of 28.826, a temperature of 20°C, and a pressure of 9 bar (Table 10.6) [10].

A primary goal of a different study is to enhance the operating conditions to maximize carbon dioxide adsorption by montmorillonite and its modified variant. The optimization process takes into account temperature, pressure, and DEA loading within a defined range. The montmorillonite adsorbent successfully achieved the desired outcome by adsorbing 219.86 mg of material per gram at a temperature of 30°C and a pressure of 9 bar [47].

The researchers conducted an optimization process using numerical optimization techniques in Design Expert software to identify the best conditions for enhancing the adsorption capacity of modified AC adsorbents [4].

10.8 STRENGTHS AND LIMITATIONS OF RSM

Strengths of RSM [50–52]:

1. Efficiency in Optimization: RSM is particularly effective in optimizing processes with multiple variables. It helps identify the optimal conditions for a response variable by exploring the experimental space efficiently.

2. Reduced Experimental Runs: RSM requires fewer experimental runs compared to traditional one-factor-at-a-time methods. This can result in significant time and cost savings, especially when dealing with complex and time-consuming experiments.
3. Model Interpretability: The mathematical models generated by RSM are often simple and interpretable. This makes it easier for researchers and practitioners to understand the relationships between input variables and the response, facilitating better decision-making.
4. Visualization of Response Surfaces: RSM provides response surface plots that offer a visual representation of the relationship between input variables and the response. These plots help in understanding the interactions between variables and identifying regions of interest.
5. Incorporation of Factor Interactions: RSM allows for the exploration and identification of interactions between factors, providing insights into the combined effects of variables on the response.
6. Adaptability to Nonlinear Responses: RSM can handle nonlinear relationships between input variables and responses, making it suitable for a wide range of applications where responses may not follow a linear trend.

Limitations of RSM [50–52]:

1. Assumption of Polynomial Models: RSM assumes that the relationships between variables can be adequately represented by polynomial models. This assumption may not always hold true, especially in cases where the true underlying relationships are highly nonlinear.
2. Sensitivity to Experimental Design: The success of RSM depends on the careful design of experiments. Inadequate or biased experimental designs can lead to inaccurate models and unreliable optimization results.
3. Limited Exploration of Experimental Space: While RSM is efficient, it may not fully explore the entire experimental space. Important regions or factors might be overlooked if the initial design is not comprehensive.
4. Model Validation Challenges: Validating the accuracy of the generated models can be challenging, especially when the available data are limited. Overfitting can be a concern, leading to models that do not generalize well to new data.
5. Assumption of Stationary Operating Conditions: RSM assumes that the relationships between variables remain constant over the experimental domain. In dynamic systems or processes with changing conditions, this assumption may not be valid.
6. Applicability to Only Continuous Variables: RSM is primarily designed for continuous variables, and its effectiveness diminishes when dealing with discrete or categorical variables.

10.9 ARTIFICIAL NEURAL NETWORKS

The neural network methodology draws inspiration from the complex structures and functions of the human brain and nervous system. Over the past two decades, it has become widely adopted in network classifiers across various fields. An artificial neural network (ANN) is a computational or mathematical model designed to emulate both the structural and operational characteristics of biological brain networks [53]. Initially employed in the early 1940s for data analysis and classification in diverse subjects, ANN gradually gained prominence [54]. However, interest waned in the 1960s and 1980s due to significant computational challenges. Seminal work by scholars such as Hopfield, Rumelhart, Widrow, and Grossberg in the 1980s laid foundational principles to overcome these challenges. Subsequent advancements successfully addressed computational limitations, positioning neural networks as a leading methodology for achieving diverse objectives [55–57]. Current research endeavors aim to deepen our understanding of human brain function, informing the development of sophisticated artificial intelligence systems grounded in principles of neuroscience [58]. Notable among the advantages of neural networks is their effectiveness in addressing complex problems within compressed time frames. When faced with data lacking discernible correlations, ANNs excel in uncovering latent connections, mirroring the cognitive processes of the human brain. This computational paradigm offers several salient features, including expedited parallel processing, nonlinear computation capabilities, adaptability, robustness against noise, error tolerance, and iterative learning mechanisms [59]. In the realm of nonlinear applications, the ANN methodology holds sway and is frequently utilized due to its versatility in nonlinear analysis, classification, optimization, and data interpretation. These attributes collectively underscore the significance and utility of neural networks as a potent computational tool for addressing multifaceted challenges across diverse disciplines.

10.9.1 Main Architectures of ANNs

As mentioned earlier, ANNs are computational models designed to utilize the structural framework of the human brain for predictive analysis of complex and nonlinear systems. In the network's structure, artificial neurons are linked across different layers, which consist of input, hidden, and output layers, with each layer serving particular functions [60].

10.9.1.1 Input Layer

This foundational layer has the role of receiving various forms of information, data, signals, or features from the external environment. These inputs, often in the form of samples or patterns, undergo normalization procedures to ensure they fall within the output range of activation functions. This normalization process is crucial for enhancing numerical accuracy in subsequent mathematical operations conducted within the network. Eq. (10.22) was utilized to normalize the input and output dataset variables, which vary from –1 to 1.

$$X_{norm} = \frac{2 \times (X - X_{min})}{X_{max} - X_{min}} - 1 \tag{10.22}$$

where X_{norm} is the normalized data, X is a raw input variable, and X_{max} and X_{min} are the maximum and minimum of the dataset, respectively [10].

10.9.1.2 Hidden Layers

These intermediate layers are crucial for identifying and extracting relevant patterns associated with the analyzed process or system. They undertake the majority of internal processing tasks within the network.

10.9.1.3 Output Layer

Comprised of neurons, the output layer is responsible for generating and presenting the final results of the network. These results signify the culmination of processing performed by neurons in the preceding layers. The primary architectures of ANNs are based on the organization and connection of neurons, incorporating various arrangements. These architectures are broadly categorized as follows:

- **Single-Layer Feedforward Networks**
  Single-layer feedforward networks represent the most basic form of ANNs, wherein the input directly connects to the output layer without any hidden layers. However, none of the provided literature directly addresses single-layer feedforward networks.
- **Multilayer Feedforward Networks**
  Multilayer feedforward networks, alternatively termed multilayer perceptrons (MLPs), exhibit greater complexity by incorporating one or more hidden layers between the input and output layers. These networks possess the capability to learn intricate nonlinear relationships [61].
- **Recurrent Networks**
  Recurrent networks are distinguished by the presence of loops within the network structure, facilitating the persistence of information. These networks are well-suited for tasks involving sequential data processing [62].
- **Convolutional Networks**
  Convolutional networks are specialized for handling data structured in a grid-like topology, such as images. These networks employ convolutional layers to autonomously and adaptively learn spatial hierarchies of features [63].

A key characteristic of ANN is their ability to learn from provided examples, mimicking the behavior of a system. Consequently, once the network identifies the relationships between inputs and outputs, it becomes adept at generating solutions that approximate the expected or desired values for given input data. To accomplish this, the neural network undergoes a training process wherein deliberate adjustments are made to align synaptic weights and neuron thresholds, enabling the network to generalize solutions from the acquired relationships.

10.9.2 Train

The network design is established by analyzing experimental data using the neural network approach, with optimization of all parameters within the network model to derive the network matrix. Consequently, a mathematical model can be developed to represent the intricate relationships between input and output parameters. At the heart of data processing within the network is the neuron, serving as the fundamental computational unit that accepts numerical inputs and produces numerical outputs. Connections within the network allow the output of one neuron to serve as input for another, with each connection assigned a weight reflecting its relative importance. A single neuron may accommodate multiple input variables and output connections. The weighted sum undergoes an activation transfer process to generate the output, followed by the application of a threshold or cutoff value to the sum. This process is demonstrated in Eq. (10.23) for the input data [49].

$$net = (\sum_{i=1}^{N} \omega_i x_i) + b \tag{10.23}$$

The model utilizes x_1, x_2,. . ., and x_n as model input variables, with n denoting the number of input nodes. The weight factor of the neuron and input node is represented by w_{ij}, while b signifies the threshold value of a neuron i and indicates the output of a neuron i. The output values, y, are computed using Eq. (10.2), where the results are processed through a transfer function f, yielding the output values y according to Eq. (10.24)

$$y = f(net) \tag{10.24}$$

This training process employs a set of directed steps known as a learning algorithm, allowing the network to discern the unique characteristics of the system under investigation from the provided examples. Typically, the entire dataset reflecting the system's behavior is divided into two subsets: the training subset and the test subset. The training subset, comprising 60–90% of random samples from the complete dataset, is specifically utilized for training, while the test subset, encompassing 10–40% of the dataset, validates the network's ability to generalize solutions to an acceptable degree. Determining the dimensions of these subsets necessitates consideration of the statistical properties of the data. Throughout the training process, wherein each example from the training set is introduced to adjust synaptic weights and thresholds, a training session occurs.

Supervised learning, denoted as the 1–3–2 strategy, entails access to desired outputs corresponding to a set of input signals. In essence, each training sample comprises both input and output signals, forming a table or property/value set that elucidates the system's behavior. The neural structure, akin to a coach, formulates a "hypothesis" concerning the investigated system based on this information [64].

Synaptic weights and network thresholds are continuously adjusted using a learning algorithm that monitors the disparity between produced and desired outputs, facilitating refinement. Once these disparities fall within an acceptable range and considering the aim of generalizing solutions, the network is deemed "trained." Supervised learning embodies an inductive inference mode wherein the network's free variables are adjusted with prior knowledge of desired system outputs. Donald Hebb introduced the first supervised learning strategy in 1949, drawing inspiration from neuropsychological observations [64].

The neural network undergoes a training phase, during which it receives input–output pairs and adjusts the connection weights between neurons using a learning algorithm. Thus, the input signals are influenced by the connection weights and layer biases. An activation function is used to alter the total of these signals, with the goal of reducing the difference between the projected output and the actual output data.

10.9.3 Activation Function

Activation functions are fundamental elements in neural networks, characterized by various shapes such as linear, sigmoid, hyperbolic tangent, or rectified linear units. These functions introduce essential nonlinearity into the network, enhancing its capacity to discern complex patterns within datasets.

Neurons within the network are organized into layers, with each layer comprising interconnected neurons. Activation functions dictate how each neuron's output is computed based on the weighted sum of its inputs.

10.9.3.1 Purelin

The Purelin activation function, also known as the linear function, computes a neuron's output as a direct weighted sum of its inputs without nonlinear transformation. It can be expressed mathematically as Eq. (10.25):

$$f(x) = x \tag{10.25}$$

Purelin is particularly useful in the output layer of the network for tasks involving linear regression or classification without specific output constraints.

10.9.3.2 Logsig

The Logsig activation function, or logistic sigmoid, applies the sigmoid function to the weighted sum of inputs. Its mathematical formulation is [64]:

$$f(x) = \frac{1}{1+e^{-n}} \tag{10.26}$$

Logsig produces output values within the [0,1] range, making it suitable for tasks such as binary classification or probabilistic outputs.

10.9.3.3 Tansig

The Tansig activation function, representing a hyperbolic tangent sigmoid, utilizes the hyperbolic tangent function to transform the weighted sum of inputs. It can be defined as Eq. (10.27) [64]:

$$f(x) = \frac{2}{1+e^{-2n}} - 1 \tag{10.27}$$

Similar to Logsig, Tansig yields output values within the [–1, 1] range, making it advantageous for tasks requiring outputs with both positive and negative values [7, 64].

These activation functions are pivotal in shaping the network's behavior and learning dynamics, facilitating effective modeling of intricate relationships within datasets.

10.9.4 Types of Networks

The radial basis function (RBF) and the MLP stand as prominent architectural paradigms within ANNs [65]. It is essential to emphasize the basic difference between these networks, which arises from how neurons operate. The RBF architecture consists of three layers: an input layer, a hidden layer, and an output layer. Neurons in the hidden layer utilize RBFs as their activation functions. This methodology facilitates determining optimal solutions through a linear optimization strategy involving weight adjustments to minimize mean square error. Unlike other ANN structures, the MLP has an input layer that precedes the output layer. The intermediate and hidden layers connect the input and output layers using different activation functions. This design allows for more complex computations and pattern recognition [66].

Furthermore, it is noteworthy to mention the significance of the support vector machine (SVM) as another influential framework in machine learning. Unlike neural networks, SVM operates as a supervised learning model, aiming to identify the optimal hyperplane for class separation within the input space. SVM is particularly adept at handling high-dimensional data and is widely employed for classification tasks across diverse domains.

10.9.4.1 MLP Training Algorithms

The algorithm's learning mechanism encompasses both data's forward propagation and errors' backward propagation. At first, raw input data enters the model via the input layer, is processed initially in the hidden layer, and then proceeds to the output layer. If the deviation between the model's predictions and the actual outputs doesn't meet specified accuracy criteria, error backpropagation begins. This entails sending the difference between the model's output and the actual output backward through the hidden layer to the input layer. Training continues until the error between the model's predictions and the actual outputs falls within acceptable bounds or until a predetermined number of training iterations is reached [67]. Six types of backpropagation algorithms include Adaptive Momentum, Self-Adaptive Learning Rate, Resilient Backpropagation, Conjugate Gradient, Quasi-Newton, and Bayesian Regularization [68].

- **Single-Layer and Multilayer Networks**

Throughout the process of neural network learning, several algorithms have been introduced to reveal the connectivity patterns within neuron weights [69]. Among these methods, the MLP technique, described in Eq. (10.28), emerges as one of the most comprehensive approaches.

$$Y = f(wx_i^k + \theta) \tag{10.28}$$

where Y denotes the output vector, x^k represents the reference vector and signifies the weighted vector of coefficients, the MLP algorithm orchestrates the acquisition, processing, and outputting of information through input and output layers. The fundamental aim of the network's training algorithm is to reduce the mean squared error (MSE) [70]. Through the training process of the neural network

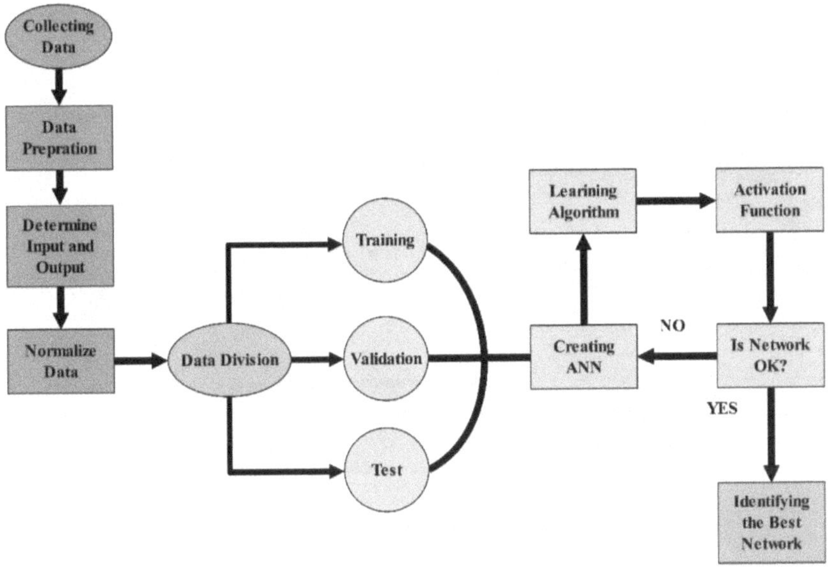

FIGURE 10.7 Schematic of the MLP structure.

feedforward propagation algorithm, the performance of the MLP algorithm is as outlined here [71]:

$$\gamma_{jk} = F_k(\sum_{i=1}^{N_{k-1}} w_{ijk}\gamma_{i(k-1)} + \beta_{jk}) \tag{10.29}$$

The neuron j's outputs from the k'slayer and the bias weight for a neuron j in a layer k are represented by γ_{jk} and β_{jk}, respectively. w_{ijk} are the link weights that were chosen at random at the start of the network training process, and F_k are the nonlinear activation transfer functions. Various functions, including the identity function, binary step function, binary, bipolar sigmoid, and, can be considered for activation. The schematic of the MLP network is presented in Figure 10.7.

10.9.4.2 Radial Basis Function

The RBF algorithm network, initially proposed by Broomhead and Lowe [72], is structured with a feedforward architecture featuring a single hidden layer. Input data from the input layers is transmitted to the hidden layer, where it undergoes processing through a transfer function, leading to nonlinear transformations. Utilizing these nonlinear transfer functions, the RBF algorithm establishes connections between the input and hidden layers. The individual hidden neurons within the network assess the distance or geometric dimension between the output vector and the weights. Eq. (10.30) elucidates the operation of the output layer within the RBF algorithm network, which relies on linear combiners [31].

$$f(x) = \sum_{i=1}^{N} w_{ij} G(\|x - c_i\| * b) \tag{10.30}$$

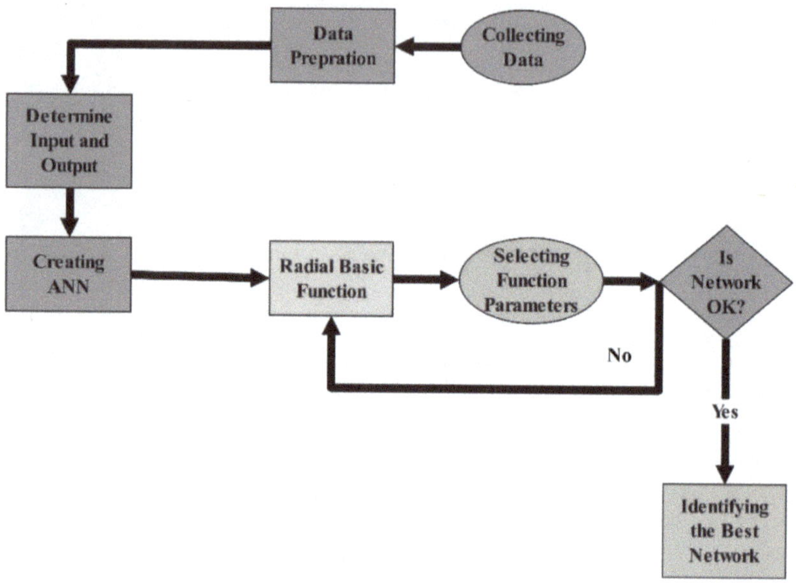

FIGURE 10.8 Schematic of the RBF structure.

N represents the number of training data sets, w_{ij} denotes the weight associated with each hidden neuron, x signifies the input vector, c_i represents the center points, and b denotes the bias. Employing an equation, Eq. (10.31) facilitates the determination of the centralized response from the hidden point [31]:

$$G(\|x - c_i\| * b) = \exp\left((-\frac{1}{2\sigma_i^2} (\|x - c_i\| * b)^2 \right) \tag{10.31}$$

σ_i represents the Gaussian function's spread. The described range denotes the extent across the input space to which the RBF neuron can react. Schematic of the RBF network is presented in Figure 10.8.

10.10 THE IMPORTANCE OF USING ANN FOR CO_2 ADSORPTION

As previously discussed, machine learning employs algorithms to identify and understand relationships between variables, including complex nonlinear interactions. ANNs are frequently applied to model and simulate adsorption processes due to their ability to capture intricate patterns in the data [73–75]. They can analyze the effects of input variables on both linear and nonlinear systems. The construction of an ANN necessitates interconnecting neurons across each layer [76]. The development of an ANN model for CO_2 adsorption prediction necessitates a systematic approach with several vital stages. Initially, data concerning CO_2 adsorption processes are gathered, encompassing variables such as temperature, pressure, CO_2 concentration, and adsorbent characteristics. This dataset is then subjected to preprocessing to rectify anomalies such as missing values or outliers, followed by normalization to ensure uniformity across variables. Subsequent to data preprocessing, the dataset is partitioned into distinct subsets for training, validation, and testing purposes. The architectural design of the model is then meticulously crafted, considering factors such as the neural network type (e.g., feedforward or convolutional), layer configuration, and choice of activation functions. The training process entails iteratively adjusting network parameters using optimization algorithms such as stochastic gradient descent to minimize the loss function. Hyperparameter optimization is conducted using the validation set to fine-tune parameters like learning rate and batch size, thereby optimizing the model's performance and guarding against overfitting. Following model training and tuning, its effectiveness is evaluated using the testing dataset, employing performance metrics such as mean absolute error and coefficient of determination for assessment. Upon achieving satisfactory evaluation results, the model is deployed for practical application, with ongoing monitoring and periodic retraining to accommodate new data and maintain optimal performance. Additionally, the trained model is analyzed to glean insights into CO_2 adsorption phenomena, including the identification of influential features. Continuous refinement and enhancement of the model are pursued through iterative iterations informed by domain expertise and advancements in neural network methodologies.

Utilizing ANNs for CO_2 adsorption by adsorbents provides several advantages. ANNs are adept at capturing complex relationships among various parameters involved in adsorption, such as temperature, pressure, adsorbent characteristics, and CO_2 concentration. This enables the development of comprehensive models that more accurately simulate real-world adsorption behavior compared to conventional methods. Additionally, ANNs excel at handling nonlinear relationships between input variables and adsorption performance, making them well-suited for predicting CO_2 adsorption behavior across diverse conditions. Furthermore, ANNs can learn from extensive datasets of experimental or simulated data, uncovering hidden patterns and correlations that traditional analytical approaches may miss. This data-driven methodology enhances prediction accuracy and facilitates the optimization of adsorption processes. ANNs also exhibit high flexibility and adaptability, allowing for the modeling of various types of CO_2 adsorbents and operational scenarios. Leveraging the predictive capabilities of ANNs enables researchers and engineers to optimize the design and operation of CO_2 adsorption systems, aiming to maximize efficiency, minimize energy consumption, and mitigate environmental impact. In summary, employing ANNs in CO_2 adsorption research and development offers a robust approach to understanding, predicting, and optimizing adsorption processes, thereby contributing to advancements in CCS technologies [49]. Some studies have investigated using ANN in understanding and analyzing adsorption processes (Table 10.7) [77–82]. Ghaemi et al. used RSM and ANN to model CO_2 adsorption and reduction by carbon molecular sieves, with the MLP network showing good agreement between experimental and predicted data [4]. Dashti et al. assessed the effectiveness of ANN in capturing CO_2 using zeolite 5A [83]. Gao et al. designed a multifactor prediction model using a backpropagation neural network to investigate how pore structure, mineral composition, and moisture content influence CO_2

TABLE 10.7
Review Studies on CO_2 Adsorption and Modeling Using ANN

Author	Variable	Structure	Activation Function	Training Algorithm	MSE	R^2	Ref
Rostami et al.	Activated carbon, pressure, and temperature	[9 4 1]	Tansig/purelin	LM	1E–5	0.9999	[88]
Khoshraftar et al.	Pressure, temperature, PEI concentration, CO_2 adsorption capacity	—	Tansig	BR	0.00011	0.9984	[10]
Khoshraftar et al.	Pressure, temperature, CO_2 adsorption capacity	[25 5]	Tansig	BR	0.00015	0.9983	[5]
Mashhadimoslem et al.	Precursors, activators, pyrolysis temperature, drop volume, adsorption pressure, adsorption temperature	[6 45 1]	Logsig/purelin	BR	0.0002	0.9716	[89]
Khoshraftar et al.	Activated carbon from Entada Africana Guill. & Perr, pressure, temperature, CO_2 adsorption capacity	[2 10 1]	Tansig/purelin	LM	9.45E–4	0.9998	[90]
Barki et al.	Specific surface area, micropore volume, temperature, pressure	[4 16 1]	Tansig/Tansig/purelin	BFGS	—	0.9965	[91]
Torkashvand et al.	Adsorption time, crosslinker ratio, adsorbent synthesis time	[3 15 10 1]	Tansig/purelin		—	0.9815	[28]
Moradi et al.	Adsorption time, synthesis time, crosslinker ratio, temperature, pressure	[6 10 8 7 1]	Tansig/purelin	BR	—	0.999	[92]
Mehrmohammadi et al.	Temperature, BET surface, volume, pressure	[45 10]	Logsig/Tansig/purelin	LM	2.6293E–5	0.9951	[93]
Norouzian et al.	Time, temperature, pressure,	[17 16]	Logsig/Tansig/purelin	LM	1.72E–04	0.998	[94]
Khalilzade et al.	Molecular weight, solution density, partial pressure of CO_2, temperature, total mass concentration	[8 7]	Tansig	LM	0.0045	0.9976	[95]
Baghban et al.	Pressure, temperature, and mass composition	100		LM	—	0.997	[96]
Mashhadimoslem et al.	Carbon biomass and surface area of the adsorbent	[45 45]	Tansig	LM	1.5E–6	0.9921	[97]
				RBF	0.0002	0.9994	
Norouzian et al.	Time, input temperature, input pressure, CO_2 adsorption capacity, and output pressure	[25 23 7 20]	Logsig/Tansig/purelin	LM	1.27E–04	0.999	[98]
				RBF	2.25E–4	0.998	
Khoshraftar et al.	Activated carbon from pistachio shells, pressure, temperature, and CO_2 adsorption capacity	—	Tansig	LM	1.14E–3	0.99905	[2]

Fathalian et al.	Graphene oxide, temperature, BET surface, volume, pressure, CO_2 adsorption capacity	[45 35 45 45]	Tansig	LM	0.024	0.9943	[75]
				RBF	0.103	0.9747	
Noroozian et al.	Temperature, pressure of the reactor, time, CO_2 adsorption capacity, and final pressure	[25 23 7 30]	Logsig/Tansig/purelin	LM	1.27E–04	0.99	[98]
Khoshraftar et al.	Temperature, pressure, and concentration	[25 5]	Purelin/trainbr	LM	0.00011	0.9984	[10]
Bahmanzadegan et al.	Temperature, BET surface, volume, time, and pressure	[30 30]	Purelin	LM	0.003021	0.9927	[49]
Moradi et al.	Adsorption time, ratio (FDA/benzene), synthesis time, pressure, temperature	[10 8 7]	Tansig/purelin	LM, BR and SCG	0.00016	0.999	[30]
				RBF	—	0.989	
Kianpour et al.	Temperature, absolute humidity, sorbent loading	[5]	Tansig	LM		1	[99]
Naeem et al.	Temperature, CO_2 vol percentage, air flow rate, time	[4,15,10,1]	Tansig	LM	—	0.983	[100]
Yulia et al.	BET surface area, pore volume, temperature, pressure, CH_4 adsorption capacity	[30]	Tansig	LM	—	0.99	[101]
Khoshraftar et al.	Temperature, pressure, and adsorption behavior of lignin	—	—	RF	0.0002199	0.9996	[9]
Khoshraftar et al.	Temperature, pressure, time, and CO_2 adsorption capacity	[25 10]	Tansig	LM	0.0004247	0.9990	[1]

LM: Levenberg–Marquardt, BR: Bayesian Regularization, BFGS: Broyden–Fletcher–Goldfarb–Shanno, RF: random forest, and SCG: scaled conjugate gradient.

adsorption in coal [84]. Raji et al. explored CO_2 capture using a variety of zeolites, such as 5A, 13X, T type, SSZ-13, and SAPO-34. They developed diverse models integrating ANN to forecast CO_2 adsorption on these zeolites [85]. Inchausti used an ANN model to predict the breakthrough behavior of a binary system in CuBTC adsorbent, with parametric studies showing the impact of temperature, feed flow rate, and pressure on breakthrough time [86]. Bahrun applied a simple ANN to predict the breakthrough time of CO_2 in a packed bed adsorption column, achieving high accuracy between actual and predicted values [87]. Fathalian et al. developed an ANN-based model to predict the CO_2 adsorption capacity of graphene oxide-based sorbents, with the MLP method showing the best performance [75].

As highlighted in the studies presented in Table 10.7, the utilization of ANN approaches underscores their significant role in elucidating and refining CO_2 adsorption processes. The table details the parameters studied, including temperature, pressure, surface area, pore volume, and the number of layers in the ANN architecture. Additionally, it presents the coefficient of determination (R^2) and MSE values derived from these models. Moreover, the table includes information about the activation functions employed, providing valuable insights into the methodology used in each study. Through ANN, researchers gain a comprehensive understanding of the impact of diverse parameters, facilitating the development of robust models for forecasting and enhancing CO_2 adsorption capacity. This advancement is pivotal in fostering the creation of effective and sustainable techniques for CCS, thereby aiding in the global effort to combat climate change. Mashhadimoslem et al. conducted a study focusing on the intricate modeling of CO_2 uptake interactions within the synthesis process of AC adsorbents sourced from diverse biomass materials. This approach aims to enhance the development of effective adsorbents tailored for CO_2 capture. The research drew upon data from a comprehensive analysis of AC synthesis experiments sourced from 35 scholarly publications to construct operability simulations employing advanced ANN methodologies. Two distinct ANN architectures, employing MLP and RBF algorithms, were devised to predict CO_2 adsorption capacity. A range of parameters, including precursor materials, activating agents, pyrolysis temperatures, pore volumes, adsorption pressures, temperatures, specific surface areas (BET), and CO_2 adsorption capacities, were carefully considered as input and output variables within the models. Detailed schematic representations of the MLP and RBF structures, showcasing the input–output configurations, are provided in Figures 10.9 and 10.10, respectively [89].

This investigation employs a range of neuron activation functions, utilizing sigmoid/Gaussian for the hidden layers and Pureline for the output layer. The selection of the optimal number of neurons during the training phase of the algorithm is based on minimizing the MSE and maximizing performance metrics. MSE was calculated as Eq. (10.32) and shown in Figure 10.11.

$$MSE = \frac{1}{N}\sum_{i=1}^{N}\left(Y_{predicted} - Y_{actual}\right)^2 \tag{10.32}$$

Mehrmohammadi et al. explored the utilization of MLP neural networks employing 13 different learning algorithms to simulate CO_2 adsorption. The input

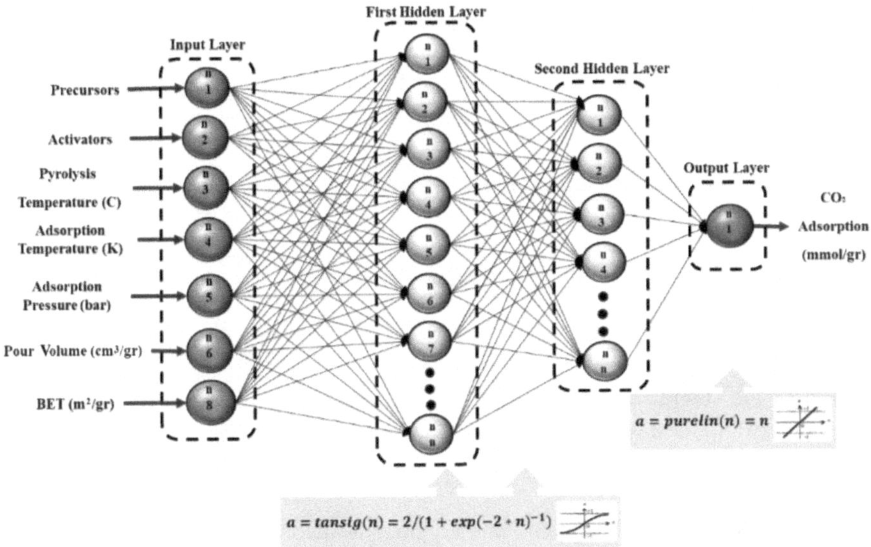

FIGURE 10.9 Schematic of MLP structure [89], copyright © 2021 American Chemical Society.

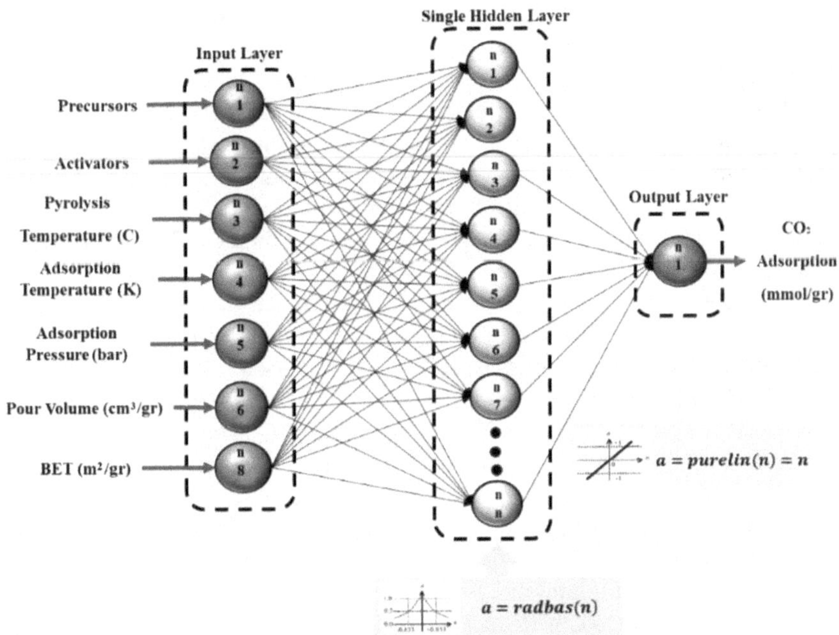

FIGURE 10.10 Schematic of RBF structure [89], copyright © 2021 American Chemical Society.

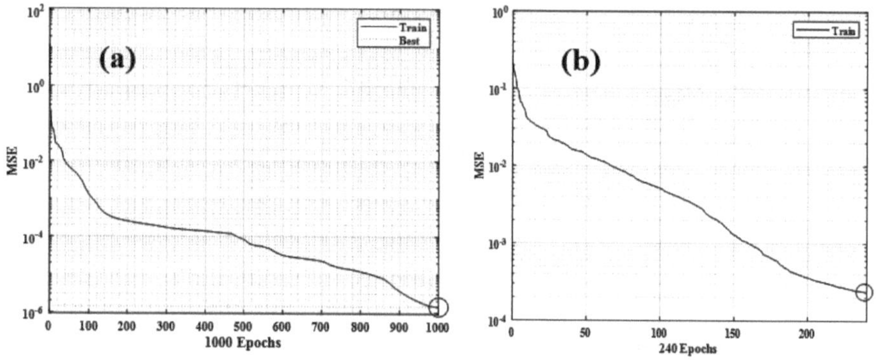

FIGURE 10.11 MSE for the MLP and RBF models: (a) CO_2 adsorption–MLP and (b) CO_2 adsorption–RBF [89], copyright © 2021 American Chemical Society.

parameters for these models encompassed various adsorbent properties such as surface area, mesopore volume, and micropore volume, along with operational factors like temperature and pressure. The examination using the Pearson correlation coefficient matrix (Figure 10.12) indicates that there isn't a notable linear correlation between the adsorbents and their CO_2 adsorption capacity. Weak associations are noticed between CO_2 adsorption capacity and both mesopore and micropore volumes. However, a significant positive correlation between micropore volume and BET surface area is identified. Additionally, CO_2 adsorption capacity demonstrates a strong positive correlation with pressure and a moderately negative correlation with temperature [93].

Notably, the LM algorithm demonstrated superior accuracy, achieving an MSE value of 2.6293E−5 and an R^2 of 0.9951. In contrast, the gestational diabetes mellitus algorithm demonstrated the least accuracy with an MSE value of 4.4192E−4 and a correlation coefficient of 0.9138. Through comparison with experimental data, depicted in Figure 10.13, the congruence between the predicted and observed data is evident.

Bahmanzadegan et al. applied ANN and RSM to forecast the performance of zeolites in CO_2 adsorption. For the hidden layers, they implemented a tangent sigmoid transfer function, while a linear transfer function was used for the output layer. The optimized MLP model yielded high predictive accuracy, with an R^2 value of 0.9927 and a low MSE of 0.003021 after 93 training epochs. A 3D plot was subsequently generated to analyze the parameter interactions and further optimize CO_2 adsorption capacity. The three-dimensional representations delineated in Figure 10.14 elucidate the response surface of an ANN model employing MLP to forecast q. These visual depictions delineate alterations in five factors while holding the remaining two constants, facilitating an in-depth analysis of their collective impact on the outcome. A comparative study between RSM and ANN outcomes reveals congruent trends, yet the ANN model demonstrates superior efficacy in predicting parameter interactions. Temperature emerges as a pivotal determinant influencing zeolite's CO_2 adsorption capacity. As depicted in Figure 10.14a and f, heightened temperatures

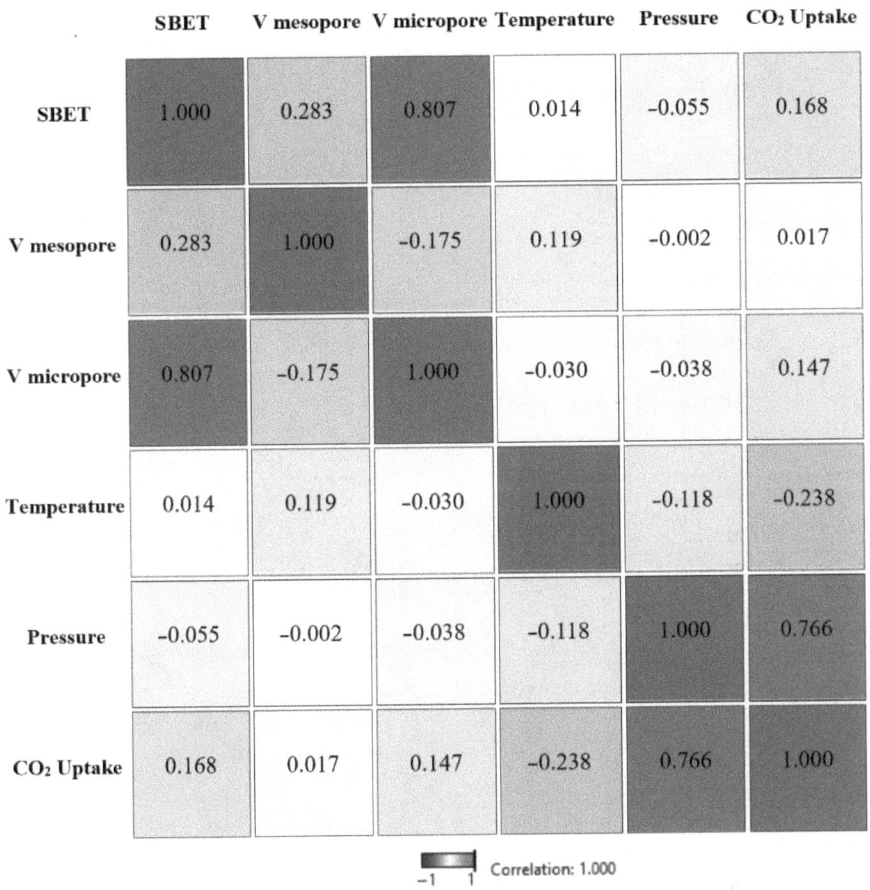

FIGURE 10.12 Pearson correlation matrix for CO_2 adsorption capacity [93], copyright © 2023 Nature.

diminish the adsorption capacity of zeolites for CO_2 owing to the heightened kinetic energy of gas molecules, thereby impeding CO_2 adsorption onto the zeolite surface. Conversely, lower temperatures facilitate CO_2 adsorption as gas molecules exhibit reduced kinetic energy, enhancing their capture by zeolites. Pressure also significantly influences CO_2 adsorption capacity, with elevated pressures augmenting zeolite's adsorption capability for CO_2. The MLP analysis, depicted in Figure 10.14a–e, reveals a positive correlation between increased pressure and CO_2 adsorption capacity in zeolites. This correlation is attributed to the greater force exerted on CO_2 molecules at higher pressures, enhancing their adsorption onto the zeolite surface. Additionally, Figure 10.14c and f emphasizes that zeolites with larger surface areas provide more adsorption sites, resulting in a significant increase in CO_2 adsorption capacity. Figure 10.14d and e further illustrates that larger pore diameters and volumes facilitate the capture of CO_2 molecules, thereby boosting the overall adsorption capacity [49].

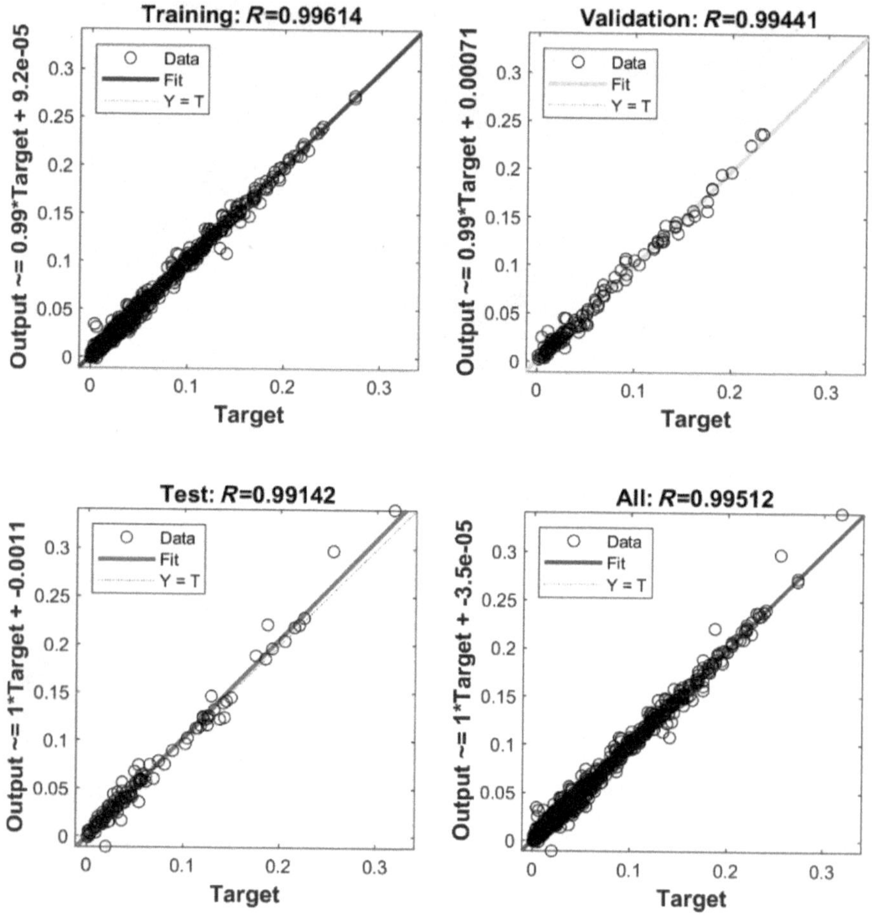

FIGURE 10.13 Evaluating the experimental data with the outcomes of the MLP neural network model across various datasets: (a) training, (b) validation, (c) test, and (d) overall data [93], copyright © 2023 Nature.

10.11 COMPRESSION OF RSM AND ANN MODELS

RSM is a statistical technique that is commonly used for modeling and optimizing processes. It relies on mathematical models that depict the correlation between input variables and output responses. RSM is typically used when the underlying relationships between variables are well understood and can be represented by simple mathematical equations. On the other hand, ANN is a more complex and flexible approach that is based on the structure and function of the human brain. It is particularly well-suited for problems that involve complex, nonlinear relationships between variables. ANN can be used to model and predict outcomes in situations where the underlying relationships are not well understood or cannot be easily represented by mathematical equations. When comparing RSM and ANN, it is important to consider

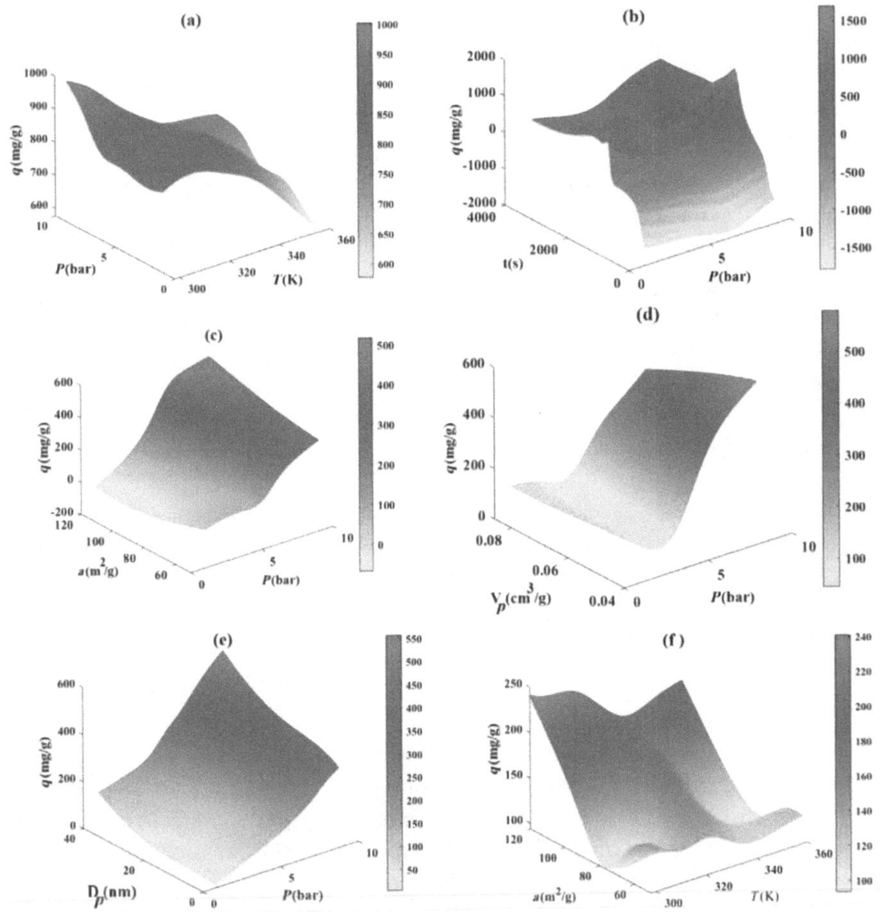

FIGURE 10.14 3D plots produced by the MLP model offer CO_2 adsorption capacity to examine the impact of (a) T and P, (b) P and t, (c) P and a, (d) P and V_p, (e) P and D_p, and (f) T and a [49], copyright (2023), Elsevier.

factors such as the complexity of the problem, the amount of available data, and the level of accuracy required for the analysis. RSM is generally more straightforward to implement and interpret, but may not be suitable for problems with highly nonlinear relationships. ANN, on the other hand, can handle complex relationships more effectively but may require more data and computational resources.

Ultimately, the choice between RSM and ANN will depend on the specific characteristics of the problem being studied and the goals of the analysis. It may be beneficial to compare the performance of both methods on a test dataset or through cross-validation to determine which approach is most appropriate for the given problem. Khoshraftar et al. compared the effectiveness of two models, RSM and ANN, in the context of compression. The performance of each model was evaluated

TABLE 10.8
Comparative Study of ANNs Models and RSM Based on MSE and R^2

Researchers	Models	MSE	R^2	Adsorbents	Ref.
Khoshraftar et al.	MLP-ANN	0.00040	0.9990	MWCNT	[1]
Khoshraftar et al.	RSM	0.02000	0.9054	MWCNT	[1]
Ghaemi et al.	MLP-ANN	0.00062	0.9978	CMS	[4]
Ghaemi et al.	RBF-ANN	0.00510	0.9497	CMS	[4]
Ghaemi et al.	RSM	0.10000	0.8958	CMS	[4]
Khoshraftar et al.	MLP-ANN	0.00115	0.9999	Pistachio-AC	[2]
Khoshraftar et al.	RSM	0.00500	0.9957	Pistachio-AC	[2]
Khoshraftar et al	RSM	0.00100	0.9916	Halloysite nanotubes modified by tetraethylenepentamine	[5]
Khoshraftar et al	MLP-ANN	0.00002	0.9995	Halloysite nanotubes modified by tetraethylenepentamine	[5]
Khoshraftar et al	RSM	0.00300	0.9868	Polyethylenimine-functionalized halloysite nanotube	[10]
Khoshraftar et al	MLP-ANN	0.00011	0.9983	Polyethylenimine-functionalized halloysite nanotube	[10]

using two key metrics: MSE and R^2. MSE quantifies the average squared difference between predicted and actual values, with lower values indicating greater accuracy. R^2 measures the proportion of variance in the dependent variable explained by the independent variables, with higher values reflecting a better model fit. In several studies, the performance of RSM and ANN models in compression was compared using these metrics. The RSM model achieved an R^2 of 0.90 and an MSE of 0.02, demonstrating a strong fit to the data. However, the MLP model performed even better with an MSE of 0.0004 and R^2 of 0.99, showing its superior predictive power and ability to explain variability in the data [1]. It was concluded that the ANN models outperformed the RSM model in comparison (Table 10.8).

10.12 CONCLUSION

This chapter has delved into the crucial realm of CO_2 adsorbent modeling and optimization, with a particular focus on the potent tools of RSM and ANN techniques. The imperative nature of addressing the escalating atmospheric CO_2 levels and their profound impact on climate change has propelled the need for effective CO_2 adsorbents, and this book endeavors to shed light on the promising potential of RSM and ANN in this pursuit. Throughout the chapters, we have navigated the foundational principles underpinning RSM and ANN, explored their application in the context of CO_2 adsorption, and provided a thorough examination of experimental methods. The comparative analysis presented in this book offers a valuable perspective on the strengths and limitations of both modeling approaches, aiding researchers and practitioners in making informed decisions based on the unique demands of their projects. As we stand at the forefront of climate change mitigation efforts, the insights derived from this comprehensive guide can pave the way for innovative and sustainable solutions. The synergy of RSM and ANN not only provides a robust framework for optimizing CO_2 adsorbents but also opens avenues for further research and development. By fostering a deeper understanding of these modelling techniques, we hope to catalyze advancements in the field and contribute meaningfully to global initiatives aimed at curbing the adverse effects of CO_2 emissions. In the future, as technology and research evolve, the integration of advanced modeling approaches with emerging materials and experimental methodologies holds the promise of yielding even more efficient and environmentally friendly CO_2 adsorbents. This chapter serves as a springboard for future investigations, encouraging a continued dialogue among researchers, fostering collaboration, and ultimately driving progress in the crucial mission of mitigating climate change through the optimization of CO_2 adsorbents.

REFERENCES

1. Z. Khoshraftar, A. Ghaemi, and A. Hemmati, "Comprehensive investigation of isotherm, RSM, and ANN modeling of CO_2 capture by multi-walled carbon nanotube," *Scientific Reports,* vol. 14, no. 1, p. 5130, 2024.
2. Z. Khoshraftar and A. Ghaemi, "Evaluation of pistachio shells as solid wastes to produce activated carbon for CO_2 capture: Isotherm, response surface methodology (RSM) and artificial neural network (ANN) modeling," *Current Research in Green and Sustainable Chemistry,* vol. 5, p. 100342, 2022.
3. Z. Khoshraftar, H. Masoumi, and A. Ghaemi, "Experimental, response surface methodology (RSM) and mass transfer modeling of heavy metals elimination using dolomite powder as an economical adsorbent," *Case Studies in Chemical and Environmental Engineering,* vol. 7, p. 100329, 2023.
4. A. Ghaemi, M. K. Dehnavi, and Z. Khoshraftar, "Exploring artificial neural network approach and RSM modeling in the prediction of CO_2 capture using carbon molecular sieves," *Case Studies in Chemical and Environmental Engineering,* vol. 7, p. 100310, 2023.
5. Z. Khoshraftar, F. S. Taheri, S. Nezami, and A. Ghaemi, "Using halloysite nanotubes modified by tetraethylenepentamine for advanced carbon capture: experimental and modeling via RSM and ANNs," *Chemical Engineering Journal Advances,* vol. 16, p. 100543, 2023.
6. Z. Khoshraftar and A. Ghaemi, "Modeling of CO_2 solubility in piperazine (PZ) and diethanolamine (DEA) solution via machine learning approach and response surface methodology," *Case Studies in Chemical and Environmental Engineering,* vol. 8, p. 100457, 2023.
7. Z. Khoshraftar and A. Ghaemi, "Modeling and prediction of CO_2 partial pressure in methanol solution using artificial neural networks," *Current Research in Green and Sustainable Chemistry,* vol. 6, p. 100364, 2023.
8. Z. Khoshraftar and A. Ghaemi, "Presence of activated carbon particles from waste walnut shell as a biosorbent in monoethanolamine (MEA) solution to enhance carbon dioxide absorption," *Heliyon,* vol. 8, no. 1, 2022.
9. Z. Khoshraftar and A. Ghaemi, "Enhanced carbon dioxide adsorption using lignin-derived and nitrogen-doped porous carbons: A machine learning approaches, RSM and isotherm modeling," *Case Studies in Chemical and Environmental Engineering,* p. 100668, 2024.

10. Z. Khoshraftar, A. Ghaemi, and F. S. Taheri, "Polyethylenimine-functionalized halloysite nanotube as an adsorbent for CO_2 capture: RSM and ANN methodology," *Current Research in Green and Sustainable Chemistry,* vol. 7, p. 100389, 2023.
11. Z. Khoshraftar and A. Ghaemi, "Prediction of CO_2 solubility in water at high pressure and temperature via deep learning and response surface methodology," *Case Studies in Chemical and Environmental Engineering,* vol. 7, p. 100338, 2023.
12. Z. Khoshraftar, A. Ghaemi, and A. H. Mohseni Sigaroodi, "The effect of solid adsorbents in triethanolamine (TEA) solution for enhanced CO_2 absorption rate," *Research on Chemical Intermediates,* vol. 47, pp. 4349–4368, 2021.
13. N. Sadeghi, S. Sharifnia, and T.-O. Do, "Optimization and modeling of CO_2 photoconversion using a response surface methodology with porphyrin-based metal organic framework," *Reaction Kinetics, Mechanisms and Catalysis,* vol. 125, pp. 411–431, 2018.
14. A. I. Khuri and S. Mukhopadhyay, "Response surface methodology," *Wiley Interdisciplinary Reviews: Computational Statistics,* vol. 2, no. 2, pp. 128–149, 2010.
15. M. Mourabet, A. El Rhilassi, H. El Boujaady, M. Bennani-Ziatni, and A. Taitai, "Use of response surface methodology for optimization of fluoride adsorption in an aqueous solution by Brushite," *Arabian Journal of Chemistry,* vol. 10, pp. S3292–S3302, 2017.
16. M. Amiri, S. Shahhosseini, and A. Ghaemi, "Optimization of CO_2 capture process from simulated flue gas by dry regenerable alkali metal carbonate based adsorbent using response surface methodology," *Energy & Fuels,* vol. 31, no. 5, pp. 5286–5296, 2017.
17. H. Pashaei, H. Mashhadimoslem, and A. Ghaemi, "Modeling and optimization of CO_2 mass transfer flux into Pz-KOH-CO2 system using RSM and ANN," *Scientific Reports,* vol. 13, no. 1, p. 4011, 2023.
18. M. Saeidi, A. Ghaemi, K. Tahvildari, and P. Derakhshi, "Exploiting response surface methodology (RSM) as a novel approach for the optimization of carbon dioxide adsorption by dry sodium hydroxide," *Journal of the Chinese Chemical Society,* vol. 65, no. 12, pp. 1465–1475, 2018.
19. B. Tyagi, C. D. Chudasama, and R. V. Jasra, "Determination of structural modification in acid activated montmorillonite clay by FT-IR spectroscopy," *Spectrochimica Acta Part A: Molecular and Biomolecular Spectroscopy,* vol. 64, no. 2, pp. 273–278, 2006.
20. D. Baş and İ. H. Boyacı, "Modeling and optimization I: Usability of response surface methodology," *Journal of Food Engineering,* vol. 78, no. 3, pp. 836–845, 2007.
21. L. A. Sarabia, M. C. Ortiz, and M. S. Sánchez, "Response surface methodology," *Comprehensive Chemometrics*, vol. 1, pp. 345–390, 2009.
22. M. Gil, M. Martínez, S. Garcia, F. Rubiera, J. Pis, and C. Pevida, "Response surface methodology as an efficient tool for optimizing carbon adsorbents for CO_2 capture," *Fuel Processing Technology,* vol. 106, pp. 55–61, 2013.
23. A. Nuchitprasittichai and S. Cremaschi, "Optimization of CO_2 capture process with aqueous amines a comparison of two simulation–optimization approaches," *Industrial & Engineering Chemistry Research,* vol. 52, no. 30, pp. 10236–10243, 2013.
24. H. Godarziani, H. Ramezanipour Penchah, and A. Ghaemi, "Triethanolamine-modified montmorillonite clay as a new adsorbent for CO_2 capture: Characterization, adsorption, and RSM modeling," *Journal of the Chinese Chemical Society,* vol. 69, no. 12, pp. 1981–1996, 2022.
25. N. K. Mohammad, A. Ghaemi, and K. Tahvildari, "Hydroxide modified activated alumina as an adsorbent for CO_2 adsorption: Experimental and modeling," *International Journal of Greenhouse Gas Control,* vol. 88, pp. 24–37, 2019.
26. F. Bahmanzadegan, M. A. Pordsari, and A. Ghaemi, "Improving the efficiency of 4A-zeolite synthesized from kaolin by amine functionalization for CO_2 capture," *Scientific Reports,* vol. 13, no. 1, p. 12533, 2023.

27. M. Khajeh and A. Ghaemi, "Exploiting response surface methodology for experimental modeling and optimization of CO_2 adsorption onto NaOH-modified nanoclay montmorillonite," *Journal of Environmental Chemical Engineering,* vol. 8, no. 2, p. 103663, 2020.
28. A. Torkashvand, H. Ramezanipour Penchah, and A. Ghaemi, "Exploring of CO_2 adsorption behavior by carbazole-based hypercrosslinked polymeric adsorbent using deep learning and response surface methodology," *International Journal of Environmental Science and Technology,* vol. 19, no. 9, pp. 8835–8856, 2022.
29. M. Khajeh and A. Ghaemi, "Strontium hydroxide-modified nanoclay montmorillonite for CO_2 capture: Response surface methodology and adsorption mechanism," *International Journal of Environmental Analytical Chemistry,* vol. 103, no. 17, pp. 5311–5336, 2023.
30. M. R. Moradi, H. Ramezanipour Penchah, and A. Ghaemi, "CO2 capture by benzene-based hypercrosslinked polymer adsorbent: Artificial neural network and response surface methodology," *The Canadian Journal of Chemical Engineering,* vol. 101, no. 10, pp. 5621–5642, 2023.
31. A. Ghaemi, H. Mashhadimoslem, and P. Zohourian Izadpanah, "NiO and MgO/activated carbon as an efficient CO_2 adsorbent: Characterization, modeling, and optimization," *International Journal of Environmental Science and Technology,* vol. 19, no. 2, pp. 727–746, 2022.
32. M. S. Shafeeyan, W. M. A. W. Daud, A. Houshmand, and A. Arami-Niya, "The application of response surface methodology to optimize the amination of activated carbon for the preparation of carbon dioxide adsorbents," *Fuel,* vol. 94, pp. 465–472, 2012.
33. D. S. Dao, H. Yamada, and K. Yogo, "Response surface optimization of impregnation of blended amines into mesoporous silica for high-performance CO_2 capture," *Energy & Fuels,* vol. 29, no. 2, pp. 985–992, 2015.
34. L. S. Tan, M. S. Azmi, L. K. Keong, and M. A. Bustam, "Application of response surface methodology to investigate CO_2 absorption column temperature rise," *Advanced Materials Research,* vol. 917, pp. 257–266, 2014.
35. M. H. El-Naas, A. F. Mohammad, M. I. Suleiman, M. Al Musharfy, and A. H. Al-Marzouqi, "Statistical analysis and optimization of a process for CO_2 capture," *International Journal of Chemical and Molecular Engineering,* vol. 10, no. 4, pp. 397–404, 2016.
36. A. F. Mohammad, M. H. El-Naas, M. I. Suleiman, and M. Al Musharfy, "Optimization of a solvay-based approach for CO_2 capture," *International Journal of Chemical Engineering and Applications,* vol. 7, no. 4, p. 230, 2016.
37. D.-M. K. Nguyen, T. Imai, T.-L. T. Dang, A. Kanno, T. Higuchi, K. Yamamoto, and M. Sekine, "Response surface method for modeling the removal of carbon dioxide from a simulated gas using water absorption enhanced with a liquid-film-forming device," *Journal of Environmental Sciences,* vol. 65, pp. 116–126, 2018.
38. M. Karimi, J. A. C. Silva, C. N. D. P. Gonçalves, J. L. Diaz de Tuesta, A. E. Rodrigues, and H. T. Gomes, "CO2 capture in chemically and thermally modified activated carbons using breakthrough measurements: experimental and modeling study," *Industrial & Engineering Chemistry Research,* vol. 57, no. 32, pp. 11154–11166, 2018.
39. A. Hemmati, H. Rashidi, A. Hemmati, and A. Kazemi, "Using rate based simulation, sensitivity analysis and response surface methodology for optimization of an industrial CO_2 capture plant," *Journal of Natural Gas Science and Engineering,* vol. 62, pp. 101–112, 2019.
40. S. M. Hosseini-Ardali, M. Hazrati-Kalbibaki, M. Fattahi, and F. Lezsovits, "Multi-objective optimization of post combustion CO_2 capture using methyldiethanolamine (MDEA) and piperazine (PZ) bi-solvent," *Energy,* vol. 211, p. 119035, 2020.

41. N. Maleki and K. Motahari, "Absorption performance of carbon dioxide in 4-Hydroxy-1-methylpiperidine+ aminoethylethanolamine aqueous solutions: Experimental measurement and modeling," *Journal of Natural Gas Science and Engineering,* vol. 56, pp. 1–17, 2018.
42. S. Babamohammadi, A. Shamiri, T. N. G. Borhani, M. S. Shafeeyan, M. K. Aroua, and R. Yusoff, "Solubility of CO_2 in aqueous solutions of glycerol and monoethanolamine," *Journal of Molecular Liquids,* vol. 249, pp. 40–52, 2018.
43. B. Morero, E. S. Groppelli, and E. A. Campanella, "Evaluation of biogas upgrading technologies using a response surface methodology for process simulation," *Journal of Cleaner Production,* vol. 141, pp. 978–988, 2017.
44. A. Nuchitprasittichai and S. Cremaschi, "Optimization of CO_2 capture process with aqueous amines using response surface methodology," *Computers & Chemical Engineering,* vol. 35, no. 8, pp. 1521–1531, 2011.
45. S. Garcia, M. Gil, J. Pis, F. Rubiera, and C. Pevida, "Cyclic operation of a fixed-bed pressure and temperature swing process for CO_2 capture: Experimental and statistical analysis," *International Journal of Greenhouse Gas Control,* vol. 12, pp. 35–43, 2013.
46. Z. Zhou, C. M. Anderson, S. K. Butler, S. K. Thompson, K. J. Whitty, T.-C. Shen, and K. J. Stowers, "Stability and efficiency of CO_2 capture using linear amine polymer modified carbon nanotubes," *Journal of Materials Chemistry A,* vol. 5, no. 21, pp. 10486–10494, 2017.
47. H. Ramezanipour Penchah, A. Ghaemi, and H. Godarziani, "Eco-friendly CO_2 adsorbent by impregnation of diethanolamine in nanoclay montmorillonite," *Environmental Science and Pollution Research,* vol. 28, no. 39, pp. 55754–55770, 2021.
48. H. Jedli, M. Almonnef, R. Rabhi, M. Mbarek, J. Abdessalem, and K. Slimi, "Activated carbon as an adsorbent for CO_2 capture: Adsorption, kinetics, and RSM modeling," *ACS Omega,* vol. 9, no. 2, pp. 2080–2087, 2023.
49. F. Bahmanzadegan and A. Ghaemi, "Exploring the effect of zeolite's structural parameters on the CO_2 capture efficiency using RSM and ANN methodologies," *Case Studies in Chemical and Environmental Engineering,* vol. 9, p. 100595, 2024.
50. R. H. Myers, D. C. Montgomery, and C. M. Anderson-Cook, *Response surface methodology: Process and product optimization using designed experiments*: John Wiley & Sons, 2016.
51. A. I. Khuri and J. A. Cornell, *Response surfaces: Designs and analyses*: CRC Press, 2018.
52. G. E. Box, J. S. Hunter, and W. G. Hunter, "Statistics for experimenters," in *Wiley series in probability and statistics*: Wiley Hoboken, NJ, 2005.
53. B. Dixon and N. Candade, "Multispectral landuse classification using neural networks and support vector machines: one or the other, or both?" *International Journal of Remote Sensing,* vol. 29, no. 4, pp. 1185–1206, 2008.
54. D. O. Hebb, *The organization of behavior: A neuropsychological theory*: Psychology Press, 2005.
55. J. J. Hopfield, "Neural networks and physical systems with emergent collective computational abilities," *Proceedings of the National Academy of Sciences,* vol. 79, no. 8, pp. 2554–2558, 1982.
56. D. E. Rumelhart, G. E. Hinton, and R. J. Williams, "Learning representations by back-propagating errors," *Nature,* vol. 323, no. 6088, pp. 533–536, 1986.
57. S. Grossberg, "How does a brain build a cognitive code?" *Studies of Mind and Brain,* pp. 1–52, 1982.
58. G. Hartwigsen, Y. Bengio, and D. Bzdok, "How does hemispheric specialization contribute to human-defining cognition?" *Neuron*, vol. 109, no. 13, pp. 2075–2090, 2021.
59. N. Siddique and H. Adeli, *Computational intelligence: Synergies of fuzzy logic, neural networks and evolutionary computing*: John Wiley & Sons, 2013.

60. D. I. Mendoza-Castillo, H. E. Reynel-Ávila, F. J. Sánchez-Ruiz, R. Trejo-Valencia, J. E. Jaime-Leal, and A. Bonilla-Petriciolet, "Insights and pitfalls of artificial neural network modeling of competitive multi-metallic adsorption data," *Journal of Molecular Liquids,* vol. 251, pp. 15–27, 2018/02/01/, doi: https://doi.org/10.1016/j.molliq.2017.12.030
61. O. Awodele and O. Jegede, "Neural networks and its application in engineering," *Sci IT,* pp. 83–95, 2009.
62. F. Shao and Z. Shen, "How can artificial neural networks approximate the brain?" *Frontiers in Psychology,* vol. 13, p. 970214, 2023.
63. B. L. Happel and J. M. Murre, "Design and evolution of modular neural network architectures," *Neural Networks,* vol. 7, no. 6–7, pp. 985–1004, 1994.
64. I. N. Silva, D. H. Spatti, R. A. Flauzino, L. Liboni, and S. Alves, *Artificial neural networks: A practical course*: Springer, 2016.
65. M. Izadi, M. Rahimi, and R. Beigzadeh, "Evaluation of micromixing in helically coiled microreactors using artificial intelligence approaches," *Chemical Engineering Journal,* vol. 356, pp. 570–579, 2019.
66. M. S. Mazloom, F. Rezaei, A. Hemmati-Sarapardeh, M. M. Husein, S. Zendehboudi, and A. Bemani, "Artificial intelligence based methods for asphaltenes adsorption by nanocomposites: Application of group method of data handling, least squares support vector machine, and artificial neural networks," *Nanomaterials,* vol. 10, no. 5, p. 890, 2020.
67. T. Lu, J. Bai, M. Demir, X. Hu, J. Huang, and L. Wang, "Synthesis of potassium bitartrate-derived porous carbon via a facile and self-activating strategy for CO_2 adsorption application," *Separation and Purification Technology,* vol. 296, p. 121368, 2022.
68. N. A. Mohd Najib, V. Sethu, S. K. Arumugasamy, and A. Selvarajoo, "Artificial neural network (ANN) modelling of palm oil mill effluent (POME) treatment with natural bio-coagulants," *Environmental Processes,* vol. 7, no. 2, pp. 509–535, 2020.
69. J. A. Richards and J. Richards, *Remote sensing digital image analysis*: Springer, 1999.
70. T. Nguyen, "Earth-return path impedances of underground cables. Part 2: Evaluations using neural networks," *IEE Proceedings-Generation, Transmission and Distribution,* vol. 145, no. 6, pp. 627–633, 1998.
71. L. V. Fausett, *Fundamentals of neural networks: Architectures, algorithms and applications*: Pearson Education India, 2006.
72. D. S. Broomhead and D. Lowe, *Radial basis functions, multi-variable functional interpolation and adaptive networks*: Royal Signals and Radar Establishment Malvern, United Kingdom, 1988.
73. S. Kolbadinejad, "Deep learning analysis of Ar, Xe, Kr, and O2 adsorption on activated carbon and zeolites using ANN approach," *Chemical Engineering and Processing—Process Intensification,* 2022, doi: https://doi.org/10.1016/j.cep.2021.108662
74. A. Arora, "Computational material screening using artificial neural networks for adsorption gas separation," *The Journal of Physical Chemistry C,* 2020, doi: https://doi.org/10.1021/acs.jpcc.0c05900
75. F. Fathalian, S. Aarabi, A. Ghaemi, and A. Hemmati, "Intelligent prediction models based on machine learning for CO_2 capture performance by graphene oxide-based adsorbents," *Scientific Reports,* vol. 12, no. 1, p. 21507, 2022.
76. M. R. Sabour and A. Amiri, "Comparative study of ANN and RSM for simultaneous optimization of multiple targets in Fenton treatment of landfill leachate," *Waste Management,* 2017, doi: https://doi.org/10.1016/j.wasman.2017.03.048
77. D. S. Dao, "Response surface optimization of impregnation of blended amines into mesoporous silica for high-performance CO_2 capture," *Energy Fuels,* 2015, doi: https://doi.org/10.1021/ef502656t
78. L. S. Tan, "Application of response surface methodology to investigate CO_2 absorption column temperature rise," *Advanced Materials Research,* 2014, doi: https://doi.org/10.4028/www.scientific.net/AMR.917.257

79. H. Pashaei, H. Mashhadimoslem, and A. Ghaemi, "Modeling and optimization of CO_2 mass transfer flux into Pz-KOH-CO2 system using RSM and ANN," *Scientific Reports*, vol. 13, p. 4011, 2023.
80. H. Derikvandi, "Comprehensive study on enhanced photocatalytic activity of heterojunction ZnS-NiS/zeolite nanoparticles: Experimental design based on response surface methodology (RSM), impedance spectroscopy and GC-MASS studies," *Journal of Colloid and Interface Science*, 2017, doi: https://doi.org/10.1016/j.jcis.2016.11.105
81. A. Mahmad, et al., "Response surface methodology and artificial neural network for prediction and validation of bisphenol a adsorption onto zeolite imidazole framework," *Groundwater for Sustainable Development*, 2023, doi: https://doi.org/10.1016/j.gsd.2023.100925
82. A. Dashti, F. Amirkhani, M. Jokar, A. Mohammadi, and K.-W. Chau, "Insights into the estimation of heavy metals ions sorption from aqueous environment onto natural zeolite," *International Journal of Environmental Science and Technology*, vol. 18, pp. 1773–1784, 2021.
83. A. Dashti, M. Raji, A. Azarafza, A. Baghban, A. H. Mohammadi, and M. Asghari, "Rigorous prognostication and modeling of gas adsorption on activated carbon and Zeolite-5A," *Journal of Environmental Management*, vol. 224, pp. 58–68, 2018.
84. F. Gao, P. Wang, D. Wang, Y. Yang, X. Zhang, and G. Bai, "Model for predicting CO_2 adsorption in coal left in goaf based on backpropagation neural network," *Energies*, vol. 16, no. 9, p. 3760, 2023.
85. M. Raji, A. Dashti, M. S. Alivand, and M. Asghari, "Novel prosperous computational estimations for greenhouse gas adsorptive control by zeolites using machine learning methods," *Journal of Environmental Management*, vol. 307, p. 114478, 2022.
86. C. Li, F. Du, P. Bénard, R. Chahine, T. Yang, and J. Xiao, "Artificial neural network prediction for breakthrough curves of H_2/CO_2 adsorption on CuBTC," in *International conference on energy engineering and environmental engineering*: Springer, 2022, pp. 37–48.
87. M. H. V. Bahrun, N. Othman, A. Bono, and M. A. A. Zaini, "Prediction of CO_2 breakthrough time in packed bed adsorption column using artificial neural network," in *International conference on computational science and technology*: Springer, 2022, pp. 367–380.
88. A. Rostami, M. A. Anbaz, H. R. E. Gahrooei, M. Arabloo, and A. Bahadori, "Accurate estimation of CO_2 adsorption on activated carbon with multi-layer feed-forward neural network (MLFNN) algorithm," *Egyptian Journal of Petroleum*, vol. 27, no. 1, pp. 65–73, 2018.
89. H. Mashhadimoslem, M. Vafaeinia, M. Safarzadeh, A. Ghaemi, F. Fathalian, and A. Maleki, "Development of predictive models for activated carbon synthesis from different biomass for CO_2 adsorption using artificial neural networks," *Industrial & Engineering Chemistry Research*, vol. 60, no. 38, pp. 13950–13966, 2021.
90. Z. Khoshraftar and A. Ghaemi, "Preparation of activated carbon from Entada Africana Guill. & Perr for CO_2 capture: Artificial neural network and isotherm modeling," *Journal of Chemical and Petroleum Engineering*, vol. 56, no. 1, pp. 165–180, 2022.
91. H. Barki, L. Khaouane, and S. Hanini, "Modelling of adsorption of methane, nitrogen, carbon dioxide, their binary mixtures, and their ternary mixture on activated carbons using artificial neural network," *Kemija u industriji: Časopis kemičara i kemijskih inženjera Hrvatske*, vol. 68, no. 7–8, pp. 289–302, 2019.
92. M. R. Moradi, H. Ramezanipour Penchah, and A. Ghaemi, "CO2 capture by benzene-based hypercrosslinked polymer adsorbent: Artificial neural network and response surface methodology," *The Canadian Journal of Chemical Engineering*, vol. 101, no. 10, pp. 5621–5642, 2023.
93. P. Mehrmohammadi and A. Ghaemi, "Investigating the effect of textural properties on CO_2 adsorption in porous carbons via deep neural networks using various training algorithms," *Scientific Reports*, vol. 13, no. 1, p. 21264, 2023.

94. M. Noroozian, S. Shahhosseini, and A. Ghaemi, "Artificial intelligence and response surface methodology to predict CO_2 capture using piperazine-modified activated alumina," *Environmental Progress & Sustainable Energy,* vol. 42, no. 3, p. e14117, 2023.
95. A. Khalilzadeh, M. Fattahi, and A. Baghban, "Modeling of carbon dioxide capture from gas stream emissions using amino acid salts blended with amine solutions," *Petroleum Science and Technology,* vol. 37, no. 16, pp. 1868–1874, 2019.
96. A. Baghban, S. Zilabi, S. Golrokhifar, and S. Habibzadeh, "Neural computations in modelling of CO_2 capture from gas stream emissions by sodium glycinate solution," *Petroleum Science and Technology,* vol. 36, no. 4, pp. 326–331, 2018.
97. H. Mashhadimoslem, M. Vafaeinia, M. Safarzadeh, A. Ghaemi, F. Fathalian, and A. Maleki, "Development of predictive models for activated carbon synthesis from different biomass for CO_2 adsorption using artificial neural networks," *Industrial & Engineering Chemistry Research,* vol. 60, no. 38, 2021.
98. M. Noroozian, A. Ghaemi, and Z. Heidari, "Potential of artificial intelligence and response surface methodology to predict CO_2 capture by KOH-modified activated alumina," *Case Studies in Chemical and Environmental Engineering,* vol. 8, p. 100442, 2023.
99. M. Kianpour, M. A. Sobati, and S. Shahhosseini, "Experimental and modeling of CO_2 capture by dry sodium hydroxide carbonation," *Chemical Engineering Research and Design,* vol. 90, no. 11, pp. 2041–2050, 2012.
100. S. Naeem, S. Shahhosseini, and A. Ghaemi, "Simulation of CO_2 capture using sodium hydroxide solid sorbent in a fluidized bed reactor by a multi-layer perceptron neural network," *Journal of Natural Gas Science and Engineering,* vol. 31, pp. 305–312, 2016.
101. F. Yulia, I. Chairina, and A. Zulys, "Multi-objective genetic algorithm optimization with an artificial neural network for CO2/CH4 adsorption prediction in metal-organic framework," *Thermal Science and Engineering Progress,* p. 100967, 2021.

11 Recyclability Study of CO_2 Adsorbents

Farzad Moazami and Ahad Ghaemi

11.1 INTRODUCTION

The recyclability of CO_2 adsorbents is essential for long-term industrial applications. Recyclability ensures that materials can undergo multiple adsorption–desorption cycles without losing effectiveness, which is critical for cost efficiency in CO_2 capture technologies. Amine-functionalized materials, such as those applied to silicas and metal-organic frameworks (MOFs), are widely studied for CO_2 capture. These materials perform well but face challenges with stability due to thermal degradation and exposure to impurities like sulfur compounds. Researches demonstrate that, with optimal conditions, amine-functionalized adsorbents can retain their performance across many cycles. However, prolonged exposure to high temperatures and oxidative conditions can degrade their amine groups, limiting recyclability [1]. MOFs are known for their high CO_2 adsorption capacity and recyclability due to their tunable structures and surface chemistries. Studies have shown that they can endure multiple adsorption–desorption cycles without losing their structural integrity. Similarly, carbon-based materials, such as AC and carbon molecular sieves, demonstrate high recyclability, maintaining CO_2 adsorption capacities across several cycles. Carbon materials, especially those activated through steam or chemical treatment, have been shown to retain their properties even after extensive use [2]. While many adsorbents show potential for long-term use, challenges remain. Maintaining chemical and thermal stability while reducing regeneration energy costs is key to improving adsorbent recyclability. Future advancements aim to create more durable materials that can withstand industrial conditions involving moisture and chemical impurities.

Carbon-based materials have long been studied for their high potential in CO_2 capture applications due to their versatile structures, surface chemistry, and physical properties. Among these materials, AC, MXene, and graphene oxide (GO) stand out for their recyclability, tunable pore structures, high surface areas, and ability to be chemically modified to enhance performance. AC is a well-established adsorbent material for CO_2 capture due to its high surface area, porosity, and cost-effectiveness. It is produced through the activation of carbonaceous precursors such as coal, coconut shells, or biomass, using physical or chemical methods. AC has been used in CO_2 capture applications for decades, with its performance often being enhanced through chemical functionalization or doping. In a study by Zhang et al. (2020), the performance of steam-ACs derived from biomass was assessed for CO_2 capture. The researchers found that these materials exhibited high adsorption capacities,

 DOI: 10.1201/9781003476191-11

particularly at low pressures, making them suitable for post-combustion CO_2 capture technologies [3]. Additionally, it was shown that chemical modification, such as amine grafting, significantly enhances CO_2 selectivity by increasing the interaction between CO_2 and nitrogen-containing functional groups on the carbon surface [4]. The recyclability of AC materials has also been a significant area of research. Studies have shown that AC can maintain its CO_2 adsorption capacity over multiple cycles of adsorption–desorption with minimal performance loss. For example, Jiang et al. (2021) reported that AC modified with nitrogen-doped porous structures retained over 95% of its initial CO_2 adsorption capacity after ten cycles of use, demonstrating its robustness under repeated use conditions [5]. However, one of the limitations of AC is its relatively low affinity for CO_2 at ambient temperatures compared to other materials, such as MOFs. Researchers are currently investigating methods to overcome this limitation by enhancing the material's surface chemistry or incorporating heteroatoms such as nitrogen, sulfur, or oxygen into the carbon framework [2].

A new class of two-dimensional CO_2 adsorbents, MXenes, a family of two-dimensional transition metal carbides and nitrides, have emerged as a promising material for CO_2 capture due to their highly tunable surface properties, large specific surface areas, and excellent electrical conductivity. Since their discovery, MXenes have attracted significant attention for applications in energy storage, catalysis, and, more recently, CO_2 adsorption. MXenes can be chemically modified by surface termination with functional groups such as hydroxyl, oxygen, and fluorine, which enhances their affinity for CO_2 molecules. As demonstrated by Khazaei et al. (2020), MXenes with oxygen and hydroxyl terminations exhibit high CO_2 adsorption capacities, especially at low pressures and ambient temperatures. The study further highlighted that these materials can be easily regenerated through thermal or vacuum treatments without significant degradation of their adsorption performance [6]. Furthermore, the layered structure of MXenes allows for intercalation of CO_2 molecules between the layers, which increases their adsorption capacity. In a study conducted by Ren et al. (2023), $Ti_3C_{2^2}$ MXenes were shown to exhibit excellent recyclability, maintaining over 90% of their CO_2 adsorption capacity after 50 adsorption–desorption cyclability [4], coupled with the material's high mechanical strength, makes MXenes an attractive candidate for long-term CO_2 capture applications in industrial settings. Despite these advantages, MXenes still face challenges related to their scalability and cost of production. Ongoing research focuses on developing cost-effective synthesis methods and improving the material's stability in humid environments, which can negatively affect their adsorption performance [4]. GO is another two-dimensional material that has garnered attention for CO_2 capture due to its high surface area, tunable surface chemistry, and ease of functionalization. GO is typically synthesized by the oxidation of graphite, resulting in a material with a large number of oxygen-containing functional groups (e.g., hydroxyl, epoxide, and carboxyl groups) that can interact with CO_2 molecules. One of the main advantages of GO is its ability to be easily modified to enhance its CO_2 capture performance. For instance, nitrogen-doped graphene oxide (N-GO) showed a significant increase in CO_2 adsorption capacity compared to pristine GO. The incorporation of nitrogen atoms into the graphene structure improved the interaction between CO_2 and the material's surface, leading to higher adsorption capacities, especially at low pressures [7]. Additionally, GO-based materials can be combined with other adsorbents, such as amines or MOFs, to further enhance their performance. In a recent study by Kim et al. (2022), GO was functionalized with amine groups, resulting in a composite material with superior CO_2 capture performance and excellent recyclability. The study demonstrated that the GO-based adsorbent retained over 95% of its CO_2 adsorption capacity after 20 cycles of adsorption–desorption [7]. GO also exhibits excellent mechanical strength and stability, making it suitable for use in harsh industrial environments. However, one of the limitations of GO is its relatively low CO_2 selectivity compared to other materials such as MOFs or amine-functionalized adsorbents. Researchers are working on improving the material's selectivity by incorporating heteroatoms or designing composite materials with enhanced surface properties [7–9]. The recyclability of carbon-based adsorbents is a critical factor for their practical application in CO_2 capture technologies. AC, MXene, and GO have all demonstrated excellent recyclability in various studies, maintaining their CO_2 adsorption capacities over multiple cycles of use. For example, as mentioned earlier, MXenes have shown the ability to retain over 90% of their initial adsorption capacity after 50 cycles, while GO composites with amine functionalization exhibit minimal loss of performance after 20 cycles [10, 11]. Generating these materials using simple thermal or vacuum treatments makes them attractive for large-scale industrial applications, where cost-effective regeneration methods are crucial. However, challenges related to material degradation, scalability, and selectivity remain, and ongoing research is focused on addressing these issues to improve the overall performance and longevity of carbon-based adsorbents. In the coming years, the development of hybrid materials that combine the advantages of multiple adsorbent types (e.g., MXene-GO composites or AC-MOF hybrids) is expected to further enhance the performance of carbon-based materials for CO_2 capture. These composite materials offer the potential to achieve higher adsorption capacities, improved selectivity, and greater stability under a wide range of operating conditions.

11.2 RECYCLABILITY OF CARBON COMPOSITE

Carbon composites are materials composed of carbon fibers reinforced with a polymer matrix, often used in high-performance applications across various industries, including aerospace, automotive, sports equipment, and renewable energy. The rising demand for lightweight, durable materials has brought carbon composites to the forefront of material science. However, their recyclability is a critical challenge due to the complex structure and bonding between the carbon fibers and the polymer matrix. Addressing this challenge is essential for minimizing environmental impact and promoting sustainable manufacturing practices [12, 13].

11.2.1 Overview of Carbon Composites and Their Applications

Carbon composites are prized for their high strength-to-weight ratio, corrosion resistance, and thermal stability, making them indispensable in industries like aerospace, where weight reduction is critical for fuel efficiency. In the automotive sector, carbon composites help achieve the lightweight design necessary to meet stricter emissions regulations. Similarly, in wind energy, carbon composites are used in turbine blades to enhance durability while reducing mass [14, 15]. Despite these advantages, the end-of-life management of carbon composites remains a significant issue. Traditional disposal methods, such as landfilling and incineration, are not sustainable, as they result in environmental pollution and loss of valuable resources. Therefore, recycling strategies are necessary to recover the carbon fibers or matrix components from discarded composite materials [16, 17].

The recyclability of carbon composites is hindered by several factors, primarily the strong chemical bonds between carbon fibers and the polymer matrix. These bonds make it difficult to separate the two components without degrading the fibers. Moreover, carbon composites often contain various additives and fillers, further complicating the recycling process. The thermoset matrix, commonly used in carbon composites, poses an additional challenge, as it does not melt upon heating, making mechanical recycling methods less effective. Instead, thermoset composites must be broken down through thermal, chemical, or mechanical means to recover the carbon fibers. Thermoplastic matrices, on the other hand, are more easily recyclable, but they are less commonly used due to their inferior mechanical properties compared to thermosets [18].

11.2.2 Mechanical Recycling

Mechanical recycling is one of the most straightforward methods for recycling carbon composites. It involves shredding or grinding the composite material into small particles or fibers, which can then be reused as fillers in new composite products. However, mechanical recycling often leads to a reduction in the mechanical properties of the recovered fibers, limiting their application to low-performance materials [17]. In a study, mechanical recycling was applied to carbon fiber-reinforced polymers (CFRPs), and the recycled fibers were used to manufacture new composite panels. While the recycled composites exhibited lower tensile strength compared to virgin composites, they still demonstrated sufficient mechanical properties for certain applications, such as automotive interiors and construction materials [18].

11.2.3 Chemical Recycling

Chemical recycling, also known as solvolysis, involves using solvents or chemicals to break down the polymer matrix while preserving the carbon fibers. This method has the advantage of operating at lower temperatures compared to pyrolysis, which reduces the risk of fiber degradation. Solvolysis is particularly suitable for recycling thermoset composites, as it can effectively dissolve the matrix while leaving the fibers intact. Recent developments in chemical recycling have focused on improving solvent efficiency and reducing environmental impact. For example, it was reported the successful use of supercritical fluids, such as supercritical water or carbon dioxide, to dissolve the polymer matrix in CFRPs. This method not only recovered the carbon fibers but also allowed for the potential recovery of valuable chemical by-products from the matrix decomposition. Another promising approach involves using ionic liquids, which can selectively target the polymer matrix without affecting the fibers [19].

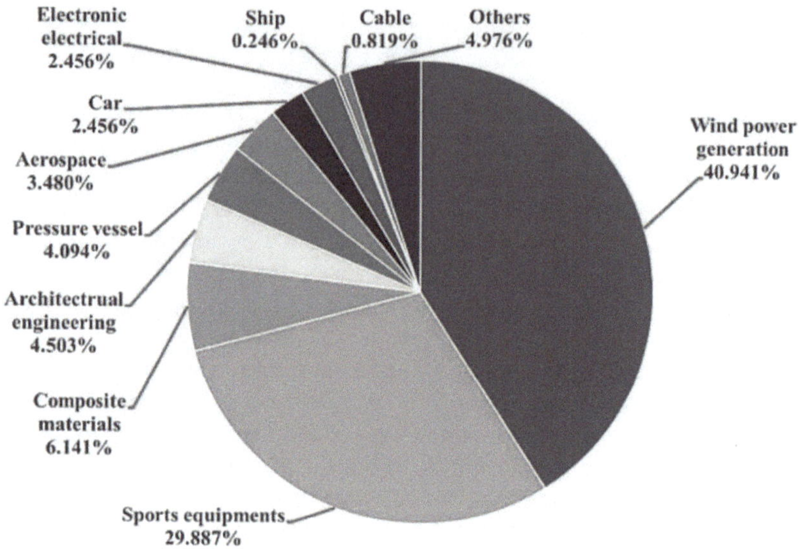

FIGURE 11.1 Percentage utilization of the (CFRP) in the different sectors [16].

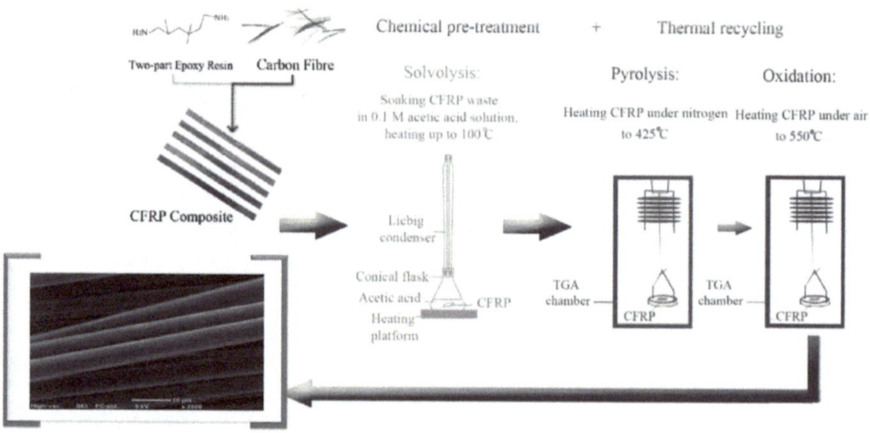

FIGURE 11.2 Thermo-chemical recycling method [20].

Chemical recycling, often referred to as solvolysis, is a highly effective method for recovering carbon fibers from the polymer matrices in CFRPs. Unlike mechanical and thermal recycling, chemical recycling offers the advantage of operating at lower temperatures, reducing the risk of degrading carbon fibers, which makes it particularly beneficial for recycling thermoset composites. These composites are otherwise difficult to recycle due to their permanent crosslinked structure that cannot be melted or reshaped [21].

11.2.4 Process and Mechanism of Solvolysis

Chemical recycling involves using specific solvents or chemical agents that dissolve the polymer matrix without damaging the embedded carbon fibers. The process typically consists of three stages.

The first stage is the application of the solvent; solvents are applied to the composite material, breaking down the resin matrix through chemical reactions. The most common solvents used in chemical recycling include supercritical fluids (such as supercritical water or CO_2), organic solvents (like acetone, ethanol, and methanol), or reactive solvents such as acids or bases. The next stage is known as dissolution and fiber separation: once the polymer matrix is dissolved, the carbon fibers are released and can be separated from the solution. At this point, the solvent–polymer solution can also be processed to recover and reuse the polymer, depending on the resin type. The fibers are then washed to remove any residue. The last stage is the post-processing of recovered fibers: after separation, the carbon fibers may undergo further treatments such as sizing (a process that applies a protective coating to improve fiber–matrix bonding), to restore some of their lost mechanical properties. The result is recycled fibers that can be reused in new composite materials, often with properties close to that of virgin fibers [19, 20].

11.2.5 Advantages of Chemical Recycling

One of the main benefits of chemical recycling is its ability to recover carbon fibers with minimal damage to their structural integrity, preserving much of their tensile strength and stiffness. This makes chemical recycling superior to mechanical recycling in terms of the quality of the recovered fibers. Chemical recycling also allows for the possibility of recovering not only the fibers but also some valuable chemicals from the resin matrix, depending on the process and the type of resin used. For instance, certain solvents can be chosen for selective dissolution, enabling the recovery of both fibers and monomers from the polymer matrix for reuse [22].

11.3 RECYCLABILITY OF POLYMERIC COMPOSITE

Polymeric composites are materials made from two or more distinct phases typically a polymer matrix and a reinforcing material such as glass or carbon fibers that exhibit superior properties compared to individual components. Due to their high strength, lightweight nature, and resistance to corrosion, polymeric composites are used extensively in various industries, including aerospace, automotive, construction, and sports equipment manufacturing [23]. However, the increasing consumption of these materials has led to a significant challenge: how to manage their end-of-life disposal. The recycling of polymeric composites is crucial to mitigate their environmental impact and conserve valuable raw materials. This review delves into the recyclability of polymeric composites, highlighting the major challenges, current recycling technologies, and emerging research trends [23, 24].

11.3.1 Composition of Polymeric Composites in CO_2 Capture

As it was mentioned before, polymeric composites are gaining attention as promising materials for CO_2 capture due to their versatility, tunable properties, and ability to incorporate various functional components that enhance their gas adsorption capacity. The composition of these composites typically includes a polymeric matrix, which serves as the structural framework, combined with functional fillers or reinforcements that drive the CO_2 adsorption mechanism. These functional components can be carbon-based materials, porous polymers, MOFs, or amine-functionalized compounds. The selection of both the polymer matrix and the functional fillers is crucial in determining the composite's efficiency, durability, and recyclability in CO_2 capture applications [25, 26].

The polymeric matrix in CO_2 capture composites can be broadly divided into two categories: thermoplastic and thermosetting polymers. *Thermoplastics* are used for their processability and flexibility, making them ideal for large-scale CO_2 capture operations. Examples include polyethyleneimine (PEI) and polyvinyl amine, which are often selected for their ability to be chemically modified with amine groups. These polymers exhibit high CO_2 affinity due to the interaction between the acidic CO_2 molecules and basic amine sites. Additionally, thermoplastics like polycarbonate and polyaniline can be easily fabricated into membranes or films, increasing the surface area for gas capture. *Thermosetting polymers*, such as epoxy resins and polyurethane, are characterized by their chemical crosslinking, which provides enhanced thermal stability and mechanical strength. These materials are less flexible than thermoplastics but are useful in CO_2 capture applications that operate under high temperatures or harsh conditions. Their rigidity offers long-term structural stability, though they are more difficult to recycle after use due to their irreversible crosslinked structure [25–28]. Functional fillers or additives are critical for improving the CO_2 capture capacity of polymeric composites. These materials provide active sites for CO_2 adsorption and can include various classes of materials, such as *carbon-based materials*, which concludes that AC, carbon nanotubes (CNTs), and GO are commonly used in polymeric composites for CO_2 capture. AC is particularly valued for its high surface area and porosity, allowing for substantial CO_2 adsorption. CNTs and GO contribute through their high conductivity and ability to disperse evenly within polymer matrices, enhancing the overall performance of the composite [29]. Amine-functionalized compound groups are widely known for their ability to chemically adsorb CO_2 through the formation of carbamate species. Polymers such as PEI are often impregnated with amine groups, or grafted with amines, to enhance their CO_2 capture efficiency. The amine-functionalized composites operate efficiently even under low CO_2 concentrations, making them ideal for post-combustion capture applications [30, 31].

The recyclability of polymeric composites used in CO_2 capture is an important consideration for sustainable implementation. Thermoplastic-based composites are more readily recyclable compared to thermosets, as they can be remelted and reprocessed. However, the presence of functional fillers can complicate recycling efforts due to the potential degradation of the fillers or the polymer matrix during processing. For instance, composites containing AC or GO may exhibit reduced performance after recycling due to changes in their structure or surface chemistry. Conversely, thermosetting composites, while more durable in high-temperature applications, are more challenging to recycle. Advanced methods such as chemical recycling through solvolysis or pyrolysis are required to recover functional components like carbon fibers or MOFs, though these processes are energy-intensive and costly [32]. Although there are challenges in recycling polymeric composites, the recyclability of polymeric composites is complicated by several factors, including the nature of the polymer matrix, the type of reinforcement used, and the bonding between the matrix and the reinforcement. Thermoset composites pose a particular challenge because they cannot be remelted or reshaped once cured, requiring alternative methods for recycling. In contrast, thermoplastic composites can be melted and reprocessed, but this process can lead to a decline in mechanical properties due to fiber damage or matrix degradation. Another significant challenge is the heterogeneous nature of polymeric composites. Unlike monolithic materials, composites consist of multiple components that need to be separated and individually processed during recycling. This complexity increases the cost and reduces the economic feasibility of recycling. Furthermore, many composites are made with additives, coatings, and fillers that complicate recycling efforts by introducing contaminants into the recycling stream [32–35].

11.3.2 Mechanical Recycling

Mechanical recycling is one of the most straightforward methods for dealing with polymeric composites, particularly thermoplastics. In this process, composite waste is shredded, ground, or milled into smaller particles or fibers, which can then be used as reinforcement in new composite products. However, mechanical recycling often results in a reduction of the mechanical properties of the recycled fibers, making them unsuitable for high-performance applications. Despite these limitations, mechanically recycled materials can still be used in lower-performance applications such as construction materials and automotive parts [36].

11.3.3 Thermal Recycling

Thermal recycling methods, including pyrolysis and fluidized bed processes, involve heating the composite material to break down the polymer matrix and recover the fibers. Pyrolysis is conducted in the absence of oxygen, preventing combustion and allowing the recovery of both carbon fibers and, in some cases, valuable chemical feedstocks from the resin. One of the main advantages of pyrolysis is its ability to recover high-quality carbon fibers, which retain a significant portion of their original mechanical properties. The fluidized bed process involves blowing hot air through a bed of particles, effectively "fluidizing" them and causing the polymer matrix to degrade or combust. This method is particularly effective for glass fiber composites, where the fibers can be recovered without significant damage. However, thermal recycling processes are energy-intensive and may generate harmful emissions if not properly controlled [37]. Yang et al. (2021) compared pyrolysis and fluidized bed methods and found that pyrolysis is more effective for carbon fiber recovery, while the fluidized bed process is better suited for glass fibers. However, both processes still face challenges in terms of energy efficiency and scalability [38].

11.3.4 Chemical Recycling

Chemical recycling, also known as solvolysis, offers a more refined approach to recycling polymeric composites, particularly thermoset-based composites. This method

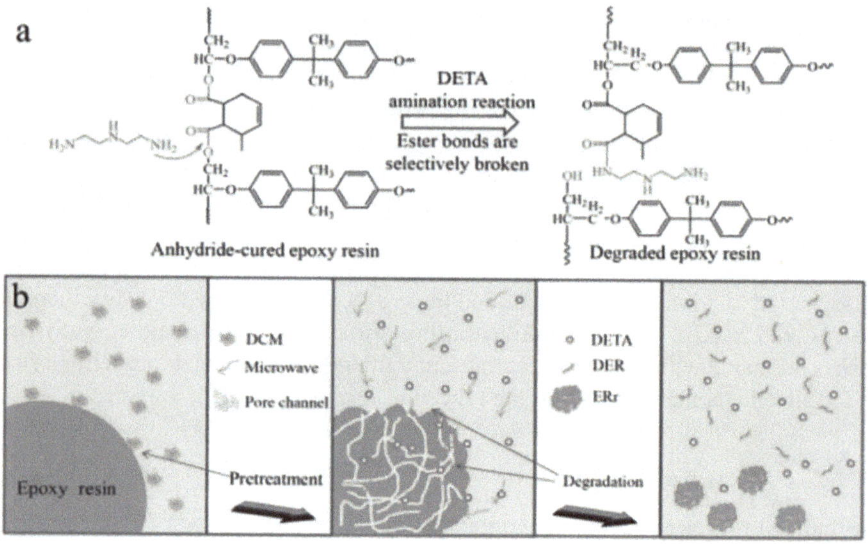

FIGURE 11.3 (a) Chemical structure and amination process of the resin and (b) pretreatment and microwave-assisted amination process of the resin [38].

involves using solvents or chemical reagents to dissolve the polymer matrix, allowing for the recovery of intact fibers. Chemical recycling can recover high-quality fibers with minimal degradation, making them suitable for high-performance applications [37, 38].

Supercritical fluids, such as supercritical CO_2, are increasingly being used as solvents in chemical recycling due to their low environmental impact and ability to penetrate the polymer matrix effectively. Research demonstrated that supercritical fluids can efficiently dissolve epoxy resins without damaging the carbon fibers, allowing for the recovery of high-quality materials. While chemical recycling has great potential, it also has limitations. The cost of solvents and the need for specialized equipment make it expensive compared to other recycling methods. Additionally, some solvents may be hazardous to the environment, requiring careful handling and disposal [39].

11.3.5 Emerging Methods in Recycling

Several emerging recycling techniques are being developed to address the limitations of current methods. Microwave-assisted pyrolysis, for example, uses microwave energy to selectively heat the polymer matrix, allowing for faster and more efficient decomposition of the resin. This method has been shown to reduce the energy consumption of traditional pyrolysis while maintaining fiber quality. Another promising technique is the use of enzymatic degradation, where specific enzymes are used to break down the polymer matrix. This approach is environmentally friendly and has the potential to target specific types of resins without damaging the fibers. However, enzymatic degradation is still in the early stages of development and has yet to be scaled up for industrial applications. Research is also being conducted on the development of self-healing and recyclable composites, which incorporate reversible bonds into the polymer matrix. These materials can be reshaped or repaired at the end of their life cycle, reducing the need for traditional recycling methods [35, 40].

11.3.6 Recyclability of Nanoadsorbents

The recyclability of nanoadsorbents is their essential feature and determines their practical application in pollution management and environmental applications. The need to use effective methods of water purification has greatly increased, due to the presence of pollutants such as heavy metals, dyes, and dangerous chemicals, thereby increasing the destruction of the environment. Nanoadsorbents, with their high surface area, unique reactivity, and exceptional pollutant adsorption capacities, have emerged as a promising solution for the adsorption and removal of pollutants from water. Apart from their effectiveness in removing pollutants, nanoadsorbents have the ability to be reused or recycled repeatedly after adsorption to minimize waste, and it is also important to reduce costs and promote sustainability. The recyclability of nanosorbents, which includes their regeneration after reaching adsorption capacity, allows them to be reused in multiple cycles (without a significant decrease in performance). Through various techniques such as thermal, chemical, and electrochemical methods, each of which is suitable for specific types of nanoadsorbents and pollutants, this regeneration can be achieved. Carbon-based nanosorbents, such as GO and CNTs, where adsorbed pollutants are removed by controlled heating, can be regenerated using thermal desorption. Also, metal oxide-based nanosorbents, especially those made of iron oxide or titanium dioxide, can be subjected to chemical regeneration processes, which increases their surface activity [38, 40]. The recycling potential of nanoadsorbents is influenced by factors such as the type of pollutant, adsorption conditions, and physical and chemical stability of the adsorbent itself. Repetition of regeneration cycles sometimes leads to the destruction of the structure, which affects the adsorption efficiency of the adsorbent. This requires the development of more robust nanosorbents and is effective in maintaining their integrity over multiple cycles. Recent advances in nanosorbent design have focused on increasing durability and ease of regeneration, with strategies including surface modification and composite formation to improve flexibility. Environmentally friendly remediation methods, including those that focus on renewable energy sources or non-toxic chemicals, are explored to minimize the environmental impact of these processes. The recyclability of nanoadsorbents for sustainable water treatment is very promising, research is of great importance to address the challenges related to degradation and loss of efficiency and thus provide scalable and environmentally friendly solutions in pollution control [40, 41].

11.3.7 Cyclic Capacity of the Sorbents

The cyclic capacity of adsorbents refers to the ability of an adsorbent to maintain a constant adsorption efficiency over multiple adsorption and desorption cycles. This is of great importance for industrial applications including carbon capture and gas purification. This property is of great importance since it determines the lifetime and economic feasibility of adsorbents in long-term use, especially in applications such as CO_2 capture from flue gases and other industrial emissions. Metal oxides such as MgO and CaO, lithium-based adsorbents, advanced materials such as perovskites, and amine-functionalized composites are among the materials that show promising cyclic capacity [42].

Lithium orthosilicate (Li_4SiO_4) has been extensively investigated (due to its initial CO_2 adsorption capacity). A large number of initial formulations showed significant capacity loss due to structural degradation and particle aggregation, which reduced the surface area. Researchers addressed these issues by synthesizing lithium orthosilicate with various silicon precursors and incorporating additives to improve the structural integrity. The use of fumed silica as a precursor has been shown to perform better with less pore collapse and less degradation of cyclic capacity over time. It has been shown in research and studies that adsorbents derived from fumed silica retain approximately 80% of their initial capacity even after 30 cycles. Another approach to improve the cyclic capacity in lithium-based adsorbents is to introduce other elements into the structure. For example, doping lithium orthosilicate with alkali metals prevents crystal growth during cycling, reducing sintering and increasing the cyclic stability of the material [41–43].

Perovskite materials have been investigated for their potential in cyclic CO_2 adsorption applications (such as those based on calcium titanate and strontium-doped oxides). Perovskites have very stable structures, even at high temperatures, which allows them to maintain adsorption efficiency and structural integrity over multiple cycles. For example, research on calcium titanate perovskites containing elements such as strontium and cobalt has shown promising results. They have also been shown to retain more than 90% of their initial adsorption capacity at elevated temperatures after several hundred cycles. These findings are particularly important for high-temperature applications, such as power plants, where adsorbents must withstand mechanical and thermal stresses over long periods. By tuning the composition of perovskite compounds, researchers are able to optimize properties such as surface area, pore size, and thermal stability, thereby increasing their cyclic performance. MgO-based adsorbents have attracted attention due to their relatively low cost as well as high theoretical CO_2 adsorption capacities. However, these MgO adsorbents face challenges related to cyclic stability. MgO tends to form a dense, impermeable carbonate layer during CO_2 adsorption, which prevents further reaction in subsequent cycles. Recent studies have investigated the use of dopants such as transition metals and alkali metals, which can modify the surface structure of MgO to prevent the formation of carbonate layers. Combining MgO with support materials, such as silica or alumina, helps maintain porosity and improves CO_2 diffusion, as well as increasing cyclic stability. With appropriate modifications, MgO-based adsorbents are able to achieve stable CO_2 capacities over dozens of cycles and are applicable on a large scale [44].

A notable development is the use of amine-functionalized materials, which provide distinct chemical mechanisms for CO_2 adsorption through chemisorption and physisorption. These materials, when combined with porous supports such as mesoporous silica or MOFs, exhibit excellent cyclic capacities. The amine groups react with CO_2 to form carbamates, which desorb at lower temperatures compared to purely physisorption materials, resulting in lower energy requirements for regeneration and increased cyclic stability of the adsorbent. Research since 2020 shows that functional materials retain CO_2 capture capacity with minimal degradation over hundreds of cycles (making them candidates for carbon capture and storage (CCS) applications in industries where frequent regeneration is required) [30, 45]. Research on cyclic capacity highlights an increasing emphasis on doping strategies, structural modifications, and composite formation to enhance the cyclic stability of adsorbents. Advances in materials synthesis have enabled researchers to create adsorbents that both efficiently capture CO_2 and withstand multiple regeneration cycles with minimal loss of performance. These innovations are of great importance to enable the deployment of carbon capture technologies on an industrial scale (the adsorbents remain efficient and cost-effective over long periods). The aim of the research progress is to develop adsorbents that combine low regeneration energy requirements, high cycle capacity, and structural flexibility to create a sustainable solution to reduce greenhouse gas emissions and climate change [30, 46, 47].

11.3.8 The Life Cycle Assessment Analysis: Environmental Management

Life cycle assessment (LCA) of CO_2 adsorbers is necessary to understand the sustainability and long-term effects of these materials throughout their lifetime, especially in the field of environmental management. Conducting an LCA for carbon dioxide removal (CDR) and adsorbent materials technologies helps to assess the environmental impacts that result from the extraction, production, use and disposal, or recycling of raw materials at the end of their life. These assessments are critical for examining the carbon footprint, resource use, energy needs, and potential emissions at each stage. This example of an environmental indicator is commonly used in LCAs to provide a comprehensive picture of life cycle impacts. The key metrics that it has include the carbon footprint that shows the greenhouse gas emissions associated with each stage of the product's life cycle and energy consumption and resource reduction metrics, as well as measuring the efficiency of energy and material consumption. These indicators help determine the stages in which environmental effects are very important and are very effective in improving production processes, material selection, and disposal methods. Studies have shown that, for some adsorbent materials, the production phase can have a greater impact due to their energy requirements, and this requires the need for cleaner production methods or a shift toward renewable energy sources [48, 49].

Recently, they investigated the life cycle impact of different CDR technologies such as direct air adsorption, biochar, and carbon storage for CO_2 adsorption. These studies emphasize the complexity of evaluating carbon removal technologies because the emissions avoided in these systems are sometimes mistakenly interpreted as negative emissions, thus underrepresenting the actual carbon removal from the atmosphere. To determine and investigate these nuances, it is recommended to use more detailed LCAs because they include multiple environmental categories and clear methodological choices. It also focuses on the sustainability of the recycling potential of CO_2 adsorbers and expands their reach. By including end-of-life scenarios in

LCA, researchers can understand the costs and environmental benefits of adsorbent regeneration, which can increase the overall sustainability of the material. Recycling processes can reduce the environmental impact of new resources; these benefits must be weighed against the energy and resource requirements of the recycling process itself. As the demand for efficient and environmentally friendly carbon sequestration solutions increases, LCA provides a fundamental approach to measure the environmental trade-offs of various CO_2 adsorbers and refine these technologies to ensure effective and sustainable climate goals are met [49].

11.3.9 Life Cycle Costing Analysis

Life cycle costing (LCC) analysis of CO_2 adsorbers is necessary to evaluate the financial feasibility of carbon adsorption technologies in their entire life cycle, and the costs of production and operation up to the stages of maintenance and end of life are included. LCC considers the costs associated with the full lifetime of the adsorber, making it a valuable tool for optimizing long-term financial and environmental results. In CO_2 adsorbers, these costs are related to energy consumption, material acquisition, routine maintenance, consumption energy, and possible recycling or disposal costs at the end of their useful life. Recently, there has been research on integrating LCA with LCC to record cost and environmental impacts, which in this research provides a comprehensive view of the sustainability profiles of adsorbents. In this combined analysis, stakeholders can make informed choices about adsorbent materials that balance economic efficiency and environmental performance. To compare different adsorbent technologies (materials with amine agent), they have used LCC, they are effective in adsorbing CO_2, but, due to energy regeneration processes, they may have higher operating costs. These predictions make it possible to choose materials that have a better trade-off between performance and life cycle costs, especially when renewable or low-energy regeneration methods are used. LCC supports decisions made regarding adsorber maintenance and replacement programs. It can be predicted that the maintenance costs of an adsorbent material will exceed its benefits when the financial impact of the degradation and regeneration cycles has been estimated, which indicates the optimum point for replacement. The analysis helps develop end-of-life strategies, taking into account that some adsorbents can be recycled or reused, which offsets future production costs and reduces waste [30, 47, 49].

Advanced software tools allow detailed LCC calculations and scenario modelling (such as regional adjustments for material and labor costs). These tools are aligned with international standards (such as ISO 15686–5) and use databases and can include calculation of life cycle costs of different adsorbents as well as utility costs, local labor rates, and material-specific maintenance requirements. LCC analysis for CO_2 adsorbers provides a comprehensive and complete method for calculating the financial and environmental aspects during the full life cycle and for the economic optimization of carbon adsorption solutions.

LCC for CO_2 adsorbers, which include material extraction, manufacturing, and any initial start-up, as well as long-term costs associated with operational energy, maintenance, and adsorber regeneration cycles, provide an in-depth and holistic view of the overall investment. It shows the need to implement and maintain carbon sequestration technologies throughout their lifetime. LCC also considers end-of-life costs, such as recycling or disposal, and provides a complete picture of financial needs at each stage of the adsorber's life cycle.

11.3.9.1 Key Phases in LCC for CO_2 Adsorbents

1. Capital costs (an initial investment that covers costs related to raw materials, manufacturing, and installation): Also, these costs for adsorbents can be according to the type of material (e.g., AC, framework metal-organic, or zeolites). The complexity involved in the synthesis of some of these materials may require extensive chemical processing, while others rely on more cost-effective natural materials.
2. Operating and maintenance costs: These costs are the largest contributors to the LCC of CO_2 adsorbers due to the energy requirements for adsorbing and regenerating the adsorbers. Some adsorbents may require high regeneration temperatures (such as amine-functionalized materials), and these types of adsorbents lead to increased energy costs. Maintenance costs include cleaning, regular inspections, and minor repairs to ensure optimal performance over time. For example, studies conducted on solidified adsorbents have concluded that some materials may suffer from a reduction in their capacity during cycles, which can require periodic replacement to maintain performance or costly maintenance.
3. Regeneration costs: Regeneration, or the process of removing CO_2 from the adsorber to restore its adsorption capacity, is a very important element in operating costs. Materials that require low energy regeneration (some solid adsorbents) have a cost advantage. Liquid-based or high-temperature adsorbers may incur significant energy costs during regeneration, increasing their life cycle costs. Among the research that is conducted with the aim of reducing these costs, it is possible to mention the improvement of regeneration efficiency and the exploration of low-energy alternatives for high-consumption adsorbents [50].
4. End-of-life costs: This step for CO_2 adsorbers can include recycling, disposal, or reuse, depending on the properties of the material. Recyclable materials provide cost savings and environmental benefits (reducing the need for new materials as well as waste disposal costs). Adsorbents with toxic or non-degradable properties may require more complex and costly disposal methods to meet environmental standards [51].

11.3.9.2 LCC in Decision-Making for CO_2 Adsorbents

LCC analysis provides information on the trade-offs between different adsorbent materials and operating methods. By integrating LCC with LCA, companies can make informed and intelligent choices that assess economic feasibility. In addition, they address environmental impacts and help avoid solutions that are financially unsustainable over time. LCC can show whether it is possible for a more expensive but more durable adsorber to ultimately cost less over the life cycle than a cheaper material that requires frequent replacement or regeneration. The results of LCC analysis are of great importance for guiding policy and regulatory frameworks on carbon

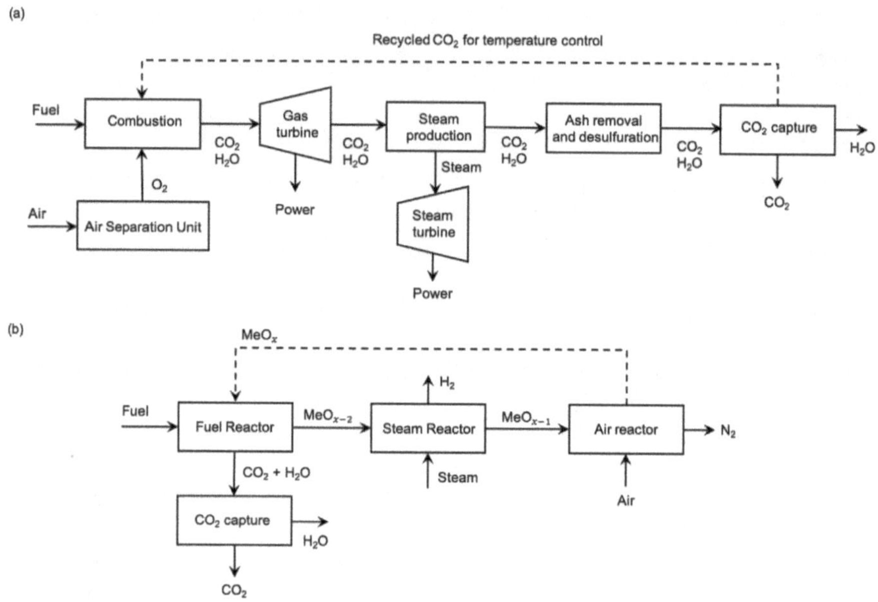

FIGURE 11.4 (a) Oxy-combustion with CO_2 capture applied to a combined cycled power plant. (b) Chemical looping combustion in a three-reactor configuration [50].

sequestration. LCC can be used by governments and industry groups to support incentive structures that support long-term investment in sustainable carbon capture solutions. This ensures that the technologies selected are cost-effective for immediate implementation as well as for long-term use and economic constraints [39, 50].

11.3.9.3 Technological Innovations and LCC Optimization

Recent advances have increased the cost-effectiveness of CO_2 adsorbers. For example, new adsorbent materials with improved selectivity and adsorption capacity help reduce the frequency of regeneration cycles, resulting in lower operating and maintenance costs. Also, compounds that have renewable energy sources can significantly reduce energy-related costs associated with LCC in regeneration processes. Technological innovations that enable the efficient recycling or reactivation of adsorbents at the end of their life will further contribute to the reduction of total life cycle costs [52].

11.4 COST–BENEFIT ANALYSIS

Due to increasing atmospheric CO_2 levels and climate change, the need for effective carbon capture technologies has increased. Among these strategies, adsorbent-based CO_2 capture has attracted attention as a cost-effective way to reduce greenhouse gas emissions. In a cost–benefit analysis (CBA), we assess the environmental and economic viability of these technologies (especially in terms of their recyclability).

This analysis considers key factors including the regeneration process, the adsorption capacity, and the overall life cycle costs of the materials involved.

Recyclability is an important consideration in the design of recycling systems as it directly affects the operational sustainability of carbon capture systems. An ideal CO_2 adsorbent exhibits high adsorption efficiency, as well as maintaining its performance over multiple adsorption and desorption cycles. MOFs, biomass-derived porous carbons (BPCs), zeolites, and amine-functionalized adsorbents are materials that have been extensively investigated for their recyclability potential. These materials differ in their economic feasibility, structural properties, and regeneration methods. The recyclability of CO_2 adsorbents affects the cost structure of carbon capture operations. These costs include initial material costs, regeneration costs, and operating parameters. For example, BPCs have a cost advantage (they are synthesized from inexpensive biomass precursors through processes such as pyrolysis or hydrothermal carbonization). They are environmentally friendly due to the use of industrial or agricultural waste, reducing the overall carbon footprint of the production process. Recent studies have shown that BPCs, if optimized under conditions, have high adsorption capacities over multiple cycles.

Regeneration is critical to the recyclability of an adsorbent. Effective regeneration ensures that the adsorbent can be reused for successive CO_2 capture cycles while maintaining its structural and functional properties. Traditional regeneration methods often involve thermal treatment, in which the adsorbent is heated to release the adsorbed CO_2. These methods can be energy-intensive and increase operating costs. Newer methods, such as microwave-assisted desorption and electrothermal regeneration, are being used as energy-efficient alternatives. These methods reduce regeneration costs, minimize structural degradation of the adsorbent, and increase its life cycle performance. MOFs are another promising class of CO_2 adsorbents, with exceptional adsorption capacities and tunable properties. Functionalized MOFs are resistant to thermal or chemical degradation, but their ability to selectively adsorb CO_2 increases recyclability. For example, functionalized MOFs induce strong interactions between amine groups and CO_2 molecules. MOF synthesis has a high initial cost, which is challenging to motivate on a large scale. Recent advances in synthesis techniques (such as scalable production methods and the use of cheaper precursors) aim to overcome these economic barriers [53–56].

In the recyclability of CO_2 adsorbent, the CBA includes several quantitative and qualitative criteria. To evaluate the net cost of using a specific adsorbent, a simple formula can be used, which is given here [56]:

$$\begin{aligned}&\text{Net Cost}\left(\text{NC}\right) = \text{Initial Material Cost}\left(\text{IMC}\right) + \\ &\left(\text{Regeneration Cost}\left(\text{RC}\right)\times \text{N}\right) - \text{Savings from Reuse}\left(\text{SR}\right)\end{aligned} \qquad (11.1)$$

In this formula, IMC represents the expenses associated with synthesizing or procuring the adsorbent, while RC denotes the energy and resources required for each regeneration cycle. The variable N reflects the total number of regeneration cycles over the adsorbent's lifespan, and SR accounts for cost savings achieved by avoiding the need for new material. This formula highlights the interplay between material performance and economic considerations, emphasizing the importance of balancing upfront costs with long-term savings.

Several studies have evaluated the recyclability of CO_2 adsorbents using this framework. For example, BPC has a low initial cost due to the abundant availability of raw materials. Since mild thermal or chemical treatment is sufficient to recover the adsorbent, it has a low regeneration cost. These materials also have high stability over multiple cycles, which increases their cost-effectiveness. Advanced adsorbents such as MOFs, while offering superior performance, require higher initial costs. Operating parameters such as the presence of impurities, temperature, and pressure play a significant role in determining the recyclability of CO_2 adsorbents. High temperatures or aggressive regeneration conditions can cause the degradation of certain adsorbents, reducing their effectiveness over time. It is possible that amine-functionalized adsorbents may become inactive during regeneration due to oxidative or thermal degradation. To address this issue, researchers have developed strategies such as incorporating stabilizing agents or optimizing regeneration protocols. These innovations reduce overall operating costs and improve the recyclability of adsorbents [52, 53, 55, 56].

Research is underway into energy-efficient regeneration techniques to increase the recyclability of CO_2 adsorbers. For example, desorption offers a rapid and uniform heating mechanism using microwaves, which reduces energy consumption. Electrothermal regeneration also uses electrical resistance to achieve localized heating, which minimizes energy consumption. These methods save costs in laboratory research and make them easy to use in industrial environments. In the regeneration processes, the integration of renewable energy sources (wind or solar energy) increases the sustainability of these technologies. However, achieving widespread adoption of recyclable CO_2 adsorbents remains challenging despite the progress made. Issues that require new solutions include high initial costs, structural degradation, and energy-intensive regeneration processes. In overcoming these challenges, the development of hybrid adsorbents that combine the properties of multiple materials, such as porous carbons and MOFs, is promising. These hybrid materials offer improved new capacity, increased stability, reduced capacity, and reduced regeneration costs, making them a viable option for future use. The scalability of recyclable CO_2 adsorbents in the future depends on continued research and development efforts. Collaboration between government agencies, academia, and industry is essential to accelerate the commercialization of these technologies. Financial credits or subsidies are policies that increase the incentive for the adoption of sustainable materials and market demand. Also, the feasibility of recyclable CO_2 adsorbers and the development of laboratory projects are facilitated by public–private partnerships.

The importance of environmental sustainability, as well as balancing economic viability, is emphasized by using a CBA of the recyclability of CO_2 adsorbents. MOFs and BPCs were among the materials that showed significant potential, but their widespread adoption presents challenges including sustainability, cost, and regeneration efficiency. Industrialists and researchers can pave the way for scalable and sustainable carbon capture solutions by leveraging advances in materials science and energy-efficient regeneration techniques. This comprehensive approach contributes to global emission reduction goals and also accelerates the transition to a low-carbon economy.

In comparing amine-functionalized adsorbents, CO_2 adsorbents, MOFs, and zeolites, cycle stability, adsorption capacity, costs, regeneration methods, and key features are reviewed. This highlights trade-offs such as selectivity, cost-effectiveness, and durability. It provides an overview of the suitability of each material for retrofitting carbon capture processes [47, 57–59].

11.5 INTRODUCTION OF TECHNO-ECONOMIC ANALYSES OF CO_2 ADSORBENTS

Due to the combustion of fossil fuels, industrial processes, and deforestation, the concentration of CO_2 in the atmosphere has increased dramatically. CCS technologies are a vital tool for reducing greenhouse gas emissions, but their economic viability remains a major challenge. The main basis for the development of CCS systems lies in the techno-economic analysis (TEA), which assesses both the technical performance and economic feasibility of these systems. Adsorbents play a pivotal role in CCS, especially in post-combustion capture. Materials like zeolites, MOFs, ACs, and emerging advanced materials such as MXenes and GO have shown promise due to their high selectivity, adsorption capacities, and regenerability. However, their real-world application depends on a balance between technical efficiency and

TABLE 11.1
Comparison of CO_2 Adsorbents

Material	Adsorption Capacity (mmol/g)	Regeneration Method	Cycles Tested	Cost Consideration	Key Features
Metal-organic frameworks (MOFs)	3.0–7.0	Thermal (120–200°C) or vacuum	>50	High material cost	High selectivity, customizable, sensitive to regeneration conditions
Amine-functionalized adsorbents	4.0–6.5	Thermal or chemical washing	>15	Moderate cost, higher energy for regeneration	High selectivity, vulnerable to oxidative degradation
Zeolites	2.0–4.0	Pressure-swing adsorption	>30	Low to moderate cost	Stable, resistant to high temperatures, limited CO_2 selectivity
Graphene	2.5–4.5	Thermal or chemical treatment	>20	Moderate cost	High surface area, tunable structure, excellent thermal stability
Graphene oxide	3.0–6.0	Thermal (<150°C)	>25	Moderate to high cost	Functional groups improve selectivity, sensitive to harsh regeneration
MXene	3.5–8.0	Thermal (<200°C) or chemical	>50	Moderate to high cost	High electrical conductivity, tunable surface chemistry, robust under cycling

cost-effectiveness. TEA focuses on identifying the cost drivers of these technologies and optimizing parameters such as capital investment, operational expenditure, and energy consumption. A holistic TEA integrates system design, material choice, and energy recovery mechanisms to ensure sustainability.

One of the main focuses in CCS, especially in post-combustion adsorption, is adsorbents. Materials such as zeolites, MOFs, ACs, and emerging advanced materials such as MXenes and GO have found wide applications due to their high selectivity, adsorption capacity, and regenerability. However, their application in the real world depends on technical efficiency, cost-effectiveness, and the balance between the two. The focus of TEA is to identify the cost drivers of these technologies and optimize parameters such as capital investment, operating costs, and energy consumption. A comprehensive TEA integrates system design, material selection, and energy recovery mechanisms to ensure sustainability [55].

11.5.1 The Cost of CO_2 Captured

In the cost analysis, the cost of CO_2 captured (CC) is a key metric for evaluating CCS systems. It quantifies the economic feasibility of capturing one unit of CO_2, usually expressed in dollars per ton (\$/ton CO_2). The calculation of CC includes the capital expenditure (CAPEX) and operating expenditure (OPEX) of the capture system, along with the system's CO_2 capture efficiency [54, 55].

$$CC = \frac{\text{CAPEX + OPEX}}{\text{Total CO}_2\text{ Captured Annually}} \tag{11.2}$$

where CAPEX refers to the initial investment in equipment, materials, and plant construction; OPEX represents ongoing costs, including maintenance, labor, and energy for regeneration; and Total CO_2 Captured annually denotes the mass of CO_2 captured in a year.

The cost of CO_2 capture (CC) is influenced by several factors, including material performance, which dictates the adsorption capacity and selectivity, thereby reducing the amount of adsorbents required and operating costs. The regeneration energy, whether thermal or pressure-based, significantly affects operating costs. Another factor is the lifespan of the adsorbent, which plays a significant role, as durable materials reduce replacement frequency and long-term costs. The impact of the scale of operation is also significant on operating efficiency, as larger plants benefit from economies of scale, while the energy source used, especially renewable options, can reduce both emissions and operating costs [56].

Example

For a plant capturing 100,000 tons of CO_2 annually, if the CAPEX is \$50 million and OPEX is \$15 million per year:

$$CC = \frac{50{,}000{,}000 + 15{,}000{,}000}{100{,}000} = 650\ {}^{\$}\!/_{\text{tonCO}_2}$$

11.5.2 Techno-Economic Analysis

A structured approach to assessing the feasibility of a CO_2 capture system is presented using TEA. TEA integrates process design, performance criteria, and economic analysis to enable decision-makers to optimize system parameters. This analysis includes modelling energy consumption, material behavior, and financial costs, taking into account environmental regulations and market dynamics.

The TEA consists of three key steps. First, the system is modelled using simulation tools for iterative measures, regeneration, and CO_2 compression. Second, it evaluates costs and discounted cash flows by categorizing fixed (CAPEX) and variable (OPEX) components to assess profitability. Economic functions, such as net present value (NPV), levelized carbon cost (LCOC), and energy intensity, are comprehensively evaluated to assess system efficiency and feasibility. Together, these steps ensure an assessment of technical and financial feasibility [51, 52, 56].

11.5.2.1 TEA Metrics

- **NPV:** represents the profitability of a project over its lifetime:

$$\text{NPV} = \sum_{t=0}^{T} \frac{R_t - C_t}{(1+r)^t} \tag{11.3}$$

where R_t represents the revenue at time t, C_t denotes the costs at time t, r is the discount rate, and T refers to the project's lifespan.

- **LCOC:** a measure of the average cost of capturing CO_2 over the system's lifespan:

$$\text{LCOC} = \frac{\text{Total Costs}}{\text{Total CO}_2\text{ Captured}} \tag{11.4}$$

Note: Energy Intensity: The energy required to capture 1 kg of CO_2, expressed in kJ/kgCO_2.

11.5.2.2 Sensitivity Analysis

How changes in key parameters are assessed in a sensitivity analysis, a critical aspect of TEA, is used to examine system performance and cost-effectiveness. It examines variables such as material costs, energy prices, adsorbent performance, and system scale and identifies those that have the greatest impact on metrics such as cost of CC, NPV, or LCOC. Risks are mitigated through system design by simulating changes within realistic ranges, ensuring economic feasibility under fluctuating market conditions, and helping to optimize CO_2 capture systems for robust and cost-effective operation. Sensitivity analysis identifies parameters that significantly influence system economics, such as:

- Energy prices
- Material cost and availability
- System efficiency and scale

A TEA study on MOF-based CO_2 adsorbers showed a 30% reduction in costs due to the optimization of regeneration energy and increased production scale. Similarly, hybrid systems combining MOFs with renewable energy sources further improved economic outcomes.

11.5.3 Sensible Heat Recovery

11.5.3.1 The Role of Heat Recovery in CCS Systems

Heat recovery is crucial in adsorption-based CO_2 capture systems, where energy recovery processes account for a significant portion of operating costs. The main focus of sensible heat recovery is on recovering waste heat to increase system efficiency and reduce OPEX. Heat recovery principles include heat integration, where waste heat from exhaust gases is used to preheat input streams, minimizing energy consumption. The use of materials such as phase change materials to capture and store excess heat for future use is part of thermal storage. In addition, advanced heat exchangers optimize heat transfer between streams and improve overall energy efficiency. Collectively, these approaches reduce operating energy demand and increase the sustainability and cost-effectiveness of industrial processes [48, 50, 51].

Efficiency of Heat Recovery (η_H)

$$\eta_H = \frac{Q_{\text{recovered}}}{Q_{\text{available}}} \times 100 \tag{11.5}$$

where $Q_{\text{recovered}}$ represents the heat recovered (kJ) and $Q_{\text{available}}$ demonstrates the total available heat (kJ).

11.5.3.2 Impact on Cost of CO_2 Captured

An important criterion in assessing the economic feasibility of CCS systems is the cost of CC. This cost is influenced by various factors such as material performance, operating parameters, energy consumption, and plant scale. Material performance plays a key role, as adsorbents with high CO_2 adsorption capacity and selectivity reduce the amount of material required, reducing initial CAPEX and ongoing replacement costs. In addition, the lifespan of the adsorbents significantly affects long-term costs. Durable materials reduce replacement frequency and minimize OPEX [47, 57].

In heat or pressure-based systems, the energy required to regenerate the adsorbent contributes significantly to OPEX. High-energy regeneration processes can increase costs, emphasizing the importance of energy-efficient materials and methods. The choice of energy source also affects costs. Another factor that can reduce both emissions and OPEX is the use of renewable energy that aligns with sustainability goals. The scale of operations affects CC through economies of scale, where larger systems achieve lower costs per unit due to the distribution of fixed costs over a larger output. Smaller plants often experience higher relative costs, highlighting the importance of scalability in project design. System design and integration, where efficient heat recovery and optimized flow systems are other issues that can reduce energy losses and associated costs, are important to consider. Policies, subsidies, and carbon pricing also play an important role, and supportive regulatory frameworks can offset some of the financial burden. CC reduction requires a multifaceted approach, integrating advanced materials, energy-efficient technologies, and optimized plant designs while providing policy incentives to achieve sustainable and economic CO_2 capture solutions [45, 48, 50, 58]. Efficient heat recovery systems can reduce regeneration energy costs by 20–40%, directly lowering CC. For example, integrating a heat exchanger that recovers 50% of the waste heat could save $10/ton CO_2.

11.6 REGULATORY AND SAFETY CONSIDERATIONS OF CO_2 ADSORBENT RECYCLING

Essential to sustainable CCS technologies is the recycling of CO_2 adsorbents, ensuring cost-effectiveness and environmental responsibility. Since this is a complex process, it must be governed by a wide range of regulatory and safety considerations to minimize environmental impacts, protect human health, and maintain operational efficiency. These considerations are influenced by national and international environmental regulations, workplace safety standards, and the unique chemical and physical properties of adsorbents such as zeolites, MOFs, AC, GO, and MXenes [59, 60].

11.6.1 Environmental Regulations and Compliance

For recycling CO_2 adsorbents, strict environmental regulations are designed to minimize emissions and waste. Regulations such as the European Union Framework Directive and the U.S. Resource Conservation and Recovery Act classify adsorbent waste based on toxicity, persistence, and environmental mobility. Adsorbents that change chemically during use may release hazardous compounds that require compliance with these regulations to prevent soil, air, or water contamination. In addition, end-of-life adsorbents must be assessed under the Basel Convention to ensure their safe handling, transportation, and disposal.

11.6.1.1 Material-Specific Challenges

Due to the different properties of different adsorbents and their chemical properties, unique challenges arise during recycling. For example, MOFs, while highly efficient at capturing CO_2, often contain metals and organic bonds that may degrade or release harmful by-products during regeneration. Proper management of these by-products is necessary to avoid regulatory violations. Similarly, the management of MXenes involves reducing the risks associated with particulate emissions, which can pose respiratory hazards (using different acids in the recycling process). Therefore, appropriate recycling protocols and environmental impact assessments are essential for each type of adsorbent.

11.6.1.2 Occupational Safety Considerations

Standards such as Occupational Safety and Health Administration control workplace safety during the recycling process in the U.S. or EU occupational safety guidelines. Exposure to dust or fumes released during regeneration or sorbent recovery is a key concern. For example, silica dust from zeolites or nano-sized particles from advanced sorbents such as GO and MXene can cause respiratory problems. Protective equipment, air filtration systems, and regular health monitoring are essential to protect workers.

11.6.1.3 Thermal and Chemical Stability

Another important safety consideration is the stability of adsorbents during recycling. Adsorbents that are exposed to high temperatures or reactive chemicals during regeneration may degrade and release potentially hazardous materials. For example, the decomposition of MOFs at high temperatures may release volatile organic compounds, requiring measures to capture and neutralize these emissions. Similarly, the thermal cycling of adsorbents must be carefully managed to avoid structural failure, which could lead to safety hazards or reduced efficiency in subsequent adsorption cycles.

11.6.1.4 Regeneration Energy and Carbon Footprint

Recycling processes must align with regulations that limit greenhouse gas emissions and energy use. The energy required for adsorbent regeneration can contribute significantly to the carbon footprint of CCS systems, potentially undermining their environmental benefits. Regulatory frameworks such as the Kyoto Protocol and the Paris Agreement incentivize the development of low-energy regeneration methods, such as pressure-swing adsorption or electric heating, to improve sustainability.

11.6.1.5 Waste Management and Disposal

Unrecoverable or degraded adsorbents must be managed as industrial waste, adhering to guidelines for hazardous materials. Adsorbents contaminated with captured CO_2 impurities, such as sulfur compounds, may require specialized treatment before disposal. Recycling facilities must also maintain documentation and traceability of waste streams to comply with legal and environmental standards.

11.6.1.6 Emerging Innovations and Policy Implications

Advancements in adsorbent recycling are introducing safer and more efficient methods, such as solvent-free regeneration techniques or closed-loop systems that minimize waste. Policymakers are increasingly recognizing the importance of integrating these innovations into regulatory frameworks to enhance the sustainability of CCS systems. Incentives for adopting greener recycling practices, such as tax benefits or subsidies, are encouraging industries to invest in sustainable technologies [47, 57, 58, 60].

The recycling of CO_2 adsorbents is a multifaceted process that requires adherence to stringent regulatory and safety standards. By addressing environmental, occupational, and technical challenges, stakeholders can ensure the safe and sustainable operation of CCS systems. Continued research, innovation, and policy support are essential to overcoming barriers and optimizing recycling practices, thereby advancing the global effort to mitigate climate change.

11.7 CONCLUSION

The recyclability of CO_2 adsorbents is a pivotal factor in ensuring the economic and environmental sustainability of carbon capture technologies. This chapter has elucidated the critical advancements and challenges in recycling various classes of adsorbents, including ACs, MXenes, MOFs, and polymeric composites. Each material demonstrates unique advantages, such as high adsorption capacity, structural stability, and efficient regeneration, which are balanced against limitations like degradation, scalability, and cost implications. ACs, with their cost-effectiveness and adaptability, exhibit significant potential for sustainable CO_2 capture, especially when chemically modified. MXenes, a novel class of materials, offer high recyclability and retention of adsorption capacity over extended cycles, though challenges in production scalability and stability under humid conditions remain. MOFs stand out for their tunable properties and high selectivity but face hurdles in high initial costs and sensitivity to harsh regeneration conditions. Meanwhile, polymeric composites and hybrid materials, which combine multiple adsorbent characteristics, emerge as promising solutions for improving performance and recyclability.

Technological innovations, such as energy-efficient regeneration methods and advanced composite designs, are crucial to overcoming the existing barriers. The integration of LCA and LCC analyses provides comprehensive insights into the economic and environmental impacts, enabling informed decision-making for large-scale implementation. Ultimately, the recyclability of CO_2 adsorbents hinges on striking a balance between material performance, economic feasibility, and environmental sustainability. Continued research, coupled with policy incentives and industrial collaboration, will be instrumental in advancing these materials. This will not only enhance their applicability but also contribute significantly to global efforts in mitigating climate change and transitioning toward a low-carbon economy.

REFERENCES

1. Jahandar Lashaki, M., S. Khiavi, and A. Sayari, Stability of amine-functionalized CO_2 adsorbents: A multifaceted puzzle. *Chemical Society Reviews*, 2019. 48(12): p. 3320–3405.
2. Liu, J., et al., Progress in adsorption-based CO_2 capture by metal–organic frameworks. *Chemical Society Reviews*, 2012. 41(6): p. 2308–2322.
3. Guo, Y., et al., Porous activated carbons derived from waste sugarcane bagasse for CO_2 adsorption. *Chemical Engineering Journal*, 2020. 381: p. 122736.
4. Ren, F. and W. Liu, Review of CO_2 adsorption materials and utilization technology. *Catalysts*, 2023. 13(8): p. 1176.
5. He, S., et al., Facile preparation of N-doped activated carbon produced from rice husk for CO_2 capture. *Journal of Colloid and Interface Science*, 2021. 582: p. 90–101.
6. Chen, J., et al., Recent progress and advances in the environmental applications of MXene related materials. *Nanoscale*, 2020. 12(6): p. 3574–3592.
7. Hiragond, C.B., et al., A novel N-doped graphene oxide enfolded reduced titania for highly stable and selective gas-phase photocatalytic CO_2 reduction into CH4: An in-depth study on the interfacial charge transfer mechanism. *Chemical Engineering Journal*, 2021. 416: p. 127978.
8. Zhou, Y., et al., Humidity-enabled ionic conductive trace carbon dioxide sensing of nitrogen-doped Ti3C2T x MXene/polyethyleneimine composite films decorated with reduced graphene oxide nanosheets. *Analytical Chemistry*, 2020. 92(24): p. 16033–16042.
9. Wang, Y., et al., N-doped porous carbon derived from rGO-incorporated polyphenylenediamine composites for CO_2 adsorption and supercapacitors. *Journal of Power Sources*, 2020. 472: p. 228610.

10. Rabiee, G., M. Behbahani, and S. Bagheri, *Functionalization of Ti3C2 MXene with Diethylenetriamine for a Column-Based Solid-Phase Extraction of Heavy Metals*. Langmuir, 2024.
11. Jimmy, J. and B. Kandasubramanian, Mxene functionalized polymer composites: Synthesis and applications. *European Polymer Journal*, 2020. 122: p. 109367.
12. Sayam, A., et al., A review on carbon fiber-reinforced hierarchical composites: mechanical performance, manufacturing process, structural applications and allied challenges. *Carbon Letters*, 2022. 32(5): p. 1173–1205.
13. Sharma, H., et al., An overview on carbon fiber-reinforced epoxy composites: Effect of graphene oxide incorporation on composites performance. *Polymers*, 2022. 14(8): p. 1548.
14. Zhang, Y., Q. Zhang, and G. Chen, Carbon and carbon composites for thermoelectric applications. *Carbon Energy*, 2020. 2(3): p. 408–436.
15. Dong, Y.-D., et al., Cellulose/carbon composites and their applications in water treatment–a review. *Chemical Engineering Journal*, 2021. 405: p. 126980.
16. Ateeq, M., A review on recycling technique and remanufacturing of the carbon fiber from the carbon fiber polymer composite: Processing, challenges, and state-of-arts. *Composites Part C: Open Access*, 2023: p. 100412.
17. Giorgini, L., et al., Recycling of carbon fiber reinforced composite waste to close their life cycle in a cradle-to-cradle approach. *Current Opinion in Green and Sustainable Chemistry*, 2020. 26: p. 100368.
18. Zhang, J., et al., Current status of carbon fibre and carbon fibre composites recycling. *Composites Part B: Engineering*, 2020. 193: p. 108053.
19. Khalil, Y., Sustainability assessment of solvolysis using supercritical fluids for carbon fiber reinforced polymers waste management. *Sustainable Production and Consumption*, 2019. 17: p. 74–84.
20. Wei, Y. and S. Hadigheh, Development of an innovative hybrid thermo-chemical recycling method for CFRP waste recovery. *Composites Part B: Engineering*, 2023. 260: p. 110786.
21. Liu, T., et al., Progress in chemical recycling of carbon fiber reinforced epoxy composites. *Macromolecular Rapid Communications*, 2022. 43(23): p. 2200538.
22. Branfoot, C., et al., Recovery of chemical recyclates from fibre-reinforced composites: A review of progress. *Polymer Degradation and Stability*, 2023: p. 110447.
23. Azwa, Z., et al., A review on the degradability of polymeric composites based on natural fibres. *Materials & Design*, 2013. 47: p. 424–442.
24. Naqvi, S., et al., A critical review on recycling of end-of-life carbon fibre/glass fibre reinforced composites waste using pyrolysis towards a circular economy. *Resources, Conservation and Recycling*, 2018. 136: p. 118–129.
25. Sattari, A., et al., The application of polymer containing materials in CO_2 capturing via absorption and adsorption methods. *Journal of CO_2 Utilization*, 2021. 48: p. 101526.
26. Choi, W., et al., Structural effects of amine polymers on stability and energy efficiency of adsorbents in post-combustion CO_2 capture. *Chemical Engineering Journal*, 2021. 408: p. 127289.
27. Dai, Y., et al., A review on the recent advances in composite membranes for CO_2 capture processes. *Separation and Purification Technology*, 2023. 307: p. 122752.
28. Friedrich, K., Polymer composites for tribological applications. *Advanced Industrial and Engineering Polymer Research*, 2018. 1(1): p. 3–39.
29. Ho, N.A.D. and C. Leo, A review on the emerging applications of cellulose, cellulose derivatives and nanocellulose in carbon capture. *Environmental Research*, 2021. 197: p. 111100.
30. Hack, J., N. Maeda, and D.M. Meier, Review on CO_2 capture using amine-functionalized materials. *ACS Omega*, 2022. 7(44): p. 39520–39530.

31. Zhao, P., et al., The latest development on amine functionalized solid adsorbents for post-combustion CO_2 capture: Analysis review. *Chinese Journal of Chemical Engineering*, 2021. 35: p. 17–43.
32. Kunalan, S. and K. Palanivelu, Polymeric composite membranes in carbon dioxide capture process: a review. *Environmental Science and Pollution Research*, 2022. 29(26): p. 38735–38767.
33. Wang, J., et al., Recent progress in porous organic polymers and their application for CO_2 capture. *Chinese Journal of Chemical Engineering*, 2022. 42: p. 91–103.
34. Kotp, M.G. and S.-W. Kuo, Selective capturing of the CO_2 emissions utilizing ecological (3-Mercaptopropyl) trimethoxysilane-coated porous organic polymers in composite materials. *Polymers*, 2024. 16(13): p. 1759.
35. Pires da Mata Costa, L., et al., Capture and reuse of carbon dioxide (CO_2) for a plastics circular economy: A review. *Processes*, 2021. 9(5): p. 759.
36. Khalid, M.Y., et al., Recent trends in recycling and reusing techniques of different plastic polymers and their composite materials. *Sustainable Materials and Technologies*, 2022. 31: p. e00382.
37. Mäntynen, L., *Use of alternative bed materials to inhibit bed agglomeration in fluidized bed combustion of biomass fuels*. Master's Thesis Tampere University Degree Programme in Environmental and Energy Engineering, 2020.
38. Chen, P.-Y., et al., Recycling and reutilization of waste carbon fiber reinforced plastics: current status and prospects. *Polymers*, 2023. 15(17): p. 3508.
39. Knappich, F., et al., Recycling process for carbon fiber reinforced plastics with polyamide 6, polyurethane and epoxy matrix by gentle solvent treatment. *Waste Management*, 2019. 85: p. 73–81.
40. Odunlami, O., et al., Advanced techniques for the capturing and separation of CO_2–a review. *Results in Engineering*, 2022. 15: p. 100512.
41. Harja, M. and G. Ciobanu, Eco-friendly nano-adsorbents for pollutant removal from wastewaters, in *Handbook of Nanomaterials and Nanocomposites for Energy and Environmental Applications*. 2021, Springer. p. 2225–2246.
42. Salaudeen, S.A., B. Acharya, and A. Dutta, CaO-based CO_2 sorbents: A review on screening, enhancement, cyclic stability, regeneration and kinetics modelling. *Journal of CO_2 Utilization*, 2018. 23: p. 179–199.
43. Albrecht, K.O., et al., Development of a CaO-based CO_2 sorbent with improved cyclic stability. *Industrial & Engineering Chemistry Research*, 2008. 47(20): p. 7841–7848.
44. Elvira, G.-B., et al., MgO-based adsorbents for CO_2 adsorption: Influence of structural and textural properties on the CO_2 adsorption performance. *Journal of Environmental Sciences*, 2017. 57: p. 418–428.
45. Jiao, J., et al., Improvement of adsorbent materials for CO_2 capture by amine functionalized mesoporous silica with worm-hole framework structure. *Chemical Engineering Journal*, 2016. 306: p. 9–16.
46. Darunte, L.A., et al., CO_2 capture via adsorption in amine-functionalized sorbents. *Current Opinion in Chemical Engineering*, 2016. 12: p. 82–90.
47. Varghese, A.M. and G.N. Karanikolos, CO_2 capture adsorbents functionalized by amine–bearing polymers: A review. *International Journal of Greenhouse Gas Control*, 2020. 96: p. 103005.
48. Ross, S. and D. Evans, Use of life cycle assessment in environmental management. *Environmental Management*, 2002. 29: p. 132–142.
49. Zhang, J., et al., Sustainable municipal waste management strategies through life cycle assessment method: A review. *Journal of Environmental Management*, 2021. 287: p. 112238.
50. Garcia, J.A., et al., Technical analysis of CO_2 capture pathways and technologies. *Journal of Environmental Chemical Engineering*, 2022. 10(5): p. 108470.

51. Wang, M., et al., Technoeconomic analysis of the electrochemically mediated amine regeneration CO_2 capture process. *Industrial & Engineering Chemistry Research*, 2020. 59(31): p. 14085–14095.
52. Li, N., et al., The overhauls technical innovation project optimization method of power grid device based on life cycle asset management. *Energy Reports*, 2020. 6: p. 1249–1254.
53. Chauvy, R. and L. Dubois, Life cycle and techno-economic assessments of direct air capture processes: An integrated review. *International Journal of Energy Research*, 2022. 46(8): p. 10320–10344.
54. Mac Dowell, N., et al., The role of CO_2 capture and utilization in mitigating climate change. *Nature Climate Change*, 2017. 7(4): p. 243–249.
55. Fu, L., et al., Research progress on CO_2 capture and utilization technology. *Journal of CO_2 Utilization*, 2022. 66: p. 102260.
56. Bochon, K. and T. Chmielniak, Energy and economic analysis of the carbon dioxide capture installation with the use of monoethanolamine and ammonia. *Archives of Thermodynamics*, 2015: p. 93–110–93–110.
57. Lai, J.Y., L.H. Ngu, and S.S. Hashim, A review of CO_2 adsorbents performance for different carbon capture technology processes conditions. *Greenhouse Gases: Science and Technology*, 2021. 11(5): p. 1076–1117.
58. Boer, D.G., J. Langerak, and P.P. Pescarmona, Zeolites as selective adsorbents for CO_2 separation. *ACS Applied Energy Materials*, 2023. 6(5): p. 2634–2656.
59. Arifutzzaman, A., et al., MXene based activated carbon novel nano-sandwich for efficient CO_2 adsorption in fixed-bed column. *Journal of CO_2 Utilization*, 2023. 68: p. 102353.
60. Dixit, F., et al., Application of MXenes for air purification, gas separation and storage: a review. *Renewable and Sustainable Energy Reviews*, 2022. 164: p. 112527.

12 Environmental Impacts and Sustainability

Fatemeh Bahmanzadegan and Ahad Ghaemi

12.1 INTRODUCTION

Carbon dioxide (CO_2) adsorbent materials represent a critical facet role in technologies for capturing and storing carbon, aiming to curb the escalating levels of CO_2 emissions responsible for climate change. These materials encompass compounds engineered to capture CO_2 molecules from industrial flue gases, natural gas processing, and other emission sources. CCS technology has the potential to decrease emissions directly at their origin by capturing CO_2 emissions from point sources, simultaneously removing other pollutants and lowering the social cost of carbon. It acts as a technological link between environmental degradation and sustainability efforts through its promotion of decarbonization. Despite its potential to decrease CO_2 emissions, there are still pending concerns about environmental risks and economic consequences (Figure 12.1) [1].

The mechanisms of CO_2 capture can vary from physical adsorption, wherein CO_2 molecules adhere to the material's surface via weak intermolecular forces, to chemical adsorption, involving covalent bonds between CO_2 and functional groups within the material.

The utilization of porous materials for physically adsorbing (physisorbed) CO_2 offers an attractive alternative due to its clean and reversible nature, coupled with lower energy requirements stemming from the decreased adsorption enthalpy compared to scrubbing methods. Industrial-scale applications of gas sorption on adsorbents, such as in O_2/N_2 separation, have largely addressed practical engineering challenges [2]. Recently, there has been an increasing focus on researching nanoporous materials designed for effective CO_2 physisorption, encompassing carbon-based materials [3], metal-organic frameworks (MOFs) [4], zeolites [5], zeolitic imidazole frameworks, porous silica [6], and their combinations. Figure 12.2 indicates the types of CO_2 adsorbents.

Although numerous groups focus on maximizing CO_2 uptake per unit weight and volume at high pressures, materials designed for selective CO_2 capture must also meet additional criteria. These include high CO_2 selectivity compared to other gases, significant CO_2 capacity during both adsorption and regeneration phases, regeneration under mild conditions using pressure or temperature changes, durable stability against impurities and moisture, and rapid adsorption rates. Figure 12.3 indicates the properties of the best adsorbent.

Besides focusing on energy-efficient regeneration, it is essential to prioritize high selectivity and working capacity to replace expensive amine scrubbing or cryogenic

DOI: 10.1201/9781003476191-12

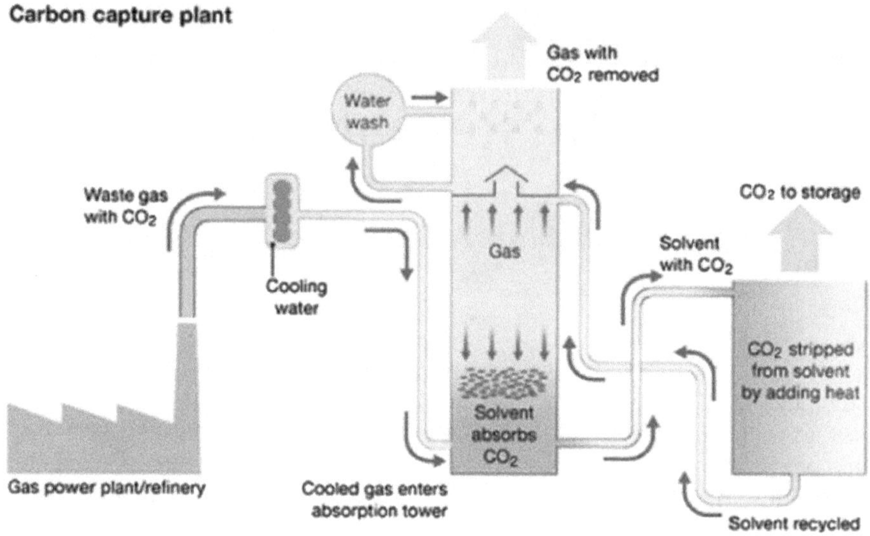

FIGURE 12.1 Schematic diagram of a possible CCS system [1], copyright (2023), Elsevier.

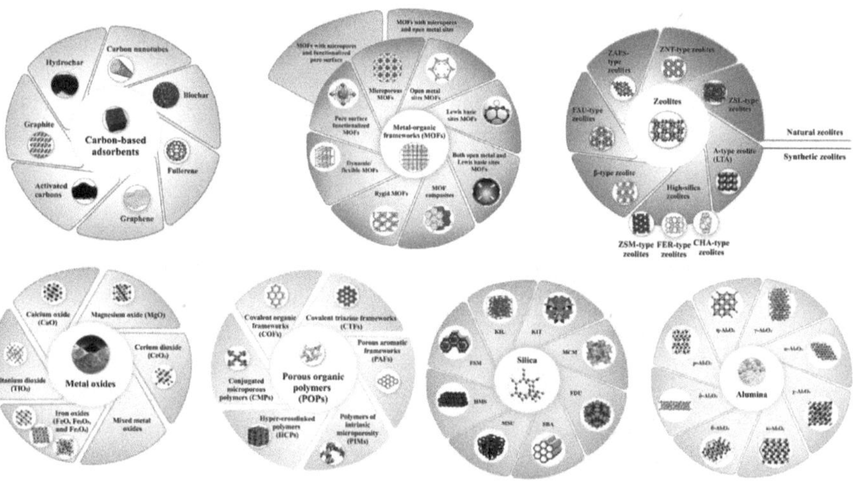

FIGURE 12.2 Types of CO_2 adsorbents [7], copyright (2021), Elsevier.

distillation with solid surface adsorption methods. While considerable advancements have been made in customizing porous materials for effective CO_2 capture through selective physisorption, achieving competitive binding of atmospheric CO_2 remains a challenging objective. Transitioning toward more adaptable multi-cycle methodologies, materials characterized by nanopores are emerging as critical components, owing to their potential to be customized to exhibit moderate regeneration energy

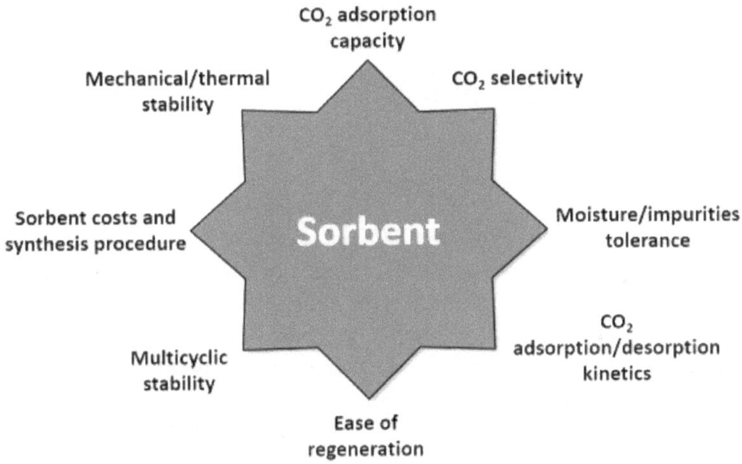

FIGURE 12.3 Schematic representation of properties of the required best adsorbent [8], copyright © 2021, American Chemical Society.

requirements while maintaining high selectivity and adsorption capacity. Despite significant strides made in this area, further efforts are essential. Particularly noteworthy is the emphasis on enhancing sorption capacity in existing research, whereas addressing sorption selectivity, although crucial for practical implementation, often receives less attention. This underscores the pressing need for sustained exploration and innovation within this sphere. Moreover, it is imperative to incorporate cost-effectiveness and sustainability considerations into the development of such materials to ensure their feasibility and environmental integrity in real-world applications. In a cyclic process, the disparity in uptake between sorption and regeneration conditions holds greater significance than achieving high uptake at higher pressures or lower temperatures. Ideally, a CO_2 adsorption isotherm that exhibits a more convex or upward-curved shape is preferred. When such a cyclic process is driven by temperature changes, a high apparent sorption enthalpy once again proves advantageous. Another critical factor in adsorption kinetics and pore diffusion is the rate of adsorption. In pores closely aligned in size to the kinetic diameter, the adsorption of CO_2 and the filling of the entire pore space tend to slow down significantly, a phenomenon well-documented, particularly in the narrow pores of zeolites used in gas filtration and catalytic processes. Efforts to address this issue typically involve introducing additional transport pores. However, it is important to recognize that, while these additional pores increase surface area to enhance transport, they can simultaneously reduce selectivity in the context of selective CO_2 capture. Industrial practices predominantly rely on amines for CO_2 capture; however, the regeneration process involves high temperatures and substantial energy demands during each adsorption/desorption cycle, resulting in material losses. To mitigate these limitations, considerable efforts have been devoted to developing solid amines characterized by high capture capacity and stability as alternative adsorbents. Despite their promise, there remains a critical gap in comprehensively assessing and discussing the environmental impacts associated with these adsorbents. Furthermore, the synthesis and consumption of these materials are closely linked to CO_2 emissions, highlighting the necessity for a more holistic approach to evaluating the sustainability of CO_2 capture technologies [9].

12.2 THE LIFE CYCLE ASSESSMENT

The life cycle assessment (LCA) of CO_2 adsorbents serves as a crucial analytical framework to assess the environmental effects of carbon dioxide capture techniques. In an era where mitigating climate change is essential, understanding the environmental implications of various CO_2 capture methods is vital for informed decision-making. The LCA of CO_2 adsorbents encompasses a thorough evaluation of their complete life cycle, from the extraction of raw materials to their manufacturing, use, and eventual disposal or recycling. This holistic approach enables researchers and policymakers to identify potential environmental hotspots and optimize processes to reduce adverse impacts.

12.2.1 Stages of LCA

The LCA of CO_2 adsorbents comprises several distinct stages that collectively offer a comprehensive evaluation of environmental impacts. These stages include raw material procurement, manufacturing, utilization, regeneration, and end-of-life considerations. Each stage encompasses specific activities and processes that must be assessed to understand the overall environmental footprint of CO_2 capture technologies. Figure 12.4 depicts a potential framework outlining the five life cycle stages of an adsorbent.

This initial stage involves sourcing the materials required for CO_2 adsorbent production, such as activated carbon (AC), zeolites, or amine-based sorbents. The environmental impact of this stage is influenced by factors such as resource extraction methods, transportation logistics, and the sustainability of material sources. Assessing the environmental footprint associated with obtaining these primary materials is critical for identifying potential areas for improvement in the supply chain. During the manufacturing stage, raw materials, including pellets, granules, or membranes, are processed and converted into adsorbent forms suitable for CO_2 capture. This phase necessitates evaluating energy consumption, emissions, and waste generation. Manufacturing processes must be optimized to minimize environmental impacts while ensuring the production of high-quality adsorbents. The efficiency and sustainability of these processes play a significant role in the overall LCA of CO_2 adsorbents. In the utilization phase, CO_2 adsorbents are deployed in capture systems, such as pre-combustion or post-combustion processes, or in industrial applications like cement production. Evaluating capture efficiency and associated energy requirements is critical in LCA assessments. Post-capture, CO_2 must be desorbed or regenerated from the adsorbent material for further use or disposal. Various regeneration methods exist, each with distinct energy inputs and emissions profiles, necessitating careful evaluation in LCAs. The final stage of the LCA involves addressing the disposal, recycling, or reprocessing of CO_2 adsorbents at the end of their operational

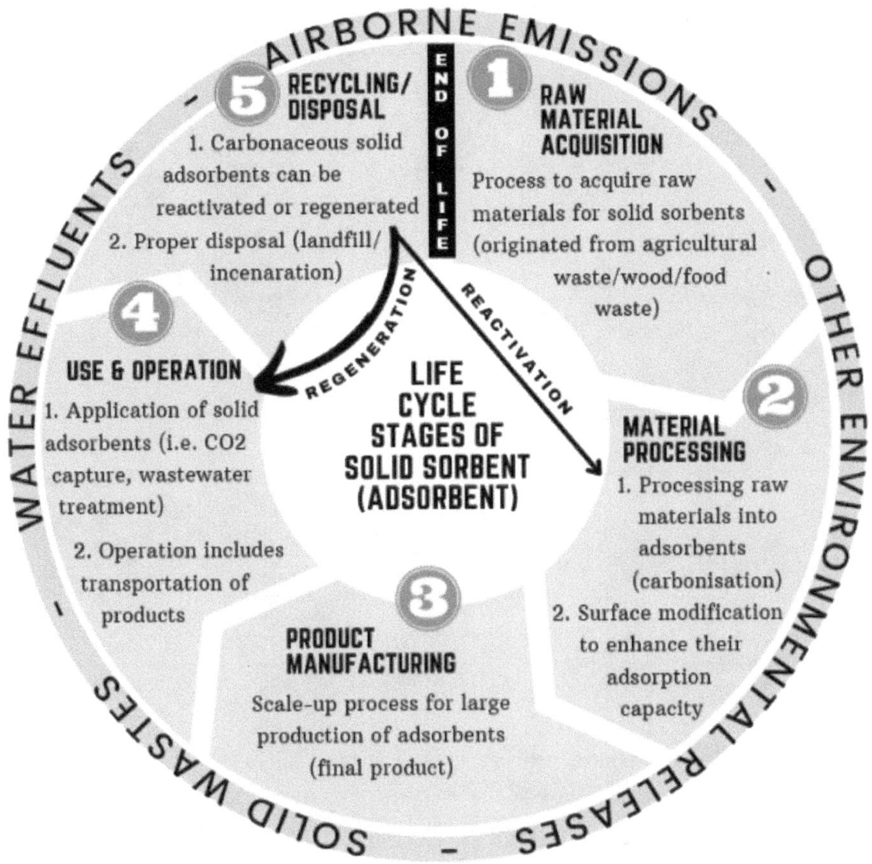

FIGURE 12.4 Conceptual framework illustrating the five life cycle stages of an adsorbent [10], copyright (2022), MDPI.

life. The environmental impact of this stage is determined by the chosen disposal method, whether it involves landfilling, incineration, or recycling. Each option has different implications for resource recovery, emissions, and environmental pollution. Assessing these factors is essential for determining the most sustainable end-of-life strategy for CO_2 adsorbents [1].

12.3 ENVIRONMENTAL IMPACTS OF CO_2 ADSORBENT PRODUCTION

Recent research has been continuously pursuing the development of new materials for capturing carbon from flue gases. Zeolites and carbon-based adsorbents have emerged as popular choices. Evaluating the environmental consequences linked to different adsorbent choices is essential to ensure the sustainability of carbon capture processes. In this context, a thorough LCA was conducted, utilizing experimental

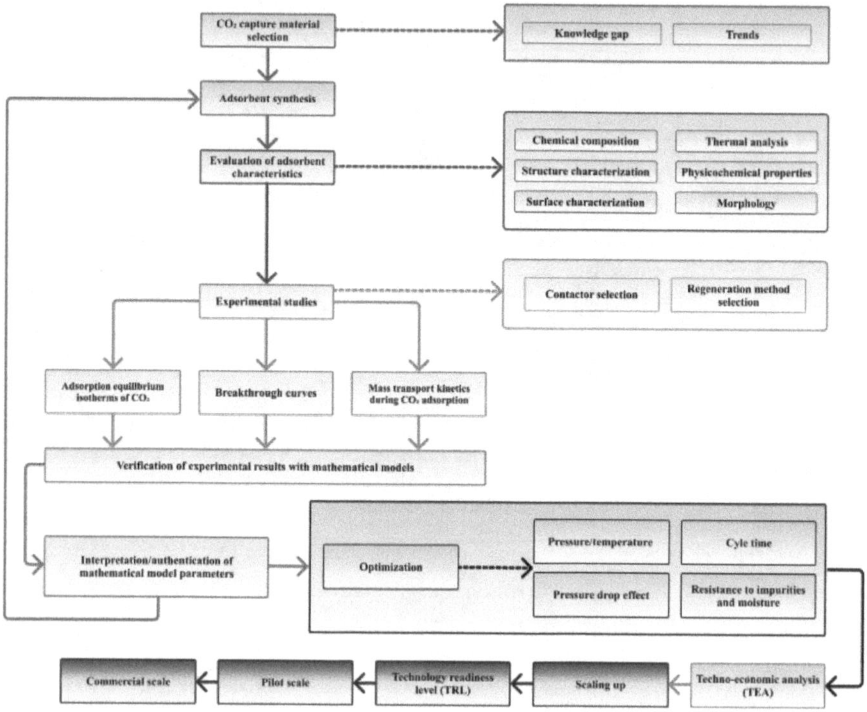

FIGURE 12.5 The assessment pathway to evaluate the suitability of materials for CO_2 capture applications [7], copyright (2021), Elsevier.

data from vacuum pressure swing adsorption (VPSA), to analyze the environmental performance of two advanced adsorbents: specifically, a zeolite (13X-APG) and a carbon molecular sieve (CMS-330).

Assessing the effectiveness of an adsorbent in capturing CO_2 involves examining the inherent characteristics of the materials produced and conducting experiments on CO_2 adsorption isotherms, breakthrough curves of the adsorbent bed, and kinetics of mass transport. Figure 12.5 shows a schematic representation of the whole process. Examination of key performance indicators revealed markedly superior metrics for 13X-APG vis-à-vis CMS-330. LCA outcomes disclosed that the zeolite configuration yielded an approximate 13% reduction in environmental burdens across all assessed categories compared to the CMS-330 counterpart. Notably, the utilization phase emerged as the primary focal point, attributing over 96% to the cumulative life cycle impacts.

A sensitivity analysis underscored that while the substitution and disposal of adsorbents were significant factors, their influence was overshadowed by the scale and manner of energy consumption during VPSA. Insights gleaned from the study suggested that the integration of renewable energy sources could substantially augment the carbon dioxide capture performance of VPSA, potentially achieving efficiency improvements of up to 96% and 84% for 13X-APG and CMS-330, respectively. In this section, an investigation compares the environmental impacts associated with producing 1 kg of adsorbents, focusing on variants derived from zeolite and carbon materials. Due to the limited availability of precise production process data for 13X-APG and CMS-330 adsorbents in the literature, data from Ecoinvent was used for this comparative analysis. Specifically, the study utilized the inventory data for zeolite powder production for the zeolite-based adsorbent and granular AC production sourced from hard coal for the carbon-based adsorbent. The findings reveal that zeolite-based adsorbents exhibit higher environmental impacts across categories such as ADP (elements), ODP, and toxicity, with an average 3.0-fold increase in environmental loads, while carbon-based counterparts show heightened impacts in other categories, averaging 1.8 times higher environmental loads. Particularly notable is that, in terms of abiotic depletion, zeolite-based adsorbents demonstrate greater impacts related to elements but lower impacts on fossil fuels compared to carbon-based adsorbents, reflecting the increased energy consumption inherent in the production of carbon-based adsorbents. This distinction is also discernible in the comparison of GWP, where carbon-based adsorbents exhibit impacts 1.8 times higher than their zeolite-based counterparts. The difference between the two materials is smallest in the case of EP, with carbon-based adsorbents scoring approximately 1.1 times that of zeolite-based adsorbents. On the other hand, the ADP (elements) comparison shows the largest gap, with the zeolite-based adsorbent outperforming its carbon-based counterpart by a factor of 42. In sum, the LCA findings delineated that employing zeolite for carbon capture entailed fewer environmental ramifications compared to CMS-330. Looking ahead, the study advocates for further exploration into alternative carbon capture adsorbents, serving as a pivotal stepping stone toward this overarching objective [11].

Gerande et al. investigated MOF materials, characterized by high surface area and remarkable design flexibility, rendering them suitable for diverse chemical processes. Although the characteristics of MOF materials have been extensively assessed in laboratory-scale powder form, where optimized synthesis produces crystalline powders, their industrial application demands efficient and scalable synthesis protocols. Moreover, to classify these materials as environmentally friendly, their synthesis must conform to sustainable practices. In this study, they conducted a cradle-to-gate LCA of four distinct synthesis protocols aimed at producing a MOF material (CPO-27-Ni, also known as MOF-74 and DOBDC), recognized as a promising adsorbent for CCS technologies. The methodology details used to produce the presented results are illustrated in Figure 12.6. Their investigation illustrates the feasibility of employing LCA methodology to devise eco-friendly design approaches for MOF materials. Through their analysis, they identified the significant impact of solvent usage during synthesis. Surprisingly, the LCA analysis showed that the use of solvents during the cleaning and purification stages has a significant impact on the environment. Therefore, it is crucial to minimize solvent usage and consider reusing solvents to reduce environmental impact wherever possible [12].

Nowrouzi et al. studied the synthesis of an environmentally friendly product by assessing its economic and environmental impacts using an LCA and a life cycle cost

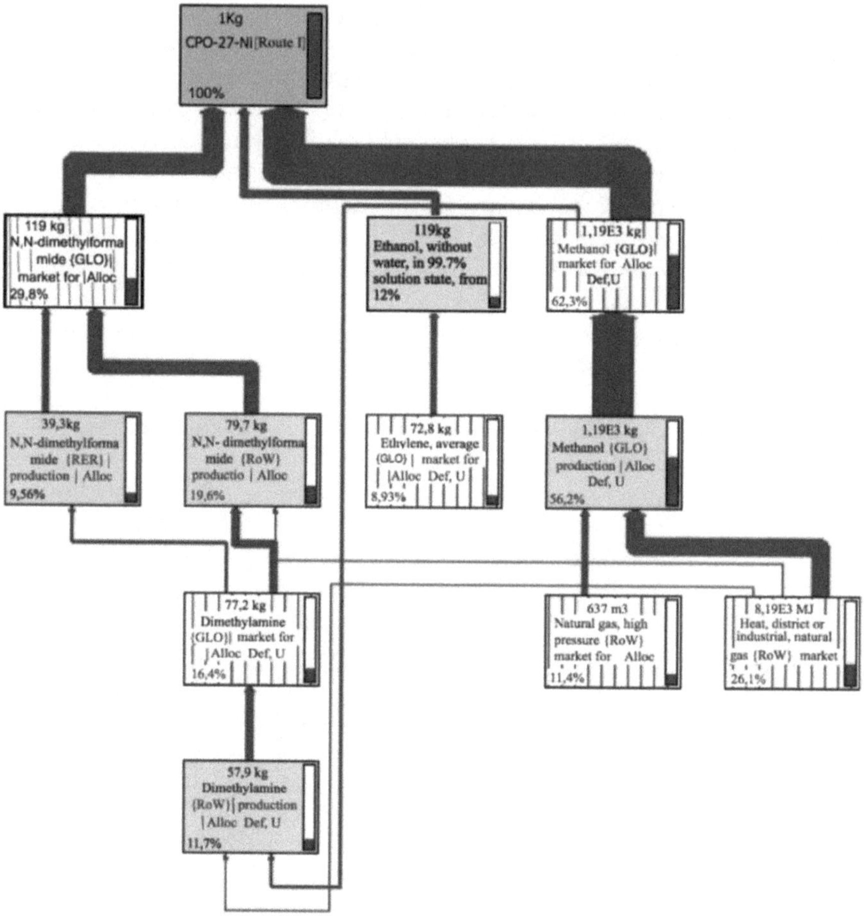

FIGURE 12.6 LCA process tree [12], copyright (2017), Elsevier.

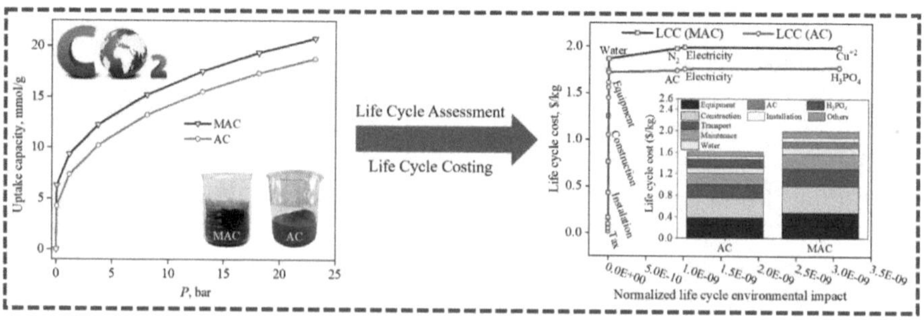

FIGURE 12.7 LCA for CO_2 adsorbent [13], copyright (2021), Elsevier.

(LCC) comparison (Figure 12.7). They focused on two highly efficient carbon-based adsorbents, AC and modified activated carbon (MAC), for adsorbing 1 kg of CO_2. They performed multiple analyses, including CML baseline 2000, cumulative energy demand (CED), ecological footprint, and greenhouse gas protocol. The research highlighted significant impacts in categories such as human toxicity (>2.63%), marine ecotoxicity (86%), and freshwater aquatic ecotoxicity (7.46%), resulting from emissions including CO_2, SO_2, NO_2, and metal releases. The CED for MAC was notably higher at 2.6 times, mainly attributed to fossil fuel supply (>91.45%). The cost of production was $1.6 per kg for AC and $2 per kg for MAC. The use of H_3PO_4 and copper ions caused significant environmental impacts, accounting for 97% of AC and 61.48% of MAC, while equipment and construction expenses represented the largest economic burden (>23%). Uncertainty analysis identified marine aquatic toxicity (19.50%) and human toxicity (16.09%) as the most sensitive categories, with a 95% confidence interval. Overall, the results from both LCA and LCC analysis indicated that MAC had more pronounced economic and environmental impacts despite its higher CO_2 adsorption capacity [13].

Wei et al. explored the production of biochar derived from various waste biomasses such as woody materials, herbaceous plants, chemical sludge, and livestock manure. This biochar, created at relatively low pyrolysis temperatures, was employed in converting CO_2 into valuable CO fuel using solar energy. Pine sawdust-derived biochar, characterized by abundant carbonyl functional groups, showed the highest yield of CO fuel due to its excellent CO_2 adsorption capacity, operating effectively without sacrificial agents, metal catalysts, or additional heating. Its performance in a 15% CO_2/N_2 mixture was particularly noteworthy, typical of industrial flue gas conditions. Additionally, a LCA indicated that biochar derived from biomass with higher cellulose and lignin content demonstrated improved CO_2 conversion efficiency and greater environmental benefits. This study significantly advances applying biomass waste-derived biochar for efficient solar-driven CO_2 conversion, contributing to sustainable energy and environmental solutions [14].

12.3.1 Amine-Based Sorbents

Amine-based sorbents are extensively recognized for their efficacy in capturing carbon dioxide through chemical adsorption processes. These materials often incorporate amine-functionalized solids interacting with CO_2, forming stable carbamate structures, making them highly effective in various carbon capture applications. From an environmental perspective, several factors must be considered. The production and disposal of amine-based sorbents involve chemical processes that can produce hazardous by-products. Careful management is required during disposal to prevent environmental contamination [15]. The regeneration of these sorbents necessitates substantial energy input, frequently derived from fossil fuels, potentially negating some environmental benefits associated with CO_2 capture. Moreover, certain amine compounds can be toxic, presenting risks to human health and the environment throughout their life cycle [16]. In terms of sustainability, amine-based sorbents offer significant advantages. These materials exhibit high stability and durability, which are crucial for their sustainable application in direct air capture technologies [17]. Utilizing renewable resources and adhering to green chemistry principles in the synthesis of amine-based sorbents can enhance their sustainability profile. Ongoing research is focused on developing more eco-friendly production methods. Advances in sorbent regeneration techniques are also improving the recyclability of amine-based sorbents, reducing the frequency of replacement and minimizing waste [16]. Amine-based adsorbents offer a highly efficient and potentially sustainable method for capturing CO_2, but it is crucial to mitigate the environmental impacts throughout their life cycle to realize their advantages fully [18]. Research indicates that despite considerable efforts aimed at improving the effectiveness and durability of CO_2 capture adsorbents, there has been a notable oversight concerning the environmental consequences linked to their manufacturing and application. This oversight is highlighted in a recent study that investigates the environmental implications of utilizing solid amine materials for carbon capture applications. This study is the first time to compare the environmental and economic trade-offs between solid amine adsorbents and traditional carbon capture methods. The findings reveal that, during the initial manufacturing phase, solid amine absorbents, particularly CNT-PEI, exhibit higher environmental costs compared to traditional MEA methods. Notably, the production of carbon nanotubes emerges as a major contributor to the overall environmental impacts throughout the life cycle. However, despite these challenges, there is potential for optimization through the adoption of sustainable energy sources, scaling up production, and implementing recycling initiatives to enhance production efficiency. It is noteworthy that despite the initial environmental and economic costs, solid amine adsorbents, such as Phy-CNT-PEI, demonstrate promising potential, with a carbon payback period of 917 cycles. This indicates a potential to reduce CO_2 costs and enhance the sustainability of existing CO_2 capture methods. However, the study highlights the necessity of conducting thorough environmental and economic impact assessments across the entire life cycle of the material. Such evaluations enable researchers and developers to identify hotspots in production processes, such as CNT and PEI syntheses, and mitigate negative environmental impacts associated with emerging adsorption technologies [19].

12.4 PRODUCTION PROCESSES

12.4.1 Material Synthesis and Activation

The synthesis and activation of materials are fundamental steps in the production of CO_2 adsorbents. These processes involve creating and enhancing materials that possess the capacity to capture and store carbon dioxide efficiently. Typically, the synthesis involves chemical reactions that produce porous materials such as zeolites, ACs, and MOFs. Activation processes often include treatments to increase the surface area and improve the adsorption properties of these materials. For instance, chemical activation using acids or bases can enhance the pore structure and reactivity of the adsorbents, thereby improving their effectiveness in capturing CO_2 from various sources.

12.4.2 Evaluate the Energy Consumption, Emissions, and Waste Generated During Production

Assessing the environmental impacts of CO_2 adsorbent production is essential, with a focus on energy use, emissions, and waste generation. The energy required for the synthesis and activation of CO_2 adsorbents can be substantial. Processes such as high-temperature activation necessitate significant energy input. The source of this energy, whether it is renewable or non-renewable, significantly affects the environmental footprint of the production process. The production of CO_2 adsorbents can result in the emission of greenhouse gases and other pollutants. For example, the synthesis of AC can release CO_2 and volatile organic compounds into the atmosphere. Additionally, if fossil fuels are the primary energy source, the energy-intensive nature of the production process can further contribute to CO_2 emissions. The production process generates waste, including chemical by-products, unreacted raw materials, and spent solvents. Effective management and disposal of these wastes are essential to minimize their environmental impact. Employing sustainable practices such as recycling and reusing by-products, along with adopting greener synthesis methods, can significantly reduce the waste footprint associated with CO_2 adsorbent production. In summary, while the synthesis and activation of CO_2 adsorbents are crucial for their performance, it is imperative to mitigate the environmental impacts of these processes through energy-efficient practices, emission controls, and sustainable waste management strategies [20]. The evaluation and application of carbon capture technologies depend heavily on the performance and energy efficiency of CO_2 adsorbents. These adsorbents are designed to selectively capture CO_2 from various gas streams, particularly from post-combustion emissions.

12.4.2.1 Performance of CO_2 Adsorbents

The effectiveness of CO_2 adsorbents is primarily measured by their adsorption capacity, selectivity, and regeneration efficiency. Carbon-based materials, such as ACs, exhibit high surface areas and porosity, which enhance their CO_2 adsorption capabilities. Zeolites, known for their well-defined pore structures, provide excellent selectivity and thermal stability. MOFs, with their tunable porosity and functionalizable surfaces, offer high adsorption capacities and are highly selective for CO_2 over other gases [21].

Additionally, hybrid materials like the CTS/GO/ZnO composite demonstrate improved adsorption efficiency due to synergistic effects between components. This composite material enhances CO_2 capture by leveraging the high surface area and reactive sites of graphene oxide, combined with the adsorption properties of chitosan and ZnO [22].

12.4.2.2 Energy Efficiency of CO_2 Adsorbents

Energy efficiency in CO_2 capture processes involves minimizing the energy required for both adsorption and regeneration. The thermodynamic properties of the adsorption process, such as the exothermic nature and spontaneity, play a crucial role. Physical adsorption processes, which involve weaker van der Waals forces, typically require less energy for regeneration compared to chemical adsorption processes, which involve stronger chemical bonds [23]. Advanced adsorbents, such as MOFs and ionic liquids, show promise in reducing energy consumption due to their low regeneration temperatures and high CO_2 affinity. Thermodynamic analyses based on carbon pump cycles suggest that optimized temperature swing adsorption (TSA) cycles can significantly improve energy efficiency by utilizing the adsorbed CO_2 as a working fluid in a closed-loop system [24].

12.4.3 Adsorption Capacity and Selectivity

Adsorption capacity refers to the quantity of CO_2 that an adsorbent can capture per unit mass or volume. This capacity is largely dependent on the properties of the adsorbent, including surface area, pore volume, and surface chemistry. Carbon-based materials, such as such as carbon nanotubes, graphene, AC, and carbon fibers, typically exhibit high adsorption capacities due to their unique properties. These materials can achieve CO_2 adsorption capacities ranging from 2.8 to 3.2 mmol/g [25]. Zeolites, with their well-defined microporous structures, also demonstrate significant adsorption capacities. The interaction between the acidic CO_2 molecules and the basic sites on zeolites enhances their adsorption performance. The adsorption capacity of zeolites can be further improved by modifying their surface properties to increase the number of active sites [26]. Selectivity is the capacity of an adsorbent to preferentially capture CO_2 over other gases, like nitrogen or methane. High selectivity is crucial for the efficient separation of CO_2 from mixed gas streams. Materials with strong interactions with CO_2, such as zeolites with high basicity, exhibit high selectivity due to the stronger binding of CO_2 compared to other gases [26]. In addition, the design of adsorbents can be tailored to enhance selectivity. For instance, the incorporation of nitrogen functional groups in carbon-based adsorbents can improve CO_2 selectivity by providing additional binding sites for CO_2 molecules. Hybrid materials, which combine different types of adsorbents, can also achieve high selectivity by leveraging the complementary properties of each component [27].

In materials of this kind, addressing the presence of H_2O poses an even more significant challenge compared to dealing with other gases present in higher concentrations. The H_2O molecule, with its dipole moment and ability to form hydrogen bonds, as well as its smaller kinetic diameter compared to CO_2, tends to hinder CO_2 binding by blocking adsorption sites in highly polar materials. This interference ultimately reduces CO_2 selectivity under practical conditions. Factors like production costs and stability against water are crucial considerations that cannot be overlooked, especially for zeolites (in terms of CO_2/H_2O selectivity) and MOFs (in terms of CO_2/H_2O selectivity and stability). These efforts aim to address the inherent thermodynamic limitations of selective CO_2 capture.

12.5 REGENERATION POTENTIAL

Regeneration entails extracting adsorbed CO_2 from the adsorbent, restoring its capacity for further adsorption cycles [28]. Several methods are employed to achieve this:

- **Thermal Swing Adsorption (TSA)**
  This method involves heating the adsorbent to desorb CO_2[29]. While effective, TSA can be energy-intensive, which may limit its practicality for large-scale operations [7].
- **Pressure Swing Adsorption (PSA)**
  In PSA, CO_2 is desorbed by reducing the pressure around the adsorbent. This method is generally less energy-intensive than TSA and is suitable for adsorbents that can regenerate at low pressures [30].

TABLE 12.1
Comparison of the Effectiveness of Various Regeneration Strategies [8]

Regeneration Mode	Reactor	Adsorbent	Scale	Recovery (%)	Energy Combustion (MJ/kg CO_2)	Reference
TSA	Fixed bed	NaUSY	Laboratory	55.5–83.6	3.40–4.50	[32]
TSA	Moving bed	Carbon sorbent	Prototype	81.3	4.35	[33]
TSA	Fixed bed	13X	Laboratory	81	3.23	[34]
TSA	Moving bed	13X	Laboratory	96	2.21	[35]
TSA	Fixed bed	Carbon honeycomb monoliths	Laboratory	85.4	3.59	[36]
TSA	Multi-stage fluidized	Amine-functionalized	Laboratory	90	2.80	[37]
TSA	Fixed bed	13X-activated carbon	Pilot	90.2	2.44	[38]

- **Chemical Regeneration**
  Some adsorbents, particularly those based on amines, can be regenerated using chemical reactions that release the adsorbed CO_2. This method is often specific to the type of adsorbent and the nature of the chemical interactions involved [31].

The energy efficiency of the regeneration process is a major concern. The energy penalty associated with regeneration can significantly impact the overall efficiency of the carbon capture system. Optimizing the regeneration process is crucial to minimize energy consumption and enhance the economic feasibility of the technology [39]. Table 12.1 presents a comparison of the performances of different regeneration strategies. Recent advancements in adsorbent materials and regeneration techniques aim to address these challenges. For instance, the development of adsorbents that regenerate at lower temperatures or pressures can significantly reduce energy requirements. Additionally, innovations in material design, such as the use of composite materials or hybrid adsorbents, offer promising avenues for improving regeneration efficiency [1].

In conclusion, the regeneration potential of CO_2 adsorbents is a pivotal aspect of carbon capture technology. Ongoing research and technological advancements continue to enhance regeneration methods, aiming to achieve more sustainable and cost-effective solutions for CO_2 capture and storage. The energy demands associated with adsorption and desorption are crucial for assessing the efficiency and practicality of carbon capture technologies. These processes involve the interaction of CO_2 molecules with an adsorbent surface, followed by the subsequent release of these molecules during desorption.

12.5.1 Adsorption Process

The activation energy required for adsorption is generally low, making the process relatively energy-efficient. The interaction between the adsorbate (CO_2) and the adsorbent surface is typically exothermic, releasing energy as the molecules attach to the surface. The adsorption energy, defined as the difference between the total energy of the adsorbate–adsorbent complex and the combined energies of the individual components, is optimized in efficient adsorbents to ensure robust yet reversible interactions with CO_2.

12.5.2 Desorption Process

Thermal desorption, the most common method for releasing adsorbed CO_2, involves heating the adsorbent. The energy required varies depending on the adsorbent material and the strength of the CO_2–adsorbent interaction [40]. Mild conditions, such as 50°C in a pure nitrogen environment, can restore the maximum adsorption capacity with relatively low energy consumption. Pressure swing desorption is another method, which reduces the surrounding pressure and is generally less energy-intensive compared to thermal desorption. The efficiency of desorption is influenced by factors like flow rate and adsorption concentration, with higher flow rates typically increasing the amount of CO_2 desorbed [41]. In summary, optimizing the energy requirements for both adsorption and desorption processes is essential for developing cost-effective and sustainable carbon capture technologies. Continuous advancements in adsorbent materials and desorption techniques enhance the efficiency of these processes, making large-scale carbon capture more viable [42].

12.6 END-OF-LIFE CONSIDERATIONS

Assessing the sustainability of CO_2 adsorbents requires a comprehensive evaluation of their end-of-life phase. This involves considering the entire life cycle of the adsorbents, from production and use to disposal, each contributing to various environmental impacts. Proper end-of-life management is crucial to mitigating negative environmental effects. Disposal methods include incineration, reactivation, and landfilling. While effective in reducing waste volume, incineration releases CO_2 and other pollutants [43], exacerbating global warming. Reactivation involves treating spent adsorbents to restore their adsorption capacity, which consumes energy but allows for reuse, thereby reducing the demand for new materials and minimizing waste [44]. Landfilling, although common, poses long-term environmental risks, such as leaching harmful substances into soil and groundwater, without addressing the carbon footprint associated with production. Environmental impacts must be considered, including the carbon footprint from production to disposal. Adsorbents that require less energy for reactivation and emit fewer pollutants during incineration are preferable for reducing overall carbon emissions. Furthermore, improper disposal and incineration can release toxic substances, affecting human health and ecosystems, while poorly managed reactivation processes also pose environmental hazards. In terms of sustainability, incorporating adsorbent reactivation into a circular economy model enhances overall sustainability. This approach decreases the need for new raw materials, lowers waste production, and reduces environmental pollution [45]. Effective end-of-life strategies, such as reactivation and sustainable disposal methods, can significantly reduce the environmental impact of CO_2

adsorbents [46]. Comprehensive LCAs are essential for identifying the most sustainable practices, promoting a circular economy, and minimizing the overall carbon footprint. Recent advancements in CO_2 capture techniques highlight the evolution toward more efficient and sustainable methods. One significant approach involves the use of adsorption technologies with solid adsorbents, which have shown promise in gas separation processes [47]. These adsorbents are pivotal in addressing greenhouse gas emissions, particularly from industries and power plants. Another innovative avenue explores biomass waste utilization for developing ACs, enhancing both environmental sustainability and CO_2 capture capabilities [9]. Such materials not only provide a renewable source for adsorbents but also contribute to reducing the carbon footprint associated with their production. In the realm of CCS, adsorption technologies play a crucial role in achieving global carbon reduction goals [8]. These technologies aim to mitigate the release of greenhouse gases into the atmosphere, thus promoting sustainable energy practices.

12.7 CONCLUSION

This chapter has elucidated the pivotal role of CO_2 adsorbents in mitigating greenhouse gas emissions and advancing sustainability objectives. Through a comprehensive LCA, the environmental impacts of various CO_2 adsorbents have been rigorously evaluated, encompassing production processes, energy consumption, emissions, and waste generation. The examination of diverse adsorbent materials and production methodologies has revealed significant disparities in energy efficiency, emissions profiles, and waste outputs. Key factors such as adsorption capacity, selectivity, and regeneration potential have been meticulously scrutinized for their implications on environmental performance. Additionally, the discussion on end-of-life considerations has underscored the imperative of adopting sustainable disposal and recycling practices. Looking ahead, optimizing CO_2 adsorbents emerges as critical for enhancing environmental sustainability and supporting global initiatives toward a low-carbon future. Continued research and innovation are essential to improve energy efficiency, minimize emissions, and maximize recyclability in CO_2 capture technologies. By doing so, we can effectively harness the potential of CO_2 adsorbents as pivotal tools in combating climate change and fostering a more sustainable global environment.

REFERENCES

1. O. O. Ayeleru, H. U. Modekwe, O. R. Onisuru, C. R. Ohoro, C. A. Akinnawo, and P. A. Olubambi, "Adsorbent technologies and applications for carbon capture, and direct air capture in environmental perspective and sustainable climate action," *Sustainable Chemistry for Climate Action,* p. 100029, 2023.
2. H. Mashhadimoslem, A. Ghaemi, A. Maleki, and A. Elkamel, "Enhancement of oxygen adsorption using biomass-based oxidized porous carbon," *Journal of Environmental Chemical Engineering,* vol. 11, no. 2, p. 109300, 2023.
3. Z. Khoshraftar, A. Ghaemi, and A. Hemmati, "Comprehensive investigation of isotherm, RSM, and ANN modeling of CO_2 capture by multi-walled carbon nanotube," *Scientific Reports,* vol. 14, no. 1, p. 5130, 2024.

4. A. Kazemi, F. Moghadaskhou, M. A. Pordsari, F. Manteghi, A. Tadjarodi, and A. Ghaemi, "Enhanced CO_2 capture potential of UiO-66-NH2 synthesized by sonochemical method: experimental findings and performance evaluation," *Scientific Reports,* vol. 13, no. 1, p. 19891, 2023.
5. F. Bahmanzadegan, M. A. Pordsari, and A. Ghaemi, "Improving the efficiency of 4A-zeolite synthesized from kaolin by amine functionalization for CO_2 capture," *Scientific Reports,* vol. 13, no. 1, p. 12533, 2023.
6. F. S. Taheri, A. Ghaemi, A. Maleki, and S. Shahhosseini, "High CO_2 adsorption on amine-functionalized improved mesoporous silica nanotube as an eco-friendly nanocomposite," *Energy & Fuels,* vol. 33, no. 6, pp. 5384–5397, 2019.
7. B. Dziejarski, J. Serafin, K. Andersson, and R. Krzyżyńska, "CO_2 capture materials: a review of current trends and future challenges," *Materials Today Sustainability,* p. 100483, 2023.
8. F. Raganati, F. Miccio, and P. Ammendola, "Adsorption of carbon dioxide for post-combustion capture: a review," *Energy & Fuels,* vol. 35, no. 16, pp. 12845–12868, 2021.
9. H. Mumtaz et al., "Biomass waste utilization for adsorbent preparation in CO_2 capture and sustainable environment applications," *Sustainable Energy Technologies and Assessments,* vol. 46, p. 101288, 2021.
10. N. S. Mohamed Hatta, M. K. Aroua, F. Hussin, and L. T. Gew, "A systematic review of amino acid-based adsorbents for CO_2 capture," *Energies,* vol. 15, no. 10, p. 3753, 2022.
11. R. Gonzalez-Olmos, A. Gutierrez-Ortega, J. Sempere, and R. Nomen, "Zeolite versus carbon adsorbents in carbon capture: A comparison from an operational and life cycle perspective," *Journal of CO_2 Utilization,* vol. 55, p. 101791, 2022.
12. C. A. Grande, R. Blom, A. Spjelkavik, V. Moreau, and J. Payet, "Life-cycle assessment as a tool for eco-design of metal-organic frameworks (MOFs)," *Sustainable materials and technologies,* vol. 14, pp. 11–18, 2017.
13. M. Nowrouzi, H. Abyar, H. Younesi, and E. Khaki, "Life cycle environmental and economic assessment of highly efficient carbon-based CO_2 adsorbents: A comparative study," *Journal of CO_2 Utilization,* vol. 47, p. 101491, 2021.
14. X. Wei, X. Zhang, L. Jin, X. Yang, W. Zou, B. Gao, and L. Dong, "Waste biomass-derived biochar in adsorption-photocatalytic conversion of CO_2 for sustainable energy and environment: Evaluation, mechanism, and life cycle assessment," *Applied Catalysis B: Environment and Energy,* vol. 351, p. 123957, 2024.
15. G. Leonzio, O. Mwabonje, P. S. Fennell, and N. Shah, "Environmental performance of different sorbents used for direct air capture," *Sustainable Production and Consumption,* vol. 32, pp. 101–111, 2022.
16. P. Priyadarshini, G. Rim, C. Rosu, M. Song, and C. W. Jones, "Direct air capture of CO_2 using amine/alumina sorbents at cold temperature," *ACS Environmental Au,* vol. 3, no. 5, pp. 295–307, 2023.
17. A. A. Al-Absi, A. M. Benneker, and N. Mahinpey, "Amine sorbents for sustainable direct air capture: long-term stability and extended aging study," *Energy & Fuels,* vol. 38, no. 10, pp. 8938–8950, 2024.
18. L. Lin et al., "Characteristics, application and modeling of solid amine adsorbents for CO_2 capture: A review," *Journal of Environmental Management,* vol. 325, p. 116438, 2023.
19. F. Wu, Z. Zhou, S. Temizel-Sekeryan, R. Ghamkhar, and A. L. Hicks, "Assessing the environmental impact and payback of carbon nanotube supported CO_2 capture technologies using LCA methodology," *Journal of Cleaner Production,* vol. 270, p. 122465, 2020.
20. R. G. Bell, "Protecting the environment during and after resource extraction," *Extractive Industries,* p. 318, 2017.
21. M. Kong, L. Song, H. Liao, S. Zhang, Y. Wang, X. Deng, and W. Feng, "A review on development of post-combustion CO_2 capture technologies: Performance of carbon-based, zeolites and MOFs adsorbents," *Fuel,* vol. 371, p. 132103, 2024.

22. F. Fathalian, H. Moghadamzadeh, A. Hemmati, and A. Ghaemi, "Efficient CO_2 adsorption using chitosan, graphene oxide, and zinc oxide composite," *Scientific Reports,* vol. 14, no. 1, p. 3186, 2024.
23. M. Ravi, S. K. Rathore, and M. Sivalingam, "Performance evaluation of CO_2 capture on using potential adsorbents in a CI engine exhaust–An experimental investigation," *Greenhouse Gases: Science and Technology,* vol. 14, no. 1, pp. 138–151, 2024.
24. S. Li, S. Deng, L. Zhao, R. Zhao, B. Liu, Y. Xu, and X. Yuan, "A journey into the energy-efficient aspect of CO_2 adsorption technologies."
25. J. Zhou, X. Wang, and W. Xing, "Carbon-based CO_2 adsorbents," in *Post-combustion Carbon Dioxide Capture Materials*, ed. Q. Wang, The Royal Society of Chemistry, pp. 1–75, 2018.
26. D. G. Boer, J. Langerak, and P. P. Pescarmona, "Zeolites as selective adsorbents for CO_2 separation," *ACS Applied Energy Materials,* vol. 6, no. 5, pp. 2634–2656, 2023.
27. Q. Yan, Y. Lin, C. Kong, and L. Chen, "Remarkable CO_2/CH_4 selectivity and CO_2 adsorption capacity exhibited by polyamine-decorated metal–organic framework adsorbents," *Chemical Communications,* vol. 49, no. 61, pp. 6873–6875, 2013.
28. L. Jiang, R. Wang, A. Gonzalez-Diaz, A. Smallbone, R. Lamidi, and A. Roskilly, "Comparative analysis on temperature swing adsorption cycle for carbon capture by using internal heat/mass recovery," *Applied Thermal Engineering,* vol. 169, p. 114973, 2020.
29. J. A. Mason, K. Sumida, Z. R. Herm, R. Krishna, and J. R. Long, "Evaluating metal–organic frameworks for post-combustion carbon dioxide capture via temperature swing adsorption," *Energy & Environmental Science,* vol. 4, no. 8, pp. 3030–3040, 2011.
30. A. Das et al., "Advancements in adsorption based carbon dioxide capture technologies-A comprehensive review." *Heliyon*, vol. 9, no. 12, 2023.
31. M. J. Bos, V. Kroeze, S. Sutanto, and D. W. Brilman, "Evaluating regeneration options of solid amine sorbent for CO_2 removal," *Industrial & Engineering Chemistry Research,* vol. 57, no. 32, pp. 11141–11153, 2018.
32. A. Ntiamoah, J. Ling, P. Xiao, P. A. Webley, and Y. Zhai, "CO_2 capture by temperature swing adsorption: use of hot CO_2-rich gas for regeneration," *Industrial & Engineering Chemistry Research,* vol. 55, no. 3, pp. 703–713, 2016.
33. M. Hornbostel, "Pilot-scale evaluation of an advanced carbon sorbent-based process for post-combustion carbon capture," SRI International, Menlo Park, CA (United States), 2016.
34. M. Clausse, J. Merel, and F. Meunier, "Numerical parametric study on CO_2 capture by indirect thermal swing adsorption," *International Journal of Greenhouse Gas Control,* vol. 5, no. 5, pp. 1206–1213, 2011.
35. G. Mondino, C. A. Grande, R. Blom, and L. O. Nord, "Moving bed temperature swing adsorption for CO_2 capture from a natural gas combined cycle power plant," *International Journal of Greenhouse Gas Control,* vol. 85, pp. 58–70, 2019.
36. M. G. Plaza, F. Rubiera, and C. Pevida, "Evaluating the feasibility of a TSA process based on steam stripping in combination with structured carbon adsorbents to capture CO_2 from a coal power plant," *Energy & Fuels,* vol. 31, no. 9, pp. 9760–9775, 2017.
37. C. Dhoke, A. Zaabout, S. Cloete, H. Seo, Y.-k. Park, L. Demoulin, and S. Amini, "Demonstration of the novel swing adsorption reactor cluster concept in a multistage fluidized bed with heat-transfer surfaces for postcombustion CO_2 capture," *Industrial & Engineering Chemistry Research,* vol. 59, no. 51, pp. 22281–22291, 2020.
38. L. Wang, Y. Yang, W. Shen, X. Kong, P. Li, J. Yu, and A. E. Rodrigues, "CO_2 capture from flue gas in an existing coal-fired power plant by two successive pilot-scale VPSA units," *Industrial & Engineering Chemistry Research,* vol. 52, no. 23, pp. 7947–7955, 2013.
39. M. Farooq, M. Soudagar, M. Imran, M. Arslan, M. Tariq, A. Pettinau, and J. Andresen, "Carbon capture for sustainable environment in developing countries," *Energy and Environmental Security in Developing Countries,* pp. 525–544, 2021.

40. H. Wang et al., "Moisture adsorption-desorption full cycle power generation," *Nature Communications,* vol. 13, no. 1, p. 2524, 2022.
41. M. Bos, S. Pietersen, and D. Brilman, "Production of high purity CO_2 from air using solid amine sorbents," *Chemical Engineering Science: X,* vol. 2, p. 100020, 2019.
42. J. E. Park, E. S. Jo, G. B. Lee, S. E. Lee, and B.-U. Hong, "Adsorption capacity and desorption efficiency of activated carbon for odors from medical waste," *Molecules,* vol. 28, no. 2, p. 785, 2023.
43. U. Khan, C. C. Ogbaga, O.-A. O. Abiodun, A. A. Adeleke, P. P. Ikubanni, P. U. Okoye, and J. A. Okolie, "Assessing absorption-based CO_2 capture: Research progress and techno-economic assessment overview," *Carbon Capture Science & Technology,* p. 100125, 2023.
44. L. J. Müller, A. Kätelhön, M. Bachmann, A. Zimmermann, A. Sternberg, and A. Bardow, "A guideline for life cycle assessment of carbon capture and utilization," *Frontiers in Energy Research,* vol. 8, p. 15, 2020.
45. A. S. Nilkar, C. J. Orme, J. R. Klaehn, H. Zhao, and B. Adhikari, "Life cycle assessment of innovative carbon dioxide selective membranes from low carbon emission sources: A comparative study," *Membranes,* vol. 13, no. 4, p. 410, 2023.
46. S. Deutz and A. Bardow, "Life-cycle assessment of an industrial direct air capture process based on temperature–vacuum swing adsorption," *Nature Energy,* vol. 6, no. 2, pp. 203–213, 2021.
47. X. Y. D. Soo, J. J. C. Lee, W.-Y. Wu, L. Tao, C. Wang, Q. Zhu, and J. Bu, "Advancements in CO_2 capture by absorption and adsorption: A comprehensive review," *Journal of CO_2 Utilization,* vol. 81, p. 102727, 2024.

13 Techno-Economic Analysis

Shanli Nezami, Farzad Moazami, and Ahad Ghaemi

13.1 INTRODUCTION

To limit greenhouse gas (GHG) emissions, the increasing urgency to mitigate climate change has made carbon capture and storage (CCS) a central strategy. Among CCS technologies, absorption-based CO_2 capture has attracted attention due to its cost-effectiveness and efficiency in various applications, including industrial emissions and direct air capture (DAC). Zeolites, activated carbon, amine-functionalized solids, and metal-organic frameworks (MOFs), are used in adsorption technologies, each with its unique advantages. For example, MOFs exhibit tunable properties to optimize adsorption capacity, and amine-based adsorbents have high selectivity for CO_2 at low concentrations. Recent advances in adsorbent synthesis (including hybrid materials such as nitrogen-enriched biochar) have increased the availability of cost-effective and scalable solutions. To evaluate CO_2 adsorbents, critical performance metrics such as absorption capacity, regeneration energy, and durability are evaluated. These parameters directly affect economic factors (such as capital investment, operating costs, and cost of avoided CO_2). Studies comparing conventional chemical adsorption methods such as monoethanolamide (MEA) with advanced adsorbents have shown significant reductions in costs and energy consumption when using solid adsorbents. Emerging technologies such as integrated carbon utilization (ICCU) and techno-economic analysis (TEA) adsorption have shown how to drive innovation. ICCU integrates CO_2 capture and utilization (CCU) in a single step, using adsorbents such as calcium oxide (CaO) and reducing agents to reduce costs and increase efficiency. ICCUs offer lower annual operating costs and improved energy efficiency compared to traditional CCUs, highlighting the potential for industrial-scale implementation. Economic assessments take into account broader factors such as market trends, scalability, and regulatory environments. The cost of capture (measured in dollars per ton of CO_2 avoided) remains an important metric. Recent TEA studies have shown that innovative adsorbent systems can achieve costs of less than \$50 to \$100 per ton, consistent with global goals for affordable decarbonization.

Challenges include the need for remediation processes, the high initial costs of developing advanced adsorbents, and the diversity of CO_2 sources. The TEA of CO_2 adsorbents provides a systematic approach to navigating the complexities of carbon capture technologies. Taking into account technical as well as cost criteria,

DOI: 10.1201/9781003476191-13

TEA identifies promising materials and processes and accelerates their commercialization path.

13.1.1 Background and Significance of CO_2 Adsorbents

With the increasing concentration of carbon dioxide (CO_2) in the atmosphere, the world is on the verge of reducing its adverse environmental impacts. For this, adsorption-based carbon capture technologies have been presented as a promising solution that offers a cost-effective, scalable alternative to traditional methods. These systems rely on adsorbents (specialized materials designed to selectively capture and store CO_2) and play an important role in the transition to sustainable practices in the industrial, environmental, and energy sectors.

13.1.1.1 Importance of CO_2 Adsorbents in Mitigation Strategies

CO_2 adsorbents address challenges in sectors such as power generation, cement production, and steel production, which contribute significantly to global emissions. When absorption technologies are integrated, these industries can meet stringent environmental regulations and greatly reduce energy consumption compared to conventional chemical absorption techniques. The flexibility that adsorbents have in capturing CO_2 from various sources, such as flue gases, biogas, and even ambient air (DAC), emphasizes their versatility and importance in comprehensive carbon management strategies [1, 2].

Modern adsorbents are classified into several categories, including:

- **Zeolites**: These crystalline aluminosilicates are renowned for their thermal stability, chemical resistance, and high adsorption selectivity, making them ideal for industrial-scale operations.
- **MOFs**: Characterized by customizable pore structures, MOFs exhibit exceptional CO_2 uptake capacity and tunability, making them a focus of advanced research.
- **Porous Organic Polymers (POPs)**: These materials combine high surface areas with chemical diversity, enabling efficient adsorption in challenging environments.
- **Activated Carbon**: Known for its cost-effectiveness, activated carbon is extensively used in large-scale applications due to its wide availability and excellent recyclability.

Research is underway to enhance adsorbent performance through molecular engineering and surface functionalization. MOFs and amine-functionalized zeolites interact with CO_2 molecules by incorporating chemical groups, improving CO_2 selectivity, developing hierarchical porous structures, increasing adsorption kinetics, and reducing the need for regenerated energy, facilitating their adoption in energy industries. Techno-economic analyses have shown that adsorbent-based systems can significantly reduce the cost of CO_2 capture, especially when deployed in pre- and post-combustion configurations. These systems enable carbon utilization pathways, such as converting the captured CO_2 into value-added products such as methanol, polymers, and biofuels. Research is focused on designing adsorbents that can operate effectively under real-world conditions (such as high humidity and variable gas compositions). To overcome these limitations, innovative materials such as hybrid adsorbents of mixed matrix membranes, which combine the properties of MOFs and activated carbon, are being investigated. The strategic development of CO_2 adsorbents is aligned with global climate goals (as outlined in agreements such as the Paris Agreement). By enabling carbon capture and efficient use, these materials help support the integration of renewable energies, reduce GHG emissions, and foster a circular carbon economy [1, 3, 4]. MXenes are a class of two-dimensional (2D) transition metal nitrides, carbides, and carbonitrides, and have attracted attention for their exceptional surface properties, such as abundant functional groups, high surface area, and electrical conductivity. The functionalized surfaces of MXenes enable selective CO_2 adsorption through chemical interactions and also enhance their adsorption capacity. Studies have shown that tuning the surface chemistry of MXenes, such as introducing hydroxyl or oxygen functional groups, significantly improves the CO_2 binding energy. The structural stability of MXenes under harsh conditions makes them ideal for industrial applications [2]. When graphene is oxidized, graphene oxide (GO) is formed. This oxide offers a highly functional surface that contains oxygen-containing groups such as hydroxyl and carboxyl groups. The functional groups provide active sites for CO_2 adsorption. The layered structure of GO facilitates gas diffusion. GO-based materials are promising in hybrid adsorbent systems, as they have high adsorption capacity combined with the advantages of other materials such as MOFs or zeolites [5]. Graphene has a one-atom-thick carbon layer and excellent mechanical and thermal properties and is widely investigated for gas adsorption applications. Its non-local π electrons facilitate weak but selective van der Waals interactions with CO_2, while doping graphene with heteroatoms such as nitrogen or boron greatly enhances its adsorption capacity. For example, nitrogen-doped graphene provides more sites for electron donation, increasing the binding energy of CO_2 molecules. Graphene-based adsorbents are lightweight and thermally stable, making them suitable for applications in the energy industry [5].

Combining MXenes, GO, or graphene with other adsorbent materials creates synergistic effects and enhances adsorption performance. For example, hybrid materials that integrate MXenes with MOFs achieve high adsorption capacity and selectivity by taking advantage of the high porosity of MOFs and the versatility of MXenes' functionality. Also, composites of GO and graphene with zeolites or activated carbon help to enhance structural stability, adsorption kinetics, and regeneration efficiency. MXenes are prone to oxidation, which can degrade their performance over time. Graphene production is expensive, which limits its economic viability. Researchers are working to improve the stability and scalability of these materials, including developing cost-effective synthesis methods for graphene and increasing the oxidation resistance of MXene. Incorporating advanced materials such as MXenes, GO, and graphene into CO_2 capture technologies represents a significant step toward achieving global decarbonization goals. These materials improve performance, while also helping to reduce the overall energy footprint of carbon capture systems

13.1.2 Purpose and Scope of the TEA

In the initial phase, the study's objective is established, encompassing the primary questions, the contextual framework, the anticipated application, the constraints, and the target audience for the analysis. This objective serves as the foundation for all subsequent components of the TEA study. Practitioners establish a preliminary objective at the outset of their work, allowing for refinement or adjustment throughout the study, albeit with careful consideration. Expanding upon this objective, practitioners outline in the assessment scope the specific elements of a product they intend to evaluate and the methodology for comparing it to rival solutions. Key tasks during the scope phase involve determining the intended application of the product, identifying the subject of analysis (product system), and establishing the dimensions for comparison with other systems (functional unit). Additionally, it includes quantifying the product in relation to other systems (reference flow), further detailing the system components (system elements), defining the inclusions and exclusions of the assessment (system boundaries), selecting appropriate systems for comparison (benchmark systems), evaluating the proximity of technologies to market entry (technology maturity), and identifying the parameters and metrics utilized (criteria and indicators). Practitioners can extract the necessary requirements for the subsequent phases of inventory, calculation, and reporting from the scope [6].

13.1.3 Objectives of the TEA

The potential methods for TEA can be delineated based on various factors, including the type of biomass and the intended products, in addition to the objectives of the process (such as economic evaluation or environmental considerations) to assess economic viability [7]. The primary objective of TEA is to assess investment variables and more clearly identify the limitations of process configurations, thereby enhancing bioenergy production and minimizing capital costs [8].

In other words, the purposes can be categorized as follows [9]:

Objectives:

- Establishing strategic investment routes for technological advancement
- Measurement of research and development performance objectives and priorities
- Recognition of essential data deficiencies
- Delivering summarized findings in a structured, succinct, and visually coherent manner

13.2 METHODOLOGY

13.2.1 Data Collection and Literature Review

Data collection must adhere to the established data requirements, chosen methodologies, and indicators, as well as the specified assessment objectives. Since TEA seeks to enhance not only the overall product system but also its individual components, it is essential to gather and document data for each identified system element. The degree of detail must align with the established processes related to the system components. Technical information includes energy and material flows, process parameters, and equipment specifications, among other factors; professionals primarily acquire this technical data from process design. Economic data includes the costs associated with equipment, the prices of inputs and outputs, and relevant market information. Like technical data, economic data can be sourced from multiple avenues, such as quotes, databases, expert opinions, and literature. However, the validity of this data is often more restricted to specific contexts, such as the location of the quote, and there can be considerable discrepancies in values between different sources, for instance, internal company pricing compared to market pricing. When gathering data from various sources, it is essential for practitioners to perform harmonization, which involves ensuring consistency and aligning assumptions, such as standardizing data to a common year and consistently utilizing lower heating values. Technical and economic data should, whenever feasible, be associated with the functional unit and reference flow. To evaluate the economic criteria, practitioners need information regarding costs, sales prices, and market volumes. Often, the cost of a facility is not readily accessible and must be approximated. To determine the sales price of substitutes, it is advisable to employ a value-based pricing strategy, while a cost-plus pricing method may provide a useful approximation. In the case of non-substitutes, a price-performance strategy is applicable for establishing the sales price. The market volume for substitutes can be inferred from the market volume of benchmark products; however, estimating the market volume for non-substitutes may present difficulties. Practitioners are required to articulate the temporal and regional context of the study, including characteristics of the value chain, along with any associated limitations and risks. They must also justify context-specific assumptions and parameters. In cases where prices or market volumes are derived from analogous studies, it is essential to ensure a reasonable alignment between the temporal and geographical conditions to avoid underestimating limitations. Government regulations can significantly differ across various regions, influencing the viability of the product system, such as through subsidies on feedstock, taxation, and environmental policies. In the context of the inventory model, input prices are essential economic data. Consequently, it is imperative to ascertain prices for input flows that traverse the system boundaries. Additionally, it may be beneficial to determine the price of a flow between system components, such as the internal production cost of 1 kg of hydrogen produced via electrolysis, to facilitate comparisons with alternative production methods. Determining an input price typically relies on three primary factors: technical specifications, assessment boundaries, and geographical location. Initially, the technical specifications of the input flow must align with the product system's requirements, which include aspects such as quantity, quality, and temporal development. Should the input flow fail to satisfy these criteria, practitioners may need to alter the source of the input flow or the production technology, adjust the system boundaries, or modify the system components. If an input flow fails to meet the required purity levels for the product system, it may be necessary to incorporate a purification step into the system. Additionally, it is essential for practitioners to determine whether the source, production, or any supplementary handling processes—such as purification, compression, or heating—are to be considered within the assessment boundaries. In this scenario, experts must factor in the additional purification step during the evaluation. It is important to note that if the source, production, and handling are included in the assessment, the flow in question does not cross the system boundaries and is not classified as an input to the product system; rather, it circulates among the system elements. Nevertheless, establishing a price for such a flow may still be relevant. Third, when practitioners identify a particular location for the input source, they can utilize the exact costs associated with transportation and storage. In cases where a specific location is not defined, employing average costs is adequate. For instance, if the production occurs within the same industrial park, experts may factor in basic transportation costs via pipeline. Practitioners must adjust the quality of price data and the number of sources in accordance with the maturity of the technology. During the initial phases of research and development, it is advisable to utilize market-average price data, as a limited number of secondary sources will generally suffice. In the development phase, experts should incorporate market-average price data that is specific to both date and location; while secondary sources may still be adequate, it is essential to include a variety of sources. In the deployment phase, practitioners are encouraged to rely on process-specific data and primary sources. The comprehensive guidelines offer additional clarification. In cases where the technology maturity is limited, specialists must incorporate learning curves and enhancements for the system components, applicable to both reference product systems and benchmark systems. Furthermore, it is essential to consider the cost reductions associated with constructing the first-of-a-kind and *n*th-of-a-kind facilities [10].

13.2.2 Selection Criteria for CO_2 Adsorbents

The world's growing need for CCS technologies in relation to climate change mitigation has increased the need for efficient and sustainable CO_2 absorbers. The selection of the appropriate absorber includes operational factors to optimize performance and evaluate various material properties, environmental sustainability, and economic feasibility [11].

- **Adsorption Capacity**
  To assess the initial efficiency of an adsorbent, we look at its CO_2 adsorption capacity, often measured in mmol/g under specific conditions. Materials such as activated carbons, MOFs, and zeolites exhibit exceptional adsorption capacity (e.g., MOFs that are functionalized with amines greatly enhance CO_2 adsorption via chemisorption mechanisms).
- **Selectivity**
- The high selectivity that CO_2 has over other gases (e.g., N_2, CH_4) is of great importance in flue gas applications. Materials such as MOFs and zeolites exhibit molecular screening effects and, because of their porous structure, have high selectivity. Functional materials, such as amine doped adsorbents, improve selectivity by strongly binding to CO_2 molecules.

- **Thermal and Chemical Stability**
  For long-term industrial use, adsorbents must exhibit thermal stability at high temperatures and chemical flexibility in the presence of moisture and contaminants. Carbon-based materials (such as graphene-based adsorbents and activated carbons) are ideal for robust operations because of their high stability.
- **Energy Requirements for Regeneration**
- To minimize the operating costs of adsorption systems, low regeneration energy is required. Solid adsorbents such as MOFs and silica-based adsorbents have been optimized for thermal adsorption and pressure swing techniques to achieve efficient regeneration. New strategies, such as electrothermal desorption, have been considered to improve energy efficiency.
- **Porosity and Surface Area**
- Porosity and surface area are very effective in CO_2 adsorption. Materials with hierarchical pore structures, such as specific MXenes and POPs, help provide more adsorption sites and enhanced diffusion pathways, facilitating high-efficiency CO_2 adsorption.
- **Hydrophobicity and Moisture Resistance**
  The presence of moisture in flue gases often causes a reduction in the performance of the adsorbent. Hydrophobic materials, including GO derivatives and specific MOFs, are formulated to resist moisture, ensuring consistent performance in humid environments.
- **Cost and Scalability**
- For an adsorbent to be economically viable, we consider the complexity of synthesis, cost of raw materials, and scalability. For large-scale applications, activated carbons and zeolites are preferred due to their low cost. MXene and graphene are also preferred due to their cost-effectiveness with high efficiency, which makes the application viable.
- **Sustainability**
- Environmentally sustainable adsorbents are important. These adsorbents are derived from biodegradable materials or renewable resources. Research is focusing on drug-active clays and bio-based polymers to achieve sustainability goals without compromising performance.

A comprehensive approach is needed in selecting CO_2 adsorbents that balance adsorption performance, operational efficiency, environmental impact, and cost. Research on advanced materials such as MOFs, carbon-based adsorbents, and MXenes is currently of great importance, addressing challenges such as regeneration efficiency, scalability, and selectivity. In the future, hybrid materials and process integration are likely to be investigated to optimize the overall carbon capture framework [1, 11–14].

13.2.3 TEA Framework

TEA serves as a systematic approach for assessing the economic feasibility of industrial processes. This methodology generally employs a modelling framework to project capital expenditures (CAPEX), operational costs, and the overall economic potential by utilizing various technical and financial input parameters to determine the commercial viability of a process or product [15]. TEA offers both qualitative and quantitative assessments regarding the financial soundness of the process. It is frequently employed as a strategy for conducting cost–benefit analyses and exploring economic viability, cash flow projections, and the evaluation of diverse technologies based on factors such as scale, efficiency, and yield for various technological applications. The application of TEA is becoming more prevalent for evaluating both existing and future technologies across various processes, allowing for comparisons based on a uniform analytical framework. It is essential to assess the current research landscape and the potential commercial feasibility of the various technologies required to design processes at an appropriate level of detail, utilizing relevant and consistent assumptions. The estimation of production costs involves analyzing the capital and operational expenses associated with emerging technological pathways for producing diverse products, while also accounting for credits related to process by-products where applicable. Figure 13.1 provides a typical methodological framework to analyze the techno-economic feasibility of the emerging technology.

The fundamental steps of a TEA include the development of a conceptual design configuration for the proposed process or product. This involves exploring various potential options for the current commercial production process. Subsequently, reliable laboratory-scale data for the identified process options is collected, supplemented

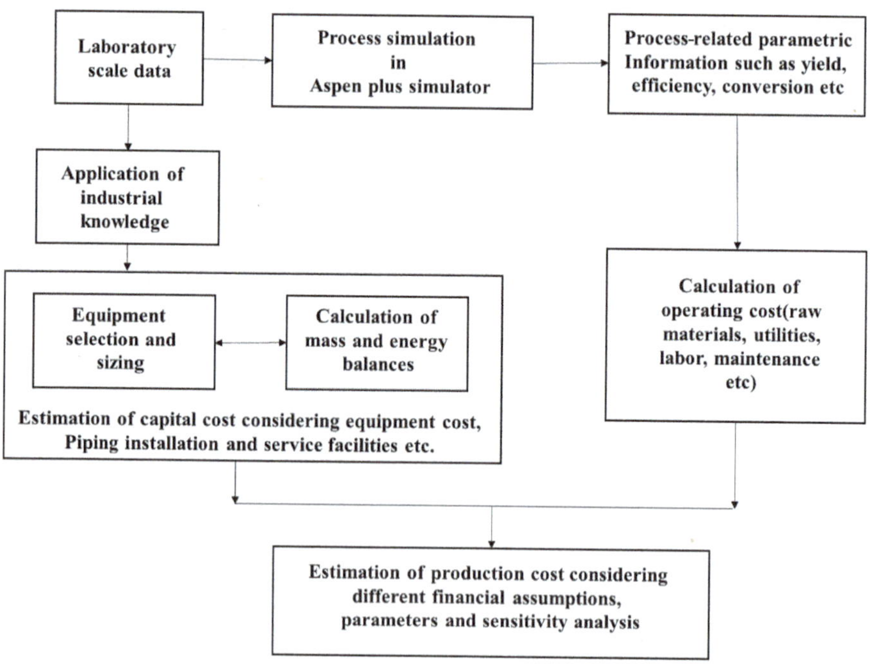

FIGURE 13.1 A comprehensive methodological framework for the techno-economic evaluation of emerging technologies [15]

by pertinent literature. Following the establishment of a conceptual process design, material and energy balance data for the proposed process is compiled using developed thermodynamic models, employing process simulation tools such as Aspen, Hysys, or CHEMCAD. These models are developed based on the findings from research and development conducted at both bench and pilot scales. In addition to the mass and energy flows of the process, various parameters such as temperature, pressure, yield, and efficiency data were extracted from the simulation. To accurately represent the existing technology, appropriate assumptions were made as needed, utilizing industrial-scale knowledge and data sourced from the literature [16]. The subsequent phase involves the assessment of capital and operational expenditures based on the data acquired from material and energy balances for various processes derived from simulations. Capital costs typically encompass the overall financial resources necessary for the establishment of an industrial facility. Working capital, representing the investment essential for initiating the plant's operations, is recouped upon the conclusion of the project's lifespan. The CAPEX associated with the proposed facility primarily encompass expenses linked to significant components such as the acquisition of equipment, land, buildings, and construction activities. The sizing of the equipment is determined based on data derived from material and energy balances, which are conducted for each process in accordance with the proposed scaling capacity [17]. The costs for equipment, as well as for instrumentation and control systems, can be approximated using information from published sources and quotations from manufacturers, taking into account the reference year. The total capital investment for a facility encompasses the fixed capital investment (FCI), start-up expenses (10% of FCI), working capital (5% of FCI), and interest accrued during the construction phase. A portion of the fixed operating costs, which includes maintenance labor, maintenance materials, overheads, and insurance, is calculated as a percentage of the FCI. Comprehensive information regarding the various components of FCI can be sourced from existing literature. These percentages serve as the foundation for the variable operating costs, which include expenses related to feedstock, electricity, heat, and catalyst costs, among others. To ascertain the production cost of the product, capital charges were incorporated into both the fixed and variable operating costs [18]. The techno-monetary analysis allows for the examination of the total duration and ongoing expenses incurred throughout the project's lifespan through various financial metrics. Continuous enhancements in the TEA are achieved by utilizing data from previous evaluations alongside newly observed information regarding the process and other system variables. As the process or technology approaches commercialization, the uncertainty associated with the TEA is expected to diminish. Additionally, sensitivity analysis is performed to evaluate and address uncertainty by varying different input parameters within the model and assessing their impact on the output parameters. Sensitivity analysis offers insights into the contributions of various parameters, including feedstock costs, plant capacity, and yields, to the variability of final outcomes and their respective impacts. TEAs are instrumental in determining which conceptual designs, along with various economic factors, possess the greatest potential for success in the near, mid, and long term for a given process or technology. Additionally, TEAs can facilitate the comparison of different process or technology pathways by utilizing technical and economic data derived from techno-economic models, thereby identifying the most promising avenues for future commercialization [19]. The supply chain of feedstock, along with factors such as location, operational scale, and technology selection, significantly influences the production costs associated with a product. A variety of techniques and metrics are employed to evaluate the economic dimensions of technologies, often based on certain assumptions. These assumptions can vary greatly from one study to another, complicating the comparison of different techno-economic assessments. Such assessments offer valuable insights for various stakeholders; for instance, in engineering research, they highlight potential enhancements that could lead to substantial cost savings. Additionally, for stakeholders like biomass suppliers, investors, governments, and energy consumers, these assessments provide critical information that aids in decision-making throughout different phases of the process [20].

- **Technical Considerations**
  - Feed flow rates
  - Cycle duration
  - Cycle planning
  - Vacuum pump performance curves
  - Mechanical engineering design
  - Thermal and work energy
  - Conversion from milligrams to kilotons' direct air capture
- **Economic Considerations**
  - IBL costing approaches
  - Costing information
  - Additional CAPEX
  - Operational expenses
  - Cost per unit of CO_2

13.2.4 Assumptions and Limitations

The development of CO_2 adsorbents is essential to mitigate climate change. To make progress in this field, we need to navigate a complex landscape of inherent assumptions and limitations. These factors influence process modelling, research designs, and the techno-economic viability of adsorption and material synthesis technologies. We will now review the major assumptions and limitations identified in the study and implementation of CO_2 adsorbents.

- **Assumptions in Material Performance**

  In CO_2 adsorbent research, the assumption is that the results obtained in the laboratory are also applicable and usable in the real world. For example, the adsorption capacity of materials such as GO, MOFs, and zeolites is reported under ideal conditions (controlled pressure, temperature, and gas composition). In operational environments, these variations are not taken into account (gases present in industrial flue gases often contain impurities and moisture, which may significantly affect the performance of the adsorbent).

Another hypothesis is that the adsorption capacity, selectivity criteria, and regeneration energy determined in controlled settings reflect their behavior over long-time scales. In many studies, it is assumed that material degradation occurs during repeated adsorption–desorption cycles (this is rarely the case in practical applications) [14].

- **Limitations of Current Adsorbents**
  Although significant advances have been made in materials science, current CO_2 adsorbents still face significant limitations, which prevent their industrial adoption. Many advanced materials, including MOFs and MXenes, exhibit poor stability under operational conditions (since they degrade in the presence of moisture or high temperatures, compromising their long-term application). Energy recovery on several adsorbents, especially those relying on chemical adsorption mechanisms with strong CO_2 binding, reduces the efficiency and cost-effectiveness of carbon capture systems. Cost and scalability are important challenges, as advanced materials such as graphene, MXene, and functional MOFs are subject to complex and expensive processes, limiting their widespread application. Activated carbons and zeolites, which are cost-effective materials, are durable compared to advanced alternatives but often perform less than optimal.
- **Assumptions in Process Modelling**
  In the case of adsorption-based carbon capture, steady-state conditions, and ideal gas behavior are often assumed in process models. These simplifications ignore dynamic variables such as pressure changes, fluctuating feed gas compositions, and thermal gradients (which are encountered in industrial processes). The effectiveness of hybrid systems that combine adsorption with other technologies (such as membrane separation or cryogenic distillation) is often predicted without fully understanding the complexities of system integration [13].
- **Environmental Impact Limitations**
  Another critical limitation is the environmental sustainability of adsorbents. While many studies focus on the potential for reducing GHG emissions, few consider the life cycle environmental impacts of adsorbent synthesis, including energy consumption, raw material extraction, and waste generation. Materials such as graphene and MOFs are energy-intensive production processes and should be optimized to offset the environmental benefits of CO_2 capture.
- **Knowledge Gaps and Research Challenges**
  Because there are a number of knowledge gaps that limit the predictive power and applicability of current findings:
  Among the key challenges in CO_2 adsorbents is the need for a deeper understanding of long-term durability, material-specific vulnerabilities, and performance under real-world conditions. The durability of adsorbents over multiple cycles has not been investigated due to the lack of standardized protocols for lifetime testing, which complicates the comparison of materials in practice. Most research has focused on CO_2 absorption in pure or binary gas mixtures, ignoring the complexities of industrial gases that

contain pollutants such as SO_2, NO_x, and particulates, which can poison or degrade the adsorbents. Furthermore, there are specific challenges with MXenes, which, despite their promise, are prone to oxidation, reducing their lifetime and absorption efficiency.

- **Operational Constraints**
  Operational limitations include reactor design, adsorbent deployment scale, and system integration. For laboratory-scale studies, fixed-bed reactors are preferred, but for large-scale operations, these reactors may not be suitable due to poor heat and mass transfer characteristics.
- **Future Directions and Overcoming Limitations**

To address the challenges of CO_2 adsorbent development, we need to focus and work on key areas. To ensure consistent and comparable evaluations across materials, it is crucial to establish standardized test protocols for regeneration energy performance, adsorption, and lifetime. At the same time, for methods such as scalable and cost-effective synthesis, advanced materials such as MXene and graphene should be prioritized to minimize environmental impacts and facilitate industrial adoption. Comprehensive TEAs that include sensitivity analyses and real-world data are needed to accurately assess the feasibility of adsorbents in different operational scenarios. Research and development into hybrid systems and advanced reactor designs can optimize the integration of adsorbents into industrial processes, expand their application, and improve efficiency [12, 13].

13.2.5 Key Performance Indicators

The objective of key performance indicators (KPIs) is to facilitate the assessment of both project and organizational performance across the construction sector (The KPI Working Group, 2000). According to Collin et al. 2016 [21], the development of KPIs necessitates the evaluation of several key factors [22].

- KPIs serve as overarching measures of performance, concentrating on essential elements of outputs or results.
- A manageable and limited set of KPIs is essential for consistent application. An excessive number of complex KPIs can lead to significant demands on time and resources.
- The consistent application of KPIs is crucial, as their effectiveness is largely dependent on their regular utilization across multiple projects. It is imperative to streamline the data collection process to enhance efficiency. A substantial sample size is necessary to mitigate the influence of project-specific variables. Consequently, KPIs ought to be formulated for application in all construction projects.
- To ensure the effectiveness of performance measurement, it is essential that the measures or indicators are recognized, comprehended, and embraced throughout the organization.
- KPIs must undergo evolution, and it is probable that a collection of these indicators will experience modifications and enhancements.

- Graphic representations of KPIs should be designed for simplicity, ease of updating, and accessibility. Considering these aspects, a comprehensive set of KPIs, encompassing both objective and subjective indicators, is established to evaluate the performance of a construction project.

In light of the previous studies conducted, a comprehensive discussion of each KPI will be presented, along with practical methodologies for their measurement. The calculation techniques for the proposed KPIs are categorized into two distinct groups. The first group employs mathematical formulas to derive the corresponding values. The formulas will be provided following comprehensive explanations of each KPI, which encompass time, cost, value, safety, and environmental performance. Conversely, the other group relies on the subjective opinions and personal judgments of stakeholders. This group encompasses aspects such as quality, building functionality, and the satisfaction levels of various stakeholders. A seven-point scale scoring system is adopted to measure these KPIs [23]. As discussed in the following paragraphs, there are nine KPI categories in total, each may include one or more measuring methods. Figure 13.2 shows a graphical representation of the KPIs.

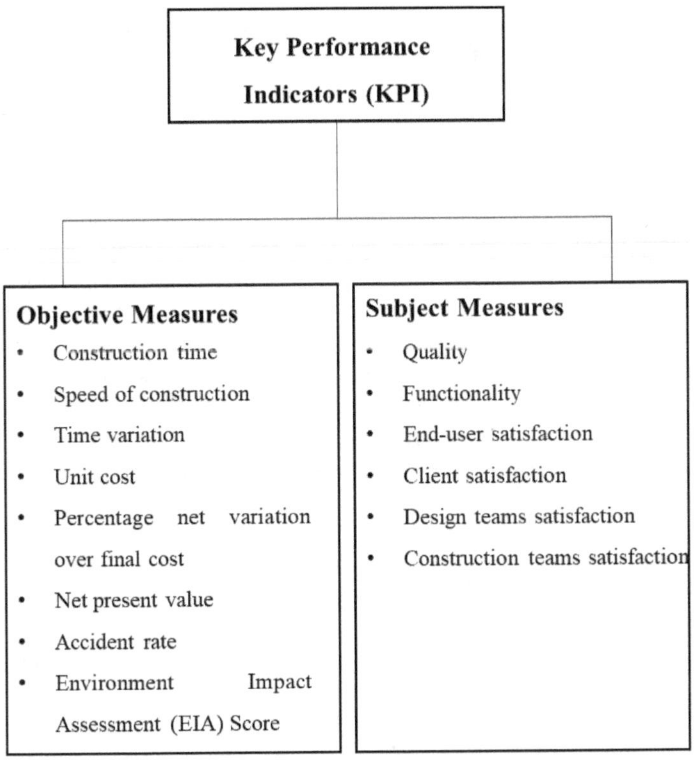

FIGURE 13.2 KPIs for project success [22].

13.3 TEA COMPONENTS

TEA is used to evaluate the economic, technical, and environmental aspects of CO_2 capture technologies. The purpose of TEA is to assess the feasibility, cost-effectiveness, and scalability of these systems, allowing informed decisions to be made for development and implementation. This includes potential revenue streams, identification and quantification of capital and operating costs, and their interaction in the context of broader technological and market factors. TEA costs include equipment procurement, land purchase, potential logistics, installation, and ancillary costs. Equipment costs, often the largest portion of the investment, include compressors, reactors, precision instrumentation, adsorbent beds, and ancillary devices necessary for CO_2 capture processes [24]. Technologies that use advanced materials such as GO, MOFs, and MXenes may require specialized equipment to match their unique physical and chemical properties. These materials require precision in operating conditions (such as temperature, pressure, and chemical compatibility), which drives up the cost of the equipment. Another critical capital component is the cost of installation and commissioning, which includes the installation of machinery, site preparation costs, calibrating systems, and running tests. For CO_2 capture systems integrated into existing industrial settings, such as power plants or refineries, these costs increase due to the complexity of retrofitting, ensuring compatibility and minimizing disruption to ongoing operations. Among the significant costs are the cost of land and infrastructure, which vary based on size, location, and proximity to resources, with scale being of paramount importance in large projects. These costs include building construction, land acquisition, as well as the creation of essential utilities such as energy, transport, and water infrastructure. Among the costs incurred during project implementation are contingency costs for uncertainties and potential surpluses. These costs are necessary to mitigate risks such as design changes, delays, material shortages, and unforeseen regulatory requirements and ensure financial stability throughout the process. Capital costs, which include environmental impact assessments (EIAs), permits, stakeholder consultations, and legal fees, are particularly relevant in the highly regulated field of CO_2 capture technologies [6, 24].

Operating costs are important in determining long-term economic viability and represent the recurring costs associated with the day-to-day operation of CO_2 capture systems. The largest operating cost is energy consumption. Energy is required in absorption-based systems to regenerate the adsorbents. One factor that affects energy costs is the choice of adsorbent and regeneration method—whether thermal, pressure swing, or electrical. Some materials, such as functionalized MOFs or graphene-based adsorbents, offer cost advantages because they require less energy (although they may require a higher initial investment). Energy costs are affected, and renewable energies offer opportunities for cost reductions and have environmental benefits. Significant operational costs include the cost of raw materials, including advanced adsorbents such as GO or MXenes, the synthesis and processing of which involve complex methods and expensive precursors. If these materials maintain their performance over long cycles, these costs are offset by recycling or regeneration of the adsorbents. Among the costs are labor costs, which include training, wages, and overhead for personnel required to set up and maintain the system. Automation reduces labor requirements, but high-tech systems require specialized skills, increasing labor costs. Emergency repairs, replacement parts, and routine inspections are among the maintenance costs. Absorption systems operating in industrial environments are subject to wear and tear and require robust maintenance protocols to ensure uninterrupted operation. Waste management costs, which relate to systems that produce by-products or inactive adsorbents, include transportation, disposal, and compliance with environmental regulations. Advanced recycling can help reduce this cost but may require additional investment. Other operating costs, including administrative costs, insurance, and unforeseen operating expenses, contribute to the overall cost structure and should be considered for an overall cost assessment [25].

In TEA, revenue streams are critical and determine the financial sustainability of CO_2 capture technologies. The value of carbon credits varies by region and regulatory framework, directly impacting the profitability of CO_2 capture systems. In enhanced oil recovery (EOR), the captured CO_2 is removed from reservoirs and used to extract additional oil, providing a significant revenue opportunity. This program has a dual benefit (creating economic returns and reducing CO_2 emissions). Captured CO_2 is also used in chemical synthesis, construction materials production, and beverage carbonation and can generate a variety of revenue streams. Its market value depends on the quality, purity, and demand in the relevant industries, as well as the availability of captured CO_2. When these revenue opportunities are integrated into TEA, it is possible to make a balanced assessment of costs and benefits [26, 27].

13.3.1 Revenue Streams

13.3.1.1 CO_2 Capture and Utilization

An innovative approach to addressing the dual challenges of climate change and sustainable development is CCU, which involves capturing carbon dioxide emissions from various sources, such as industrial facilities, power plants, and even directly from the atmosphere, and using the captured CO_2 for long-term storage or to create valuable products. The global goals of this approach are to transition to a circular carbon economy (where waste emissions are converted into cost-effective economic outputs), reduce GHG emissions, and so on. The CCU framework encompasses a wide range of technologies and applications (biological, chemical, and mineral pathways) and offers opportunities to reduce environmental impacts and create economic value [28].

CO_2 separation from gas streams in CCU absorption uses technologies such as direct air absorption (DAC), pre-combustion, oxy-fuel combustion, and post-combustion. Post-combustion absorption is the best-known absorption method and involves the removal of CO_2 from flue gases produced during the combustion of fossil fuels or biomass. It uses techniques such as chemical absorption using solvents such as amines, physical absorption using materials such as activated carbon or zeolites, and membrane separation. Post-combustion absorption faces challenges such as operational degradation management and high energy for solvent regeneration. Pre-combustion absorption removes CO_2 before combustion (during synthesis gas production in processes such as integrated gasification combined cycle systems). In the combustion-based absorption method, CO_2 is absorbed at higher concentrations and pressures and is more efficient but requires significant investment in terms of cost. In oxy-fuel combustion, the fuel burns in pure oxygen, resulting in a concentrated CO_2 stream, simplifying capture but requiring advanced air separation technologies. DAC represents the cutting edge of CO_2 capture, which directly removes carbon dioxide from the atmosphere using liquid solvents or solid adsorbents. Although DAC has the potential to address both diffuse and diffuse emissions, it is currently limited by high energy consumption and low atmospheric CO_2 concentrations [26, 28].

After capture, CO_2 has multiple applications, converting it into more valuable products and reducing its impact as a GHG. In chemical conversion processes, there is significant potential to use CO_2 as a feedstock for the production of chemicals, fuels, and polymers. For example, CO_2 can be hydrogenated as a versatile chemical to produce methanol (which serves as a precursor for acetic acid, formaldehyde, and synthetic hydrocarbons). Catalysts such as copper–zinc–alumina play a key role in the conversion process, improving efficiency and selectivity. CO_2 can also be converted into carbonates (such as polycarbonates and cyclic carbonates, which are widely used in the plastics and coatings industries). Other applications include the synthesis of urea, a key fertilizer produced by reacting CO_2 with ammonia, which is one of the largest industrial applications of captured CO_2. CO_2 can also be converted into fuel and combined with hydrogen derived from renewable energy sources, providing a renewable energy cycle. Synthetic methane can be produced via the Sabatier reaction, and liquid fuels, such as gasoline and diesel, can be synthesized via Fischer–Tropsch processes [29].

Another promising route for using CO_2 is mineralization, in which CO_2 reacts with alkaline earth metals such as magnesium or calcium to form stable carbonates. This approach offers a permanent, environmentally friendly storage option while simultaneously producing valuable building materials such as concrete and aggregates. Mineralization has the dual benefit of reducing GHG emissions from industries such as cement production and creating durable, CO_2-sequestering products. One approach with significant potential is the biological use of CO_2. This method uses natural processes such as photosynthesis to convert CO_2 into biomass. Microalgae have the potential to use CO_2 as a carbon source for the production of animal feed, biofuels, and bioplastics. This approach offers additional benefits such as wastewater treatment and nutrient recovery and integrates well with industrial processes.

CCU faces challenges including regulatory, technical, and economic challenges. The most important obstacles include the high energy required to capture and convert CO_2. The cost of advanced materials can limit scalability (such as high-performance catalysts or functionalized adsorbents), and material degradation under operational conditions poses challenges to long-term efficiency. Market conditions, regulatory incentives, and demand for CO_2-derived products are factors that will determine the economic viability of CCU technologies. Revenue streams such as EOR, carbon credits, and sales of CO_2-based chemicals play an important role in offsetting costs. In addition, the development of robust regulatory frameworks and standardized protocols is essential to ensure the safe and effective deployment of CCU technologies [26, 28, 29].

A transformative approach to addressing climate change is CCU, which turns waste emissions into economic opportunities. Advances in capture technologies, including efficient solvents and scalable DAC systems and improved adsorbents, are expanding the feasibility of capturing CO_2 from a variety of sources. Fuel production, chemical synthesis, and mineralization provide new avenues for integrating CO_2 into industrial and energy systems. Sustained research, cross-sector collaboration, and supportive policies are needed to overcome challenges such as energy demand, cost, and regulatory complexity. As CCU technologies evolve, they have the potential to become a cornerstone of sustainable development, bridging the gap between economic growth and environmental stewardship [29].

13.3.1.2 Carbon Offsetting and Trading

A carbon offset refers to a decrease in the emissions of carbon dioxide or other GHGs, implemented to counterbalance or mitigate emissions produced in a different location. One ton of carbon offset signifies the reduction of one ton of carbon dioxide or its equivalent in other GHGs [30]. There exist two distinct markets for carbon offsets: (1) the more extensive compliance market, in which corporations, governmental bodies, or other organizations acquire carbon offsets to adhere to regulations limiting their total carbon dioxide emissions; and (2) the comparatively smaller voluntary market, where individuals, businesses, or governments opt to purchase carbon offsets to counterbalance their own GHG emissions resulting from activities such as transportation and electricity consumption [31]. Typically, carbon offsets are associated with initiatives aimed at reducing GHG emissions, either in the short term or over a longer duration. A prevalent category of projects involves renewable energy initiatives, including wind farms, biomass energy, and hydroelectric dams. Additional project types encompass energy efficiency improvements, the elimination of industrial pollutants or agricultural waste, the mitigation of landfill methane emissions, as well as land use, land use change, and forestry initiatives, and efforts aimed at reducing emissions from deforestation and forest degradation (REDD), among others. Krasovskii et al. were driven by the uncertainties surrounding future CO_2 pricing in relation to the carbon emissions offsetting practices of regulated entities. Inquiries were made regarding the feasibility of mitigating these financial risks. The authors of this study examined the bilateral relationship between a REDD supplier and a GHG-emitting energy producer within an incomplete emissions offsets market. Their findings indicated that the flexibility of flobsion offered benefits over a conventional option, assisting GHG-emitting energy producers in managing their compliance risks while simultaneously promoting the advancement of REDD programs. Additionally, carbon trading represents another method of carbon pricing within cap-and-trade frameworks. Cap-and-trade represents a strategy for managing and ultimately decreasing carbon emissions. The government establishes a national cap on carbon emissions and subsequently restricts the volume of carbon dioxide that companies are permitted to emit. Companies that are able to reduce their carbon emissions more effectively can sell their surplus permits in the emissions market to those companies that find it challenging to lower their emissions. Consequently, carbon trading markets are created. The prevalence of emissions trading systems globally is on the rise. In conjunction with the EU emissions trading system, various national or subnational systems are currently operational or being developed in countries such as Canada, China, Japan, New Zealand, South Korea, Switzerland, and the United States [19]. To examine the effects of emission reduction policies on the production and economic performance of the steel industry, Duan et al. [32] developed a two-stage dynamic game model, which facilitated an analysis of the influence of different emission reduction policies on both the steel industry and individual enterprises. The findings of the research demonstrated that, as the emission reduction target rises (ranging from 15% to 30%) and the price of carbon quota trading increases (from 12.65 to 137.59 Yuan), both social welfare and producer surplus exhibit an upward trend, while macro losses associated with emissions display a downward trend.

13.3.2 Financial Analysis

13.3.2.1 Net Present Value

The net present value (NPV) is an indicator used to evaluate the profitability of investment projects by summing all inflows and outflows of cash over the project's lifetime. A positive NPV indicates that the project earnings surpass the expected costs, meaning that the project is profitable, while a negative NPV will event in a net loss. To assess the value of the project, financed by a combination of debt and equity, the discount rate applied to the cash flows in NPV calculation corresponds to an overall cost of capital from a weighted average cost of all capital sources invested in the project. The NPV considers this discount rate (calculated as the weighted average cost of capital (WACC)) over the project lifetime, giving the annual cash flows in present values [33–36]. Hence, the NPV can be equated to Eq. (13.1):

$$NPV(i,N) = \sum_{t=0}^{N} \frac{C_t}{(1+i)^t} \tag{13.1}$$

where i is the financial discount rate (WACC), C_t the annual cash flow in the year, and N the total number of years. Time period t¼ 0 relates to the investment during the design and construction phase. The WACC is calculated based on the cost of equity and the cost of debt (borrowed capital), which allows determining the discount rate applicable to cash flows to calculate the NPV. The WACC is given by Eq. (13.2):

$$WACC = \left(\frac{E}{E+D}\right)K_e + \left(\frac{D}{E+D}\right)K_d(1-T) \tag{13.2}$$

where E is the market value of the equity, D is the market value of the debt, K_e cost of equity, K_d cost of debt, and T marginal tax rate.

13.3.2.2 Internal Rate of Return

The internal rate of return (IRR) stands as an indicator used to measure the profit ability of an investment project. The higher the IRR, the greater the profitability of the project will be. Additionally, the IRR is the discount rate that makes the NPV of all cash flows equal to zero, determining the minimum rate of return to make the project viable. If the IRR is higher than the discount rate (WACC), the project is feasible [36, 37]. The IRR is calculated by Eq. (13.3):

$$NPV(IRR, N) = \sum_{t=0}^{N} \frac{C_t}{(1+IRR)^t} = 0 \quad (13.3)$$

13.3.2.3 Payback Period

The payback period (PBP) is the time required to reclaim the initial capital investments. The shorter the PBP, the stronger is the financial feasibility of the project [36, 38]. In this analysis PBP is calculated by finding the year in which the cumulative NPV cash flow becomes positive and is formulated as Eq. (13.4):

$$PBP = A + \frac{B}{C} \quad (13.4)$$

where A is the last year with a negative cumulative NPV, B is the absolute value of cumulative NPV at the end of year A, and C is the total annual cash flow during the year after A.

13.3.2.4 Sensitivity Analysis

Sensitivity analysis plays an important role in financial analysis, especially for the technical-economic evaluation of emerging technologies (e.g., CO_2 capture). This analytical method examines how changes in key parameters will affect the financial and technical viability of a project. For CO_2 capture and sequestration, optimization of designs and sensitivity analysis are essential for identifying risks, adopting new capture technologies, and understanding the economic consequences. This structured approach ensures that investors, stakeholders, and policymakers make informed decisions based on robust, scenario-based assessments.

TEA integrates technical performance criteria with economic feasibility assessments to arrive at a comprehensive evaluation of a technology or process. For CO_2 absorbers, TEA focuses on reducing energy requirements, costs, and efficiency in CO_2 capture, as well as broader economic considerations including operating costs, capital costs, and market conditions. Sensitivity analysis increases TEA by examining how deviations in specific variables—such as energy prices, material costs, or process efficiency—affect overall financial performance. This is of particular importance for CO_2 absorbers, as their ability to be adopted depends on balancing performance criteria such as regeneration efficiency, absorption capacity, and selectivity with consideration of cost [26, 28].

In TEA, the cost of adsorbent materials for CO_2 capture technologies is a key parameter and affects the overall viability of the project. Materials such as zeolites, amines, or advanced 2D structures such as Ti_3C_2 MXenes are evaluated based on their scalability, synthesis, and availability costs. Sensitivity analysis assesses how changes in these factors affect the total cost of ownership for a CO_2 capture system.

Sensitivity analysis can examine whether a technology is economically feasible and also predict and determine increases in CAPEX. This analysis helps identify cost-reduction opportunities, such as alternative synthesis routes or economies of scale.

Among the conditions that affect the economics and performance of a CO_2 adsorbent are operating conditions such as pressure and temperature. How these parameters change and their impact on CO_2 absorption capacity and energy requirements is evaluated by sensitivity analysis. For example, materials such as Ti_3C_2 MXene exhibit different absorption efficiencies at different temperatures. Sensitivity analysis identifies optimal ranges by simulating scenarios under different operating conditions, which maximizes absorption capacity and minimizes energy consumption, helping to balance operational efficiency with economic sustainability.

Another important parameter during CO_2 capture and removal is energy consumption, which is evaluated in sensitivity analysis and TEA. Among the factors that affect the overall operating costs of a CO_2 capture system is the energy required for processes such as pressure swing adsorption or temperature swing adsorption. Sensitivity analysis examines and analyses how changes in energy prices or process energy demand will affect profitability. For example, a 20% increase in energy prices could highlight the need for energy-efficient materials and processes and make certain adsorbents economically unviable. The analysis would drive innovation and increase the development of adsorbents with lower regeneration energy requirements as well as alternative adsorption mechanisms that would reduce energy dependency [39].

Among the technical parameters are absorption capacity and selectivity, which directly affect the feasibility of a CO_2 absorber. These criteria determine the amount of CO_2 absorbed per unit mass of the absorber and the ability to preferentially absorb CO_2 over other gases such as nitrogen. In sensitivity analysis, it is evaluated how changes in these parameters affect the cost per ton of CO_2 absorbed. For example, a 15% reduction in adsorption capacity may require increased operating costs, increased material usage, larger equipment, or increased capital investment. This analysis provides insights into the trade-off between material performance and economic feasibility, helping to guide the development and selection of high-performance adsorbents.

The pricing of CO_2 emissions, or market conditions, plays a key role in the economic viability of CO_2 capture technologies. Sensitivity analysis examines how fluctuations in external factors affect financial results. For example, providing subsidies for carbon capture projects as well as changes in carbon tax rates can have a significant impact on the return on investment. Modeling different market scenarios helps sensitivity analysis to prepare for regulatory changes and align strategies to evolving policy landscapes. The dynamics of the carbon credit or penalty market can directly impact adoption rates. The basic economic components of TEA include CAPEX and operating expenditure (OPEX). In sensitivity analysis, we examine how changes in costs affect the overall feasibility of CO_2 capture projects. CAPEX in capture-based technologies includes the costs of materials, equipment, and installation, while OPEX includes the costs of materials, maintenance, and energy. Cost drivers are identified by sensitivity analysis, and their impact on financial performance is assessed. For example, for a 25% increase in maintenance costs, sensitivity analysis predicts how the NPV or PBP will be affected. This information allows stakeholders to efficiently identify resources and prioritize cost-saving measures [26, 28, 29].

Scale-up is an important consideration in the techno-economic evaluation of CO_2 absorbers. In the transition from laboratory-scale to industrial-scale production, there is often uncertainty in cost and performance. In sensitivity analysis, uncertainties are addressed by modelling the impact of scale-up factors, including increased material demand and process efficiency losses on financial results. For example, with a 10% reduction in the scale-up of the capture efficiency, the resulting increase in CO_2 capture costs can be estimated by sensitivity analysis. This approach is done by proactively reducing risks and ensuring the viability of the technology during commercialization.

In financial analysis for technical-economic evaluations of CO_2 absorbers, sensitivity analysis is a vital tool. Sensitivity analysis provides a systematic assessment of how changes in key parameters affect technical and financial results, in terms of risk, performance, and feasibility. This analytical approach to CO_2 absorption technologies ensures that economic considerations are aligned with environmental objectives and leads to the development and proposal of cost-effective solutions.

Sensitivity analysis for financial and techno-economic evaluations uses formulas to quantify the relationship between variables and their impact on the results. The following are some of the key formulas used in sensitivity analysis:

- **Sensitivity Coefficient**
  The *sensitivity coefficient* (Eq. 13.5) measures the percentage change in an outcome relative to a percentage change in a specific input variable [29, 40]:

$$(S)=\frac{\frac{\Delta Y}{Y}}{\frac{\Delta X}{X}} \tag{13.5}$$

  In this relation, ΔY represents the change in the input variable (e.g., cost of adsorbent material) and ΔY represents the change in the output variable (e.g., NPV). Also, X and Y are the initial value of the input variable and the initial value of the output variable, respectively.
  A high sensitivity coefficient indicates that the output is strongly influenced by changes in the input.
- **Levelized Cost of CO_2 Capture (LCOC)**
  LCOC (Eq. 13.6) evaluates the cost per ton of CO_2 captured, factoring in CAPEX, OPEX, and CO_2 captured:

$$\text{LCOC}=\frac{\text{Total Annualized Cost}}{\text{TotalCO}_2\text{ Captured}\left(\frac{\text{tons}}{\text{year}}\right)} \tag{13.6}$$

  In Eq. (13.6), the total annualized cost is formulated as Eq. (13.7):

$$\text{Total Annualized Cost} = \text{CAPEX}\times\text{CRF}+\text{OPE} \tag{13.7}$$

The capital recovery factor of Eq. (13.7) is obtained using Eq. (13.8):

$$CRF = \frac{r(1+r)^n}{(1+r)^n - 1} \tag{13.8}$$

In Eq. (13.8), n and r represent the plant lifetime in years and the discount rate, respectively. Sensitivity analysis explores how changes in CAPEX, OPEX, or adsorbent performance influence the LCOC.

- **Break-Even Analysis**
  Sensitivity analysis often determines the break-even point (Eq. 13.9) for variables such as material costs or energy prices:

$$\text{Break-Even Point}=\frac{\text{Fixed Cost}}{\text{Unit Price-Variable Cost per Unit}} \tag{13.9}$$

  This helps estimate the minimum performance or cost targets required for the project to remain viable.
- **Percentage Impact Formula**

The recent impact formula (Eq. 13.10) is used to quantify how a particular input variable affects the outcome:

$$\text{Percentage Impacy} = \frac{\Delta Y}{Y} \times 100 \tag{13.10}$$

For example, if increasing energy costs by 10% results in a 5% increase in operating costs, the percentage impact reflects the sensitivity of operating costs to energy prices.

These formulas, combined with scenario modeling, enable a quantitative evaluation of how variations in parameters like adsorbent costs, energy requirements, and market conditions affect financial and technical outcomes [39–41].

13.3.3 Environmental Impact Assessment

EIA involves a systematic process for predicting, identifying, and evaluating the potential environmental impacts of a proposed project or development. It is a vital tool for ensuring sustainable development and helps to understand the environmental implications before the project begins. The EIA process typically includes a baseline for collecting data on current conditions, scoping to identify significant environmental concerns, and studies to predict potential impacts. Mitigation strategies are developed to minimize adverse impacts. EIA in industrial projects, such as those involving CO_2 capture technologies, assesses factors such as land use, air and water quality, resource consumption, and biodiversity. EIA supports decisions that balance economic growth with environmental protection.

13.3.3.1 Life Cycle Assessment

Life cycle assessment (LCA) is a comprehensive framework for assessing the environmental impacts of a product, process, or service over its entire life cycle. This includes stages such as raw material extraction, use, transport, manufacturing, and disposal or recycling at the end of life. LCA provides a cradle-to-grave analysis and quantifies impacts such as energy consumption, GHG emissions, waste generation, and water consumption. For example, LCA for CO_2 absorbers determines how much the environmental benefits of CO_2 absorption outweigh the impacts of producing and regenerating the absorbers. This method includes defining the impact assessment, the purpose and scope, interpreting the results, analyzing the inventory, assessing the impact, and so on. LCA helps to identify critical points where environmental impacts are concentrated and also enables targeted improvements to reduce the carbon footprint or resource consumption. It is a valuable tool for comparing alternative technologies or materials, ensuring that sustainability goals are met without unintended consequences [42].

13.3.3.2 GHG Emissions

GHGs are emitted from industrial processes, energy production, transportation, and agriculture. Methane, CO_2, and nitrous oxide are the primary GHGs. Emissions of these gases (GHGs) contribute significantly to climate change and are of great importance in environmental assessments. Quantifying and managing emissions is essential to reduce global warming and achieve net-zero goals. In projects involving CCS, a thorough analysis of the GHG emissions associated with the entire system is carried out. To reduce atmospheric CO_2 levels, GHG LCAs help determine the overall effectiveness of carbon capture technologies. Regulatory frameworks such as cap-and-trade, cap-and-trade, and carbon pricing guide the assessment and management of GHG emissions.

13.3.3.3 Waste Generation and Management

Waste generation and management in industrial projects are integral to environmental impacts. Wastes include solid, liquid, and gaseous by-products and require specific management and disposal methods. Promoting recycling, minimizing the volume of waste generated, reusing, and ensuring safe disposal to prevent environmental pollution are the objectives of waste management strategies. Waste materials in CO_2 capture processes include spent adsorbents, process residues, or wastewater. To assess the environmental impact of these wastes, we assess their potential for soil, air, and water pollution. One of the things that helps with the circular economy is life cycle approaches, which help identify steps in the process where waste generation can be reduced or eliminated. Adopting innovative technologies and complying with waste management regulations (such as biodegradable materials or waste-to-energy systems) increases the sustainability of industrial operations [42, 43].

13.3.3.4 Ecological Footprint

The ecological footprint measures the amount of land and water required to sustain biological production (to sustain human activities) and provides a quantitative indicator of environmental impact. It calculates the resources consumed and waste generated and converts them into equivalent land areas. The ecological footprint is a valuable tool for assessing the sustainability of a project or process, helping to identify whether resource use exceeds the earth's regenerative capacity. Ecological footprints for industrial projects, including CO_2 capture technologies, can quantify impacts on water consumption, land use, and biodiversity. When ecological footprint analysis is integrated into project planning, organizations can develop strategies to reduce their environmental burden (such as using renewable energy, optimizing resource efficiency, or increasing habitat restoration efforts). In this comprehensive approach, global sustainability goals are aligned with industrial practices.

13.3.3.5 Social Impact

Social Impact Assessment: This assesses how a project or process will affect the well-being of individuals and communities. These assessments consider factors such as public health, job opportunities, social equity, and community displacement. Projects that have significant environmental components (such as social impacts, carbon sequestration, and impacts) may involve changes in local labor markets, access to clean air and water, or the distribution of benefits and burdens across communities. Engaging stakeholders through participatory approaches ensures that diverse perspectives are included in decision-making, promotes social acceptance, and addresses potential conflicts. Social impact assessment is also considered (evaluating how today's projects will affect future generations, ethical dimensions, such as intergenerational equity). By incorporating social considerations into EIAs, projects can strike a balance between economic development, technological progress, and societal well-being, fostering inclusive and sustainable growth [40, 43].

13.4 CASE STUDIES: A POWER PLANT INTEGRATED WITH CO_2 CAPTURE TECHNOLOGY UTILIZING AN INNOVATIVE ABSORBENT

13.4.1 Process Design

Numerous studies have been conducted on the enhancement of CO_2 absorbents, as improvements in their performance lead to a reduction in the energy consumption associated with the CO_2 capture process. This section aims to provide a conceptual understanding of how innovative absorbents influence the techno-economics of the capture-integrated process. A case study is presented focusing on a novel absorbent known as modulated amine blend (MAB), which was developed by the research teams at the Korea Carbon Capture & Sequestration R&D Center. The regeneration energy required for MAB is estimated to be approximately 2.2 GJ/tCO_2, taking into account the full integration and optimization of the CO_2 capture process within a power plant. The design of the commercial-scale absorption process utilizing the MAB absorbent is based on thermodynamic models and the physical characteristics of MAB, as established by Kim and Lee. Additionally, a simulation of the commercial-scale absorption process has been conducted, allowing for an analysis of the process economics and sensitivities derived from the simulation results [44, 45]. A front-end engineering design (FEED) study has been conducted by the engineering firm as part of the research and development project, ensuring comprehensive technical feasibility while significantly reducing uncertainties associated with process design and scale-up. The power plant under consideration for the implementation of CCS is an ultra-supercritical (USC) coal-fired facility with a net power generation capacity of 522.5 MW. A new construction scenario is proposed, wherein the gross power output of the plant is increased to sustain the same net power production following the integration of CO_2 capture technology. A process optimization study has been undertaken to assess various design alternatives for potential enhancements, which include the incorporation of intercooling, the use of rich solvent split flow, and the reconfiguration of the heat exchanger network. A comprehensive evaluation of various operating conditions, including stripper pressure and lean loading, is conducted rigorously.

The process design analysis indicates a regeneration energy requirement of 2.17 GJ/tCO_2 for the MAB system. The CO_2 compression process is engineered to compress the captured CO_2 to 152.7 bar(a), which is necessary for its subsequent transportation and storage. The economic framework and principal assumptions of the TEA are detailed in Tables 13.1 and 13.2. The equipment costs associated with the TEA model have been revised to align with the findings from the FEED study, and the costs related to CO_2 capture and avoidance for the integrated process utilizing the innovative MAB absorbent have been calculated.

TABLE 13.1
Cost Foundation and Essential Premises for TEA [45]

Economic Assumptions	**Financial Base Year**		**2015**
	Chemical engineering plant cost index (CEPCI)	2015 annual index	556.8
		2011 annual index	585.7
		2007 annual index	525.4
	Capacity factor (%)		85
	Plant lifetime (yr)		30
Total plant cost (TPC)	EPC and home office cost (% of BEC)		10
	Process contingency	CO_2 capture process (% of BEC)	20
		Coal-fired power plant (% of BEC)	0
	Project contingency	CO_2 capture process and compression process (% of BEC)	22
		Coal-fired power plant (% of BEC)	8–22
Total overnight cost (TOC)	Preproduction cost	Number of months of all labor	6
		Number of months of maintenance materials cost	1
		Number of months non-fuel consumables cost	1
		Percentage of one-month fuel cost at 100% CF	25
		miscellaneous (% of TPC)	2

(Continued)

TABLE 13.1 (*Continued*)
Cost Foundation and Essential Premises for TEA [45]

Economic Assumptions	Financial Base Year		2015
	Inventory capital	Number of days of supply of fuel and consumables at 100% CF	60
		Spare parts (% of TPC)	0.5
	Land costs (106 USD2011)		0.9
	Other owner's costs (% of TPC)		15
	Financing costs (% of TPC)		2.7
Fixed operating costs	Operating labor rate (base, USD2011/h)		39.7
	Operating labor requirements per shift (operator, skilled operator, foreman, etc.)		14.0 (w/o CCS)16.3 (w/ CCS)
	Administrative and support labor (% of O&M labor cost)		25
	Property taxes and insurance (% of TPC)		2
Variable operating costs	Maintenance material costs (% of TPC)		1
	Unit costs of water, chemicals, waste disposal, and other		—

Note: CO_2 transportation and storage cost is not considered.

TABLE 13.2
Economic Foundation and Principal Assumptions for TEA of the CO_2 Capture and Compression Process [45]

Economic Assumptions	Financial Cost Base Year		2015
	Chemical engineering plant cost index (CEPCI)	2015 annual index	556.8
		2009 annual index	521.9
		2006 annual index	499.6
		2002 annual index	395.6
		2001, May–September	397.0
		2000, January	391.1
Bare erected cost (BEC)	Installation materials and labor cost (% of equipment purchased cost)		118 (capture)
			50 (compression)
Variable operating cost	MEA (USD2006/kg)		2.09
	Unit costs of consumable		—

TABLE 13.2 (*Continued*)
Economic Foundation and Principal Assumptions for TEA of the CO_2 Capture and Compression Process [45]

Economic Assumptions	Financial Cost Base Year		2015
Purchased equipment cost	Units of capacity	Correlation	
Column (absorber and stripper) (USD2001)	Diameter, *D* (m)	*H* × (100.5633 × (log *D*)²+1.0566 × (log *D*) + 3.8057)	
	Column height, *H* (m)		
Column packing (USD2009)	Packed volume, *V* (m³)	1700 × V	
Heat exchanger (USD2000)	Area, *A* (m²)	94,093 + 1127 × A0.98	
Pump (USD2001)	Flow rate, *F* (m³/h)	100.2468 × (log *F*)2– 0.5966 × log *F* + 3.9213	
Turbine (USD2002)	Power, *P* (kWe)	100.7191 × log *P* + 3.0574	
Compressor (USD2002)	Power, *P* (kWe)	100.9518 × log *P* + 2.9184	
Blower (USD2002)	Flow rate, *F* (m³/h)	100.6126 × log *F* + 4.6614	

TABLE 13.3
The Expense Associated with the Integrated Process Utilizing Innovative Monoclonal Antibody (MAB) Technology for Carbon Dioxide Capture [45]

Power Plant	Pulverized-Coal-Fired USC Power Plant	Pulverized-Coal-Fired USC Power Plant
CO_2 capture process	Adsorption-based CO_2 capture with MAB absorbent	w/o CO_2 capture
CO_2 capture rate (%)	90.0	—
Gross power (MWe)	608.0	550.0
Net power (MWe)	522.5	522.5
Annualized CAPEX (TOC) (106 USD2015/y)	270.2	158.3
OPEX (106 USD2015/y)	206.4	155.1
Cost of electricity (USD2015/MWh)	122.6	80.6
CO_2 capture cost (USD2015/tCO_2)	47.2	—
CO_2 avoidance cost (USD2015/tCO_2)	60.3	—

A sensitivity analysis has also been performed to assess the economic implications of cost parameters on the expenses associated with CO_2 capture and avoidance. Key process data and the calculated cost outcomes are presented in Table 13.3. The cost of CO_2 capture is determined to be 47.2 USD_{2015}/tCO_2, reflecting a reduction of 14.8% from the 55.4 USD_{2015}/tCO_2 reported for the CANSOLV® system by DOE/NETL.

Meanwhile, the cost of CO_2 avoidance is calculated at 60.3 USD_{2015}/tCO_2, which is 15.7% lower than the CANSOLV system's cost of 71.5 USD_{2015}/tCO_2.

13.4.2 Sensitivity Analysis and Localization

Economic assumptions and the cost framework are primarily derived from the baseline study conducted by DOE/NETL, which encapsulates the practices and conditions prevalent in process engineering and project management within the United States. Given that the industrial landscape for project implementation is highly dependent on location, it is essential to localize various factors and/or cost parameters utilized in the TEA. For instance, economic indicators such as capacity factor, interest rate, plant lifespan, and contingency levels exhibit variability across different regions; similarly, costs associated with equipment, consumables, fuel, and labor can vary significantly by locality. In conjunction with the sensitivity analysis, the TEA outcomes are revised using localized economic parameters. For the sensitivity analysis, six economic parameters are identified, with their base case levels presented in Table 13.4 [45].

a Average project contingency of the entire integrated process.
b Value of operating labor rate.

The costs of electricity (COE), CO_2 capture, and avoidance are recalculated as each parameter is adjusted within a range of −15% to +15%, with the results illustrated in Figure 13.3 [45] and presented in Table 13.5.

The cost of CO_2 capture for the MAB-based process is determined to be 25.7 USD2015 per ton of CO_2 when localized. To enable a fair comparison of CO_2 capture costs, the cost associated with the CANSOLV system has been recalibrated based on localized parameters, resulting in a figure of 30.7 USD2015/tCO_2. This represents a 44.6% decrease relative to the cost derived from the DOE/NETL framework. In addition, the CO_2 capture cost of the innovative MAB-based system is found to be 16.3% lower than that of the CANSOLV system, indicating a comparable reduction when assessed against the DOE/NETL cost basis. Moreover, the economic implications of CO_2 emissions are taken into account, and their effects on the COE, as well as the costs associated with CO_2 capture and avoidance, are examined, as illustrated in Figure 13.4 [45]

TABLE 13.4
Economic Variables Chosen for Sensitivity Analysis [45]

Parameter	Value (Base Case Scenario)
Capacity	85.0%
Discount	0.120
Plant life time	30 years
Project contingency[a]	16.6%
Fuel cost	71.8 USD2015/ton
O & M labor cost[b]	37.7 USD2015/h

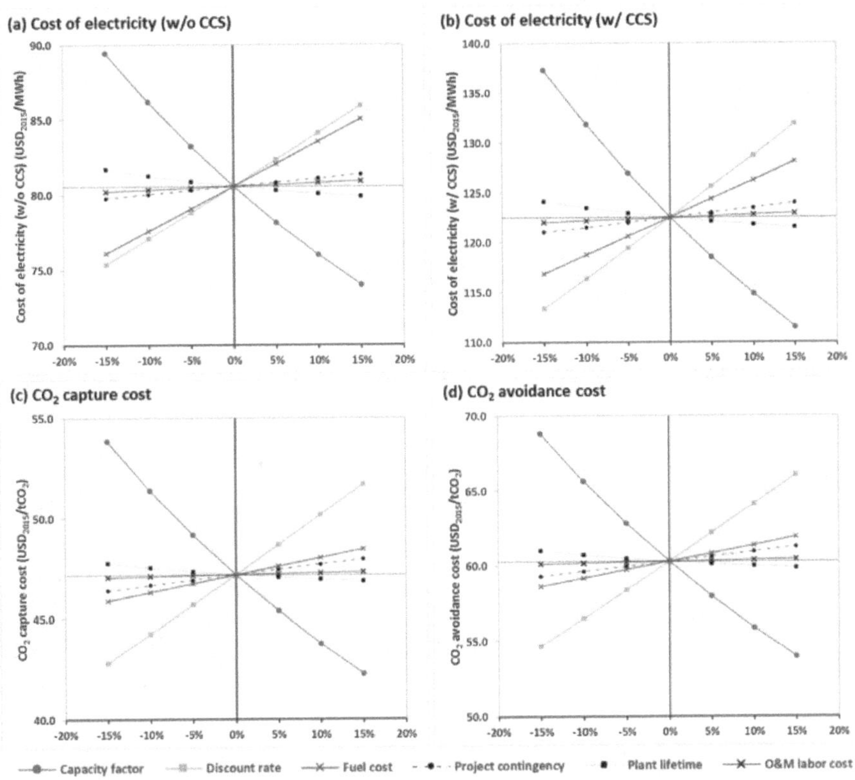

FIGURE 13.3 The outcomes of the sensitivity analysis are as follows: (a) cost of energy (COE) in the absence of carbon capture and storage (CCS); (b) COE with CCS; (c) cost associated with CO_2 capture; and (d) cost related to CO_2 avoidance [45] are significantly affected by the capacity factor (CF) and the discount rate, whereas the influence of the other four parameters is considerably less pronounced. To incorporate the effects of localization on the TEA results, the economic assumptions and cost basis have been modified to align with the engineering context of this study, specifically within the Republic of Korea. The findings are presented in Table 13.5 [45].

The DOE/NETL analysis indicates that the MAB system becomes economically feasible when a CO_2 emission penalty exceeds 58.3 USD2015 per ton of CO_2. In contrast, the CANSOLV system is deemed economically appealing when the penalty for CO_2 emissions reaches 69.2 USD2015 per ton of CO_2. This underscores the significance of advancements in CO_2 capture technology concerning its economic viability.

13.4.3 The Economic Implications Arising from the Possible Improvement of Carbon Dioxide Capture

Understanding the economic advantages that can be derived from potential advancements in CO_2 capture technologies is essential for systematically identifying priority areas for further research and development. In the case study discussed

TABLE 13.5
TEA Outcomes Based on DOE/NETL and Localized Assessments [45]

		DOE/NETL Basis			Localized Basis		
CO_2 capture technology		MAB	CANSOLV	MEA (30 wt%)	MAB	CANSOLV	MEA (30 wt%)
Gross power (MWe)		608.0	641.3	687.1	608.0	641.3	687.1
Net power (MWe)		522.5	550.0	549.5	522.5	550.0	549.5
w/o CCS	Annualized CAPEX (TOC) (106 USD2015/y)	158.3	152.0	155.3	81.7	78.9	80.5
	OPEX (106 USD2015/y)	155.1	168.4	173.6	146.0	160.4	164.1
	Cost of electricity (USD2015/ MWh)	80.6	78.3	80.3	54.1	54.0	55.2
w/ CCS	Annualized CAPEX (TOC) (106 USD2015/y)	270.2	281.1	323.3	135.1	141.1	162.7
	OPEX (106 USD2015/y)	206.4	237.6	268.8	188.7	216.9	243.2
	Cost of electricity (USD2015/ MWh)	122.5	126.7	144.7	76.9	80.8	91.6
CO_2 capture cost (USD2015/ tCO_2)		47.2	55.4	62.8	25.7	30.7	35.5
CO_2 avoidance cost (USD2015/ tCO_2)		60.3	71.5	97.7	32.8	39.6	55.3

in the preceding section, the potential economic benefits are evaluated based on three specific enhancements in absorbent-based CO_2 capture processes: separation efficiency, corrosiveness, and degradation rate. Due to the corrosive nature of amine components, stainless steel (SS) and alloy steel with a corrosion allowance have been chosen as the materials for the absorption-based CO_2 capture process. Typically, equipment constructed from SS is approximately 1.5 to 2.5 times more costly than that made from carbon steel (CS) [45, 46]. By improving the corrosiveness of the absorbent, it becomes feasible to substitute the primary equipment of

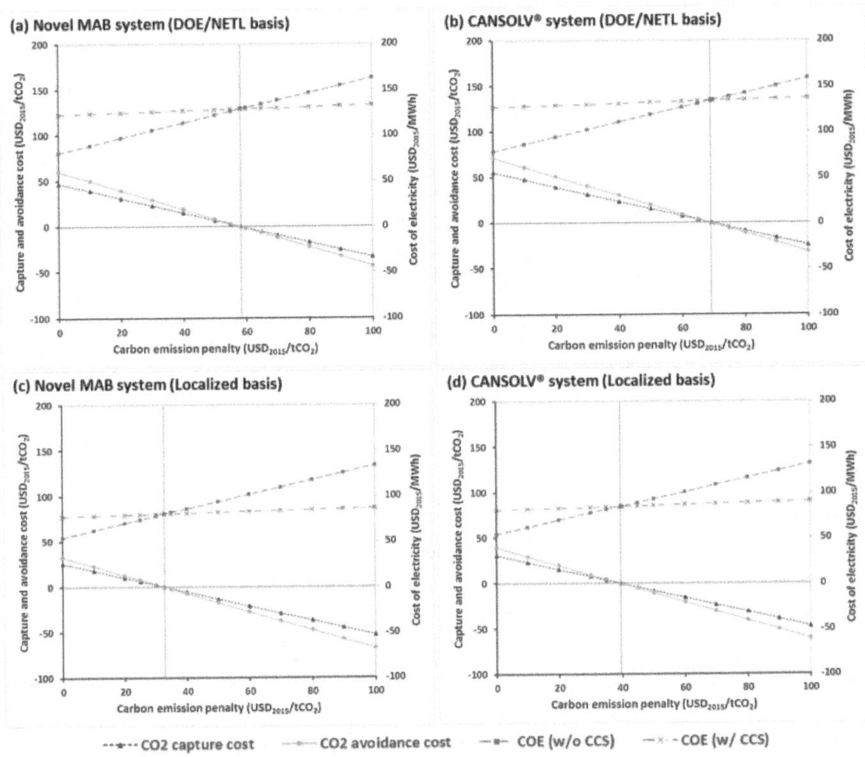

FIGURE 13.4 The responsiveness of carbon emission penalties in relation to capture costs and the cost of electricity (COE) is examined through the following systems: (a) an innovative MAB system based on DOE/NETL standards; (b) the CANSOLV system also aligned with DOE/NETL standards; (c) an innovative MAB system utilizing a localized framework; and (d) the CANSOLV system employing a localized framework [45]

the CO_2 capture process from SS to CS, with the relevant TEA results presented in Table 13.6 [45].

The materials for the primary equipment, specifically the absorber, stripper, lean/rich heat exchanger, and amine cooler should be substituted with carbon steel (CS), so the cost of CO2 capture would be reduced to 44.8 USD2015 per ton of CO2, representing a 5% decrease compared to the baseline scenario. Furthermore, if CS were utilized for all equipment, excluding the package-based units, the CO_2 capture cost would further decline to 44.2 USD2015 per ton of CO_2, which is 6% lower than the baseline case. Additionally, this analysis examines how improvements in absorption performance can lead to reductions in CO_2 capture and avoidance costs. Enhanced absorbents with superior separation capabilities increase the absorption rate and the necessary contact time for flue gas treatment, resulting in a decrease in

TABLE 13.6
Properties of TEA Are Altered due to Enhancements in Its Absorbent Corrosivity [45]

Improvement of Absorbent Corrosivity	DOE/NETL Basis			Localized Basis		
	CS for Entire System	CS for Major Equipment	Base Case	CS for Entire System	CS for Major Equipment	Base Case
Material for major equipment of CCS process	CS	CS	Design material	CS	CS	Design material
Material for rest of CCS process	CS	CS	Design material	CS	CS	Design material
w/ CCS Annualized CAPEX (TOC)(106 USD2015/y)	261.0	262.9	270.2	130.0	131.1	135.1
w/ CCS OPEX (106 USD2015/y)	205.1	205.4	206.4	188.5	188.5	188.7
w/ CCS Cost of electricity (106 USD2015/y)	119.8	120.4	122.5	75.6	75.9	76.9
CO_2 capture cost (USD2015/tCO_2)	44.2	44.8	47.2	24.3	24.6	25.7
CO_2 avoidance cost (USD2015/tCO_2)	56.4	57.2	60.3	31.0	31.4	32.8

TABLE 13.7
TEA Undergoes Modifications due to Enhancements in Absorbent Efficiency and Advancements in Equipment Technology [45]

Improvement of Adsorption Performance and Equipment Sophistication	DOE/NETL Basis			Localized Basis		
	Colum Diameter and Height Decreased	Colum Height Decreased	Base Case	Colum Diameter and Height Decreased	Colum Height Decreased	Base Case
Relative diameter	66.7%	100%	100%	66.7%	100%	100%
Relative height	66.7%	100%	100%	66.7%	100%	100%
w/ CCS Annualized CAPEX (TOC)(106 USD2015/y)	264.5	267.5	270.2	132.0	133.6	135.1
w/ CCS OPEX (106 USD2015/y)	205.6	206.0	206.4	188.5	188.6	188.7
w/ CCS Cost of electricity (106 USD2015/y)	120.8	121.7	122.5	76.1	76.5	76.9
CO_2 capture cost (USD2015/tCO_2)	45.3	46.3	47.2	24.8	25.3	25.7
CO_2 avoidance cost (USD2015/tCO_2)	57.9	59.2	60.3	31.7	32.3	32.8

TABLE 13.8
TEA Undergoes Modifications due to Enhancements in Absorption Loss and Degradation [45]

	DOE/NETL Basis			Localized Basis		
Improvement of Absorbent Loss and Degradation	**No Degradation and Loss**	**50% Degradation and Loss**	**Base Case**	**No Degradation and Loss**	**50% Degradation and Loss**	**Base Case**
Relative absorbent makeup rate	0%	50%	100%	0%	50%	100%
w/ CCS annualized CAPEX (TOC) (106 USD2015/y)	270.1	270.1	270.2	135.1	135.1	135.1
w/ CCS OPEX (106 USD2015/y)	203.8	205.1	206.4	185.8	187.3	188.7
w/ CCS cost of electricity (106 USD2015/y)	121.8	122.1	122.5	76.2	76.5	76.9
CO_2 capture cost (USD2015/tCO_2)	46.4	46.8	47.2	24.9	25.3	25.7
CO_2 avoidance cost (USD2015/tCO_2)	59.3	59.8	60.3	31.8	32.2	32.8

the dimensions of both the absorber and the stripper. Consequently, size reductions in these components, relative to the baseline scenario, are assessed to determine the economic implications of improved absorber performance on capture costs, as detailed in Table 13.7.

When the height of both the absorber and the stripper is set at 66.7% of the baseline scenario, the cost of CO_2 capture is lowered to 46.3 USD2015 per ton of CO_2. Furthermore, if the diameter of both columns is also reduced by 66.7%, the cost further decreases to 45.3 USD2015 per ton of CO_2, representing a 4% reduction compared to the baseline scenario. Table 13.8 illustrates the potential economic advantages gained from enhancing the absorbent, taking into account degradation and loss rates, with two scenarios analyzed: a 50% and a 100% reduction from the baseline. The resulting CO_2 capture costs range from 46.4 to 46.8 USD2015 per ton of CO_2. Consequently, the CO_2 capture cost can decrease by as much as 6.4% relative to the baseline, which is approximately 45 USD2015 per ton of CO_2, if all possible enhancements are implemented alongside a reduction in the regeneration energy required for absorbents.

13.5 CONCLUSION

This study provides a robust TEA of CO_2 adsorbents, with a particular focus on revenue streams, financial analysis, EIA, and real-world case studies. These facets collectively underscore the transformative potential of CO_2 capture technologies in addressing climate change and advancing sustainable industrial practices.

Revenue streams are pivotal in ensuring the financial viability of CO_2 capture technologies. The monetization of captured carbon through utilization pathways such as EOR, chemical synthesis, and carbon credits creates diverse economic opportunities. EOR serves as a dual-purpose mechanism, generating economic returns while sequestering CO_2. Similarly, the conversion of CO_2 into valuable chemicals like methanol and polymers not only offsets costs but also drives circular carbon economies. The integration of carbon trading frameworks, including cap-and-trade systems, further enhances the financial appeal of these technologies. These systems reward organizations for reducing emissions, making CO_2 capture an integral part of decarbonization strategies. In the financial analysis, metrics such as NPV, IRR, and PBP demonstrate the economic feasibility of CO_2 capture projects. The use of innovative adsorbents, such as MAB, has shown significant cost reductions compared to traditional methods. For example, the localized analysis of MAB technology revealed CO_2 capture costs as low as \$25.7 per ton under optimized conditions, emphasizing its economic competitiveness. Sensitivity analyses highlight how variables like energy costs, plant capacity, and carbon pricing influence overall profitability. These insights inform stakeholders about risk mitigation and resource allocation, ensuring the adaptability of these technologies to changing market and policy landscapes. EIAs complement the financial analysis by evaluating the sustainability of CO_2 capture systems. LCAs reveal the trade-offs between environmental benefits and the ecological footprint of adsorbent production. While advanced materials like MOFs and MXenes offer high efficiency, their synthesis processes require optimization to minimize resource consumption and waste generation. Additionally, GHG emission assessments demonstrate that CO_2 capture technologies significantly reduce net emissions, contributing to global climate goals. Waste management strategies and ecological footprint analyses further reinforce the importance of integrating sustainability into the design and implementation of these systems. The case studies examined within this analysis provide real-world evidence of the efficacy of CO_2 capture technologies. For instance, the integration of MAB technology into a coal-fired power plant demonstrated a CO_2 capture rate of 90% while maintaining economic viability. Process optimizations, such as intercooling and heat exchanger reconfiguration, reduced energy requirements, and operating costs, set benchmarks for future industrial applications. Sensitivity analyses conducted on these case studies identified key factors influencing performance and cost, offering a roadmap for scaling up these technologies while addressing operational challenges. The synergy between revenue generation, financial viability, environmental stewardship, and practical implementation defines the success of CO_2 capture technologies. This study underscores the importance of advancing innovative adsorbents, refining TEA methodologies, and fostering cross-sector collaborations to achieve widespread adoption. By integrating economic, environmental, and technical considerations, CO_2 capture systems can transition from emerging innovations to indispensable tools for global decarbonization, aligning with the overarching goals of sustainable development and climate resilience.

REFERENCES

1. Pan, H., et al., Emerging porous materials for carbon dioxide adsorptive capture: progress and challenges. *Materials Chemistry Frontiers*, vol. 7, no. 24, pp. 6463–6482, 2023.
2. Ren, F. and W. Liu, Review of CO_2 adsorption materials and utilization technology. *Catalysts*, 2023. 13(8): p. 1176.
3. Ochedi, F.O., Y. Liu, and Y.G. Adewuyi, State-of-the-art review on capture of CO_2 using adsorbents prepared from waste materials. *Process Safety and Environmental Protection*, 2020. 139: p. 1–25.
4. Hynes, N.R.J., et al., Modern enabling techniques and adsorbents based dye removal with sustainability concerns in textile industrial sector-A comprehensive review. *Journal of Cleaner Production*, 2020. 272: p. 122636.
5. Chowdhury, S., G.K. Parshetti, and R. Balasubramanian, Post-combustion CO_2 capture using mesoporous TiO2/graphene oxide nanocomposites. *Chemical Engineering Journal*, 2015. 263: p. 374–384.
6. Zimmermann, A.W., et al., Techno-economic assessment guidelines for CO_2 utilization. *Frontiers in Energy Research*, 2020. 8: p. 5.
7. Swanson, R.M., et al., Techno-economic analysis of biomass-to-liquids production based on gasification. *Fuel*, 2010. 89: p. S11–S19.
8. Wright, M.M., et al., Techno-economic analysis of biomass fast pyrolysis to transportation fuels. *Fuel*, 2010. 89: p. S2–S10.
9. Mirkouei, A., et al., A review and future directions in techno-economic modeling and optimization of upstream forest biomass to bio-oil supply chains. *Renewable and Sustainable Energy Reviews*, 2017. 67: p. 15–35.
10. Van der Spek, M., S. Roussanaly, and E.S. Rubin, Best practices and recent advances in CCS cost engineering and economic analysis. *International Journal of Greenhouse Gas Control*, 2019. 83: p. 91–104.
11. Shi, Y., R. Ni, and Y. Zhao, Review on multidimensional adsorbents for CO_2 capture from ambient air: Recent advances and future perspectives. *Energy & Fuels*, 2023. 37(9): p. 6365–6381.
12. Lu, S., et al., Three birds with one stone: Designing a novel binder-free monolithic zeolite pellet for wet VOC gas adsorption. *Chemical Engineering Journal*, 2022. 448: p. 137629.
13. Miglio, V., *Optimization of porous materials for environmental purposes*. PhD thesis, University of Eastern Piedmont, Italy, 2023.
14. Low, M.-Y.A., et al., Analytical review of the current state of knowledge of adsorption materials and processes for direct air capture. *Chemical Engineering Research and Design*, 2023. 189: p. 745–767.
15. Mandade, P. and Y.M. Nimdeo, Techno-economic assessment of biofuel production using thermochemical pathways, in *Biofuels and Bioenergy*. 2022, Elsevier. p. 653–671.
16. Doanh, N.K., L. Do Dinh, and N.N. Quynh, Tea farmers' intention to participate in Livestream sales in Vietnam: The combination of the Technology Acceptance Model (TAM) and barrier factors. *Journal of Rural Studies*, 2022. 94: p. 408–417.
17. Bridgwater, A., A. Toft, and J. Brammer, A techno-economic comparison of power production by biomass fast pyrolysis with gasification and combustion. *Renewable and Sustainable Energy Reviews*, 2002. 6(3): p. 181–246.
18. Elliott, C.D., G. Murray, and L. Pearson, *Differential Ability Scales*. Administration and scoring manual. San Antonio, TX: The Psychological Corporation, 1990.
19. Krasovskii, A., et al., Flexible options for greenhouse gas-emitting energy producer. *Energies*, 2019. 12(19): p. 3792.

20. Van Dael, M., et al., A techno-economic evaluation of a biomass energy conversion park. *Applied Energy*, 2013. 104: p. 611–622.
21. Collins, A.J., et al., An improvement selection methodology for key performance indicators. *Environment Systems and Decisions*, 2016. 36: p. 196–208.
22. Chan, A.P. and A.P. Chan, Key performance indicators for measuring construction success. *Benchmarking: An International Journal*, 2004. 11(2): p. 203–221.
23. Perrenoud, A., et al., Using best-value procurement to measure the impact of initial risk-management capability on qualitative construction performance. *Journal of Management in Engineering*, 2017. 33(5): p. 04017019.
24 Van der Spek, M., et al., Uncertainty analysis in the techno-economic assessment of CO_2 capture and storage technologies. Critical review and guidelines for use. *International Journal of Greenhouse Gas Control*, 2020. 100: p. 103113.
25. Chauvy, R. and L. Dubois, Life cycle and techno-economic assessments of direct air capture processes: An integrated review. *International Journal of Energy Research*, 2022. 46(8): p. 10320–10344.
26. Mac Dowell, N., et al., The role of CO_2 capture and utilization in mitigating climate change. *Nature Climate Change*, 2017. 7(4): p. 243–249.
27. Fu, L., et al., Research progress on CO_2 capture and utilization technology. *Journal of CO_2 Utilization*, 2022. 66: p. 102260.
28. Goren, A.Y., et al., Recent developments on carbon neutrality through carbon dioxide capture and utilization with clean hydrogen for production of alternative fuels for smart cities. *International Journal of Hydrogen Energy*, 2024. 79: p. 551–578.
29. Lv, Z., et al., Recent progress and perspective on integrated CO_2 capture and utilization. *Current Opinion in Green and Sustainable Chemistry*, 2023. 40: p. 100771.
30. Shogren, J., *Encyclopedia of Energy, Natural Resource, and Environmental Economics*. 2013, Newnes.
31. Tsai, W.-H., *Carbon Emission Reduction—Carbon Tax, Carbon Trading, and Carbon Offset*. 2020, MDPI. p. 6128.
32. Duan, Y., et al., Research on the impact of various emission reduction policies on china's iron and steel industry production and economic level under the carbon trading mechanism. *Energies*, 2019. 12(9): p. 1624.
33. Bochon, K. and T. Chmielniak, Energy and economic analysis of the carbon dioxide capture installation with the use of monoethanolamine and ammonia. *Archives of Thermodynamics*, 2015: p. 93–110.
34. Jouny, M., W. Luc, and F. Jiao, General techno-economic analysis of CO_2 electrolysis systems. *Industrial & Engineering Chemistry Research*, 2018. 57(6): p. 2165–2177.
35. Rajabi, M.M., et al., Probabilistic net present value analysis for designing techno-economically optimal sequential CO_2 sequestration and geothermal energy extraction. *Journal of Hydrology*, 2022. 612: p. 128237.
36. Sukor, N.R., et al., Techno-economic analysis of CO_2 capture technologies in offshore natural gas field: Implications to carbon capture and storage in Malaysia. *Processes*, 2020. 8(3): p. 350.
37. Cardoso, J., V. Silva, and D. Eusebio, Techno-economic analysis of a biomass gasification power plant dealing with forestry residues blends for electricity production in Portugal. *Journal of Cleaner Production*, 2019. 212: p. 741–753.
38. Zhu, Y., et al., Techno-economic analysis of solar tower aided coal-fired power generation system. *Energies*, 2017. 10(9): p. 1392.
39. Lau, H.C., The contribution of carbon capture and storage to the decarbonization of coal-fired power plants in selected Asian countries. *Energy & Fuels*, 2023. 37(20): p. 15919–15934.
40. Law, L.C., et al., A thermodynamics model for the assessment and optimisation of onboard natural gas reforming and carbon capture. *Emission Control Science and Technology*, 2024. 10(1): p. 52–69.

41. Mohd Tamidi, A., et al., Numerical modeling and economic analysis of ultrasonic-assisted CO_2 absorption process for offshore application. *Processes*, 2023. 11(11): p. 3089.
42. Choe, C., et al., Mitigating climate change for negative CO_2 emission via syngas methanation: Techno-economic and life-cycle assessments of renewable methane production. *Renewable and Sustainable Energy Reviews*, 2023. 185: p. 113628.
43. Li, S., et al., Production of sustainable biofuels from microalgae with CO_2 bio-sequestration and life cycle assessment. *Environmental Research*, 2023. 227: p. 115730.
44. Kim, H., S.J. Hwang, and K.S. Lee, Novel shortcut estimation method for regeneration energy of amine solvents in an absorption-based carbon capture process. *Environmental Science & Technology*, 2015. 49(3): p. 1478–1485.
45. Yun, S., S.-Y. Oh, and J.-K. Kim, Techno-economic assessment of absorption-based CO_2 capture process based on novel solvent for coal-fired power plant. *Applied Energy*, 2020. 268: p. 114933.
46. Peters, M.S. and K.D. Timmerhaus, *Plant Design and Economics for Chemical Engineers*. 1968, McGraw-Hill, Inc.

14 Irradiation Effects on CO_2 Adsorbents

Shanli Nezami and Ahad Ghaemi

14.1 INTRODUCTION

One of the most vital greenhouse gases that lead to global warming is CO_2. The biggest source of CO_2 gas production is fossil fuels, which supply almost 80% of the world's energy. Considering the environmental effects that CO_2 greenhouse gas emissions bring to the environment, various ideas have been proposed for the necessity of reducing the amount of CO_2 from the earth's atmosphere [1]. Among the different technologies, adsorption has been more interesting due to its cyclability, fast kinetics, low energy consumption, and high selectivity. In the adsorption process, the engineered functional materials possess the capability to selectively adsorb or isolate particular molecules and ions. It is important to recognize that the performance of these materials is influenced not only by their structural and morphological characteristics but also by the surrounding environmental conditions, such as temperature and the medium in which they operate [2, 3]. Despite the development of advanced adsorbents that demonstrate exceptional efficacy in treating complex mixtures, their practical application remains constrained by the lengthy production processes and inconsistent material properties associated with industrial manufacturing. This limitation arises from the slow heating rates of conventional methods, such as electro-thermal heating and steam, which create a temperature gradient from the equipment's wall to the bulk liquid. Furthermore, the traditional adsorption approach for separating polar and nonpolar substances is fraught with several drawbacks, including elevated costs, sluggish processing speeds, and significant energy consumption [4]. The introduction of microwave technology holds significant promise for addressing the aforementioned challenges. This is due to the interaction between microwaves and polar materials via dipole polarization-relaxation, which facilitates the direct transfer of electromagnetic energy to specific substances, resulting in rapid and uniform heating of the material [5]. In the presence of a swiftly changing electric field (EV), molecules possessing a permanent dipole are realigned in accordance with the EV's direction, while intermolecular collisions result in the production of thermal energy. Consequently, microwaves will preferentially heat polar substances that exhibit high dielectric loss, while having minimal interaction with nonpolar substances that demonstrate low dielectric loss. This method of selective heating, which is predicated on the variance in dielectric properties, offers a viable approach for enhancing the separation of polar and nonpolar mixtures. Furthermore, the rapid and uniform heating characteristics of microwaves facilitate effective and straightforward management of synthetic

DOI: 10.1201/9781003476191-14

conditions during the synthesis of adsorbents [6]. In the last few decades, solid adsorbents, especially chemical adsorbents for CO_2 adsorption, have attracted the attention of researchers [7–9]. Usually, chemical adsorbents are stabilized by having amine functional groups and performing physical or chemical interactions on hybrid or organic solid substrates. But after several steps of adsorption/desorption with amine washing, the physically impregnated amine functional groups lose their effectiveness. In other words, despite the high adsorption capacity and stability resulting from the covalent bonding of amine functional groups to the absorbent substrate, the adsorption kinetics and efficiency of amines are reduced [10, 11]. Among amine-containing adsorbents, microporous polymers and microfibers have good performance in absorbing CO_2 gas with thermal stability, mechanical strength, high chemical resistance, and amine functional groups compared to other adsorbents [12, 13]. However, in this field, production cost and work capacity are among the things that need more development. The capacity and performance of these adsorbents increased the surface area by reducing the fiber size from micro to nanoscale and creating a three-dimensional structure with porous morphology. Among the various methods such as template synthesis and phase separation, electrospinning is the most common and diverse method of nanofiber production [14, 15]. The process of producing microfibers and nanofibers from polymer solutions under pressure, atmospheric temperature, and high EV is called electrospinning [16]. To promote the advancement of radiation technology, specifically in microwave and polymerization applications for industrial-scale adsorption, this chapter outlines the preparation of adsorbents utilizing radiation and emphasizes the enhanced mechanisms involved in the production of these materials. Initially, the fundamentals of microwave-assisted adsorption are discussed, followed by an exploration of microwave-assisted preparation techniques. Furthermore, a summary of various adsorbents developed through microwave and polymerization methods as documented in the existing literature is provided. Lastly, this review proposes future research avenues in the domain of radiation-enhanced adsorption, informed by the current state of research.

14.2 FUNDAMENTALS OF MICROWAVE-ASSISTED ADSORPTION

The straightforward design of microwave-assisted material preparation and microwave-enhanced adsorption separation necessitates a comprehensive understanding of the interactions between microwaves and materials [17]. When microwaves interact with materials during transmission, three phenomena occur, as illustrated in Figure 14.1: reflection, adsorption, and penetration. Materials are typically classified into four categories based on their response to microwave irradiation: conductors, insulators, dielectrics, and magnetic materials [18].

When microwaves interact with metal conductors, a minimal portion of the microwave energy penetrates the metal, resulting in an exponential decay of the microwaves on the surface of the conductor, a phenomenon known as skin depth. Consequently, metal conductors are frequently employed to direct and reflect microwave signals. In contrast, when microwaves encounter insulating materials, the majority of the microwaves traverse the material, with only a negligible fraction of

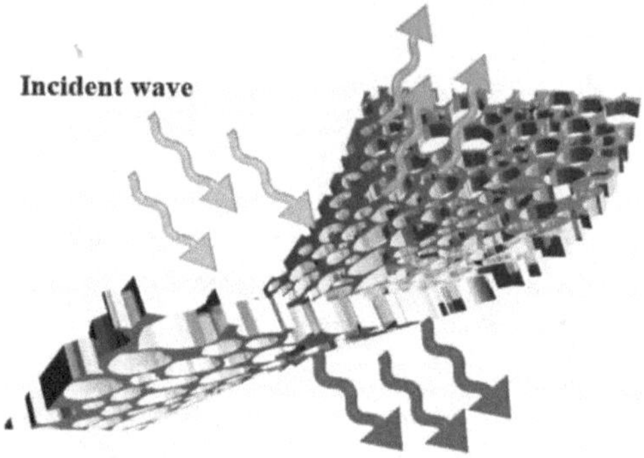

FIGURE 14.1 A schematic representation illustrating the interactions between microwave radiation and a material [17].

the energy being absorbed. Upon interacting with dielectric materials, microwaves are partially reflected, partially absorbed, and partially transmitted, with the specific mode of interaction being dependent on the dielectric characteristics of the material. The behavior of magnetic materials in response to microwaves is analogous to that of dielectric materials; however, the distinction lies in the fact that the effect of dielectric materials on microwaves is associated with the EV, whereas the influence of magnetic materials is linked to the magnetic field [19, 20]. The ongoing research regarding microwave-assisted adsorbent preparation primarily emphasizes adsorptive materials, specifically dielectric and magnetic substances. When subjected to a microwave field, these reactive components transform electromagnetic energy into thermal energy, which subsequently initiates a system response characterized by a rapid temperature increase, an accelerated reaction rate, and reduced heat loss [21]. Electromagnetic wave irradiation exerts a distinct influence on condensed matter, which is contingent upon the polarization relationship between the electromagnetic field and dielectric molecules. This includes various types of polarization such as electronic polarization, atomic polarization, dipole polarization, orientation polarization, and interface polarization. The relaxation times for electronic and atomic polarization are approximately 10^{-16} and 10^{-13} seconds, respectively, whereas the oscillation period of the microwave alternating EV spans from 10^{-9} to 10^{-12} seconds.

Electromagnetic waves at this frequency do not induce electron or atomic polarization; instead, they align with the frequency associated with dipole and interface polarization. Currently, it is understood by researchers that the interaction between microwave radiation and the majority of chemical substances is primarily governed by dipolar polarization mechanisms (Figure 14.2). In particular, when a material containing polar molecules is subjected to a high-frequency electromagnetic field, these polar molecules undergo continuous reorientation in response to the fluctuations of the electromagnetic field [17]. In this process, polar molecules engage in hundreds of millions of rapid movements each second. These molecules must swiftly navigate the interference and barriers posed by neighboring molecules to maintain

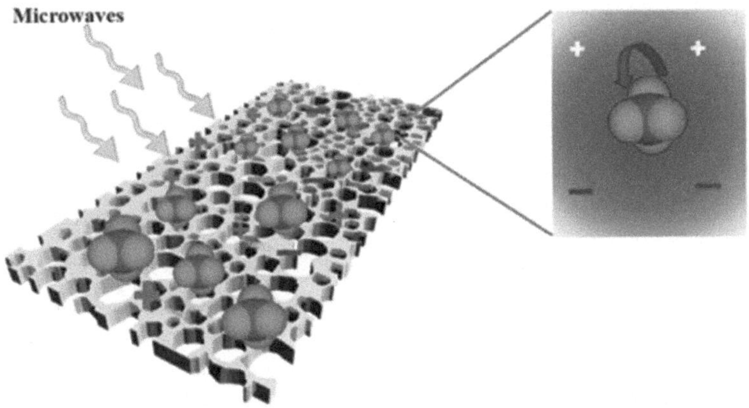

FIGURE 14.2 The reorganization of polar molecules within the electromagnetic field [17].

their regular and periodic motions. The activity of polar molecules generates a thermal effect akin to molecular friction, resulting in an increase in the temperature of the polar medium. Consequently, the microscopic magnetic product transforms the electromagnetic energy from microwaves into the kinetic energy of polar substances, which is observed macroscopically as the heating of these substances within the electromagnetic field [2, 22, 23]. The varied reactions of materials to microwave irradiation result in the selective heating property of microwaves, which are more effective on strongly polar substances compared to nonpolar or weakly polar ones. This understanding elucidates the beneficial impact of microwaves on the preparation of adsorbents and the adsorption process. Microwaves possess the ability to penetrate barriers and uniformly influence small-volume systems, as evidenced by their consistent heating in larger contexts. In hybrid systems, materials exhibiting high dielectric losses are preferentially heated, creating "hot spots," which facilitate rapid heating of the entire system through heat transfer [24]. Rapid overheating is advantageous for the growth of crystals. Additionally, consistent heating can enhance the uniformity of the porous material during its formation. In the processes associated with the synthesis of the adsorbent material, the swift heating provided by microwaves can expedite the reactions significantly. As knowledge of microwaves expands, researchers from diverse fields have observed phenomena that cannot be accounted for solely by the thermal effects of microwave irradiation [25, 26]. These phenomena are referred to as nonthermal effects. In particular, the nonthermal effect of microwaves encompasses various influences beyond thermal effects, including electric, magnetic, and chemical effects [27]. The nonthermal effect is understood as the direct interaction of microwaves with the internal energy of molecules rather than their transformation into thermal energy. This phenomenon can be seen as the interaction between the EV and specific polar molecules present in the reaction medium. Since its identification, the potential for nonthermal effects has been explored in a range of fields, including chemical reactions and materials science, encompassing areas such as the Maillard reaction, bacterial mutagenesis, protein denaturation, and polymer solubility [28,

29]. The prevailing theories regarding "nonthermal effects" primarily focus on the molecular polarization and vibration induced by the alternating EV. In the following text, we provide a summary of these theories.

- The rearrangement of the reactant molecules
- The breaking of bonds among the molecules
- The creation of an electric gradient force within the reactant (Figure 14.3)

Rapid synthetic techniques offer significant advantages for researchers, enabling them to conduct a greater number of experiments within the same timeframe, thereby enhancing their scientific creativity in the development of innovative adsorbents. Additionally, these techniques are advantageous for manufacturers, facilitating the production of highly efficient and cost-effective adsorbents for industrial use. In this context, microwave-assisted preparation has demonstrated considerable promise in achieving these objectives, as this method of remote heating directly supplies energy to the reactants. This approach minimizes the adverse effects of temperature gradients in the bulk liquid during synthesis, which can lead to improved yields in material production. With the enhancement of microwave irradiation during the reaction, certain materials exhibit a tendency toward improved performance, while the variety and yields of synthesized materials also rise [30]. The viscosity of polymer raw materials is typically high, and their thermal conductivity is limited. As a result, uneven heating frequently occurs during the synthesis of polymer materials, which can result in inconsistent properties in the final polymer product. Additionally, the adhesive reactants that cling to the walls of reactors pose challenges for cleaning. In the case of most polymerization reactions, the reaction rate is notably enhanced under microwave conditions, which also brings benefits such as minimized side reactions, increased yield, and enhanced properties of the polymer. In the process of

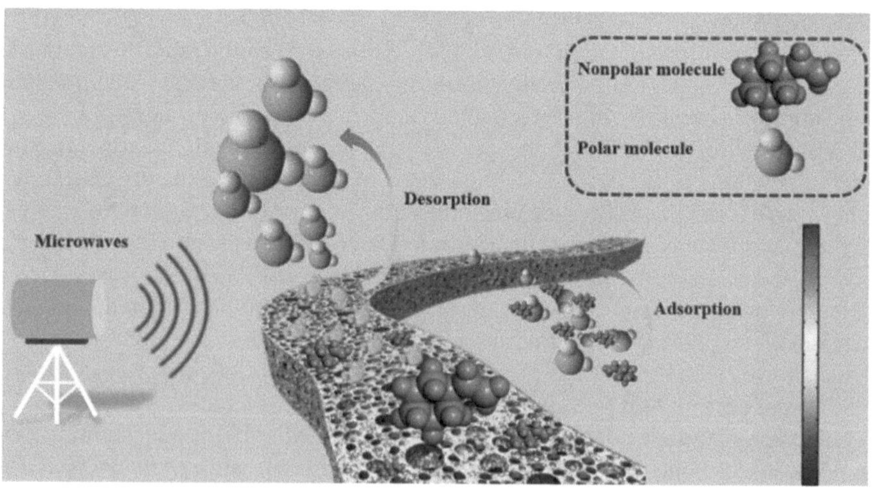

FIGURE 14.3 Steps for preparing the adsorbent with microwave radiation [17]

preparing polymer adsorbents, microwaves serve as a supplementary method to heat the reaction system and enhance the rate of the reaction [31–34].

14.3 THE SURFACE MODIFICATION BY RADIATION-INDUCED GRAFTING

Various techniques are used to modify the appropriate structure in adsorption processes. One of these techniques is radiation-induced graft copolymerization (RIG), which is used to modify the physical and chemical properties of polymeric materials without changing the intrinsic and structural properties. When polymer substrates are exposed to high-energy radiation, the bonds of the copolymerization process begin.

During this process, radicals are produced that have the ability to start a reaction with monomer molecules in different environments such as solvents and emulsions. In addition to the effect of the absorbent surface on the adsorption performance, the effect of the surface chemistry on the adsorption power cannot be ignored. The ability to absorb a molecule with an adsorbent is increased by the functional groups of the adsorbent [36, 37]. CO_2 adsorption increases with the presence of amine groups in the adsorbent. In addition to the effect of the absorbent surface on the adsorption performance, the effect of the surface chemistry on the adsorption power cannot be ignored. The ability to absorb a molecule with an adsorbent is increased by the functional groups of the adsorbent [38, 39]. The CO_2 adsorption increases with the presence of amine groups in the adsorbent. Therefore, this method is considered a cost-effective modification, does not produce toxic products, and does not destroy the environment [40].

The four main techniques for bond-forming polymerization, shown in Figure 14.4, include the following:

- UV treatment
- Electron beam
- Plasma treatment
- Gamma radiation

The degree of grafting (DG) is the desired effect of radiation on polymerization grafting reactions and is defined as the criterion of grafting efficiency. The level of

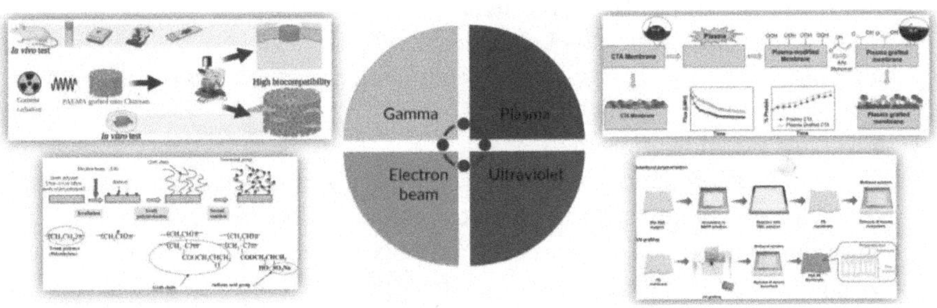

FIGURE 14.4 The various techniques of radiation-induced grafting that are frequently employed [35].

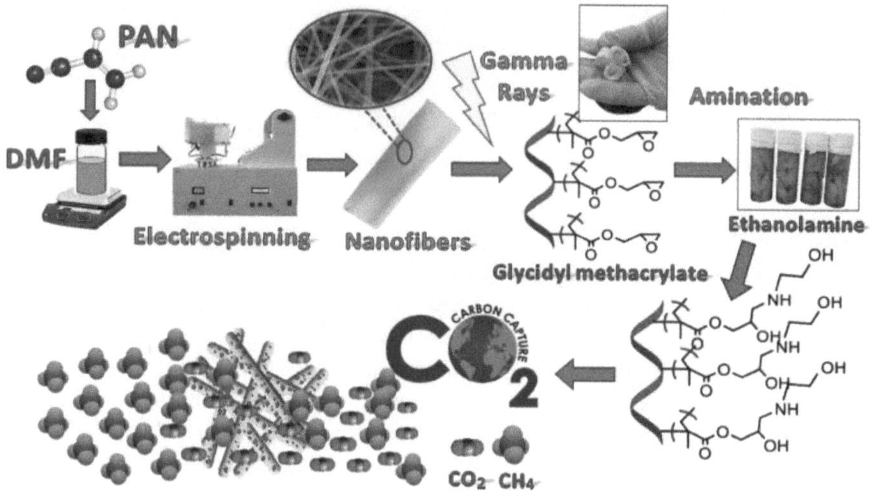

FIGURE 14.5 Schematic method for the arrangement of the polymeric adsorbents [42].

polymerization graft response is determined by the grafting efficiency. High bonding efficiency, in addition to the wide presence of monomer functions in the polymer structure, shows the superiority of functional materials [35].

- **Polymerization Technique**
  In the gamma radiation method, the cobalt-60 radiation source works successfully to form bonds and change the performance of polymers. In other words, the formation of bonds with high efficiency and the absence of catalytic contamination and other destructive initiators can be seen with the formation of instantaneous and homogeneous active radical sites as a result of bond formation due to gamma radiation. Therefore, the use of gamma radiation in the field of biomedicine is more suitable [41]. Since the radioactive element is used in the discussion of gamma radiation, the disadvantages of gamma radiation include the production of radioactive and toxic wastes that are very dangerous and toxic to human health and the environment [35]. In Figure 14.5, A.A. Tourzani et al. schematically show the synthesis of polymer adsorbents, which includes polyacrylonitrile (PAN) electrospinning, nanofiber radiation bonding, and modification method, and experimental tests with CO_2/CH_4 gases [42].
- **Electron Beam Radiation**
  Electron beam radiation is one of the effective methods to improve the performance of polymers. The formation of active sites that easily react with monomers leads to the formation of a copolymer bond as a result of electron beam irradiation to the surface of the polymer. The copolymer bond formed is introduced as a branched copolymer from the main chain [43]. One of the advantages of the electron beam method is the production of more stable products. However, the gamma-ray method has a significant advantage over

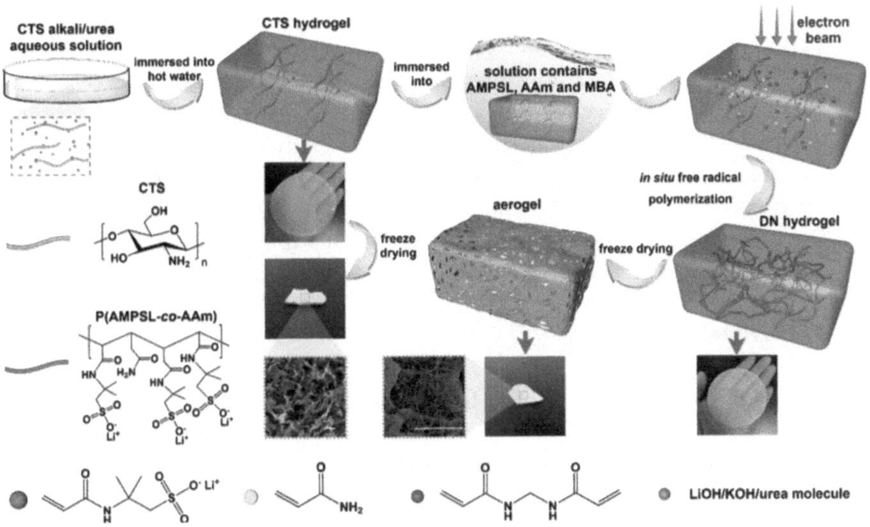

FIGURE 14.6 A schematic representation of the synthesis process for chitosan/P(AMPSL-co-AAm) hydrogel and aerogel initiated by electron beam radiation [45].

the electron beam method in improving the performance of polymers due to the high-efficiency bonding based on the copolymerization of natural polymer and polystyrene [44]. Zhiyan Liu and colleagues synthesized a novel high-strength chitosan/lithium sulfonate double network hydrogel by electron beam irradiation (Figure 14.6). Since the electron beam has a wide irradiation area and certain penetrating power, free radical polymerization can be initiated uniformly and rapidly in the hydrogel. The network structure of the hydrogel prepared by radiation polymerization is more uniform than that prepared by conventional chemical initiator polymerization. Meanwhile, the introduction of a second network to construct the double network structure does not reduce the surface area of the aerogel, which is different from the conventional method of grafting or impregnating modified porous materials. Furthermore, the produced aerogels exhibit commendable physical and chemical stability. The frozen aerogels possess a porous architecture and demonstrate the ability to adsorb CO_2, attributed to their CO_2-philic double network structure. The utilization of inexpensive raw materials and a straightforward irradiation process allows for a reduction in the cost of CO_2 adsorbents, presenting promising applications in the domain of solid CO_2 adsorbents. The chitosan hydrogel is regenerated from an aqueous alkali/urea solution, while the secondary lithium sulfonate network is initiated through in situ free radical polymerization using electron beam irradiation. The frozen aerogel is capable of CO_2 absorption [45].

- **Plasma Irradiation**

  In the last decade, the use of plasma irradiation methods to modify the functional surface of polymers has attracted the attention of researchers. In this method, modification is done without the use of solvents and in a short period of time using environmentally friendly active functional groups [35]. In 2022, Chen Gu and his colleagues successfully demonstrated, inspired by nature, the enhancement of CO_2 absorption capacity with metal-organic adsorbents under the influence of plasma irradiation technology and temperature fluctuation (Figure 14.7). In Figure 14.7a, it is shown that, during adsorption, the radiative cooling coating on the top of the adsorbents facing the sun causes the adsorbents to absorb CO_2 at a low temperature. And Figure 14.7b shows that, in the case of desorption, the adsorbents are exposed to solar radiation for solar heating to release CO_2 [46].

- **Ultraviolet Irradiation**

Based on their spectrum, ultraviolet (UV) radiation is known in three forms: UVB (290–320 nm), UVA (320–400 nm), and UVC (320–400 nm) [47]. UVC is considered the most harmful ultraviolet radiation due to its stronger energy, shorter wavelength,

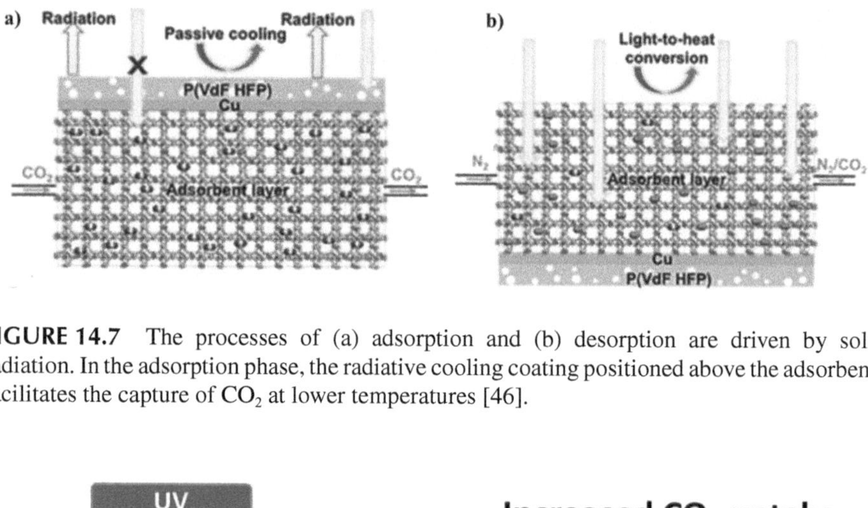

FIGURE 14.7 The processes of (a) adsorption and (b) desorption are driven by solar radiation. In the adsorption phase, the radiative cooling coating positioned above the adsorbents facilitates the capture of CO_2 at lower temperatures [46].

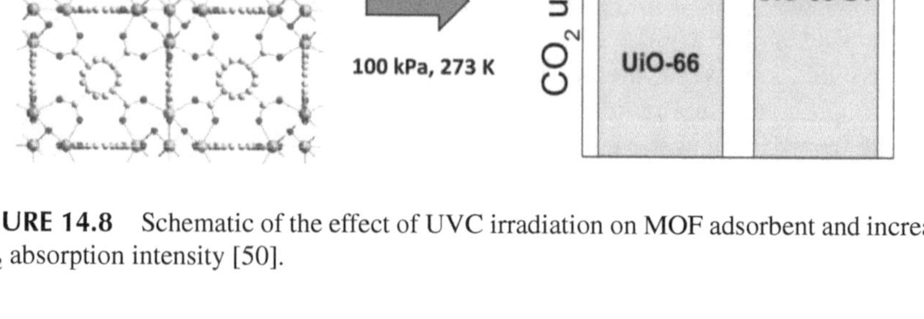

FIGURE 14.8 Schematic of the effect of UVC irradiation on MOF adsorbent and increasing CO_2 absorption intensity [50].

and the severity of damage to human life. Among these three radiations, UVB is the most appropriate choice for breaking bonds and photo destruction. Among the advantages of the ultraviolet radiation technique for modifying the properties of polymers compared to other methods are the ease and speed of operation, slow reactions, cost-effectiveness, and the possibility of being used on an industrial scale [48, 49].

J.B. Jeffries and colleagues reported in 2005 that the CO_2 adsorption capacity was increased by simple modification of the adsorbent (MOF) UiO-66 using UVC. Figure 14.8 shows that, as a result of UVC irradiation, the overall structure and crystallinity of the MOF were preserved and the modified adsorbent had a more structural complexity with a more open framework and new interaction sites, resulting in increased CO_2 adsorption [50].

14.4 GRAFTING YIELD

The grafting efficiency is reported as the ratio of monomers attached to the polymer structure. The DG percentage value is used to determine grafting efficiency. The parameters affecting DG during the polymerization process include temperature, absorbed dose, irradiation distance, irradiation time, and monomer concentration. Ochoa-Segundo and his colleagues treated poly[(R)-3-hydroxybutyric acid] with n-hydroxyethylacrylamide monomer by gamma irradiation (Cobalt-60) modified. Ethanol (EtOH), hexane (Hex), ethyl acetate (EtOAc), chloroform ($CHCl_3$), acetone (Ace), sodium acetate, silica nanoparticles, and acetic acid are used in this report. They reported that the grafting efficiency of 83.06% for ethyl acetate solvent compared to other solvents is due to the ability of ethyl acetate to produce functional active sites during the polymerization process, which is very effective on the degree of gravity grafting. Irradiation temperature and reaction time have been reported as other parameters affecting grafting efficiency. Aoki and others investigate the effect of reaction time parameters and irradiation temperature at constant solvent concentration on the formation of N-vinyl pyrrolidone (NVP) antimicrobial bonds on polyolefin fabric and iodine adsorption in the electron beam irradiation method. In this work, the authors reported that the highest grafting efficiency (98%) was obtained at an irradiation temperature of 337°C and a reaction time of 30 min [51]. In a similar work, Selambakkannu and co-workers showed that high reaction temperature accelerates the reaction kinetics and forms higher degrees of crosslinks. On the other hand, the ease of connection of activated sites on the surface will be created by cross-diffusion inside the solution due to temperature conditions [52]. Finally, based on the investigations, it can be said that among the four techniques of plasma radiation, electron beam, gamma radiation, and ultraviolet radiation that are used to improve the surface performance of polymers, gamma radiation will have a higher performance than the other methods. Also, electron beam radiation has received more attention from researchers due to the lack of radioactive waste generation in the last few decades, and UV and plasma radiation methods have been investigated by some researchers. Also, the degree of bonding is the most important parameter in evaluating the bonding during polymerization, and the performance of the formed bonds is greatly influenced by parameters such as temperature, time, distance, dose of radiation, types of solvents, and concentration of monomers. Therefore, to maximize the radiation-induced bonding efficiency, the optimal conditions for these parameters were investigated.

14.5 EXPERIMENTAL CHARACTERIZATION METHODS

14.5.1 Overview of Characterization Techniques Used to Assess the Effects of Irradiation on CO_2 Adsorbents

In a detailed report, Zahra Imanian et al. modified PAN nanofibers from industrial PAN materials using electrospinning and glycidyl methacrylate (GMA) monomer bond formation on the surface of PAN using beam bonding method, and the structural results of this absorbent and its efficiency checked in the structural examination by scanning electron microscopy (SEM) images, absorbent substrate, and synthetic nanoabsorbents have been presented in Figure 14.4.

In this report, the diameters were randomly measured using the basic software associated with the device, and in Figure 14.9a, the PAN nanofiber substrate shows an average diameter from 227 to 229.25 nm. Following the crosslinking reaction with GMA during PAN irradiation, the average fiber diameter increased to 574.4 nm, as shown in Figure 14.9b. This increase in diameter is indicative of successful GMA grafting, with the most significant diameter expansion of 323.45% DG occurring at 30 kG in the presence of 25% GMA monomer. In Figure 14.9c, the average diameter of GMA-modified PAN increased to 1 μm after an amination reaction with ethanolamine (EA). This change indicates the effective implementation of the amination reaction and the establishment of amino group layers on NF-PAN/Rad/GMA/EA. The experimental results for the optimal adsorbent (NF-PAN/Rad/GMA/EA) show the maximum DG% equal to 323.45% and the degree of amination (DA%) equal to 170.88%. Also, in this report, the highest adsorption capacity with NF-PAN/Rad/GMA/EA adsorbent is equal to 3.23 mmol/g in environmental conditions, a flow rate of 100 ml/min, and a CO_2 concentration of 5% [13].

In a related study, Ali Abbasi and his team developed a novel adsorbent featuring nanofibrous amine for CO_2 capture. This was achieved through the electrospinning of syndiotactic polypropylene (s-PP), followed by crosslinking using GMA radiation and subsequent amination with EA to facilitate crosslink formation. To analyze the structural changes resulting from radiation, they conducted various assessments, including Fourier-transform infrared (FTIR) spectroscopy and SEM. FTIR results

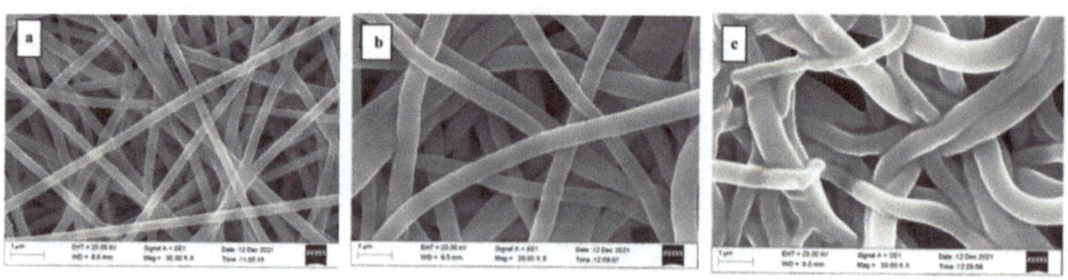

FIGURE 14.9 SEM images (a) NF-PAN; (b) NF-PAN/Rad/GMA; and (c) NF-PAN/Rad/GMA/EA [13].

validated the incorporation of GMA into the electrospun s-PP nanofibers and the subsequent modification with EA. The data illustrated (Figure 14.10) the emergence of new adsorption bands at 1725, 1255, 905, and 845 cm^{-1} post-grafting. Literature indicates that the prominent band at 1725 cm^{-1}, and the broad, weaker band at 1255 cm^{-1} correspond to the stretching vibrations of –C=O and CO-acrylate, while the smaller peaks at 905 and 845 cm^{-1} are associated with the epoxy ring. Notably, the intensity of the carbonyl band of poly(GMA) side chains at 1725 cm^{-1} remained unchanged during the amination process. However, as the DA increased from 30% (c) to 60% (d) and subsequently to 94% (e), the intensities of the bands at 905 and 845 cm^{-1}, linked to the epoxy groups of poly(GMA) side chains, diminished significantly, with these peaks nearly vanishing at 94%. Additionally, a new broadband emerged in the range of 3500–3000 cm^{-1}, attributed to the O–H stretching of the hydroxyl group around 3400 cm^{-1}, which overlaps with the N–H stretching of the secondary amine group near 3300 cm^{-1}. The C–N stretching vibration of aliphatic amines is observed between 1020 and 1250 cm^{-1}; however, this spectral region is congested with multiple peaks characteristic of the s-PP base polymer, obscuring the C–N peak. These findings align with the SEM images, confirming the successful amination of the grafted samples, which was accompanied by the opening of the polylinked epoxy (GMA) rings [12].

Anish Mathai Varghese and his team developed and examined a UV-irradiated graphene oxide foam (UV-GOF) for its capacity to absorb CO_2, focusing on the structural, morphological, and surface characteristics of both the standard GOF and the UV-treated variant. X-ray diffraction analysis revealed that the characteristic diffraction peak of GOF shifted to lower angles following UV treatment for up to 10 hours, after which the trend reversed with prolonged exposure. This initial shift

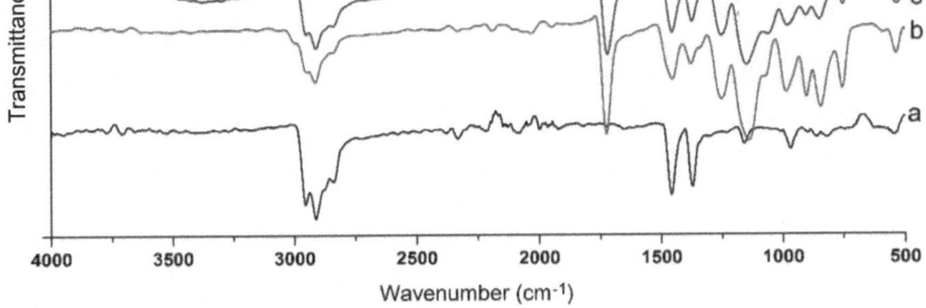

FIGURE 14.10 FTIR spectra of pristine s-PP nanofibers (a) compared with 300% grafted nanofibers (b) and fibers with 30% (c), 60% (d), and 94% (e) DA using EA [12].

suggests an increase in interlayer spacing due to oxidation and the incorporation of additional oxygen functional groups, as confirmed by FTIR and energy-dispersive spectroscopy analyses presented in this study. Conversely, the later trend indicates a partial reduction of active groups with extended treatment duration. Notably, characteristic diffraction peaks were observed at 10.04° and 10.13° for the 10UV-GOF and 15UV-GOF samples, corresponding to interlayer spacings of 8.80 and 8.73 Å, respectively. This research highlights that the structural modifications induced by UV radiation significantly enhance adsorption performance, resulting in a selectivity increase of 30 times and a CO_2 adsorption capacity that rises up to sevenfold after five hours of UV exposure [53].

Gabriela Duran-Jiménez and her team conducted a study on CO_2 adsorption using activated carbons generated through microwave processing. Their findings revealed that the hybrid carbon produced at a low temperature of 600°C exhibited the highest CO_2 adsorption capacity, reaching 225 mg/g. This outcome underscores the synergistic effect of hybrid heating in markedly improving CO_2 adsorption. Furthermore, the analysis of BET structural data indicated a linear relationship among surface area, micropore volume, adsorption intensity, and the effects of radiation on the absorbent [54].

14.6 IRRADIATION EFFECTS ON CO_2 ADSORBENTS

Adsorption of an effective technology in the laboratory and industry is in the discussion of separation and removal. One of the most important parameters affecting the adsorption performance is temperature. The intensity of adsorption of solid and liquid molecules on the adsorbent increases with increasing temperature. Therefore, the process that is observed due to microwave radiation on the absorber and the increase in adsorption intensity is as follows: In the first stage, due to the activation of the absorbent material, due to the microwave radiation, the adsorption performance is affected. In the next step, the probability of collision between the adsorbent and the adsorbed molecules increases as a result of the radiation intensifying the molecular motion of the adsorbed substances. Finally, with the structural change of absorbent materials, for example, the porosity of the material and its specific surface due to microwave radiation, the adsorption capacity increases. In several reports, the effect of microwaves on adsorption capacity has been investigated. In 2010, Yang and colleagues demonstrated the use of an amine fiber adsorbent (PAN-AF) developed through pre-irradiation copolymerization for CO_2 capture. An adsorption capacity of 6.22 mmol/g was achieved by employing a 50% concentration of allylamine monomer in conjunction with a redox system consisting of $(NH_4)_2S_2O_8$ and $NaHSO_3$. The reconstitution of this adsorbent (PAN-AF) occurs at a temperature of 100°C for a duration of 30 minutes. It is important to highlight that the allylamine monomer poses significant toxicity and environmental hazards [55]. In 2013, Zhuang and colleagues employed a PP polymer adsorbent for carbon dioxide (CO_2) capture utilizing a pre-irradiation technique involving GMA monomer, followed by modification with triethylenetetramine (TETA). At a CO_2 concentration of 20%, the adsorption capacity achieved was 4.72 mmol/g, with the reduction process conducted at 100°C for a duration of 20 minutes. Additionally, Nasef and his team demonstrated that a polyethylene (PE)/PP nonwoven polymer adsorbent, treated with electron beams, exhibited the highest adsorption capacity of 4.52 mol/kg. This adsorbent was synthesized using GMA monomer and triethylamine (TEA) [56]. In a separate investigation, polyamide 6 modified with a polyethylene imine functional group was employed for the adsorption of CO_2, achieving an adsorption capacity of 1.16 mmol/g. This capacity was diminished when subjected to a temperature of 105°C for a duration of 40 minutes [57]. In a seminal report by Mohamed Mahmoud Nasef et al. in 2015, they used glycidimethacrylate (GMA) on a PE/PP nonwoven fabric to synthesize a number of adsorbents with different applications using radiation-induced cross-linking (RIG). The functionalization of the grafted fabric was carried out by one of three functionalization reactions, modification with (1) N-dimethyl-D-glucamine, (2) triethylamine, and (3) triethylamine and alkalization with KOH. The structure and physical and chemical changes of the adsorbents were analyzed by FTIR and SEM. Figure 14.11 shows functionalization and bond formation on the absorber due to radiation.

By examining the adsorption power of these adsorbents in absorbing CO_2 from CO_2/N_2 mixtures, it has been shown that this adsorbent can be a potential alternative in industrial applications instead of granular resin, which requires further development [58]. In 2017, Young-Jung Heo et al. synthesized magnesium oxide (MgO)-modified carbon adsorbents using a nitrogen-enriched carbon precursor by microwave-assisted irradiation for CO_2 adsorption. The transformation of the spherical structure of the original material into a hybrid structure with MgO crystalline particles in a carbon matrix was performed by changing the microwave reaction time. Structural analyzes were performed to investigate the evolution of the morphology and properties of the prepared materials. The authors reported that under optimal conditions, the synthesized adsorbent shows a high adsorption of 1.22 mmol/g

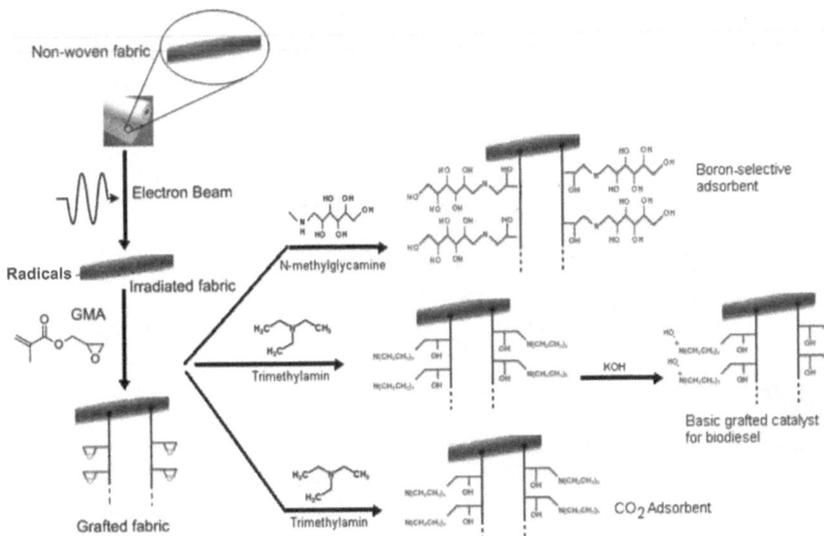

FIGURE 14.11 Functionalization and bond formation on the absorber due to radiation [58].

of CO_2 from a CO_2/N_2 mixture with 15% of CO_2. Finally, the authors report that the structural changes of MgO by microwave irradiation are the main factor in the effective adsorption of CO_2 [59]. In 2018, research was conducted on the grafting of polyethyleneimine (PEI) onto poly(acrylic acid)-polyethylene membranes. This adsorbent demonstrated a significant capacity for CO_2 capture and hydrogenation. The findings indicated that the linkages of poly(acrylic acid) increased as the density of amino groups rose [60]. In 2019, Abbasi et al. synthesized a new type of nanofibrous amine-containing adsorbent for CO_2 adsorption by electrospinning s-PP and then by UV radiation, the bond caused by GMA radiation and amination with EA was created. The adsorbents obtained were used to absorb CO_2 from CO_2/N_2 containing 5–15% CO_2 in a fixed bed adsorption column at atmospheric pressure. N. Yasandag reported the maximum adsorption capacity of 2.87 mmol/g for the sample with 300% grafting degree and 94% amination degree at 15% feed concentration at 30°C. This adsorption capacity was accompanied by high mechanical stability and excellent amine efficiency at high temperature, which indicates that the obtained fibrous adsorbent has a high potential for CO_2 adsorption [12]. In 2020, Anish Mathai Varghes et al. produced a UV-irradiated UV-GOF for CO_2 adsorption. They showed that the UV-modified GOF adsorbent, with structural and surface modifications associated with the modification, exhibits high CO_2 adsorption performance in terms of capacity, selectivity, stability, and hydrophilicity. They also showed that the selectivity at 100 mbar with 5-hour UV modification increased nearly 30 times and the adsorption capacity increased nearly 7 times. Finally, the authors have concluded that the modification of graphene oxide absorbers with ultraviolet rays can increase their adsorption performance for carbon dioxide and be exploited on an industrial scale as a low-cost technology. Figure 14.12 summarizes the modification of graphene oxide with ultraviolet rays and the results of CO_2 adsorption with that adsorbent [53].

In 2023, Imanian et al. synthesized PAN nanofibers from industrial PAN materials using the electrospinning method and then successfully created GMA monomer grafting on the surface of PAN using the radiation grafting method. Also, the modification

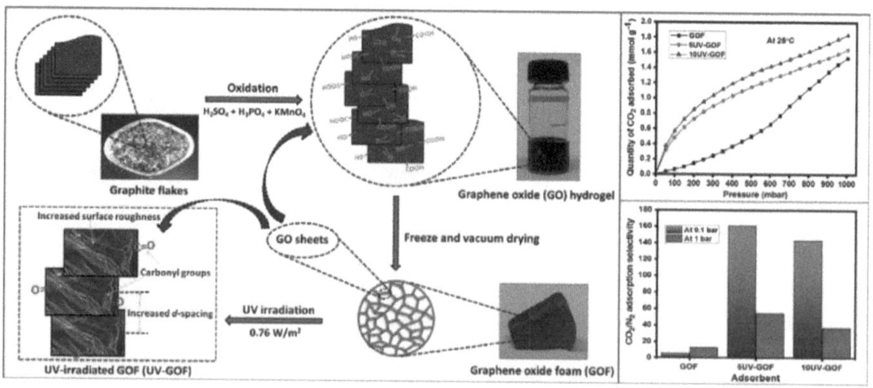

FIGURE 14.12 Summary of UV modification of graphene oxide and CO_2 adsorption with that sorbent [53].

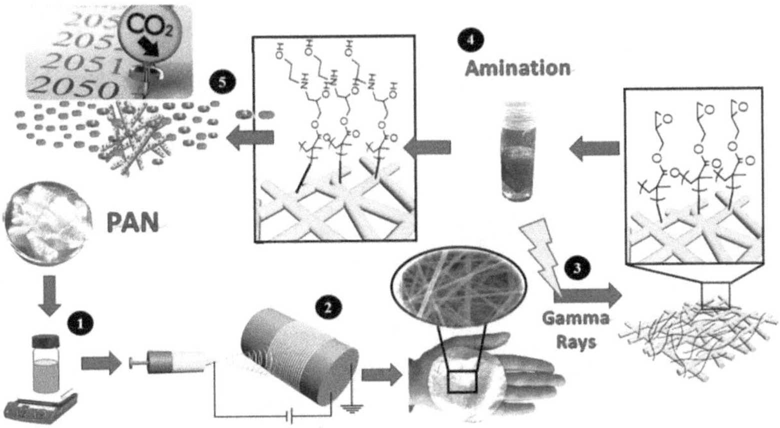

FIGURE 14.13 CO_2 adsorption by NF-PAN/Rad/GMA/EA [13].

method was performed using the amination approach with EA. Figure 14.13 shows the steps of synthesis, adsorbent modification and CO_2 adsorption [13].

The authors showed that the optimal adsorbent (NF-PAN/Rad/GMA/EA) has the maximum DG% equal to 323.45% and the DA% equal to 170.88%. Also, in this work, the highest adsorption capacity with NF-PAN/Rad/GMA/EA adsorbent is reported to be 3.23 mmol/g in ambient conditions, flow rate 100 ml/min and 5% CO_2 concentration [13]. In 2023, Ali Ahmadizadeh Tourzani et al. synthesized poly(acrylonitrile) nanofibers by electrospinning and bonded them to the surface by gamma irradiation in a fixed bed column for CO_2 adsorption. In this work, the authors modified the nanofibers with EA, diethylamine, and TEA to increase the adsorption capacity. This modified adsorbent showed the maximum adsorption capacity of 2.84 mmol/g with 172.26% DG in 10 kg for CO_2 adsorption from a CO_2/N_2 mixture with a CO_2 feed concentration of 5% M [42]. Based on the collected information and a review of the reports, it is shown that the use of microwaves to increase the CO_2 adsorption capacity has not yet been properly and fully investigated. It should be noted that designing and setting up CO_2 adsorption units is one of the most expensive industrial activities. Therefore, it is very important to design and manufacture adsorbents that have selectivity, permeability, proper performance, and high stability in CO_2 adsorption. To date, amine adsorbents are one of the most widely used adsorbents for CO_2 adsorption by performing reversible reactions with CO_2. On the other hand, solid adsorbents with high stability and cost-effectiveness can be a suitable substitute for liquid amine adsorbents. Porous adsorbents functionalized with amine compounds have been widely investigated in CO_2 adsorption. These adsorbents are synthesized by one-covalent or chemical bonding with amine compounds (they have low stability), two-solid polymerization of amine monomers in porous structures, and three-stabilization of amine polymers in porous materials. The use of microwaves in the modification of solids synthesized by the first two methods is one of the most appropriate methods. At room temperature, the high penetrating power

TABLE 14.1
CO_2 Adsorption Capacity in Different Types of Polymer Adsorbents [62]

Polymer	Monomer	Irradiation Method	Functional Agent	Type Adsorbent	CO_2 %	Adsorption Capacity (mmol/g)	Ref
PE	Acrylic acid	Pre-irradiation UV-grafting	PEI	Membrane	—	—	[60]
PE-PP	N-Vinylformamid	Pre-irradiation Electron bean	Hydrolysis with NaoH	Nonwoven fiber	40	1.36	[63]
UHMWPE	N-Vinylformamid	Pre-irradiation Electron bean	Hydrolysis with NaoH	Film	10	0.0486	[64]
PP	GMA	Pre-irradiation Electron bean	EA	Nanofiber	15	2.87	[12]
PE-PP	GMA	Pre-irradiation Electron bean	TEA	Nonwoven fiber	—	4.52	[56]
PAN	Allyamine	Pre-irradiation γ-ray	$(NH_4)_2S_2O_8$/ $NaHSO_3$	Fiber	15	6.22	[55]
PP	GMA	Pre-irradiation γ-ray	TETA	Fiber	—	4.72	[65]
PAN	GMA	Simultaneous irradiation	EA	Nanofiber	5	2.95	[62]

of gamma rays is very effective in the polymerization of vinyl monomers adsorbed on any solid substrate. In the radiation chemistry of polymers, the creation of bonds due to radiation exposure is one of the most important aspects studied. Bonding of monomers on solids, even metals, can be created by irradiation [17, 61]. In 2024, A.A. Tourzani and his team successfully synthesized NF-PAN nanofibers utilizing two cost-effective polymers: industrial PAN and industrial polyurethane. They employed the electrospinning technique for modification. Following their corrections, the authors reported a maximum CO_2 absorption capacity of 2.95 mmol/g CO_2 of absorbent, achieved at a radiation dose of 28.03 kg, with 25.77% of GMA and 66.45% of EA. Furthermore, a notable enhancement in absorption capacity was recorded in the irradiated absorber, which exhibited an increase of 73.15% compared to the non-irradiated counterpart. Additionally, this report includes a concise overview of the conducted activities, as detailed in Table 14.1.

14.7 CONCLUSION

In this chapter, the application of the methods of using microwaves in the development of absorbers and the increase of CO_2 adsorption capacity by them were investigated. Studies show that the emerging process of developing absorbents using microwaves with the aim of increasing adsorption capacity compared to traditional heating methods is an effective and efficient method. However, this method is in the early stages and requires a lot of research and analysis by scientists and chemists. In the first step, it is necessary to investigate and analyze the mechanism of change of absorbers due to microwave radiation on materials. In the next step, it is necessary to check and measure the temperature in the process of microwave radiation in absorbent materials with appropriate methods. Also, for the simulation of microwave waves, it is necessary that the developed mathematical models are closer to the real conditions. The last point is that to expand the use of microwaves in increasing the capacity of absorbers, it is necessary to conduct joint research in several fields and theories so that they can be used in specific industrial applications. It is also necessary for scientists and chemists to direct the world toward the use of this technology in CO_2 removal with convincing scientific evidence about the use of microwaves in increasing the absorbent capacity in CO_2 adsorption and its economic benefits.

REFERENCES

1. Gibbins, J. and H. Chalmers, Carbon capture and storage. *Energy Policy*, 2008. 36(12): p. 4317–4322.
2. Chen, K., et al., Graphene adsorption and separation functional materials. *Chemical Engineering & Technology*, 2019. 42(2): p. 266–286.
3. Sen, T., Y. Kawajiri, and M.J. Realff, Integration of material and process design for kinetic adsorption separation. *Industrial & Engineering Chemistry Research*, 2021. 60(6): p. 2536–2546.
4. Liu, S., et al., Cascade adsorptive separation of light hydrocarbons by commercial zeolites. *Journal of Energy Chemistry*, 2022. 72: p. 299–305.
5. Paramasivan, B., Microwave assisted carbonization and activation of biochar for energy-environment nexus: A review. *Chemosphere*, 2022. 286: p. 131631.
6. Li, H., et al., Fundamentals and applications of microwave heating to chemicals separation processes. *Renewable and Sustainable Energy Reviews*, 2019. 114: p. 109316.
7. Zhang, G., et al., Amine-modified SBA-15 (P): A promising adsorbent for CO_2 capture. *Journal of CO_2 Utilization*, 2018. 24: p. 22–33.
8. Santos, K.M., et al., Natural gas dehydration by adsorption using MOFs and silicas: A review. *Separation and Purification Technology*, 2021. 276: p. 119409.
9. Muchan, P., C. Saiwan, and M. Nithitanakul, Investigation of adsorption/desorption performance by aminopropyltriethoxysilane grafted onto different mesoporous silica for post-combustion CO_2 capture. *Clean Energy*, 2020. 4(2): p. 120–131.
10. Mahmoodi, N.M. and M.H. Saffar-Dastgerdi, Clean Laccase immobilized nanobiocatalysts (graphene oxide-zeolite nanocomposites): From production to detailed biocatalytic degradation of organic pollutant. *Applied Catalysis B: Environmental*, 2020. 268: p. 118443.
11. Oveisi, M., M.A. Asli, and N.M. Mahmoodi, Carbon nanotube based metal-organic framework nanocomposites: synthesis and their photocatalytic activity for decolorization of colored wastewater. *Inorganica Chimica Acta*, 2019. 487: p. 169–176.
12. Abbasi, A., et al., Amine functionalized radiation induced grafted polyolefin nanofibers for CO_2 adsorption. *Radiation Physics and Chemistry*, 2019. 156: p. 58–66.
13. Imanian, Z., et al., CO_2 adsorption-desorption in batch and continuous conditions by using a nano-fibrous polyacrylonitrile radiation grafted adsorbent under response surface methodology. *Journal of CO_2 Utilization*, 2023. 73: p. 102506.
14. Ibrahim, H.M. and A. Klingner, A review on electrospun polymeric nanofibers: Production parameters and potential applications. *Polymer Testing*, 2020. 90: p. 106647.
15. Abbasi, A., et al., Highly flexible method for fabrication of poly (Glycidyl Methacrylate) grafted polyolefin nanofiber. *Radiation Physics and Chemistry*, 2018. 151: p. 283–291.

16. Bhardwaj, N. and S.C. Kundu, Electrospinning: A fascinating fiber fabrication technique. *Biotechnology Advances*, 2010. 28(3): p. 325–347.
17. Kou, Z., et al., A review of adsorption intensified by microwave irradiation from absorbent preparation to separation processes. *Chemical Engineering and Processing-Process Intensification*, 2023. 184: p. 109300.
18. Kowalski, S.J., J. Banaszak, and A. Rybicki, Damage analysis of microwave-dried materials. *AIChE Journal*, 2012. 58(7): p. 2097–2104.
19. Nüchter, M., et al., Microwave assisted synthesis–a critical technology overview. *Green Chemistry*, 2004. 6(3): p. 128–141.
20. Kingston, H.M. and L.B. Jassie, *Introduction to Microwave Sample Preparation: Theory and Practice*, Vol. 61. 1988, Washington, DC: American Chemical Society.
21. Zhao, D.C. and L.S. Zhang, Biological effects of electromagnetic radiation and protection. *Applied Mechanics and Materials*, 2014. 513: p. 3313–3316.
22. Mauge-Lewis, K., et al., Metal-assisted and microwave-accelerated evaporative crystallization: proof-of-principle application to proteins. *Crystal Growth & Design*, 2015. 15(7): p. 3212–3219.
23. Zhao, Z., et al., Microwave-induced enhancement of distillation separation based on dielectric properties difference. *Chemical Industry and Engineering Progress*, 2020. 39(6): p. 2275.
24. Zhao, Z., et al., Watching microwave-induced microscopic hot spots via the thermosensitive fluorescence of europium/terbium mixed-metal organic complexes. *Angewandte Chemie International Edition*, 2022. 61(6): p. e202114340.
25. Gou, D., et al., Influence of weak microwaves on spatial collision and energy distribution of water molecules. *Chemical Physics*, 2021. 540: p. 110977.
26. Wang, N., et al., Study and application status of the nonthermal effects of microwaves in chemistry and materials science–a brief review. *RSC Advances*, 2022. 12(27): p. 17158–17181.
27. Kubo, M.T., et al., Non-thermal effects of microwave and ohmic processing on microbial and enzyme inactivation: A critical review. *Current Opinion in Food Science*, 2020. 35: p. 36–48.
28. Chandrasekaran, S., S. Ramanathan, and T. Basak, Microwave material processing—a review. *AIChE Journal*, 2012. 58(2): p. 330–363.
29. Gut, J.A.W., Response to "Non-thermal microwave effects: Conceptual and methodological problems". *Food Chemistry*, 2022. 390: p. 133216.
30. Adnađević, B., et al., Comparative study on isothermal kinetics of fullerol formation under conventional and microwave heating. *Chemical Engineering Journal*, 2008. 140(1–3): p. 570–577.
31. Dawood, K.M. and A. Kirschning, Combining enabling techniques in organic synthesis: solid-phase-assisted catalysis under microwave conditions using a stable Pd (II)-precatalyst. *Tetrahedron*, 2005. 61(51): p. 12121–12130.
32. Marimuthu, A. and G. Madras, Continuous distribution kinetics for microwave-assisted oxidative degradation of poly (alkyl methacrylates). *AIChE Journal*, 2008. 54(8): p. 2164–2173.
33. Rusanov, A.L., et al., Preparation and characterization of sulfonated polyphenylquinoxalines. *High Performance Polymers*, 2008. 20(6): p. 627–641.
34. Zhang, C., L. Liao, and S.S. Gong, Recent developments in microwave-assisted polymerization with a focus on ring-opening polymerization. *Green Chemistry*, 2007. 9(4): p. 303–314.
35. Seman, N., et al. Modified surface polymer prepared by using radiation-induced grafting co-polymerization method and its application: A mini review, in *IOP Conference Series: Earth and Environmental Science*. 2022, IOP Publishing.
36. Nasef, M.M. and E.-S.A. Hegazy, Preparation and applications of ion exchange membranes by radiation-induced graft copolymerization of polar monomers onto non-polar films. *Progress in Polymer Science*, 2004. 29(6): p. 499–561.

37. Torkaman, R., et al., Assessing the radiation-induced graft polymeric adsorbents with emphasis on heavy metals removing: A systematic literature review. *Journal of Water Process Engineering*, 2021. 44: p. 102371.
38. Ueki, Y., N. Seko, and Y. Maekawa, Machine learning approach for prediction of the grafting yield in radiation-induced graft polymerization. *Applied Materials Today*, 2021. 25: p. 101158.
39. Nasef, M.M. and O. Güven, Radiation-grafted copolymers for separation and purification purposes: Status, challenges and future directions. *Progress in Polymer Science*, 2012. 37(12): p. 1597–1656.
40. Thinkohkaew, K., et al., Radiation induced graft polymerization of fluorinated methacrylate onto polypropylene spunbond nonwoven fabric. *Surfaces and Interfaces*, 2021. 24: p. 101125.
41. Nisar, S., et al., γ-Radiation induced L-glutamic acid grafted highly porous, pH-responsive chitosan hydrogel beads: A smart and biocompatible vehicle for controlled anticancer drug delivery. *International Journal of Biological Macromolecules*, 2021. 182: p. 37–50.
42. Tourzani, A.A., et al., Effective CO_2 capture by using poly (acrylonitrile) nanofibers based on the radiation grafting procedure in fixed-bed adsorption column. *Scientific Reports*, 2023. 13(1): p. 6173.
43. Lim, S.J. and I.H. Shin, Graft copolymerization of GMA and EDMA on PVDF to hydrophilic surface modification by electron beam irradiation. *Nuclear Engineering and Technology*, 2020. 52(2): p. 373–380.
44. Khamplod, T., et al., Electron beam radiation grafting of styrene on natural rubber using Taguchi's design. *Polymer*, 2015. 79: p. 135–145.
45. Liu, Z., et al., Radiation-initiated high strength chitosan/lithium sulfonate double network hydrogel/aerogel with porosity and stability for efficient CO_2 capture. *RSC Advances*, 2021. 11(33): p. 20486–20497.
46. Gu, C., et al., Solar-radiation-induced adsorption/desorption system for carbon dioxide capture. *Cell Reports Physical Science*, 2022. 3(11).
47. Chen, X., et al., Ultraviolet B radiation down-regulates ULK1 and ATG7 expression and impairs the autophagy response in human keratinocytes. *Journal of Photochemistry and Photobiology B: Biology*, 2018. 178: p. 152–164.
48. Chen, Y., et al., Preparation of antifouling poly (ether ether ketone) hollow fiber membrane by ultraviolet grafting of polyethylene glycol. *Materials Today Communications*, 2021. 27: p. 102326.
49. Lee, J.-Y., et al., Surface-attached brush-type CO_2-philic poly (PEGMA)/PSf composite membranes by UV/ozone-induced graft polymerization: Fabrication, characterization, and gas separation properties. *Journal of Membrane Science*, 2019. 589: p. 117214.
50. Lozano, L.A., et al., Enhancement in the CO_2 uptake of UiO-66 by a simple exposure to ultraviolet light. *Materials Today Communications*, 2022. 31: p. 103540.
51. Aoki, S., et al., Antimicrobial fabric adsorbed iodine produced by radiation-induced graft polymerization. *Radiation Physics and Chemistry*, 2013. 84: p. 242–245.
52. Selambakkannu, S., et al., Preparation and optimization of thorium selective ion imprinted nonwoven fabric grafted with poly (2-dimethylaminoethyl methacrylate) by electron beam irradiation technique. *Journal of Environmental Chemical Engineering*, 2020. 8(3): p. 103737.
53. Varghese, A.M., et al., Performance enhancement of CO_2 capture adsorbents by UV treatment: the case of self-supported graphene oxide foam. *Chemical Engineering Journal*, 2020. 386: p. 124022.
54. Durán-Jiménez, G., et al., Simultaneous conventional and microwave heating for the synthesis of adsorbents for CO_2 capture: Comparative study to pristine technologies. *Chemical Engineering Journal*, 2022. 438: p. 135549.

55. Yang, Y., et al., Preparation and characterization of a solid amine adsorbent for capturing CO_2 by grafting allylamine onto PAN fiber. *Langmuir*, 2010. 26(17): p. 13897–13902.
56. Nasef, M.M., A. Abbasi, and T. Ting, New CO_2 adsorbent containing aminated poly (glycidyl methacrylate) grafted onto irradiated PE-PP nonwoven sheet. *Radiation Physics and Chemistry*, 2014. 103: p. 72–74.
57. Zainab, G., et al., Free-standing, spider-web-like polyamide/carbon nanotube composite nanofibrous membrane impregnated with polyethyleneimine for CO_2 capture. *Composites Communications*, 2017. 6: p. 41–47.
58. Nasef, M.M., et al., Radiation grafted adsorbents for newly emerging environmental applications. *Radiation Physics and Chemistry*, 2016. 118: p. 55–60.
59. Heo, Y.-J. and S.-J. Park, Facile synthesis of MgO-modified carbon adsorbents with microwave-assisted methods: effect of MgO particles and porosities on CO_2 capture. *Scientific Reports*, 2017. 7(1): p. 5653.
60. Wang, Y., et al., Polyethylenimine-modified membranes for CO_2 capture and in situ hydrogenation. *ACS Applied Materials & Interfaces*, 2018. 10(34): p. 29003–29009.
61. Güven, O. and B. Han, The prospects for radiation technology in mitigating carbon footprint. *Radiation Physics and Chemistry*, 2022. 198: p. 110282.
62. Tourzani, A.A., et al., Enhancing the formation of eco-friendly CO_2 adsorbent through utilizing the RSM technique and implementation of PAN/PU electrospinning and radiation-grafted approaches. *Journal of Cleaner Production*, 2024. 434: p. 140213.
63. Zubair, N.A., et al., Kinetic studies of radiation induced grafting of N-vinylformamide onto polyethylene/polypropylene fibrous sheets and testing its hydrolysed copolymer for CO_2 adsorption. *Radiation Physics and Chemistry*, 2020. 171: p. 108727.
64. Rojek, T., et al., Polyvinylamine-containing adsorbent by radiation-induced grafting of N-vinylformamide onto ultrahigh molecular weight polyethylene films and hydrolysis for CO_2 capture. *Industrial & Engineering Chemistry Research*, 2017. 56(20): p. 5925–5934.
65. Zhuang, L., et al., Preparation of a solid amine adsorbent based on polypropylene fiber and its performance for CO_2 capture. *Journal of Materials Research*, 2013. 28(20): p. 2881–2889.

15 Industrial Applications and Case Studies

Hassan Pashaei, Hossein Alizadeh, and Ahad Ghaemi

15.1 INTRODUCTION

In recent years, warming environmental issues, because of the diffusion of greenhouse gases, have become a global problem [1, 2]. According to the different climate models, global warming will increase the temperature of the earth closely by 1.4–5.8°C until the first of the 22nd century [3]. To fulfill the global important objectives, it is necessary to develop a technically practical CO_2 absorption process that allows for the decrease of greenhouse gases from the flue gases [4].

With increasing worldwide demand for energy, fossil fuels will retain their dominance as a global energy source because of their availability and relatively low cost. Therefore, reducing CO_2 emissions from fossil fuel use is urgently required to reduce their potential negative impact on the environment [5]. CO_2 capture and storage (CCS) is one strategy that is able to alleviate and control the CO_2 emissions from large exhaust sources such as fossil fuel power plants, cement industries, iron and steel mills, and other industry sectors, which account for about 60% of total CO_2 emissions [6]. Typically, a minimum CO_2 concentration of 95% is required for the deep geological sequestration of CO_2 based on technical and economic feasibility studies [7]. Within the field of CCS, a range of technology options for CO_2 capture are being evaluated and developed, and the number of demonstration-scale projects in operation or planned is growing [8].

Since 1992, many studies (bench, pilot, and demonstration scales) have been conducted worldwide in the field of CO_2 capture by adsorption, especially in Japan and Korea. The Japanese power industry started investigating flue gas CO_2 capture with pilot plants using physical adsorption at their power stations in the early 1990s [9].

Generally, the main approaches to the removal of CO_2 are restricted to cryogenic distillation, membrane purification, absorption with solutions, and adsorption using solids [10–14]. Although cryogenic distillation is widely used in gas separation, this technique is likely to be ruled out in the case of CO_2 due to the high energy demand involved. Membranes are an efficient mass-separating agent for bulk separations. However, when CO_2 is a minor component as in the flue gases, membranes have reduced efficiency and are difficult to scale up [15–18]. Absorption with chemical solutions is the most mature technology for large-scale CO_2 capture using aqueous alkylamine solutions like monoethanolamine (MEA) and diethanolamine or other

DOI: 10.1201/9781003476191-15

fluids with the basic character [19–23]. However, there are some problems associated with this technique, such as high equipment corrosion, high energy consumption in regeneration, oxidative degradation of absorbents, and loss of effectiveness over time due to low thermal stability and foaming in the gas–liquid interface [24–27]. Because of the limitation of these technologies, solid adsorption has been offered due to its plenteous benefits such as low energy requirement for renewal, better CO_2 adsorption capacity, great selectivity, low equipment cost, and easy of handling. Lower energy penalty in the adsorption–desorption procedure is owed to the deficiency in water and solid adsorbent low heat capacity [28]. The adsorption process is one of the new concepts for CO_2 capture. For various circumstances, numerous adsorbents have been employed (Table 15.1).

Adsorption in porous solid materials such as activated carbon is an attractive alternative technology to overcome the techno-economical limitations of the aforementioned technologies. Adsorption is a widely used technology for gas treatment due to its versatility and efficiency. Overall, adsorption of CO_2 by solid adsorbents is an applied method due to its high CO_2 adsorption capacities, suitable for batch and continuous processes, high possibility of regeneration and potential reuse, and low energy

TABLE 15.1
CO_2 Adsorption with Different Adsorbent

Sorbent	Temperature (K)	Capacity (mmol/g)	Process	Ref
Dry NaOH	581	15.6	Fixed-bed reactor	[29]
Polyethyleneimine	785	3.02	Thermogravimetric adsorption	[30]
Monoethanolamine	30	1.0	Thermogravimetric adsorption	[31]
Molecular sieve 13X	298	2.8–3.6	Pressure swing adsorption	[32]
Active carbon–KOH	298	—	Pressure swing adsorption	[33]
NaOH-granular coconut shell AC	332	27.1	Fixed-bed reactor	[34]
NaOH/CaO	350	3	Temperature swing adsorption	[35]
Graphene	195	0.8	—	[36]
Na_2CO_3	50–70	2.3	Thermogravimetric adsorption	[37]
DTO	313	8.07	—	[38]
Granular activated carbon	303	1.63	—	[39]
NaOH-modified coconut shell–activated carbon	308	27.1	Fixed-bed reactor	[34]

requirement [34, 40–42]. Various adsorbents such as inorganic-porous materials, activated carbons, and basic oxide and amine-based systems have been employed in the catalysis and adsorption of CO_2. But some of these adsorbents like the amine-based ones require high energy and possess degradation threat through oxidation that results in corrosion [28, 43–46]. Carbon-based materials have high chemical and thermal stability as well as high adsorption capacities. Their low cost and recyclability make them ideal for gas pollutions prevention. Activated carbon as a highly microporous material with a large surface area has been recognized as one of the carbon-based materials for removing gas phase pollutants [42]. In recent years, activated carbon was modified with KOH, NaOH, and phosphoric acid for functionalization and improvement of the surface area. Impregnation with sodium hydroxide (NaOH) is more effective in the modification of activated carbon because NaOH is inexpensive, minimally corrosive, and environmentally friendly. Tan et al. (2014) investigated CO_2 adsorption capacity on modified coconut shell–activated carbon at different adsorption temperatures, including 35°C, 45°C, and 55°C and showed that a 32% NaOH concentration with a 3 h dwelling time provided the best CO_2 adsorption capacity [45]. Another study by Buczek (2016) showed that process reactivation of carbon changes its particle size as well as density properties and increases by nearly twice the amounts of CH_4 and CO_2 adsorbed under high-pressure conditions [33]. Guo et al. (2006) studied the adsorption of CO_2 on raw activated carbon A and three modified activated carbon samples B, C, and D at temperatures ranging from 29.85°C to 59.85°C and showed that the active ingredients impregnated in the carbon samples show significant influence on the adsorption for CO_2. The volumes adsorbed on modified carbon samples B, C, and D are all larger than that on the raw carbon sample A. On the other hand, the physical parameters such as surface area, pore volume, and micropore volume of carbon samples show no influence on the adsorbed amount of CO_2 [47]. Shahkarami et al. (2015) studied the effects of different activation techniques on CO_2 adsorption performance at temperature range of 25°C to 65°C. The results revealed that KOH-activated carbon with a total micropore volume of 0.62 cm^3/g and a surface area of 1400 m^2/g had the highest CO_2 adsorption capacity due to its microporous structure [48]. Another study by Sreńscek-Nazzal et al. (2016) demonstrated that the AC modified with KOH had the highest SBET, V_{tot}, and V_{mic} values of 2063 m^2/g, 1.13 cm^3/g, and 0.67 cm^3/g, respectively, and the maximum CO_2 adsorption was 14.44 mmol/g for DTO/KOH-modified carbon, whereas 8.07 mmol/g of CO_2 was adsorbed at DTO [49].

15.2 POST-COMBUSTION CAPTURE

CCS technologies have gained significant attention in recent years as a means to reduce greenhouse gas emissions and combat climate change. CCS consists of the separation of fossil CO_2 from industrial and energy-related sources and transport to a storage location, where it can be permanently stored (Figure 15.1); captured CO_2 can be alternatively used in a variety of ways, including enhanced oil recovery (EOR), improving the growth of plants and algae, or as a raw material in the production of fuels and chemicals.

To date, dependent on the process or plant application in question, there are four main approaches to capture CO_2 generated from primary fossil fuel (coal, natural gas,

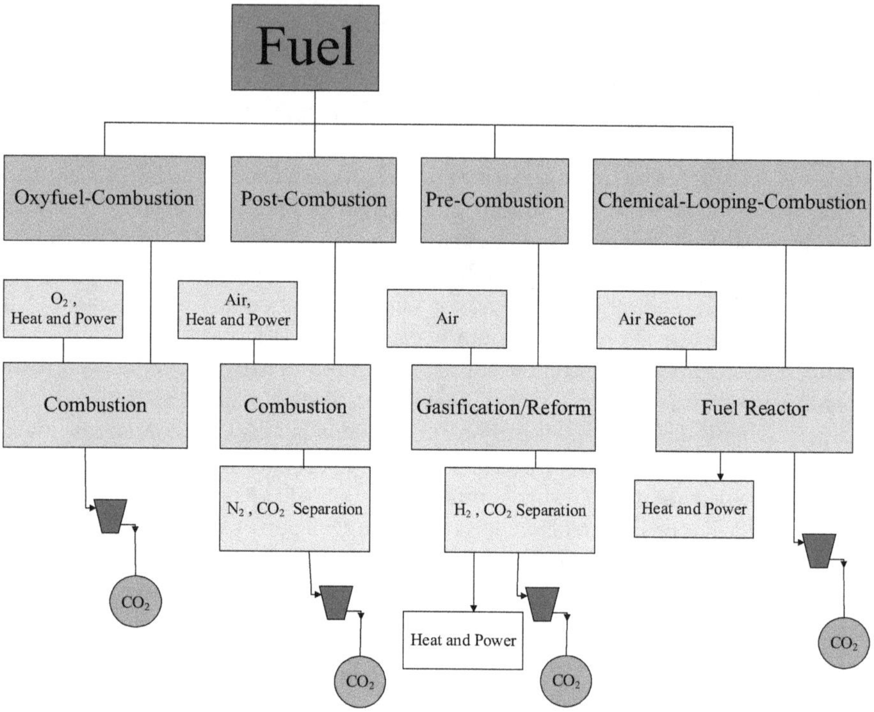

FIGURE 15.1 Schematic representation of capture systems.

or oil), biomass, wastes, or mixtures of these fuels: post-combustion, pre-combustion, oxy-fuel, and chemical looping processes (Figure 15.1). In post-combustion systems, CO_2 is separated from the flue gases produced by the combustion of the primary fuel in the air, thus providing the main advantage of being able to be adapted/retrofitted to any existing plants, many of which were built recently and could operate for decades to come. In pre-combustion systems, the primary fuel is processed in a reactor with steam and air or oxygen to produce a mixture consisting mainly of CO and H_2 ("synthesis gas"). Additional H_2, together with CO_2, is produced by reacting CO with steam in a second reactor (a "shift reactor"). The resulting mixture of H_2 and CO_2 can then be separated; CO_2 is stored, whereas H_2 is a carbon-free energy carrier that can be combusted to generate power and/or heat or used to produce fuels/chemicals. The high concentrations of CO_2 produced by the shift reactor (typically up to 60% by volume on a dry basis) and the high pressures often encountered in these applications make the CO_2 separation easier than in post-combustion; however, the initial fuel conversion steps are more elaborate and costly than in post-combustion systems. In oxy-fuel combustion systems, O_2 is used instead of air for combustion of the primary fuel to produce a flue gas that is mainly water vapor and CO_2. This results in a flue gas with high CO_2 concentrations (greater than 80% by volume). The water vapor is then removed by cooling and compressing the gas stream. However, oxy-fuel combustion requires the upstream separation of oxygen from the air, with a purity of 95–99% oxygen assumed in most current designs. The main idea of chemical looping combustion is to split the combustion of a hydrocarbon or carbonaceous fuel into separate oxidation and reduction reactions. In particular, a solid oxygen carrier (generally a metal oxide) is used to transfer oxygen from the air to the fuel. The advantage of this technique compared to normal combustion is that CO_2 and H_2O are inherently separated from the other components of the flue gas, namely, N_2 and unreacted O_2, and thus no extra energy is needed for CO_2 separation. However, the development of a suitable oxygen carrier, being able to provide a high fuel conversion ratio, high oxygen transport capacity, and good stability, still represents an open challenge. The potential of these technologies is summarized in Table 15.2.

Among all of these possible solutions, post-combustion CO_2 capture is a straightforward approach and forms the basis of the current infrastructure in CCS. Despite the distinct advantages of both pre-combustion and oxy-fuel capture, these methods will unlikely replace post-combustion capture (PCC) on a global scale. On the contrary,

TABLE 15.2
The Summary of Each Technology Potential

Post-Combustion	Pre-Combustion	Oxy-Fuel Combustion
Applicable to the majority of existing coal-fired plants	Generally higher CO_2 concentration than for post-combustion capture	Very high CO_2 concentration in flue gas
Low CO_2 partial pressure	Higher driving force for CO_2 separation	Combustors would be fairly conventional
Significantly higher performance or circulation volume is required for high-capture levels	Fuel processing is needed	Large cryogenic oxygen production requirements may be cost prohibitive
CO_2 produced at low pressure compared to sequestration requirements	Potential for reduction in compression costs/loads	Large cryogenic oxygen production requirement may be cost prohibitive
Retrofit technology option	For coal plants, efficiency and cost penalties are generally lower than for post-combustion capture	Need to recycle large quantities of flue gas to avoid excessively high combustion temperatures
Improved solvents may reduce energy losses	Barriers to commercial application of gasification are common in pre-combustion capture	CO_2 is recycled to the compressor to provide the expansion medium, instead of air
Significant cost savings are possible	Extensive supporting systems requirements	Novel turbine cycles have been proposed, which eliminate shift reaction/fuel gas CO_2 separation
		Potential for advanced oxygen separation membranes with lower energy consumption
		Decreased process efficiency

PCC remains the only solution capable of delivering significant emission reductions from existing large stationary sources, essentially power stations and large industrial plants. Moreover, in combination with bioenergy or through direct air capture, it can also generate "negative emissions," a challenging option for limiting future temperature increases to 2°C or below. In this framework, the main difficulty of post-combustion capture is represented by the need to produce a highly concentrated CO_2 stream matching the purity requirement for transportation and storage from the flue gas stream, where CO_2 is highly diluted: between 4% for the natural gas combined cycle (NGCC) and 15% for pulverized coal (PC). One more explanation for the slow deployment of fully integrated commercial post-combustion capture schemes is the considerable cost of the capture phase, which represents approximately two-thirds of the total cost for CCS. Therefore, the development of an efficient and cost-effective CO_2 capture technique is considered to be one of the highest priorities in the field of CCS. Dependent on the principle of the capture process, there are different types of separation techniques to separate CO_2 from the flue gas stream: absorption by means of solvents, membranes, cryogenics, and adsorption by solid materials, and calcium looping cycle. Among these, amine-based absorption is the most mature option for post-combustion and is actually applied for the separation of CO_2 in real industrial processes (e.g., natural gas sweetening and production of hydrogen and ammonia). However, when applied for post-combustion applications, absorption is characterized by serious shortcomings, such as high energy demands to regenerate the sorbent, corrosion problems, amine losses as a result of evaporation, thermal/chemical degradation of the amines caused by the presence of oxygen, and environmental concerns about amine disposal. In this framework, adsorption, relying on the ability of gases (the adsorbate) to be adsorbed on a solid surface (the adsorbent), which can be subsequently regenerated by acting on either the temperature (temperature swing) or the pressure (pressure swing) of the system, has been receiving great research interest because of its favorable characteristics over all of the other alternative solutions. In particular, it can provide remarkable energy savings with respect to the amine-based absorption approach. Besides that, it can be quite easily retrofitted to existing plants offering a wide flexibility of capturing CO_2 from different industrial CO_2 sources as a result of different available sorbent regeneration modes and reactor types. Thus far, research efforts have been focused primarily on the development of the sorbent materials aiming at reducing the energy penalty by minimization of the heat of adsorption and maximization of the adsorption capacity. However, it is important to point out that other aspects, affecting the performance/competitiveness and overall cost of the CO_2 adsorption process, must be taken into account for adsorption to become one of the leading capture techniques. Indeed, even though this gas separation technology is mature in a number of large-scale industrial applications, its application to post-combustion CO_2 capture still needs very intense research effort to develop, beyond suitable materials, also innovative designs for an improved gas–solid contact system and appropriate regeneration strategies.

There are several post-combustion gas separation and capture technologies being investigated: (a) absorption, (b) cryogenic separation, (c) membrane separation, (d) microalgal bio-fixation, and (e) adsorption. Figure 15.2 summarizes various technology options for post-combustion CO_2 capture.

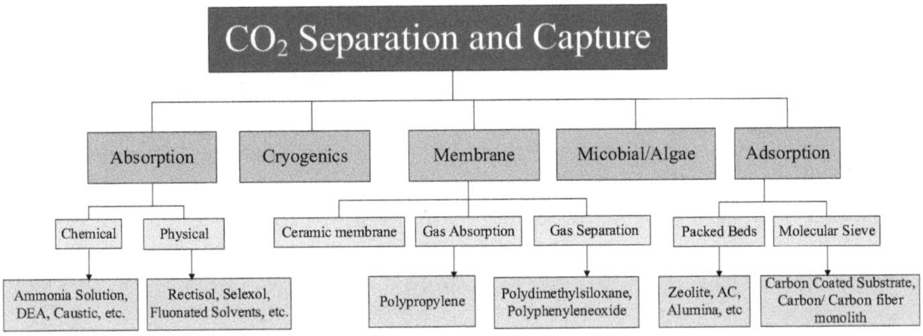

FIGURE 15.2 Different technologies for CO_2 removal.

Post-combustion capture (PCC) refers to the separation of CO_2 from flue gas derived from combusting fossil fuels—coal, natural gas, or oil—in the air. In the case of coal-based power, coal is combusted in air and the liberated heat is converted to electricity by steam-driven turbines connected to generators. The combustion results in a flue gas mixture consisting of N_2, CO_2, H_2O, O_2, and a host of compounds such as SO_x, NO_x, and heavy metals among others. Some of these are removed using existing technologies such as selective catalytic reduction, electrostatic precipitation, and flue gas desulphurization. A PCC process then aims to selectively separate CO_2 from the remaining gas mixture.

- PCC can be typically built into existing industrial plants and power stations (known as retrofitting) without significant modifications to the original plant.
- PCC can be integrated into new plants to achieve a range of greenhouse gas intensity reductions near-zero emissions.
- In contrast to competing technologies, PCC offers high operational flexibility (partial retrofit, zero to full capture operation) and it can match market conditions for both existing and new power stations.
- PCC offers a lower technology risk compared to competing technologies; this is further enhanced by the ability for staged implementation, which is not possible with competing technologies.
- Renewable technologies can be integrated into the PCC process; in particular, PCC allows the use of low-cost solar thermal collectors to provide the necessary heat to separate CO_2 from sorbents, effectively reducing the loss of electrical output caused by capture.
- PCC can be applied to capture CO_2 from natural gas-fired power stations and other large stationary sources of CO_2, for instance, smelters, cement kilns, and steelworks.

In particular, separation techniques for CO_2 capture such as pressure swing adsorption (PSA), amine absorption, and cryogenic separation have been widely used in industry for CO_2 separation in chemical and petrochemical applications. For example, the amine-based absorption with an aqueous MEA solution is the most common technology used for PCC, since it is capable of achieving a high level of CO_2 capture (more than 90%) from flue gas owing to fast kinetics and a strong chemical reaction. Nevertheless, the amines are corrosive, are predisposed to degradation by trace flue gas constituents such as sulfur oxides, and require a considerable amount of energy, mainly in the regeneration step [50].

Cryogenic CO_2 capture is constituted by many steps; some of them reported here:

1. Cooling of CO_2-laden flue gas to desublimation temperatures (from −100°C to −135°C)
2. Separation of solid CO_2 formed from flue gas
3. Use of the cold products to cool incoming gases in a recuperative heat exchanger
4. Compression of the solid/liquid CO_2 to final pressures (100–200 atm; 10.13–20.27 MPa)
5. Delivering of a compressed CO_2 stream separated from an atmospheric pressure light-gas stream

The overall energy and economic costs appear to be at least 30% lower than most competing processes involving air separation units (ASUs), solvents, or similar technologies. In addition, the cryogenic CO_2 capture process shows several benefits (Bolland and Undrum, 2003), including the following:

1. No chemical absorbents are required.
2. It can be operated at atmospheric pressure.
3. It provides highly efficient removal of many pollutants (Hg, SO_x, NO_2, HCl, etc.).
4. It has possible energy storage capacity.
5. It has potential water savings.

Cryogenic separation is widely used commercially for streams that have high CO_2 concentrations (typically > 90%), but it is not used for more dilute CO_2 streams. In fact, the major disadvantage of CO_2 cryogenic separation is the amount of energy required to provide the refrigeration necessary for the process, particularly for dilute gas streams.

15.2.1 Post-Combustion in Thermal Power Plant

Post-combustion CO_2 capture methods are based on removing carbon dioxide from flue gas. The capture unit is placed after the purification systems, such as desulfurization, denitrogenation, and dedusting installations. Figure 15.3 shows a general block diagram of the PCC technique.

In existing conventional power units, post-combustion technologies are the most frequently considered [51]; nevertheless, there is one main barrier to using these methods. Since the partial pressure of CO_2 in the flue gas is low (flue gas is under atmospheric pressure and the concentration of carbon dioxide is within 13–15%), the driving force for CO_2 is also low [52]. Post-combustion technologies can be divided

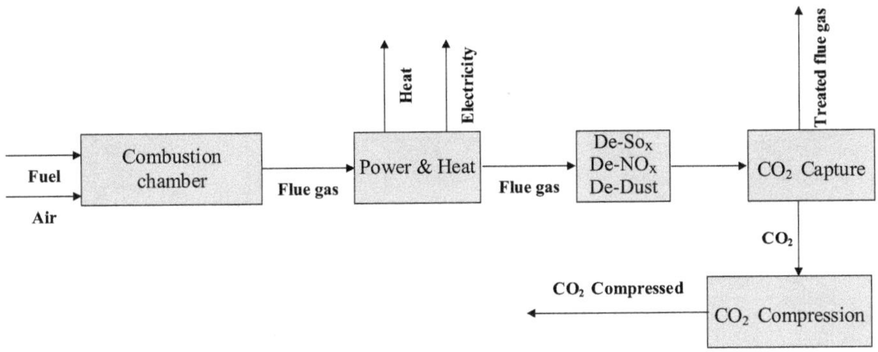

FIGURE 15.3 Block diagram of electricity generation and heat production through the use of the post-combustion CO_2 capture method.

up according to the type of process used for capturing carbon dioxide, such as absorption solvent-based methods, membrane separation, chemical looping combustion and calcium looping process (CLP), cryogenic method, converting CO_2 into value-added chemicals, and adsorption–physical separation.

Adsorption is a process that uses a solid surface to remove carbon dioxide from a mixture. Physical separation relies on adsorption, absorption, and cryogenic separation methods. It can be physical (van der Waals forces for adhesion CO_2—physisorption) or chemical (covalent bonding between compounds—chemisorption) [53]. Physical adsorption uses various porous materials (such as activated carbon, alumina, metallic oxides, or zeolites) to absorb carbon dioxide. Activated carbons contain amorphous carbon, and it is a low-cost material with the advantage of having a large surface area and the possibility of modifying its pore structure. However, the weak binding energy with carbon dioxide causes this material to be highly microporous to be useful for carbon capture [51]. Zeolites (crystalline aluminosilicates) have good adsorption properties for CO_2 capture, but they are hydrophilic. The presence of water weakens these properties by reducing the strength of interactions between coupled compounds [8]. A new approach is to use metal–organic frameworks (MOFs) in adsorption processes. MOFs consist of metal ions or ion clusters linked by organic ligands and bridges that create strong coordination bonds. On account of this, MOFs are characterized as having benefits such as ease of design and synthesis, a high porosity, and tailored pore properties [54]. One of the other adsorption materials is silica. Silicas are non-carbonaceous substances with a large surface area and pore size, and they are highly mechanically stable. Mesoporous silica materials use amine-based substances for CO_2 capture [51, 53]. The methods of adsorption are as follows: PSA, temperature swing adsorption (TSA), vacuum swing adsorption (VSA), and pressure–temperature swing adsorption (PTSA).

15.3 PRE-COMBUSTION CAPTURE

The pre-combustion carbon capture technique mainly focuses on extracting carbon dioxide from fossil fuel or biomass fuel before the combustion processes generates

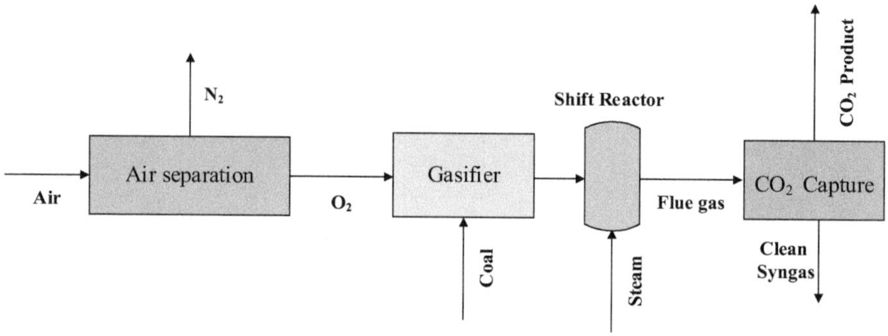

FIGURE 15.4 Process diagram carbon capture and compression for an IGCC power plant.

energy. Generally, this pre-combustion capture technology is used in the process of gasification of coal, natural gas, and biomass, which can generate syngas and in natural gas power plants [55]. Typically, a pre-combustion carbon capture system starts with the gasification of fossil or biomass fuels to form syngas, which contain mainly hydrogen, carbon monoxide, and carbon dioxide. Usually, the syngas is used to power the turbine generator and produce electricity, so pre-combustion carbon capture technology usually extracts carbon dioxide out of the syngas before the gas is combusted in the turbine [55]. To achieve this, syngas is shifted to water-gas-shift (WGS) reactions, which will create extra hydrogen and convert the carbon monoxide in syngas into extra carbon dioxide. The increasing density of carbon dioxide after the WGS reaction will make the carbon-capturing process more effective. The whole procedures are shown in Figure 15.4. The post-combustion capture, in general, has four aspects to study, including solvent-based carbon capture, sorbent-based carbon capture, membrane-based carbon capture, and novel concepts [55].

In natural gas processing, natural gas reforming, gasification, and IGCC, the ultimate capture of the CO_2 is currently accomplished under pressure by an acid gas removal process of absorption in a solvent followed by regenerative stripping of the rich solvent to release the CO_2, which with subsequent compression can be sent to sequestration or supplied for EOR. There are two major generic types of "acid gas" (i.e., CO_2, H_2S, COS) removal solvents—chemical and physical.

In the pre-combustion method, the fuel (coal, gas, and biomass) is not completely combusted in the reactor but is converted into a mixture of CO and H_2 in the reforming or gasification process. Subsequently, using the water–gas shift, CO_2 and H_2 are produced. Pre-combustion capture is used, for example, in an integrated gasification combined cycle (IGCC). Carbon dioxide is removed after the gasification process. Figure 15.5 shows a block diagram of the pre-combustion CO_2 capture method in a power plant.

IGCC technology utilizes pre-combustion CO_2 capture to enhance electricity generation efficiency. This process involves gasifying fossil fuels to produce syngas, which are primarily composed of hydrogen and carbon monoxide. During this stage, CO_2 is removed before combustion, typically using methods like the Selexol or Purisol processes, achieving high-capture efficiencies (up to 95%) due to the high

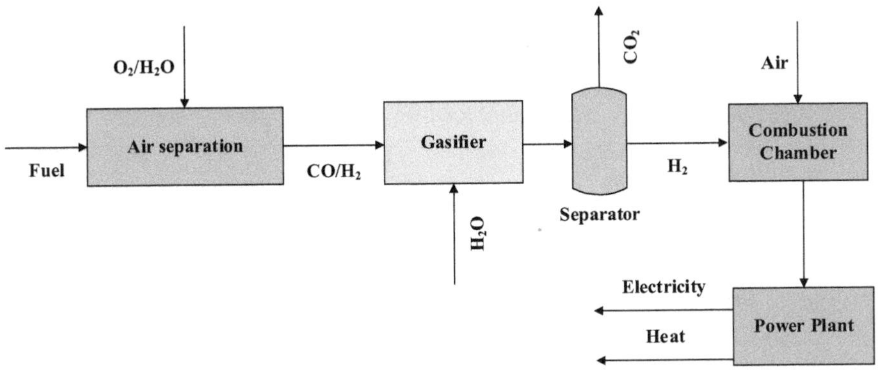

FIGURE 15.5 Block diagram of electricity generation heat production with the use of the pre-combustion CO_2 capture method.

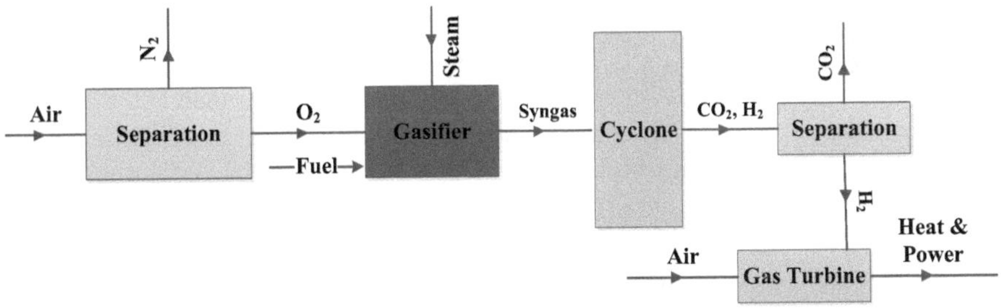

FIGURE 15.6 Scheme of IGCC for electricity generation using a gas turbine using the pre-combustion CO_2 capture method.

concentration of CO_2 in syngas. This technology not only reduces carbon emissions but also allows for the production of blue hydrogen, aiding the transition to cleaner energy sources. An example of a typical process for power and heat generation in a gas turbine with pre-combustion capture is shown in Figure 15.6.

In this process, steam and oxygen are provided to the gasifier to produce syngas enriched in hydrogen and carbon monoxide. Then, syngas is sent to a cyclone separator, where it is filtered to remove ash. After, the conversion of syngas and steam to CO_2 and H_2 occurs in the water–gas shift reactor. The received gas needs to be purified of sulfur in the desulfurization unit. Subsequently, CO_2 is captured in the CO_2 separator and is sent for storage or utilization. Received hydrogen is provided to the gas turbine as fuel [56]. Pre-combustion methods are very effective in CO_2 separation on the grounds of the high concentration of CO_2 in fuel before combustion. On the other hand, these processes are expensive due to the need for a gasification unit.

The CLP is the method of pre-combustion CO_2 capture, where CO_2 capture is achieved effectively at a low cost. This method involves the sorption of CaO with CO_2 and the desorption of $CaCO_3$ to release CO_2 at an optimal temperature. This cycling process repeats multiple times, and waste heat from the gasifier is used to reduce the heat consumption of the CO_2 capture process. Implementing pre-combustion carbon capture in IGCC power plants faces several challenges [57]:

1. High Capital Costs: The technology is primarily suitable for new plants, making retrofitting existing facilities economically unfeasible due to integration complexities.
2. Technological Maturity: Pre-combustion capture methods are less developed compared to post-combustion technologies, limiting their large-scale application.
3. Energy Penalties: The process requires significant energy for cooling syngas and regenerating solvents, which can reduce overall efficiency.
4. Hydrogen Fuel Issues: Utilizing hydrogen-rich fuels presents operational challenges, including decay issues and heat transfer limitations.

The pre-combustion carbon capture demo plant in Port Arthur, United States, has successfully captured 1 million tons of CO_2 since it started operations, without problems. This plant proved that, using the dual PSA technology method, purification of hydrogen >99.9% and a high-efficiency CO_2 capture can be achieved. When the streaming gas has a low pressure, hydrogen purification is performed and the tail gas is sent to undergo vacuum pressure swing adsorption (VPSA) to separate purified CO_2. If the streaming gas has high pressure, CO_2 capture is achieved without VPSA first, and hydrogen purification is achieved from the exiting gas [58].

15.4 OXY-FUEL COMBUSTION

Oxy-combustion carbon capture occurs after the combustion process in an oxygen atmosphere by separating CO_2 generated during the oxy-combustion process. Oxy-combustion is a carbon capture technology where fuels are burned in a mixture of pure oxygen and recycled flue gas instead of air, producing flue gas primarily composed of CO_2. This method enhances CO_2 capture efficiency, achieving up to 100% capture due to higher CO_2 concentrations, which facilitate easier separation and storage. Oxy-combustion also significantly reduces nitrogen oxide (NO_x) emissions compared to traditional air combustion, making it a promising solution for reducing greenhouse gas emissions in power generation. However, the high energy demand for oxygen production remains a challenge for commercialization. The general scheme for the process using the oxy-combustion method is presented in Figure 15.7. Oxy-fuel combustion demonstrates superior energy efficiency compared to other CCS technologies. It requires significantly less energy, particularly at high CO_2 removal rates, achieving energy savings of up to 84% compared to air combustion systems. While oxy-fuel systems incur an energy penalty due to the ASU, their higher CO_2 concentrations enhance mass transfer efficiency, leading to lower overall energy demands during capture processes. In contrast, traditional post-combustion methods typically exhibit higher energy penalties and auxiliary power consumption.

Application of oxy-combustion technology mainly concerns solid fuel-fired boilers, including PC boilers or circulating fluidized bed boilers, but more and more consideration is being given to the possibility of using oxy-combustion in energy

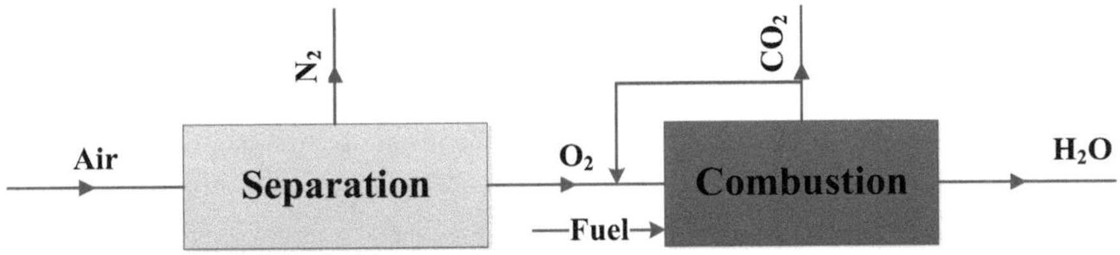

FIGURE 15.7 Block diagram of electricity generation heat production with the oxy-combustion CO_2 capture method.

systems with gas turbines. Oxy-combustion using CO_2 capture via adsorption presents a promising approach for reducing carbon emissions in power generation. This method involves burning fossil fuels in pure oxygen, resulting in flue gas with a high concentration of CO_2, facilitating efficient capture. Key applications and benefits include:

1. High CO_2 Concentration: The flue gas can contain up to 90% CO_2, significantly enhancing the efficiency of adsorption processes compared to traditional air combustion methods.
2. Reduced Energy Consumption: Oxy-combustion systems require less energy for CO_2 capture, with studies indicating energy savings of up to 84% compared to conventional methods.
3. No Nitrogen Oxides Emissions: The absence of nitrogen in the combustion process eliminates NO_x emissions, contributing to cleaner air quality.

Adsorbents such as MOFs and ionic liquids are being explored for their effectiveness in capturing CO_2 under these conditions, further optimizing the process.

The strengths of oxy-combustion are nitrogen oxide (NO_x) reduction, boiler dimension reductions, a simplified CO_2 capture method compared to other technologies, the possibility of applying in existing technologies, and less mass flow rate of exhaust gases (about 75% less compared to combustion in the air). The weaknesses of oxy-combustion are the high material requirements because of the high temperatures, an efficiency decrease (oxygen production process is energy-consuming), and a high capital cost.

Oxy-combustion methods are mainly used at the laboratory scale and for pilot installations, among which are the callide power station [59] and compostilla thermal power. A 30 MW_e experimental unit in the callide power station started operation in 2012. The nominal flow of 98% pure oxygen, which was supplied to the boiler, was 19,200 m_n^3/h. Various types of fuel have been the object of investigation (Callide coal, Minerva coal, etc.). Daily production of CO_2 based on cryogenic capture technology was 75 t. The unit was closed after a successful demonstration in 2016. A pilot power plant unit with CO_2 capture, based on oxy-combustion technology, was developed as part of the OXYCFB 300 project at compostilla thermal power. The CO_2 capture installation was equipped with a circulating fluidized bed boiler with a thermal power of 30 MW_{th} and a PC-fired boiler with a thermal power of 20 MW_{th}. The flue gas stream was 800 m^3/h. The daily capacity of the separation process, which was carried out using the cryogenic method, was 3–5 tons of CO_2.

The development of oxy-combustion technology is not only connected with solid-fired fuel units. It also concerns systems equipped with gas turbines, where combustion in an oxygen-enriched atmosphere takes place. Within the framework of the project Negative CO_2 Emission Gas Power Plant [60], the concept of a negative CO_2 emission gas power plant based on oxy-combustion combined with CO_2 capture from flue gas was developed. Application of CO_2-neutral fuel, such as sewage sludge in combination with oxy-combustion and CO_2 capture, will allow negative emission levels [61]. The CO_2 capture process will be possible, among other things, through the use of a prototype spray-ejector condenser (SEC). The main task of the SEC will be to condense the water vapor from the exhaust gases.

Among other CO_2 capture technologies, oxy-combustion carbon capture does not require many process modifications. Oxygen is used for the combustion process instead of air to eliminate the nitrogen content in flue gas, which leaves flue gas with carbon dioxide and water vapor as the major contents. This flue gas does not require high energy consumption for CO_2 separation. The boiler's temperature is controlled by again sending part of the flue gas (about 70%) to the boiler. The ASU, which separates oxygen from air, is the most energy-consuming part of the oxy-fuel combustion process. An industrial-scale cryogenic ASU consumes up to 200–225 kWh/t on an industrial scale [62]. To improve efficiency and CO_2 capture, oxy-fuel combustion carbon capture is combined with moderate and intense low-oxygen dilution (MILD). MILD oxy-combustion carbon capture holds many benefits, such as improving the efficiency of the plant, improving the purity of the CO_2, and reducing energy consumption [63]. The oxy-fuel combustion CCS technology used in the Allam cycle shows a higher efficiency of 55–59%, which is higher when compared to that of a combined cycle power plant with a carbon capture unit. In the Allam cycle, the heat generated from the ASU is sent to the regenerator to heat up the CO_2 to 400°C, which is then reused in the combustion chamber, improving the cycle efficiency [64].

15.5 INDUSTRIAL APPLICATIONS OF CO_2 CAPTURE AND STORAGE

CO_2 adsorption has significant industrial applications, particularly in sectors like cement, steel, and natural gas processing. It utilizes porous materials such as MOFs and zeolites, which enhance CO_2 capture efficiency and reduce energy consumption during regeneration. PSA is favored for its cost-effectiveness and operational flexibility, enabling effective CO_2 separation from flue gases. Furthermore, integrating CO_2 capture with utilization technologies can convert captured CO_2 into valuable chemicals, contributing to sustainability goals.

15.5.1 Power Generation

CO_2 concentrations in the atmosphere have increased by 40% compared to pre-industrial times and recently passed the 400 ppm milestone. One possible mitigation measure to stabilize atmospheric CO_2 concentrations while continuing to exploit fossil fuel resources is CCS. CCS involves separating CO_2 from large anthropogenic point sources such as thermal power plants, compressing it for transportation, and permanently storing it in underground geological formations.

The industrial application of CO_2 adsorption in power generation focuses on capturing carbon emissions effectively during energy production processes [65]. Key areas include:

1. Pre-combustion Capture: In IGCC plants, CO_2 is captured from syngas before combustion. Technologies like the Selexol process utilize physical solvents to absorb CO_2 efficiently, achieving up to 95% capture rates under high-pressure conditions.
2. PCC: Adsorption technologies are employed to capture CO_2 from flue gases in conventional power plants. Techniques such as rotary adsorption and advanced sorbents enhance the efficiency of CO_2 removal, contributing to lower emissions.
3. Cost Reduction: Ongoing research aims to develop more affordable and efficient adsorbents, such as MOFs and zeolites, which can operate effectively at various temperatures and pressures, thereby reducing overall carbon capture costs.

These applications are critical for meeting climate goals by minimizing greenhouse gas emissions from power generation. Many different techniques have been proposed for capturing CO_2. These include chemical or physical absorption, adsorption, reactive solids, membranes, and cryogenic processes [66, 67]. All these various processes are neither equivalent nor at the same stage of development. Cryogenic separation does not appear to be a contender, because it requires too much energy and is too expensive when the volume of CO_2 in the exhaust gases is too low and is at atmospheric pressure. However, in the case of a high-pressure gas mixture with a high content of CO_2, it would be possible to liquefy it by cooling it without using too much energy, to separate it from other gases. Separation by membranes is attractive, particularly in the case in which membranes are associated with a chemical solvent for which they act as a selective contactor with CO_2 (combination of filtration and chemical absorption), but there are also many problems with the development of this technique: impact of dust, steam, physical degradation of membranes. To date, all commercial CO_2 capture plants are based on absorption for separating CO_2 [68], as it is the most mature and well-understood technology. However, its large-scale deployment is hindered by the large power consumption, which negatively affects the energy efficiency of the plant. That, combined with other concerns related to solvent toxicity and the potentially high corrosion rate, makes it advisable to investigate alternatives. The PSA process is analyzed as an option for post- and pre-combustion CO_2 capture. During the PSA adsorption step, the CO_2 present in the feed gas stream is fixed on the surface of the selected adsorbent. Then, the regeneration of the bed is carried out by a pressure swing operation. The potential advantage connected to this process is the absence of any thermal energy duty during the regeneration step. Adsorption processes have been successfully employed for CO_2 removal from synthesis gas for hydrogen production [69]. With regard to CCS applications, PSA process suitability has to be proven yet. A large number of studies have been done to assess PSA processes operating in the conditions typical of post- and pre-combustion [70, 71] applications. A significant lack was found in the analysis of more comprehensive systems, where the PSA process is integrated with the rest of the plant. Few works deal with the understanding of such complex arrangements. In post-combustion applications, only preliminary studies have been carried out, whose results can be considered partial [72] and/or focusing on a particular side of the topic (e.g., economic considerations) [73]. In pre-combustion applications, more thorough analyses have been performed. Liu and Green [74] evaluated the applicability of PSA as CO_2 removal technology in an IGCC. They simulated a warm PSA process based on a tailored adsorbent, able to perform at elevated temperatures. The results achieved are in line with those of a Selexol absorption process. Other studies investigated the performance of sorption-enhanced water gas shift, an innovative CO_2 capture process for pre-combustion applications, applied to both IGCC [75] and NGCC [76]. In either case the outcome appears to be extremely promising. The objective of this section is to provide a full-plant analysis of coal-fired plants implementing CO_2 capture by a cold PSA process, meaning that the process takes place at temperature levels suitable for many of the most common adsorbents.

In the current section, the suitability of the PSA process for CO_2 capture in coal-fired power plants has been assessed. The effectiveness of PSA is evaluated on three different levels: CO_2 separation performance, energy efficiency, and footprint of the technology. A post- and a pre-combustion scenario have been considered.

The model of the power plant was developed by Thermoflow Inc. products: STEAM PRO, GT PRO, and THERMOFLEX. The focus is on coal-fired power plants, since the combustion of coal produces high specific emissions of CO_2 per unit of electricity generated. Accordingly, two thermal power plant layouts were conducted to represent the most common systems for coal-based power generation. These systems are an advanced super critical (ASC) PC plant and an IGCC plant, respectively constituting the basis for the post- and pre-combustion CO_2 capture scenario.

I ASC Plant with CO_2 Capture by PSA

PSA is an effective method for CO_2 capture in ASC coal-fired power plants. It operates by integrating a PSA unit to separate CO_2 from flue gases post-combustion. Studies show that PSA can achieve competitive CO_2 separation performance and energy efficiency, often outperforming traditional absorption methods like amine-based systems. The technology utilizes specific adsorbents, such as zeolites, tailored for optimal performance in various operational scenarios, making it a promising option for reducing greenhouse gas emissions in power generation. The integration of a PSA unit in the ASC plant is not affecting much the general layout. The additional units are all downstream of the flue gas treatment units and consist of a water removal section, a PSA process, and a compression stage for CO_2 transport. The plant upstream remains basically unchanged. The resultant plant layout is represented in

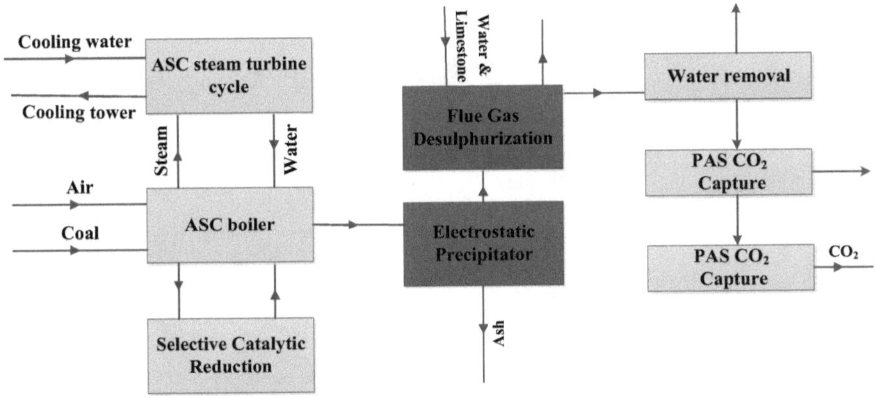

FIGURE 15.8 ASC plant with integrated PSA unit for CO_2 capture and a CO_2 compression unit.

Figure 15.8. The application of PSA for CO_2 capture in ASC power plants focuses on efficiently separating carbon dioxide from flue gases. Key aspects include:

1. Technology Suitability: PSA is evaluated as a viable option for CO_2 separation in coal-fired power plants, demonstrating competitive performance compared to traditional absorption methods, particularly in terms of energy efficiency and separation effectiveness.
2. Process Design: The PSA process involves multiple cycle steps, including feed, heavy reflux, and depressurization, which optimize CO_2 recovery rates to over 90% with high purity levels.
3. Economic Viability: Recent studies indicate that the cost of CO_2 capture via PSA can be lower than conventional absorption techniques, making it an attractive option for power generation facilities aiming to reduce emissions.
4. Flexibility and Integration: PSA systems can be tailored for various operational conditions and are suitable for retrofitting existing plants, enhancing their carbon capture capabilities without significant infrastructure changes.

II **IGCC Plant with CO_2 Capture by PSA**

IGCC plants can utilize PSA technology for effective CO_2 capture. PSA enhances CO_2 separation from hydrogen, improving overall efficiency and reducing energy consumption. Various configurations, such as stripping and dual reflux PSA, are explored to optimize performance. Studies indicate that, while PSA may not outperform absorption methods, it offers a viable option for near-zero emission power generation by facilitating the CCS, thus mitigating climate change impacts.

The addition of a PSA unit to the IGCC plant requires a higher degree of integration compared to the post-combustion scenario. A major difference is that the CO_2-lean gas stream leaving the PSA process (i.e., the H_2-rich gas stream) is further

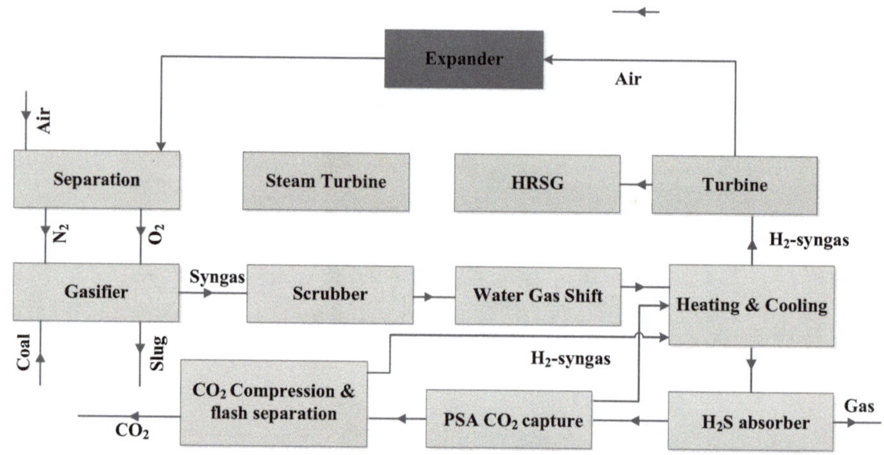

FIGURE 15.9 IGCC plant with integrated a PSA unit for CO_2 capture and a CO_2 compression unit.

processed in the plant, constituting the fuel for the gas turbine. The additional units, with respect to the reference IGCC plant [77], consist of a WGS section, a PSA process, and a compression stage for CO_2 transport. The plant layout is presented in Figure 15.9.

The WGS converts CO and H_2O into CO_2 and H_2, providing a beneficial effect on the following CO_2 separation due to the increase in the CO_2 partial pressure. COS hydrolysis is also carried out in the WGS process. The syngas is then cooled down. During the cooling process, condensing water is removed. Thanks to the relatively high pressure, water presence is drastically decreased (≈0.6%). The syngas stream at an appropriate temperature is fed to the H_2S removal unit and successively to the PSA unit. The outputs of the PSA process are a CO_2-rich stream and an H_2-rich stream. The latter is the fuel for the gas turbine cycle and is preheated by the syngas leaving the WGS process. Since the CO_2-rich gas stream does not achieve the requirements for being processed and transported, a further purification step is implemented.

In the post-combustion scenario a PSA process is integrated with an ASC PC plant. The outputs of the full-plant analysis are compared to those of a similar plant implementing a state-of-the-art absorption process for capturing CO_2. A two-stage PSA process is necessary to achieve satisfactory characteristics of the CO_2-rich stream to be transported and stored.

The application of a PSA process in a pre-combustion scenario returns promising results. The PSA process is integrated into an IGCC plant. The outputs of the full-plant analysis were compared to those of a similar plant implementing a state-of-the-art absorption process for capturing CO_2. Using PSA for CO_2 capture in IGCC plants offers several advantages:

1. Energy Efficiency: PSA can achieve lower energy penalties compared to traditional absorption methods, particularly in configurations that avoid tail gas compression.

2. Flexibility: PSA systems can be designed to operate in various configurations, allowing for adaptable performance based on operational needs.
3. Purity and Recovery: PSA can provide high purity levels of captured CO_2, with certain designs achieving purities above 95%.
4. Reduced Operational Costs: The absence of chemical solvents in PSA reduces costs associated with solvent management and handling, leading to lower capital and operational expenditures. New adsorbents are being developed specifically for CO_2 capture in IGCC plants using PSA. Notably, TDA research has patented the AMS-19 activated carbon adsorbent, which is being optimized for pre-combustion CO_2 capture. This development includes various cycle designs that incorporate steps like pressure equalization and steam purge to enhance performance. These innovations aim for high CO_2 purity (>95%) and recovery (>90%), showing promise for improving the efficiency of CO_2 capture processes in IGCC applications [78].

15.5.2 Cement Industry

The global production of cement reached 4.2 Gt in 2019[79]. CO_2 adsorption in the cement industry is crucial for reducing greenhouse gas emissions, as cement production accounts for about 8% of global CO_2 emissions. Technologies like amine-based carbon capture and oxy-fuel combustion are being developed to enhance CO_2 capture efficiency, achieving up to 90% removal rates in pilot projects. Additionally, captured CO_2 can be utilized in processes like methanol production, contributing to a circular economy. Despite high costs and logistical challenges, advancements in carbon capture technology offer significant opportunities for sustainable practices in the cement sector [80]. The market is expected to grow by 12–23% in the coming decades through 2050, as a result of the increasing global population and urbanization trends, coupled with infrastructure development needs [81]. Over the years the cement industry has significantly reduced CO_2 emissions per ton of cement by increasing energy efficiency, replacing fossil fuels with wastes that may be considered "carbon neutral" and increasing the cement: clinker ratio by increasing the use of additives. The scope for further reductions is limited by these tools, but there is a growing need to reduce CO_2 emissions to avoid contributing to anthropogenic climate change. CCS is one of the opportunities to significantly reduce greenhouse gas emissions. In many ways the cement industry is a good opportunity for CCS, as cement plants are relatively large point sources of CO_2, the concentration of CO_2 in cement plant flue gas is relatively high, and over 60% of total CO_2 emissions from a modern cement plant are from mineral decomposition where the resulting CO_2 emissions cannot be avoided by the use of alternative energy sources. Scaling up CO_2 capture technology in cement plants faces several significant challenges:

1. High Costs: The initial investment for carbon capture systems is substantial, making retrofitting existing plants economically challenging.
2. Infrastructure Limitations: There is a lack of established infrastructure for transporting and storing captured CO_2, complicating logistics and increasing costs.

3. Technical Complexity: Each cement plant has unique emissions profiles, requiring customized solutions that complicate the development of standardized carbon capture systems.
4. Regulatory and Policy Issues: Inadequate incentives and inconsistent policies hinder investment in carbon capture technologies across the industry.
5. Energy Requirements: Significant energy is needed for the capture process, which can impact overall efficiency and operational costs.

Figure 15.10 shows an advanced cement manufacturing process. Limestone is ground with clay and other minor components and fed into the preheater, where the raw meal comes into contact with the hot kiln exhaust gases in a series of vertical cyclones and is heated to about 900°C. Below the preheater, in the calciner, the limestone decomposes to form calcium oxide and release CO_2 (see Reaction 15.1). These are referred to as process emissions, as they are inherent to the clinker production process. Most of the CO_2 is usually released from the limestone raw materials in the preheater and calciner [82]. The pre-calcined meal is heated to 1450°C in a rotary kiln, where calcination is complete and the calcium oxide reacts with silica, alumina, and iron oxides to form calcium silicates, aluminates, and ferrites that make up the clinker. At the kiln outlet, the clinker is rapidly cooled in the grate cooler using incoming combustion air. Finally, the cooled clinker is mixed with gypsum and ground into a powder to produce Portland cement, or blended cement is produced with additional components—such as slag, fly ash, or limestone—replacing part of the clinker. Each cement product has its own unique composition of raw materials, which depends heavily on the location of the cement plant and the availability of raw materials.

$$CaCO3 \leftrightarrow CaO + CO2 \tag{15.1}$$

The cement sector, due to its size and the inherent characteristics of its production process, is one of the main sources of anthropogenic CO_2, accounting for 8% of global emissions [83]. Up to 0.95 tons of CO_2 is released for every ton of cement produced.

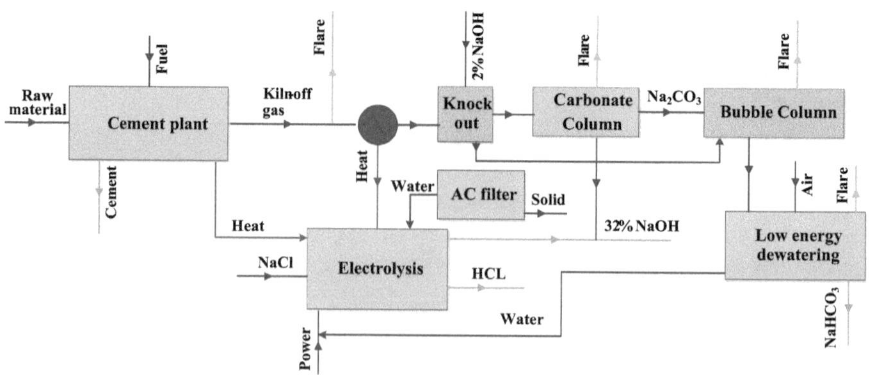

FIGURE 15.10 Scheme of a state-of-the-art cement manufacturing process.

Adsorption is a surface process in which adsorbent molecules, initially present in the bulk fluid phase, tend to concentrate on the surface of a solid adsorbent. When the phenomenon is not limited to a solid surface, the term "adsorption" is preferred. Solid adsorption-based separation processes are driven by selective attractions between a specific adsorbent, present in the bulk fluid mixture, and a solid adsorbent. There are two main mechanisms for gas adsorption by solids: physical adsorption (or physisorption) and chemical adsorption (or chemisorption). These are distinguished based on the nature of the interaction between the sorbate and the adsorbent: intermolecular (van der Waals) forces or the formation of chemical bonds, respectively. Low-temperature solid-state adsorption Candidate physisorbants for CO_2 adsorption include carbonaceous materials, zeolites, aluminophosphates, aluminum-silicophosphates, or MOFs [84]. Chemical adsorbents generally consist of an active phase, such as amines or carbonates, deposited on a porous support. Adsorbent development seeks to achieve higher CO_2 absorption capacities, faster kinetics, higher selectivity toward CO_2, milder regeneration conditions, better stability, impurity tolerance, and lower costs. For continuous operation, the adsorbent can be regenerated by reducing the pressure, in PSA processes, by increasing its temperature in TSA processes, or by using a purge gas in concentration swing adsorption processes or with a combination of the previously mentioned regeneration methods [85]. Since the cement plant flue gas is at near atmospheric pressure, the adsorbent can be regenerated at subatmospheric pressures in VSA processes. Process development seeks to reduce the overall energy consumption of capture processes, while meeting purity and recovery constraints. The main advantage of adsorption-based CO_2 capture processes is their negligible environmental impact, no emissions or toxic waste, and no corrosion problems. Furthermore, adsorption offers a context for reducing the energy penalty of the benchmark technology, amine adsorption, although it has a lower development stage. Research Triangle Institute (RTI) solid adsorbent technology was evaluated against Aker's ACCTM soluble amine technology, membrane separation, and GE's reductive calcium cycle in the Norcem CO_2 capture project (2014–1017). The RTI technology uses a high-capacity chemical adsorbent (35 wt% polyethyleneimine, PEI, loaded on silica) that flows with the flue gas in a multistage fluidized moving bed adsorber in a TSA process, where the adsorbent is circulated (see Figure 15.10). This type of reactor provides efficient heat management, which is crucial for the chemical adsorption process, as any increase in temperature in the adsorbent will reduce the CO_2 capacity of the adsorbent. RTI evaluated the sorbent exposure to real cement flue gas using the automated absorbent test rig installed at the Brevik cement plant, in Norway, during the first phase of the project. The slipstream from the flue gas was cooled and the condensate was removed. This proved to be an effective method for reducing contaminants that could affect the RTI adsorbent, as most of them are removed with the condensate. Tests showed that exposure to 100 ppm SO_2 caused a 30% reduction in CO_2 adsorption capacity. A prototype of an RTI, with an adsorbent inventory of 75 kg, was operated at the Berwick Cement Plant between September and November 2016. A consistent CO_2 capture rate of 80–90% was demonstrated with an energy consumption of 2.4 GJ/t CO_2 avoided (without heat recovery). According to the technical and economic assessments carried out, the cost of CO_2 capture, for a minimum capture rate of 85%, without heat integration, is

45.8 €/t of CO_2 avoided. For a minimum absorption rate of 85%, but using the waste heat available in the cement plant, €40.7/t of carbon dioxide is avoided. By using all available low-grade heat, without a minimum absorption rate restriction, the cost is reduced to 38.6 €/t CO_2[86].

Adsorption-based technology might be near to profit-oriented scale demonstration in the cement sector: in January 2020, Svante (formerly Inventys), LafargeHolcim, Oxy Low Carbon Ventures, and Total launched a joint study to assess the viability and design of a commercial-scale carbon capture facility at the LafargeHolcim's Portland Cement Plant in Florence, Colorado, United States [87]. This initial scoping study was successfully completed in June 2020, and in September 2020, the U.S. Department of Energy's National Energy Technology Laboratory (DOE-NETL) awarded $1.5 million of federal funding to support the advancement of the LH CO_2MENT Colorado Project. This is the largest demonstration in the cement sector with a capacity of up to 2 million tons of CO_2 captured annually [87]. The carbon capture facility uses Svante's CCS technology, which is an intensified TSA process that uses a patented structured adsorbent architecture (spacer sheets), and a proprietary process cycle design, the Veloxotherm™ process, in which a rotating adsorbent is used for utilization.

The carbon capture facility would use Svante's CCS technology, which is an intensified TSA process that uses a patented structured adsorbent architecture (spaced sheets), and a proprietary process cycle design, the Veloxotherm process, where a rotary adsorber is used to capture the CO_2 from the flue gas and release it by displacement with steam (and vacuum, VTCSA), regenerating the adsorbent in a continuous manner and in a single unit (see Figure 15.11). Svante's process claims to be material agnostic: it can use a variety of adsorbents, from carbon to functionalized silicas or MOFs. The latter has great potential due to the absorption and rejection of pressure and temperature fluctuations, leading to lower parasitic energy loads, faster kinetic speeds, and higher work capacities than conventional adsorbents, although they are

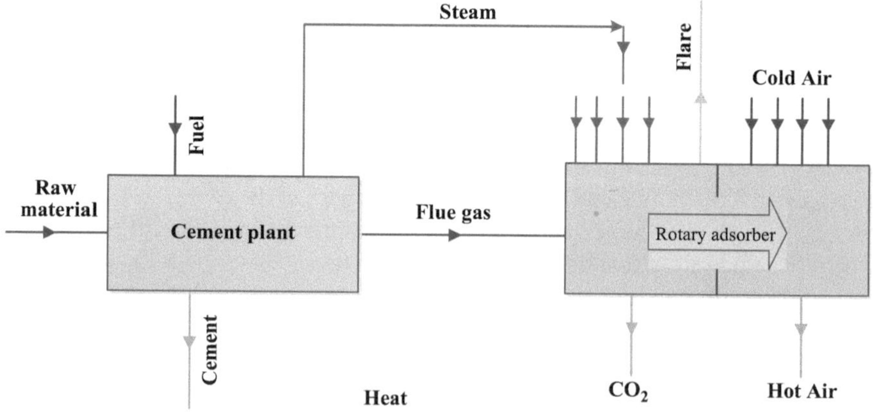

FIGURE 15.11 Simplified flow diagram of Svante's Veloxotherm process integrated into a cement plant.

still under development. As advanced adsorbents are developed, cost reductions can be expected [88]. According to Svante's CEO, the abatement cost for the cement plant is approximately \$50 per ton of CO_2. This fact, combined with the 45Q tax credit, which is \$35 per ton of CO_2 saved through EOR, and the current price of CO_2, around 20 tons of CO_2, makes the project economically feasible [89].

Recent advancements in cement manufacturing focus on sustainability and efficiency. The new Mitchell cement plant in Indiana, the second-largest in North America, exemplifies this trend with a capacity of 2.4 million tons per year and significant reductions in CO_2 emissions through advanced technology and alternative fuels. Other notable facilities, like Holcim Ste Genevieve plant, emphasize environmentally efficient operations and high production capabilities. The industry is also exploring innovative fuel mixes, including biomass and hydrogen, to achieve net-zero emissions by 2050.

The Mitchell plant significantly contributes to Heidelberg Materials' sustainability goals by focusing on low-carbon production and carbon capture technologies. It aims to reduce CO_2 emissions from clinker production by nearly 30% through natural gas usage and is set to implement a Carbon Capture, Utilization, and Storage (CCUS) project, targeting a 95% capture rate of emissions by 2030. This facility also produces EcoCem PLC™, a lower-carbon cement, enhancing the company's product offerings while addressing U.S. supply chain constraints and promoting sustainable practices in the cement industry. The Mitchell cement plant reduces energy consumption and emissions through several advanced strategies: (1) Natural Gas Utilization: It operates primarily on natural gas, leading to a nearly 30% reduction in carbon dioxide emissions per ton of clinker produced. (2) Carbon Capture Technology: The facility plans to implement a CCUS project aiming to capture 95% of its CO_2 emissions by 2030, potentially reducing emissions by about 2 million tons annually. (3) Advanced Equipment: The plant features high-efficiency equipment, including an X AF calciner and optimized kiln systems, which enhance energy efficiency and allow for the use of alternative fuels. The Mitchell plant is significant for the American Midwest market as it represents a major investment of over \$600 million, enhancing local cement production capacity to 2.4 million tons annually. This facility is crucial for alleviating supply constraints in the region and is expected to create approximately 50 new full-time jobs. Additionally, it incorporates advanced technologies aimed at reducing CO_2 emissions by nearly 30% and supports Heidelberg Materials' sustainability goals, including a commitment to carbon capture and low-carbon products, positioning it as a leader in sustainable cement manufacturing in North America.

15.5.3 Natural Gas Processing and Storage

CO_2 adsorption plays a crucial role in natural gas processing, particularly in CCS. It addresses CO_2 emissions from natural gas combustion, which are less concentrated than those from coal. Efficient adsorbents are needed to capture CO_2 at low concentrations (≤3.9%) and elevated temperatures (≥40°C) to meet regulatory targets. Technologies like PSA are utilized for CO_2 removal, enhancing the purity of natural gas and reducing environmental impact. Innovations in adsorbent materials and processes can improve capture rates and operational efficiency. The main challenges in using CO_2 adsorption for natural gas processing include:

1. Low CO_2 Concentration: Natural gas flue emissions contain only about 3.9% CO_2, making capture more energetically demanding compared to coal. Adsorbents must have high capacities (>2–3 mmol/g) to be effective at these low concentrations.
2. High Regeneration Temperatures: Effective desorption of strongly-bound CO_2 requires high temperatures (>120°C), which can lead to thermal degradation of adsorbents and increased energy costs.
3. Water Saturation: Natural gas emissions often have high water content (8.4%), complicating CO_2 capture as adsorbents must selectively capture CO_2 in humid conditions while maintaining stability.
4. Economic Viability: Current adsorption technologies, such as PSA, have higher operational costs and lower efficiencies compared to traditional methods like amine scrubbing, limiting their commercial application.

I General

CO_2 separation technology has been used in some industrial sectors, such as natural gas processing, for decades. These separation technologies were developed to remove unwanted CO_2 from mixtures with other gases. In other industries, CO_2 is a by-product of chemical processes used to produce materials such as hydrogen or fertilizer. Natural gas is a naturally occurring hydrocarbon gas. It is composed mainly of methane and a certain percentage of carbon dioxide (CO_2), hydrogen sulfide (H_2S), nitrogen, and varying amounts of higher alkanes. Natural gas is typically associated with other fossil fuels in coal beds. Removing carbon dioxide from natural gas has recently become a global issue. In addition, natural gas for liquefied natural gas (LNG) production needs to meet gas and pipeline specifications and effectively liquefy natural gas processes. Reducing CO_2 emissions into the atmosphere is essential to combat global warming. The most common method of CO_2 removal is amine absorption. PSA, distillation, and membranes are emerging techniques for CO_2 capture.

The first project began in 1972 using a waste stream of CO_2 by-products from several natural gas processing facilities in the Val Verde region of South Texas. Carbon dioxide previously separated from the natural gas stream at Valverde gas plants was compressed, rather than discharged, and transported through the first large-scale, long-distance CO_2 pipeline to an oil field several hundred kilometers away elsewhere in Texas. The CO_2 was then injected into SACROC. Techniques for removing CO_2 from raw natural gas streams to meet sales gas needs have been used since the 1930s. Available methods for removing CO_2 from natural gas include physical solvents, chemical solvents, absorption processes, and membrane contactors. A combination of chemical and physical solvents is widely used in LNG processing facilities. The main processes used to purify certain types of gases are in accordance with the gas specifications and the healthy environment and are classified as chemical and physical adsorption, solid surface adsorption, membrane separation, cryogenic separation, and a combination of physical and chemical solvents (mixed solution). Factors such as the type and concentration of impurities in the feed gas, the characteristics of the final product and sales gas, the composition of the feed gas, and the feed gas conditions affect the selection process. The two main factors are the sour feed gas composition and the operating conditions. Increased physical solvent efficiency is achieved with a higher CO_2 partial pressure in the purified gas stream. In contrast, high amounts of heavy hydrocarbons reduce the use of physical adsorbents.

The uptake of acidic gas components by a solid adsorbent is known as the adsorption process [90, 91]. The main industrial adsorption materials are ionic oxide, zinc oxide, and molecular sieve (zeolite) [92]. The microporous structure of the adsorbent selectively retains the adsorbed components. When the solid adsorbent particles become saturated by the remaining acid gas, the particles are regenerated or the adsorbents are removed from the container for regeneration purposes [93]. Regeneration is usually accomplished by blowing hot sweet gas through a molecular sieve bed. Zeolite is a suitable adsorbent used to absorb carbon dioxide from natural gas. Zeolite is suitable for cases where the CO_2 concentration in natural gas is low. The advantages of adsorption systems include the absence of mechanical damage, easy operation of the power plant, no influence from pressure fluctuations, simultaneous removal of acid gas, and the possibility of gas dehydration. However, the adsorption process is a batch system and hence is limited to small amounts of gas. The process drawbacks are overcome by providing two absorption units: while one unit is running, the other is on standby or in the regeneration phase (see Figure 15.12) [94]. In adsorption systems, purification and separation of gas mixtures is carried out by selective adsorption of a component of the gas mixture onto porous adsorbents. This process can be used to separate pollutants from various industrial processes. In the natural gas purification industry, molecular sieves are used to dehydrate natural gas. Zeolite is used to remove CO_2 from natural gas. High pressure is required to introduce impurities into porous solid particles.

Natural gas processing often involves CO_2 adsorption to meet purity specifications. PSA is a promising method for CO_2 removal, particularly in high-pressure

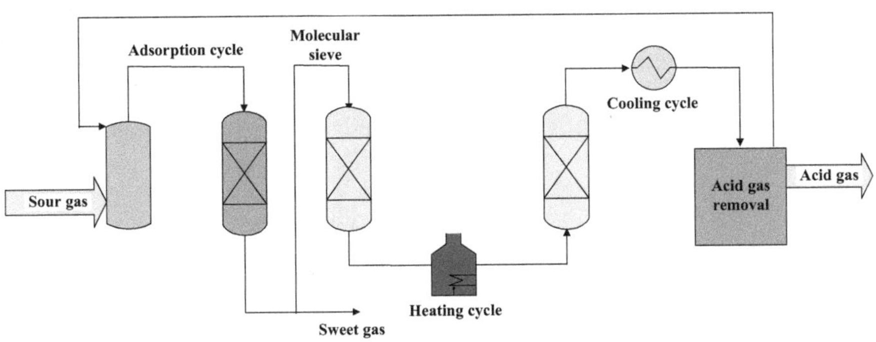

FIGURE 15.12 Schematic diagram for the CO_2 adsorption process on a solid-molecular sieve.

environments typical of natural gas production. Studies indicate that, while PSA is not yet commercially viable, it shows potential for improving CO_2 capture efficiency compared to traditional methods like amine absorption. Additionally, facilities like Norway's Sleipner-T demonstrate successful CCS, injecting about 1 million tons annually into deep saline aquifers. Overall, advancements in adsorption technologies are essential for effective carbon management in natural gas processing. CO_2 adsorption in natural gas processing typically employs PSA or vacuum pressure swing adsorption (VPSA) techniques. These methods utilize materials like zeolite 13X for selective CO_2 removal, achieving high selectivity ratios at ambient temperatures. The PSA process aims to produce pipeline-quality natural gas with CO_2 levels below 2.5%. Challenges include optimizing adsorbent materials and minimizing methane losses during the CO_2 recovery process, which can significantly impact operational costs. Scaling up VSA–PSA processes for industrial use faces several challenges:

1. Pressure Drops: Large pressure drops can occur in scaled-up systems, leading to increased energy consumption and operational costs.
2. Uneven Fluid Distribution: Achieving uniform fluid distribution across the adsorbent bed is difficult, which can affect separation efficiency.
3. Complexity of Control Systems: The dynamic nature of PSA processes requires advanced control strategies, which can be complicated and may introduce delays in system response.
4. Cost of Adsorbents: High capital costs associated with adsorbent materials and system infrastructure pose economic barriers to scaling.
5. Optimization Needs: The lack of systematic methodologies for optimizing large-scale PSA cycles complicates effective design and operation.

II. CO_2CRC Otway site

In the case of natural gas transportation through pipelines, the CO_2 specification depends on each country but is around 2–4%, so if the natural gas exceeds this percentage, the CO_2 must be removed to meet the specification. CO_2 absorption from NG can be accomplished by several techniques such as solvent washing, membranes, or cryogenic distillation. CO_2 capture by the absorption process has the potential to reduce energy requirements and operating costs due to lower energy consumption and low maintenance requirements. Physical adsorbents and the PSA process are considered suitable for CO_2 absorption at higher CO_2 partial pressures. The PSA technology for this software is not commercial and is only in the research and development phase. The high pressures and high flow rates used in this application are not similar to other applications in which PSA is widely used.

The CO_2CRC Otway Project is a pioneering CCS initiative located in southwestern Victoria, Australia. It aims to demonstrate the safety and efficacy of CO_2 storage through extensive monitoring and verification programs. The project has undergone multiple stages since its inception in 2004, including the injection of over 95,000 tons of CO_2 into various geological formations. It serves as a global benchmark for future CCS efforts, focusing on reducing greenhouse gas emissions and ensuring environmental safety. Currently, the most mature and practical gas sweetening process is the solvent-based absorption process and the combination of the absorption and membrane system, but neither is effective enough to deal with a gas source with high pressure at the gas wellhead. Adsorbent-based processes may be applicable to the separation of CO_2 from natural gas in high-CO_2 fields. Several previous studies have shown that higher CO_2 concentrations facilitate CO_2 separation through absorption processes [95]. Furthermore, the high pressure that occurs in typical natural gas fields, which inherently provides the driving force for separation, can potentially make the absorption process a low-energy process. Absorption processes are often used in the areas of air separation and natural gas dehydration, but in the case of applications for CO_2 removal from natural gas, they are still in the research and development phase. When implementing adsorption-based gas separation, there are two critical factors: designing an effective PSA loop process and selecting the appropriate adsorbents.

Santos et al. conducted a biogas upgrading process on a two-plate PSA system and studied the effect of recycling pressure equalization and purification streams on performance. They showed that the process with a pressure equalization step outperformed the other processes [96]. Kang et al. used a four-bed PSA experimental device to separate carbon dioxide from biogas (54.9% CH_4 and 45.1% CO_2) [97]. By comparing with process simulation results, they found several general rules to guide the design of future CO_2/CH_4 adsorption separation using PSA processes. For both of the aforementioned studies, the highest pressure during the process was about 4 bar, which was much lower than the practical pressure for natural gas sweetening application.

Grande and his colleagues have recently completed studies on high-pressure CO_2/CH_4 separation using PSA processes. They designed a complex PSA cycle to achieve the separation performance required for natural gas sweetening [98]. According to their simulation results, a well-designed 12-bed PSA system can treat a natural gas feed stream with 10% CO_2 at a flow rate of 500,000 m^3/h and achieve a purity of 97.8% CH_4 in the hydrocarbon product stream, which is just suitable for piping transportation, and a purity of 84.5% CO_2 in the CO_2 product stream. As the authors pointed out, however, the need for very large footage for a 12-column PSA plant, as well as the loss of methane in the CO_2 product stream, become major drawbacks compared to a conventional absorption-based process. This study is the most relevant study of PSA technology for CO_2 separation from natural gas, but no corresponding test rig was built to follow up on the research findings.

The adsorbents used to remove CO_2 from methane include activated carbon [98], carbon molecular sieves (CMSs) [96, 99], and zeolites [100]. Recently, highly ordered silica [101], zeolitic imidazolate frameworks [102], and MOFs [103] were studied as novel adsorbents to achieve CO_2/CH_4 separation, but few of these studies were applied to high pressure gas adsorptive separation. Grande et al. chose traditional adsorbents such as CMS and activated carbon for high pressure process evaluation; however, the separation performance was not satisfactory [104]. MOFs showed extraordinary selectivity and capacity of CO_2 to CH_4 at high pressure. However, due to their vulnerable structures under high humidity and temperature, as well as their extremely high cost, MOFs are not yet feasible in large-scale CO_2/CH_4 PSA separation processes. MCM-41, a periodic mesoporous silicate material, showed promising properties for natural gas sweetening. Its CO_2 selectivity over CH_4 at 45 bar 298 K can reach around 8 and with increase with increasing total pressure [101].

Grande et al. [104] have evaluated the feasibility of using a PSA process for the removal of carbon dioxide from natural gas. This process is not commercially available yet and thus faces several challenges regarding the utilization of proper materials and efficient regeneration schemes. The study has considered the effect of different PSA process operation modes. The PSA unit was designed to upgrade natural gas with a composition of 83% CH_4, 10% CO_2, and 7% C_2H_6 available at 70 bars at a temperature of 313 K and with an inlet flow rate of 500,000 Sm^3/h. The analysis also considered several process variables. Due to the lack of existing data at the pressure levels, unpublished adsorption equilibrium and kinetic data of CMS available in SINTEF were used for designing the PSA unit. A PSA unit with 12 columns using a multi-feed principle (three columns processing the feed gas at the same time) with 15 steps including a rinse of CO_2 was necessary to produce pipeline-quality natural gas and CO_2 purity with a purity of 84.5%. It has been observed that, to minimize gas losses with the CO_2 stream, the number of pressure equalizations have to be significantly increased (a maximum of four was used). A cost comparison between the existing amine scrubbing process and the designed PSA unit indicates that the cost of CO_2 removal (including transport and storage) is 77 €/tCO_2 captured compared with 50 €/tCO_2 captured obtained in the amine scrubbing process. The main contributor for the higher cost of the PSA unit is the amount of natural gas losses with the CO_2 stream. In the case that a PSA unit can significantly reduce the natural gas losses, costs can be reduced to around 46 €/tCO_2 captured. Since this is the first study of its kind, there is a lot of room to improve process performance and that adsorption processes for this application should not only be limited to very low concentrations.

In industry, one of the early patents assigned by Air Products & Chemicals [105] is to remove CO_2 from natural gas using carbon-based porous materials in a pressure-vacuum swing adsorption process at a pressure of up to 35 bar (500 psia). Engelhard (now BASF Catalysts) applied their molecular gate technology initially for nitrogen rejection from natural gas using PSA processes and later successfully for CO_2 separation as well. The proposed adsorbents for this technology were a type of synthetic titanosilicate ETS^{-4}. Xebec Adsorption Inc. developed a new approach named Rapid Cycle PSA (RCPSA) [106], which significantly reduced the total footage of the separation units. Instead of bulky piping and valves, RCPSA used two rotary valves in a compact and integrated system, which allowed similar throughput but only in 1/20 of the conventional PSA process footprint. ExxonMobil refinery has built and operated a prototype of RCPSA units for obtaining high purity hydrogen production, and theoretically it could be repurposed for natural gas sweetening.

CO_2CRC's National Research Facility at the Otway site has contained 400,000 tons of in-situ CO_2 and methane mixture, which could serve as a prefect source for long-term testing of different natural gas sweetening processes at scale. The design of their Capture Facility has provided a feed gas composition from 5% up to 79% CO_2, spanning the range of most naturally occurring natural gas wells. The CO_2 capture demonstration skid at the Otway site is shown in Figure 15.13 and will test new membranes and adsorbents for novel CO_2 capture applications. The capture skid includes a two-bed adsorption unit, with provision for running a general adsorption cycle.

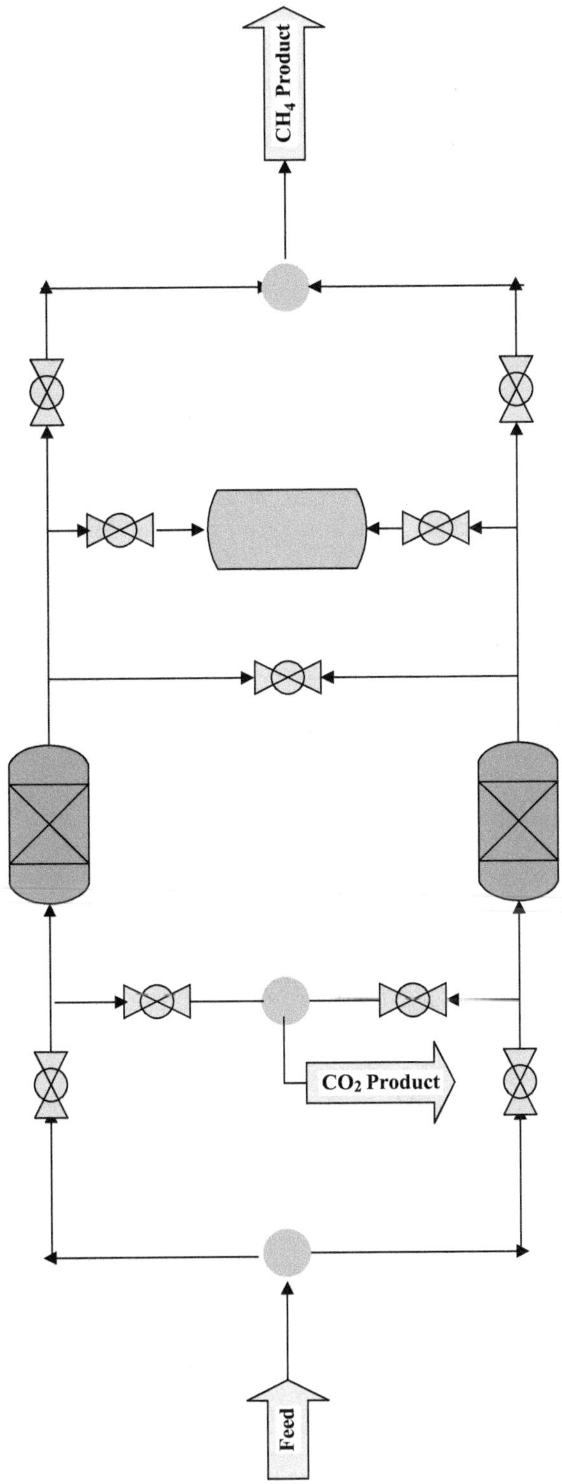

FIGURE 15.13 Scheme of CO_2CRC Otway Capture skid.

Two months of experimental data from May 2017 to June 2017 were collected showing that CO_2 was successfully separated from the feed mixture gas using a commercial adsorbent. With a feed gas of 30% CO_2/70% CH_4, the CO_2 concentration in the CO_2-rich product stream reached 50% and 12% of CO_2 concentration in the CH_4-rich product. The results of 60% CO_2 and 80% CH_4 products could also be observed from the experiment with a 50% CO_2/50% CH_4 feed gas. However, due to the limitation of the design, ideal separation performance was not obtained on the rig due to the high desorption pressure. This problem is now being rectified. Simulation results with an in-house simulation tool MINSA were in good agreement with the experimental data.

Based on their preliminary PSA operating results, simulation predictions revealed that if the desorption pressure could decrease to 2 bar instead of 5 to 7 bar in the actual process, the CO_2 capture performance could be significantly improved.

The CO_2CRC Otway Project has achieved several key milestones:

1. Successful CO_2 Injection: Over 95,000 tons of CO_2 has been injected into both depleted gas reservoirs and saline formations, demonstrating safe storage methods since 2008.
2. Advanced Monitoring Techniques: The project developed innovative monitoring technologies, including permanent pressure monitoring and distributed acoustic sensing, which significantly reduce monitoring costs by up to 85%.
3. Regulatory Frameworks: It has contributed to establishing regulatory standards for CCS in Australia, influencing future CCS projects globally.

These achievements position the Otway Project as a leader in CCS technology and research.

15.5.4 Petrochemical Industry

In petrochemical industrial processes, necessary care is taken to prevent environmental pollution, but, in some cases, pollution is inevitable. One of the important points is controlling greenhouse gas emissions. Water vapor contributes the most to the greenhouse effect, followed by carbon dioxide, methane, ozone, and chlorofluorocarbons. Of course, it is not possible to accurately determine the contribution of each of these gases. These pollutants are harmful to biological systems, change atmospheric temperatures, and cause global warming. They can also react chemically with minerals, changing their physical and chemical properties. Energy economics, laws and policies, environmental protection, and global temperature change are among the most important issues in controlling greenhouse gases. The most important factor in greenhouse gas emissions is the consumption of fossil fuels. CO_2 control is one of the most important issues in controlling greenhouse gases. One of the major problems in the strategy to reduce carbon dioxide emissions is the high cost of sequestration. Currently, separation is only cost-effective for thick, high-pressure, high-volume flows, while most flows are not. Therefore, improving efficiency and reducing separation costs to more effectively implement CO_2 utilization to reduce its emissions into the atmosphere is an important research topic. Bahrain Gulf Petrochemical Industries Company has launched a carbon dioxide recovery unit in collaboration with Italy's Tecnoment Company and Japan's Mitsubishi Heavy Industries Company. This is the first petrochemical company in the Middle East to use environmentally friendly technology. This technology reduces the amount of carbon dioxide emitted into the environment to prevent global temperature changes. In this process, carbon dioxide is separated from the gas stream exiting the methanol unit and sent to the relevant unit as feedstock in the production of urea and methanol. This is expected to reduce more than 100,000 tons of carbon dioxide emissions. The $52 million unit produces 450 tons of carbon dioxide per day.

15.6 CONCLUSION

This chapter examines various industrial applications of CO_2 capture, along with case studies that highlight successful implementations and lessons learned. The following CO_2 capture methods: pre-combustion and post-combustion based on chemical absorption, physical separation, membrane separation, chemical loop combustion, and oxy-combustion are discussed in this chapter. Post-combustion scenario analysis shows that PSA can be competitive with respect to separation performance and energy. PSA is able to meet CO_2 separation requirements, and the relative energy penalty is slightly lower than amine adsorption. Pre-combustion technology for CO_2 adsorption involves capturing CO_2 from syngas before combustion in power generation. PSA in the pre-combustion scenario returns encouraging results, approaching the results achieved with absorption in terms of both CO_2 separation performance and plant energy efficiency. The footprint, even if significantly larger, seems reasonable for real-world implementation.

CO_2 adsorption technologies are increasingly applied in various industries for effective carbon capture and mitigation of greenhouse gas emissions. Key applications include:

1. Power Plants: Adsorption processes like TSA and PSA are utilized to capture CO_2 from flue gases, offering lower energy requirements compared to traditional methods. This chapter has provided the first assessment of the suitability of PSA as a viable option for CCS in coal-fired power plants. A complete analysis of an ASC plant and an IGCC plant operating with a PSA unit is presented.
2. Cement Industries: These sectors employ adsorption techniques to reduce CO_2 emissions during production processes. To achieve climate goals, implementing CCUS in the cement industry is a necessity, not an option. Although no adsorption technology has yet reached commercial scale in the cement sector, much progress has been made in the past decade. This chapter provides an overview of CO_2 capture technologies that have been evaluated in the cement industry at pilot scale so far, as well as current plans for future commercial demonstration.
3. Natural Gas Processing: Natural gas is one of the fossil fuels that have the lowest carbon dioxide emissions per unit of energy produced. Hence, removing

CO_2 from natural gas is crucial to promote its use. To this end, CO_2CRC Ltd., the leading organization in CCS research and development in Australia, has designed and implemented a carbon capture demonstration slide at Australia's Otway National Research Center. The skid includes a PSA rig designed to remove CO_2 from a natural gas resource with high CO_2 content in the Otway basin. The operation results showed that the PSA rig could successfully capture CO_2 from high-pressure natural gas with a recovery of 66%. However, the purity of the final CH_4 product has not been sufficient to meet the public demand for natural gas (only 80% instead of 95%).

4. Petrochemical Industry: The petrochemical industry is increasingly adopting carbon capture and utilization (CCU) technologies to mitigate CO_2 emissions. This involves capturing CO_2 from flue gases using methods like adsorption, absorption, and membrane separation, which prevent its release into the atmosphere. Captured CO_2 can then be converted into valuable products such as methanol or polymers, contributing to a circular economy. While CCU presents significant environmental benefits, challenges remain in improving technology efficiency and establishing supportive regulatory frameworks. Advanced reforming processes also enhance CO_2 capture efficiency in syngas production.

Overall, adsorption methods are favored for their efficiency, lower costs, and adaptability to existing industrial systems. CO_2 adsorption technology involves capturing carbon dioxide from gas streams using solid adsorbents. Key methods include PSA and TSA, which are favored for their energy efficiency and lower operational costs compared to traditional absorption techniques. Advanced materials, such as carbonaceous and MOFs, are being developed to enhance adsorption capacity and reduce regeneration energy requirements. Recent advancements focus on optimizing pore structures and integrating various technologies to improve performance in industrial applications, particularly for post-combustion CO_2 capture.

REFERENCES

1. López, A., et al., Characterization of MIPA and DIPA aqueous solutions in relation to absorption, speciation and degradation. *Journal of Industrial and Engineering Chemistry*, 2015. 21: p. 428–435.
2. Ramazani, R., et al., Characterization of monoethanolamine+ potassium lysinate blend solution as a new chemical absorbent for CO_2 capture. *International Journal of Greenhouse Gas Control*, 2016. 51: p. 29–35.
3. Bajpai, A. and M.K. Mondal, Equilibrium solubility of CO_2 in aqueous mixtures of DEA and AEEA. *Journal of Chemical & Engineering Data*, 2013. 58(6): p. 1490–1495.
4. Stec, M., et al., Density of unloaded and CO_2-loaded aqueous solutions of piperazine and 2-amino-2-methyl-1-propanol and their mixtures from 293.15 to 333.15 K. *Physics and Chemistry of Liquids*, 2016. 54(4): p. 475–486.
5. Mondal, M.K., H.K. Balsora, and P. Varshney, Progress and trends in CO_2 capture/separation technologies: A review. *Energy*, 2012. 46(1): p. 431–441.
6. Kheshgi, H., H. de Coninck, and J. Kessels, Carbon dioxide capture and storage: Seven years after the IPCC special report. *Mitigation and Adaptation Strategies for Global Change*, 2012. 17: p. 563–567.

7. Delgado, J.A., et al., Numerical analysis of CO_2 concentration and recovery from flue gas by a novel vacuum swing adsorption cycle. *Computers & Chemical Engineering*, 2011. 35(6): p. 1010–1019.
8. Boot-Handford, M.E., et al., Carbon capture and storage update. *Energy & Environmental Science*, 2014. 7(1): p. 130–189.
9. Wakamura, O., K. Shibamura, and K. Uenoyama, Development of PSA type equipment for generating carbon dioxide from combustion waste gas. Nensho hai gas karano PSA ni yoru tansan gas seizo setsubi no kaihatsu. *Shinnettetsu Giho (Japan)*, 1992. 345.
10. Pashaei, H., A. Ghaemi, and M. Nasiri, Modeling and experimental study on the solubility and mass transfer of CO_2 into aqueous DEA solution using a stirrer bubble column. *RSC Advances*, 2016. 6(109): p. 108075–108092.
11. Hajilary, N. and M. Rezakazemi, CFD modeling of CO_2 capture by water-based nanofluids using hollow fiber membrane contactor. *International Journal of Greenhouse Gas Control*, 2018. 77: p. 88–95.
12. Norouzbahari, S., S. Shahhosseini, and A. Ghaemi, Modeling of CO_2 loading in aqueous solutions of piperazine: Application of an enhanced artificial neural network algorithm. *Journal of Natural Gas Science and Engineering*, 2015. 24: p. 18–25.
13. Ghaemi, A., V. Hashemzadeh, and S. Shahhosseini, An experimental investigation of reactive absorption of carbon dioxide into an aqueous NH3/H_2O/NaOH solution. *Iranian Journal of Oil and Gas Science and Technology*, 2017. 6(3): p. 55–67.
14. Ghaemi, A., S. Shahhosseini, and M. Ghannadi Maragheh, Experimental investigation of reactive absorption of ammonia and carbon dioxide by carbonated ammonia solution. *Iranian Journal of Chemistry and Chemical Engineering (IJCCE)*, 2011. 30(2): p. 43–50.
15. Shirazian, S., A. Marjani, and M. Rezakazemi, Separation of CO_2 by single and mixed aqueous amine solvents in membrane contactors: Fluid flow and mass transfer modeling. *Engineering with Computers*, 2012. 28: p. 189–198.
16. Rezakazemi, M., M. Sadrzadeh, and T. Matsuura, Thermally stable polymers for advanced high-performance gas separation membranes. *Progress in Energy and Combustion Science*, 2018. 66: p. 1–41.
17. Zhang, Z., et al., Modeling of a CO_2-piperazine-membrane absorption system. *Chemical Engineering Research and Design*, 2018. 131: p. 375–384.
18. Rezakazemi, M., A. Khajeh, and M. Mesbah, Membrane filtration of wastewater from gas and oil production. *Environmental Chemistry Letters*, 2018. 16: p. 367–388.
19. Norouzbahari, S., S. Shahhosseini, and A. Ghaemi, Chemical absorption of CO_2 into an aqueous piperazine (PZ) solution: development and validation of a rigorous dynamic rate-based model. *RSC Advances*, 2016. 6(46): p. 40017–40032.
20. Razavi, S.M.R., et al., Simulation of CO_2 absorption by solution of ammonium ionic liquid in hollow-fiber contactors. *Chemical Engineering and Processing: Process Intensification*, 2016. 108: p. 27–34.
21. Norouzbahari, S., S. Shahhosseini, and A. Ghaemi, CO_2 chemical absorption into aqueous solutions of piperazine: Modeling of kinetics and mass transfer rate. *Journal of Natural Gas Science and Engineering*, 2015. 26: p. 1059–1067.
22. Gao, J., et al., Experimental study of a hybrid solvent MEA-Methanol for post-combustion CO_2 absorption in an absorber packed with three different packing: Sulzer BX500, Mellapale Y500, Pall rings 16× 16. *Separation and Purification Technology*, 2016. 163: p. 23–29.
23. Rezakazemi, M., I. Heydari, and Z. Zhang, Hybrid systems: Combining membrane and absorption technologies leads to more efficient acid gases (CO_2 and H2S) removal from natural gas. *Journal of CO_2 Utilization*, 2017. 18: p. 362–369.
24. Wang, L., et al., Amine-modified ordered mesoporous silica: The effect of pore size on CO_2 capture performance. *Applied Surface Science*, 2015. 324: p. 286–292.

25. Rezakazemi, M., et al., CFD simulation of natural gas sweetening in a gas–liquid hollow-fiber membrane contactor. *Chemical Engineering Journal*, 2011. 168(3): p. 1217–1226.
26. Díez, N., et al., CO_2 adsorption capacity and kinetics in nitrogen-enriched activated carbon fibers prepared by different methods. *Chemical Engineering Journal*, 2015. 281: p. 704–712.
27. Yu, C.-H., C.-H. Huang, and C.-S. Tan, A review of CO_2 capture by absorption and adsorption. *Aerosol and Air Quality Research*, 2012. 12(5): p. 745–769.
28. Samanta, A., et al., Post-combustion CO_2 capture using solid sorbents: a review. *Industrial & Engineering Chemistry Research*, 2012. 51(4): p. 1438–1463.
29. Saeidi, M., et al., Exploiting response surface methodology (RSM) as a novel approach for the optimization of carbon dioxide adsorption by dry sodium hydroxide. *Journal of the Chinese Chemical Society*, 2018. 65(12): p. 1465–1475.
30. Xu, X., et al., Novel polyethylenimine-modified mesoporous molecular sieve of MCM-41 type as high-capacity adsorbent for CO2 capture. *Energy & Fuels*, 2002. 16(6): p. 1463–1469.
31. Jadhav, P., et al., Monoethanol amine modified zeolite 13X for CO_2 adsorption at different temperatures. *Energy & Fuels*, 2007. 21(6): p. 3555–3559.
32. Siriwardane, R.V., et al., Adsorption of CO_2 on molecular sieves and activated carbon. *Energy & Fuels*, 2001. 15(2): p. 279–284.
33. Buczek, B., Preparation of active carbon by additional activation with potassium hydroxide and characterization of their properties. *Advances in Materials Science and Engineering*, 2016. 2016(1): p. 5819208.
34. Tan, Y., et al., Adsorption of carbon dioxide by sodium hydroxide-modified granular coconut shell activated carbon in a fixed bed. *Energy*, 2014. 77: p. 926–931.
35. Siriwardane, R.V., et al., Novel regenerable sodium-based sorbents for CO_2 capture at warm gas temperatures. *Energy & Fuels*, 2007. 21(4): p. 2088–2097.
36. Ghosh, A., et al., Uptake of H2 and CO_2 by graphene. *The Journal of Physical Chemistry C*, 2008. 112(40): p. 15704–15707.
37. Lee, J.B., et al., Sodium-based dry regenerable sorbent for carbon dioxide capture from power plant flue gas. *Industrial & Engineering Chemistry Research*, 2008. 47(13): p. 4465–4472.
38. Młodzik, J., et al., Activated carbons from molasses as CO_2 sorbents. *Acta Physica Polonica A*, 2016. 129(3): p. 402–404.
39. Shafeeyan, M.S., et al., Ammonia modification of activated carbon to enhance carbon dioxide adsorption: Effect of pre-oxidation. *Applied Surface Science*, 2011. 257(9): p. 3936–3942.
40. Foroutan, R., et al., Adsorption behavior of Cu (II) and Co (II) using chemically modified marine algae. *Environmental Technology*, 2018. 39(21): p. 2792–2800.
41. Dali, A.M., A.S. Ibrahem, and A. Hadi, General study about activated carbon for adsorption carbon dioxide. *Journal of Purity, Utility Reaction and Environment*, 2012. 1(5): p. 236–251.
42. Somy, A., et al., Adsorption of carbon dioxide using impregnated activated carbon promoted by Zinc. *International Journal of Greenhouse Gas Control*, 2009. 3(3): p. 249–254.
43. Lee, S.-Y. and S.-J. Park, A review on solid adsorbents for carbon dioxide capture. *Journal of Industrial and Engineering Chemistry*, 2015. 23: p. 1–11.
44. Mirzaeian, M. and J. Peter, Thermodynamical studies of irreversible sorption of CO_2 by Wyodak coal. *Iranian Journal of Chemistry and Chemical Engineering (IJCCE)*, 2008. 27(2): 59–68.
45. Nazari Kudahi, S., A. Noorpoor, and N.M. Mahmoodi, Adsorption performance indicator for power plant CO_2 capture on graphene oxide/TiO2 nanocomposite. *Iranian Journal of Chemistry and Chemical Engineering (IJCCE)*, 2019. 38(3): p. 293–307.

46. Auta, M., et al., Fixed-bed column adsorption of carbon dioxide by sodium hydroxide modified activated alumina. *Chemical Engineering Journal*, 2013. 233: p. 80–87.
47. Guo, B., L. Chang, and K. Xie, Adsorption of carbon dioxide on activated carbon. *Journal of Natural Gas Chemistry*, 2006. 15(3): p. 223–229.
48. Shahkarami, S., et al., Breakthrough CO_2 adsorption in bio-based activated carbons. *Journal of Environmental Sciences*, 2015. 34: p. 68–76.
49. Sreńscek-Nazzal, J., et al., Modification of commercial activated carbons for CO_2 adsorption. *Acta Physica Polonica A*, 2016. 129(3): p. 394–401.
50. Maneeintr, K., et al., Comparative mass transfer performance studies of CO_2 absorption into aqueous solutions of DEAB and MEA. *Industrial & Engineering Chemistry Research*, 2010. 49(6): p. 2857–2863.
51. Nord, L.O. and O. Bolland, *Carbon Dioxide Emission Management in Power Generation*. 2020, John Wiley & Sons.
52. Wang, Y., et al., A review of post-combustion CO_2 capture technologies from coal-fired power plants. *Energy Procedia*, 2017. 114: p. 650–665.
53. Osman, A.I., et al., Recent advances in carbon capture storage and utilisation technologies: a review. *Environmental Chemistry Letters*, 2021. 19(2): p. 797–849.
54. Li, J.-R., et al., Carbon dioxide capture-related gas adsorption and separation in metal-organic frameworks. *Coordination Chemistry Reviews*, 2011. 255(15–16): p. 1791–1823.
55. Spliethoff, H. and H. Spliethoff, Carbon capture and storage (CCS), in *Power Generation from Solid Fuels, Power Systems*. Berlin, Heidelberg: Springer. 2010. p. 629–667.
56. Theo, W.L., et al., Review of pre-combustion capture and ionic liquid in carbon capture and storage. *Applied Energy*, 2016. 183: p. 1633–1663.
57. Wu, F., P.A. Dellenback, and M. Fan, Highly efficient and stable calcium looping based pre-combustion CO_2 capture for high-purity H2 production. *Materials Today Energy*, 2019. 13: p. 233–238.
58. Grande, C.A., et al., Experimental results of pressure swing adsorption (PSA) for pre-combustion CO_2 capture with metal organic frameworks. *Energy Procedia*, 2017. 114: p. 2265–2270.
59. Safina, U.F. and G. Jones, A review of local economic development and negative externalities generated by a coal-fired power station in regional Queensland: Recommendations for future actions. *Refereed Proceedings*, 2019, p. 133.
60. Ziółkowski, P., et al., Negative CO_2 emission gas power plant as technology for utilization of sewage sludge, production of electrical energy, and CO_2 capture-case of chamber under transpiration cooling, in *Turbo Expo: Power for Land, Sea, and Air*. 2024, American Society of Mechanical Engineers.
61. Ziółkowski, P., et al., A novel concept of negative CO_2 emission power plant for utilization of sewage sludge. *Proceedings of the 6th International Conference on Contemporary Problems of Thermal Engineering CPOTE*, 2020.
62. Cormos, C.-C., Oxy-combustion of coal, lignite and biomass: A techno-economic analysis for a large scale carbon capture and storage (CCS) project in Romania. *Fuel*, 2016. 169: p. 50–57.
63. Gładysz, P., et al., Thermo-ecological evaluation of an integrated MILD oxy-fuel combustion power plant with CO_2 capture, utilisation, and storage–a case study in Poland. *Energy*, 2018. 144: p. 379–392.
64. Kindra, V., et al., Techno-economic analysis of the oxy-fuel combustion power cycles with near-zero emissions. *Energies*, 2021. 14(17): p. 5358.
65. Metz, B., et al., *IPCC Special Report on Carbon Dioxide Capture and Storage*. 2005, Cambridge: Cambridge University Press.
66. Romeo, L.M., I. Bolea, and J.M. Escosa, Integration of power plant and amine scrubbing to reduce CO_2 capture costs. *Applied Thermal Engineering*, 2008. 28(8–9): p. 1039–1046.

67. Hwang, H.T., et al., A membrane-based reactive separation system for CO_2 removal in a life support system. *Journal of Membrane Science*, 2008. 315(1–2): p. 116–124.
68. Force, C.A.T., Advanced post-combustion CO_2 capture. Freund, H., J. Bauer, T. Zeiser and G. Emig (2005)." Detailed simulation of transport processes in fixed-beds." *Industrial & Engineering Chemistry Research*, 2009. 44(16): p. 6423–6434.
69. Ribeiro, A.M., et al., A parametric study of layered bed PSA for hydrogen purification. *Chemical Engineering Science*, 2008. 63(21): p. 5258–5273.
70. Tlili, N., G. Grévillot, and C. Vallières, Carbon dioxide capture and recovery by means of TSA and/or VSA. *International Journal of Greenhouse Gas Control*, 2009. 3(5): p. 519–527.
71. Casas, N., et al., A parametric study of a PSA process for pre-combustion CO_2 capture. *Separation and Purification Technology*, 2013. 104: p. 183–192.
72. Panowski, M., R. Klainy, and K. Sztelder. Modelling of CO_2 adsorption from exhaust gases, in *Proceedings of the 20th International Conference on Fluidized Bed Combustion*. 2010, Springer.
73. Ho, M.T., G.W. Allinson, and D.E. Wiley, Reducing the cost of CO_2 capture from flue gases using pressure swing adsorption. *Industrial & Engineering Chemistry Research*, 2008. 47(14): p. 4883–4890.
74. Liu, Z. and W.H. Green, Analysis of adsorbent-based warm CO_2 capture technology for integrated gasification combined cycle (IGCC) power plants. *Industrial & Engineering Chemistry Research*, 2014. 53(27): p. 11145–11158.
75. Gazzani, M., E. Macchi, and G. Manzolini, CO_2 capture in integrated gasification combined cycle with SEWGS–Part A: Thermodynamic performances. *Fuel*, 2013. 105: p. 206–219.
76. Manzolini, G., et al., Integration of SEWGS for carbon capture in natural gas combined cycle. Part B: Reference case comparison. *International Journal of Greenhouse Gas Control*, 2011. 5(2): p. 214–225.
77. Alstom, U., *European Best Practice Guidelines for Assessment of CO_2 Capture Technologies*. 2011, Trondheim, Norway: SINTEF.
78. Subraveti, S.G., et al., Cycle design and optimization of pressure swing adsorption cycles for pre-combustion CO_2 capture. *Applied Energy*, 2019. 254: p. 113624.
79. Farfan, J., M. Fasihi, and C. Breyer, Trends in the global cement industry and opportunities for long-term sustainable CCU potential for Power-to-X. *Journal of Cleaner Production*, 2019. 217: p. 821–835.
80. Ige, O.E., et al., Environmental impact analysis of Portland cement (CEM1) using the midpoint method. *Energies*, 2022. 15(7): p. 2708.
81. IEA, C., *Technology Roadmap Low-Carbon Transition in the Cement Industry*. 2018, France/WBCSD. Geneva, Switzerland; Paris: IEA.
82. Barker, D., et al., CO_2 capture in the cement industry. *Energy Procedia*, 2009. 1(1): p. 87–94.
83. Brun, M. and S.H.B. Granstrøm, *Assessment of Norway's Potential for a New CCS Industry*. 2019, Norway: University of Stavanger.
84. Sayari, A., Y. Belmabkhout, and R. Serna-Guerrero, Flue gas treatment via CO_2 adsorption. *Chemical Engineering Journal*, 2011. 171(3): p. 760–774.
85. Ruthven, D.M., *Principles of Adsorption and Adsorption Processes*. 1984, John Wiley & Sons.
86. Nelson, T.O., et al., RTI's solid sorbent-based CO_2 capture process: technical and economic lessons learned for application in coal-fired, NGCC, and cement plants. *Energy Procedia*, 2017. 114: p. 2506–2524.
87. González Plaza, M., S. Martínez Gómez, and F. Rubiera González, CO_2 capture, use, and storage in the cement industry: state of the art and expectations. *Energies*, 2020. 13(21): 5692.
88. Rezaei, S., A. Liu, and P. Hovington, Emerging technologies in post-combustion carbon dioxide capture & removal. *Catalysis Today*, 2023: p. 114286.

89. Moore, A., *US 45Q Tax Credit Key to Developing Carbon Capture Facility in Colorado*. 2020, S&P Global Platts.
90. Miller, B., Greenhouse gas–carbon dioxide emissions reduction technologies. *Fossil Fuel Emissions Control Technologies*, 2015: p. 367–438.
91. Bohloul, M., et al., CO_2 absorption using aqueous solution of potassium carbonate: Experimental measurement and thermodynamic modeling. *Fluid Phase Equilibria*, 2017. 447: p. 132–141.
92. Mofarahi, M. and F. Gholipour, Gas adsorption separation of CO_2/CH4 system using zeolite 5A. *Microporous and Mesoporous Materials*, 2014. 200: p. 1–10.
93. Chaffee, A.L., et al., CO_2 capture by adsorption: materials and process development. *International Journal of Greenhouse Gas Control*, 2007. 1(1): p. 11–18.
94. Nie, L., et al., Recent developments and consideration issues in solid adsorbents for CO_2 capture from flue gas. *Chinese Journal of Chemical Engineering*, 2018. 26(11): p. 2303–2317.
95. Ling, J., et al., Effects of feed gas concentration, temperature and process parameters on vacuum swing adsorption performance for CO_2 capture. *Chemical Engineering Journal*, 2015. 2(65): 47–57.
96. Santos, M.n.P., C.A. Grande, and A.r.E. Rodrigues, Pressure swing adsorption for biogas upgrading. Effect of recycling streams in pressure swing adsorption design. *Industrial & Engineering Chemistry Research*, 2011. 50(2): p. 974–985.
97. Kim, Y.J., Y.S. Nam, and Y.T. Kang, Study on a numerical model and PSA (pressure swing adsorption) process experiment for CH4/CO_2 separation from biogas. *Energy*, 2015. 91: p. 732–741.
98. Grande, C.A., et al., High-pressure separation of CH4/CO_2 using activated carbon. *Chemical Engineering Science*, 2013. 89: p. 10–20.
99. Rocha, L.A., K.A. Andreassen, and C.A. Grande, Separation of CO_2/CH4 using carbon molecular sieve (CMS) at low and high pressure. *Chemical Engineering Science*, 2017. 164: p. 148–157.
100. Gholipour, F. and M. Mofarahi, Adsorption equilibrium of methane and carbon dioxide on zeolite 13X: Experimental and thermodynamic modeling. *The Journal of Supercritical Fluids*, 2016. 111: p. 47–54.
101. Belmabkhout, Y., R. Serna-Guerrero, and A. Sayari, Adsorption of CO_2 from dry gases on MCM-41 silica at ambient temperature and high pressure. 1: Pure CO_2 adsorption. *Chemical Engineering Science*, 2009. 64(17): p. 3721–3728.
102. McEwen, J., J.-D. Hayman, and A.O. Yazaydin, A comparative study of CO_2, CH4 and N2 adsorption in ZIF-8, Zeolite-13X and BPL activated carbon. *Chemical Physics*, 2013. 412: p. 72–76.
103. Furukawa, H., et al., Ultrahigh porosity in metal-organic frameworks. *Science*, 2010. 329(5990): p. 424–428.
104. Grande, C.A., et al., CO_2 capture in natural gas production by adsorption processes. *Energy Procedia*, 2017. 114: p. 2259–2264.
105. Kumar, R., Adsorptive process for producing two gas streams from a gas mixture. United States Patent, No. US 5,026,406, 11 p. 1991.
106. Alizadeh-Khiavi, S., et al., Rapid cycle syngas pressure swing adsorption system. 2010, Google Patents.

16 International Laws Government Policies and Regulations

Mohsen Shayanmehr and Ahad Ghaemi

16.1 INTRODUCTION

Carbon dioxide is one of the most important gases leading to global warming. They are also the driving force behind climate change, especially carbon emissions. This is because it traps heat in the earth's atmosphere. Whenever there is an increase in CO_2 levels, more heat gets retained, thereby raising the temperature of the earth's surface. Moreover, this has been going on linearly, which means whenever we have the total amount of emitted carbons then it will correspond with the average global temperature change. Controlling the amount of carbon dioxide will lower temperatures around the world, thus preventing severe impacts caused by extreme weather conditions. Additionally, it is necessary that we stabilize our concentration levels for carbon dioxide and not go beyond certain limits to avoid dangerous effects on the climate and keep them within manageable bounds. Global CO_2 emissions from fossil fuels, industry, and land-use change from 1960 to 2023 are shown in Figure 16.1, with a steady rise in emissions from fossil fuels and industry, while land-use change remains relatively stable.

As is generally known, China is the world's highest emitter of greenhouse gases (GHGs). But this was not always so. In the last few decades, the country's GHG emissions have risen owing to rapid economic growth and industrialization. The timeline of the CO_2 emissions from transport by income level and sector is illustrated in Figure 16.2.

Different energy sources contribute to the amount of CO_2 emitted harmfully into the atmosphere at the global level. This is attributed to their carbon intensity and global energy consumption prevalence. Figure 16.3 shows possible future CO_2 emissions from global transport from 2010 to 2050, based on different models and scenarios. Each point represents an estimate from a specific study, with the solid red line showing the average trend and the dashed lines marking the highest and lowest projections. The range highlights how changes in technology, policy, and economic factors could impact future emissions.

Worldwide governments have introduced different laws and regulations to counteract the effect of industrial operations on the environment. In doing so, the laws are majorly centered on improving energy efficiency and reducing the emission of GHGs notably carbon dioxide (CO_2), which is the most common GHG emitted through

DOI: 10.1201/9781003476191-16

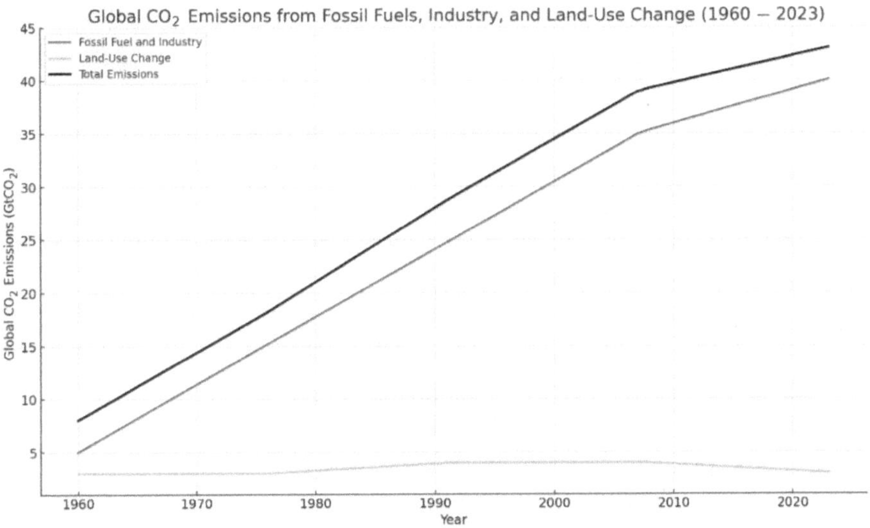

FIGURE 16.1 Global CO_2 emissions from fossil fuels, industry, and land-use change (1960–2023) [1].

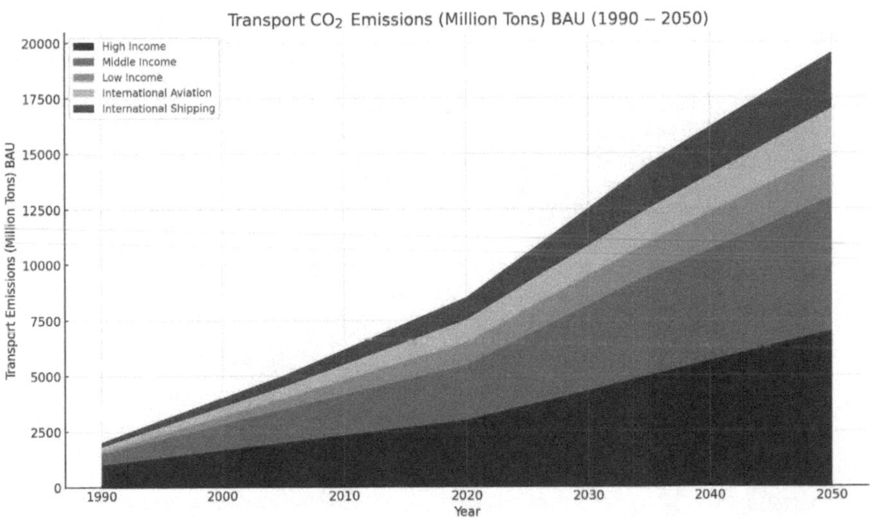

FIGURE 16.2 Future CO_2 emissions from transport by income level and sector (1990–2050) [1].

human activities. Such emissions result in persistent dry spells, polar ice meltdown, high sea levels, and rapid changes in temperature among others, which arise from a lack of enough water sources.

To solve these difficulties, conferences and agreements have been set up under United Nations sponsorship through international cooperation frameworks. Stabilizing GHG concentrations so as not to harm the climate system is their target. To this end,

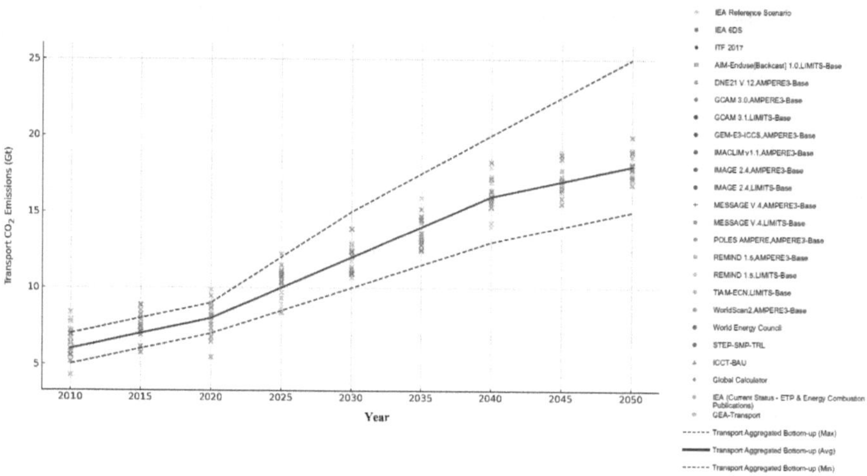

FIGURE 16.3 Projected global CO_2 emissions from transport (2010–2050) across different scenarios, showing possible pathways and overall trends in red [1].

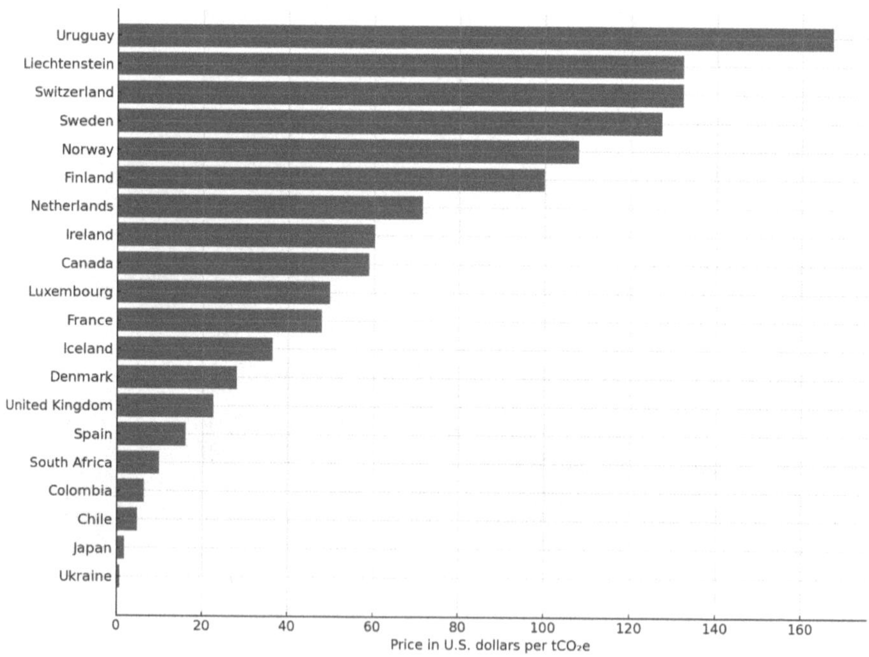

FIGURE 16.4 Carbon tax rates in various regions worldwide as of April 2024 [2].

they have decided to adopt the United Nations Framework Convention on Climate Change. This document was adopted in 1992 at the Rio Summit and has been ratified by 197 countries. It is the basis for their activities in this field. The United Nations Framework Convention on Climate Change (UNFCCC) objective is to stabilize the atmospheric concentration of GHGs at a level that would not cause dangerous anthropogenic harm to the climate system. It should also promote natural ecosystems' ability to adapt and support sustainable economic development globally. In this regard, countries have implemented a range of policy tools, including national and international energy efficiency standards, regulations, and carbon pricing mechanisms such as carbon taxes. Landmark agreements like the Kyoto Protocol and the Paris Agreement further reinforce these commitments. Figure 16.4 illustrates carbon tax rates in various regions around the world [2].

16.2 FRAMEWORKS FOR INTERNATIONAL COOPERATION

More people are increasingly acknowledging that climate change is a serious global problem caused by human activity that poses threats to human civilization. A rise in GHG emissions from human activity especially since the industrial revolution is the primary driver of climate change and global warming. Long-term droughts, the melting of polar ice, sea-level rise, the depletion of water supplies and reserves, abrupt temperature swings, and more are some of the major effects of these climatic changes [3]. In response, various regions of the world have hosted major international conferences governed by legal frameworks to address climate change and develop strategies for mitigation and adaptation. The United Nations has played a key role in organizing these gatherings. Table 16.1 provides an overview of several major UN-coordinated world assemblies focused on climate change and their outcomes [4, 5].

16.2.1 United Nations Framework Convention on Climate Change

The purpose of the UNFCCC is to address the environmental effects of global warming. It was adopted during the Rio Summit in Brazil in 1992 and ratified by 197 countries. The main goal of UNFCCC is to stabilize GHG concentrations in the atmosphere to prevent any dangerous anthropogenic interference with the earth's climate. It also provides for a united response by nations to climate change aiming at keeping these gases within states' control systems. Parties meet often for "COP" talks on climate topics. There, they check progress, swap ideas, and work as a team to fight climate change better. These sessions connect nations globally for this huge shared challenge [6, 7].

16.2.1.1 Overview of the UNFCCC and Its Role in Addressing Climate Change

Because it deals with climate change worldwide, the UNFCCC is very important. A fundamental strategy for cooperating globally to address air pollution is the UNFCCC. This should also help people accept the fact that weather patterns are becoming unpredictable; hence, they need to change their way of life. Besides, when it comes to areas like energy or manufacturing, cleanliness should not be compromised under any circumstances because that will only lead us back into the same problem; we're trying so hard to fix. The UNFCCC's main aim is to stabilize the levels of GHGs in the atmosphere so that no harmful human interference with the

TABLE 16.1
A Review of Several Important World Assemblies on Climate Change Coordinated by the United Nations [4, 5]

Row	Name of the Conference	Year of Holding	Event Place	Description
1	United Nations Conference on the Human Environment	1972	Stockholm	• First major environmental conservation conference involving 115 countries • Issued the "Declaration on the Human Environment," outlining principles for environmental protection
2	Earth Summit, Rio de Janeiro	1992	Rio de Janeiro	• Led to the establishment of the UNFCCC and adoption of the Rio Declaration • Key event for global environmental cooperation and development
3	Johannesburg Conference	2002	Johannesburg	• Emphasized economic efficiency, climate justice, and environmental conservation • Adopted an action plan for sustainable development goals
4	Bali Conference	2007	Bali	• Focused on setting emission reduction goals for developed and developing nations • Highlighted the importance of global cooperation in tackling climate change
5	Copenhagen Climate Conference	2009	Copenhagen	• Convened representatives from 192 countries to address climate change • Aimed to develop strategies for reducing greenhouse gas emissions
6	Cancun Conference	2010	Cancun	• Made decisions to combat climate change, reduce emissions, and promote clean energy markets • Advanced the global climate agenda with accountability measures
7	Durban Conference	2011	Durban	• Aimed to reach a new international agreement on environmental protection • Continued discussions on global climate action and cooperation
8	Water and Climate Festival	2014	New York	• Focused on protecting forests, promoting smart agriculture, and adapting to climate change • Highlighted the interconnectedness of water and climate issues
9	Katowice Conference	2018	Katowice	• Reached an agreement on implementing the Paris Agreement and controlling temperature rise • Marked progress in global efforts to combat climate change
10	Madrid Conference	2019	Madrid	• 66 countries committed to achieving carbon neutrality by 2050 • Demonstrated growing consensus on the urgency of climate action
11	Ambition Summit	2020	United Kingdom	• Leaders declared bold commitments and immediate actions to address climate change • Fostering momentum for global climate action despite challenges like the COVID-19 pandemic
12	Glasgow Conference	2021	Glasgow	• Reached agreements on support, finance, and emission reduction efforts • Significant step forward in global climate cooperation and transition to low-emission energy

climate system is made; hence, the endangerment is averted. This is based on the fact that the increase in GHGs causes the intensification of the natural greenhouse effect, resulting in global warming and consequently climate change. By keeping the GHG levels constant, ecological systems can adjust themselves to these variations by natural selection as well as ensuring there is no threat posed on food production, while allowing for sustainable economic development. This objective emphasizes the importance of striking a delicate balance between human actions, especially those linked to the utilization and production of energy and natural processes of the earth [8]. One of the main goals set by UNFCCC is to foster sustainable development. It means stimulating economic growth and development as well as guaranteeing that natural resources are utilized responsibly and sustainably. According to the Convention, economic development is necessary for addressing climate change since it provides adaptation resources and mitigation capacities. However, it states that such a development should not undermine the ability of future generations to meet their own needs. Thus, within UNFCCC sustainable development implies shifting to economies with low carbon dioxide emissions, improving energy efficiency, and promoting renewable sources of energy [9]. The UNFCCC is based on equity, which means everyone should be treated fairly. It also includes the idea of CBDR-RC. According to this principle, all countries must take part in solving climate change problems even if they are not responsible for creating them equally. It is because some countries may lack the necessary resources to deal with such issues while others have more than enough means to do so. Developed nations are supposed to cut their emissions first since they have released a lot of GHGs into the atmosphere before and also help develop ones in their fight against climate change. This goal shows how fair and just the world's reaction to global warming should be. The UNFCCC's main goal is to help poor countries adjust to climate change and move toward sustainable development. This involves giving them money, technology, and training. The Convention states that developing nations suffer most from global warming and have the least amount of money for dealing with it. Consequently, these countries must be helped internationally to adapt themselves to its impacts, follow strategies for growth which produce few emissions, and take part effectively in worldwide efforts at reducing them [10]. The UNFCCC has long been concerned with helping countries cope with the adverse effects of climate change. This includes efforts to deal with such events as rising sea levels, extreme weather conditions, and alterations in levels of precipitation. Methods of adaptation can be diverse, from constructing barriers against floods and creating crops that are resistant to drought all the way to establishing systems which would give warnings about natural disasters well ahead of time. Moreover, the Convention strives toward increasing people's ability to adjust; making them more robust and less susceptible to these environmental changes while placing special emphasis on assisting nations or groups thereof considered most at risk [11]. One of the main objectives of UNFCCC is to provide international climate targets and objectives to guide countries on how they should reduce their emissions of GHGs, which are leading to global warming. Countries use different platforms such as the Conference of the Parties, where they negotiate for limitations in the production of carbon IV oxide as well as increasing their ability to withstand harsh weather conditions caused by climate change. With time, nations need to keep updating and making their promises to cut down on carbon dioxide release stronger since this organization allows them to create new targets. The UNFCCC also acts as a meeting point among nations for joint efforts in fighting climate change. Through its universal participation, the UNFCCC facilitates discussions, negotiations, and the exchange of good practices in climate action. It is important to have this worldwide organization to bring all countries together to respond to climate change as a whole [12]. To respond to climate change, the UNFCCC deals with both cutting emissions and helping communities adapt to new situations. Efforts to reduce emissions—such as increasing the use of renewable energy and improving energy efficiency—can be complemented by enhancing carbon dioxide sequestration and implementing other mitigation strategies. Resilience should be built against impacts such as sea-level rise, extreme weather conditions, or changing patterns of rainfall, which are brought about by adaptation measures among others. Nations are also supposed to report regularly on how far they have gone with implementing what they agreed on about their roles toward meeting these shared challenges. They must regularly tell others verbally how much progress each state is making in achieving their goal under this agreement. Furthermore, countries' advancements toward fulfilling their nationally determined contributions (NDCs) should be regularly published under the UNFCCC. This transparency and accountability framework helps track global emissions trends, assess the effectiveness of climate policies, and identify areas where additional action is needed, and finally it can be said that the UNFCCC plays a critical role in advancing climate science and research by synthesizing and disseminating the latest scientific findings on climate change. This scientific knowledge informs policy decisions and helps countries understand the urgency and magnitude of the climate challenge [13].

16.2.1.2 Stabilizing GHG Concentrations to Prevent Dangerous Anthropogenic Interference with the Climate System

GHGs trigger the earth's climate system. These gases keep heat, leading to global warming and climate change. It is essential to stop the rise of GHG levels. This aim is very important for the planet, human well-being, and economies all over the world. The major GHGs include carbon dioxide, methane, and nitrous oxide. Their levels greatly affect the weather conditions. If we can stabilize these gases, then we will be able to avoid dangerous impacts of people on the earth's climate system. Due to the many effects that it has on different parts of the world, this problem has become a matter of global concern.

According to the U.S. National Academy of Sciences, there are links between changes in atmospheric CO_2 concentration, global temperature, and cumulative CO_2 emissions (Figure 16.1a and b). Total CO_2 discharges have been found to have a near-linear relationship with variations in worldwide average temperature [14]. Figure 16.5 shows the connection between global mean temperature change and (a) atmospheric CO_2 concentrations and (b) cumulative CO_2 emissions, demonstrating a near-linear relationship that underscores the urgency of emission reductions.

There are three different CO_2 stabilization scenarios that have been developed by the International Energy Agency (IEA) to address different levels of global warming. These scenarios aim at controlling the concentration of CO_2 in the atmosphere to

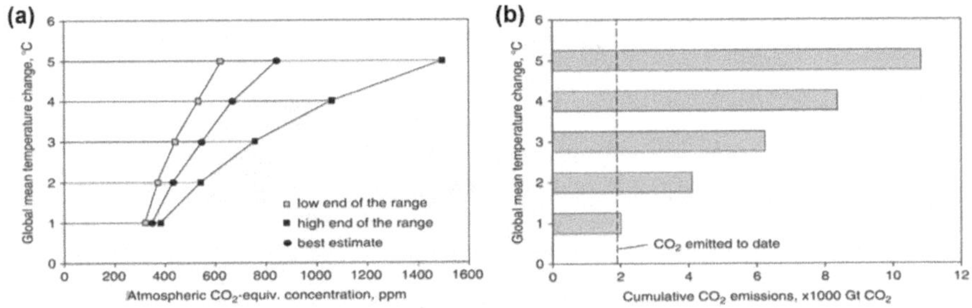

FIGURE 16.5 Connection between global mean temperature change and atmospheric CO_2 concentrations (a) and cumulative CO_2 emissions (b) [14].

limit the rise of global temperatures. We are now going to give detailed descriptions for each scenario:

1. 6°C Scenario (6DS): This is basically doing nothing in terms of cutting down CO_2 admission. According to 6DS, the global temperature is expected to rise to up to 6°C before 2100. A scenario assumes that more fossil fuels will be used and less renewable energy will be embraced. The consequences of this scenario would be felt as harsh climatic changes, leading to things like very high ocean levels, very severe storms, or tornadoes among others, disrupting many lives—both human beings' health and animals' habitats [15].
2. 4°C Scenario (4DS): The 4DS presents a medium departure from current policies. It includes some attempts at controlling carbon dioxide (CO_2) emissions, but they are insufficient to keep global warming under 2°C. This scenario is designed to hold CO_2 concentration to the point where the temperature of the entire planet will rise by about 4°C. It also presupposes wider use of renewable energy sources, certain enhancements in energy efficiency as well as carbon capture and storage (CCS) technologies on a limited scale. Compared with 6DS, this one is better; however, it still has considerable consequences for climate change such as higher chances of droughts, floods, and extinction of species [16].
3. 2°C Scenario (2DS): The 2DS is an ambitious plan that aims to halve energy-related carbon dioxide emissions by 2050 compared with 2009 levels. It is intended to guarantee that, in compliance with international climate targets, the average world temperature rises no more than 2°C over pre-industrial levels. For the 2DS to be achieved, significant adjustments must be made to the global energy system. This entails using CCS technology; expanding the use of renewable energy sources; reducing the use of coal, oil, and gas; and making significant advancements in energy efficiency. Moreover, the 2DS gives prominence to decarbonizing electricity generation and increasing the application of low-carbon fuels in industry and transportation. Also, it is important to note that, while this is a plausible scenario for averting dangerous anthropogenic interference with the climate system, it cannot be realized except through robust policy measures, technological

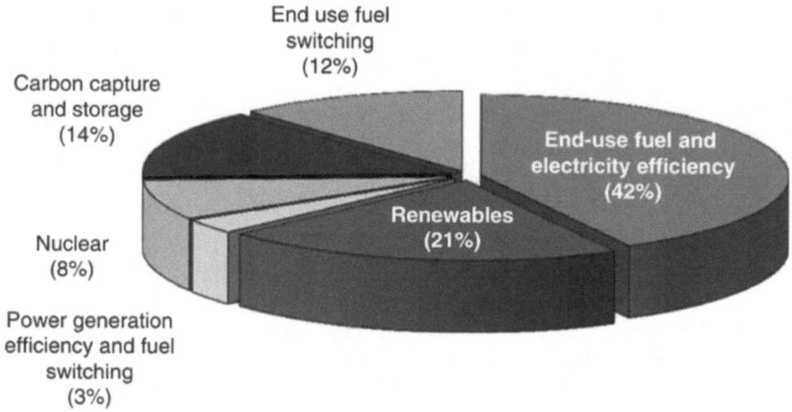

FIGURE 16.6 Contributions of technologies to carbon reductions in the 2DS pathway versus 6DS (IEA, 2012) [17].

breakthroughs, and international collaboration [16]. Figure 16.6 illustrates how various technologies contribute to carbon reductions in the 2°C Scenario (2DS) compared to the 6°C Scenario (6DS), highlighting the scale of transformation required under each pathway.

To combat climate change and avoid dangerous interference, which is primarily due to human activities, the 2DS scenario plays a very important role. The following are the most important things about it: keeping global average temperature rise below 2°C. It is called for short for the "2°C" target. The 2DS scenario tries to not go beyond 2 degrees higher than what it was like before people started using machines that burn fuels to do work because many scientists say that going past this point would have serious and irreversible effects on the weather. Hence, we can state that one of the goals of the 2DS scenario is to prevent the most severe consequences of climate change caused by not keeping temperature increase limited within its bound. Stabilizing atmospheric concentrations of GHGs. At present there is too much carbon dioxide, methane, nitrous oxide, and other gases which keep heat captured in the earth's atmosphere, so they need to be controlled if we want to achieve our goal according to the 2DS scenario—stabilization of GHG amounts at some point above pre-industrial levels but below those corresponding with 2°C warming. The situation of the 2 degrees scenario mainly refers to the global temperature control measures can still let the temperature rise to 2 and it will not exceed 2 energy world, GHG emissions, fossil fuel use and the proportion of renewable energy use, power supply system, the development of new technology and new energy perspectives, supply and demand to balance the multiple pathways of the energy transition, as well as bypassing the framework convention on climate change issues such as equity and financial support. This strategy is different and spreads the effort to reduce the seriousness of something throughout various technologies and economic sectors so that no one area has to deal with it all. These gases are becoming more concentrated because of two primary reasons. One is that people have been burning fossil fuels like oil, coal, and natural gas for energy and transportation, which has caused a lot of additional CO_2 to be released into the air since the Industrial Revolution. The other reason is deforestation; when trees are cut down, there's less carbon stored naturally, so this also contributes. Furthermore, the concentration of these same gases does rise due to natural processes like volcanoes, solar activity, or oceanic current patterns as well. The scientific evidence is clear: more than half of the global temperature increase since the 1950s has been caused by people putting GHGs—mostly from burning fossil fuels—into the atmosphere. Though natural factors do influence climate, they are not its main driver at present; humanity is. Therefore, coordinated strong measures need to be taken to abate emissions of these heat-trapping chemicals on a scale consistent with targets such as capping average planetary warming at 2°C above pre-industrial levels. It also serves to stabilize "the amount of GHGs" as an important and necessary step which has numerous benefits for environmental protection, social development, and economic prosperity [18]. The fundamental concern is how to keep GHG emissions at a steady state. This may be accomplished by implementing a variety of plans and initiatives that lower emissions and strengthen the natural systems that remove these pollutants from the environment. Table 16.2 lists strategies that help maintain stable concentrations of greenhouse gases, including both emission reduction measures and enhancement of natural carbon sinks.

TABLE 16.2
Ways to Maintain GHG Level Concentrations [19]

No.	Strategies	Description
1	Reducing emissions	• Transitioning to renewable energy sources such as solar, wind, and hydropower to reduce reliance on fossil fuels • Improving energy efficiency in industries, transportation, and buildings to lower overall energy consumption • Implementing carbon capture and storage technologies to capture and store CO_2 emissions from industrial processes and power plants
2	Afforestation and reforestation	• Planting trees and restoring forests to absorb CO_2 from the atmosphere through photosynthesis • Protecting existing forests from deforestation to maintain their role as carbon sinks
3	Enhancing carbon sequestration	• Promoting sustainable agricultural practices that enhance soil carbon sequestration • Restoring degraded lands to increase their capacity to absorb and store carbon
4	Reducing methane and nitrous oxide emissions	• Implementing measures to reduce methane emissions from livestock, rice paddies, and landfills • Improving agricultural practices to reduce nitrous oxide emissions from fertilizers and manure
5	International agreements and policies	• Participating in international agreements like the Paris Agreement to set emission reduction targets and promote global cooperation on climate action • Implementing carbon pricing mechanisms such as carbon taxes or cap-and-trade systems to incentivize emission reductions
6	Research and innovation	• Investing in research and development of new technologies for clean energy, carbon capture, and sustainable land-use practices • Supporting innovation in climate-resilient agriculture and forestry practices

Doing this all over the world together will help even more if we work against climate change by decreasing or stabilizing the amount of GHGs in our air. "This planet could be saved from destruction in future generations if different methods were used" [20].

16.3 THE KYOTO PROTOCOL

Today's environmental crisis is mainly due to global warming and climate change. This is primarily brought about by the release of GHGs when burning coal, oil, and gas. The establishment of different environmental conventions also indicates the significance of this problem. Adopted in 1997 in Kyoto, Japan, and entered into force in 2005 [21], the Kyoto Protocol represents a milestone in international environmental regulation. It has been agreed on through the UNFCCC, its main aim being to combat global warming by cutting down on GHG emissions [22–24]. The Protocol stands as one of the first major steps taken by the international community toward addressing climate change at a worldwide level. In this protocol, developed countries are the ones that are charged with the responsibility of reducing GHG emissions and they do not require developing countries to do so for some time. There are two categories of states as per the treaties brought about by this protocol: category one is made up of industrialized or Annex I nations, which should cut down their emissions. Some of the leading manufacturers under Annex I are the United States, Japan, Russia, Germany, Italy, Spain, Poland, Canada, England, Australia, Ukraine, and France. Category two consists of developing countries outside the Annex. (These countries have been given exemptions during the initial stage of the Kyoto Protocol implementation.) There are also two types of developing countries. The first group is made up of countries that are economically advanced among the less industrialized nations, which are supposed to supply money and technology for cutting down on GHG emissions. Known to be several biggest emitters among them are China, Brazil, India, Mexico, Indonesia, Thailand, South Africa, and South Korea. The other group comprises those that have no obligation toward reduction of the amount of carbon dioxide being released into the atmosphere; these nations include Iran since it falls under the category "developing country," which didn't assume any commitments with regard to cutting GHGs under provisions laid down in conventions such as Climate Change Convention and the Kyoto Protocol, among others. However, it may join in clean development projects or take national assessments of GHG emissions on a voluntary basis. The examination of the GHG emissions mechanism discloses that the industrial, transport, residential, commercial, and public service sectors have been identified as the major sources with a higher percentage [25]. Figure 16.7 shows the amount of CO_2 emitted into the atmosphere from 1920 to 2020, measured in billion tons.

The Kyoto Protocol is made up of three principles. The first principle employs the rule of Joint Common but Differentiated Responsibilities (CBDR) is based on the fact that developed countries have over the past 150 years engaged in industrial activities, which have led to the release of high amounts of GHGs into the atmosphere, thus, those countries are held chiefly responsible for the situation. This was known as the CBDR [27] principle. Second, each participating developed country

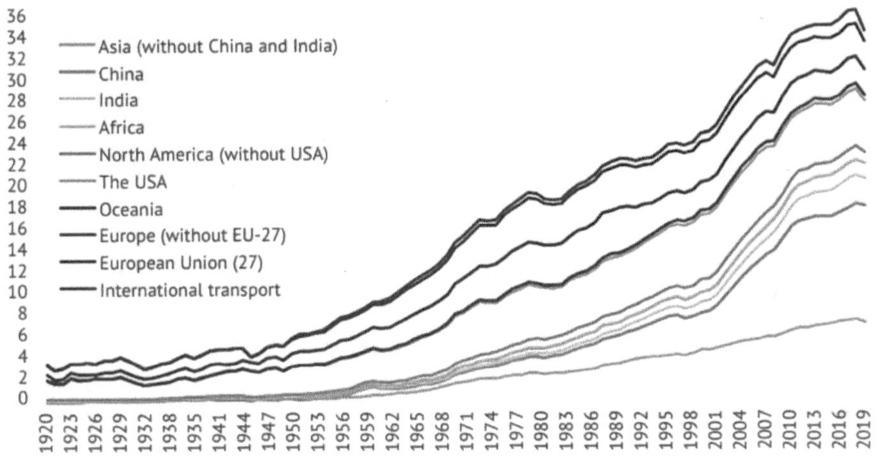

FIGURE 16.7 The amount of CO_2 emitted into the atmosphere from 1920 to 2020, billion tons [26].

was to have specific targets for emissions reduction contained in its publications. These targets were supposed to vary to take into account differing national capabilities and different levels of responsibility according to respective abilities. The third principle adopts flexible mechanisms. Therefore, the Kyoto Protocol brought in three types of market instruments that would allow countries to meet their emission limitations at lower costs than would otherwise be possible. Nations with a surplus of emissions cuts under their own targets could sell the excess to other countries authorized under the system known as Joint Implementation. For developed nations, the Clean Development Mechanism was a way to invest in emission reduction projects located in less developed regions where costs might be lower while earning Certified Emission Reduction credits from it. International Emissions Trading (IET) lets countries buy or sell GHG units among themselves as per an agreement called the Kyoto Protocol [28].

Despite its innovative approach, the Kyoto Protocol faced several challenges and criticisms. And not all countries ratified the protocol, while some major emitters like the United States withdrew from or did not ratify it. Moreover, critics argued that the emission reduction targets were not significant enough to mitigate global warming adequately. Finally, the protocol did not set emission reduction targets for developing countries, which raised concerns about increased emissions from such regions as well. The data shows that the implementation of the environmental agreement according to the Kyoto Protocol caused a decrease in nitrous oxide and methane, two GHGs that contribute less than carbon dioxide to GHG emissions, in all three country groups. However, when it comes to carbon dioxide—the gas that makes up the highest percentage of greenhouse emissions—emission cuts have only taken place in developed countries. To meet their obligations under the Kyoto Protocol during its first commitment period, affluent nations sought to adopt different economic and environmental measures with varying degrees of success. Nonetheless, there has still been no significant reduction in the overall amount of these gases despite some progress being made by the industrialized world in curbing emissions. Therefore, it can be said that there was little success achieved by implementing this agreement toward its set objectives for cutting down on targeted pollutants. The first term of the Kyoto Protocol came to an end in 2012. A second commitment period which is called the Doha Amendment was agreed upon in 2012 that would extend the agreement to 2020. Nevertheless, the document most vastly resembling the Kyoto Protocol is the Paris Agreement, adopted in 2015, which is based on that agreement and commits to limiting the temperature rise of global warming to below 2 degrees, if possible, at 1.5°C compared to pre-industrial time [29–30].

16.4 PARIS AGREEMENT

The Hellenic Accord would be a landmark international treaty planning to address the urgent issue of global warming by bringing together all nations to make pledges to cut down on their GHG emissions and also to adapt to climate change. Since its establishment in 2015, it has greatly contributed toward shaping worldwide efforts against this environmental threat. To get to the history of the Paris Agreement one needs to go back past prior international climate pacts like the Kyoto Protocol. The Kyoto Protocol was among the first major global agreements, which required developed countries to limit their emission of GHGs [31]. It was put into effect in 1997. Adversely, it excluded some countries from the mandatory reductions such as those that are still developing. This caused it to face backlash because it was claimed that the only way to effectively address the issue of climate change is through an agreement that involves all nations and is more ambitious on a global level. Although the Paris Agreement is based on prior pacts which were made including those like the Kyoto Protocol, it differs by bringing together all countries for joint activities while enabling each country to come up with its own nonbinding domestic action plan called an NDC. A major feature of this accord is its general objective of keeping the increase in average global temperature below 2°C during the pre-industrial period, with aspirations of further capping it at 1.5°C. This ambitious target reflects growing scientific consensus on the urgent need for decisive action to avoid catastrophic climate impacts [32]. Besides its temperature objectives, adaptation and mitigation efforts are another significant part of the Paris Agreement. Apart from reducing emissions, countries are also required to bolster their ability to adjust to new weather patterns and become more resilient against future impacts. This all-encompassing strategy acknowledges that the issue of climate change cannot be resolved by simply decreasing emissions at their origin points but rather by making communities ready for unavoidable transformations. Under the Paris Agreement, countries' involvement and contributions vary greatly depending on their individual circumstances and capacities. Developed nations must take the lead not only in cutting down on GHG production but also in offering financial aid as well as transferring technology know-how so that developing countries can switch over to low-carbon development pathways. Some key elements contained in the Paris Agreement are outlined in Table 16.3.

TABLE 16.3
Key Provisions of the Paris Agreement

Row	Topic	Description	Source [33]
1	Nationally Determined Contributions (NDCs)	Each country must submit plans detailing emission reduction targets and actions to achieve them.	Paris Agreement
2	Transparency and Accountability	The agreement establishes a framework ensuring transparent and accurate reporting of countries' actions, subject to review.	Paris Agreement
3	Global Stocktake	Conducted every five years, this assessment evaluates collective progress toward the agreement's goals, informing future actions.	Paris Agreement
4	Financial Support	Developed nations are required to financially aid developing countries in climate change mitigation and adaptation efforts.	Paris Agreement
5	Technology Transfer	The agreement promotes international cooperation in transferring environmentally sound technologies, especially to support developing nations.	Paris Agreement

16.5 GOVERNMENT POLICIES AND REGULATIONS

16.5.1 Carbon Pricing Mechanisms

One way that people try to slow down global warming is by using carbon pricing. This means that they charge businesses and individuals for every ton of GHGs released into the atmosphere. Since CO_2 is the most common GHG, the "price" is often measured in terms of CO_2eqs. Putting a price on carbon is supposed to create an incentive for everyone to reduce their emissions. The idea is that if it costs more to release carbon, then people will do less of it. While the primary goal of carbon pricing is environmental—that is, getting individuals and companies alike to shrink their carbon footprints—it also aims at achieving economic and social ends. This encourages organizations and people to decrease their carbon impression by methods for energy productivity, selection of cleaner advancements, and moving to inexhaustible wellsprings of energy. The monetary results of carbon pricing are such that it advances development in low-carbon innovations. Businesses are incentivized to invest in research and development of innovations that reduce environmental emissions, including the development of cleaner fuels and other sustainable technologies. In addition, carbon taxes, along with auctioning emission, permits under cap-and-trade systems create large amounts of income. These funds can be allocated to support renewable energy projects, improve energy efficiency, and finance climate change adaptation strategies. Such investments would help countries manage the impacts of extreme weather events and other climate-related challenges. However, assessing the value of these interventions may not be suited to simple models like quadratic equations or linear regression, as more complex, context-specific analytical methods are often required. Although it may seem as though the concept of carbon pricing is limited to a national or regional arena, this provides a template for global action. Coordinating global efforts to prevent climate change becomes feasible by setting an example to follow and maybe connecting various carbon markets. Similarly, you can collect the money in such a way that it does not affect industries and disadvantaged populations much so it supports equitable transitioning toward a low carbon economy. These measures may involve specific refunds, financial incentives for using renewable energy technology, and support schemes for staff moving out of coal-based industries. In conclusion, carbon pricing systems intend to reduce GHG emissions at an affordable cost; foster the creation of new clean sources of power; and generate income for climate financing and energy programs while involving a wider range of people in activities aimed at addressing climate change [34]. Each carbon pricing approach has its contradictions and factors to consider. The choice of the mechanism is often based on political feasibility, economic efficiency, and environmental effectiveness. Table 16.4 compares different types of carbon pricing methods with their advantages and disadvantages.

Because they are more economically efficient, feasible, and environmentally friendly, carbon taxes and cap-and-trade systems represent the key carbon pricing systems and will be discussed at length in the following section.

16.5.1.1 Carbon Taxes and Cap-and-Trade Systems, Implemented by Governments to Incentivize Emissions Reductions

Politicians everywhere have implemented two main market-based methods for stimulating the reduction of GHG emissions (which are primarily carbon dioxide) to tackle climate change—cap-and-trade systems and carbon taxes. Both of these tactics work toward pricing carbon emissions so that people and firms who release them into the environment can be encouraged economically to cut their carbon footprint. However, they are built on different foundations.

- **Carbon Taxes**

Since the world's industrialization became widespread, climate change has been a big issue, and one of the biggest levels of response for this issue is the Kyoto Protocol. Therefore, an essential part of dealing with global warming problems is controlling the emissions of GHGs. A new method to reduce the emission of carbon dioxide, which is the most common GHG, was introduced by the carbon trade market that was put in place following this agreement. Carbon trading is fundamentally composed of carbon pricing mechanisms. It can send signals for policy making as well as restricting activities in industries that emit a lot of carbon dioxide if the prices of these permits fluctuate. A fee imposed on the burning of fuels such as coal, oil, or gas known as a carbon tax serves as a direct and simple way to price carbon. A company or individual is taxed for each ton of carbon dioxide they emit into the atmosphere. The purpose of the carbon tax is to minimize GHG release through creating a fixed cost for it, promoting sustainable research and development (R&D) investments, and also funding devices that lower emissions. A sentiment of hopefulness toward study methods that could be used in the future is established when projects record minimal

TABLE 16.4
Different Types of Carbon Pricing Mechanisms [35–37]

Mechanisms	Definition	Mechanism	Advantages	Disadvantages
Carbon taxes	Carbon taxes are a direct fee imposed on the carbon content of fossil fuels. The tax is typically levied per ton of carbon dioxide (CO_2) emitted when the fuel is burned.	Companies or individuals that emit carbon dioxide are required to pay a tax based on the amount of emissions produced. The tax incentivizes polluters to reduce their emissions to avoid higher costs.	Provides price certainty for carbon emissions	May disproportionately impact low income households
Cap-and-trade systems	Cap-and-trade systems, as explained earlier, involve setting a cap on total emissions and allowing companies to trade emissions allowances.	Companies are allocated a certain number of emissions allowances, which they can buy, sell, or trade with other entities. The total emissions are capped, and the cap decreases over time to reduce overall emissions.	Allows the market to determine the price of emissions reductions	Complex to implement and administer
Baseline and credit systems	Baseline and credit systems, also known as offset programs, allow entities to earn credits by reducing emissions below a predetermined baseline level. These credits can then be sold to other entities to offset their emissions.	Companies or projects that reduce emissions below a specified baseline can generate carbon credits. These credits represent the amount of emissions reduced and can be traded in a market.	Incentivizes emission reductions through offset projects	Challenges in ensuring additional offset projects
Fee bate Systems	Fee bate systems combine elements of both carbon taxes and subsidies to incentivize the adoption of cleaner technologies. They impose fees on high-emission products and provide rebates or subsidies for low-emission alternatives.	Products or activities that generate high emissions are subject to a fee, while those with lower emissions receive a rebate or subsidy. This creates a financial incentive for consumers to choose cleaner options.	Shifts consumer behavior toward cleaner alternatives	Potential for market manipulation and gaming
Emissions trading systems	Emissions trading systems (ETSs), also known as cap-and-trade systems, are market-based mechanisms that set a cap on total emissions and allow entities to buy and sell emission allowances.	Under an ETS, a government sets a cap on the total amount of emissions allowed and issues a corresponding number of allowances. Companies can buy and sell these allowances in a market, providing flexibility in meeting emission reduction targets.	Provides flexibility in meeting emission reduction targets	Dependent on the effectiveness of the cap set by the government
Carbon offsetting	Carbon offsetting involves compensating for emissions by investing in projects that reduce or remove greenhouse gas emissions elsewhere.	Companies or individuals can purchase carbon offsets from projects that reduce emissions, such as renewable energy projects, reforestation initiatives, or methane capture projects.	Compensates for emissions by investing in offset projects	Verification and credibility issues with offset projects

success. Through pricing, the carbon tax indirectly influences levels of pollution but cannot ensure any specific reduction amount unless it is sufficiently high. It is difficult to find the right balance between lowering the quantity of emissions that cause global warming without harming economic growth too much by determining the best level at which this should be set. As an example of this, different industries will be affected by a carbon tax in various ways; for instance, it will have a particularly significant impact on companies involved in electricity generation from fossil fuels or those engaged in transportation, agriculture, and forestry which are considered. People with low incomes are impacted the most by not having enough money to spend on things that would save energy or keep them warm during winter. Because they can't afford this, their living expenses rise without being able to change how much power they use in their homes. However, getting a carbon tax passed is hard due to the fact that many citizens need to support it.

The commonness of the carbon content in fossils makes tax computations simple. Additionally, the carbon tax can be applied anywhere along the supply chain of fossil fuels from extraction to burning, thus including a wide range of sources for emissions. Figure 16.8. illustrates explicit carbon prices around the world in 2023, showing how various countries have adopted this mechanism to address greenhouse gas emissions. Another important benefit is that it creates income for the government. Even though many carbon pricing schemes are designed as "revenue-neutral" to compensate for the costs of the tax, some economists have suggested using these funds to reduce budget deficits. To illustrate, receipts from the carbon tax could go toward paying off the national debt or funding cuts in existing rates of other taxes. Such charges explicitly put a price on carbon, which enables firms to plan better for their investments in clean technologies. The government may be able to earn a lot of money through carbon tax. This money can be used to pay for renewable energy projects, reduce other taxes, or give it back to people to help them pay for the increased electricity cost. If there were no taxes imposed, then it would cost a lot more to find ways of cutting down on GHG emission. This way, businesses are given the freedom

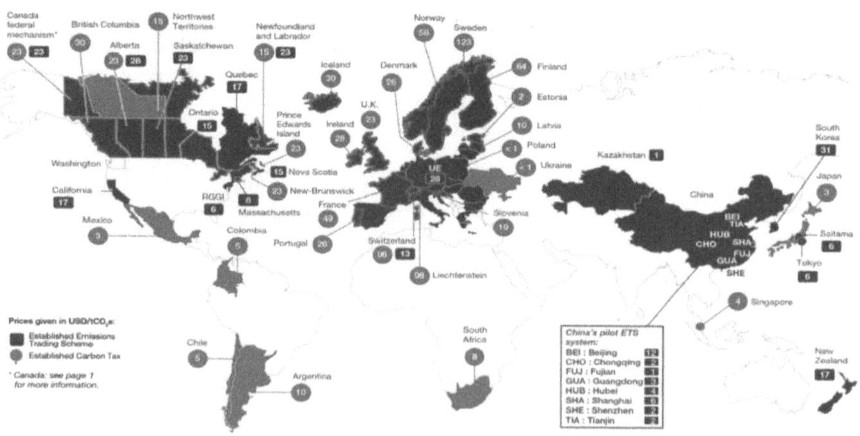

FIGURE 16.8 Map of explicit carbon prices around the world in 2023 [39].

to reduce their emissions at the lowest possible expense. With uniform pricing of carbon emitting, all companies will be able to plan for future investments because they have an idea of what will happen. Nevertheless, some individuals especially those with low income may not benefit from these duties hence they should not cause more problems than they solve. Introducing such a tax can be difficult politically. Companies may resist due to increased costs while customers fear higher prices for energy and goods. It is important to set the right amount for this charge. A low one does not encourage cutting down on emissions that cause global warming. High rates could slow economic growth [38].

- **Cap-and-Trade Systems**

The following is the next regulation that we are going to discuss, which is the cap-and-trade systems. Cap-and-trade systems are also known as carbon trading mechanism that is designed to control the total amount of carbon emissions by setting upper limit or ceiling. Figure 16.9 illustrates different carbon pricing options—low, medium, and high caps—based on 2017 emissions data and 2021 green premium estimates. In this system, the limit of total emission is not set but rather carbon tax is imposed which means that there is cost on emitting carbon; these systems further limit total allowed amount of GHG release into atmosphere and allow businesses purchase emission permits within those limits buying and selling roofs. After some time, the cap is lowered step by step so as to reduce the overall emissions. Unlike carbon tax, advantages involve firm limits on emission reduction goals achievement. Companies have flexibility to reduce emissions or buy allowances from others who've reduced emissions more affordably, leading to cost-effective reductions. The

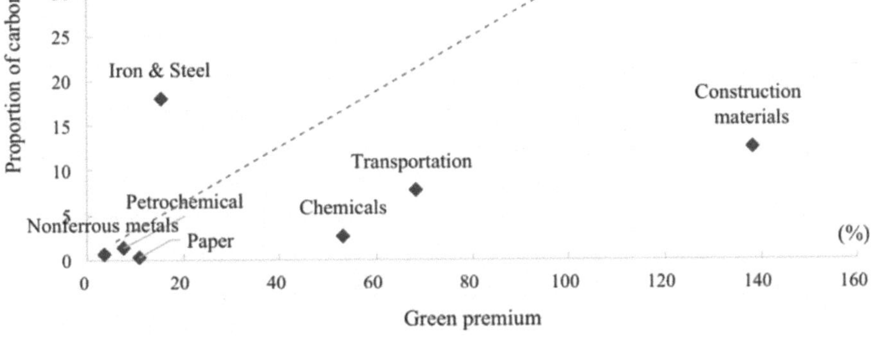

FIGURE 16.9 Different Carbon Pricing Options: Low, Medium, and High Caps Based on 2017 Emissions Data and 2021 Green Premium Estimates [41].

system encourages innovative clean technologies, as companies can profit by selling extra allowances if they cut emissions below their allocation. Like the carbon tax, this mechanism has challenges. It can be complex to design and monitor, requiring robust tracking to ensure compliance. The price of allowances can fluctuate, making it difficult for businesses to plan emissions reduction investments. How the allowances are distributed (auctioned or given for free) can significantly impact the system's effectiveness and fairness [40].

Carbon taxes and cap-and-trade systems are two ways by which the level of carbon dioxide in the atmosphere can be reduced. It will all depend on what is required, either economically or politically, in a specific part of the world such as a country or even a region within that country. Sectors that could be looked at might include agriculture, among others; this should also take into consideration what needs to be achieved regarding climate change locally versus globally. Some nations have gone further to combine these methods to get more benefits out of them [42]. In general, while cap-and-trade systems provide a structured and predictable approach to emissions reduction, they require careful management of market dynamics and price volatility to ensure long-term effectiveness and stability.

16.5.1.2 The Effectiveness and Challenges Associated with These Market-Based Approaches

Carbon pricing mechanisms have a multifaceted impact on the market. The concept of a unified carbon price is scrutinized, as there is a distinction between the "social cost of carbon" and the "net social cost." This suggests that a unified carbon price may not be suitable for all industries due to variations in net social costs. Calculating differentiated carbon costs is a complex and time-consuming process compared to a unified calculation. Complicating matters further is the introduction of the "green premium" as a means of computing differential carbon prices. This idea takes into account both the social cost and net social cost viewpoints, recognizing the difficulties in determining differentiated costs in addition to the ambiguities in determining a single carbon cost based on social cost. The adoption of different carbon pricing mechanisms is made more complicated by the numerous issues they bring.

First, calculating carbon costs is difficult due to the existence of various carbon price concepts with different meanings, leading to ambiguity in discussions across contexts. There are disagreements over the fundamental data and calculation approaches used in mainstream carbon cost estimation methods. This is further complicated by vastly different estimates of the social cost of carbon by prominent economists like Nordhaus and Stern, who have divergent views on the discount rate. The assumption of a unified carbon cost across all sectors by mainstream methods is also questioned, as it may not accurately capture the varying net social costs across different industries. The idea of unified carbon pricing faces scrutiny as it fails to consider industry-specific net social costs and the complexities in calculating differentiated carbon costs. Debates continue over the efficiency and fairness of differentiated carbon pricing, with valid arguments on both sides. Implementing such mechanisms poses challenges, requiring complex operational designs within carbon markets or alternative mechanisms like carbon taxes. The proposal of a "parity carbon cost" suggests a more practical approach, based on the green premium, which may better reflect market realities compared to the theoretically discounted social cost of carbon. Addressing these challenges is crucial for the effective implementation of market-based carbon pricing mechanisms and the realization of their intended environmental and economic objectives [43].

16.5.2 Renewable Energy Policies

Renewable energy comes from virtually limitless natural resources like the sun, wind, water, and earth's heat. Fighting climate change, reducing GHG emissions, and a greener and more sustainable energy system are all pegged on the use of renewable sources. Figure 16.10 shows the growth in installed renewable energy capacity worldwide from 2010 to 2020. One major type of this power is solar, which comes from the sun. Solar thermal or photovoltaic systems are used to trap solar radiation from sun rays. Photovoltaic cells unlike solar thermal systems do not require sunlight to heat a liquid so that it produces steam for turning turbines, which generate electricity, but they change light directly into electric current using sunlight. Solar energy is abundant and clean and can be used on either small scales such as rooftop solar panels or large scales like solar farms. Wind power harnesses the wind's kinetic energy and turns it into electrical power—typically through spinning blades on a turbine generator system. Land-based wind farms are installed where there are strong winds, while offshore facilities are built in the ocean or sea areas with high-speed currents. Large dams, river systems, or pumped storage facilities can all be used to accomplish this. A dependable and adaptable renewable energy source that contributes to grid stabilization and stable power supply is hydropower. The heat found deep under the earth's surface is the source of geothermal energy. To create electricity, power plants

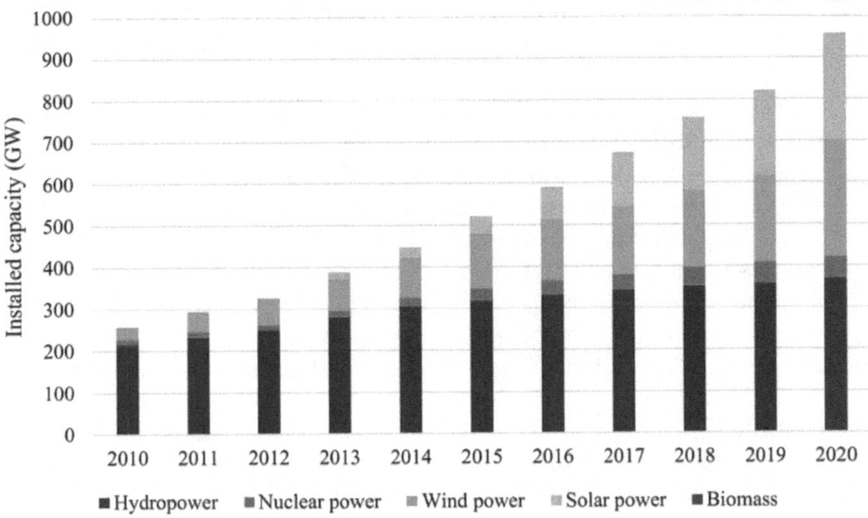

FIGURE 16.10 The capacity of installed renewable energy in 2010–2020 [45].

can use the heat from steam or hot water in geothermal reservoirs. With little GHG emissions, geothermal energy is a dependable and sustainable energy source. Natural resources including wood, agricultural waste, and rubbish are the source of biomass energy. It can be immediately burnt to provide heat or power, or it can be processed to create biogas or biofuels like ethanol and biodiesel's. Biomass energy can help cut waste, is flexible, and doesn't release carbon into the atmosphere. Utilizing the energy of ocean waves and tides, tidal and wave energy produces electricity. Wave energy employs converters, but tidal energy requires barrages or turbines. While still in their infancy, these technologies have the potential to provide sustainable energy on a regular basis [44].

Using renewable energy instead of fossil fuels has problems. Various technological, financial, and regulatory challenges related to the shift from traditional to renewable sources of energy are targeted by renewable energy policies. The following is a detailed analysis of each policy and law that has been put in place by governments across the globe to promote the utilization of renewable energy which aims at giving a broader view:

1. Renewable Portfolio Standards (RPS): RPS regulations force energy providers to make sure a certain portion of the power they sell originates from renewable sources. The demand for renewable energy is increased by this strategy, which incentivizes utilities to invest in or buy renewable energy. Trading renewable energy certificates, which serve as evidence that power was produced using renewable energy sources, is a common way to assure compliance [46].
2. Feed-in Tariffs, or FiTs, are government regulations that, frequently above market rates, establish a predefined price for renewable energy provided to the grid for a certain period of time. This lowers investment risk and promotes the growth of renewable energy projects by ensuring a steady income for producers of renewable energy. FiTs have been very effective in hastening the installation of wind and solar power in a number of nations.
3. Tax Credits and Incentives: For the purpose of reducing the financial load on renewable energy projects, tax credits, exemptions or deductions are being provided by different governments. A tax credit is offered per kilowatt hour (kWh) of electricity produced from specific energy resources through production tax credits, whereas certain investment costs can be deducted from taxes under investment tax credits.
4. Policies Regarding Net Metering: Under net metering, customers that produce their own power using solar panels or other renewable energy sources can resell extra energy to the grid. This adds to the supply of energy overall and offers a financial incentive for people and companies to participate in renewable energy.
5. Government Grants and Loans: Direct financial assistance in the form of grants and low-interest loans can assist in removing the first financial obstacles that come with initiatives involving renewable energy. These monies may be used for infrastructure initiatives, new technology commercialization, or R&D.

6. Public Procurement of Renewable Energy: By obtaining renewable energy for their own needs, governments may set an exemplary example. This demonstrates a commitment to sustainable practices and highlights the need for renewable energy. Also, through economies of scale, public procurement can help increase the renewable energy market and lower prices.
7. Carbon Pricing Mechanisms: When governments put a price on carbon emissions with taxes or cap-and-trade systems, they make dirty fossil fuels more expensive than clean alternatives. With this, businesses are encouraged to cut down their emission levels while consumers are motivated to adopt energy efficiency measures as well as invest in renewables.
8. R&D Funding: One of the most important things is to invest in renewable energy technology with an aim at advancing them by making sure that they are efficient enough besides being reliable and cost effective. Early-stage research funding which could be too risky for private sector investors may be supported through government grants leading to realization breakthroughs in this area.
9. Public Awareness Campaigns: Educating the public about the benefits of renewable energy and how to implement renewable solutions can increase acceptance and demand. Awareness campaigns can also inform consumers about incentives and programs available to them, further encouraging the adoption of renewable energy.
10. International Agreements and Cooperation: Participation in international agreements, such as the Paris Agreement, commits countries to reducing GHG emissions and supports the global transition to renewable energy. The sharing of technology, best practices, and financial resources may be facilitated by international collaboration, particularly to encourage the growth of renewable energy in underdeveloped nations.
11. Building Codes and Standards: Implementing building codes that require or encourage the integration of renewable energy systems in new constructions can significantly increase the uptake of technologies like solar panels and geothermal heating. Energy efficiency standards for buildings and appliances can also reduce overall energy demand, complementing the shift toward renewable energy.
12. Land Use and Zoning Policies: By designating specific areas for renewable energy projects and streamlining the permitting process, governments can reduce barriers to the development of renewable energy infrastructure. Zoning policies can help balance the need for renewable energy projects with environmental protection and community interests.

When properly thought out and executed, these measures and statutes can immensely hasten the move toward a future where energy comes from inexhaustible sources, therefore aiding in the fight against climate change, ensuring there is energy security as well as fostering economic growth [46–48].

The adoption of renewable energy is met with several challenges in different sectors such as economics, technology, regulations, society, and education. For instance, economically there are high initial costs that may deter investors who also face hard times penetrating markets already dominated by cheaper fossil fuels. Technical and infrastructure wise, grid integration becomes an issue because power generated from wind or solar panels cannot be stored for use when there is no sunshine or wind blowing; hence, energy storage systems need to be installed while considering the maturity levels of various renewable energy technologies among other things. Additionally, there needs to be international coordination, but this becomes difficult due to many countries having their own regulations regarding renewables, which also affects policy certainty since permits may take a long time before being issued domestically let alone internationally where many more steps must be followed than just getting one permit approved within national borders, and so forth. The workforce's skill gaps, especially in developing nations, and the requirement for knowledge transfer to facilitate the broad implementation of renewable energy solutions round out the educational and workforce problems. A comprehensive strategy that incorporates financial incentives, technical innovation, international cooperation, and supporting legislation is needed to address these difficulties.

16.5.3 Energy Efficiency Standards

16.5.3.1 Government Regulations and Standards Aimed at Improving Energy Efficiency in Industry

Energy efficiency criteria are the minimum level of energy usage that is allowed in equipment as well as in industries set by the government. They are put in place to provide for energy conservation which will lead reduction of GHGs being emitted into the atmosphere thus making sustainable industrial operations. This is achieved by ensuring that industrial enterprises embrace practices and technologies of using power effectively by coming up with the lowest performance standards. In return, this plays a role in saving money for companies due to reduced wastage among other things like cutting down costs on the environmental impact of their activities related to power consumption. It involves a broad spectrum of areas including but not limited to the manufacturing sector, transport industry, and construction field among many others depending on what each particular field requires in terms of energy sources or devices connected thereto. Industries should not breach specific levels otherwise they will face penalties based on the energy performance standards provided. Adopting these rules comes with various advantages like better performance, cuts on power bills, becoming more competitive, and positive environmental impacts. Putting money in devices and ways that save power lowers expenses in the long run while meeting sustainability objectives. Officials collaborate very closely with business players in the field, energy specialists, and ecological bodies in creating and implementing these laws by employing stakeholder involvement, open forums, and regular assessments so that they remain pertinent and efficient. There are instances where these directives may be in line with international norms and treaties. This promotes uniformity, thereby fostering worldwide partnerships for conserving energy. Such measures are aimed at simplifying policies as well as propagating good ideas beyond national boundaries. In general, boosting efficiency should be seen as key to a more sustainable industry one that is also energy efficient. These rules push for invention in the sector, reducing power demand and ultimately working toward a greener tomorrow. In America, the Environmental Protection Agency's Energy Star program stipulates what levels different industrial products and machines should meet to save on power usage; thus, companies using such equipment will save energy too and be more environmentally friendly. Additionally, the ISO 50001 standard for energy management can help organizations improve performance by establishing systems whereby they identify areas where heat is lost or gains are possible; this helps them use what they have effectively while minimizing waste. This needs member states to ensure energy is saved and used efficiently in industrialization. California's Energy Efficiency Standards Title 24 regulates buildings to save energy, which includes industries. Lighting, HVAC (Heating, Ventilation, and Air Conditioning) systems, and the design of the building for the reduction of energy consumption and GHG emissions are part of these regulations. In Japan the Top Runner Program has set efficiency levels based on the best products available. Similarly, it is important for the manufacturers to comply with the energy efficiency thresholds of the best-performing products. This encourages creativity and advancement of energy-saving technologies within the sector. By advocating for efficient methods, promoting innovation, and reducing ecological effects, these criteria have been instrumental in cutting down on industrial energy consumption. Adherence to them has resulted in substantial declines in global manufacturing energy usage and GHG emissions [49].

16.5.3.2 Analyze the Impact of Energy Efficiency Policies on CO_2 Emissions Reduction

In the battle against GHGs, mainly carbon dioxide (CO_2), which is the leading cause of climate change, saving power is important. It can lead to decreased energy consumption in various sectors such as transport, industry, and buildings, thus reducing CO_2 emissions as a result. Efforts toward improving efficiency are made within the transport industry too. These include but are not limited to the development of new technologies that enhance fuel economy, promoting the use of public transport systems and non-motorized means to cut down on individual car usage levels and switching over to electric cars powered by renewable sources, which produce less CO_2 when compared to others. Similarly, the commercial field employs different methods aimed at conserving energy like optimizing manufacturing processes and utilizing waste heat recovery systems among others, while ensuring effective monitoring and control of electricity consumption through energy management systems. Buildings must achieve energy efficiency. This will involve energy-efficient designs, materials, and technologies in new constructions as well as the retrofitting of existing ones. Besides, a change in occupants' behavior toward saving energy can also help a lot. It is estimated that if transportation sector energy efficiency policies are pursued to the latter, they can reduce CO_2 emissions by 10% to 30% and even more depending on such policies' specificity and implementation. The success of these measures depends on various factors, including the composition of each industry, the availability of alternative technologies, and broader economic and sociocultural conditions. Therefore, it is essential to develop inclusive, context-specific strategies when designing such policies to ensure their effectiveness. This will maximize the impact of energy efficiency measures on lowering CO_2 emissions and support the growth of sustainable transportation systems [50, 51].

Combining energy-saving policies with green power drives could cut down CO_2 releases to a large extent. There are many different strategies that can be used to achieve this kind of integration to make the most impact possible. The first thing is to have supporting legislation that is designed to work hand in hand with one another. When overall energy consumption is reduced by efficiency drives, it becomes cheaper to meet the remaining demand using renewable sources, therefore hastening the shift toward a low-carbon emitting energy system. The next important step lies in planning and implementation through an inclusive approach, which should be taken seriously. It calls for synchronization among various departments, ensuring that measures aimed at saving power are not only put up but also supported by projects of generating electricity from such resources as wind or solar; thus, an example could be making it mandatory to include them when constructing energy-efficient buildings. This enhances the sharing of information and correct apportionment of funds to come up with better ways of reducing CO_2 emissions. If these comprehensive methods are used, leaders and communities concerned can utilize the advantages brought about by energy efficiency and renewable sources of power. As a result, there will be significant long-term cuts in carbon dioxide production, while still promoting economic development, energy independence, and living standards [52].

16.6 INTERNATIONAL EMISSIONS TRADING

Carbon trading, commonly known as IET of GHGs, is a plan for controlling pollution that uses economic incentives to reduce the amount of pollutants in the atmosphere. This is one of the main methods suggested for dealing with global climate change. Under this system, "cap and trade," a country or company can buy or sell permits to emit carbon dioxide or other GHGs. This will allow them to make maximum cost-effective reductions in their emissions. The concept of this system as an economic solution to environmental problems came up some years back. The idea gained popularity toward the end of the 20th century, especially after the U.S. Acid Rain Program became successful using similar tactics to decrease sulfur dioxide releases. After this, it was also enforced for GHG discharges on a worldwide level through what is referred to as the Kyoto Protocol in 1997. This is the international agreement that was adopted in 1992 where the countries have pledged to reduce the emissions of gases like carbon dioxide, which result in global warming [53].

16.6.1 IET Systems

IET systems are instruments created for minimizing GHG emissions affordably. Such systems enable the trading of emission permits or credits between countries or companies. Through this kind of commerce, the market can identify the most efficient ways of reducing emissions. One common technique employed in emissions trading is known as "cap-and-trade." The participating entities have a specified limit on the total amount of certain GHGs they can emit within this system. This ceiling is successively reduced over time to bring about overall emission reductions. In these schemes, entities are allocated emission allowances, which permit them to emit specific quantities of different gases with warming potential. Nevertheless, the aggregate number of authorizations issued should not exceed the set limit. The trading component comes into play when an organization manages to lower its emissions below the amount of allowances it holds. In such cases, it can sell its extra allowances to other organizations that are struggling to meet their emission reduction targets. This system provides flexibility to organizations in how they achieve their emissions reduction goals, thereby encouraging cost-effective reductions and fostering innovation in emissions mitigation strategies. Some key examples of international ETSs include the EU ETS and various regional or national cap-and-trade programs implemented by countries globally. For example, the European Union Emission Trade System was established in 2005. Under this system, there are more than 11,000 power plants and factories across the EU as well as airlines are included. This system is operated based on the cap-and-trade mechanism where the total amount of emission is limited and companies can trade emission allowances. It has significantly cut down on emissions within the EU and served as a model for many similar schemes globally. Another case in point is the Regional Greenhouse Gas Initiative (RGGI), which began in 2009 involving states from northeastern and mid-Atlantic U.S. regions. RGGI aims at auctioning off carbon dioxide permits to lower emissions from the electricity industry. The RGGI has been successful at lowering CO_2 emissions in its member states. To achieve this goal, California established the cap-and-trade program, which is one component of its plan to reduce GHG levels back to 1990 by 2020. It targets major sources such as power plants and oil refineries, while also linking up with other carbon markets, thereby amplifying its effectiveness. The largest national carbon market so far was created when China set up its own ETS whose capacity surpasses any other scheme globally since 2021. The original plan was for the project to cover eight main industries. At present, it is still in its early stages, but the ETS will be crucial in helping China cut down on its emissions. These systems are based on emissions trading, which takes different forms, thus highlighting how much can be achieved through market-driven means that support efforts toward reducing global GHGs that cause climate change [54].

Furthermore, an introduction should be made about ETSs. Environmentally, it has been found that ETS helps in the reduction of GHGs. This is made possible when ETS sets limits on the amount of emissions that can be made, leading to a gradual decrease over time. This means that there is always control over the total emissions released. Additionally, the implementation of ETS leads to an increase in prices for carbon-emitting fuels, thereby encouraging companies to adopt renewable sources of energy generation, which are cheaper to use. Moreover, awareness about carbon footprint reduction is created among business enterprises through the establishment of ETS, thus fostering a sustainable culture within them as well as among the general public. On the whole, ETS presents a well-rounded approach. It provides environmental benefits and economic ones. ETSs are economical. These systems are designed to be efficient by allowing companies that have lower costs associated with reducing emissions to sell off their surplus allowances to those with higher costs, thus making an overall saving. Also, financial inducements will be provided for companies to invest in clean technology R&D; through this, they will be able to find cheaper ways of reducing their GHG releases into the atmosphere. Emitters may choose between different options under the flexibility of the ETS such as meeting their targets internally, buying permits from others, or even investing in offset projects. Auctioning permits can also serve as a means through which governments make revenues from charging for carbon emissions, which are then plowed back into schemes aimed at curbing global warming while assisting people in adapting to its effects. This indicates that a strong economy can still grow while emissions are significantly reduced. ETSs also enable nations to trade carbon allowances with one another, thereby fostering global cooperation in addressing the climate crisis. This facilitates the achievement of more ambitious worldwide environmental targets [55].

16.6.1.1 European Union ETS

The European Union initiated the ETS in 2005. It is an essential part of the EU's solution to global warming and a cost-effective means to reduce GHG emissions. This system represents the world's first big step toward controlling carbon pollution and still remains the largest. In addition, it includes more than 11,000 power plants and factories from 31 countries as well as airlines [57]. Being a central element in the European Union's effort for GHG reduction under the Kyoto Protocol, the EU ETS sought to create a market-based mechanism that would incentivize polluters to cut down their emissions of specified heat-trapping gases [58]. The EU ETS is based on the cap-and-trade principle. The cap or limit is the total amount of these GHGs that are able to be emitted by the installations covered by the system. This cap gets lower over the years; therefore, the total emissions are constantly reduced. The EU ETS requires large carbon emitters within the EU to monitor and report their CO_2 emissions annually. These companies must also surrender enough emission allowances to match their yearly emissions. These allowances can be traded, allowing companies that can reduce emissions more cheaply to sell their extra allowances to other companies for whom emission reduction is more expensive. The EU ETS has gone through multiple stages, each aimed at

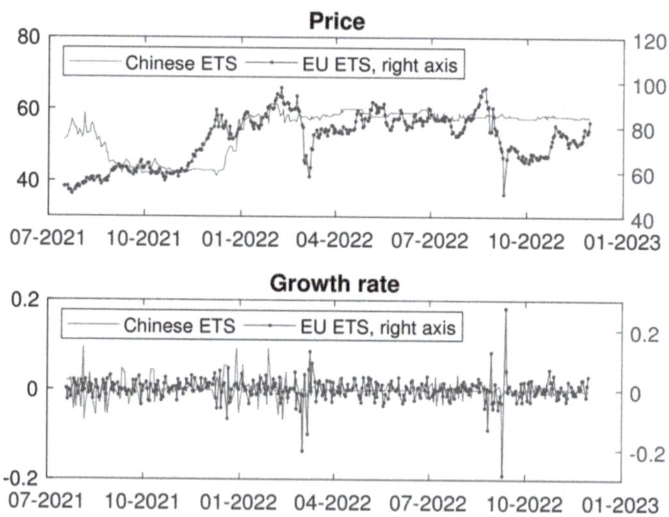

FIGURE 16.11 The prices and the price growth rates of the Chinese ETS and European ETS. The price units for the Chinese and European ETS are the RMB and Euro, respectively [56].

making the system stronger, more efficient, and broader in scope. These phases reflect the EU's increasing determination to tackle climate change, as well as its attempts to refine the system based on prior experiences and hurdles.

Phase I (2005–2007): The Pilot Phase

In the early stages, what they were trying to do was see if trading carbon in the European Union was possible. This part of the test was used to find out how well a system like this could work over a large area. It involved CO_2 emissions from power stations, heat plants, and some types of factories that use a lot of energy. By looking at many different parts of the economy, it was possible to study what might happen if people traded carbon and how good an idea it would be. When it came to handing out allowances, most were given away for nothing. This was mainly based on what had been done before but also called "grandfathering." However, it was not easy sailing during this period. It was marked by a number of hurdles such as too many allowances being given out, thus causing the price of carbon in the market to fall. In as much as there were some drawbacks experienced, there were also some good things that happened during the first phase. One of them is that the scheme built a base for future programs related to the trade of carbon dioxide by putting in place important systems. Moreover, it provided very important information, which if used well would ensure that later phases become more effective and successful in dealing with EU's GHGs [59, 60].

Phase II (2008–2012): Kyoto Protocol Commitment Period

EU ETS under the first commitment period of the Kyoto Protocol was developed to achieve the union emission reduction objectives outlined in the Protocol. This indicated that the EU was in line with international efforts to combat climate change. The scheme grew to include more sectors as well as GHGs with a view of realizing comprehensive cuts across different industries. Although the majority of allowances were still being issued for free, some countries started selling them through auctioning, thus introducing market-based mechanisms into the scheme. In spite of the progress made in cutting emissions, there were challenges at the implementation stage due to the economic recession, which resulted in surplus permits. The surplus led to the persistence of low prices in carbon markets, highlighting the complex interplay between economic considerations and environmental policy goals [61, 62]

Phase III (2013–2020): Toward Auctioning and Harmonization

In this particular stage, the objective was to make the system more efficient and stimulate investment in clean technologies that support a shift toward sustainability and environmental friendliness. To achieve this, the scope of the system was widened to cover other sectors and types of GHGs based on the fact that such gases are behind global warming. Notably, how emission permits were distributed changed significantly by allowing more permits to be sold through auction as opposed to free allocation thereby showing that there was a departure from the past methods which means that more market-oriented approaches are coming into play within this scheme. Additionally, with a view to implementing this move toward auctioning, efforts were made to align different approaches used in the member states through the development of common rules for the allocation of permits at the EU level, thus doing away with previous national allocation plans. The main goal of this move was to make the system more equitable and regular across nations. The implementation of the market stability reserve (MSR) in 2019 was a crucial measure in dealing with an excess of allowances and increasing its resilience. There were good outcomes at this point; emissions fell in all covered sectors, while carbon prices rose steadily toward the end of the period, which indicates that reforms made were successful in cutting down emissions as well as stabilizing the EU ETS market [63].

Phase IV (2021–2030): Strengthening the EU ETS

This step's goal is to harmonize the EU ETS with the increased climate targets of the EU for 2030 and the ultimate goal of net-zero emissions by 2050. A significant reduction in GHG emissions is required to achieve this challenging task. The emissions cap will be tightened more steeply to drive a 43% cut in GHG emissions from the covered sectors by 2030 compared to 2005 levels. Furthermore, the proportion of allowances to be auctioned within the system is set to rise further as a way of spurring cuts, while stimulating investments into low-carbon technologies. This phase also contains measures aimed at promoting innovation and modernization in energy through earmarked funds. The innovation and modernization funds are meant to assist in coming up with new technologies that release less carbon into the atmosphere, which can then be used on a widespread basis. They are also intended to update existing systems to make them more environmentally friendly. Another thing being done is trying to solve the excess of permits in the carbon market. This will be achieved through reinforcing the MSR, thus making sure that there is high stability and resilience within the carbon market. Additionally, there has been a proposal to introduce a carbon border adjustment mechanism. The main reason for this kind of arrangement is to avoid what is known as "carbon leakage" whereby companies transfer their production to countries with laxer emission standards leading to an overall increase in emissions globally. It would entail imposing a charge on imports of certain products from countries outside the EU based on the amount of carbon dioxide attributed to their production [64]. This shall create fairness among nations and act as an encouragement toward reducing GHG emissions universally.

EU ETS's every new phase is a product of the past's learning. It has been progressively widening its reach and effectiveness. The current steps and those to come have been crafted with the primary goal of keeping EU ETS as one of the main pillars of EU climate policy. This system is meant to cut down on carbon dioxide discharge besides spurring growth through technology.

EU Directive 2003/87/EC established the European Union Emission Trading Scheme (EU ETS), and, since its launch in 2005, it has played a critical role in cutting GHG emissions across member states. The latter event marked an important shift in the EU's climate agenda, which had already seen another significant milestone achieved by this time. It managed to lower covered sector discharges by 35% within 15 years compared to their level in 2005.

The system has encouraged innovation in low-carbon technology, improved pricing efficiency in the market through carbon valuation, and established a worldwide benchmark in terms of carbon trading. In the future, the EU plans to tighten emission limits further, guarantee the stability of the market with the help of the MSR, and may also include new areas like maritime transport. It remains effective as a means of reducing GHGs; hence, there is a need for more linkage with other such schemes across different parts of the world, coupled with investment toward innovation and modernization. The EU ETS is said to also contribute toward achieving tough climate goals while becoming a model for pricing carbon and trading emissions at the international level [65].

16.6.1.2 China's National Carbon Market

The China Emissions Trading Scheme is the largest national carbon market in the world, launched by China in a significant move to address global warming and reduce its GHG emissions. It was established in 2021 and covers more carbon dioxide than any other system, including the EU ETS [66]. The main objective of this project relates to supporting the country's pledge toward achieving net zero by 2060 and peaking before 2030 [67]. This strategy works by putting a price on emissions, thereby making it expensive for organizations that release large amounts of CO_2 into the atmosphere. The China ETS follows a cap-and-trade methodology. The government determines the maximum amount of GHGs that can be emitted by all participating firms. Firms obtain or buy emission rights within this threshold. They are then able to exchange these rights among themselves, which gives them a reason to lower emissions so that they can sell any extra ones and make a profit. The success of China's future national carbon market is not only important for the country's environment and economy but also for global efforts against climate change. It will have a big impact on worldwide strategies concerning carbon pricing and the wider switch to economies that use less carbon since it is the biggest market in the world in terms of emissions covered. The electricity sector is the first focus of the ETS in China. However, plans also have been suggested to expand it by incorporating other industries with high emissions such as steel, cement, chemicals, and aviation. This will help to have a wider impact on the overall GHG production in the country [68]. For the ETS to be in line with China's climate targets, it may need to apply more strict limits on emissions or increase the reduction goals over time. Consequently, the quantity of emission allowances available for sale should gradually reduce. Consequently, businesses would be forced to employ cleaner technologies and practices. The effectiveness of the China ETS would be increased by technological breakthroughs like blockchain systems that track carbon transactions and emissions data as well as artificial intelligence, coupled with big data analytics for improved projection and identification of opportunities for cutting down on releases [69]. An opportunity to connect with carbon trading markets across the globe means global systems of carbon pricing could be made to work together and allow for affordable emissions cuts worldwide. But doing so demands delicate talks that cater to differences in market structures and regulations among nations. For the future growth of its ETS, China must have strong policy backing as well as clear guidelines on joining, trading within, adhering to, and being penalized under them. For public popularity of carbon markets and their aims stakeholder involvement is vital. This includes making the public understand why carbon should be priced as well as having the industrial sector, NGOs, and international organizations fine-tune ETSs. China's national cap-and-trade system will largely depend on how ambitious its global warming policies are and what role it plays in international efforts against climate change. Other countries can learn much from the experiences and models offered by this market toward perfecting their own carbon pricing mechanisms or even introducing new ones altogether [70].

16.7 CONCLUSION

Global efforts to reduce CO_2 emissions analysis shows that combating climate change requires extensive cooperation among nations and strong policies. Environmental problems such as droughts, melting of polar ice caps, increasing sea levels, and rapid temperature changes are due to the growing CO_2 emission into the atmosphere through industries and other human activities. To control the GHG concentration countries must join hands with each other under international agreements and frameworks like the UNFCCC, which gave the world the possibility of concerted steps and coordination in their efforts. The Kyoto Protocol and the Paris Agreement not only set the strictest targets for emissions reduction but also encouraged the production and development of low-carbon energy sources and energy efficiency. The analysis of different methods to price carbon such as cap-and-trade systems, carbon taxes, baseline and credit systems, ETSs, and feebate reveals their importance as tools that provide economic incentives for reducing emissions. Carbon tax, alongside cap-and-trade systems, features most prominently due to its wide acceptance and ability to promote economic efficiency as well as environmental welfare gains. Temperature projections show that, according to these policies' effective execution and compliance, the increase in global temperature can be significantly slowed down. The Paris Agreement has outlined that, through the use of efficient carbon pricing mechanisms as well as strategies for reducing emissions, we might keep the world from getting warmer than 1.5–2°C higher than it was before industrialization. Such stabilization of heat is necessary if we want to lessen severe consequences like more frequent high temperatures, storms, or disruptions in agriculture caused by changing environments around us. In conclusion, though each method of carbon pricing is beset with problems, their successful execution can greatly diminish the release of GHGs and help regularize worldwide temperatures. Continuous international cooperation, imaginative policy initiatives, and technological breakthroughs are necessary to accomplish the objectives set forth by climate pacts. It is imperative to stabilize concentrations of GHGs so as not to disrupt the climate system, thus shielding the planet and ensuring human welfare as well as sustainable development globally. The journey toward an economy that utilizes little carbon may be difficult, but it must be taken to mitigate against climate change, stabilize temperature levels worldwide, and secure a future that is sustainable for all.

REFERENCES

1. Gota, Sudhir, et al. "Decarbonising transport to achieve Paris Agreement targets." *Energy Efficiency* 12.2 (2019): 363–386.

2. Jansson, Torbjörn, et al. "Carbon taxes and agriculture: the benefit of a multilateral agreement." *Climate Policy* 24.1 (2024): 13–25.
3. Nikendei, Christoph, et al. "Klimawandel: Ursachen, Folgen, Lösungsansätze und Implikationen für das Gesundheitswesen." *Zeitschrift für Evidenz, Fortbildung und Qualität im Gesundheitswesen* 156 (2020): 59–67.
4. Tawfeeq, Saif Nussrat, Karrar Noori Hammed, and Jumana Khaldon Saadoun. "The role of US financial institutions in the international economic sanctions mechanism." *Tikrit Journal For Political Science* 4.26 (2022): 121–147.
5. Maarouf, Habib. *The Issue of Global Climate Change between the Failures of Kyoto and Paris Expectations*, 1st edition (Beirut, Lebanon: Dar Al-Farabi, 2016).
6. Boon, Foo Kim. "The third session of the conference of the parties to the United Nations framework convention on climate change-Kyoto, Japan 1–10 December 97." *Singapore Journal of International and Comparative Law* 2 (1998): 191.
7. Mace, Mary Jane. "Adaptation under the UN framework convention on climate change: The international legal framework." In *Fairness in Adaptation to Climate Change*, W. Neil Adger, Jouni Paavola, Saleemul Huq, and M. J. Mace (2006). MIT Press. DOI: https://doi.org/10.7551/mitpress/2957.001.0001 ISBN electronic: 9780262266819.
8. Cadman, Tim. "The United Nations framework convention on climate change." *The Palgrave Handbook of Contemporary International Political Economy* (2019): 359–375.
9. Krause, D. "Sustainable development in times of climate change." *Policy Innovations for Transformative Change: UNRISD Flagship Report* (2016): 141–166.
10. Hassan, Alaa Mohammed, and Saif Nussrat Tawfeeq. "the role of the united nations in mitigating global climate change." *Russian Law Journal* 11.9S (2023): 521–536.
11. Ebi, Kristie L., and Jan C. Semenza. "Community-based adaptation to the health impacts of climate change." *American Journal of Preventive Medicine* 35.5 (2008): 501–507.
12. Kuyper, Jonathan, Heike Schroeder, and Björn-Ola Linnér. "The evolution of the UNFCCC." *Annual Review of Environment and Resources* 43 (2018): 343–368.
13. Prasad, Ravi S., and Ridhima Sud. "The pivotal role of UNFCCC in the international climate policy landscape: A developing country perspective." *Global Affairs* 7.1 (2021): 67–78.
14. National Research Council, et al. *Climate Stabilization Targets: Emissions, Concentrations, and Impacts Over Decades to Millennia* (National Academies Press, 2011).
15. Freitas, Isabel Maria Bodas, Eva Dantas, and Michiko Iizuka. "The Kyoto mechanisms and the diffusion of renewable energy technologies in the BRICS." *Energy Policy* 42 (2012): 118–128.
16. International Energy Agency. *Energy Technology Perspectives. Pathways to a Clean Energy System* (Paris: IEA/OECD, 2012).
17. Van de Graaf, Thijs. "Obsolete or resurgent? The International Energy Agency in a changing global landscape." *Energy Policy* 48 (2012): 233–241.
18. International Energy Agency. *World Energy Outlook 2009 Edition* (Paris, 2009). ISBN: 978-92-64-06130-9.
19. International Energy Agency. *Energy Technology Perspectives. Pathways to a Clean Energy System* (Paris: IEA/OECD, 2012).
20. Muradov, Nazim, and Nazim Muradov. "Stabilization of atmospheric CO_2: Prospects and implications." *Liberating Energy from Carbon: Introduction to Decarbonization* (2014): 91–115.
21. Wang, Chien-Ho, Ming-Hui Ko, and Wan-Jiun Chen. "Effects of Kyoto protocol on CO_2 emissions: A five-country rolling regression analysis." *Sustainability* 11.3 (2019): 744.
22. Yost, Freya R. "Sharing the data: The information policies of NOAA and EUMETSAT." *IFLA Journal* 42.1 (2016): 5–15.
23. Howarth, Candice, et al. "Enhancing the contribution and role of practitioner knowledge in the Intergovernmental Panel on Climate Change (IPCC) Working Group (WG) II process: Insights from UK workshops." *Climate Services* 5 (2017): 3–10.

24. Wang, Chien-Ho, Ming-Hui Ko, and Wan-Jiun Chen. "Effects of Kyoto protocol on CO_2 emissions: A five-country rolling regression analysis." *Sustainability* 11.3 (2019): 744.
25. Iwata, Hiroki, and Keisuke Okada. "Greenhouse gas emissions and the role of the Kyoto protocol." *Environmental Economics and Policy Studies* 16 (2014): 325–342.
26. Miller, A., A. Atakhanov, M. Guliyev, T. Azizov, and Kh. Huseynova. "The economic effect of the measures provided for by the Kyoto Protocol by region (as of the 2020s)." *Scientific Horizons* 26.4 (2023): 136–145.
27. Grubb, Michael. "Kyoto and the future of international climate change responses: From here to where." *International Review for Environmental Strategies* 5.1 (2004): 15–38.
28. Ge, Jiamin, and Boqiang Lin. "Does the Kyoto protocol as an international environmental policy promote Forest carbon sinks?." *Journal of Global Information Management (JGIM)* 30.6 (2021): 1–22.
29. UNFCCC. *What is the Kyoto Protocol?* https://unfccc.int/kyoto_protocol (accessed on 15 March 2021).
30. Balogh, Jeremiás Máté, and Tamás Mizik. "Global impacts of climate policy and trade agreements on greenhouse gas emissions." *Agriculture* 13.2 (2023): 424.
31. Cahill-Webb, Finn. *International Environmental Governance and the Paris Agreement on Climate Change: The Adoption of the "Pledge and Review" Governance Approach.* No. 99/2018. Working Paper, 2018.
32. Zhizheng Zhou. "Evaluation of the effectiveness of Paris agreement—an angle from the determining of normative framework." *Lecture Notes in Education Psychology and Public Media*, 2023.
33. https://unfccc.int/sites/default/files/resource/parisagreement_publication.pdf.
34. Kossoy, Alexandre. *State and Trends of Carbon Pricing 2015* (World Bank Publications, 2015).
35. Borodin, Alex, et al. "Mechanisms for tax regulation of CO_2-equivalent emissions." *Energies* 15.19 (2022): 7111.
36. Xu, Ke, et al. "Carbon price prediction based on multiple decomposition and XGBoost algorithm." *Environmental Science and Pollution Research* 30.38 (2023): 89165–89179.
37. Buckley, Neil J. "Baseline-and-credit emission permit trading: Experimental evidence under variable output capacity." In *Environmental Economics, Experimental Methods* (Routledge, 2007): 31–50.
38. Andersson, Julius J. "Carbon taxes and CO_2 emissions: Sweden as a case study." *American Economic Journal: Economic Policy* 11.4 (2019): 1–30.
39. Kalirajan, Kaliappa. "Low-carbon energy systems: seizing market-based business opportunities across borders." *Journal of Social and Economic Development* (2024): 1–15.
40. Bealby, Drew. *Policy analysis of emissions cap and trade: The United State and the European Union.* Master's Theses, 3175, 2021.
41. CICC Research, CICC Global Institute Zipeng. Zhou@ cicc. com. cn. "The capacity of carbon pricing mechanism." In *Guidebook to Carbon Neutrality in China: Macro and Industry Trends under New Constraints* (Singapore: Springer Singapore, 2022): 45–76.
42. Yu, Jinhan. "Research on carbon cap-and-trade policies and corporate low-carbon value creation." *Frontiers in Business, Economics and Management* 6.2 (2022): 58–60.
43. Cosma, Maria. *Capping Carbon: A Survey of Carbon Pricing Mechanisms and their Efficacy.* Diss. Pennsylvania State University, 2016.
44. Saravanan, K. K., N. Stalin, and S. T. Jayasuthahar. "Review of renewable energy resources in clean green environment." *International Journal of Emerging Trends in Electrical and Electronics* 1.1 (2013): 54–58.
45. Wei, Wei, et al. "The effectiveness and trade-offs of renewable energy policies in achieving the dual decarbonization goals in China: A dynamic computable general equilibrium analysis." *International Journal of Environmental Research and Public Health* 19.11 (2022): 6386.

46. Mischke, Peggy, and Kenneth B. Karlsson. "Modelling tools to evaluate China's future energy system–a review of the Chinese perspective." *Energy* 69 (2014): 132–143.
47. Wang, Ye, et al. "How to design renewable energy support policies with imperfect carbon pricing?." *Frontiers in Environmental Science* 10 (2022): 958979.
48. Dwivedi, Chesta. "Influence of production and investment tax credit on renewable energy growth and power grid." In *2018 IEEE Green Technologies Conference (GreenTech)* (IEEE, 2018).
49. Li, Ming-Jia, and Wen-Quan Tao. "Review of methodologies and polices for evaluation of energy efficiency in high energy-consuming industry." *Applied Energy* 187 (2017): 203–215.
50. Price, Lynn. "Technologies and policies to improve energy efficiency in industry." In *AIP Conference Proceedings*, Vol. 1044, No. 1 (American Institute of Physics, 2008).
51. Zhang, Feng, Yongming Huang, and Xiaoli Nan. "Advanced research methods and their applications on the nexus of energy efficiency and environment: evidence from five RCEP economies." *Economic research-Ekonomska istraživanja* 35.1 (2022): 5676–5698.
52. del Río, Pablo. "Why does the combination of the European Union Emissions Trading Scheme and a renewable energy target makes economic sense?." *Renewable and Sustainable Energy Reviews* 74 (2017): 824–834.
53. Li, Chenxing, et al. "An authenticated and secure accounting system for international emissions trading." *Climate Policy* 22.9–10 (2022): 1333–1342.
54. Han, Jing, Weilin Zhu, and Chaofan Chen. "Identifying emissions reduction opportunities in international bilateral emissions trading systems to achieve China's energy sector NDCs." *International Journal of Environmental Research and Public Health* 20.2 (2023): 1332.
55. Crespo, Antonio Marcio Ferreira, and Chun Wang. "European Union emissions trading scheme: Design evolution and effectiveness analysis." In *Handbook of Research on Interdisciplinary Approaches to Decision Making for Sustainable Supply Chains* (IGI Global, 2020): 189–210.
56. Cheng, Qiyun, et al. "Price dynamics and interactions between the Chinese and European carbon emission trading markets." *Energies* 16.4 (2023): 1624.
57. Ciesielska-Maciagowska, D., and D. Klimczak. "Skrzek-Lubasi nska, M. Central and Eastern European CO_2 market—challenges of emissions trading for energy companies." *Energies* 14 (2021): 1051.
58. Aldy, Joseph E., and Robert N. Stavins. "Introduction: International policy architecture for global climate change." In *Architectures for Agreement: Addressing Global Climate Change in the Post-Kyoto World* (Cambridge University Press, 2007).
59. Bel, Germà, and Stephan Joseph. "Emission abatement: Untangling the impacts of the EU ETS and the economic crisis." *Energy Economics* 49 (2015): 531–539.
60. Gronwald, Marc, and Beat Hintermann, eds. *Emissions Trading as a Policy Instrument: Evaluation and Prospects* (MIT Press, 2015).
61. Creti, Anna, Pierre-André Jouvet, and Valérie Mignon. "Carbon price drivers: Phase I versus Phase II equilibrium?." *Energy Economics* 34.1 (2012): 327–334.
62. Hintermann, Beat, Sonja Peterson, and Wilfried Rickels. "Price and market behavior in phase II of the EU ETS: A review of the literature." *Review of Environmental Economics and Policy* 10.1 (2016).
63. Ellerman, A. Denny, Claudio Marcantonini, and Aleksandar Zaklan. "The EU ETS: Eight years and counting." *Robert Schuman Centre for Advanced Studies Research Paper* 2014/04 (2014).
64. Knopf, Brigitte, et al. *The European Emissions Trading System (EU ETS): Ex-Post Analysis, the Market Stability Reserve and Options for a Comprehensive Reform.* Nota di Lavoro 79.2014 (Milan, Italy: Fondazione Eni Enrico Mattei, 2014).
65. Mandaroux, Rahel, Chuanwen Dong, and Guodong Li. "A European emissions trading system powered by distributed ledger technology: an evaluation framework." *Sustainability* 13.4 (2021): 2106.

66. Zhang, Weijie, Ning Zhang, and Yanni Yu. "Carbon mitigation effects and potential cost savings from carbon emissions trading in China's regional industry." *Technological Forecasting and Social Change* 141 (2019): 1–11.
67. Liu, Jing, et al. "Research on the spillover effect of China's carbon market from the perspective of regional cooperation." *Energies* 16.2 (2023): 740.
68. Zhang, ZhongXiang. "China's carbon market: Development, evaluation, coordination of local and national carbon markets, and common prosperity." *Journal of Climate Finance* 1 (2022): 100001.
69. Pan, Chunyu, et al. "A linkage framework for the china national emission trading system (CETS): Insight from key global carbon markets." *Sustainability* 13.13 (2021): 7459.
70. Duan, Chener, Yiran Mi, and Ziye Lan. "Analysis models for China's carbon trading market: comparison and outlook." In *IOP Conference Series: Earth and Environmental Science*, Vol. 1011, No. 1 (IOP Publishing, 2022).

Index

FSC
www.fsc.org
MIX
Papier | Fördert gute Waldnutzung
FSC® C083411

Kolibri 360 GmbH
Ferdinand-Jühlke-Straße 7
99095 Erfurt, Deutschland
produktsicherheit@kolibri360.de